HUMAN BIOLOGY

TENTH EDITION

HUMAN BIOLOGY

TENTH EDITION

Cecie Starr

Beverly McMillan

BROOKS/COLE
CENGAGE Learning®

Australia • Brazil • Japan • Korea • Mexico • Singapore • Spain • United Kingdom • United States

BROOKS/COLE
CENGAGE Learning

***Human Biology*, Tenth Edition**
Cecie Starr and Beverly McMillan

Publisher: Yolanda Cossio

Senior Acquisitions Editor, Life Sciences:
Peggy Williams

Assistant Editor: Shannon Holt, Cynthia Ashton

Editorial Assistant: Sean Cronin

Media Editor: Lauren Oliveira

Brand Manager: Nicole Hamm

Market Development Manager: Tom Ziolkowski

Content Project Manager: Jennifer Risden

Art Director: John Walker

Manufacturing Planner: Karen Hunt

Rights Acquisitions Specialist: Dean Dauphinais

Production Service: Lachina Publishing Services

Text Designer: Yvo Riezebos

Photo Researcher: Bill Smith Group

Text Researcher: Karyn Morrison

Copy Editor: Barbara Armentrout

Illustrator: Lachina Publishing Services

Cover Designer: Riezebos Holzbaur/
Andrei Pasternak

Cover Image: AFP/Getty Images

Compositor: Lachina Publishing Services

For product information and technology assistance, contact us at
Cengage Learning Customer & Sales Support, 1-800-354-9706

For permission to use material from this text or product,
submit all requests online at **www.cengage.com/permissions**
Further permissions questions can be emailed to
permissionrequest@cengage.com

Library of Congress Control Number: 2012943394

ISBN-13: 978-1-133-59916-6

ISBN-10: 1-133-59916-8

Brooks/Cole
20 Davis Drive
Belmont, CA 94002-3098
USA

Cengage Learning is a leading provider of customized learning solutions with office locations around the globe, including Singapore, the United Kingdom, Australia, Mexico, Brazil, and Japan. Locate your local office at **www.cengage.com/global**

Cengage Learning products are represented in Canada by Nelson Education, Ltd.

To learn more about Brooks/Cole, visit **www.cengage.com/brookscole**

Purchase any of our products at your local college store or at our preferred online store **www.cengagebrain.com**

Printed in China
2 3 4 5 16 15 14 13

BRIEF CONTENTS

CONTENTS

17 Development and Aging 331

18 Cell Reproduction 355

This edition of *Human Biology* continues our commitment to provide an accessible, up-to-date introduction to well-established concepts and principles that underpin the study of human biology. At the same time, we know that today's students demand relevance more than ever before, and so we continue to focus on and expand our coverage of real-world applications of the topics we discuss, from health concerns and diseases and disorders to the many ways in which human activities can alter the environment and threaten both human health and the biodiversity of our planet. Throughout the book, we reviewed and reworked text and diagrams as needed to make difficult concepts easier for students to grasp. Because today's students, especially non-science majors, learn best when the written text is integrated with dynamic visuals, we've brightened graphic elements and diagrams, and included a wealth of new photographs.

Highlights of This Edition

Section-Based Glossary As with the previous edition, in addition to a full glossary of terms at the end of the book, each section has a section-based glossary that includes all boldfaced terms for that section. A student can simply glance at this glossary for a quick check of a term's meaning.

Section-Based Links to Earlier Concepts Bulleted *Links to Earlier Concepts* appear at the start of most sections and facilitate review of material that a student may not have fully mastered.

Section-Based Concept Review Each text section closes with a focused *Take-Home Message*. This review restates the section's main concepts in the form of a simple question that is followed by short, bulleted explanations.

Visual Chapter Summary To further enhance visual learning, we've added graphics to each section of the chapter summary.

Expanded Self-Quizzes In addition to revising questions for clarity, in most chapters we have increased the number of Self-Quiz questions to provide students with a more complete review of key facts and concepts.

Enhancement Features We have expanded the number of our well-received *Think Outside the Book* features. Another popular feature, *Explore on Your Own*, now appears before the chapter summary. We've updated many of the *Your Future* topics that close each chapter with a glimpse of emerging discoveries and health applications in the realm of human biology.

Chapter-Specific Changes

Chapter 1 Text and art revisions to clarify explanations of concepts related to critical thinking, energy flow, and cycling of materials in the biosphere.

Chapter 2 Revised chapter introduction with new photograph; expanded visuals to support discussions of properties of water, functional groups, acids, and acid deposition as an environmental concern; clarifying labels added to illustrations of structure of cholesterol and ATP.

Chapter 3 Updated discussion of lactate fermentation in muscle cells; updated and expanded *Focus on Health* features on mitochondrial disorders and the cholera epidemic in Haiti.

Chapter 4 Revised diagram of epithelium types; revised discussion of exocrine and endocrine glands, now with explanatory diagrams; new diagrams of stem cell functions and skin structure; illustrated discussion of non-shivering heat production added to the section on homeostasis and thermoregulation.

Chapter 5 Revised section on the types of joints, including a new overview illustration; the illustration of muscle attachments in the knee joint has been moved to the fuller discussion of muscle attachments in Chapter 6.

Chapter 6 Updated introductory essay on strength training; revised Section 6.2 includes a diagram of muscle attachments in a knee joint; a more thorough discussion of muscle fatigue notes how glycogen depletion and calcium leaks contribute to the phenomenon; a new visual strengthens the discussion of muscle twitches and functioning of whole muscles; an updated *Focus on Health* discussion of muscle health cross-references research on the "exercise hormone" irisin (Chapter 15).

Chapter 7 Updated overview of the cardiovascular system now includes an image of branching blood vessels; a new diagram better illustrates the direction of blood flow in capillary beds and the location and functioning of precapillary sphincters.

Chapter 8 A new chapter introduction focuses on blood transfusions; the chapter also includes an enhanced discussion and illustrations in sections on blood typing and in the *Focus on Health* dealing with information revealed by blood tests.

Chapter 9 Fuller discussion of defensive roles of the lymph vascular system; updated introduction to innate immunity with expanded visual on inflammation; expanded treatment of immunoglobulins and uses of monoclonal antibodies in research and health applications; expanded discussion of allergic reactions; the section on HIV/AIDS has an updated discussion of anti-retroviral therapy.

Chapter 10 More fully illustrated section on partial pressures and principles of gas exchange; new *Think Outside the Book* points students to resources for learning more about the Heimlich maneuver; a new, large diagram illustrates shifting partial pressure gradients of blood gases; an improved diagram more clearly shows arterial and brain sensors that monitor blood pH and concentrations of carbon dioxide and oxygen.

Chapter 11 Revised overview of the small intestine; for clarity, the main discussion and illustration of brush border cells is now in the section on nutrient absorption in the small intestine; expanded coverage and visuals of liver cirrhosis in the discussion of digestive system disorders; the chapter's nutrition sections include an updated discussion of USDA nutrition guidelines with the new "plate" graphic, information on phytonutrients, and a clearer description of gastric banding methods.

Chapter 12 A revised introductory section now includes an image of a glomerulus; the section on urine formation features a new illustration that more clearly shows the major events of filtration, reabsorption, and secretion; a new diagram shows the location of the internal and external urethral sphincters; a photograph of a polycystic kidney has been added to the discussion of kidney disorders.

Chapter 13 A clarified introduction to neurons features a revised diagram of a motor neuron and of ion flows across the neuron plasma membrane; photographs added to several sections integrate with text discussions of functions of the PNS and of the limbic system; updated discussion of nervous system disorders includes Guillain-Barré syndrome, and research on brain cancer and cell phone use.

Chapter 14 Numerous new visuals and revisions for clarity throughout; a new *Your Future* box discusses gene therapy efforts to treat patients with a genetic form of blindness.

Chapter 15 Revisions for clarity throughout; the section on other hormone sources (heart, gut, brain) includes a discussion of the newly identified hormone irisin produced by skeletal muscles in response to exercise.

Chapter 16 Discussion of female reproductive hormones revised for clarity; updated discussions of sexually transmitted diseases include HPV and cancers, anti-HPV vaccination programs, and the newly discovered strain of gonococcus that is resistant to all known antibiotics.

Chapter 17 Modified organization discusses extraembryonic membranes in two sequential sections; Section 17.5 describes the roles of the yolk sac, amnion, allantois, and chorion; Section 17.6 is devoted to the structure and functions of the placenta. An updated discussion of implantation includes a photograph of an implanting human blastocyst; also updated is the discussion on structural and physiological effects of aging; a new *Think Outside the Book* raises the issue of rising rates of births by Cesarean section.

Chapter 18 New and revised visuals include the cell cycle and the four stages of mitosis; a new *Think Outside the Book* asks questions about gender testing of female athletes.

Chapter 19 The discussion of segregation of chromosomes during meiosis has been revised for clarity, as has the text describing transmission of the chin dimple trait as an example of Mendelian inheritance; the section on varying effects of single genes features a new diagram of sickle cell disease impacts; discussion of the genetic underpinnings of ABO blood types includes a new Punnett square diagram showing possible gene combinations.

Chapter 20 Updated visuals include the chapter opener on cystic fibrosis and photographs accompanying the discussion of achondroplasia and the disorder aniridia (lack of an iris in the eye) that is due to a chromosome deletion; the section on X-linked traits and disorders cross-references an expanded Appendix VI, which now contains information students can use to explore Internet resources on human genetic disorders.

Chapter 21 The introduction to DNA now includes a photograph of human DNA; the discussion of transcription has been revised for clarity and notes thalassemia as an mRNA-splicing disorder; new diagrams illustrate translation of mRNA and basic processes of gene therapy; the section on genetically modified organisms other than humans now includes the example of GM corn.

Chapter 22 Updated discussion of carcinogenesis with new illustration summarizing the basic steps; new visuals include the image and story of a young woman who developed malignant melanoma from heavy use of tanning beds; updated American Cancer Society statistics on cancer incidence; the section on cancer screening and diagnosis touches on the use of personal genetic profiling to gauge risk for breast cancers.

Chapter 23 Updated section on fossils and biogeography as evidence for evolution; revised and updated discussions and visuals related to extinction and adaptive radiation, using early humans as the prime example; a new diagram presents branching of the primate lineage.

Chapter 24 An updated introduction to principles of ecology includes new visuals, including coral reef and boreal forest ecosystems; the diagram of a simple food chain is now a numbered figure for added impact and easier reference.

Chapter 25 Chapter has been revised throughout to reference timely issues of human impacts on the biosphere; the introduction to principles of population biology includes an updated graphic on human population growth; the section on air pollution includes new visuals and an updated discussion of polar ozone thinning; the discussion of greenhouse gases includes updated information on global climate change; the discussion of nuclear power notes issues raised by reactor meltdowns at the Fukushima Dai-ichi power plant in Japan.

Student and Instructor Resources

Biology CourseMate Cengage Learning's Biology CourseMate brings course concepts to life with interactive learning, study, and exam preparation tools that support the printed textbook or the included eBook. With CourseMate, professors can use the included Engagement Tracker to assess student preparation and engagement. Use the tracking tools to see progress for the class as a whole or for individual students. New to this edition of the CourseMate are new 3D animations, videos from BBC Motion Gallery, crossword puzzles, and new interactive activities.

Test Bank Thousands of test questions ranked according to difficulty and consisting of multiple-choice (organized by section heading), matching, labeling, and short-answer exer-

cises. Includes selected images from the text. Also included in Microsoft® Word format on the PowerLecture DVD.

ExamView® Create, deliver, and customize tests (both print and online) in minutes with this easy-to-use assessment and tutorial system. Each chapter's end-of-chapter material is also included.

Instructor's Resource Manual Includes chapter outlines, objectives, key terms, lecture outlines, suggestions for presenting the material, classroom and lab enrichment ideas, discussion topics, possible answers to critical thinking exercises, and more. Also included in Microsoft® Word format on the PowerLecture DVD.

Student Interactive Workbook Labeling exercises, self-quizzes, review questions, and critical thinking exercises help students with retention and better test results.

PowerLecture™ This convenient tool makes it easy for you to create customized lectures. Each chapter includes the following features, all organized by chapter: lecture slides, all chapter art and photos, bonus photos, animations, videos, Instructor's Resource Manual, Test Bank, ExamView testing software, and JoinIn™ polling and quizzing slides. Available online or on disc, PowerLecture places all of the media resources at your fingertips.

WebTutor™ for WebCT™ and Blackboard® Jump start your course with customizable, rich, text-specific content. Whether you want to Web-enable your class or put an entire course online, WebTutor delivers. WebTutor offers a wide array of resources including media assets, quizzing, web links, exercises, flash cards, and more.

CengageNOW Save time, learn more, and succeed in the course with CengageNOW, an online set of resources (including Personalized Study Plans) that gives you the choices and tools you need to study smarter and get the grade. You will have access to hundreds of animations that clarify the illustrations in the text, videos, and quizzes to test your knowledge.

eBook This complete online version of the text is integrated with multimedia resources and special study features, providing the motivation that so many students need to study and the interactivity they need to learn.

Brooks/Cole Biology Video Library featuring BBC Motion Gallery Looking for an engaging way to launch your lectures? The Brooks/Cole series features short, high-interest segments, including new videos featuring information on improvements on artificial hearts, bionic hands, the threats to coral reefs and mountain glaciers and their impact on people, and the intelligence of ants.

Influential Reviewers

Sheila Prabhakar Abraham
Southeastern University

John Alcock
Arizona State University

Donald K. Alford
Metropolitan State College of Denver

Venita F. Allison
Southern Methodist University

D. Andy Anderson
Utah State University

David R. Anderson
Pennsylvania State University, Fayette

Dorothy J. Anthony
Keystone College

Peter Armstrong
University of California, Davis

Tami L. Asplin
North Dakota State University

Amanda Attryde
Arizona State University, West

Leigh Auleb
San Francisco State University

Kirat Baath
University of Southern Indiana

Caryn Babaian
Bucks County Community College

Marne Bailey
Lewis University

Aimee H. Bakken
University of Washington, Seattle

Tamatha R. Barbeau
Francis Marion University

Susan R. Barnum
Miami University

Sally M. Bauer
Hudson Valley Community College

Edmund E. Bedecarrax
City College of San Francisco

Jack Bennett
Northern Illinois University

Sheila Bennett
University of Maine, Augusta

David F. Bohr
University of Michigan, Ann Arbor

Charles E. Booth
Eastern Connecticut State University

Laurie Bradley
Hudson Valley Community College

J. D. Brammer
North Dakota State University

Sharon T. Broadwater
College of William and Mary

Melvin K. Brown
Erie Community College, City Campus

Linda D. Bruslind
Oregon State University

Alfred B. Buchanan
Santa Monica College

Douglas J. Burks
Wilmington College

Victor Chow
City College of San Francisco

Ann Christensen
Pima Community College

A. Kent Christensen
University of Michigan

Joe Connell
Leeward Community College

George W. Cox
San Diego State University

Jerry Coyne
University of Chicago

Richard M. Crosby
Treasure Valley Community College

Lisa Danko
Mercyhurst College

Fred Delcomyn
University of Illinois, Urbana-Champaign

Melvin Denner
University of Southern Indiana

Katherine J. Denniston
Towson State University

Tom E. Denton
Auburn University, Montgomery

Dean Dluzen
Northeastern Ohio University – College of Medicine

Gordon J. Edlin
University of Hawaii

Inga Eley
Hudson Valley Community College

Michael Emsley
George Mason University

Gina Erickson
Highline Community College

Mary Erwin
San Francisco City College

Daniel J. Fairbanks
Brigham Young University

Richard H. Falk
University of California, Davis

Laurine J. Ford
Inver Hills Community College

Glenn M. Fox
Jackson Community College

John E. Frey
Mankato State University

Jeffery Froehlich
University of New Mexico

Larry M. Frolich
Yavapai College

David E. Fulford
Edinboro University of Pennsylvania

James G. Garner
Long Island University, C.W. Post Campus

Saul M. Genuth
Case Western Reserve University

Edmund A. Gianferrari
Keene State College of the University of New Hampshire

H. Maurice Goodman
University of Massachusetts – Medical School

Sheldon R. Gordon
Oakland University

Kenneth W. Gregg
Winthrop University

John C. Grew
New Jersey City University

Martin Hahn
William Paterson College

N. Gail Hall
Trinity College

John P. Harley
Eastern Kentucky University

Aslam S. Hassan
University of Illinois

Alan I. Hecht
Hofstra University

Michael Henry
Santa Rosa Junior College

Paul E. Hertz
Barnard College

Merrill B. Hille
University of Washington

Eileen Ternove Hinks
Virginia Military Institute

Stanton F. Hoegerman
College of William and Mary

Ronald W. Hoham
Colgate University

Tara Jo Holmberg
Northwestern Connecticut Community College

Howard L. Hosick
Washington State University

Perry Huccaby
Elizabethtown Community College

Celeste Humphrey
Dalton State College

Madelyn D. Hunt
Lamar University

Eugene W. Hupp
Texas Woman's University

Mitrick A. Johns
Northern Illinois University

Leonard R. Johnson
University of Tennessee – College of Medicine

Ted Johnson
St. Olaf College

Vincent A. Johnson
St. Cloud State University

Carolyn K. Jones
Vincennes University

Jann Joseph
Grand Valley State University

Peter Kareiva
University of Washington

Gordon I. Kaye
Albany Medical College

Ronald Keiper
Valencia Community College

Kenneth A. R. Kennedy
Cornell University

Dean H. Kenyon
San Francisco State University

Jack Keyes
Linfield College, Portland Campus

Keith K. Klein
Mankato State University

David T. Krohne
Wabash College

Barbara Krumardt
Des Moines Area Community College, Urban Campus

Charles E. Kupchella
Southeast Missouri State University

Howard Kutchai
University of Virginia

Dale Lambert
Tarrant County College

John M. Lammert
Gustavus Adolphus College

Robert Lapen
Central Washington University

William E. Lassiter
University of North Carolina, Chapel Hill – School of Medicine

Lee H. Lee
Montclair State University

Matthew N. Levy
Mt. Sinai Hospital

Robert C. Little
Medical College of Georgia
Cran Lucas
Louisiana State University, Shreveport
Elizabeth A. Machunis-Masuoka
University of Virginia
Alan Mann
University of Pennsylvania
Nancy J. Mann
Cuesta College
Philip I. Marcus
University of Connecticut
Joseph V. Martin
Rutgers University, Camden
James N. Mathis
West Georgia College
Patricia Matthews
Grand Valley State University
Charles Mays
DePauw University
John F. McCue
St. Cloud State University
Karen A. McMahon
University of Tulsa
Shirley A. McManus
Fresno City College
F. M. Anne McNabb
Virginia Polytechnic Institute and State University
Leah M. Meek
Kent State University, Salem Campus
John L. A. Mitchell
Northern Illinois University
Robert B. Mitchell
Pennsylvania State University
David E. Mohrman
University of Minnesota, Duluth
Hylan Moises
University of Michigan
David Mork
St. Cloud State University
David Morton
Frostburg State University
Michael I. Mote
Temple University
Rod Mowbray
University of Wisconsin, LaCrosse

Chantilly Munson
San Francisco City College
Heather Murdock
San Francisco State University
Richard A. Murphy
University of Virginia – Health Sciences Center
Donald L. Mykles
Colorado State University
David O. Norris
University of Colorado, Boulder
William Parson
University of Washington
Lewis Peters
Northern Michigan University
Joel B. Piperberg
Millersville University
Harold K. Pleth
Fullerton Community College
Robert Kyle Pope
Indiana University, South Bend
Shirley Porteous-Gaffard
Fresno City College
G. Pozzi-Galluzi
Dutchess Community College
David Quadagno
Florida State University
Jill D. Reid
Virginia Commonwealth University
Maren Reiner
University of Richmond
David Reznick
University of California, Riverside
Caroline Clark Rivera
Tidewater Community College
Jane C. Roberts
Creighton University
Jan C. Rogers
State University of New York, Morrisville
Susan Rohde
Triton College
Elaine Rubenstein
Skidmore College
Andrew M. Scala
Dutchess Community College
Maureen Scott
Norfolk State University

Lois Sealy
Valencia Community College

Gordon M. Shepherd
Yale University – School of Medicine

John W. Sherman
Erie Community College, North

Lauralee Sherwood
West Virginia University

Richard H. Shippee
Vincennes University

Eric M. Sikorski
University of Tampa

Om V. Singh
University of Pittsburgh

Roger D. Sloboda
Dartmouth College

Barbara W. Smigel
Community College of Southern Nevada

Robert L. Smith
West Virginia University

Terrill C. Smith
City College of San Francisco

Ron Sorochin
State University of New York – College of Technology at Alfred

Ruth Sporer
Rutgers University, Camden

Craig W. Steele
Edinboro University of Pennsylvania

Fred E. Stemple Jr.
Tidewater Community College

Patricia M. Steubing
University of Nevada, Las Vegas

Gregory J. Stewart
State University of West Georgia

Analee G. Stone
Tunxis Community College

John D. Story
Northwest Arkansas Community College

Robert J. Sullivan
Marist College

Eric Sun
Macon State College

Rosalyn M. Sweeting
Saginaw Valley State University

David Tauck
Santa Clara University

William Thieman
Ventura College

Chad Thompson
State University of New York – Westchester Community College

Ed W. Thompson
Winona State University

Sandra L. W. Thornton
Eastern Shore Community College

Ian Tizard
Texas A&M University

William Trotter
Des Moines Area Community College – Sciences Center

Jeremy B. Tuttle
University of Virginia – Health Sciences Center

James W. Valentine
University of California, Berkeley

Kent M. Van De Graaff
Weber State University

Pete Van Dyke
Walla Walla Community College

Alexander Varkey
Liberty University

Bruce Walsh
University of Arizona

Margaret R. Warner
Purdue University

Dan R. Weisbrodt
William Paterson University

Mark L. Weiss
Wayne State University

Brian J. Whipp
St. George's Hospital Medical School

Susan Whittemore
Keene State College

Diane Wilkening Fritz
Gateway Community and Technical College

Roberta B. Williams
University of Nevada, Las Vegas

Morgan Wilson
Hollins University

Mary Wise
Northern Virginia Community College

Stephen L. Wolfe
University of California, Davis (Emeritus)

Shanna Yonenaka
San Francisco State University

LEARNING ABOUT HUMAN BIOLOGY

LINKS TO EARLIER CONCEPTS

Each chapter in this book builds on previous ones. Bullets and cross-references will link you to sections in earlier chapters where you can review related topics.

KEY CONCEPTS

The Nature of Life

Living things share basic features, including the genetic material DNA and the need to maintain a state of internal stability called homeostasis. A cell is the smallest unit of life. Sections 1.1, 1.8

Life's Organization and Diversity

Nature is organized from simple to complex, starting with nonliving atoms. The most inclusive level of life's organization is the whole living world, the biosphere. Sections 1.2–1.3

Studying Life

Critical thinking is the foundation for scientific study of the living world. It also is valuable in many life decisions. Sections 1.4–1.7

Top: Daniel McDonald/The Stock Shop/Black Star Middle: David De Lossy/Photodisc/Jupiterimages Bottom: © Raymond Gehman/Corbis

What's happening in your world today?

You may already have checked your favorite social media sites or seen the latest news on your Web browser. Headlines, blog posts, and messages from friends mingle news about wars or political wrangles with tips for managing your love life or choosing food supplements. There may be an alert about a threatening infectious disease like the H1N1 "swine flu" or the devastation and suffering caused by a powerful storm or natural catastrophe. More and more these days we see reports about human activities having major—often negative—impacts on nature. We hear more and more about global climate change.

But these are exciting times, too. We humans can study nature, including ourselves, in ways that may help us better understand the natural world and our place in it. We can examine the world around us, come up with ideas, and find ways to test them. Gradually we can learn a great deal about factors that affect our health, the environment, and other issues. That's what this book is for—to give you a fuller understanding of how your body works and where you fit in the larger world.

This chapter begins our survey of human biology, starting with the basic features of all forms of life. It paves the way for a brief survey of the chemical foundations of life and how our body cells are built and operate. We will then explore how the body's tissues, organs, and organ systems function. You will also learn about the effects of many diseases and disorders, how parents pass traits to their children, and about basic concepts of evolution and ecology.

© istockphoto.com/Alberto Pomares

1.1 The Characteristics of Life

■ Several basic characteristics allow us to distinguish between living things and nonliving objects.

cell An organized unit that can survive and reproduce by itself, using energy, necessary raw materials, and DNA instructions.

homeostasis A state of overall internal chemical and physical stability that is required for survival of cells and the body as a whole.

Living and nonliving things are all alike in some ways. For instance, both are made up of atoms, which are the smallest units of nature's fundamental substances. On the other hand, wherever we look in nature we find that all living things share some features that nonliving ones don't have. There are five basic characteristics of life.

1. **Living things take in and use energy and materials.** Like other animals, and many other kinds of organisms, we humans take in energy and materials by consuming food (Figure 1.1). Our bodies use the energy and raw materials in food to build and operate their parts in ways that keep us alive.

2. **Living things sense and respond to changes in the environment.** For example, a plant wilts when the soil around its roots dries out, and you might put on a sweater on a chilly afternoon.

Figure 1.1 Humans take in energy by eating food. This boy's body will extract energy and raw materials from the food and use them for processes that are required to keep each of his cells, and his body as a whole, alive.

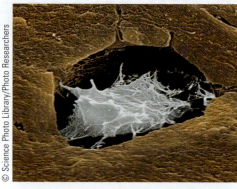

Figure 1.2 Cells are the basic units of life. A bone cell looks white and delicate in this picture. Like other types of body cells, it contains DNA and uses ATP energy.

3. **Living things consist of one or more cells.** A **cell** is an organized unit that can live and reproduce by itself, using energy, the required raw materials, and instructions in DNA. Figure 1.2 shows a living bone cell. Cells are the smallest units that can be alive. The energy for all cell activities comes from another special chemical found only in living things, ATP.

4. **Living things maintain homeostasis.** Changes inside and outside of organisms affect the ability of cells and body parts built from them to carry out their activities. Living things compensate for these changes through mechanisms that maintain an overall internal state of chemical and physical stability. This overall internal stability, called **homeostasis** (hoe-me-oh-STAY-sis), is necessary for the survival of cells and the body as a whole. *Homeostasis* means "staying the same." You will learn much more about it in later chapters.

5. **Living things reproduce and grow.** Organisms can make more of their own kind, based on instructions in DNA, the genetic material. Only living things have DNA. Guided by the instructions in their DNA, most organisms develop through a series of life stages. For us humans, the basic life stages are infancy, childhood, adolescence, and adulthood.

TAKE-HOME MESSAGE

WHAT CHARACTERISTICS SET LIVING ORGANISMS APART FROM NONLIVING OBJECTS?

- Living things take in and use energy and materials, and they sense and can respond to changes in their environment.
- Living things can reproduce and grow, based on instructions in DNA.
- The cell is the smallest unit that can be alive.
- Organisms maintain homeostasis by way of mechanisms that keep conditions inside the body within life-supporting limits.

Our Place in the Natural World

■ Human beings arose as a distinct group of animals during an evolutionary journey that began billions of years ago.

Humans have evolved over time

The term "evolution" means change over time. Chapter 23 of this textbook explains how populations of organisms may evolve by way of changes in DNA. This biological evolution is a process that began billions of years ago on the Earth and continues today. In the course of evolution, major groups of life forms have emerged.

Figure 1.3 provides a snapshot of how we fit into the natural world. Humans, apes, and some other closely related animals are **primates** (PRY-mates). Primates are mammals, and mammals make up one group of "animals with backbones," the **vertebrates** (VER-tuh-braytes). Of course, we share our planet with millions of other animal species, as well as with plants, fungi, countless bacteria, and other life forms. Biologists classify living things according to their characteristics, which in turn reflect their evolutionary heritage. Notice that Figure 1.3 shows three domains of life. Animals, plants, fungi, and microscopic organisms called protists are assigned to kingdoms in a domain

Left: Rich Buzzelli/Tom Stack & Associates
Right: bilderlounge/Jupiter Images

Figure 1.4 Humans are related to Earth's other organisms. Bonobos (*left*) are one of four species of apes, our closest primate relatives. Like us, they walk upright and use tools.

called Eukarya (you-KARE-ee-uh). The other two domains are reserved for bacteria and some other single-celled life forms. Some biologists prefer different schemes. For example, for many years all living things were simply organized into five kingdoms—animals, plants, fungi, protists, and bacteria. The key point is that despite the basic features all life forms share, evolution has produced a living world of incredible diversity.

Humans are related to all other living things—and they have some distinctive characteristics

Due to evolution, humans are related to every other life form and share characteristics with many of them. For instance, we and other mammals are the only vertebrates that have body hair. We share the most features with apes, our closest primate relatives (Figure 1.4). We humans also have some distinctive features that appeared as evolution modified traits of our primate ancestor. For example, we have great manual dexterity due to the way muscles and bones in our hands are arranged and how our nervous system has become wired to operate them. Even more astonishing is the human brain. This extraordinarily complex organ gives us the capacity for sophisticated language and analysis, for developing advanced technology, and for a huge variety of social behaviors.

primates A distinct group of mammals that includes humans, apes, and their close relatives.

vertebrate An animal that has a backbone.

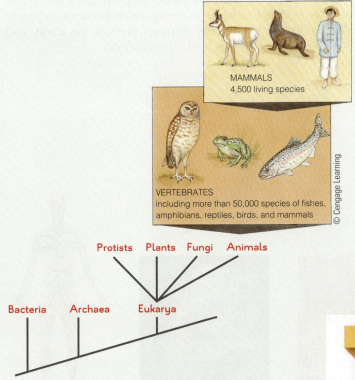

© Cengage Learning

MAMMALS
4,500 living species

VERTEBRATES
including more than 50,000 species of fishes, amphibians, reptiles, birds, and mammals

Protists Plants Fungi Animals

Bacteria Archaea Eukarya

Figure 1.3 Animated! Organisms are classified into groups according to their characteristics. Humans are one of more than a million species in the animal kingdom, which is part of the domain Eukarya. Plants, fungi, and some other life forms make up other kingdoms in Eukarya. The domains Bacteria and Archaea contain vast numbers of single-celled organisms.

TAKE-HOME MESSAGE

WHY IS EVOLUTION A BASIC CONCEPT IN HUMAN BIOLOGY?

- Like all life forms, humans arose through evolution.
- Evolution has produced the features that set humans apart from other complex animals. These characteristics include sophisticated verbal skills, analytical abilities, and extremely complex social behavior.

Life's Organization

■ Nature is organized on many levels, from nonliving materials to the entire living world.

Nature is organized on many levels

Nature is organized on eleven basic levels, which you see summarized in Figure 1.5. At the most basic level are atoms. Next come molecules, which are combinations of atoms. Atoms and molecules are the nonliving components from which cells are built. In humans and other multicellular organisms, cells are organized into tissues—muscle, the epithelium of your skin, and so forth. Different kinds of tissues make up organs, and systems of organs make up whole complex organisms.

We can study the living world on any of its levels. Many courses in human biology focus on organ systems, and a good deal of this textbook explores their structure and how they function.

Nature's organization doesn't end with individuals. Each organism is part of a population, such as the Earth's whole human population. In turn, populations of different organisms interact in communities of species occupying the same area. Communities interact in ecosystems. The most inclusive level of organization is the **biosphere**. This term refers to all parts of the Earth's waters, crust, and atmosphere in which organisms live.

biosphere All parts of the Earth's waters, crust, and atmosphere in which organisms live.

Organisms are connected through the flow of energy and cycling of materials

Organisms take in energy and materials to keep their life processes going. Where do these essentials come from?

Energy flows into the biosphere from the sun (Figure 1.6). This solar energy is captured by "self-feeding" life forms such as plants, which use a sunlight-powered process called photosynthesis to make fuel for building tissues, such as a grain of wheat. Raw materials such as carbon that are needed to build the wheat plant come from air, soil, and water. Thus self-feeding organisms are the living world's basic food producers.

Animals, including humans, are the consumers: When we eat plant parts, or feed on animals that have done so, we take in materials and energy to fuel our body functions. You tap directly into stored energy when you eat bread made from grain, and you tap into it indirectly when you eat the meat of an animal that fed on grain. Organisms such as bacteria and fungi obtain energy and materials when they decompose tissues, breaking them down to substances that can be recycled back to producers. This one-way flow of energy through organisms, and the cycling of materials among them, means that all parts of the living world are connected.

Because of the interconnections among organisms, it makes sense to think of ecosystems as webs of life. With this perspective, we can see that the effects of events in one part of the web will eventually ripple through the whole and may even affect the entire biosphere. For example, we see evidence of large-scale impacts of human activities in the loss of biodiversity in many parts of the world, acid rain, climate change, and other concerns.

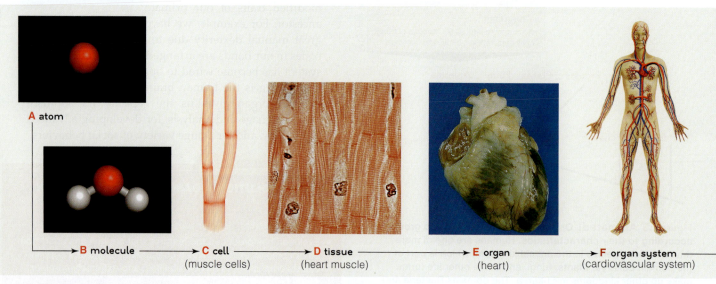

A atom

B molecule ────→ **C** cell ────→ **D** tissue ────→ **E** organ ────→ **F** organ system
(muscle cells) (heart muscle) (heart) (cardiovascular system)

Figure 1.5 Animated! An overview of the levels of organization in nature. (A, B, C, F: © Cengage Learning D: Ed Reschke/Peter Arnold E: L. Bassett/Visuals Unlimited, Inc.)

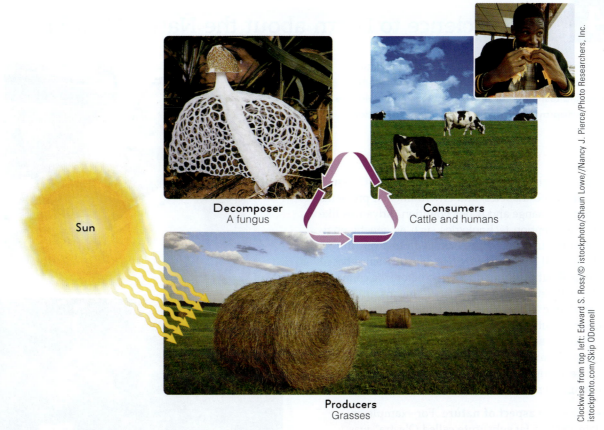

Decomposer
A fungus

Consumers
Cattle and humans

Sun

Producers
Grasses

Clockwise from top left: Edward S. Ross/© istockphoto/Shaun Lowe//Nancy J. Pierce/Photo Researchers, Inc. istockphoto.com/Skip ODonnell

Figure 1.6 The flow of energy and the cycling of materials maintain nature's organization. The bottom photograph shows producers—grass plants used for hay that feeds livestock. The plants obtained the energy to make their roots, seeds, and other parts from the sun. They obtained nutrients for their growth from soil and air. Consumers include animals, such as cattle and humans, and decomposers include organisms such as fungi and bacteria.

TAKE-HOME MESSAGE

HOW IS NATURE ORGANIZED?

• Nature is organized in levels that are sustained by a flow of energy and cycling of materials.

• Energy flows into the biosphere from the sun. Raw materials cycle within the biosphere as consumers obtain food from producers, and decomposers break down tissues to substances that help nourish producers.

• Because living things are interconnected, ecosystems are webs of life in which all the parts are linked.

G organism
(human)

H population

I community

J ecosystem

K the biosphere

G: Yuri Arcurs/Shutterstock.com H: Darko Sikman/Shutterstock.com I: David De Lossy/Photodisc/Jupiterimages J: Ed Reschke/Peter Arnold K: NASA

■ Science basically is a way of thinking about the natural world. Scientists try to explain natural phenomena by making and testing predictions. They search for evidence that may disprove or support a proposed explanation.

Science is a systematic study of nature

Antibiotics. Insights into genetic disorders, health issues such as cancer and diabetes, and environmental problems such as climate change and water pollution. Advances like these—not to mention technologies such as genetic engineering and the Internet—have changed our lives. In this textbook you will be learning a great deal of science-based information about the human body. So before we continue, let's look briefly at what "doing science" means.

We can define "science" as a systematic way of getting information about the natural world. This system is sometimes called the **scientific method**, but there is no single script for it. Researchers can pursue their work in the laboratory or in the field, using a variety of tools (Figure 1.7). The following steps are common.

1. **Observe some aspect of nature.** For example, in the late 1990s, a fat substitute called Olestra® was approved for use in foods. Made from vegetable oil and sugar, Olestra is indigestible and seemed to be a dieter's dream. When potato chips and corn chips made with Olestra were marketed, however, some consumers reported intestinal gas, cramps, and diarrhea.

2. **Ask a question about the observation or identify a problem to explore.** Olestra's manufacturer hired an independent clinical research firm to investigate the intestinal upsets Olestra users were reporting. Was Olestra causing the problems?

3. **Develop a hypothesis.** A **hypothesis** is a proposed explanation for an observation or how some natural process works. With a scientific hypothesis, there must be some objective way of testing it, such as experiments. The research firm hypothesized that Olestra can cause digestive upsets and organized a study to test the hypothesis.

4. **Make a prediction.** As a first step in testing the hypothesis, the study was based on a prediction: Eating food containing Olestra is likely to produce intestinal side effects. As in this example, a prediction states what you should observe about the question or problem if the hypothesis is valid.

5. **Test the prediction.** To test their prediction, the study team recruited 3,181 volunteers aged 2 to 89 and gave them bags of potato or corn chips labeled as containing Olestra. In fact, however, half the snacks were Olestra-

Centers for Disease Control and Prevention

© Raymond Gehman/Corbis

Figure 1.7 **Scientists do research in the laboratory and in the field. A** At the Centers for Disease Control and Prevention, Mary Ari testing a sample for the presence of dangerous bacteria. **B** Making field observations in an old-growth forest.

free controls. Participants could eat as many chips as they liked for six weeks, keeping a record of any gastrointestinal symptoms. About 38 percent of those who ate "Olestra" chips reported intestinal problems—but so did nearly 37 percent in the "regular" group. When scientists at Johns Hopkins University devised a similar, one-time test using 1,100 people, they also found no evidence that Olestra caused digestive problems. The experiments were not failures, because a properly designed scientific test is supposed to reveal flaws. If the findings don't support the initial prediction, then some factor that influenced the test may have been overlooked, or the hypothesis may have been wrong.

6. **Repeat the tests or develop new ones**—the more the better. Hypotheses that are supported by the results of repeated testing are more likely to be correct.

7. **Analyze and report the test results and conclusions.** Scientists typically publish their findings in scientific

Hypothesis
Olestra® causes intestinal cramps.

↓

Prediction
People who eat potato chips made with Olestra will be more likely to get intestinal cramps than those who eat potato chips made without Olestra.

↓

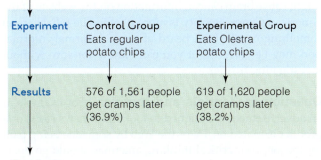

Experiment	**Control Group** Eats regular potato chips	**Experimental Group** Eats Olestra potato chips
Results	576 of 1,561 people get cramps later (36.9%)	619 of 1,620 people get cramps later (38.2%)

↓

Conclusion
Percentages are about equal. People who eat chips made with Olestra are just as likely to get intestinal cramps as those who eat chips made without Olestra. These results do not support the hypothesis.

Figure 1.8 The Olestra study followed steps used in many scientific experiments.

journals, with a detailed description of their methods so that other researchers can try the same test and see if they get a similar result. This is what happened in both our Olestra examples.

Many scientists conduct experiments

Experimenting is a time-honored way to test a scientific prediction. An **experiment** is a test that is carried out under controlled conditions that the researcher can manipulate. Figure 1.8 shows the typical steps followed, using the Olestra study as an example. To get meaningful test results, experimenters use safeguards. They begin by reviewing information that may bear on their project. This step includes considering any previous studies on the topic. Then the researchers design a controlled experiment, one that will test only a single prediction of a hypothesis at a time. In both the Olestra studies, researchers predicted that people who consume Olestra have a greater chance of developing intestinal side effects.

Most aspects of the natural world are the result of interacting variables. A **variable** is a factor that can change with time or in different circumstances. Researchers design experiments to test one variable at a time. They also set up a **control group** to which one or more experimental groups

can be compared. The control groups in both Olestra studies were identical to the experimental ones except for the variable being studied—chips containing Olestra. Identifying possible variables, and eliminating unwanted ones, is extremely important if the results of an experiment are to be reliable. For instance, if any of the participants were already eating foods made with Olestra, it would have been impossible for the experimenters to determine if any reported side effects were due to the test chips or to long-term use.

Scientists usually can't observe all the individuals in a group they want to study. In studies of a food additive such as Olestra it would be hard to include all possible consumers around the world. Results obtained from a subset of test subjects may differ from results obtained from the whole group. This sort of distortion is called **sampling error**. It often happens when a sample size is too small. To avoid that source of sampling error, researchers use a test group that is large enough to be representative of the whole. That is why the Olestra studies recruited so many participants. You can learn firsthand about sampling error in the *Explore on Your Own* exercise at the end of this chapter.

Science never stops

A scientist must draw logical conclusions about any findings. That is, the conclusion cannot be at odds with the evidence used to support it. Interestingly, in the years since Olestra was first developed, the U.S. Food and Drug Administration (FDA) has received more than 20,000 consumer complaints alleging problems, and Olestra has been reformulated to reduce certain side effects. Today, although it is used in a variety of processed foods, some advocates say more research is needed.

control group In an experiment, a group to which one or more experimental groups can be compared.

experiment A test carried out under controlled conditions that the researcher can manipulate.

hypothesis A proposed explanation for an observation or how a natural process works.

sampling error Distortion of experimental results, often because the sample size is too small.

scientific method Any systematic way of obtaining information about the natural world.

variable A factor that can change over time or under different circumstances.

■ To think critically, it is important to evaluate information before accepting it.

critical thinking Using systematic, objective strategies to judge the quality of information; evidence-based thinking.

fact Verifiable information, not opinion or speculation.

opinion A subjective judgment.

Have you ever tried a new or "improved" product and been disappointed when it didn't work as expected? Everyone learns, sometimes the hard way, how useful it can be to cast a skeptical eye on advertising claims or get an unbiased evaluation of, say, a used car you are considering buying. This objective evaluation of information is called *evidence-based* or **critical thinking**.

Scientists use critical thinking in their own work and to review findings reported by others. Anyone can make a mistake, and there is always a chance that pride or bias will creep in. Critical thinking is a smart practice in everyday life, too, because so many decisions we face involve scientific information. Will an herbal food supplement really boost your immune system? Is it safe to eat irradiated food? Table 1.1 gives guidelines for evidence-based, critical thinking.

© Corbis

Evaluate the source of information

An easy way to begin evaluating information is to notice where it is coming from and how it is presented. Here are two simple strategies for assessing sources.

Let credible scientific evidence, not opinions or hearsay, do the convincing For instance, if you are concerned about reports that heavy use of a cell phone might cause brain cancer, information on the website of the American Cancer Society is more likely to be reliable than something cousin Fred heard at work. Informal information may be correct, but you can't know for sure without investigating further.

Question credentials and motives For example, if an advertisement is designed to look like a news story, or a product is touted on TV or the Web by someone being paid for the job, your critical thinking antennae should go up. Is the promoter simply trying to sell a product with the help of "scientific" window dressing? Can any facts presented be checked out? Responsible scientists try to be cautious and accurate in discussing their findings and are willing to supply the evidence to back up their statements.

Evaluate the content of information

Even if information seems authoritative and unbiased, it is important to be aware of the difference between the cause of an event or phenomenon and factors that may only be correlated with it. For example, studies show that recirculation of air in an airplane's passenger cabin increases travelers' exposure to germs coughed or sneezed out by others. An "airplane cold," however, is caused directly by infection by a virus.

Also keep in mind the difference between facts and opinions or speculation. A **fact** is verifiable information, such as the price of a loaf of bread. An **opinion**—whether the bread tastes good—can't be verified because it involves a subjective judgment. Likewise, a marketer's prediction that many consumers will favor a new brand of bread is speculation, at least until there are statistics to back up the claim.

TABLE 1.1   A Critical Thinking Guide and Checklist

To think critically about any subject:

✔ **Do** gather information or evidence from reliable sources.

✗ **Don't** rely on hearsay.

✔ **Do** look for facts that can be checked independently and for signs of obvious bias (such as paid testimonials).

✗ **Don't** confuse *cause* with *correlation*.

✔ **Do** separate *facts* from *opinions*.

Once you have formed your opinion:

Be able to state clearly your view on a subject.

Be aware of the evidence that led you to hold this view.

Ask yourself if there are alternative ways to interpret the evidence.

Think about the kind of information that might make you reconsider your view.

If you decide that nothing can ever persuade you to alter your view, recognize that you are not being objective about this subject.

THINK OUTSIDE THE BOOK

Controversy swirls around claims that an extract from berries of the acai plant can produce rapid, easy weight loss. Using reputable sources such as the National Institutes of Health, do some Web research on this topic. What is the fuss all about?

TAKE-HOME MESSAGE

WHAT IS CRITICAL THINKING?

- Critical thinking is an objective, evidence-based evaluation of information.
- Critical thinking is required for doing science. It also is a smart strategy in many aspects of daily life.

1.6 Science in Perspective

■ **A scientific theory explains a large number of observations.**

We know that the practice of science can yield powerful ideas, like the theory of evolution, that explain key aspects of life. At the same time, we also know that science is only one part of human experience.

It is important to understand what the word *theory* means in science

You've probably said, "I've got a theory about that!" This expression usually means that you have an untested idea about something. A **scientific theory** is the opposite: It is an explanation of a broad range of related natural events and observations that is based on repeated, careful testing of hypotheses. Table 1.2 lists some major scientific theories related to biology. Before scientific research established one of them, the germ theory of disease, some people tried to appease malevolent spirits they blamed for outbreaks of infectious disease (Figure 1.9).

A hypothesis usually becomes accepted as a theory only after years of testing by many scientists. Then, if the hypothesis has not been disproved, scientists may feel confident about using it to explain more data or observations. The theories of evolution and natural selection—topics we will look at in Chapter 23—are prime examples of "theories" that are supported by tens of thousands of scientific observations.

Science demands critical thinking, so a theory can be modified, and even rejected, if results of new scientific tests call it into question. It's the same with other scientific ideas. Today, for instance, advances in technology are giving us a new perspective on subjects such as the links between emotions and health. Some "facts" in this textbook one day will likely be revised as we learn more about various processes. This willingness to reconsider ideas as new information comes to light is a major strength of science.

Science has limits

Because science requires an objective mindset, scientists can only do certain kinds of studies. No experiment can explain the "meaning of life," for example, or why each of us dies at a certain moment. Such questions have *subjective* answers that are shaped by our experiences and beliefs. Every culture and society has its own standards of morality and esthetics, and there are probably thousands of different sets of religious beliefs. All guide their members in deciding what is important and morally good and what is not. By contrast, the external world, rather than internal conviction, is the only testing ground for scientific views.

TABLE 1.2 Examples of Scientific Theories

CELL THEORY	All organisms consist of one or more cells, the cell is the basic unit of life, and all cells arise from existing cells.
GERM THEORY	Germs cause infectious diseases.
THEORY OF EVOLUTION	Change can occur in lines of descent.
THEORY OF NATURAL SELECTION	Variation in heritable traits influences which individuals of a population reproduce in each generation.

© Bettmann/Corbis

Figure 1.9 In the 1300s, people tried all sorts of strategies to ward off the bubonic plague epidemic—the Black Death—that may have killed half the people in Europe.

Because science does not involve value judgments, it sometimes has been or can be used in controversial pursuits. For example, some people worry about issues such as the use of animals in scientific research and possible negative consequences of genetic modification of food plants. There has been great debate over the causes of global climate change and the use of "industrial" fishing methods on the high seas. Meanwhile, whole ecosystems are being altered by technologies that each year allow millions of a forest's trees to be cut and hundreds of millions of fishes to be taken from the sea. These are matters we can't leave to the scientific community alone to resolve. That responsibility also belongs to us.

scientific theory A thoroughly tested explanation of a broad range of natural events and observations.

TAKE-HOME MESSAGE

WHAT ARE THE STRENGTHS AND LIMITS OF SCIENTIFIC STUDY?

- Science applies to questions and problems that can be tested objectively.
- A scientific theory remains open to tests, revision, and even rejection if new evidence comes to light.
- Responsibility for the wise use of scientific information must be shared by all.

■ **Every chapter in this textbook contains one or more** *Focus* **features that give you added insight into major health and medical topics, environmental problems such as water pollution, or social concerns such as genetic profiling. We begin in this chapter with a** *Focus on Health* **that introduces one of the most pressing modern public health issues, the global threat of infectious disease.**

antibiotic A substance that can kill microorganisms.

emerging disease A disease caused by a pathogen that until recently did not infect humans, or did so only rarely.

Humans have always lived with countless health threats, but today people everywhere are locked in an escalating global battle with pathogens—bacteria, viruses, and parasites that can cause disease. Most pathogens are invisible to the naked eye. Figure 1.10 gives you an idea of what some of these foes look like under the microscope. You will learn much more about disease pathogens in many of the chapters to follow.

Emerging diseases present new challenges

Today health officials worry especially about **emerging diseases**. These diseases are caused by pathogens that until recently did not infect humans or were present only in limited areas. Many are caused by viruses. This group includes the encephalitis caused by West Nile virus and the severe respiratory disease caused by the SARS virus (Figure 1.10C). Other examples are "hemorrhagic fevers" that cause massive bleeding. In this latter group are dengue fever and the illness caused by the Ebola virus. You have probably heard of Lyme disease, which is a major emerging disease

in the United States. It is caused by the bacterium *Borrelia burgdorferi*, which is transmitted by ticks when they suck blood.

Why is all this happening? A few factors stand out. For one, there are simply many more of us on the planet, interacting with our surroundings and with each other. Each person is a potential target for pathogens. Also, more people are traveling, carrying diseases along with them. Another important factor is the misuse and overuse of antibiotics.

Antibiotics are a major weapon against infectious disease

Antibiotics were developed in the 1940s, when much of the world was engulfed in World War II, and they were soon harnessed to fight disease (Figure 1.11). An **antibiotic** (literally, "against a living thing") can destroy living organisms such as bacteria and some other microorganisms, or prevent them from growing. Most antibiotics are natural substances produced by bacteria and fungi. The penicillins and some other antibiotics kill microbes by interfering with different cell processes that you will read about in Chapter 3.

You may already know that antibiotics don't work against viruses, which are not cells and so do not have "life processes." Some of the body defenses you will read about in Chapter 9 may prevent certain viruses from multiplying inside cells. Antiviral drugs interfere with the viral "life cyle" in some way.

Antibiotics can have side effects. For example, some trigger an allergic response in susceptible people. Others can discolor teeth or reduce the effectiveness of birth control pills.

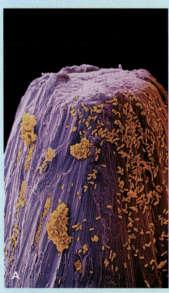

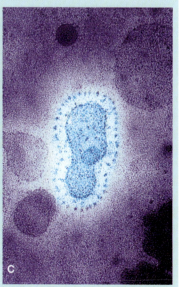

Figure 1.10 A wide variety of pathogens may live on or in the human body. A Bacteria on the tip of a pin. **B** *Trypanosoma brucei*, microscopic protozoan that causes African sleeping sickness—a disease that afflicts millions of people in Africa. It is shown next to a red blood cell. **C** The SARS virus, which causes an emerging respiratory disease.
A: © Dr. Tony Brain & David Parker/Photo Researchers, Inc. B: © Eye of Science/Photo Researchers, Inc. C: © Sercomi/Photo Researchers, Inc.

Increasing resistance to antibiotics is a major public health Issue

Today antibiotic resistance is a major and growing public health problem. Drug-resistant bacteria already include strains that cause some cases of tuberculosis, strep throat, sexually transmitted diseases (STDs) such as syphilis and gonorrhea, childhood middle-ear infections, urinary tract infections, and infections of surgical wounds.

Several factors have contributed to the emergence of bacteria that are genetically resistant to antibiotics that might otherwise have killed them. Pressure from patients led some doctors to prescribe antibiotics for people who had viral illnesses. In nations where antibiotics are not prescription drugs, people may take the drugs without getting medical advice. Some patients stop taking an antibiotic when they start to feel better, not finishing the full recommended course of treatment. As a result, more-resistant microbes may not be killed. Antibiotics have long been fed to livestock used for human food and have been added to consumer products like soaps and wipes.

The future will bring increased efforts to stem the tide of improper antibiotic use, especially in poorer nations, where antibiotic resistance is growing the fastest. Researchers will try to develop new therapies using a "cocktail" of several drugs instead of a single one. And pressure will mount for pharmaceutical companies to develop new antibiotics, even if that effort is not where the biggest financial profits lie.

Figure 1.11 Penicillin saved the lives of many soldiers in World War II. This ad is from a 1944 issue of *Life* magazine.

1.8 Homeostasis

CONNECTIONS

Section 1.1 introduced the concept of homeostasis—the state of chemical and physical stability inside the body that must exist if cells, and the whole body, are to stay alive. Homeostasis is one of the most important concepts in this textbook. Figure 1.12 is a visual summary of the main ideas, using the muscular system as an example. Each body system you will study during your human biology course performs functions that contribute to homeostasis in other systems. Those chapters conclude with a *Connections* section that summarizes each system's key contributions to homeostasis.

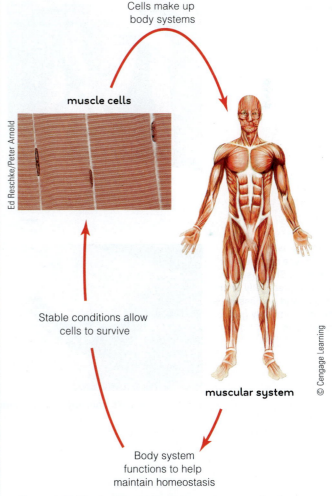

Figure 1.12 The body's survival depends on mechanisms that maintain internal homeostasis.

EXPLORE
ON YOUR OWN

As you read in Section 1.4, having a sample of test subjects or observations that is too small can skew the results of experiments. This phenomenon is called *sampling error*. To demonstrate this for yourself, all you need is a partner, a blindfold, and a jar containing beans of different colors—jelly beans will do just fine (Figure 1.13). Have your partner stay outside the room while you combine 120 beans of one color with 280 beans of the other color in a bowl. This will give you a ratio of 30 to 70 percent. With the bowl hidden, blindfold your partner; then ask him or her to pick one bean from the mix. Hide the bowl again and instruct your friend to remove the blindfold and tell you what color beans are in the bowl, based on this limited sample. The logical answer is that all the beans are the color of the one selected.

Next repeat the trial, but this time ask your partner to select 50 beans from the bowl. Does this larger sample more closely approximate the actual ratio of beans in the bowl? You can do several more trials if you have time. Do your results support the idea that a larger sample size more closely reflects the actual color ratio of beans?

A Natalie, blindfolded, randomly plucks a jelly bean from a jar of 120 green and 280 black jelly beans, a ratio of 30 to 70 percent.

C Still blindfolded, Natalie randomly picks 50 jelly beans from the jar and ends up with 10 green and 40 black ones.

B The jar is hidden before she removes her blindfold. She observes a single green jelly bean in her hand and assumes the jar holds only green jelly beans.

D The larger sample leads her to assume one-fifth of the jar's jelly beans are green and four-fifths are black (a ratio of 20 to 80 percent). Her larger sample more closely approximates the jar's green-to-black ratio. The more times Natalie repeats the sampling, the greater the chance she will come close to knowing the actual ratio.

All Photos: © Cengage Learning/Gary Head

Figure 1.13 Here's one way you can demonstrate sampling error.

SUMMARY

Section 1.1 Humans have the characteristics found in all forms of life, as listed in Table 1.3.

Section 1.2 All life on Earth has come about through a process of evolution. The defining features of humans include a large and well-developed brain, great manual dexterity, sophisticated skills for language and mental analysis, and complex social behaviors.

Section 1.3 The living world is highly organized. Atoms, molecules, cells, tissues, organs, and organ systems make up whole, complex organisms. Each organism is a member of a population, populations live together in communities, and communities form ecosystems. The biosphere is the most inclusive level of biological organization. A continual flow of energy and cycling of raw materials sustains the organization of life.

Section 1.4 Science is an approach to gathering knowledge. There are numerous versions of the scientific method. Table 1.4 lists elements that are important in all of them. Reputable scientists must draw conclusions that are not at odds with the evidence used to support them.

Section 1.5 Critical thinking skills include scrutinizing information sources for bias, seeking reliable opinions, and separating the causes of events from factors that may only be associated with them.

Sections 1.6, 1.7 A scientific theory is a thoroughly tested explanation of a broad range of related phenomena. Science does not address subjective issues, such as religious beliefs and morality.

REVIEW QUESTIONS

1. You are a living organism. Which characteristics of life do you exhibit?

2. Why is the concept of homeostasis meaningful in the study of human biology?

3. What is meant by biological evolution?

TABLE 1.3 Summary of Life's Characteristics

1. Living things take in and use energy and materials.
2. Living things sense and respond to changes in their surroundings.
3. Living things consist of one or more cells.
4. Living things maintain the internal steady state called homeostasis.
5. Living things reproduce and grow based on information in DNA.

TABLE 1.4 Scientific Method Review

HYPOTHESIS	Possible explanation of a natural event or observation
PREDICTION	Proposal or claim of what testing will show if a hypothesis is correct
EXPERIMENT	Controlled procedure to gather observations that can be compared to prediction
CONTROL GROUP	Standard to compare test group against
VARIABLE	Aspect of an object or event that may differ with time or between subjects
CONCLUSION	Statement that evaluates a hypothesis based on test results

4. Study Figure 1.5. Then summarize what biological organization means.

5. Define and distinguish between:
    a. a hypothesis and a scientific theory
    b. an experimental group and a control group

SELF-QUIZ *Answers in Appendix V*

1. Instructions in _____ govern how organisms are built and function.

2. A _____ is the smallest unit that can live and reproduce by itself using energy, raw materials, and DNA instructions.

3. _____ is a state in which an organism's internal environment is maintained within a tolerable range.

4. Humans are _____ (animals with backbones); like other primates, they also are _____.

5. Starting with cells, nature is organized on at least _____ levels.

6. A scientific approach to explaining some aspect of the natural world includes all of the following except _____.
   a. a hypothesis
   b. testing
   c. faith-based views
   d. systematic observations

7. A controlled experiment should have all the following features except _____.
   a. a control group
   b. a test subject
   c. a variable
   d. many testable predictions

8. A related set of hypotheses that collectively explain some aspect of the natural world makes up a scientific _____ .
   a. prediction
   b. test
   c. theory
   d. authority
   e. observation

9. Which of the following is not a feature of a scientific theory?
   a. It begins as a hypothesis.
   b. It eventually is accepted as absolute truth.
   c. It requires critical thinking.
   d. It is is not accepted as a theory until it has been tested repeatedly.

10. The diagram below depicts the concept of _____ .
    a. evolution
    b. reproduction
    c. levels of organization
    d. energy transfers in the living world

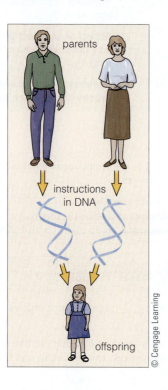

parents

instructions in DNA

offspring

© Cengage Learning

CRITICAL THINKING

1. The diagram to the right shows how tiles can be put together in different ways. How does this example relate to the role of DNA as the universal genetic material in organisms?

Your Future

Every day, scientists around the world are looking for answers to questions relating to human health, medicine, environmental issues, or other concerns that may affect your life. At the end of each chapter *Your Future* gives you a quick preview of what the future is likely to bring with respect to an issue or concern related to the chapter's content.

© Chris Knapton/Alamy

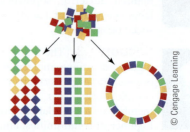

© Cengage Learning

2. Court witnesses are asked "to tell the truth, the whole truth, and nothing but the truth." Research shows, however, that eyewitness accounts of crimes often are unreliable because even the most conscientious witnesses misremember details of what they observed. What other factors that might affect the "truth" a court witness presents?

3. Design a test (or series of tests) to support or refute this hypothesis: People who have no family history of high blood pressure (hypertension) but who eat a diet high in salt are more likely to develop high blood pressure than people with a similar family history but whose diet is much lower in salt.

4. In a popular magazine article the author reports health benefits attributed to a particular dietary supplement. What kinds of evidence should the article cite to help you decide whether the information is likely to be accurate?

5. Researchers studied 393 patients in a hospital's coronary care unit. In the experiment, volunteers were asked to pray daily for a patient's rapid recovery and for the prevention of complications and death.

   None of the patients knew if he or she was being prayed for. None of the volunteers or patients knew each other. The research team categorized how each patient fared as "good," "intermediate," or "bad." They concluded that "prayed for" patients fared a little better than other patients—the experiment having documented results that seemed to support the prediction that prayer might have beneficial effects for seriously ill patients.

   The results brought a storm of criticism, mostly from scientists who cited bias in the experimental design. For instance, the patients were categorized after the experiment was over, instead of as they were undergoing treatment, so the team already knew which ones had improved, stayed about the same, or gotten worse. Why do you suppose the experiment generated a heated response from many in the scientific community? Can you think of at least one other variable that might have affected the outcome of each patient's illness?

CHEMISTRY OF LIFE

KEY CONCEPTS

Atoms and Elements

Atoms are the basic units of matter. Each chemical element consists of a single type of atom. Bonds between atoms form molecules. Sections 2.1–2.4

Water and Body Fluids

Life depends on properties of water. Substances dissolved in the water of body fluids have major effects on all body functions. Sections 2.5–2.7

Biological Molecules

Biological molecules include carbohydrates, lipids, proteins, and nucleic acids. All contain atoms of the element carbon. Sections 2.8–2.13

Top: © Cengage Learning Middle: somchaij/Shutterstock.com
Bottom: © Photos.com/JupiterImages

Trans fats long were a key ingredient in some vegetable oils used to prepare store-bought baked goods, fast-food French fries, microwave popcorn, fried chicken, and so on. Then they started to get a bad reputation. It was well deserved. Many research studies were showing that trans fats raise the level of a harmful form of cholesterol in human blood more than any other fat. We now know that they also alter the structure of our blood vessels in harmful ways, increasing the risk of heart disease. Today, government regulations require manufacturers of prepared foods to list the trans fat content on food labels. Some restaurants, including popular fast-food outlets, do the same. As a result, consumers who want to "eat healthy" can avoid trans fats much more easily.

In studying human biology, it's useful to understand some basic chemistry, in part because chemical reactions explain the effects—harmful or helpful—of substances we take into our bodies. Unlike trans fats, many substances we consume are indispensable in the chemical events that build cell parts and allow them to function properly. In this chapter you will learn some simple chemical basics that will help you understand topics of later chapters, such as how certain substances serve as vital nutrients while some others are health hazards.

Exactostock/SuperStock

Homeostasis Preview

In this chapter we discuss two topics that bear directly on the body's ability to maintain the internal stability of homeostasis. These are the properties of water and changes in the chemical makeup of body fluids.

2.1 | Atoms and Elements

- Pure substances called elements are the basic raw material of living things.
- Each element consists of one type of atom.
- The parts of atoms determine how the molecules of life are put together.
- Link to Life's organization 1.3

Elements are pure substances

Like all else on Earth, your body consists of chemicals, some of them solids, others liquid, still others gases. Each of these chemicals consists of one or more elements. An **element** is a pure substance that cannot be broken down to another substance by ordinary physical or chemical techniques. There are more than ninety natural elements on Earth, and scientists have created many other artificial ones.

atom The smallest unit having the properties of a given element.

element A pure substance that cannot be broken down to another substance by ordinary chemical or physical techniques.

Overall, organisms consist mostly of four elements: oxygen, carbon, hydrogen, and nitrogen. The human body also contains some calcium, phosphorus, potassium, sulfur, sodium, and chlorine, plus many different trace elements (Figure 2.1). A trace element is one that makes up less than 0.01 percent of body weight. Trace elements still are vital, however. For example, your red blood cells can't carry oxygen without the trace element iron. The body's chemical makeup is finely tuned. Many trace elements found in our tissues—such as arsenic, selenium, and fluorine—are toxic in amounts larger than normal.

Atoms of the same or different elements can combine into molecules—the first step in biological organization. Molecules in turn can combine to form larger structures, as described shortly.

Atoms are composed of smaller particles

An **atom** is the smallest unit that has the properties of a given element. A million could fit on the period at the end of this sentence. In spite of their tiny size, however, all atoms are composed of more than one hundred kinds of subatomic particles. The ones we are concerned with in this book are protons, electrons, and neutrons, illustrated in Figure 2.2.

All atoms have one or more protons, which carry a positive charge, marked by a plus sign (p^+). Atoms also have one or more neutrons, which have no charge. Neutrons and protons make up the atom's core, the atomic nucleus. Electrons move around the nucleus, in the space that occupies 99.99 percent of the atom's volume. Electrons have a negative charge, which we write as e^-. An atom usually has equal numbers of electrons and protons.

Each element is assigned an "atomic number," which is the number of protons in its atoms. Elements also have a "mass number"—the sum of the protons and neutrons

Human		Earth's crust	
Oxygen	65	Oxygen	46.6
Carbon	18	Silicon	27.7
Hydrogen	10	Aluminum	8.1
Nitrogen	3	Iron	5.0
Calcium	2	Calcium	3.6
Phosphorus	1.1	Sodium	2.8
Potassium	0.35	Potassium	2.6
Sulfur	0.25	Magnesium	2.1
Sodium	0.15	Other elements	1.5
Chlorine	0.15		
Magnesium	0.05		
Iron	0.004		
Iodine	0.0004		

Cape Verde National Institute of Meteorology and Geophysics and the U.S. Geological Survey

Figure 2.1 Everything in the biosphere, from humans to the Earth's crust, is made of elements.

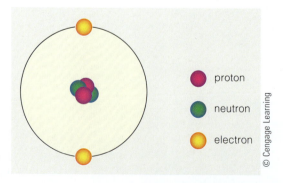

proton

neutron

electron

© Cengage Learning

Figure 2.2 Animated! **Atoms consist of subatomic particles.** This model does not show what an atom really looks like. Electrons travel in spaces located around a nucleus of protons and neutrons. These spaces are about 10,000 times larger than the nucleus.

in the nucleus of their atoms. Appendix II of this textbook has charts of the elements and of the atomic numbers of the common elements in living things.

Isotopes are varying forms of atoms

All atoms of a given element have the same number of protons, but they may *not* have the same number of neutrons. When an atom of an element has more or fewer neutrons than the most common number, it is called an **isotope** (EYE-so-tope). For instance, while a "standard" carbon atom will have six protons and six neutrons, the isotope called carbon 14 has six protons and *eight* neutrons. These two forms of carbon atoms also can be written as ^{12}C and ^{14}C. The prefix *iso-* means "same," and all isotopes of an element interact with other atoms in the same way. Most elements have at least two isotopes. Cells can use any isotope of an element for their metabolic activities, because the isotopes behave the same as the standard form of the atom in chemical reactions.

Have you heard of radioactive isotopes? A French scientist discovered them in 1896, after he had set a chunk of rock on top of an unexposed photographic plate in a desk drawer. The rock contained isotopes of uranium, which emit energy. This unexpected chemical behavior is what we today call radioactivity.

The nucleus of a **radioisotope** is unstable, but it stabilizes itself by emitting energy and certain types of particles. This process, called radioactive decay, takes place spontaneously, and it transforms a radioisotope into an atom of a different element. The decay process happens at a known rate. For instance, over a predictable time span, potassium 40 becomes argon 40. Scientists can use radioactive decay rates to determine the age of very old substances, such as ancient rocks and fossils.

isotope An atom of an element that has a different number of neutrons than the most common, standard number.

radioisotope An isotope with an unstable nucleus that becomes stable by emitting energy and particles, a process known as radioactive decay.

tracer Molecule with a detectable substance such as a radioisotope attached to it.

2.2 PET Scanning—Using Radioisotopes in Medicine

SCIENCE COMES TO LIFE

Emissions from radioisotopes can reveal the activity of body cells. As a result, they are useful tools in medicine, because they permit physicians to diagnose disease, or track its course, without doing surgery.

The technology called PET (short for Positron Emission Tomography) is a prime example. Figure 2.3A shows a PET scan from a cancer patient. The patient was injected with a **tracer**—a molecule in which radioisotopes have been substituted for some atoms. The cells in a cancerous tumor are more active than normal body cells, so they take up the tracer faster. A scanner then detects radioactivity that becomes concentrated in the tumors. Figure 2.3B shows a PET scan from two subjects, one of them a smoker and one a nonsmoker. Researchers did the scan to obtain an image of how smoking may change the activity of a substance that is important to normal functioning of body organs such as the brain, heart, and kidneys.

Radioisotopes also are used to help treat

some cancers. For safety's sake, such treatments use only radioisotopes that decay quickly into a different, more stable element.

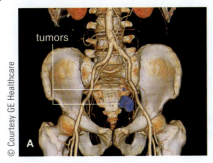
© Courtesy GE Healthcare

tumors

A

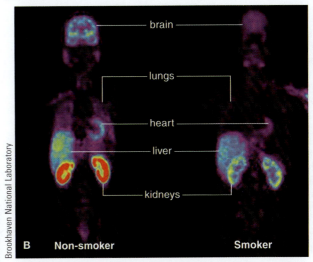
Brookhaven National Laboratory

brain

lungs

heart

liver

kidneys

B   Non-smoker   Smoker

Figure 2.3 Radioisotopes have important medical uses. A PET image showing tumors (*blue*) in and near the bowel of a cancer patient. **B** PET scans showing the activity of a substance in a nonsmoker (*left*) and a smoker (*right*). The activity is color-coded from red (highest) to purple (lowest).

- Atoms may share, give up, or gain electrons.
- Whether an atom will interact with other atoms depends on how many electrons it has.
- Chemical bonds between atoms form molecules.

Atoms interact through their electrons

chemical bond A union between the electron structures of atoms.

compound A molecule containing atoms of two or more elements in proportions that are always the same.

mixture A substance in which two or more kinds of molecules mingle in proportions that may vary.

molecule The structure that is formed when chemical bonding joins atoms.

By way of their electrons, atoms of many elements interact with other atoms. Electrons may be shared, one atom may donate one or more electrons to another atom, or an atom may receive electrons from other atoms. Which of these events takes place depends on how many electrons a given atom has and how the electrons are arranged.

You've probably heard that like charges (++ or −−) repel each other and unlike charges (+−) attract. Electrons carry a negative charge, so they are attracted to the positive charge of protons. On the other hand, electrons repel each other. In an atom, electrons respond to these pushes and pulls by moving around the atomic nucleus in "shells" (Figure 2.4). A shell has three dimensions, like the space inside a balloon, and the electron or electrons inside it travel in "orbitals." Each orbital is like a room that can hold no more than two occupants. This means that in an atom, a maximum of two electrons can occupy an orbital. Recall from Section 2.1 that atoms of different elements differ in how many electrons they have. They also differ in how many of their "rooms" are filled.

Hydrogen is the simplest atom. It has one electron in a single shell (Figure 2.4A). In atoms of other elements, the first shell holds two electrons. Any additional electrons are in shells farther from the nucleus.

The shells around an atom's nucleus are equivalent to energy levels. The shell closest to the nucleus is the lowest energy level. Each shell farther out from the nucleus is at a progressively higher energy level. Because the atoms of different elements have different numbers of electrons, they also have different numbers of shells that electrons can occupy (Figure 2.4B, C). A shell can have up to eight electrons, but not more. This means that larger atoms, which have more electrons than smaller ones do, also have more shells.

A The first shell corresponds to the first energy level, and it can hold up to 2 electrons. Hydrogen has one proton, so it has 1 electron and 1 vacancy. A helium atom has 2 protons, 2 electrons, and no vacancies. The number of protons in each model is shown.

first shell

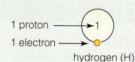
1 proton
1 electron
hydrogen (H)

helium (He)

B The second shell corresponds to the second energy level, and it can hold up to 8 electrons. Carbon has 6 protons, so its first shell is full. Its second shell has 4 electrons and four vacancies. Oxygen has 8 protons and two vacancies. Neon has 10 protons and no vacancies.

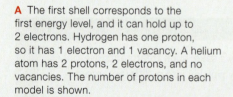
second shell

carbon (C)

oxygen (O)

neon (Ne)

C The third shell, which corresponds to the third energy level, can hold up to 8 electrons. A sodium atom has 11 protons, so its first two shells are full; the third shell has one electron. Thus, sodium has seven vacancies. Chlorine has 17 protons and one vacancy. Argon has 18 protons and no vacancies.

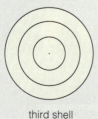
third shell

sodium (Na)

chlorine (Cl)

argon (Ar)

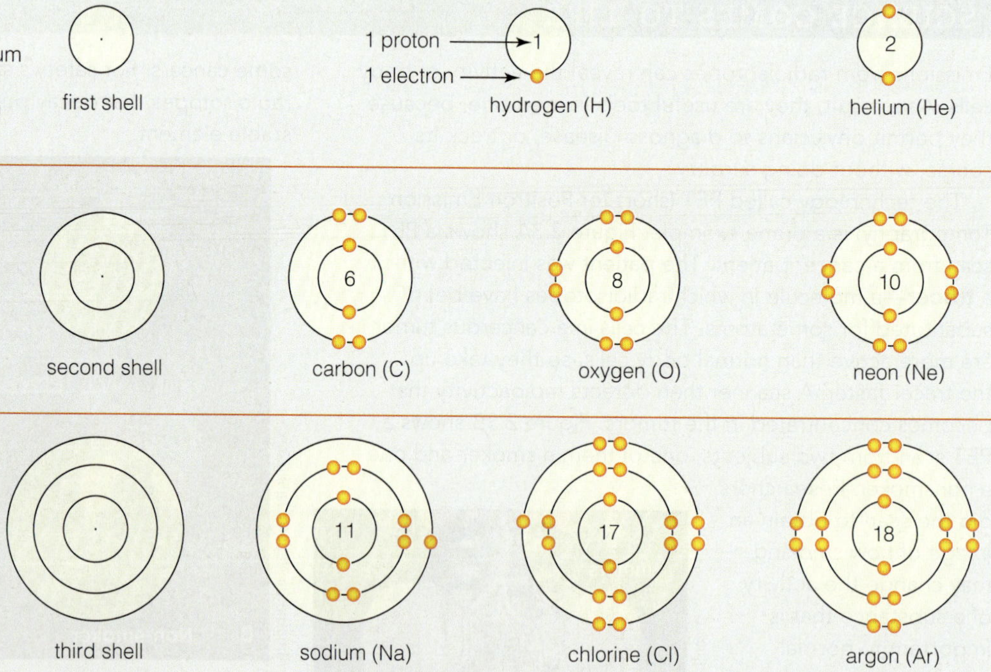

Figure 2.4 Animated! **The shell model helps you visualize the vacancies in an atom's outer orbitals.** Each circle represents all of the orbitals on one energy level. The larger the circle, the higher the energy level. (© Cengage Learning)

Chemical bonds join atoms into molecules

When the electron structures of atoms unite, the union is called a **chemical bond**. This chemical bonding joins atoms into a new type of structure, a **molecule** (Table 2.1).

Bonds form because an atom is most stable when its outer shell is filled. For atoms that have too few electrons to fill their outer shell, chemical bonding with other atoms can provide stability. As shown in Figure 2.4A, hydrogen and helium atoms have a single shell. It is full when it contains two electrons. Some other kinds of atoms that have unfilled outer shells tend to form chemical bonds that fill vacant "slots" in their outer shell so that it has a full set of eight electrons. Atoms of oxygen, carbon, hydrogen, and nitrogen—the most abundant elements in the body—are in this category. Look for electron vacancies in an atom's outer shell and you will always have a clue as to whether the atom will bond with others.

In Figure 2.4 you can count the electron vacancies in the outer shell of each of the atoms pictured. Atoms like helium, which have no vacancies, are said to be *inert*. They usually don't take part in chemical reactions.

Molecules may contain atoms of a single element or of different elements

Many molecules contain atoms of only one element. Molecular nitrogen (N_2), with its two nitrogen atoms, is an example. Many other molecules are **compounds**—they combine two or more elements in proportions that never vary. For example, water is a compound. No matter where water molecules are—in a lake or your bathtub—each one always has one oxygen atom bonded to two hydrogen atoms. Figure 2.5 explains how to read the notation used in representing chemical reactions that occur between atoms and molecules.

In a **mixture**, two or more kinds of molecules simply mingle. The proportions may or may not be the same. For example, the sugar sucrose is a compound of carbon, hydrogen, and oxygen. If you swirl together molecules of sucrose and water, you'll get a mixture—sugar-sweetened water. If you keep the same amount of water but add more sucrose you will still have a mixture—just an extremely sweet one, such as syrup.

TABLE 2.1 Different Ways to Represent the Same Molecule

COMMON NAME	Water	Familiar term.
CHEMICAL NAME	Hydrogen oxide	Describes the elements making up the molecule.
CHEMICAL FORMULA	H_2O	Indicates proportions of elements. Subscripts show number of atoms of an element per molecule. There is no subscript when only one atom is present.
STRUCTURAL FORMULA	H—O—H	Represents a bond as a single line between atoms. The bond angles also may be represented.
STRUCTURAL MODEL		Shows the positions and relative sizes of atoms.
SHELL MODEL		Shows how pairs of electrons are shared.

We use symbols for elements when writing *formulas*, which identify the composition of compounds. For example, water has the formula H_2O. Symbols and formulas are used in *chemical equations*, which are representations of reactions among atoms and molecules.

In written chemical reactions, an arrow means "yields." Substances entering a reaction (reactants) are to the left of the arrow. Reaction products are to the right. For example, the reaction between hydrogen and oxygen that yields water is summarized this way:

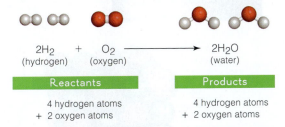

$$2H_2 + O_2 \longrightarrow 2H_2O$$
(hydrogen)   (oxygen)   (water)

Reactants	Products
4 hydrogen atoms + 2 oxygen atoms	4 hydrogen atoms + 2 oxygen atoms

Note that there are as many atoms of each element to the right of the arrow as there are to the left. Although atoms are combined in different forms, none is consumed or destroyed in the process. The total mass of all products of any chemical reaction equals the total mass of all its reactants. All equations used to represent chemical reactions, including reactions in cells, must be balanced this way.

Figure 2.5 Symbols are a shorthand way to describe chemical reactions. (© Cengage Learning)

2.4 Important Bonds in Biological Molecules

■ The characteristics of atoms determine which types of bonds form in biological molecules.

An ionic bond joins atoms that have opposite electrical charges

Overall, an atom carries no charge because it has as many electrons as protons. That balance can change if an atom has a vacancy—an unfilled orbital—in its outer shell. For example, a chlorine atom has one vacancy and therefore can gain one electron. A sodium atom, on the other hand, has a single electron in its outer shell, and that electron can be knocked out or pulled away. When an atom gains or loses an electron, the balance between its protons and its electrons shifts so that it has a positive or negative charge. An atom or other particle that has a charge is called an **ion**.

It's common for neighboring atoms to accept or donate electrons among one another. When one atom loses an electron and one gains, both become ionized. Depending on conditions inside the cell, the ions may separate, or they may stay together as a result of the mutual attraction of their opposite charges. An association of two ions that have opposite

biological molecule A molecule that contains carbon and that is formed in a living organism.

covalent bond A bond in which the atoms share two electrons.

hydrogen bond A weak attraction between a covalently bound hydrogen atom and an electronegative atom in the same or a different molecule.

ion A particle that has a charge, either positive or negative.

ionic bond An association between two ions that have opposite charges.

charges is called an **ionic bond**. Figure 2.6 shows how sodium ions (Na^+) and chloride ions (Cl^-) interact through ionic bonds, forming NaCl, or table salt.

The process in which an atom or molecule loses one or more electrons to another atom or molecule is known as *oxidation*. It's what causes a match to burn and an iron nail to rust, and it is part of all kinds of important metabolic events in body cells.

In a covalent bond, atoms share electrons

In a **covalent bond**, atoms *share* two electrons (Figure 2.7). The bond forms when two atoms each have a lone electron in their outer shell and each atom's attractive force "pulls" on the other's unpaired electron. The tug is not strong enough to pull an electron away completely, so the two electrons occupy a shared orbital. Covalent bonds are extremely stable.

As you saw in Table 2.1, in structural formulas a single line between two atoms means they share a single covalent bond. Molecular hydrogen, a molecule that consists of two hydrogen atoms, has this kind of bond and can be written as H—H. In a *double* covalent bond, two atoms share two electron pairs, as in an oxygen molecule (O=O). In a *triple* covalent bond, two atoms share three pairs of electrons. A nitrogen molecule (N≡N) is this way. All three examples are gases. When you breathe, you inhale H_2, O_2, and N_2 molecules.

In a *nonpolar* covalent bond, the two atoms pull equally on electrons and so share them equally. The term "nonpolar" means there is no difference in charge at the two ends

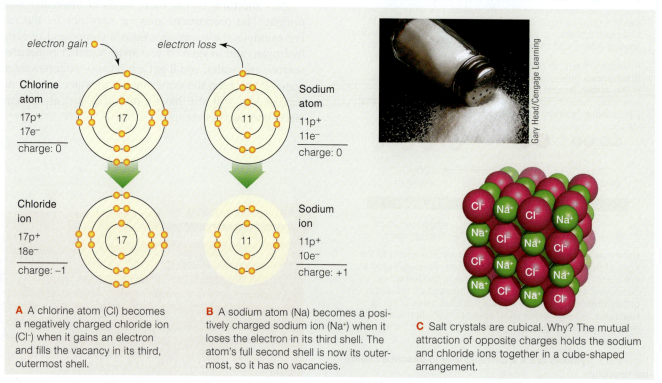

A A chlorine atom (Cl) becomes a negatively charged chloride ion (Cl^-) when it gains an electron and fills the vacancy in its third, outermost shell.

B A sodium atom (Na) becomes a positively charged sodium ion (Na^+) when it loses the electron in its third shell. The atom's full second shell is now its outermost, so it has no vacancies.

C Salt crystals are cubical. Why? The mutual attraction of opposite charges holds the sodium and chloride ions together in a cube-shaped arrangement.

Figure 2.6 Animated! An ionic bond may form between two oppositely charged atoms. (© Cengage Learning)

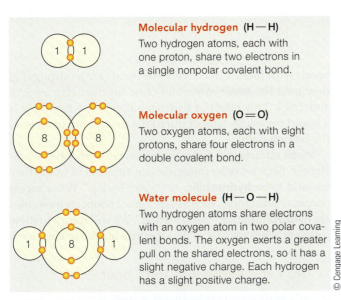

Molecular hydrogen (H—H)
Two hydrogen atoms, each with one proton, share two electrons in a single nonpolar covalent bond.

Molecular oxygen (O=O)
Two oxygen atoms, each with eight protons, share four electrons in a double covalent bond.

Water molecule (H—O—H)
Two hydrogen atoms share electrons with an oxygen atom in two polar covalent bonds. The oxygen exerts a greater pull on the shared electrons, so it has a slight negative charge. Each hydrogen has a slight positive charge.

© Cengage Learning

Figure 2.7 Animated! Shared electrons make up covalent bonds. Two atoms with unpaired electrons in their outer shell become more stable by sharing electrons. Two electrons are shared in each covalent bond. When the electrons are shared equally, the covalent bond is nonpolar. If one atom exerts more pull on the shared electrons, the covalent bond is polar.

("poles") of the bond. Molecular hydrogen is a simple example. Its two hydrogen atoms, each with one proton, attract the shared electrons equally.

In a *polar* covalent bond, two atoms do not share electrons equally. The atoms are of different elements, and one has more protons than the other. The one with the most protons pulls more, so its end of the bond ends up with a slight negative charge. We say it is "electronegative." The atom at the other end of the bond ends up with a slight positive charge. For instance, a water molecule (H—O—H) has two polar covalent bonds. The oxygen atom carries a slight negative charge, and each of the two hydrogen atoms has a slight positive charge.

A hydrogen bond is a weak bond between polar molecules

A **hydrogen bond** is a weak attraction that has formed between a covalently bound hydrogen atom and an electronegative atom in a different molecule or in another part of the same molecule (Figure 2.8A). The dotted lines in Figure 2.8B represent this link.

Individual hydrogen bonds are weak, so they form and break easily. Despite this property, hydrogen bonds are vital in **biological molecules**—molecules that contain carbon and that are formed in living things. For example, the genetic material DNA is built of two parallel strands of chemical units, and the strands are held together by hydrogen bonds, as shown in Figure 2.8B. In Section 2.5 you will learn how hydrogen bonds between water molecules contribute to properties of water that make it essential for life.

Table 2.2 summarizes the basic characteristics of hydrogen bonds and the other main chemical bonds in biological molecules.

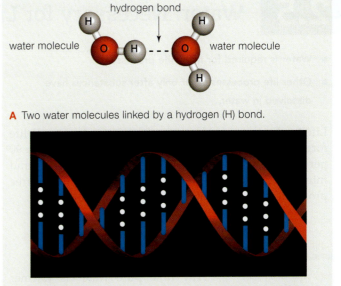

A Two water molecules linked by a hydrogen (H) bond.

B Hydrogen bonds (*white dots*) hold the two coiled-up strands of a DNA molecule together. Each H bond is weak, but collectively they are strong.

Figure 2.8 Animated! Hydrogen bonds can form when a hydrogen atom is already covalently bonded in a molecule. The hydrogen's slight positive charge weakly attracts an atom with a slight negative charge that is already covalently bonded to something else. **A** The hydrogen atoms of water molecules interact in hydrogen bonds. **B** In large molecules such as DNA, the many hydrogen bonds help stabilize the molecule's shape.
(© Cengage Learning)

TABLE 2.2 Major Chemical Bonds in Biological Molecules

Bond	Characteristics
IONIC	Joined atoms have opposite charges.
COVALENT	Strong; joined atoms share electrons. In a *polar* covalent bond one end is slightly positive, the other slightly negative.
HYDROGEN	Weak; joins a hydrogen (H⁺) atom in one polar molecule with an electronegative atom in another polar molecule.

TAKE-HOME MESSAGE

WHAT ARE THE MAIN TYPES OF CHEMICAL BONDS THAT OCCUR IN BIOLOGICAL MOLECULES?

- Biological molecules are formed mainly by ionic bonds, covalent bonds, and hydrogen bonds.
- In an ionic bond, ions of opposite charge attract each other and stay together.
- In a covalent bond, atoms share electrons. If the electrons are shared equally, the bond is nonpolar. If the sharing is not equal, the bond is polar—slightly positive at one end, slightly negative at the other.
- In a hydrogen bond, a covalently bound hydrogen atom attracts a small, negatively charged atom in a different molecule or in another part of the same molecule.

- Water is required for many life processes.
- Other life processes occur only after substances have dissolved in water.

Life on Earth probably began in water, and for all life forms it is indispensable. Human blood is more than 90 percent water, and water helps maintain the shape and internal structure of our cells. As described next, three properties of water suit it for its key roles in the body.

hydrophilic Chemically attracted to water.

hydrophobic Chemically repelled by water.

Hydrogen bonding makes water liquid

Any time pure water is warmer than about 32°F or cooler than about 212°F, it is a liquid. Therefore it is a liquid at body temperature; our watery blood flows, and our cells have the fluid they need to maintain their structural integrity and to function properly. What keeps water liquid? You may recall that while a water molecule has no net charge, it does carry charges that are distributed unevenly. The water molecule's oxygen end is slightly negative and its hydrogen end is a bit positive (Figure 2.9A). This uneven distribution of charges makes water molecules "polar." Because they are polar, the molecules can attract other water molecules and form hydrogen

bonds with them. Collectively, the bonds are so strong that they hold the water molecules close together (Figure 2.9B and 2.9C). This effect of hydrogen bonds is why water is a liquid unless its temperature falls to freezing or rises to the boiling point.

Water attracts and hydrogen-bonds with other polar substances. Because polar molecules are attracted to water, they are said to be **hydrophilic**, or "water loving." Water repels nonpolar substances, such as oils. Hence nonpolar molecules are **hydrophobic**, or "water fearing." We will return to these concepts when we look at the structure of cells in Chapter 3.

Water can absorb and hold heat

Water's hydrogen bonds give it a high *heat capacity*—the ability to absorb a great deal of heat energy before water warms significantly or evaporates. This is because it takes a large amount of heat to break the many hydrogen bonds in a quantity of water. Water's ability to absorb a lot of heat before becoming hot is the reason it was used to cool automobile engines in the days before alcohol-based coolants became available. In a similar way, water helps stabilize the temperature inside cells, which are mostly water. The chemical reactions in cells produce heat, yet cells must stay fairly cool in order for their proteins to function properly.

When water absorbs enough heat energy, hydrogen bonds between water molecules break apart. Then liquid

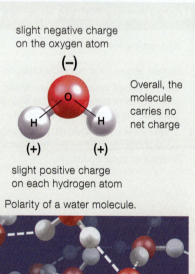

slight negative charge on the oxygen atom

(−)

Overall, the molecule carries no net charge

(+) **(+)**

slight positive charge on each hydrogen atom

A Polarity of a water molecule.

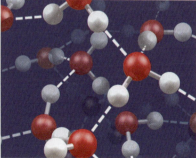

B Hydrogen bonds between molecules in liquid water (*dashed lines*).

C Water's cohesion. When water flows over a high ledge, the fall (gravity) pulls molecules away from the surface. The individual water molecules don't scatter every which way, however, because hydrogen bonds pull inward on those at the surface. As a result, the molecules tend to stay together in droplets and streams. *Right*: Cohesion gives water a skinlike surface that can support lightweight objects, such as a leaf or an insect.

Figure 2.9 Animated! Water is essential for life. (© Cengage Learning)

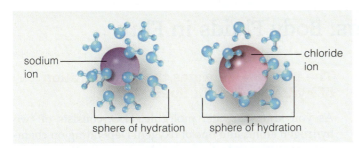

sodium ion

chloride ion

sphere of hydration   sphere of hydration

Figure 2.10 Animated! Charged substances dissolve easily in water. This diagram depicts water molecules clustered around a sodium ion and a chloride ion. The clusters are called "spheres of hydration." (© Cengage Learning)

water evaporates: Molecules at its surface begin to escape into the air. Heat is lost when a large number of water molecules evaporate. This is why sweating helps cool you off on a hot, dry day. Your sweat is 99 percent water. When it evaporates from the millions of sweat glands in your skin, heat leaves with it.

Water is a solvent

Water also is a superb **solvent**, which means that ions and polar molecules easily dissolve in it. In chemical terms a dissolved substance is called a **solute** (SAHL-yoot). When a substance dissolves, water molecules cluster around its individual molecules or ions and form "spheres of hydration."

This is what happens to solutes in blood and other body fluids. Most chemical reactions in the body occur in water-based solutions.

Figure 2.10 shows what happens to table salt (NaCl) when you pour some into a glass of water. After a while, the salt crystals separate into Na^+ and Cl^-. Each Na^+ attracts the negative end of some of the water molecules while each Cl^- attracts the positive end of others.

antioxidant Substance that gives up an electron to a free radical.

free radical An unstable molecule that includes an atom with an electron vacancy in its outer shell.

solute A dissolved substance.

solvent Water-based solution in which polar molecules and ions easily dissolve.

TAKE-HOME MESSAGE

WHAT ARE THE CHEMICAL PROPERTIES OF WATER THAT HELP SUPPORT LIFE?

• A water molecule is polar. Its oxygen atom is slightly positive and its hydrogen atoms are slightly negative.

• Polarity allows water molecules to form hydrogen bonds with one another and with other polar (hydrophilic) substances.

• Water molecules tend to repel nonpolar (hydrophobic) substances.

• The hydrogen bonds in water help it stabilize temperature in body fluids and allow it to dissolve many substances.

2.6 How Antioxidants Protect Cells

FOCUS ON HEALTH

The oxidations that go on in our cells (Section 2.4) also release highly unstable molecules called **free radicals**. A free radical (such as O_2^-) lacks a full complement of electrons in its outer shell. To fill the empty slot, a free radical can easily "steal" an electron from a stable molecule. This theft disrupts the structure and functioning of the affected molecule.

When free radicals are present in large numbers, they pose a serious threat to many cell molecules, including DNA. Cigarette smoke and the ultraviolet radiation in sunlight produce additional free radicals in the body.

An **antioxidant** can give up an electron to a free radical before the rogue damages DNA or some other important cell component. The body makes some antioxidants, including the hormone melatonin (Chapter 15), but this homegrown chemical army isn't enough to balance the ongoing production of free radicals. Many nutritionists recommend adding antioxidants to the diet by eating lots of the foods that contain them, using supplements only in moderation (Figure 2.11).

Ascorbic acid—vitamin C—is an antioxidant, as is vitamin E. So are some carotenoids, such as alpha carotene, which are pigments in orange and leafy green vegetables, among other foods. Antioxidant-rich foods typically also are low in fat and high in fiber.

Background: iStockphoto.com/xefstock Foreground: Sue Hartzell

Figure 2.11 Antioxidants help counter free radicals.

Acids, Bases, and Buffers: Body Fluids in Flux

acid A substance that donates protons (as H⁺) to other solutes or to water molecules when it dissolves in water.

base A substance that accepts H⁺ when it dissolves in water.

hydrogen ion A proton, H⁺.

hydroxide ion The negatively charged molecule OH⁻.

pH scale A measure of the concentration of H⁺ in a fluid.

- Ions such as H⁺ dissolved in the fluids inside and outside cells influence cell functions.

- Buffer systems help maintain proper ion balance.

Every instant of every day, chemical reactions in or outside your cells add or remove substances from your body fluids. Homeostasis—and our health—depend on the body's ability to manage these changes.

The pH scale indicates the concentration of hydrogen ions in fluids

As you know, a water molecule, H_2O, consists of two hydrogen atoms and one of oxygen. Depending on chemical conditions, a water molecule can naturally separate into two ions—a proton, also called a **hydrogen ion**, or H⁺—and a **hydroxide ion** (OH^-). These ions are the basis for the **pH scale** (Figure 2.12). This numerical scale represents the concentration (relative amount) of H⁺ in water, blood, and other fluids. There are huge numbers of hydrogen ions in the body and they can have major effects on body functions.

Pure water (not rainwater or tap water) always has equal numbers of H⁺ and OH⁻ ions. This state is *neutrality*, or pH 7, on the pH scale. Each unit of change away from neutrality corresponds to a tenfold increase or decrease in the concentration of H⁺.

The watery fluid inside most body cells is about 7 on the pH scale. Blood and the watery fluids outside cells usually have a slightly higher pH, ranging between 7.3 and 7.5. These facts are relevant because proteins and many other biological molecules can function properly only within a narrow pH range. Even small changes in pH can drastically affect life processes.

Acids give up H⁺ and bases accept H⁺

An **acid** donates protons (as H⁺) to other solutes or to water molecules when it dissolves in water. A **base** accepts H⁺ when it dissolves in water. When either an acid or a base dissolves, OH⁻ then forms in the solution as well. *Acidic* solutions, such as black coffee and lemon juice, release more H⁺ than OH⁻; their pH is below 7. *Basic* solutions, such as household bleach and dissolved baking soda, release more OH⁻ than H⁺. Basic solutions are also called *alkaline* fluids; their pH is above 7.

Most acids are classed as either weak or strong. Weak acids, such as acetic acid, don't readily donate H⁺. Depending on the pH, they just as easily accept H⁺ as give it up, so they alternate between acting as an acid and acting as a base. On the other hand, strong acids totally give up H⁺ when they dissociate in water. The hydrochloric acid (HCl) in your stomach and sulfuric acid (H_2SO_4) are examples.

High concentrations of strong acids or strong bases can be helpful in the stomach. For instance, when you eat, cells in your stomach secrete HCl, which separates into H⁺ and Cl⁻ in water. The H⁺

pH		
0 —	10^0	battery acid
1 —	10^{-1}	gastric fluid
2 —	10^{-2}	acid rain / lemon juice / cola / vinegar
3 —	10^{-3}	
4 —	10^{-4}	orange juice / tomatoes, wine / bananas
5 —	10^{-5}	beer / bread / black coffee / urine, tea, typical rain
6 —	10^{-6}	corn / butter / milk
7 —	10^{-7}	pure water
8 —	10^{-8}	blood, tears / egg white / seawater
9 —	10^{-9}	baking soda / phosphate detergents / Tums
10 —	10^{-10}	toothpaste / hand soap / milk of magnesia
11 —	10^{-11}	household ammonia
12 —	10^{-12}	hair remover
13 —	10^{-13}	bleach / oven cleaner
14 —	10^{-14}	drain cleaner

more acidic

more basic

Figure 2.12 The pH scale indicates the acidity of a solution. (Figure by Lisa Starr)

ions make stomach fluid more acidic, and the increased acidity switches on enzymes that can chemically break down food. The acid also helps kill harmful bacteria. Eating too much of certain kinds of foods can lead to "acid stomach." Antacids are strong bases. For example, milk of magnesia releases magnesium ions and OH^-, which combines with excess H^+ in your stomach fluid. This chemical reaction raises the fluid's pH, and your acid stomach goes away.

Strong acids or bases can also be harmful. For example, many drain cleaners and other household products can cause severe chemical burns. So can sulfuric acid in car batteries. Smoke from fossil fuels and motor vehicle exhaust releases strong acids that alter the pH of rain (Figure 2.13). This "acid rain" is an environmental threat discussed in Chapter 25.

A salt releases other kinds of ions

Salts are compounds that release ions *other than* H^+ and OH^- in solutions. Salts and water often form when a strong acid and a strong base interact. Depending on a solution's pH value, salts can form and dissolve easily. Many salts dissolve into ions that have key functions in cells. For example, nerve impulses depend on ions of sodium, potassium, and calcium.

Buffers protect against shifts in pH

Because shifts in pH can seriously disrupt body functions, there must be homeostatic mechanisms to counteract them. Fortunately, body fluids usually stay at a consistent pH because they are stabilized by **buffers**—substances that can compensate for pH changes by donating or accepting H^+. Pairs of buffers, often a weak acid or a base and its salt, operate as a balancing system that can keep the pH of a solution stable.

For example, when a base is added to a fluid, OH^- is released. However, if the fluid is buffered, the weak acid partner gives up H^+. The H^+ combines with the OH^-, forming a small amount of water that does not affect pH. So, a buffered fluid's pH stays constant even when a base is added.

A key point to remember is that the action of a buffer can't make new hydrogen ions or eliminate those that already are present. It can only bind or release them.

Carbon dioxide forms in many reactions in the body and it takes part in an important buffer system in the blood. In this system it combines with water to form the compounds carbonic acid and bicarbonate. When the acidity of blood starts to drop (that is, its pH starts to rise) due to other factors, the carbonic acid neutralizes the excess OH^- by releasing H^+:

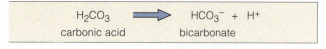

$$H_2CO_3 \longrightarrow HCO_3^- + H^+$$
carbonic acid   bicarbonate

Figure 2.13 **Acids produced by human activities affect the environment.** In this photograph, camera lens filters reveal otherwise invisible sulfur dioxide emissions from a coal-burning power plant. Sulfur dioxide is a major component of acid rain. The eroded statue pictured here is evidence of its damage.

When the blood becomes more acidic, the bicarbonate absorbs excess H^+ and thus shifts the balance of the buffer system toward carbonic acid.

$$HCO_3^- + H^+ \longrightarrow H_2CO_3$$
bicarbonate   carbonic acid

Together these reactions usually keep the blood pH slightly basic, between 7.3 and 7.5, but a buffer system can neutralize only so many ions. Even slightly more than that limit causes the pH to swing widely.

A buffer system failure in the body can be disastrous for homeostasis. If blood's pH (7.3–7.5) declines to even 7, a person will fall into the deep state of unconsciousness called a *coma*. In *acidosis*, carbon dioxide builds up in the blood, too much carbonic acid forms, and blood pH plummets. The condition called *alkalosis* is an abnormal increase in blood pH. Untreated, acidosis or alkalosis can cause death.

buffer A chemical that can stabilize the pH of a solution by donating or accepting hydrogen ions (H^+).

salt A compound that releases ions other than H^+ and OH^- in a solution.

TAKE-HOME MESSAGE

HOW DO ACIDS, BASES, SALTS, AND BUFFERS AFFECT THE MAKEUP OF BODY FLUIDS?

- Cell processes produce large numbers of hydrogen ions (H^+), which are chemically active and make body fluids more acidic.
- The pH scale represents the relative amount of hydrogen ions in body fluids.
- Acids release H^+ ions, and bases accept them.
- Salts release ions other than H^+ and OH^-.
- Buffer systems counteract potentially harmful shifts in the pH of body fluids.

Michael Grecco Photography/Workbook Stock/Getty Images; inset: iStockphoto.com/Linda Steward

2.8 Molecules of Life

- Biological molecules are built on atoms of the element carbon.

Biological molecules contain carbon

There are four main kinds of biological molecules: carbohydrates, lipids, proteins, and nucleic acids. Each one is an **organic compound**: It contains the element carbon and at least one hydrogen atom. Chemists once defined organic substances as those obtained from animals and vegetables, as opposed to "inorganic" ones from minerals.

Carbon's key feature is versatile bonding

The human body consists mostly of oxygen, hydrogen, and carbon (Figure 2.1). The oxygen and hydrogen are mainly in the form of water. Carbon makes up more than half of what is left.

Carbon's importance to life starts with its versatile bonding behavior. As you can see in the sketch below, each carbon atom can share pairs of electrons with as many as four other atoms. The covalent bonds are fairly stable, because the carbon atoms share pairs of electrons equally. This type of bond links carbon atoms together in chains. The chains form a backbone to which atoms of hydrogen, oxygen, and other elements can attach.

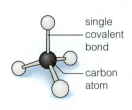

single covalent bond

carbon atom

The angles of the covalent bonds help produce the shapes of organic compounds. A chain of carbon atoms, bonded covalently one after another, forms a backbone from which other atoms can project:

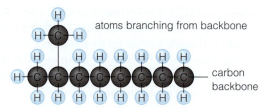

atoms branching from backbone

carbon backbone

A carbon backbone with only hydrogen atoms attached to it is a hydrocarbon. The backbone also may form a ring, like this:

or

carbon rings

Functional groups affect the chemical behavior of organic compounds

Biological molecules also have parts called functional groups. A **functional group** is an atom or cluster of atoms that are covalently bonded to carbon. The kind, number, and arrangement of these groups determine specific properties of molecules, such as polarity or acidity.

Figure 2.14 shows some functional groups. Sugars and other organic compounds classified as alcohols have one or more hydroxyl groups (—OH). Water forms hydrogen bonds with hydroxyl groups, which is why sugars can dissolve in water. The backbone of a protein forms by reactions between amine groups and carboxyl groups.

Human sex hormones illustrate the importance of where a functional group attaches to a biological molecule. Estrogen and testosterone account for many differences

Group	Found In	Structure
hydroxyl	amino acids; sugars and other alcohols	—OH
methyl	fatty acids, some amino acids	$-\overset{H}{\underset{H}{C}}-H$
carbonyl	sugars, amino acids, nucleotides	$-\overset{}{\underset{O}{C}}-H$ (aldehyde) $\overset{}{\underset{O}{C}}-$ (ketone)
carboxyl	amino acids, fatty acids, carbohydrates	$-\overset{}{\underset{O}{C}}-OH$ $-\overset{}{\underset{O}{C}}-O^-$ (ionized)
amine	amino acids, some nucleotide bases	$-\overset{}{\underset{H}{N}}-H$ $-\overset{H}{\underset{H}{N}}H^+$ (ionized)
phosphate	nucleotides (e.g., ATP); DNA and RNA; many proteins; phospholipids	$-O-\overset{O^-}{\underset{O}{P}}-O^-$ (P) ion
sulfhydryl	many cellular molecules	—SH —S—S— (disulfide bridge)

Figure 2.14 Functional groups help determine the properties of biological molecules.

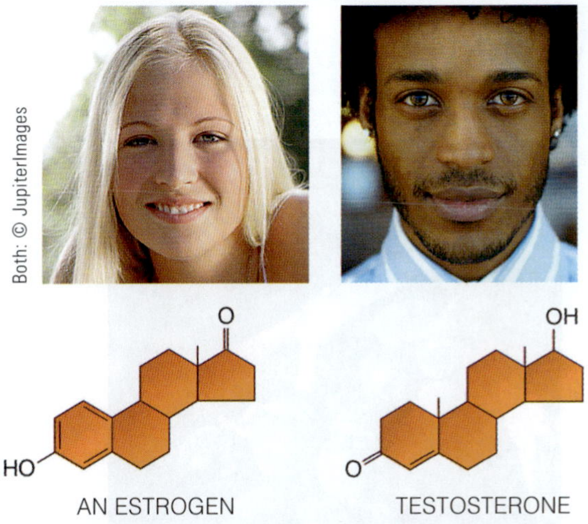

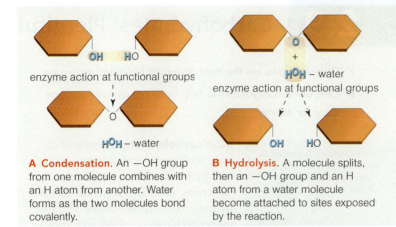

AN ESTROGEN   **TESTOSTERONE**

Both: © JupiterImages

Figure 2.15 The location of functional groups determines the difference between the sex hormones estrogen and testosterone.

A Condensation. An —OH group from one molecule combines with an H atom from another. Water forms as the two molecules bond covalently.

B Hydrolysis. A molecule splits, then an —OH group and an H atom from a water molecule become attached to sites exposed by the reaction.

Figure 2.16 **Animated!** Metabolic reactions build, rearrange, and break apart most biological molecules. (© Cengage Learning)

between males and females. The hormones have the same functional groups, but the groups are in different places, as you can see in Figure 2.15.

Cells have chemical tools to assemble and break apart biological molecules

How do cells make the organic compounds they need for their structure and functioning? To begin with, whatever happens in a cell requires energy, which is provided by a compound called ATP that you will learn more about shortly. Chemical reactions in cells also require a class of proteins called **enzymes**, which make reactions take place faster than they would on their own. Table 2.3 lists the ways cells build, rearrange, or split apart organic compounds. Two important types of reactions are called condensation and hydrolysis.

Condensation reactions   As a cell builds or changes organic compounds, a common step is the **condensation reaction**. Often in this kind of reaction, enzymes remove a hydroxyl group from one molecule and an H atom from another, then speed the formation of a covalent bond between the two molecules (Figure 2.16A). The discarded

hydrogen and oxygen atoms may combine to form a molecule of water (H_2O). Because this kind of reaction often forms water as a by-product, condensation is sometimes called *dehydration* ("un-watering") *synthesis*. Cells can use condensation reactions to assemble polymers. *Poly-* means "many," and a **polymer** is a large molecule built of three to millions of subunits. The subunits, called **monomers**, may be the same or different.

Hydrolysis reactions   Hydrolysis is like condensation in reverse (Figure 2.16B). In a first step, enzymes that act on particular functional groups split molecules into two or more parts. Then they attach an —OH group and a hydrogen atom from a molecule of water to the exposed sites. With hydrolysis, cells can break apart large polymers into smaller units when these are required for building blocks or energy.

condensation reaction Chemical reaction that covalently bonds two molecules into a larger one. Water often forms as a by-product.

enzyme Type of protein that speeds up chemical reactions.

functional group Atom or atoms bonded to carbon in a molecule and that helps determine the molecule's chemical properties.

hydrolysis reaction Chemical reaction that splits a large molecule into smaller parts, often using a water molecule in the process.

monomer A small subunit of a larger molecule (a polymer).

organic compound Compound that contains carbon and at least one hydrogen atom.

polymer A large molecule built of monomer subunits.

TABLE 2.3	What Cells Do to Organic Compounds
Class of Reaction	**What Happens**
CONDENSATION	Two molecules covalently bond into a larger one.
CLEAVAGE	A molecule splits into two smaller ones, as by hydrolysis.
FUNCTIONAL GROUP TRANSFER	One molecule gives up a functional group, and a different molecule immediately accepts it.
ELECTRON TRANSFER	One or more electrons from one molecule are donated to another molecule.
REARRANGEMENT	Moving internal bonds converts one type of organic compound to another.

Carbohydrates: Plentiful and Varied

- Carbohydrates are the most abundant biological molecules.

- Cells use carbohydrates to help build cell parts or package them for energy.

carbohydrate A biological molecule built of carbon, hydrogen, and oxygen atoms, usually in a 1:2:1 ratio.

monosaccharide The simplest class of carbohydrate, consisting of a single sugar monomer. A glucose molecule is an example.

oligosaccharide A carbohydrate that consists of a short chain of sugar units. Sucrose is an example.

polysaccharide A complex carbohydrate that consists of straight or branched chains of sugar monomers. Cellulose is an example.

Most **carbohydrates** consist of carbon, hydrogen, and oxygen atoms in a 1:2:1 ratio. Due to differences in structure, chemists separate carbohydrates into three major classes: monosaccharides, oligosaccharides, and polysaccharides.

Simple sugars are the simplest carbohydrates

Saccharide comes from a Greek word meaning "sugar." A **monosaccharide**, meaning "one monomer of sugar," is the simplest carbohydrate. It has at least two —OH groups joined to the carbon backbone plus an aldehyde or a ketone group. Monosaccharides usually taste sweet and dissolve easily in water. The most common ones have a backbone of five or six carbons; for example, there are five carbon atoms in deoxyribose, the sugar in DNA. The simple sugar glucose is the main energy source for body cells.

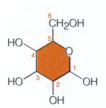

Each glucose molecule (at left) has six carbons, twelve hydrogens, and six oxygens. (Notice how it meets the 1:2:1 ratio noted above.) Glucose is a building block for larger carbohydrates. It also is the parent molecule (precursor) for many compounds, such as vitamin C, which are derived from sugar monomers.

Grapes, a natural source of sucrose in the diet.

David M. Phillips/Visual Unlimited, Inc.

Oligosaccharides are short chains of sugar units

Unlike the simple sugars, an **oligosaccharide** is a short chain of two or more sugar monomers that are joined by dehydration synthesis. (*Oligo-* means "a few.") The type known as *di*saccharides consists of just two sugar units. Lactose, sucrose, and maltose are examples. Lactose (a glucose and a galactose unit) is a milk sugar. Sucrose, the most plentiful sugar in nature, consists of one glucose and one fructose unit (Figure 2.17). You consume sucrose

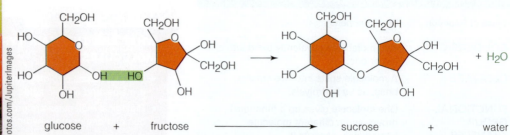

© Photos.com/JupiterImages

glucose + fructose ⟶ sucrose + water

Figure 2.17 Sucrose, or table sugar, is a disaccharide formed from glucose and fructose. As you can see in this diagram, the synthesis of a sucrose molecule is a condensation reaction, which forms water as a by-product. (© Cengage Learning)

A Cellulose occurs only in plants. Chains of glucose units stretch side by side and hydrogen-bond at many —OH groups. The hydrogen bonds stabilize the chains in tight bundles that form long fibers, such as cotton fibers humans use for clothing.

B In amylose, one type of starch, glucose units are monomers that form a coiling polymer chain. Plants store starch in their roots, stems, leaves, seeds, and fruits, such as apples.

C Glycogen. This polysaccharide functions as an energy reservoir. The liver and muscles of active animals, including people, store large amounts of it.

Maridav/Shutterstock.com

Figure 2.18 Animated! **Complex carbohydrates are chains of many sugar monomers.** This diagram shows the structure of **A** cellulose, **B** starch, and **C** glycogen. Glucose is the basic building block of all three of these carbohydrates. (© Cengage Learning)

when you eat fruit, among other plant foods. Table sugar is sucrose crystallized from sugar cane and sugar beets.

Proteins and other large molecules often have oligosaccharides attached as side chains to their carbon backbone. Some chains have key roles in activities of cell membranes, as you will read in Chapter 3. Others are important in the body's defenses against disease.

Polysaccharides are sugar chains that store energy

The "complex" carbohydrates, or **polysaccharides**, are straight or branched chains of sugar monomers. Often thousands are joined by dehydration synthesis. The many chemical bonds in polysaccharides store a great deal of energy. That energy is released to cells when the digestive system breaks these sugars down. Polysaccharides make up most of the carbohydrates humans eat. The most common ones—glycogen, starch, and cellulose—consist only of glucose.

Plants store a large amount of glucose in the form of cellulose (Figure 2.18A). Humans don't have digestive enzymes that can break down the cellulose in whole grains, vegetables, fruits, and other plant tissues. We do

benefit from it, however, as undigested "fiber" that adds bulk and so helps move wastes through the lower part of the digestive tract.

Many plant-derived foods are rich in starch, which is one form in which plants store glucose. In starch the glucose subunits form a string, as with the starch amylose illustrated in Figure 2.18B.

The polysaccharide glycogen is one form in which animals store sugar, most notably in muscles and the liver (Figure 2.18C). When a person's blood sugar level falls, liver cells break down glycogen and release glucose to the blood. When you exercise, your muscle cells tap into their glycogen stores as a quick source of energy.

TAKE-HOME MESSAGE

WHAT ARE CARBOHYDRATES?

- Carbohydrates range from simple sugars such as glucose to molecules composed of many sugar units.
- From simple to complex, the three major types of carbohydrates are monosaccharides, oligosaccharides, and polysaccharides.
- Cells use carbohydrates for energy or as raw materials for building cell parts.

Lipids: Fats and Their Chemical Relatives

■ Cells use lipids to store energy, as structural materials, and as signaling molecules.

Oil and water don't mix. Why? Oils are a type of lipid, and a **lipid** is a nonpolar hydrocarbon. A lipid's large nonpolar region makes it hydrophobic, so it does not dissolve easily in water. Lipids do easily dissolve in other nonpolar substances. For example, you can dissolve melted butter in olive oil. Here we are interested first in fats and phospholipids, both of which have chemical "tails" called fatty acids. We will also consider sterols, which have a backbone of four carbon rings.

chicken fat. The fatty acid tails of *unsaturated* fats have one or more double covalent bonds. Such strong bonds make rigid kinks that prevent unsaturated fats from packing tightly. Most vegetable oils such as canola oil, peanut oil, corn oil, and olive oil are unsaturated. They stay liquid at room temperature.

Butter, lard, oils, and other dietary fats consist mostly of **triglycerides**. These fats have three fatty acid tails attached to a glycerol backbone (Figure 2.20). Triglycerides are the most common lipids in the body as well as its richest source of energy. Compared to complex carbohydrates, they yield more than twice as much energy when they are broken down. This is because triglycerides have more removable electrons than do carbohydrates—and energy is released when electrons are removed. In the body, cells of fat-storing tissues stockpile triglycerides as fat droplets.

Some unsaturated fats, like the trans fats described at the beginning of this chapter, are unhealthy. A double bond in "cis" fatty acids keeps them kinked, but in trans fatty acids a double bond keeps them straight (Figure 2.21). Some trans fatty acids occur naturally in beef, but most of those in human food are formed by a manufacturing process (called hydrogenation) that is used to solidify vegetable oils for solid margarines and shortenings that are used in many prepared foods. A diet high in trans fatty acids increases the risk of heart disease.

fat A lipid molecule that has up to three fatty acid tails.

fatty acid A chemical compound with a backbone of carbon atoms bonded to a carboxyl group.

lipid A nonpolar hydrocarbon.

phospholipid A complex lipid that has a phosphate functional group.

sterol A type of lipid that has no fatty acid tail. Sterols include cholesterol and steroid hormones.

triglyceride A fat that has three fatty acid tails attached to a glycerol backbone.

Fats are energy-storing lipids

The lipids called **fats** have as many as three fatty acids, all attached to glycerol. Each **fatty acid** has a backbone of up to thirty-six carbons and a carboxyl group (—COOH) at one end. Hydrogen atoms occupy most or all of the remaining bonding sites. A fatty acid typically stretches out like a flexible tail (Figure 2.19).

In *saturated* fats, the fatty acid backbones have only single covalent bonds. Animal fats are saturated and solid at room temperature. Examples are butter, lard, or

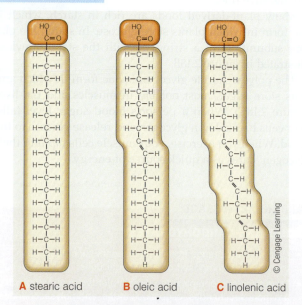

Figure 2.19 Animated! Fats are built from fatty acids. **A** Stearic acid has a carbon backbone fully saturated with hydrogens. **B** Oleic acid, with a double bond in the carbon backbone, is unsaturated. **C** Linolenic acid, with three double bonds, is a polyunsaturated fatty acid.

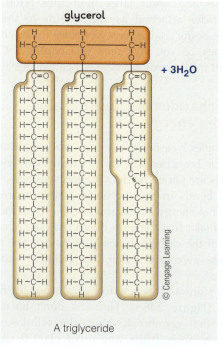

Figure 2.20 Animated! Triglycerides have three fatty acid tails attached to glycerol.

Figure 2.21 **Some foods contain unhealthy trans fats.** French fries cooked in certain types of vegetable oil contain a great deal of trans fatty acids. It is the arrangement of carbon atoms around the carbon-carbon double bond (*orange arrow*) in the middle of a trans fatty acid that makes it a very unhealthy food.

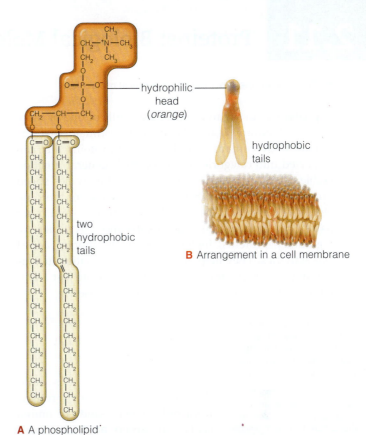

A A phospholipid

B Arrangement in a cell membrane

Figure 2.22 **Phospholipids contain a phosphate atom. A** Structural formula and **B** a simple diagram of a common phospholipid in human cell membranes. (© Cengage Learning)

Phospholipids are basic building blocks of cell membranes

A **phospholipid** has a glycerol backbone, two fatty acid tails, and a hydrophilic "head" with a phosphate group—a phosphorus atom bonded to four oxygen atoms—and another polar group (Figure 2.22A). Phospholipids are the main materials of cell membranes, which have two layers of lipids (Figure 2.22B). The heads of one layer are dissolved in the cell's fluid interior, while the heads of the other layer are dissolved in the surroundings. Sandwiched between the two are all the fatty acid tails, which are hydrophobic.

Cholesterol and steroid hormones are built from sterols

Sterols are among the lipids that have no fatty acid tails. Sterols differ in the number, position, and type of their functional groups, but they all have a rigid backbone of four fused-together carbon rings (Figure 2.23A). Many people associate the sterol cholesterol (Figure 2.23B) with heart disease. However, normal amounts of this sterol are essential in the body. For instance, the sterol cholesterol is a vital component of membranes of every cell in your body. Important derivatives of cholesterol include vitamin D (essential for bone and tooth development), bile salts (which help with fat digestion in the small intestine), and steroid hormones such as estrogen and testosterone. In later chapters we will discuss how steroid hormones influence reproduction, development, growth, and many other functions.

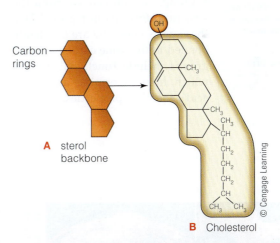

A sterol backbone

B Cholesterol

Figure 2.23 **Cholesterol is the most common sterol in the human body.** Notice the carbon rings in its backbone.

TAKE-HOME MESSAGE

WHAT ARE LIPIDS?

- Lipids are hydrophobic greasy or oily compounds.
- The type called triglycerides are major reservoirs of energy.
- Phospholipids are the main components of cell membranes.
- Sterols (such as cholesterol) are components of membranes and precursors of steroid hormones and other vital molecules.

2.11 Proteins: Biological Molecules with Many Roles

■ Proteins are the most diverse biological molecules.

A **protein** is an organic compound built of one or more chains of amino acids. Biochemists estimate that the human body contains roughly 30,000 proteins that can be sorted into categories based on their general function (Table 2.4). For instance, proteins called enzymes speed up chemical reactions. Structural proteins are building blocks of cells and tissues in bones, muscles, and other parts. Transport proteins move substances. A variety of regulatory proteins, including some hormones, adjust cell activities. They help make possible activities such as waking, sleeping, and engaging in sex, to cite just a few. Other proteins are important in body defenses.

amino acid Any of the small organic compounds that are the building blocks of proteins.

peptide bond Covalent bond that joins the amino group of one amino acid to the carboxyl group of a second amino acid.

polypeptide chain A chain of three or more amino acids joined by peptide bonds.

primary structure Of a protein, the particular sequence of amino acids that makes up the protein.

protein An organic compound composed of one or more chains of amino acids.

Proteins are built from amino acids

Amazingly, our body cells build thousands of different proteins from only twenty kinds of amino acids. An **amino acid** is a small organic compound that consists of an amino group, a carboxyl group (an acid), an atom of hydrogen, and one or more atoms called its R group. As you can see from the structural formula in Figure 2.24A, these parts generally are covalently bonded to the same carbon atom. R groups include functional groups, which help determine an amino acid's chemical properties.

TABLE 2.4 Some Roles Proteins Play in the Body	
Type of Protein	**Examples**
STRUCTURAL	Serves as building materials for cells and tissues. Examples: Girderlike support fibers inside cells; collagen fibers that strengthen skin.
ENZYME	Speeds up chemical reactions. Example: Digestive enzymes that speed the breakdown of complex carbohydrates, fats, and dietary proteins in the digestive system.
TRANSPORT	Carries substances in body fluids or moves them into or out of cells. Examples: The protein hemoglobin, which carries oxygen to cells; proteins that pump ions into or out of cells.
MOVEMENT	Produces movements of cells and cell parts. Examples: Contraction of muscle cells; swimming by sperm cells.
REGULATOR	Adjusts cell activities. Examples: Hormones such as sex hormones that govern puberty; insulin, which regulates blood sugar.
RECEPTOR	Binds molecules to or inside cells. Example: Receptors that bind hormones to target cells.
DEFENSE	Assists in immune responses and other bodily defenses. Examples: Antibodies that attach to invading organisms and molecules; proteins that identify cells as "self" (belonging to a given person's body).

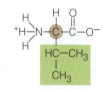

B valine (val)

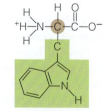

tryptophan (trp)

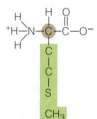

methionine (met)

Amino Group   Carboxyl Group

R Group (20 kinds, each with distinct properties)

A

Figure 2.24 All amino acids have the same basic chemical parts. As you can see in **A**, these building blocks are an amino group, a carboxyl group, an R group, and a hydrogen atom, all connected to a carbon atom by covalent bonds. A variety of foods provide these small organic compounds. **B** shows structural formulas for three common amino acids human cells use. (© Cengage Learning)

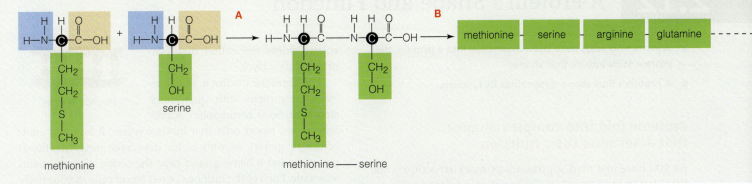

Figure 2.25 Animated! **A protein is built as peptide bonds form between amino acids. DNA determines the order of amino acids in a polypeptide chain. A When the chain starts to form, a peptide bond forms between the first two amino acids—here, methionine and serine. Notice that the bond forms between the carboxyl group of the methionine and the amino group of the serine. Peptide bonds are formed during condensation reactions, so as each one joins amino acids, water forms as well. B More amino acids are added to the chain, with a peptide bond linking each one to the next in line.** (© Cengage Learning)

The sequence of amino acids is a protein's primary structure

When a cell makes a protein, amino acids become linked, one after the other, by **peptide bonds**. As Figure 2.25 shows, this is the type of covalent bond that forms between one amino acid's amino group (NH_3^+) and the carboxyl group ($—COO^-$) of the next amino acid.

When peptide bonds join two amino acids together, we have a dipeptide. When they join three or more amino acids, we have a **polypeptide chain**.

Each type of polypeptide chain, and therefore each type of protein, has its own unique sequence of amino acids. The sequence forms as different amino acids are added in a specific order, one at a time, from the twenty kinds available to body cells. Figure 2.24B gives you an idea of how different amino acids can vary in their chemical structure.

As a later chapter describes, DNA determines the order in which amino acids are added to the growing chain. Every kind of protein in the body will have its own sequence of amino acids, linked one to the next like the links of a chain. This sequence is called the **primary structure** of a protein. This is a representation of the primary structure of a small but vital protein in humans, the hormone insulin, which consists of just fifty-one amino acids:

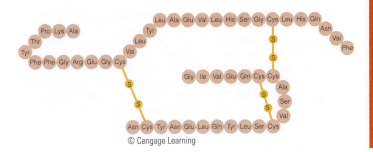

© Cengage Learning

A large number of amino acids can be linked up this way. The primary structure of the largest known protein, which is a building block of human muscle, is a string of 27,000 amino acids!

THINK OUTSIDE THE BOOK

Dietary supplements of some amino acids sometimes are promoted to consumers as helpful for "curing" herpes infections or providing relief from depression or some other health problem. Visit the website of the federal Food and Drug Administration (FDA) at www.fda.gov and check out information there on food supplements. Does the FDA regulate dietary supplements? Who is responsible for ensuring the safety of amino acid supplements?

TAKE-HOME MESSAGE

WHAT ARE PROTEINS?

- Proteins are organic molecules built of one or more chains of amino acids.
- DNA determines the order of amino acids in the chain. The sequence is unique for each type of protein.
- The sequence of amino acids that makes up a protein is the protein's primary structure.
- Proteins perform many functions in the body. Some are building blocks, others transport substances, and still others regulate body functions or serve in defense.

2.12 | A Protein's Shape and Function

- When amino acids have been assembled into a protein, the protein folds into its final shape.
- A protein's final shape determines its function.

Proteins fold into complex shapes that determine their function

As you have just read, a protein's primary structure is the first step in the formation of a functioning protein (Figure 2.26A). Secondary structure emerges as the chain twists, bends, loops, and folds. These shape changes occur as hydrogen bonds form between different amino acids in different parts of the chain (Figure 2.26B). Even though the primary structure of each protein is unique, similar patterns of coils, sheets, and loops occur in most proteins.

The coils, sheets, and loops of a protein fold up even more, much like an overly twisted rubber band. This is the third level of organization, or *tertiary* structure, of a protein (Figure 2.26C). Tertiary structure is what makes a protein a molecule that can perform a particular function. For instance, some proteins fold into a hollow "barrel" that provides a channel through cell membranes.

A protein may have more than one polypeptide chain

glycoprotein A protein that has a sugar, such as an oligosaccharide, attached to it.

lipoprotein A protein that has a lipid attached to it.

Some proteins are built of more than one polypeptide chain. This type of protein has *quaternary* structure (Figure 2.26D). Interactions between its polypeptide chains (such as hydrogen bonds) hold the chains together. In some cases the links include covalent bonds between sulfur atoms of R groups. These bonds between two

Disulfide bridges

sulfur atoms are called disulfide bridges (*di* = two).

The hormone insulin is an example of a protein with quaternary structure. So is hemoglobin, a protein in red blood cells that binds oxygen. It has four molecules of globin, as well as an iron-containing functional group (called a heme group) near the center of each globin molecule. Each of the millions of red blood cells in your body is transporting a billion molecules of oxygen, bound to some 250 million molecules of hemoglobin. You will learn more the function of hemoglobin in Chapter 8.

Hemoglobin and insulin are globular proteins. So are most enzymes. Many other proteins with quaternary structure are fibrous—like heavy-duty thread, they are elongated and strong. An example is collagen, the most common protein in the body. Your skin, bones, corneas, and other body parts depend on its strength. Multiple polypeptide chains of some proteins may be organized into coils or sheets. Keratin, a structural protein of hair, is like this (Figure 2.27).

Glycoproteins have sugars attached and lipoproteins have lipids

Some proteins have other organic compounds attached to their polypeptide chains. For example, **lipoproteins** form when certain proteins circulating in blood combine with cholesterol, triglycerides, and phospholipids that were consumed in food. Most **glycoproteins** (from *glukus*, the Greek word for "sweet") have oligosaccharides bonded to them. Most of the proteins found at the surface of cells are glycoproteins, as are many proteins in blood and those that cells secrete (such as protein hormones).

| lysine | glycine | glycine | arginine |

A The primary structure of a protein is its linear sequence of amino acids. This string of amino acids is a polypeptide chain.

B Secondary structure comes about as a polypeptide chain twists or folds. Hydrogen bonds hold the molecule in this shape.

C More folding of the chain produces a protein's tertiary structure—its overall three-dimensional shape. The folding results in pockets or crevices that establish how a protein will function chemically.

D In proteins with quaternary structure, bonds and other forces hold two or more polypeptide chains together in one molecule. This example shows how hemoglobin, which consists of four chains (*here colored green or blue*). A pocket in each chain holds a heme group (*red*) that contains an iron atom.

Figure 2.26 Proteins can have up to four levels of organization. (© Cengage Learning)

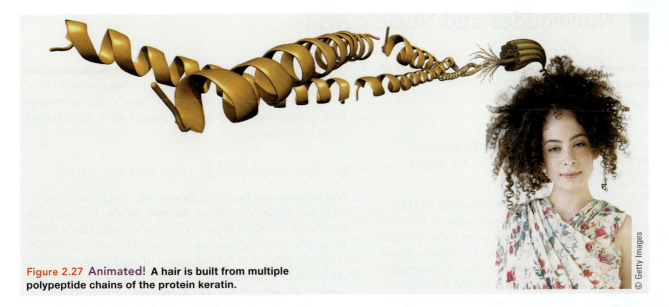

Figure 2.27 Animated! A hair is built from multiple polypeptide chains of the protein keratin.

© Getty Images

Disrupting a protein's shape prevents it from functioning normally

When a protein or any other large molecule loses its normal three-dimensional shape, it is *denatured*. For example, hydrogen bonds are sensitive to increases or decreases in temperature and pH. If the temperature or pH exceeds a protein's tolerance, its hydrogen bonds break, polypeptide chains unwind or change shape, and the protein no longer functions. The chemicals used in a permanent wave break hydrogen bonds in disulfide bridges in the keratin chains in hair. After the hair is wrapped around curlers that hold polypeptide chains in new positions, a second chemical causes disulfide bridges to form between different sulfur-bearing amino acids. The rearranged bonding locks the hair in curls (Figure 2.28A). Cooking an egg destroys weak bonds that contribute to the three-dimensional shape of the egg white protein albumin. Some denatured proteins can resume their shapes when normal conditions are restored, but not albumin. There is no way to uncook a cooked egg white (Figure 2.28B). In some cultures people enjoy uncooked dishes made of raw shrimp or other seafood soaked in lemon or lime juice. The acid in the citrus juice "cooks" the bits of fish by denaturing the proteins they contain.

TAKE-HOME MESSAGE

HOW DOES A PROTEIN GET ITS FINAL SHAPE?

- After amino acids have been linked to form a protein, the protein folds into its secondary structure, a coil or an extended sheet.
- More folding produces the third level of protein structure, which dictates how the protein will function.
- Proteins with more than one polypeptide chain have a fourth level of organization called quaternary structure.
- If some factor disrupts a protein's final shape, it becomes denatured and cannot function properly.

Figure 2.28 Animated! Changes in the chemical structure of a protein may show up in changes in the structure or functioning of body parts. **A** Actress Nicole Kidman's hair changed shape after a structural protein, keratin, was exposed to the chemicals that create a permanent wave. **B** The heat of cooking denatures the protein albumin in egg white. (Left: Ron Davis/Shooting Star; Middle: © Frank Trapper/Sygma/Corbis; Right: © Matthew Farruggio)

- The fourth and final class of biological molecules consists of nucleotides and nucleic acids.
- Link to Life's characteristics 1.1

ATP Adenosine triphosphate, a nucleotide that has phosphate groups attached and that serves as an energy carrier in cells.

coenzyme An "enzyme helper" molecule that moves hydrogen atoms and electrons to the sites of chemical reactions in cells.

DNA Deoxyribonucleic acid, which contains the sugar deoxyribose; the genetic material.

nucleic acid A single- or double-stranded molecule built of nucleotides. DNA is a double-stranded nucleic acid.

nucleotide A molecule built of a sugar (deoxyribose or ribose), a nitrogen-containing base, and one or more phosphate groups.

RNA Any of several ribonucleic acids, all of which contain the sugar ribose; RNAs help build cell proteins.

Nucleotides are energy carriers and have other roles

A **nucleotide** (NOO-klee-oh-tide) is composed of one sugar, at least one phosphate group, and one nitrogen-containing base. The sugar—ribose or deoxyribose—has a five-carbon ring structure. Ribose has two oxygen atoms attached to the ring, and deoxyribose has one. The bases have a single or double carbon ring structure.

The nucleotide **ATP** (for adenosine triphosphate), has a row of three phosphate groups attached to its sugar (Figure 2.29). In cells, ATP links chemical reactions that *release* energy with other reactions that *require* energy. This connection is possible because ATP can transfer a phosphate group to many other molecules in the cell, providing the acceptor molecules with the energy they need to enter into a reaction.

Some nucleotides are part of **coenzymes**, or "enzyme helpers." They move hydrogen atoms and electrons from one reaction site to another. Some other nucleotides act as chemical messengers inside and between cells. One of these is a nucleotide called cAMP (for cyclic adenosine monophosphate). It is extremely important in the action of some hormones.

Nucleic acids include DNA and the RNAs

Nucleotides are building blocks for single- or double-stranded molecules called **nucleic acids**. In a strand's backbone, covalent bonds join each nucleotide's sugar to a phosphate group of the neighboring nucleotide (Figure 2.30A). In this book you will read often about the nucleic acid **DNA** (deoxyribonucleic acid), which contains the sugar deoxyribose. DNA consists of two strands of nucleotides, twisted together in a double helix (Figure 2.30B). Hydrogen bonds between the nucleotide bases hold the strands together, and the sequence of bases encodes genetic information. Unlike DNA, **RNA** (short for ribonucleic acid) is usually a single strand of nucleotides. There are several kinds of RNA, but all have the sugar ribose. RNAs have crucial roles in processes that use genetic information to build proteins in cells.

> **TAKE-HOME MESSAGE**
>
> **WHAT IS A NUCLEIC ACID?**
> - A nucleic acid is a single or double-stranded molecule built of nucleotides.
> - Nucleic acids include DNA and RNAs.

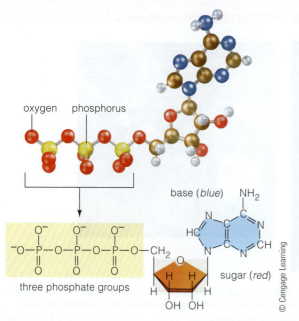

oxygen phosphorus

base (*blue*) NH₂

sugar (*red*)

three phosphate groups

Figure 2.29 **ATP is the energy-carrying nucleotide in cells.**

© Cengage Learning

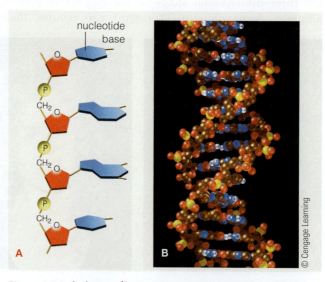

nucleotide base

A **B**

© Cengage Learning

Figure 2.30 **Animated! Chains of nucleotides form nucleic acids. A** Bonds between the bases in nucleotides. **B** Model of DNA, a nucleic acid with two strands of nucleotides joined by hydrogen bonds and twisted into a double helix.

Food Production and a Chemical Arms Race

If you eat like most people in developed countries, a variety of agricultural chemicals help provide your daily supply of organic compounds. For example, the lettuce for your salad most likely grew in fertilized cropland, and the grower may well have used herbicides to eradicate weeds, insecticides to kill unwanted insects, and fungicides against harmful molds and other destructive fungi.

Pesticides are quite useful in some applications. Many research studies show that modern pesticides used properly increase food supplies and profits for farmers. They also save lives by killing disease-causing insects and other pathogens. And despite understandable worries of consumers, for now there is little evidence that the usual amounts of pesticides in or on food pose a significant health risk.

On the other hand, pesticides are powerful chemicals. Some kill natural enemies of the targeted pest, and others harm wildlife such as birds. Some, such as DDT, stay active for many years. (DDT is banned in the United States, although not in many other countries.) And when people are exposed to unsafe doses, either by accident or misuse, some pesticides can trigger rashes, hives, headaches, asthma, and joint pain. According to some authorities, young children who are exposed to pesticides applied to keep a lawn thick and green may be at risk of developing learning disabilities and other problems. Manufacturers dispute these claims, but it is worth noting that according to the U.S. Environmental Protection Agency, homeowners in the United States use ten times more pesticides on their lawns than farmers do in agriculture.

At the beginning of this chapter we discussed how scientific studies of trans fats have led to stricter regulations on their use. In the future we can expect to see expanded research on the health effects of several chemical compounds that find their way into the drinking water supplies of millions of people around the globe.

One of these compounds is the weed killer atrazine. In 2009 the U.S. Environmental Protection Agency announced new efforts to study possible health impacts of atrazine, which has been used widely in agriculture and lawn care products for more than forty years (Figure 2.31). As a result of this long-term use, water supplies in many parts of the United States contain atrazine, although at levels that were thought to be safe for human consumption.

Recently, though, new research has suggested that atrazine may be associated with reproductive problems in humans. The issues include premature births, newborns with abnormally low birth weights, and menstrual problems in women. Some studies suggest that atrazine may be implicated in some cases of prostate cancer. The jury is still out on all these questions, and the manufacturers of atrazine maintain that their product is safe when applied properly. The hope is that new research will clarify whether current levels of atrazine in drinking water are safe, or if more stringent regulations on its use are in order.

Figure 2.31 **Vegetables and fruits may contain residues of agricultural chemicals.** The photograph at right shows a low-flying crop duster raining pesticides in an agricultural field. Atrazine is applied this way to cotton and some other crops.

EXPLORE
ON YOUR OWN

It's easy to demonstrate the practical consequences of differences between hydrophilic and hydrophobic molecules. Just try this simple kitchen experiment. Take two identical clean plates. Smear one with grease (such as margarine or lard) and pour syrup over the other.

Next run comfortably warm water over both plates for thirty seconds and observe the results. Which plate got cleaner, and why? The companies that make dishwashing detergents manipulate them chemically so that their molecules have both hydrophobic and hydrophilic regions. Given what you know about the ability of water by itself to dissolve hydrophilic and hydrophobic substances, why might this be?

Courtesy Barry N. Williams

SUMMARY

Section 2.1 An element is a fundamental substance that cannot be broken down to other substances by ordinary chemical means. The four main elements in the body are oxygen, carbon, hydrogen, and nitrogen.

An atom is the smallest unit that has the properties of an element. Atoms are composed of protons, neutrons, and electrons. An element's atoms may vary in how many neutrons they contain. These variant forms are isotopes. The number and arrangement of an atom's electrons determine its interactions with other atoms.

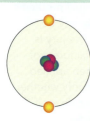

Section 2.3 Electrons move in orbitals within a series of shells around an atom's nucleus. An atom with one or more unfilled orbitals in its outer shell is likely to take part in chemical bonds.

A chemical bond is a union of the electron structures of atoms. Bonds join atoms into molecules. A chemical compound consists of atoms of two or more elements in unchanging proportions. In a mixture, two or more kinds of molecules mingle in variable proportions.

Section 2.4 Atoms generally have no net charge. An atom that gains or loses one or more electrons becomes an ion with a positive or negative charge.

In an ionic bond, positive and negative ions stay together by the mutual attraction of their opposite charges. In a covalent bond, atoms share one or more electrons. A hydrogen bond is a weak bond between polar molecules.

Section 2.5 Water is vital for the physical structure and chemical activities of cells. Hydrogen bonds between its molecules give water special properties, such as the ability to resist temperature changes and to dissolve other polar substances. A dissolved substance is a solute. Polar molecules are hydrophilic (attracted to water). Nonpolar substances, such as oils, are hydrophobic (repelled by water).

Section 2.7 The pH scale measures the concentration of hydrogen ions in a fluid. Acids release hydrogen ions (H^+), and bases release hydroxide ions (OH^-) that can combine with H^+.

At pH 7, the H^+ and OH^- concentrations in a solution are equal; this is a neutral pH. A buffer system maintains pH values of blood, tissue fluids, and the fluid inside cells. A salt is a compound that releases ions other than H^+ and OH^-.

Section 2.8 Carbon atoms bonded together in linear or ring structures are the backbone of organic compounds. Functional groups help determine the chemical and physical properties of many compounds.

Cells assemble and break apart most organic compounds by way of five kinds of reactions: transfers of functional groups, electron transfers, internal rearrangements, condensation reactions (dehydration synthesis), and cleavage reactions such as hydrolysis. Enzymes speed all these reactions. A polymer is a molecule built of three or more subunits; each subunit is called a monomer.

Cells have pools of dissolved sugars, fatty acids, amino acids, and nucleotides. These are small organic compounds with no more than about twenty carbon atoms. They are building blocks for the larger biological molecules—the carbohydrates, lipids, proteins, and nucleic acids (Table 2.5).

Section 2.9 Cells use carbohydrates for energy or to build cell parts. Monosaccharides, or single sugar units, are the simplest ones. Chains of sugars linked by covalent bonds are oligosaccharides; common ones, such as glucose, are disaccharides built of two sugar units. Polysaccharides are longer chains that store energy in the bonds between the sugar units (Table 2.5).

Section 2.10 The body uses lipids for energy, to build cell parts, and as signaling molecules. The most important dietary fats are triglycerides. Phospholipids are building blocks of cell membranes; sterols also are constituents of membranes and various key molecules.

TABLE 2.5 Summary of the Main Organic Molecules in Living Things

Category	Main Subcategories	Some Examples and Their Functions	
CARBOHYDRATES . . . contain an aldehyde or a ketone group and one or more hydroxyl groups.	**MONOSACCHARIDES** Simple sugars	Glucose	Energy source
	OLIGOSACCHARIDES Short-chain carbohydrates	Sucrose (a disaccharide)	Most common form of sugar; the form transported through plants
	POLYSACCHARIDES Complex carbohydrates	Starch, glycogen	Energy storage
		Cellulose	Structural roles
LIPIDS . . . are mainly hydrocarbon; generally do not dissolve in water but do dissolve in nonpolar substances, such as alcohols and other lipids.	**GLYCERIDES** Glycerol backbone with one, two, or three fatty acid tails (e.g., triglycerides)	Fats (e.g., butter), oils (e.g., corn oil)	Energy storage
	PHOSPHOLIPIDS Glycerol backbone, phosphate group, another polar group, and often two fatty acids	Lecithin	Key component of cell membranes
	STEROLS Four carbon rings; the number, position, and type of functional groups differ among sterols	Cholesterol	Component of animal cell membranes; precursor of many steroids and vitamin D
PROTEINS . . . are one or more polypeptide chains, each with as many as several thousand covalently linked amino acids.	**MOSTLY FIBROUS PROTEINS** Long strands or sheets of polypeptide chains; often strong, water-insoluble	Keratin	Structural component of hair, nails
		Collagen	Structural component of bone
		Myosin, actin	Functional components of muscles
	MOSTLY GLOBULAR PROTEINS One or more polypeptide chains folded into globular shapes; many roles in cell activities	Enzymes	Great increase in rates of reactions
		Hemoglobin	Oxygen transport
		Insulin	Control of glucose metabolism
		Antibodies	Immune defense
NUCLEIC ACIDS . . . are chains of units (or individual units) that each consist of a five-carbon sugar, phosphate, and a nitrogen-containing base.	**ADENOSINE PHOSPHATES**	ATP	Energy carrier
		cAMP	Messenger in hormone regulation
	NUCLEOTIDE COENZYMES	NAD^+, $NADP^+$, FAD	Transfer of electrons, protons (H^+) from one reaction site to another
	NUCLEIC ACIDS Chains of nucleotides	DNA, RNAs	Storage, transmission, translation of genetic information

Sections 2.11, 2.12 Proteins are built of amino acids and each one's function depends on its structure. Linked amino acids form a polypeptide chain. The linear sequence of the amino acids is a protein's primary structure. A protein's final shape comes about as the polypeptide chain bends, folds, and coils. Many proteins consist of more than one polypeptide chain. Some have other organic compounds bonded to them; examples are glycoproteins, which have oligosaccharides attached, and lipoproteins, which have lipids attached. A protein becomes denatured when some factor changes its usual three-dimensional shape.

Section 2.13 Nucleic acids such as DNA and RNA consist of nucleotides. A nucleotide is composed of one sugar (such as deoxyribose, the sugar in DNA), one or more phosphate groups, and a nitrogen-containing base. The nucleotide ATP transfers energy that powers chemical reactions in cells.

REVIEW QUESTIONS

1. Distinguish between an element, an atom, and a molecule.

2. Explain the difference between an ionic bond and a covalent bond.

3. Ionic and covalent bonds join atoms into molecules. What do hydrogen bonds do?

4. Name three vital properties of water in living cells.

5. Which small organic molecules make up carbohydrates, lipids, proteins, and nucleic acids?

6. Which of the following is the carbohydrate, the fatty acid, the amino acid, and the polypeptide?
   a. $^+NH_3$—CHR—COO^- c. $(glycine)_{20}$
   b. $C_6H_{12}O_6$ d. $CH_3(CH_2)_{16}COOH$

7. Describe the four levels of protein structure. How do a protein's side groups influence its interactions with other substances? What is denaturation?

8. Distinguish among the following:
   a. monosaccharide, polysaccharide, disaccharide
   b. peptide bond, polypeptide
   c. glycerol, fatty acid
   d. nucleotide, nucleic acid

SELF-QUIZ *Answers in Appendix V*

1. Fill in the blanks: The backbone of organic compounds forms when _____ atoms are covalently bonded.

2. A carbon atom can form up to _____ bonds with other atoms.
   a. four   c. eight
   b. six   d. sixteen

3. All of the following except _____ are building blocks or energy sources in cells.
   a. fatty acids   d. amino acids
   b. simple sugars   e. nucleotides
   c. lipids

4. Which of the following is not a carbohydrate?
   a. glucose molecule   c. margarine molecule
   b. simple sugar   d. polysaccharide

5. _____, a class of proteins, make metabolic reactions proceed much faster than they would on their own.
   a. Nucleic acids   c. Fatty acids
   b. Amino acids   d. Enzymes

6. Examples of nucleic acids are _____.
   a. polysaccharides   c. proteins
   b. DNA and RNA   d. simple sugars

7. Which phrase best describes what a functional group does?
   a. assembles large organic compounds
   b. influences the behavior of organic compounds
   c. splits molecules into two or more parts
   d. speeds up metabolic reactions

8. In _____ reactions, small molecules are linked by covalent bonds, and water can also form.
   a. hydrophilic   c. condensation
   b. hydrolysis   d. ionic

9. Match each type of molecule with its description.
   ____ chain of amino acids   a. carbohydrate
   ____ energy carrier   b. phospholipid
   ____ glycerol, fatty acids,   c. protein
        phosphate   d. DNA
   ____ chain of nucleotides   e. ATP
   ____ one or more sugar units

10. What kinds of bonds often control the shape (or tertiary form) of large molecules such as proteins?
    a. hydrogen   d. inert
    b. ionic   e. single
    c. covalent

CRITICAL THINKING

1. The pH of black coffee is 5, and that of milk of magnesia is 10. Is the coffee twice as acidic as milk of magnesia?

2. Draw a shell model of an uncharged nitrogen atom. Hint: Nitrogen has seven protons.

3. A store clerk says that vitamin C from rose hips is healthier than synthetic vitamin C. Based on what you know of the structure of organic compounds, does this claim seem credible? Why or why not?

4. Use the Web to find three examples of acid rain damage and efforts to combat the problem. You might start with the U.S. Environmental Protection Agency's acid rain home page.

5. Manufacturers make carbonated drinks by forcing pressurized carbon dioxide gas into flavored water. A chemical reaction between water molecules and some of the CO_2 molecules creates hydrogen ions (H^+) and bicarbonate, which is a buffer. In your opinion, is this reaction likely to raise the pH of a soda above 7, or lower it? Give your reasoning.

CELLS AND HOW THEY WORK

3

LINKS TO EARLIER CONCEPTS

The living cell is one of the first levels of organization in nature (1.3).

This chapter explains how lipids are organized to form cell membranes (2.10). You will learn where DNA and RNA are found in cells (2.13) and which cell parts build large molecules from carbohydrates and amino acids (2.9, 2.11, 2.12).

The chapter explains principles that govern the movement of water and solutes into and out of cells (2.5). It also considers how cells make and use the nucleotide ATP to fuel their activities (2.13).

KEY CONCEPTS

Basic Cell Features

All cells have an outer plasma membrane, and they contain cytoplasm and DNA. Most cells can only be seen with a microscope. Cells of all complex organisms contain compartments called organelles. Sections 3.1–3.4

Cells and Their Parts

Cell organelles have specialized functions. The plasma membrane controls the movement of substances into and out of the cell. The nucleus is a control center that contains the cell's DNA. Sections 3.6–3.12

Energy for Cell Activities

Organelles called mitochondria use organic compounds to make a molecule called ATP. ATP, a nucleotide, is the main fuel for cell activities. Sections 3.13–3.17

Top: G.L. Decker, Baylor College of Medicine; Middle: © Cengage Learning; Bottom: © Professors Pietro M. Motta & Tomonori Naguro/Photo Researchers, Inc.

Ethyl alcohol, the form in alcoholic beverages, is a powerful drug. In the stomach it triggers the release of acid that irritates cells in the stomach lining. Even moderate drinkers may develop ulcers and be at higher risk for cancers of the mouth, throat, and esophagus. In the brain, alcohol slows cell operations. Long-term, heavy use may damage memory, reflexes, and other functions. Binge drinking—five or more drinks in a brief period—can be deadly because the flood of alcohol can stop the heart. Liver cells detoxify 95 percent of the alcohol a person drinks, but in the long run this "detox" damages them too. Its legacy may be alcohol-related hepatitis and cirrhosis. Every day we make choices about which substances to put into our bodies. And in one way or another, everything that enters the body affects the ability of our cells to carry out a wide range of basic and specialized tasks.

© BananaStock/SuperStock

Homeostasis Preview

This chapter discusses how cells bring in some substances, keep others out, and make and release others. These activities constantly change the chemical and physical conditions in which cells operate.

3.1 What Is a Cell?

- From its size and shape to the structure of its parts, a cell is built to carry out life functions efficiently.

- Links to Life's characteristics 1.1, 1.3, Phospholipids 2.10

cell theory Scientific theory stating that cells are the smallest units of life, all organisms consist of one or more cells, and all cells come from pre-existing ones.

cytoplasm The contents of a cell between the outer plasma membrane and the nucleus.

cytosol The jellylike fluid portion of a cell's cytoplasm.

eukaryotic cell A cell that has a nucleus containing its DNA.

organelle Any of the compartments and sacs in a cell.

plasma membrane Covering that encloses a cell's internal parts.

prokaryotic cell A cell in which the DNA is not contained inside a nucleus; bacteria are prokaryotic cells.

There are trillions of cells in your body, and each one is a highly organized bit of life. A desire to understand cells led early biologists to develop the **cell theory**:

1. **Every organism is composed of one or more cells.**

2. **The cell is the smallest unit having the properties of life.**

3. **All cells come from pre-existing cells.**

In addition to these basics, today we know that chemical reactions occur in cells, and that cells contain and can pass on the hereditary material DNA.

All cells are alike in three ways

All living cells have three things in common. They have an outer **plasma membrane**, they contain DNA, and they contain cytoplasm.

The plasma membrane This outer covering encloses the cell's internal parts, so that cell activities can go on apart from events that may be taking place outside the cell. The plasma membrane doesn't completely isolate the cell's interior. Substances still can move across the membrane, as you will read later in this chapter.

Manfred Cage/Peter Arnold, Inc.

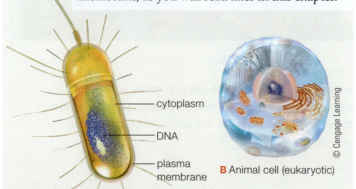

cytoplasm

DNA

plasma membrane

A Bacterial cell (prokaryotic)

B Animal cell (eukaryotic)

© Cengage Learning

Figure 3.1 Animated! There are two basic types of cells.
A A prokaryotic cell. **B** A eukaryotic cell, which has many types of organelles, including a nucleus.

TABLE 3.1 Eukaryotic and Prokaryotic Cells Compared

	Eukaryotic	Prokaryotic
PLASMA MEMBRANE	yes	yes
DNA-CONTAINING REGION	yes	yes
CYTOPLASM	yes	yes
NUCLEUS INSIDE A MEMBRANE	yes	no

DNA Cells contain DNA. Cells also contain molecules that can copy or "read" the genetic instructions DNA carries.

Cytoplasm Cytoplasm (sigh-toe-plaz-um) is everything between the plasma membrane and the region of DNA. It consists of a thick, jellylike fluid, the **cytosol**, and various other components.

There are two basic kinds of cells

Cells are classified into two basic kinds, depending on how they are organized internally (Table 3.1). In a **prokaryotic cell** (*prokaryotic* means "before the nucleus") nothing separates the cell's DNA from other internal cell parts. Bacteria, like the one diagrammed in Figure 3.1A, are prokaryotic cells.

Cells that have their DNA inside a nucleus are called **eukaryotic cells** ("true nucleus"). The nucleus is one of numerous **organelles** ("little organs") in eukaryotic cells (Figure 3.1B). Organelles are compartments or sacs we will discuss more fully in Section 3.2.

Most cells have a large surface area compared to their volume

A few cells—including the yolks of chicken eggs—can be seen with the unaided eye, but most cells are so small that they can only be seen with a microscope. For instance, a

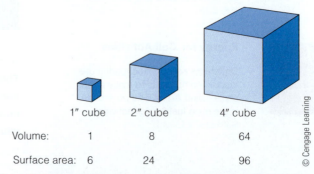

	1″ cube	2″ cube	4″ cube
Volume:	1	8	64
Surface area:	6	24	96

© Cengage Learning

Figure 3.2 The relationship of surface to volume influences the size of cells. Here boxes represent cells. If the linear dimensions of a box double, the volume increases 8 times but the surface area increases only 4 times. As in the text example, if the linear dimensions increase by 4 times, the volume is 64 times greater but the surface area is only 16 times larger.

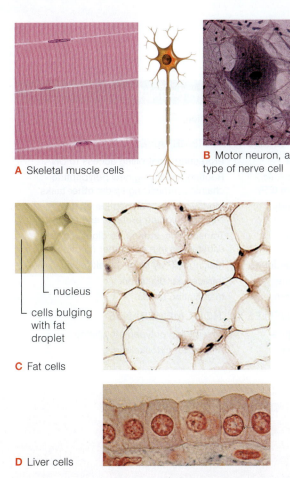

A Skeletal muscle cells

B Motor neuron, a type of nerve cell

nucleus

cells bulging with fat droplet

C Fat cells

D Liver cells

Figure 3.3 Human cells come in many shapes and sizes. A Cells of skeletal muscles are long and slender. **B** A motor neuron, a type of nerve cell, has slender extensions. **C** The cells that make up body fat are rounded and contain whitish lipid molecules. **D** These boxy-looking liver cells are shown in cross section. Each cell's nucleus looks reddish because it has been stained with dye. (Top Left: Ed Reschke/Peter Arnold; Top right: Carolina Biological Supply/Phototake; Middle: (c) University of Cincinnati, Raymond Walters College, Biology; Bottom: G.L. Decker, Baylor College of Medicine)

human red blood cell is so tiny that you could line up 2,000 of them across your thumbnail.

The **surface-to-volume ratio** is responsible for the small size of cells. This ratio is a physical relationship. It dictates that as the linear dimensions of a three-dimensional object increase, the volume of the object increases faster than its surface area does (Figure 3.2). For instance, if a round cell grew like an inflating balloon so that its diameter increased to 4 times the starting girth, the volume inside the cell would be 64 times more than before, but the cell's surface would be just 16 times larger. The cell would not have enough surface area to allow nutrients to flow inward rapidly, or for wastes or cell products to move rapidly outward. A large, round cell also would have trouble moving materials through its cytoplasm. In short order the cell would die.

In small cells, though, random motions of molecules easily distribute materials. If a cell isn't small, it likely is long and thin or has folds that increase its surface area relative to its volume. The smaller or narrower or more frilly the cell, the more efficiently materials can cross its surface and disperse inside it. Figure 3.3 shows four of the many

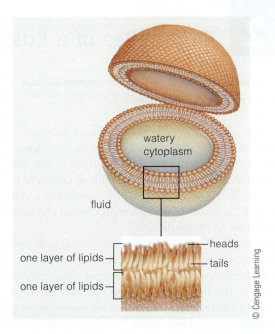

watery cytoplasm

fluid

heads

one layer of lipids

tails

one layer of lipids

© Cengage Learning

Figure 3.4 Animated! In cell membranes, phospholipids are arranged in a bilayer.

shapes of cells in your own body. Figure 3.3A depicts long, slender cells in a type of muscle called skeletal muscle. In the biceps of your upper arm they are many inches long—as long as the muscle itself.

Membranes enclose cells and organelles

Membranes enclose a eukaryotic cell and its organelles. Most molecules in cell membranes are phospholipids (Section 2.10). You may remember that a phospholipid has a hydrophilic (water-loving) head and two fatty acid tails, which are hydrophobic (water-fearing). In cell membranes, phospholipids organize into two layers with all the hydrophobic tails sandwiched between all the heads (Figure 3.4). This heads-out, tails-in arrangement is called a **lipid bilayer**. All cell membranes have the lipid bilayer structure. The hydrophilic heads of the phospholipids are dissolved in the watery fluids inside and outside cells.

lipid bilayer The structure of the plasma membrane, in which two parallel layers of phospholipids form with their heads facing outward and their tails facing inward.

surface-to-volume ratio The physical relationship by which the volume of a growing three-dimensional object increases faster than its surface area does.

TAKE-HOME MESSAGE

WHAT BASIC COMPONENTS DO ALL LIVING CELLS SHARE?

- All living cells have an outer plasma membrane, cytoplasm inside the membrane, and the genetic material DNA.
- Eukaryotic cells make up complex organisms such as humans. A eukaryotic cell's DNA is contained in the nucleus, an organelle.
- Prokaryotic cells, such as bacteria, have DNA but no nucleus.
- The surface-to-volume ratio limits cell size.
- A cell's membranes consist mainly of phospholipids arranged in a lipid bilayer.

3.2 Organelles of a Eukaryotic Cell

■ The interior of a cell is divided into organelles, each with one or more special functions.

In every eukaryotic cell, at any given moment, a vast number of chemical reactions are going on. Many of the reactions would conflict if they occurred in the same cell compartment. For example, a molecule of fat can be built by some reactions and taken apart by others, but a cell gains nothing if both sets of reactions proceed at the same time on the same fat molecule.

In eukaryotic cells organelles solve this problem. Table 3.2 lists those in animal cells. Most of them have an outer membrane that separates the inside of the organelle from the rest of the cytoplasm. It also controls the types and amounts of substances that enter or leave the organelle. For example, organelles called lysosomes contain enzymes that break down various unwanted substances. If the enzymes escaped from the organelle, they could destroy the entire cell. A membrane is not present in the organelles called ribosomes and centrioles.

Organelles also may serve as "way stations" for operations that occur in steps. Proteins are assembled and modified in steps involving several organelles.

Figure 3.5 shows where organelles and some other structures might be located in a body cell. This is only a general picture of cells. There are major differences in the structures and functions of cells in different tissues.

TABLE 3.2 Organelles of Animal Cells

Name	Function
ORGANELLES WITH MEMBRANES	
Nucleus	Protecting, controlling access to DNA
Endoplasmic reticulum (ER)	Routing, modifying new polypeptide chains; synthesizing lipids; other tasks
Golgi body	Modifying new polypeptide chains; sorting, shipping proteins and lipids
Vesicles	Transporting, storing, or digesting substances in a cell; other functions
Mitochondrion	Making ATP by sugar breakdown
Lysosome	Intracellular digestion
Peroxisome	Inactivating toxins
ORGANELLES WITHOUT MEMBRANES	
Ribosomes	Assembling polypeptide chains
Centriole	Anchor for cytoskeleton

TAKE-HOME MESSAGE

WHY ARE ORGANELLES IMPORTANT IN CELLS?

- Organelles isolate and physically organize chemical reactions in cells.
- Nearly all organelles have an outer membrane that separates the inside of the organelle from the cytosol and the rest of the cytoplasm.
- Organelles also provide separate locations for activities that occur in a sequence of steps.

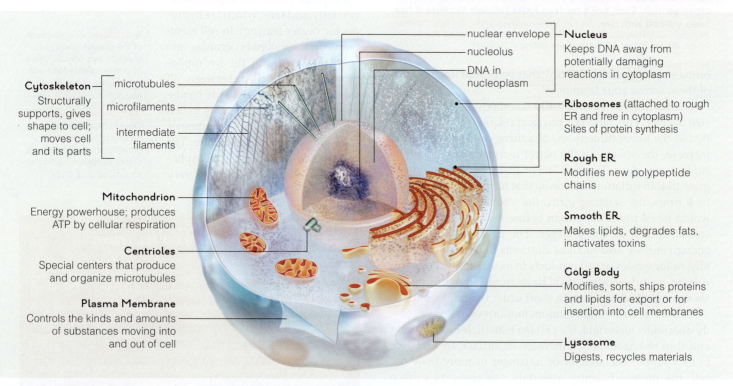

Cytoskeleton
Structurally supports, gives shape to cell; moves cell and its parts
— microtubules
— microfilaments
— intermediate filaments

Mitochondrion
Energy powerhouse; produces ATP by cellular respiration

Centrioles
Special centers that produce and organize microtubules

Plasma Membrane
Controls the kinds and amounts of substances moving into and out of cell

— nuclear envelope
— nucleolus
— DNA in nucleoplasm

Nucleus
Keeps DNA away from potentially damaging reactions in cytoplasm

Ribosomes (attached to rough ER and free in cytoplasm)
Sites of protein synthesis

Rough ER
Modifies new polypeptide chains

Smooth ER
Makes lipids, degrades fats, inactivates toxins

Golgi Body
Modifies, sorts, ships proteins and lipids for export or for insertion into cell membranes

Lysosome
Digests, recycles materials

Figure 3.5 Animated! An animal cell has a variety of internal parts. (© Cengage Learning)

3.3 How Do We See Cells?

The use of microscopes, called **microscopy**, has allowed us to learn a great deal about cells . A photograph of an image formed by a microscope is called a **micrograph**.

The micrographs in Figure 3.6 compare the sorts of detail three different types of microscopes can reveal. The red blood cells in Figure 3.6A were viewed with a compound light microscope. It has two or more glass lenses that bend (refract) incoming light rays to form an enlarged image of a specimen. With this method, the cell must be small or thin enough for light to pass through, and its parts must differ in color or optical density from their surroundings. Unfortunately, most cell parts are nearly colorless and they have about the same density. For this reason, before viewing cells through a light microscope, cells often are treated with dyes that react with some cell parts but not with others. Light microscopes only provide sharp images when the diameter of the object being viewed is magnified by 2,000 times or less.

Electron microscopes use magnetic lenses to bend beams of electrons. They reveal smaller details than even the best light microscopes can. There are several types, with new innovations occurring often.

With a scanning electron microscope, a beam of electrons is directed back and forth across a specimen thinly coated with metal. The metal emits some of its own electrons, and then the electron energy is converted into an image of the specimen's surface on a television screen. Most of the images have fantastic depth (Figure 3.6B).

A transmission electron microscope (Figure 3.6C) uses a magnetic field as the "lens" that bends a stream of electrons and focuses it into an image.

> **micrograph** The photograph of an image formed by a microscope.
>
> **microscopy** The use of a microscope to view objects, including cells, that are not visible to the unaided eye.

A Compound light microscope

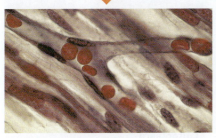

This image shows red blood cells inside a small blood vessel, as revealed by a light microscope. You may use this type of microscope in your biology class laboratory.

B Scanning electron micrographs

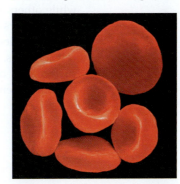

This scanning electron micrograph (SEM) with color added shows the "doughnut without a hole" shape of red blood cells.

This colored SEM is of a cancer cell. The fuzzy-looking white balls around it are white blood cells, part of the body's immune defenses.

C Transmission electron microscope

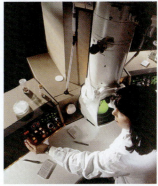

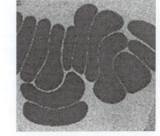

In this transmission electron micrograph (TEM) we see hemo-globin packed inside red blood cells. Hemoglobin is a protein that carries oxygen in the blood.

A, top: © Jupiterimages A, bottom: Science Source/Photo Researchers, Inc. B, top: Dr. Fred Hossler/Visuals Unlimited, Inc. B, bottom: Lennart Nilsson/ScanPix C, top: George Musil/Visuals Unlimited C, bottom: Lennart Nilsson/Scanpix

Figure 3.6 Different types of microscopes reveal different kinds of details about cells or their parts.

- The plasma membrane controls the movement of substances into and out of cells.
- Links to Polar molecules 2.4, Enzymes 2.8, Phospholipids 2.10

The plasma membrane is a mix of lipids and proteins

The plasma membrane isn't a solid, rigid wall between a cell's cytoplasm and the fluid outside. If it were, needed substances couldn't enter the cell and wastes couldn't leave it. Instead, the plasma membrane has a fluid quality, something like cooking oil. The membrane also is extremely thin. A thousand stacked like pancakes would be about as thick as this page.

In Figure 3.4 you've already seen a simple picture of a plasma membrane lipid bilayer with its "sandwich" of phospholipids. This structure often is described as a "mosaic" of proteins and different kinds of lipids. These include phospholipids, glycolipids, and, in the cells of humans and other animals, the lipid we call cholesterol. Plasma membrane proteins are embedded in the bilayer or attach to its outer or inner surface.

What makes the membrane fluid? A key factor is the movement of the molecules in it. Most phospholipids can spin on their long axis like a chicken on a rotisserie. They also move sideways and their tails flex. These movements help keep neighboring molecules from packing into a solid layer.

Proteins carry out most of the functions of cell membranes

The proteins that are embedded in or attached to a lipid bilayer carry out most of a cell membrane's functions (Figure 3.7). Many of these proteins are enzymes; you may recall from Chapter 2 that enzymes speed chemical reactions in cells. Other membrane proteins serve a range of functions. Some are channels through the membrane, while others are transporters that move substances across it. Still others are receptors; they are like docks for signaling molecules, such as hormones, that trigger changes in cell activities. Recognition proteins that sit like flags on the surface of a cell are chemical "fingerprints" that identify the cell as being of a specific type.

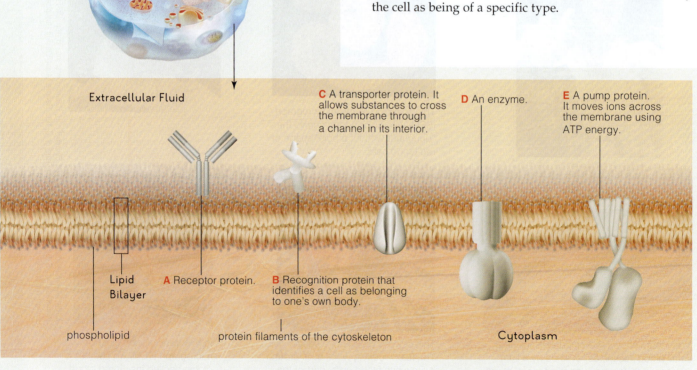

Extracellular Fluid

C A transporter protein. It allows substances to cross the membrane through a channel in its interior.

D An enzyme.

E A pump protein. It moves ions across the membrane using ATP energy.

Lipid Bilayer

A Receptor protein.

B Recognition protein that identifies a cell as belonging to one's own body.

phospholipid

protein filaments of the cytoskeleton

Cytoplasm

Figure 3.7 Animated! **A cell's plasma membrane consists of lipids and proteins.** Most of the lipids are phospholipids. This diagram also shows examples of membrane proteins. Biologists refer to the membrane's mix of lipids and proteins as a "mosaic."
(© Cengage Learning)

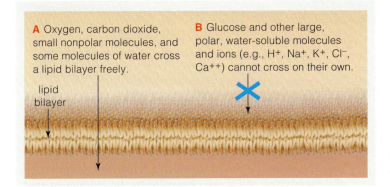

A Oxygen, carbon dioxide, small nonpolar molecules, and some molecules of water cross a lipid bilayer freely.

lipid bilayer

B Glucose and other large, polar, water-soluble molecules and ions (e.g., H+, Na+, K+, Cl−, Ca++) cannot cross on their own.

much crosses at a given time. Lipids in the bilayer are mostly nonpolar, so they let small, non-polar molecules such as carbon dioxide and oxygen slip across. Water molecules are polar, but some can move through gaps that briefly open up in the bilayer. Ions and large polar molecules (such as the blood sugar glucose) cross the bilayer through the interior of its transporter proteins. You will read more about this topic in Section 3.10.

selective permeability A property of the cell plasma membrane, in which the membrane allows only certain substances to cross it.

Figure 3.8 Animated! Cell membranes are selectively permeable. (© Cengage Learning)

The plasma membrane is "selective"

You have just read that a cell's plasma membrane is a bilayer containing lipids and proteins. These molecules give the membrane **selective permeability**. They allow some substances but not others to enter and leave a cell (Figure 3.8). They also control *when* a substance can cross and how

TAKE-HOME MESSAGE

HOW DOES THE PLASMA MEMBRANE'S STRUCTURE RELATE TO ITS FUNCTION?

- The plasma membrane is a lipid bilayer. It is a mix of various lipids and proteins and has a fluid quality.
- Proteins of the bilayer carry out most of the membrane's functions.
- The structure of the plasma membrane makes it selectively permeable. Some substances can cross it but others cannot.

3.5 A Watery Disaster for Cells

FOCUS ON HEALTH

Soon after the massive earthquake that devastated the island nation of Haiti in 2010, health officials called upon to help cope with the disaster began worrying about water. Specifically, they were concerned about waterborne diseases such as cholera, which can pose serious problems in places where public sanitation is poor. A bacterium, *Vibrio cholerae*, causes cholera. The microbe is found in coastal seas and in contaminated public water supplies. It can also taint shellfish such as oysters. It produces a toxin that affects pump proteins in the plasma membranes of cells in the small intestine. The toxin causes cells to pump out various ions, and other dissolved substances follow. As these substances leave, cells lose their water by osmosis, a process that is described in Section 3.10.

According to the Centers for Disease Control (CDC), a single glass of contaminated water or a few bites of tainted seafood may contain enough cholera bacteria to make a person seriously ill. Mildly infected people may not show symptoms, but they still can pass infectious bacteria in their feces for as long as ten days.

Treating cholera patients is a two-step task. Most urgently, patients need to be rehydrated with fluids that replenish both lost water and ions. Once this rehydration therapy is underway (either orally or intravenously), the next step is to administer an antibiotic that kills *V. cholerae*.

Vibrio cholerae bacteria, which cause cholera. This image is a scanning electron micrograph with color added.

Cholera's main symptom is massive watery diarrhea. In severe cases it can literally drain a person's body of water in less than a day. It is a common health threat in parts of Africa, Asia, and South America. So as the disaster unfolded in Haiti, getting bottled water to survivors became an urgent priority (Figure 3.9). Despite these efforts, several hundred thousand Haitians developed cholera, and more than 5,000 died.

Figure 3.9 Getting clean drinking water to Haiti earthquake survivors was an urgent priority in order to prevent disease.

3.6 The Nucleus

- Like a master control center, the nucleus contains and protects the cell's DNA, the genetic material.

chromatin A cell's DNA molecules and proteins attached to them.

chromosome An individual DNA molecule and attached proteins.

nuclear envelope A double membrane that separates the inside of the nucleus from the cytoplasm. It has many pores.

nucleolus A cluster of the RNA and proteins used to assemble ribosomes from their subunits.

nucleus Organelle that encloses a eukaryotic cell's DNA.

The **nucleus** encloses the DNA of a eukaryotic cell. DNA contains instructions for building a cell's proteins. Those proteins in turn determine a cell's structure and function. In a human cell there are forty-six DNA molecules that together would be more than 6 feet long if they were stretched out end to end.

Figure 3.10 shows the basic structure of the nucleus. The nucleus has several key functions. To begin with, it prevents DNA from getting tangled up with structures in the cytoplasm. When a cell divides, its DNA molecules must be copied so that each new cell receives a full set. If the DNA is separate, it is easier to copy and organize these hereditary instructions. Also, outer membranes of the nucleus are a boundary where the movement of substances to and from the cytoplasm can be controlled.

A nuclear envelope encloses the nucleus

Unlike the cell itself, the nucleus has two outer lipid bilayers, one pressed against the other. This double-membrane

system is called a **nuclear envelope** (Figure 3.11). The envelope surrounds the fluid part of the nucleus (the nucleoplasm), and many proteins are embedded in its layers. The outer portion of the nuclear envelope merges with the membrane of ER, an organelle in the cytoplasm that you'll read about in Section 3.7.

Threadlike bits of protein attach to the inner surface of the nuclear envelope. They anchor DNA molecules to the envelope and help keep them organized.

Proteins that span both bilayers have a wide variety of functions. Some are receptors or transporters. Others form pores, as you can see in Figure 3.11B. The pores are passageways. They allow small ions and molecules dissolved in the watery fluid inside and outside the nucleus to cross the nuclear membrane.

The nucleolus is where cells make the parts of ribosomes

As a cell grows, one or more dense masses appear inside its nucleus. Each mass is a **nucleolus** (noo-KLEE-oh-luhs), a construction site where some proteins and RNAs are combined to make the parts of ribosomes. These subunits eventually will cross through nuclear pores to the cytoplasm. There, they will briefly join up to form ribosomes. These organelles are "workbenches" where amino acids are assembled into proteins.

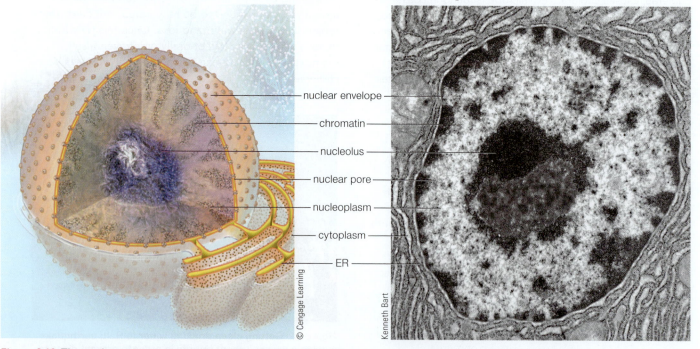

nuclear envelope

chromatin

nucleolus

nuclear pore

nucleoplasm

cytoplasm

ER

© Cengage Learning

Kenneth Bart

Figure 3.10 The nucleus of an animal cell contains the cell's DNA. The microscope image on the right shows the nucleus of an animal pancreas cell.

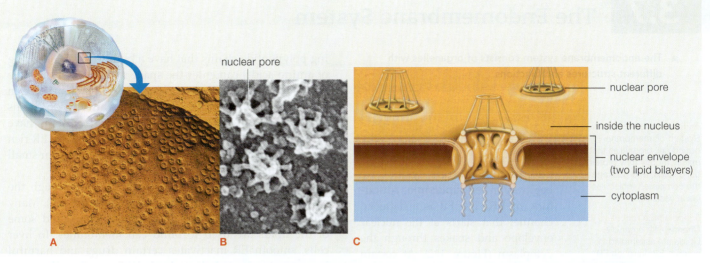

nuclear pore

nuclear pore

inside the nucleus

nuclear envelope
(two lipid bilayers)

cytoplasm

A   B   C

Figure 3.11 **Animated!** **The nuclear envelope is a double membrane with pores.** **A** This view of a cell's nuclear envelope shows pores that form channels through it. **B** This micrograph image reveals that each pore is a cluster of membrane proteins. They selectively allow certain substances to move into and out of the nucleus. The sketch of the nuclear envelope in **C** shows the envelope's structure. (A: Dr. Donald Fawcett/Visuals Unlimited, Inc B: A.C. Faberge, Cell and Tissue Research, 151:403-415, 1974 C: © Cengage Learning)

DNA is organized in chromosomes

When a eukaryotic cell is not dividing, you cannot see individual DNA molecules, nor can you see that each consists of two strands twisted together. The nucleus just looks grainy, as in Figure 3.10. When a cell is preparing to divide, however, it copies its DNA so that each new cell will get all the required hereditary instructions. Soon the duplicated DNA molecules are visible as long threads. They then fold and twist into a compact structure.

Early microscopists named the grainy-looking substance in the nucleus *chromatin*, and they called the compact structures *chromosomes* ("colored bodies"). Today we define **chromatin** as the cell's DNA along with the proteins associated with it. Sections of chromatin make up each **chromosome**—a double-stranded DNA molecule that carries genetic information. A chromosome looks grainy or compact depending on whether the cell is dividing or is in another part of its life cycle:

Events that begin in the nucleus continue in the cell cytoplasm

Outside the nucleus, new polypeptide chains for proteins are assembled on ribosomes. Many of them are used at once or stockpiled in the cytoplasm. Others enter the endomembrane system. As you'll read in the next section, this system includes various structures. It is where many proteins get their final form and where lipids are assembled and packaged.

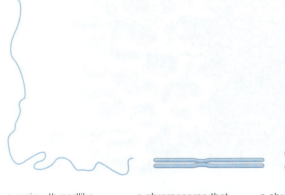

a grainy, threadlike molecule of DNA (two strands, with proteins)
© Cengage Learning

a chromosome that has been duplicated (two DNA molecules with proteins)

a chromosome that has been duplicated, then twisted and folded

3.7 The Endomembrane System

■ The endomembrane system consists of organelles with different structures and functions.

ER is a protein and lipid assembly line

To understand the functions of a cell's **endomembrane system,** we begin with **endoplasmic reticulum**, or **ER**. The ER is a flattened channel that starts at the nuclear envelope and snakes through the cytoplasm (Figure 3.12). At certain points inside the channel, lipids are assembled and "raw" polypeptide chains are processed into final proteins. In different places the ER looks rough or smooth, depending mainly on whether the organelles called ribosomes are attached to the side of the membrane that faces the cytoplasm. A **ribosome** is a platform for building a cell's proteins. Its parts were synthesized in the nucleolus, as described in Section 3.6.

Rough ER is studded with ribosomes (Figure 3.12A). Newly form-ing polypeptide chains that have a built-in signal (a string of amino acids) can enter the space inside rough ER or be incorporated into ER membranes. Once the chains are in rough ER, enzymes in the channel may attach side chains to them. Body cells that secrete finished proteins have extensive rough ER. For example, in your pancreas, ER-rich gland cells make and secrete enzymes that enter your small intestine and help you digest the food you eat.

Smooth ER has no ribosomes and curves through the cytoplasm like flat connecting pipes (Figure 3.12B). Many cells assemble most lipids inside these pipes, and some of the lipids are used to build cell membranes. In liver cells, smooth ER inactivates certain drugs and harmful by-products of metabolism. In skeletal muscle cells a type of smooth ER called sarcoplasmic reticulum stores and releases calcium ions essential for muscles to contract.

Golgi bodies "finish, pack, and ship"

A **Golgi body** is a series of flattened sacs that are often compared to a stack of cupped pancakes (Figure 3.12C). Enzymes in the sacs put the finishing touches on proteins and lipids, then package the completed molecules in vesicles for shipment to specific locations. A **vesicle** is a tiny sac that moves through the cytoplasm or takes up positions in it. For example, an enzyme in one Golgi region might attach a phosphate group to a new protein and then "pack"

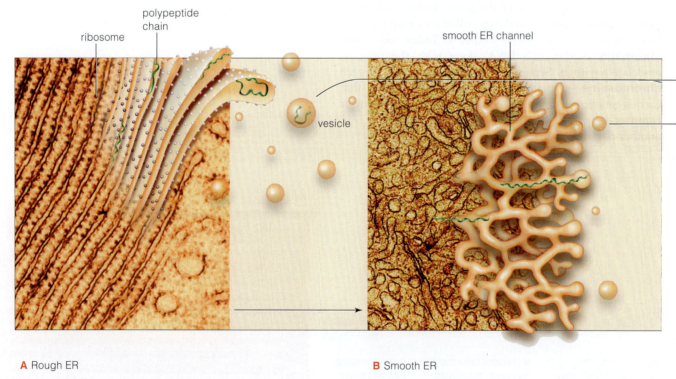

polypeptide chain

ribosome

smooth ER channel

vesicle

A Rough ER　　　　　　　　　　　**B** Smooth ER

Figure 3.12 The endomembrane system builds lipids and modifies many cell proteins. These molecules are sorted and shipped to other cell parts or to the plasma membrane to be exported out of the cell.

the protein into a vesicle, thereby giving it a "mailing tag" to its proper destination. The pancake at the top of a Golgi body is the organelle's "shipping gate" for molecules to be exported. Vesicles form there as patches of the membrane bulge out and then break away into the cell's cytoplasm.

Vesicles have a range of roles in cells

Vesicles may have other specialized roles in cells. An example is the lysosome, a type of vesicle that buds from the membranes of Golgi bodies (Figure 3.12D). A **lysosome**'s function is to chemically digest (break down) substances. It contains a potent stew of enzymes that speed the breakdown of proteins, some lipids, complex sugars, and nucleic acids. Lysosomes may even digest whole cells or cell parts. Often, lysosomes fuse with other vesicles

that have formed at a cell's plasma membrane and that contain bacteria or other undesirable items that attach to the plasma membrane. White blood cells of the immune system dispose of foreign material in the vesicles.

Vesicles called **peroxisomes** are sacs of enzymes that break down fatty acids and amino acids. The reactions produce hydrogen peroxide, a potentially harmful substance. But before hydrogen peroxide can injure the cell, another enzyme in peroxisomes converts it to water and oxygen or uses it to break down alcohol. When someone drinks alcohol, peroxisomes of liver and kidney cells are able to break down about half of it.

> **lysosome** Vesicle in which enzymes digest (break down) unwanted molecules.
>
> **peroxisome** Vesicle in which enzymes break down fatty acids and amino acids.

THINK OUTSIDE THE BOOK

In genetic conditions called lysosomal storage diseases, an enzyme is missing from cell lysosomes. The substance the missing enzyme would break down builds up instead. As it accumulates, it interferes with cell activities. Briefly research this topic. What is the most common lysosomal storage disease in humans? How does the disease affect a person who is born with it?

One good information source is the Lysosomal Storage Disease Center at the University of California San Francisco Children's Hospital (www .ucsfhealth.org/clinics/lysosomal_storage_disease_ center/index.html).

TAKE-HOME MESSAGE

WHAT DO ENDOMEMBRANE SYSTEM ORGANELLES DO?

- Ribosomes on rough endoplasmic reticulum (ER) build new cell proteins.
- In the smooth ER and Golgi bodies, lipids are assembled and many proteins are modified into their final form.
- Some vesicles move substances into or around cells or transport them to the outside.
- The vesicles called lysosomes and peroxisomes break down unwanted material.

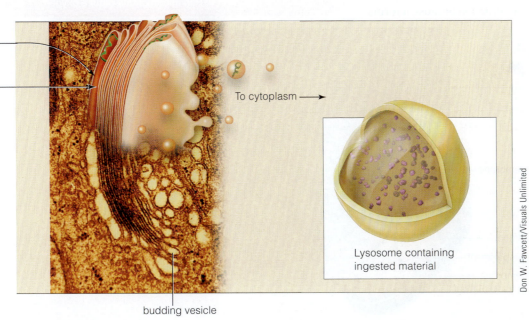

To cytoplasm →

Lysosome containing ingested material

Don W. Fawcett/Visuals Unlimited

budding vesicle

C Golgi body   **D** A lysosome

3.8 Mitochondria: The Cell's Energy Factories

- The energy for cell activities comes from ATP made in the cell's mitochondria.

- Link to ATP 2.13

Mitochondria make ATP

Section 2.13 introduced ATP, the main energy carrier in cells. Because ATP can deliver energy to nearly all the sites where chemical reactions occur in a cell, ATP is the fuel for most cell activities. ATP forms during reactions that break down organic compounds to carbon dioxide and water. These reactions occur in a **mitochondrion** (my-toe-KON-dree-ahn; plural: mitochondria).

mitochondrion Organelle that produces ATP, the main cell fuel.

Only eukaryotic cells contain mitochondria. The one shown in Figure 3.13 gives you an idea of their structure. The ATP-forming reactions that occur in mitochondria extract far more energy from organic compounds than can be obtained any other way. The reactions can't be completed without an ample supply of oxygen. Every time you inhale, you are taking in oxygen mainly for the mitochondria in your cells.

ATP forms in an inner compartment of the mitochondrion

A mitochondrion has a double-membrane system. As shown in the sketch in Figure 3.13, the outer membrane faces the cell's cytoplasm. The inner one generally folds back on itself, accordion-fashion. This membrane system is the key to the mitochondrion's function because it forms two separate compartments inside the organelle. In the outer one, enzymes and other proteins stockpile hydrogen ions. As Section 3.14 will explain, energy from electrons fuels this process.

Mitochondria have intrigued biologists because they are about the same size as bacteria and function like them in many ways as well. Mitochondria even have their own DNA (called mtDNA) and some ribosomes, and they divide independently of the cell they are in. Many biologists believe mitochondria evolved from ancient bacteria that were consumed by another ancient cell, yet did not die. If they became protected, permanent residents in the host cell, they might have lost structures and functions required for independent life while they were evolving into mitochondria, the ATP-producing organelles without which we humans could not survive.

Joe McBride/Stone/Getty Images

TAKE-HOME MESSAGE

WHAT DO MITOCHONDRIA DO?

- Mitochondria are the ATP-producing powerhouses of cells.
- ATP is produced by reactions that take place in the inner compartment formed by a mitochondrion's double-membrane system.
- Reactions that form ATP in mitochondria require oxygen.

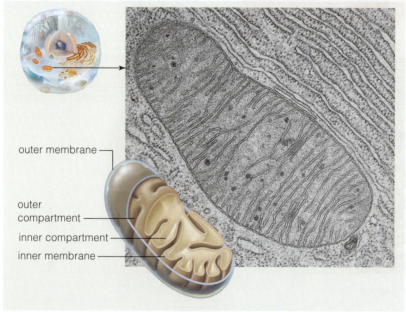

outer membrane

outer compartment

inner compartment

inner membrane

Micrograph, Keith R. Porter

Figure 3.13 Animated! **Mitochondria form ATP.** Sketch and transmission electron micrograph of a mitochondrion. Reactions inside mitochondria produce ATP, the major energy carrier in cells.

3.9   The Cell's Skeleton

- A cell's structural framework is called the cytoskeleton. The cytoskeleton's elements assemble and disassemble as needed for cell activities.

The **cytoskeleton** is a system of interconnected fibers, threads, and lattices in the cytosol (Figure 3.14A). Different proteins form these parts, which collectively give cells their shape, organization, and ability to move.

centrioles Cell structures that give rise to microtubules.

cilia Short, bendable structures built of microtubules.

cytoskeleton The cell's internal structural framework.

flagella Whiplike structures built of microtubules.

intermediate filaments Cytoskeleton filaments that anchor proteins (actin and myosin) in the cytosol and add strength to it.

microfilaments Filaments in the cytoskeleton that reinforce or anchor cell parts.

microtubules The largest elements of the cytoskeleton.

Microtubules are the largest cytoskeleton elements. They spatially organize the interior of the cell, and also help move cell parts. **Microfilaments** often reinforce some part of a cell, such as the plasma membrane. They also anchor some membrane proteins.

Some kinds of cells also have **intermediate filaments** that add strength much as steel rods strengthen concrete pillars. Intermediate filaments also anchor the filaments of two proteins, called actin and myosin, which interact in muscle cells and enable the muscle to contract. Chapter 6 explains how they function.

Some types of cells move about by **flagella** (singular: flagellum) or have moving **cilia** (singular: cilium) (Figures 3.14B and 3.14C). In both structures nine pairs of microtubules ring a central pair. A system of spokes and links holds this "9 + 2 array" together (Figure 3.15). The flagellum or cilium bends when microtubules in the ring slide over each other. Whiplike flagella propel human sperm.

Cilia are shorter than flagella, and there may be more of them per cell. In your respiratory tract, thousands of ciliated cells whisk out mucus laden with dust or other undesirable material. The microtubules of cilia and flagella arise from **centrioles**, which remain at the base of the completed structure as a "basal body." Centrioles also have a major role to play when a cell divides (Chapter 18).

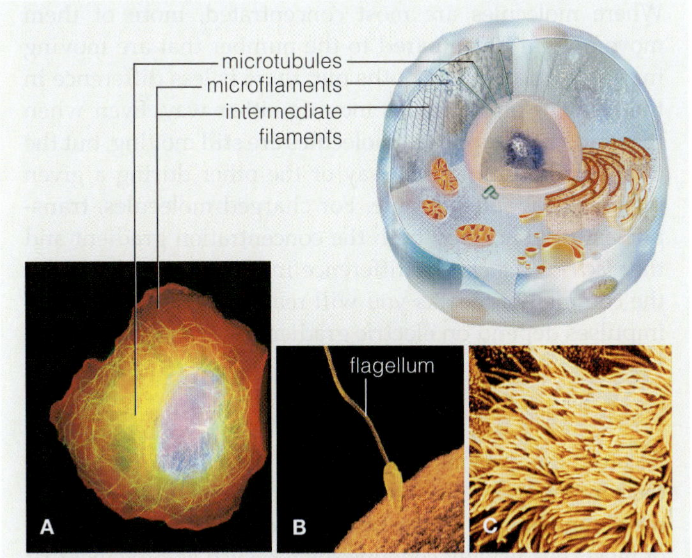

TAKE-HOME MESSAGE

WHAT IS THE ROLE OF THE CYTOSKELETON?

- The cytoskeleton gives a cell its shape, internal structure, and capacity for movement.
- The main elements of the cytoskeleton are microtubules, microfilaments, and intermediate filaments.
- Some types of cells have flagella or cilia, which move by way of microtubules.

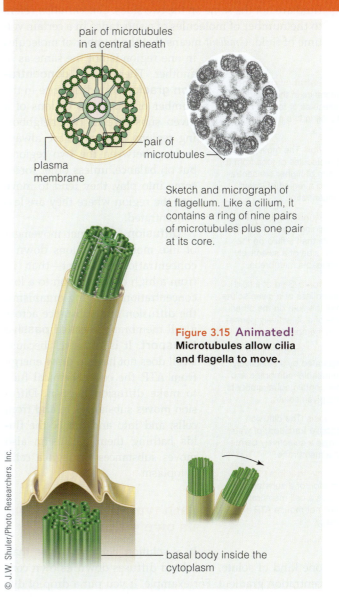

pair of microtubules in a central sheath

plasma membrane

pair of microtubules

Sketch and micrograph of a flagellum. Like a cilium, it contains a ring of nine pairs of microtubules plus one pair at its core.

Figure 3.15 Animated! **Microtubules allow cilia and flagella to move.**

basal body inside the cytoplasm

microtubules
microfilaments
intermediate filaments

flagellum

A   **B**   **C**

Figure 3.14 Animated! **Microtubules and filaments make up the cytoskeleton. A The cytoskeleton of a pancreas cell. B The flagellum of a sperm cell. C Cilia in an airway in the lungs.** (Left: © Jennifer C. Waters/Photo Researchers, Inc. Center: © Don W. Fawcett/Photo Researchers, Inc. Right: © J.W. Shuler/Photo Researchers, Inc.)

© J.W. Shuler/Photo Researchers, Inc.

3.10 | How Diffusion and Osmosis Move Substances across Membranes

- A cell takes in and expels substances across its plasma membrane. Diffusion and osmosis are the major means for accomplishing these tasks.

- Phospholipids 2.10, Protein function 2.12

As you now know, a cell's plasma membrane is selectively permeable. It allows only certain kinds of substances to enter and leave the cell. Why does a solute move one way or another at any given time? The answer starts with concentration gradients.

In diffusion, a dissolved molecule or ion moves down a concentration gradient

There is fluid on both sides of a cell's plasma membrane, but the kinds and amounts of dissolved substances in the fluid are not the same on the two sides. *Concentration* refers to the number of molecules of a substance in a certain volume of fluid. *Gradient* means that the number of molecules in one region is not the same as in another. Therefore, a **concentration gradient** is a difference in the number of molecules or ions of a given substance in two neighboring regions. Molecules are always moving between the two regions, but on balance, unless other forces come into play, they tend to move into the region where they are less concentrated.

Diffusion is the net movement of like molecules or ions down a concentration gradient—that is, from a high concentration to a low concentration. In living organisms, the diffusion of a substance across a cell membrane is called **passive transport**. It is "passive" because a cell does not have to draw energy from ATP, the cell's chemical fuel, to make diffusion happen. Diffusion moves substances to and from cells, and into and out of the fluids bathing them. Diffusion also moves substances through a cell's cytoplasm.

Each type of solute follows its own gradient

If a solution contains more than one kind of solute, each kind diffuses down its own concentration gradient. For example, if you put a drop of dye

concentration gradient A difference in the number of molecules or ions of a substance in two neighboring regions.

diffusion The movement of molecules or ions from a region of higher concentration to a region of lower concentration.

hypertonic Said of a fluid containing more of a given solute than a fluid on the other side of a selectively permeable membrane.

hypotonic Said of a fluid having less of a given solute than the fluid on the other side of a selectively permeable membrane.

isotonic Said of fluids separated by a selectively permeable membrane and that contain equal amounts of a given solute.

osmosis The diffusion (passive transport) of water across a selectively permeable membrane.

passive transport The diffusion of a substance across a cell membrane; does not require ATP energy.

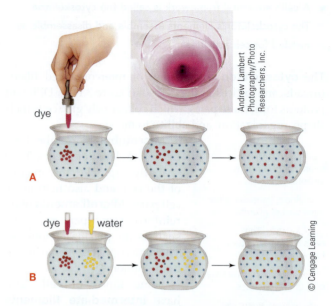

Andrew Lambert Photography/Photo Researchers, Inc.

© Cengage Learning

Figure 3.16 Substances diffuse down a concentration gradient. A A drop of dye enters a bowl of water. Gradually the dye molecules disperse evenly through the molecules of water. **B** The same thing happens with the water molecules. If red dye and yellow dye are added to the same bowl, each substance will move (diffuse) down its own concentration gradient.

in one side of a bowl of water, the dye molecules diffuse to the region where they are less concentrated. Likewise, the water molecules move in the opposite direction, to the region where *they* are less concentrated (Figure 3.16).

Molecules diffuse faster when the gradient is steep. Where molecules are most concentrated, more of them move outward, compared to the number that are moving in. As the gradient smooths out, there is less difference in the number of molecules moving either way. Even when the gradient disappears, molecules are still moving, but the total number going one way or the other during a given interval is about the same. For charged molecules, transport is influenced by both the concentration gradient and the *electric gradient*—a difference in electric charge across the cell membrane. As you will read in Chapter 13, nerve impulses depend on electric gradients.

Water crosses membranes by osmosis

Because the plasma membrane is selectively permeable, the concentration of a solute can increase on one side of the membrane but not on the other. For example, the cytoplasm of most cells usually contains solutes (such as proteins) that cannot diffuse across the plasma membrane. When solutes become more concentrated on one side of the plasma membrane, the resulting solute concentration gradients affect how water diffuses across the membrane. **Osmosis** (OSS-MOE-sis)

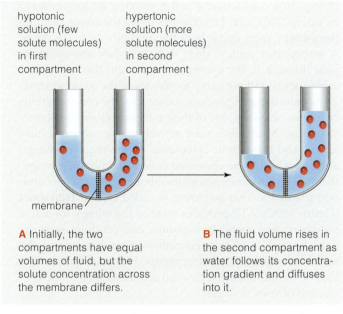

hypotonic solution (few solute molecules) in first compartment

hypertonic solution (more solute molecules) in second compartment

membrane

A Initially, the two compartments have equal volumes of fluid, but the solute concentration across the membrane differs.

B The fluid volume rises in the second compartment as water follows its concentration gradient and diffuses into it.

Figure 3.17 Animated! **The concentration of a solute affects the movement of water by osmosis.** (© Cengage Learning)

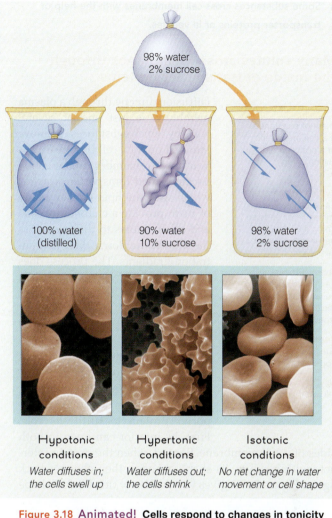

98% water
2% sucrose

100% water (distilled)

90% water 10% sucrose

98% water 2% sucrose

Hypotonic conditions

Water diffuses in; the cells swell up

Hypertonic conditions

Water diffuses out; the cells shrink

Isotonic conditions

No net change in water movement or cell shape

All three images: M. Sheetz, R. Painter and S. Singer, Journal of Cell Biology, 70:193 (1976), by copyright permission of the Rockefeller University Press

Figure 3.18 Animated! **Cells respond to changes in tonicity of body fluids.** In the sketches, membrane-like bags that allow water but not sucrose to cross are placed in hypotonic, hypertonic, and isotonic solutions. Arrow width represents the relative amount of water movement in each container. Red blood cells cannot actively take in or expel water. The micrographs show what happens to them when they are placed in solutions like those in the sketches. Red blood cells in a hypotonic solution quickly explode.

is the name for the diffusion of water across a selectively permeable membrane in response to solute concentration gradients. Figure 3.17 is a simple diagram of this process.

Tonicity is the concentration of solutes in a solution. When solute concentrations in the fluids on either side of a cell membrane are the same, the fluids are **isotonic** (*iso-* means "same") and there is no net flow of water in either direction across the membrane. When the solute concentrations are not equal, one fluid is **hypotonic**—it has fewer solutes. The other has more solutes and it is **hypertonic**. Figure 3.18 shows how the tonicity of a fluid affects red blood cells. A key point to remember is that water always tends to move from a hypotonic solution to a hypertonic one because it always moves down its concentration gradient.

If too much water enters a cell by osmosis, in theory the cell will swell up until it bursts. This is not a danger for most body cells because they can selectively move solutes out—and as solutes leave, so does water. Also, the cytoplasm exerts pressure against the plasma membrane. When this pressure counterbalances the tendency of water to follow its concentration gradient, osmosis stops.

Every moment, cell activities and other events change the factors that affect the solute concentrations of body fluids and water movements between them. Chapter 12 explains how osmotic water movements help maintain the body's water balance.

3.11 Other Ways Substances Cross Cell Membranes

■ Some substances cross cell membranes with the help of transporter proteins or in vesicles.

Many solutes cross membranes through the inside of transporter proteins

active transport
Movement of substances across a cell membrane against a concentration gradient, using energy from ATP.

endocytosis Process by which a cell takes in a large molecule or particle by forming a vesicle that encloses it and moves it into the cell cytoplasm.

exocytosis Process in which a vesicle encloses and moves a large molecule or particle to the cell surface and expels it.

facilitated diffusion Diffusion assisted by a transporter protein.

phagocytosis Endocytosis of a cell or other organic matter.

Diffusion directly through a plasma membrane is just one of several ways by which substances can move into and out of a cell (Figure 3.19A). You may remember that Section 3.4 mentioned transporter proteins, which span the lipid bilayer. Many of them provide a channel for ions and other solutes to diffuse across the membrane down their concentration gradients. The process does not require ATP energy, so it is a form of passive transport (Figure 3.19B). This type of passive transport sometimes is called **facilitated diffusion** because the transporter proteins "facilitate" the movement by providing a route for the solute that is crossing the cell membrane.

Two features allow a transporter protein to fulfill its role. First, its interior can open to both sides of a cell membrane. Second, when the protein interacts with a solute, its shape changes, then changes back again. Figure 3.19D gives you an idea of this kind of shape change in a case where a cell is moving an ion outward by active transport. The changes move the solute through the protein, from one side of the lipid bilayer to the other. A transporter protein does not allow just any solute to pass through it. For example, the protein that transports amino acids will not carry molecules of the sugar glucose.

As cells use and produce substances, the concentrations of solutes on either side of their membranes are constantly changing. A cell also must actively move certain solutes in, out, and through its cytoplasm. Action requires energy, and so cells have mechanisms called "membrane pumps" that move substances across membranes *against* concentration gradients. This pumping is called **active transport** (Figure 3.19C). ATP provides most of the energy for active transport, and membrane pumps can continue working until the solute is *more* concentrated on the side of the membrane where it is being pumped. This difference lays the chemical foundation for vital processes such as the contraction of your muscles.

Vesicles transport large solutes

Transporter proteins can only move small molecules and ions into or out of cells. To bring in or expel larger molecules or particles, cells use vesicles that form through endocytosis and exocytosis (Figure 3.20).

In **endocytosis** ("coming inside a cell"), a cell takes in substances next to its surface. A small indentation forms at the plasma membrane, balloons inward, and pinches off. The resulting vesicle transports its contents or stores them in the cytoplasm (Figure 3.20A). When endocytosis

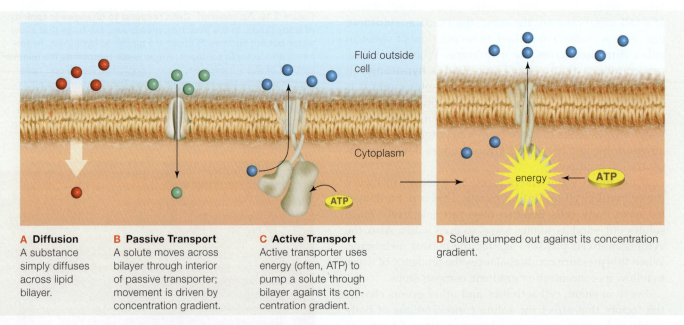

A Diffusion
A substance simply diffuses across lipid bilayer.

B Passive Transport
A solute moves across bilayer through interior of passive transporter; movement is driven by concentration gradient.

C Active Transport
Active transporter uses energy (often, ATP) to pump a solute through bilayer against its concentration gradient.

D Solute pumped out against its concentration gradient.

Fluid outside cell

Cytoplasm

energy ← ATP

Figure 3.19 Animated! **Substances cross cell membranes in a variety of ways.** Notice that **A** diffusion and **B** passive transport do not require the cell to invest energy. **C** Active transport uses ATP energy, whether a cell is moving a needed substance inward or **D** releasing a substance to the outside. (© Cengage Learning)

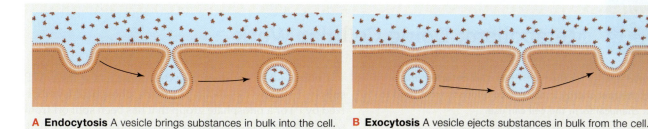

A **Endocytosis** A vesicle brings substances in bulk into the cell.

B **Exocytosis** A vesicle ejects substances in bulk from the cell.

© Cengage Learning

Figure 3.20 **Animated!** In **A** endocytosis and **B** exocytosis, vesicles move large molecules or particles across the plasma membrane.

brings organic matter into the cell, the process is called **phagocytosis**, or "cell eating."

In **exocytosis** ("moving out of a cell"), a vesicle moves to the cell surface and the protein-studded lipid bilayer of its membrane fuses with the plasma membrane (Figure 3.20B). Its contents are then released to the outside.

3.12 When Mitochondria Fail

FOCUS ON HEALTH

In as many as 1 in every 4,000 babies born in the United States each year, mitochondria, the cell energy factories, don't function properly. Other children will develop a mitochondrial disease before the age of ten. These inborn conditions usually are caused by gene changes, or mutations. In other cases, toxins—including impurities in illicit drugs—damage mitochondria. Regardless, cells with "sick" mitochondria don't have enough ATP energy to fuel normal operations. Mitochondrial disorders tend to have the most serious effects in cells in the heart, brain, and hard-working muscles because those organs require a great deal of energy.

A mitochondrial disorder called Luft's syndrome was the first disease to be directly linked to a malfunctioning cell organelle. The brother and sister pictured in Figure 3.21 were diagnosed with Friedreich's ataxia, an inherited condition in which changes to mitochondria eventually kill the affected organelles. Symptoms include a loss of muscle coordination (ataxia), weak muscles, and serious heart problems. Many affected people die in early adulthood.

Unfortunately, diagnosing mitochondrial disorders can be a tricky business. Any of several hundred different gene mutations may be at fault and symptoms can vary widely. And until a disorder is properly diagnosed, it may not be properly treated. Researchers at the Seattle Children's Research Institute and at the University of Washington have teamed up to change this bleak picture. They are working to develop a high-tech gene screening method that will speed up diagnosis by quickly pinpointing the precise genetic problem in sick children whose symptoms suggest that a mitochondrial disorder is the culprit.

To date, more than 40 different mitochondrial disorders have been identified. Some affect many body organs, others only a single one. None of these disorders can be cured, but early diagnosis and proper treatment may help curb symptoms and prevent progression of the disease. Some patients live full and reasonably active lives for decades.

© Louise Chalcraft-Frank and FARA

Figure 3.21 **Leah and Joshua both inherited the mitochondrial disorder called Friedreich's ataxia.**

- Cells need energy for their activities. Cell mitochondria convert the raw energy in organic compounds from food to ATP—a chemical form the cell can use.
- Links to Organic compounds 2.8, Energy carriers 2.13

active site Area on the surface of an enzyme where the enzyme and its substrate can interact.

anabolism Metabolic activity that builds large molecules from smaller ones.

ATP/ADP cycle A cycle in which a phosphate attaches to ADP, forming ATP, which then transfers a phosphate elsewhere, becoming ADP again.

catabolism Metabolic activity that breaks down large molecules into smaller ones.

metabolism The chemical reactions in cells.

substrate The particular kind of molecule that interacts with a given enzyme.

ATP is the cell's energy currency

The chemical reactions in cells are called **metabolism**. Some reactions release energy and others require it. ATP links the two kinds of reactions, carrying energy from one reaction to another. You may remember from Section 2.13 that ATP is short for adenosine triphosphate, one of the nucleotides. A molecule of ATP consists of the five-carbon sugar ribose to which adenine (a nucleotide base) and three phosphate groups are attached (Figure 3.22A). ATP's stored energy is contained in the bond between the second and third phosphate groups.

Enzymes can break the bond between the second and third phosphate groups of the ATP molecule. The enzymes then can attach the released phosphate group to another molecule. When a phosphate group is moved from one molecule to another, stored energy goes with it.

Cells use ATP constantly, so they must renew their ATP supply. In many metabolic processes, phosphate (symbolized by P_i) or a phosphate group that has been split off from some substance is attached to ADP, adenosine diphosphate (the prefix *di-* indicates that *two* phosphate groups are present). Now the molecule, with three phosphates, is ATP. And when ATP transfers a phosphate group elsewhere, it reverts to ADP. In this way it completes the **ATP/ADP cycle** (Figure 3.22B).

Like money earned at a job and then spent to pay your expenses, ATP is earned in reactions that produce energy and spent in reactions that require it. That is why textbooks often use a cartoon coin to symbolize ATP.

There are two main types of metabolic pathways

At this moment thousands of reactions are transforming thousands of substances inside each of your cells. Most of these reactions are part of metabolic pathways, steps in which reactions take place one after another. There are two main types of metabolic pathways, called anabolism and catabolism.

In **anabolism**, small molecules are put together into larger ones. In these larger molecules, the chemical bonds hold more energy. Anabolic pathways assemble complex carbohydrates, proteins, and other large molecules. The energy stored in their bonds is a major reason why we can use these substances as food.

In **catabolism**, large molecules are broken down to simpler ones. Catabolic reactions disassemble complex

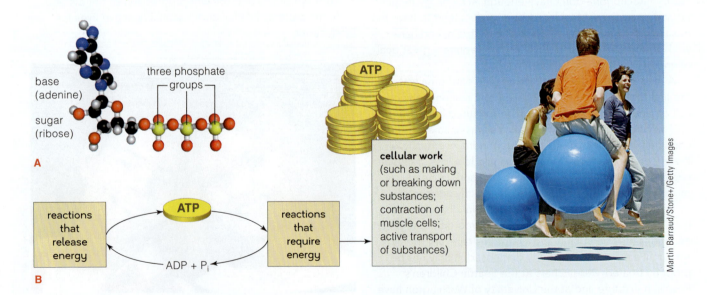

base (adenine)

three phosphate groups

sugar (ribose)

A

ATP

cellular work (such as making or breaking down substances; contraction of muscle cells; active transport of substances)

reactions that release energy

ATP

reactions that require energy

ADP + P_i

B

Martin Barraud/Stone+/Getty Images

Figure 3.22 ATP provides energy for cell activities. A Structure of ATP. **B** ATP connects energy-releasing reactions with energy-requiring ones. In the ATP/ADP cycle, the transfer of a phosphate group turns ATP into ADP, then back again to ATP.

carbohydrates, proteins, and similar molecules, releasing their components for use by cells. For example, when a complex carbohydrate is catabolized, the reactions release the simple sugar glucose, the main fuel for cells.

Any substance that is part of a metabolic reaction is called a *reactant*. A substance that forms between the beginning and the end of a metabolic pathway is an *intermediate*. Substances present at the end of a reaction or a pathway are the *products*.

Many metabolic pathways advance step by step from reactants to products:

reactant $\xrightarrow{\text{enzyme 1}}$ intermediate $\xrightarrow{\text{enzyme 2}}$ intermediate $\xrightarrow{\text{enzyme 3}}$ product

In other pathways the steps occur in a cycle, with the products serving as reactants to start things over.

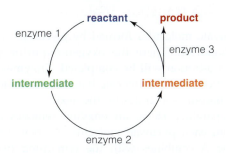

Enzymes are essential in metabolism

Metabolic reactions require *enzymes,* which you first read about in Section 2.8. Most enzymes are proteins, and all are catalysts: They speed up chemical reactions. In fact, enzymes generally make reactions occur hundreds to millions of times faster than would be possible otherwise. Enzymes are not used up in reactions, so a given enzyme molecule can be used over and over.

Each kind of enzyme can only interact with specific kinds of molecules, which are called its **substrates**. The enzyme can chemically recognize a substrate, bind it, and change it in some way. An example is thrombin, one of the enzymes required to clot blood. It only recognizes a side-by-side set of two particular amino acids in a protein. When thrombin "sees" this arrangement, it breaks the peptide bond between the amino acids.

An enzyme and its substrate interact at a surface crevice on the enzyme. This area is called an **active site**. Figure 3.23 shows how enzyme action can combine two substrate molecules into a new, larger product molecule.

Powerful as they are, enzymes only work well within a certain temperature range. For example, if a person's body temperature rises too high, the increased heat energy breaks bonds holding an enzyme in its three-dimensional shape. The shape changes, substrates can't bind to the active site as usual, and chemical reactions do not occur as normal. For this reason people usually die if their internal temperature reaches 44°C (112°F).

Enzymes also function best within a certain pH range—in the body, from pH 7.35 to 7.4. Above or below this range most enzymes cannot operate normally.

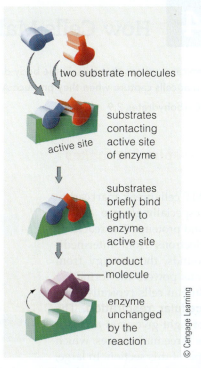

Figure 3.23 Animated! **Enzymes and substrates fit together physically.** When substrate molecules contact an enzyme's active site, they bind to the site for a brief time and a product molecule forms. When the product molecule is released, the enzyme goes back to its previous shape. The reaction it catalyzed does not change it in any way.

Section 2.13 mentioned that organic molecules called *coenzymes* assist with many reactions. Many coenzymes are derived from vitamins, which is one reason why vitamins are important in the diet.

The body controls the activity of enzymes

Controls may boost the action of enzymes, slow it down, or adjust how fast new enzyme molecules are made—and thus how many are available for a given metabolic pathway. For example, when you eat, food entering your stomach causes gland cells there to secrete the hormone gastrin into your bloodstream. Stomach cells with receptors for gastrin respond in a variety of ways, such as secreting the ingredients of "gastric juice"—including enzymes that break down food proteins.

TAKE-HOME MESSAGE

WHAT ARE METABOLIC PATHWAYS?

- Metabolic pathways are sequences of chemical reactions that build or dismantle molecules in cells.
- Pathways of anabolism require ATP energy to build large molecules from smaller ones. Pathways of catabolism release ATP energy as large molecules are broken down to smaller ones.
- Enzymes speed the rate of chemical reactions. Each type acts only on specific substrates.
- A given enzyme acts only on specific substrates. All enzymes function best within certain ranges of temperature and pH.

- The chemical reactions that sustain the body depend on energy that cells capture when they produce ATP.
- Link to Carbohydrates 2.9

Cellular respiration makes ATP

To make ATP, cells break apart carbohydrates, especially glucose, as well as lipids and proteins. The reactions remove electrons from intermediate compounds, then energy from the electrons powers the formation of ATP. Human cells typically form ATP by **cellular respiration**. The diagram at right gives you an overview of its three main stages, which are the topic of this section. In large, complex organisms like ourselves, all but the first stage of this process usually is aerobic—it uses oxygen. Glucose is the most common raw material for cellular respiration, so it will be our example here.

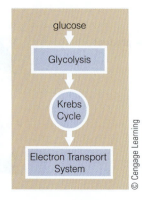

glucose

Glycolysis

Krebs Cycle

Electron Transport System

© Cengage Learning

Step 1: Glycolysis breaks glucose down to pyruvate

Cellular respiration starts in the cell's cytoplasm, in a set of reactions called **glycolysis**—literally, "splitting sugar." You may recall that glucose is a simple sugar. Each glucose molecule consists of six carbon atoms, twelve hydrogens, and six oxygens, all joined by covalent bonds. During glycolysis, a glucose molecule is broken into two molecules of a compound called pyruvate. As shown in Figure 3.24, each pyruvate molecule has three carbons.

When glycolysis begins, two ATPs each transfer a phosphate group to glucose, donating energy to it. This kind of transfer is called **phosphorylation**. It adds enough energy to glucose to begin the energy-releasing steps of glycolysis.

The first energy-releasing step breaks the glucose into two molecules of PGAL (for phosphoglyceraldehyde), which are converted to intermediates. These molecules then each donate a phosphate group to ADP, forming ATP. The same thing happens with the next intermedi-

cellular respiration The overall aerobic (oxygen-using) process by which cells break down organic molecules to make ATP.

electron transport system The chain of reactions in mitochondria that uses energy from electrons to generate many ATP molecules.

glycolysis Process that breaks apart glucose molecules, forming pyruvate, in the first stage of cellular respiration.

Krebs cycle Process that produces energy-rich compounds (NADH and FADH$_2$) that deliver electrons to electron transport systems in mitochondria. The cycle also produces a small amount of ATP.

phosphorylation The transfer of a phosphate group to a molecule.

ate in the sequence, and the end result is two molecules of pyruvate and four ATP. However, because two ATP were invested to start the reactions, the final, *net* energy yield is only two ATP.

Notice that glycolysis does not use oxygen. If oxygen is not available for the following aerobic steps of cellular respiration, for a short time a cell can still form a small amount of ATP by a process of fermentation, which also does not use oxygen. You will read more about this "back-up" process for forming ATP later in the chapter.

Step 2: The Krebs cycle produces energy-rich transport molecules

The pyruvate molecules formed by glycolysis move into a mitochondrion. There the oxygen-requiring phase of cellular respiration will be completed. Enzymes catalyze each reaction, and the intermediate molecules formed at one step become substrates for the next.

In preparatory steps, an enzyme removes a carbon atom from each pyruvate molecule. A coenzyme called coenzyme A combines with the remaining two-carbon

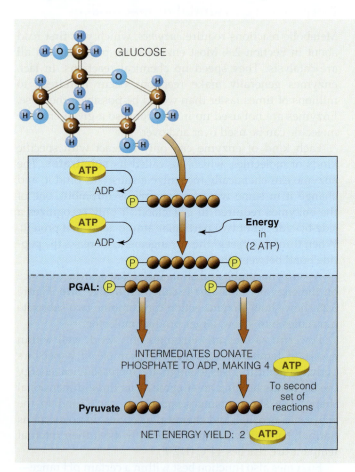

GLUCOSE

ATP
ADP
P

ATP
ADP
P

Energy in (2 ATP)

P — P

PGAL: P

P

INTERMEDIATES DONATE PHOSPHATE TO ADP, MAKING 4 ATP

To second set of reactions

Pyruvate

NET ENERGY YIELD: 2 ATP

Figure 3.24 Animated! Glycolysis splits glucose molecules and forms a small amount of ATP. (© Cengage Learning)

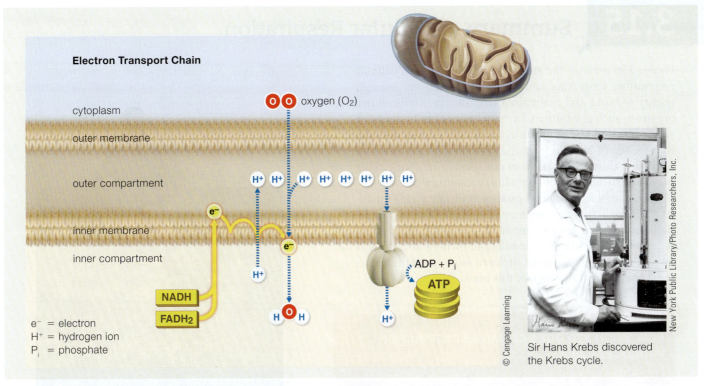

Electron Transport Chain

cytoplasm

outer membrane

outer compartment

inner membrane

inner compartment

oxygen (O₂)

ADP + Pᵢ

ATP

NADH

FADH₂

e⁻ = electron
H⁺ = hydrogen ion
Pᵢ = phosphate

© Cengage Learning

Sir Hans Krebs discovered the Krebs cycle.

New York Public Library/Photo Researchers, Inc.

Figure 3.25 **Animated!** **Electron transport forms large amounts of ATP.**

fragment and becomes a compound called acetyl-CoA. This substance enters the **Krebs cycle**. For each turn of the cycle, six carbons, three from each pyruvate, enter and six also leave, in the form of carbon dioxide. The bloodstream then transports this CO_2 to the lungs where it is exhaled.

Reactions in mitochondria before and during the Krebs cycle have three important functions. First, they produce two molecules of ATP. Second, they regenerate intermediate compounds required to keep the Krebs cycle going. And in a third, crucial step, a large number of the coenzymes called NAD^+ and FAD pick up H^+ and electrons, in the process becoming NADH and $FADH_2$. Loaded with energy, NADH and $FADH_2$ will now move to the site of the third and final stage of reactions that make ATP.

Step 3: Electron transport produces many ATP molecules

ATP production increases during the last stage of cellular respiration. Now, chains of reactions capture and use energy released by electrons. Each chain is called an **electron transport system**. It includes enzymes inside the membrane that divides the mitochondrion into two compartments (Figure 3.25). As electrons flow through the system, each step transfers a small amount of energy to a molecule that briefly stores it. This gradual releasing of energy reduces the amount of energy that is lost (as heat) while a cell is making ATP.

As shown at the lower left of Figure 3.25, an electron transport system uses electrons and hydrogen ions that are provided by NADH and $FADH_2$. The electrons are

transferred from one molecule of the transport system to the next in line. The yellow "bouncing" line in Figure 3.25 represents this process. When molecules in the chain accept electrons and then donate them, they also pick up hydrogen ions in the inner compartment, then release them to the outer compartment. At the end of an electron transport system, oxygen accepts electrons in a reaction that forms water (H_2O).

As the system moves hydrogen ions into the outer compartment of a mitochondrion, an H^+ concentration gradient develops. As the ions become more concentrated in the outer compartment, they follow the gradient back into the inner compartment, crossing the inner membrane through the interior of enzymes that can catalyze the formation of ATP from ADP and phosphate (P_i). This step is shown at the far right of Figure 3.25.

TAKE-HOME MESSAGE

HOW DOES CELLULAR RESPIRATION MAKE ATP?

- ATP forms as the stages of cellular respiration take place in the cell cytoplasm and mitochondria.

- Glycolysis occurs in the cytoplasm and does not require oxygen. Glycolysis breaks down a carbohydrate such as glucose, with a net yield of two ATP molecules.

- A second set of reaction steps, now in mitochondria, require oxygen. Enzymes acting on pyruvate molecules from glycolysis strip away carbon atoms that end up in carbon dioxide. The rest of the molecule enters the Krebs cycle, which produces two more ATP.

- Much more ATP forms in mitochondria as electrons and H^+ move through transport systems in which enzymes add a phosphate group to ADP.

3.15 Summary of Cellular Respiration

Figure 3.26 reviews the steps and ATP yield from cellular respiration. Only this aerobic pathway delivers enough energy to build and maintain a large, active, multicellular organism such as a human. In many types of cells, the third stage of reactions forms thirty-two ATP. When we add these to the final yield from the preceding stages, the total harvest is thirty-six ATP from one glucose molecule. This is a very efficient use of our cellular resources!

While aerobic cellular respiration typically yields thirty-six ATP, the actual amount may vary, depending on conditions in a cell at a given moment—for instance, if a cell requires a particular intermediate elsewhere and pulls it out of the reaction sequence. To learn more about this topic, see Appendix I at the back of this book.

© Jim Cummins/Corbis

TAKE-HOME MESSAGE

HOW MUCH ATP DOES CELLULAR RESPIRATION PRODUCE?

- From start to finish, cellular respiration typically nets thirty-six ATP for every glucose molecule.
- By far the most ATP is produced during the aerobic pathway that occurs in a cell's mitochondria.

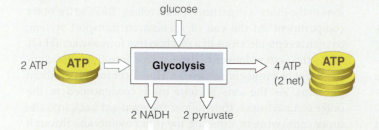

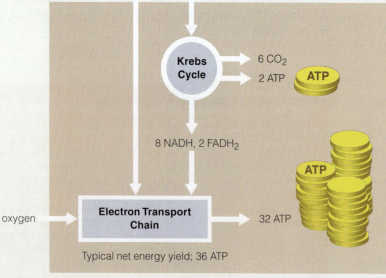

Cytoplasm

A The first stage, glycolysis, occurs in the cell's cytoplasm. Enzymes convert a glucose molecule to 2 pyruvate for a net yield of 2 ATP. During the reactions, 2 NAD⁺ pick up electrons and hydrogen atoms, so 2 NADH form.

Mitochondrion

B The second stage, the Krebs cycle and a few steps before it, occurs inside mitochondria. The 2 pyruvates are broken down to CO_2, which leaves the cell. During the reactions, 8 NAD⁺ and 2 FAD pick up electrons and hydrogen atoms, so 8 NADH and 2 $FADH_2$ form. 2 ATP also form.

C The third and final stage, the electron transport chain, occurs inside mitochondria. 10 NADH and 2 $FADH_2$ donate electrons and hydrogen ions at electron transfer chains. Electron flow through the chains sets up H⁺ gradients that drive ATP formation. Oxygen accepts electrons at the end of the chains.

Figure 3.26 Animated! This diagram summarizes aerobic cellular respiration. (© Cengage Learning)

3.16 Other Energy Sources

■ Carbohydrates, fats, and proteins all can supply needed raw materials for making ATP.

Glucose from carbohydrates is the body's main energy source, but fats and proteins also can supply this sugar. If you consume more glucose than your cells need for the moment, an intermediate of glycolysis is diverted into an anabolic pathway that makes a storage sugar called glycogen. This switch occurs often in muscle and liver cells, which store most of the body's glycogen.

Other kinds of cells tend to store excess glucose as fat, mostly in the form of triglycerides. These lipids build up in the cells of body fat (called *adipose tissue*), which occurs in the buttocks and other locations beneath the skin. Between meals or during exercise, the body may tap triglycerides as alternatives to glucose. Enzymes in fat cells break them into glycerol and fatty acids, which enter the bloodstream. Both can enter pathways of cellular respiration—glycerol in glycolysis (in the liver), and fatty acids as raw materials for the Krebs cycle.

The body doesn't store excess proteins but dismantles them into amino acids. A cell may use leftover carbons to make fats or carbohydrates. Alternatively, electrons removed from them may be used to help make ATP in the electron transport systems of the cell's mitochondria.

Sudden, strenuous exercise may call on cells in skeletal muscles (which attach to bones) that use an ATP-forming mechanism called *lactate fermentation* (Figure 3.27). The process converts pyruvate from glycolysis to lactic acid. It does not use oxygen and produces ATP quickly but not for long. Muscles feel sore when lactic acid builds up in them.

TAKE-HOME MESSAGE

WHAT NONCARBOHYDRATES CAN PROVIDE ENERGY FOR CELLS?

• If need be, cells also can use fats and proteins to make ATP.

Figure 3.27 A gymnast's muscle cells can briefly make ATP by lactate fermentation.

3.17 No Thanks to Arsenic

FOCUS ON OUR ENVIRONMENT

The element arsenic is a powerful poison. When arsenic atoms enter cells, they disrupt the Krebs cycle and electron transport chain. ATP production stops cold, and the affected cell then dies. This effect has made arsenic a useful ingredient (in carefully regulated amounts) in some anticancer drugs, wood preservatives, and insecticides. On the other hand, murderers have used killer doses of arsenic to dispatch their victims for thousands of years. More often, people are exposed to low doses of arsenic in tainted food, water, or industrial emissions.

Arsenic occurs naturally in soil and rock in many areas of the world, including parts of the western United States and in Bangladesh, where some 19 million people drink arsenic-laced well water. With enough exposure, arsenic can cause cancer, disfiguring skin disorders, and severe damage to many internal organs.

Dr. Abul Hussam, a chemist at George Mason University in Virginia, was born in Bangladesh. When he learned that his own family's well was contaminated with naturally occurring arsenic, he and his brothers built a device that costs about $35 and that can filter the arsenic from roughly 130 gallons of water per day—enough to meet the needs of several families

(Figure 3.28). In 2008 the National Academy of Engineering awarded Dr. Hussam a $1 million prize for his work. He pledged $250,000 to help fund more arsenic research. Nearly all the remaining prize money will go to buy filters for poor Bangladeshi families.

Because arsenic is a problem in the United States as well, the Environmental Protection Agency recently announced it will sponsor research on improved technologies for limiting arsenic pollution of water and soil.

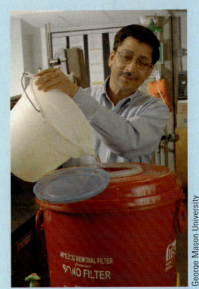

Figure 3.28 Dr. Abul Hussam's low-cost water filtering device effectively removes arsenic from drinking water.

EXPLORE
ON YOUR OWN

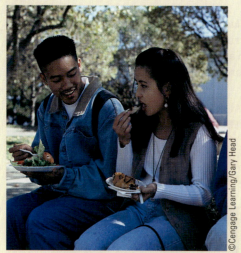

©Cengage Learning/Gary Head

Figure 3.29 Enzymes digest the different kinds of biological molecules in foods.

In this chapter you learned that an enzyme can act only on certain substrates. Because your saliva contains enzymes that can use some substances as substrates but not others, you can easily gain some insight into practical impacts of this concept (Figure 3.29). Start by holding a bite of plain cracker in your mouth for thirty seconds, without chewing it. What happens to the cracker, which is mostly starch (carbohydrate)? Repeat the test with a dab of butter or margarine (lipid), then with a piece of meat, fish, or even scrambled egg (protein). Based on your results, what type of biological molecules do your salivary enzymes act upon?

SUMMARY

Sections 3.1, 3.2 A plasma membrane surrounds the inner region of cytoplasm of a living cell. In a eukaryotic cell, including human cells, membranes divide the cell into organelles, compartments that separate metabolic reactions in the cytoplasm.

Section 3.4 Cell membranes consist mainly of phospholipids and proteins. The phospholipids form a lipid bilayer. Various kinds of proteins in or attached to the membrane perform most of its functions.

Some membrane proteins are transporter proteins. Others are receptors. Still others have carbohydrate chains that serve as a cell's identity tags.

Section 3.6 The largest organelle is the nucleus, which contains the genetic material DNA. The nucleus is surrounded by a double membrane, the nuclear envelope. Pores in the envelope help control the movement of substances into and out of the nucleus.

A cell's DNA and proteins associated with it are called chromatin. Each chromosome in the nucleus is one DNA molecule with its associated proteins.

Section 3.7 The endomembrane system includes the endoplasmic reticulum (ER), Golgi bodies, and various vesicles. In this system new proteins are modified into final form and lipids are assembled. Unwanted materials may be broken down in lysosomes and peroxisomes.

Sections 3.8, 3.9 Mitochondria carry out the oxygen-requiring reactions that make ATP, a nucleic acid that is the cell's energy currency. These reactions occur in the inner compartment of mitochondria.

The cytoskeleton gives a cell its shape and internal structure. It consists mainly of microtubules and microfilaments; some types of cells also have intermediate filaments. Microtubules are the structural framework for cilia or flagella, which develop from centrioles and are used in movement.

Section 3.10 A cell's plasma membrane is selectively permeable—only certain substances may cross it, by way of transport mechanisms.

In diffusion, substances move down their concentration gradient. *Osmosis* is the name for the diffusion of water across a selectively permeable membrane in response to a concentration gradient, a pressure gradient, or both. In passive transport, a solute moves down its concentration gradient through a membrane transporter protein.

Section 3.11 In active transport, a solute is pumped through a membrane protein *against* its concentration gradient. Active transport requires an energy boost, as from ATP.

Cells use vesicles to take in or expel large molecules or particles. In exocytosis, a vesicle moves to the cell surface and fuses with the plasma membrane. In endocytosis, a vesicle forms at the surface and moves inward. In phagocytosis, an endocytic vesicle brings organic matter into a cell.

Section 3.13 The chemical reactions in a cell are collectively called its metabolism. A metabolic pathway is a stepwise sequence of chemical reactions catalyzed by enzymes—catalytic molecules that speed up the rate of metabolic reactions. Each enzyme interacts only with a specific substrate, linking with it at one or more active sites.

Anabolism builds large, energy-rich organic compounds from smaller molecules. Catabolism breaks down molecules to smaller ones. Most anabolic reactions run on energy from ATP, which is replenished by way of the ATP/ADP cycle.

Sections 3.14, 3.15 In human cells, cellular respiration produces ATP molecules. This pathway releases chemical energy from glucose and other organic compounds. Cellular respiration begins with glycolysis (in the cytoplasm). which makes a little ATP. Then in mitochondria the Krebs cycle and electron transport generate a large amount of ATP in steps that require oxygen. These steps are the aerobic pathway of cellular respiration, in which oxygen is the final acceptor of electrons removed from glucose. The typical net energy yield of cellular respiration is thirty-six ATP.

Section 3.16 In cells, complex carbohydrates are broken down to the simple sugar glucose, the body's main metabolic fuel. Alternatives to glucose include fatty acids and glycerol from triglycerides and sometimes amino acids from proteins.

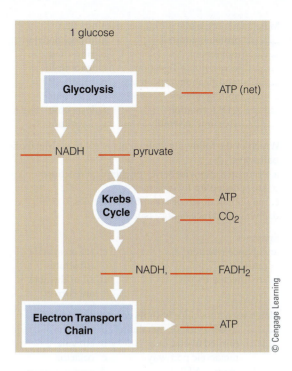

REVIEW QUESTIONS

1. Describe the general functions of the following in a eukaryotic cell: the plasma membrane, cytoplasm, DNA, ribosomes, organelles, and cytoskeleton.

2. Which organelles are in the endomembrane system?

3. Distinguish between the following pairs of terms:
   a. diffusion; osmosis
   b. passive transport; active transport
   c. endocytosis; exocytosis

4. What do enzymes do in metabolic reactions?

5. In aerobic cellular respiration, which reactions occur only in the cytoplasm? Which ones occur only in a cell's mitochondria?

6. For the diagram of the aerobic pathway shown in the next column, fill in the number of molecules of substances formed at each stage.

SELF-QUIZ *Answers in Appendix V*

1. The plasma membrane _____.
   a. surrounds the cytoplasm
   b. separates the nucleus from the cytoplasm
   c. separates the cell interior from the environment
   d. both a and c are correct

2. Fill in the blanks: The _____ is responsible for a eukaryotic cell's shape, internal organization, and cell movement.

3. Cell membranes consist mainly of a _____.
   a. carbohydrate bilayer and proteins
   b. protein bilayer and phospholipids
   c. phospholipid bilayer and proteins

4. _____ carry out most membrane functions.
   a. Proteins   c. Nucleic acids
   b. Phospholipids   d. Hormones

5. The passive movement of a solute through a membrane protein down its concentration gradient is an example of _____.
   a. osmosis   c. endocytosis
   b. active transport d. diffusion

6. Match each organelle with its correct function.
    _____ protein synthesis   a. mitochondrion
    _____ movement   b. ribosome
    _____ intracellular digestion   c. smooth ER
    _____ modification of proteins   d. rough ER
    _____ lipid synthesis   e. nucleolus
    _____ ATP formation   f. lysosome
    _____ ribosome assembly   g. flagellum

7. Which of the following statements is *not* true? Metabolic pathways _____.
    a. occur in a stepwise series of chemical reactions
    b. are speeded up by enzymes
    c. may break down or assemble molecules
    d. always produce energy (such as ATP)

8. Enzymes _____.
    a. enhance reaction rates   c. act on specific substrates
    b. are affected by pH   d. all of the above are correct

9. Match each substance with its correct description.
    _____ a coenzyme or metal ion   a. reactant
    _____ formed at end of a metabolic pathway   b. enzyme
        c. cofactor
    _____ mainly ATP   d. energy carrier
    _____ enters a reaction   e. product
    _____ catalytic protein

10. Cellular respiration is completed in the _____.
    a. nucleus   c. plasma membrane
    b. mitochondrion   d. cytoplasm

11. Match each type of metabolic reaction with its function.
    _____ glycolysis   a. many ATP, NADH, $FADH_2$, and CO_2 form
    _____ Krebs cycle
    _____ electron transport   b. glucose to two pyruvate molecules and some ATP
        c. H^+ flows through channel proteins, ATP forms

12. In a mitochondrion, where are the electron transport systems and enzymes required for ATP formation located?

CRITICAL THINKING

1. Using Section 3.3 as a reference, suppose you want to observe the surface of a microscopic section of bone. Would the best choice for this task be a compound light microscope or an electron microscope?

2. Jogging is considered aerobic exercise because the cardio-vascular system (heart and blood vessels) can adjust to supply the oxygen needs of working cells. In contrast, sprinting the 100-meter dash might be called "anaerobic" (lacking oxygen) exercise, and golf "nonaerobic" exercise. Explain these last two observations.

3. Section 3.17 mentions that arsenic poisons human cells because it halts the production of ATP. This happens because the structure of arsenic atoms closely resembles that of phosphorus atoms—so close, in fact, that arsenic can take the place of phosphorus in chemical reactions. Why would the substitution of arsenic atoms for phosphorus atoms prevent the formation of ATP?

4. The cells of your body never use nucleic acids as an energy source. Can you suggest a reason why?

TISSUES, ORGANS, AND ORGAN SYSTEMS

LINKS TO EARLIER CONCEPTS

This chapter focuses on the tissue, organ, and organ system levels of biological organization (1.3).

You will also get a look at some of the variations on basic cell structure (3.1–3.9) that occur in your body. The variations remind us that cells that perform specialized functions must be built to carry out those tasks.

KEY CONCEPTS

Types of Body Tissues

Epithelium, connective tissue, muscle tissue, and nervous tissue are the basic types of body tissues.
Sections 4.1–4.7

Organs and Organ Systems

Combinations of tissues form organs. The body's organ systems consist of interacting organs.
Sections 4.8–4.9

Homeostasis

A feedback mechanism works to maintain homeostasis—stable operating conditions in the body.
Sections 4.10–4.11

Top: Ed Reschke/Peter Arnold; Middle: © Cengage Learning; Bottom: © Jupiterimages

Each year tens of thousands of people develop a disease or suffer an injury that severely damages an organ or tissues. If only it were possible to replace those body parts! Actually, that is the dream of researchers who study stem cells, like Junying Yu, pictured below.

All body cells "stem" from stem cells, which are the first cells to form in an embryo. In theory, an embryonic stem cell can produce every kind of cell in the body. This is why many scientists are keen to try to use embryonic stem cells to develop therapies that can replace damaged tissues and organs. Other people believe that using embryonic stem cells for any reason is unethical, because doing so destroys or may seriously harm the embryo.

Adults also have stem cells, which are less controversial. Although adult stem cells are more limited than embryonic ones, active types replace dead or damaged cells, including blood and skin cells. In the laboratory, adult stem cells have shown promise for regenerating tissues such as cartilage and heart muscle damaged by a heart attack. Recently, several patients suffering from cancer of the trachea (windpipe) have received "new" tracheas grown from their own stem cells.

This chapter focuses on the basic types of cells and tissues that make up the body. It launches our study of body structures and functions—that is, of human anatomy and physiology. *Anatomy* refers to the body's parts and how they are put together. *Physiology* refers to how body parts function.

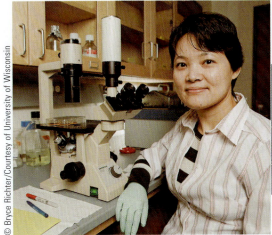
© Bryce Richter/Courtesy of University of Wisconsin

Embryonic stem cells
(© Juergen Berger, Max-Planck Institute/Photo Researchers, Inc.)

Homeostasis Preview

In this chapter you will learn about a major mechanism of homeostasis called negative feedback and see how it regulates internal body temperature.

Epithelium: The Body's Covering and Linings

- Epithelial tissues cover the body surface or line its cavities and tubes.

- Link to the Cell cytoskeleton 3.9

basement membrane A noncellular membrane positioned between an epithelium and the underlying tissue.

endocrine gland Gland that makes a hormone, releasing it directly into the fluid outside the gland.

epithelium A sheetlike tissue that has one free surface. Epithelia line body cavities, ducts, and tubes or protect an underlying tissue.

A **tissue** is a group of similar cells that perform a certain function. The tissue called **epithelium** (plural: epithelia) has a sheetlike structure, and one of its surfaces faces an internal body fluid or the outside environment (Figure 4.1A). The other surface rests on a **basement membrane** that is sandwiched between it and the tissue below. A basement membrane is densely packed with proteins and polysaccharides. It does not have cells.

There are two basic types of epithelium

Simple epithelium is the thinnest type of epithelium. It has just one layer of cells. It lines the body's cavities, ducts, and tubes—for example, the chest cavity, tear ducts, and the tubes in the kidneys where urine is formed (Figure 4.1B–D). In general, the cells in a simple epithelium function in the diffusion, secretion, absorption, or filtering of substances across the layer.

Some single-layer epithelia look stratified in a side view because the nuclei of neighboring cells don't line up. Most of the cells also have cilia. This type of simple epithelium is termed *pseudostratified* (*pseudo-* means false). It lines the throat, nasal passages, reproductive tract, and other sites in the body where cilia sweep mucus or some other fluid across the tissue's surface.

Stratified epithelium has more than one layer of cells, and it usually has a protective function. For example, this is the

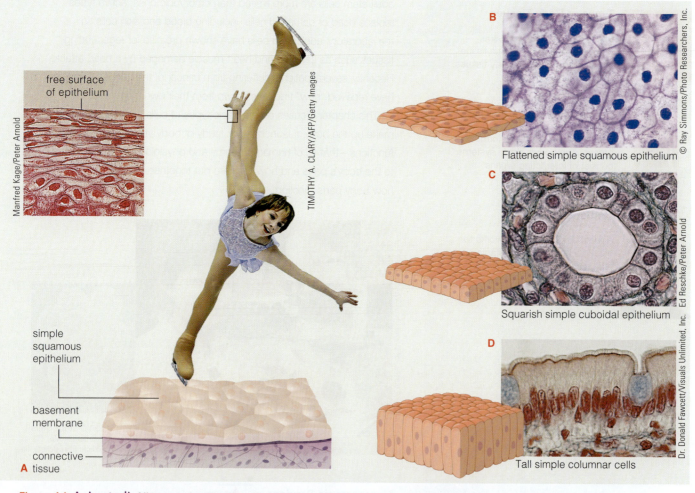

free surface of epithelium

Manfred Kage/Peter Arnold

TIMOTHY A. CLARY/AFP/Getty Images

simple squamous epithelium

basement membrane

connective tissue

A

Flattened simple squamous epithelium

© Ray Simmons/Photo Researchers, Inc.

B

Squarish simple cuboidal epithelium

Ed Reschke/Peter Arnold

C

Tall simple columnar cells

Dr. Donald Fawcett/Visuals Unlimited, Inc.

D

Figure 4.1 Animated! All types of epithelium share basic characteristics. All epithelia have a free surface that faces either the outside environment or an internal body fluid. **A** Squamous epithelium of skin consists of several layers of cells that flatten as they near the free surface. The basement membrane is sandwiched between the lower epithelial surface and underlying connective tissue. The diagram shows simple epithelium, a single layer of cells. **B–D** Examples of simple epithelium, showing the three basic cell shapes in this type of tissue.

Exocrine gland

parotid gland
(secretes saliva)

parotid duct (delivers
saliva to mouth)

A

Endocrine gland

blood vessel   cell that secretes hormone

B

thyroid gland (secretes hormones into blood)

iStockphoto.com/Flashon Studio

© Cengage Learning

Figure 4.2 **Glands make and release substances.** This illustration shows examples of an exocrine and an endocrine gland. Chapter 15 discusses the endocrine system, which includes a variety of glands and cells that release hormones.

tissue at the surface of your skin, which is exposed to nicks, bumps, scrapes, and so forth.

The two basic types of epithelium are categorized by the shape of cells at the tissue's free surface (Table 4.1). The cells of *squamous epithelium* are flattened, while they are cube-shaped in *cuboidal epithelium* and elongated in *columnar epithelium*. Each shape correlates with a given function. For instance, oxygen and carbon dioxide easily diffuse across the thin simple squamous epithelium that makes up the walls of fine blood vessels (Figure 4.1B). Cells of cuboidal and columnar epithelia may secrete or absorb substances.

Glands develop from epithelium

A **gland** makes and releases products such as saliva or mucus. Some glands consist of a single cell, while others are more complex. All glands develop from epithelial tissue and often stay connected to it. Mucus-secreting goblet cells,

for instance, are embedded in epithelium that lines the trachea (your windpipe) and other tubes leading to the lungs. The stomach's epithelial lining contains gland cells that release mucus and digestive juices.

Glands may be classified by how their products reach the place where they are used (Figure 4.2). **Exocrine glands** release substances through ducts or tubes (Figure 4.2A). Mucus, saliva, oil, earwax, milk, and digestive enzymes are in this group. Many exocrine glands simply release the substance they make; salivary glands and most sweat glands are like this. In other cases, a gland's secretions include bits of the gland cells. For instance, milk from a nursing mother's mammary glands contains bits of the glandular epithelial tissue. In still other cases, such as sebaceous (oil) glands in your skin, whole cells full of material are shed into the duct, where they burst and their contents spill out.

Endocrine glands do not release substances through tubes or ducts. They make hormones that are released directly into the fluid bathing the glands (Figure 4.2B).

exocrine gland Gland that releases the substance it makes through a duct or tube.

gland A structure built of one or more cells that makes and releases products such as saliva, milk, mucus, or oil.

tissue A group of similar cells that perform a specific function.

TABLE 4.1 Major Types of Epithelium		
Type	**Shape**	**Typical Locations**
SIMPLE	Squamous	Linings of blood vessels, air sacs of lungs (alveoli)
	Cuboidal	Glands and their ducts, ovary surfaces, iris of eye
	Columnar	Stomach, intestines, uterus
PSEUDOSTRATIFIED	Columnar	Throat, nasal passages, sinuses, trachea, male genital ducts
STRATIFIED	Squamous	Skin, mouth, throat, vagina
	Cuboidal	Ducts of sweat glands
	Columnar	Male urethra, salivary gland ducts

Connective Tissue: Binding, Support, and Other Roles

- Connective tissue connects, supports, and anchors the body's parts.

- Links to Lipids 2.10, Structural proteins 2.11

Connective tissue provides support and protection for cells, tissues, and organs. It makes up more of your body than any other tissue. In most kinds of connective tissue, the cells secrete fiberlike structural proteins and a "ground substance" of polysaccharides. These ingredients form a **matrix** around the cell. The matrix can range from hard to liquid, and it gives each kind of connective tissue its specialized properties (Table 4.2).

Fibrous connective tissues are strong and stretchy

The various kinds of **fibrous connective tissue** all have cells and fibers in a matrix, but in different proportions that make each kind of fibrous connective tissue well suited to perform its special function.

The different forms of *loose connective tissue* have few fibers and cells that are loosely arranged in a jellylike ground substance. This structure makes loose connective tissue flexible. The example in Figure 4.3A wraps many organs and helps support the skin. A "reticular" (netlike) form of loose connective tissue is the framework for soft organs such as the liver, spleen, and lymph nodes.

TABLE 4.2 Connective Tissues at a Glance

Fibrous Connective Tissues

LOOSE	Collagen and elastin loosely arranged in ground substance; quite flexible and fairly strong
DENSE	Mainly collagen; strong and somewhat flexible. Its collagen fibers are aligned in parallel in tendons and ligaments
ELASTIC	Mainly elastin; easily stretches and recoils

Special Connective Tissues

CARTILAGE	Mainly collagen in a watery matrix; resists compression
BONE	Very strong, mineral-hardened matrix
ADIPOSE TISSUE	Mainly cells filled with fat; soft matrix
BLOOD	Matrix is the fluid blood plasma, which contains blood cells and other substances

Dense connective tissue has more collagen than does loose connective tissue, so it is less flexible but much stronger. The form pictured in Figure 4.3B helps support the skin's lower layer, the dermis. It also wraps around muscles and organs that do not need to stretch much, such as kidneys.

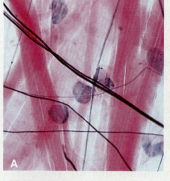
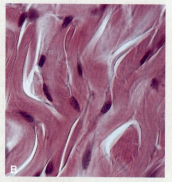
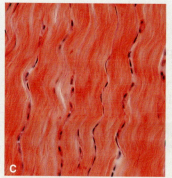
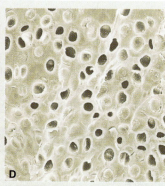

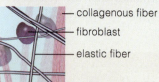
collagenous fiber
fibroblast
elastic fiber

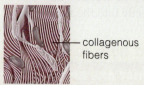
collagenous fibers

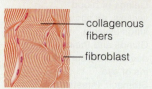
collagenous fibers
fibroblast

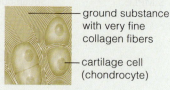
ground substance with very fine collagen fibers
cartilage cell (chondrocyte)

Type Loose connective tissue
Description Fibroblasts, other cells, plus fibers loosely arranged in semifluid matrix
Common Locations Under the skin and most epithelia
Function Elasticity, diffusion

Type Dense, irregular connective tissue
Description Collagenous fibers, fibroblasts, less matrix
Common Locations In skin and capsules around some organs
Function Support

Type Dense, regular connective tissue
Description Collagen fibers in parallel bundles, long rows of fibroblasts, little matrix
Common Locations Tendons, ligaments
Function Strength, elasticity

Type Cartilage
Description Cells embedded in pliable, solid matrix
Common Locations Ends of long bones, nose, parts of airways, skeleton of embryos
Function Support, flexibility, low-friction surface for joint movement

Figure 4.3 Animated! Connective tissues connect, support, and anchor. (A–C: Ed Reschke/Peter Arnold; D: © Science Photo Library/Photo Researchers, Inc.)

Another version of this tissue has bundles of collagen fibers aligned in the same plane (Figure 4.3C). It is found in tendons, which attach many skeletal muscles to bones, and in ligaments, which attach bones to one another. The tissue's structure allows a tendon to resist being torn, and in ligaments the tissue's elastic fibers allow the ligament to stretch so bones can move at joints such as the knee.

Elastic connective tissue is a form of dense connective tissue in which most of the fibers are the protein elastin. As a result, this tissue is elastic and is found in organs that must stretch, such as the lungs, which expand and recoil as air moves in and out.

Special connective tissues include cartilage, bone, adipose tissue, and blood

Like rubber, **cartilage** is solid, pliable, and not easily compressed. Its matrix is a blend of collagen and elastin fibers in a rubbery ground substance. The result is a tissue that can withstand considerable physical stress. The collagen-producing cells are trapped inside small cavities in the matrix (Figure 4.3D). If you have ever torn a cartilage, you know that injured cartilage heals slowly. This is because cartilage lacks blood vessels.

Most cartilage in the body is whitish, glistening *hyaline cartilage* (*hyalin* = "glassy"). Hyaline cartilage at the ends of bones reduces friction in movable joints. It also makes up parts of your nose, windpipe (trachea), and ribs. An early embryo's skeleton consists of hyaline cartilage.

Elastic cartilage has both collagen and elastin fibers. It occurs where a flexible yet rigid structure is required, such as in the flaps of your ears. Sturdy *fibrocartilage* is

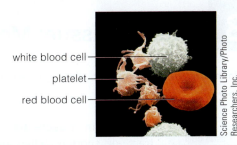

Figure 4.4 **Blood is an unusual connective tissue that transports substances.** Here you see some components of human blood. This tissue's liquid matrix (plasma) is mostly water in which various substances are dissolved.

white blood cell

platelet

red blood cell

packed with thick bundles of collagen fibers. It can withstand a lot of pressure, and it forms the cartilage "cushions" in joints such as the knee and in the disks between the vertebrae in the spinal column.

Bone tissue is the main tissue in bones. It is hard because its matrix includes not only collagen fibers and ground substance but also calcium salts (Figure 4.3E). Bones serve the body in ways described in Chapter 5.

Adipose tissue stores fat—the way the body deals with carbohydrates and proteins that are not immediately used for metabolism. It is mostly cells packed with fat droplets, with just a little matrix between them (Figure 4.3F). Most of our adipose tissue is located just beneath the skin, where it provides insulation and cushioning.

Blood is classified as connective tissue even though it does not "connect" or bind other body parts. Instead blood's role is transport. Its matrix is the fluid plasma, which contains proteins (blood's "fibers") as well as a variety of blood cells and cell fragments called platelets (Figure 4.4). Chapter 8 discusses this complex tissue.

adipose tissue Tissue that stores fat in adipose (fat) cells.

bone tissue The hard main tissue in bones, mineralized with calcium salts.

cartilage Pliable tissue with a matrix consisting of collagen and elastin fibers in a rubbery ground substance.

connective tissue Tissue that connects and supports body parts; all types consist of cells and fibers in a matrix.

fibrous connective tissue Connective tissue having a matrix that makes it strong and stretchy; varying characteristics of the matrix result in loose, dense, and elastic forms.

matrix The blend of cells, fibers, and ground substance that gives each type of connective tissue its specialized properties.

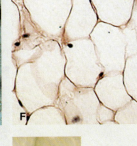

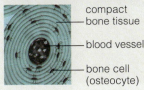

compact bone tissue

blood vessel

bone cell (osteocyte)

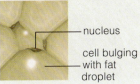

nucleus

cell bulging with fat droplet

Type Bone tissue
Description Collagen fibers, matrix hardened with calcium
Common Locations Bones of skeleton
Function Movement, support, protection

Type Adipose tissue
Description Large, tightly packed fat cells occupying most of matrix
Common Locations Under skin, around heart, kidneys
Function Energy reserves, insulation, padding

TAKE-HOME MESSAGE

WHAT IS CONNECTIVE TISSUE?

- Connective tissue binds together and supports other body tissues and organs.
- The differing types of fibrous connective tissues have different amounts and arrangements of collagen and elastin fibers in their matrix.
- Cartilage, bone, blood, and adipose tissue are specialized connective tissues. Cartilage and bone are structural materials. Blood transports substances. Adipose tissue stores energy.

(E: Ed Reschke; F: © University of Cincinnati, Raymond Walters College, Biology)

4.3 Muscle Tissue: Movement

■ Cells in muscle tissue can contract, allowing muscle to move body parts.

muscle tissue Tissue built of cells that can contract.

The cells in **muscle tissue** contract, or shorten, when they are stimulated by an outside signal. Then they relax and lengthen. Muscle tissue has long, cylindrical cells lined up in parallel. This shape is why muscle cells are often called "muscle fibers." Muscle layers and muscular organs contract and relax in a coordinated way. This is how the action of muscles maintains and changes the positions of body parts, movements that range from leaping to blinking your eyes. The three types of muscle tissue are skeletal, smooth, and cardiac muscle tissues.

Skeletal muscle is the main tissue of muscles that attach to your bones (Figure 4.5A). Skeletal muscle cells are unusual in that they have more than one nucleus. In a typical muscle, the cells line up in parallel bundles and look striped, or *striated*. The bundles, called fascicles, are enclosed by a sheath of dense connective tissue. This arrangement of muscle and connective tissue makes up the organs we call "muscles." Because we can exert conscious control over our skeletal muscles, their contractions are said to be "voluntary." The structure and functioning of skeletal muscle tissue are topics we consider in Chapter 6.

Smooth muscle cells taper at both ends (Figure 4.5B). They are bundled inside a connective tissue sheath. This type of muscle tissue is specialized for ongoing contraction. It is found in the walls of internal organs—including blood vessels, the stomach, and the intestines. The contraction of smooth muscle is "involuntary" because we usually cannot make it contract just by thinking about it (as we can with skeletal muscle).

Cardiac muscle (Figure 4.5C) is found only in the wall of the heart and its sole function is to pump blood. As you will read in Chapter 7, special junctions fuse the plasma membranes of cardiac muscle cells. In places, communication junctions allow the cells to contract as a unit. When one cardiac muscle cell is signaled to contract, the cells around it contract, too.

TAKE-HOME MESSAGE

WHAT IS MUSCLE TISSUE?

• Muscle tissue helps move the body and its parts. It is built of cells that can contract.
• Muscle tissue can contract (shorten) when it is stimulated by an outside signal.
• Skeletal muscle attaches to bones.
• Smooth muscle is found in internal organs.
• Cardiac muscle makes up the walls of the heart.

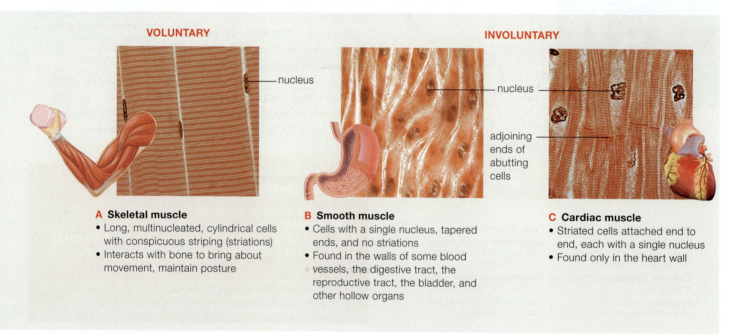

VOLUNTARY   **INVOLUNTARY**

nucleus

nucleus

adjoining ends of abutting cells

A Skeletal muscle
• Long, multinucleated, cylindrical cells with conspicuous striping (striations)
• Interacts with bone to bring about movement, maintain posture

B Smooth muscle
• Cells with a single nucleus, tapered ends, and no striations
• Found in the walls of some blood vessels, the digestive tract, the reproductive tract, the bladder, and other hollow organs

C Cardiac muscle
• Striated cells attached end to end, each with a single nucleus
• Found only in the heart wall

Figure 4.5 Animated! All types of muscle tissue consist of cells that can contract. (A, C: Ed Reschke; B: Biophoto Associates/Photo Researchers, Inc.)

4.4 Nervous Tissue: Communication

■ **Nervous tissue makes up the nervous system.**

The body's **nervous tissue** consists mostly of **neurons**, the "nerve cells," and support cells. Tens of thousands of neurons occur in the brain and spinal cord, and millions more are present throughout the body. Neurons make up the body's communication lines. The signals they carry are often called nerve impulses.

Neurons carry messages

Like other kinds of cells, a neuron has a cell body that contains the nucleus and cytoplasm. It also has two types of extensions, or cell "processes." Branched processes called **dendrites** receive incoming messages. Processes called **axons** conduct outgoing messages. Depending on the type of neuron, its axon may be very short, or it may be as long as three or four feet. In the image below (right), you can see

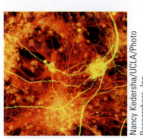
Nancy Kedersha/UCLA/Photo Researchers, Inc.

A glial cell

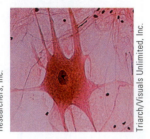
Triarch/Visuals Unlimited, Inc.

A motor neuron

the cell processes of a motor neuron, which carries signals to muscles and glands.

Neuroglia are support cells

About 90 percent of the cells in the nervous system are **glial cells** (also called *neuroglia*). The word *glia* means "glue," and glial cells were once thought to simply be the "mortar" that physically supported neurons. Today we know that they have various functions. In the central nervous system, glia help supply nutrients to neurons, provide physical support, and remove unwanted material. Outside the brain and spinal cord glial cells provide insulation—a function that helps speed nerve impulses through the body, as described in Chapter 13.

> **axon** The neuron extension that carries outgoing messages.
>
> **dendrite** A neuron extension that receives incoming messages.
>
> **glial cell** Any of the large number of cells in the nervous system that support neurons physically or in other ways.
>
> **nervous tissue** Tissue made up of neurons and glial cells.
>
> **neuron** A nerve cell.

TAKE-HOME MESSAGE

WHAT IS NERVOUS TISSUE?

- Nervous tissue contains neurons, which are the body's communication cells.
- Support cells called glia (neuroglia) make up most of the body's nervous tissue.

4.5 Healing with Stem Cells and Lab-Grown Tissues

FOCUS ON HEALTH

Stem cells can be manipulated to produce various types of specialized cells (inset). The hope is that stem cell research may lead to therapies that can help patients with health problems such as Parkinson's disease, type 2 diabetes, sickle cell anemia, and paralysis due to spinal cord injury.

Scientists at the National Institutes of Health (NIH) are making progress toward a stem-cell-based cure for sickle cell anemia. In this genetic disease, faulty stem cells in bone marrow produce defective red blood cells. Working with ten adult patients, NIH doctors used radiation to kill the malfunctioning stem cells and then replaced them with healthy donated stem cells that began producing normal red blood cells. Two years later, nine of the ten patients had no sign of the disease.

Other technologies focus on growing replacement tissues. The replacement tracheas mentioned in the chapter introduction were grown from adult stem cells seeded on a scaffold of synthetic material. Only a few patients have undergone the procedure, which is still considered experimental. A cultured skin substitute (Figure 4.6) is a more common option for burn victims and people with chronic wounds. The tissue is grown from cells extracted from foreskins removed when infant boys are circumcised.

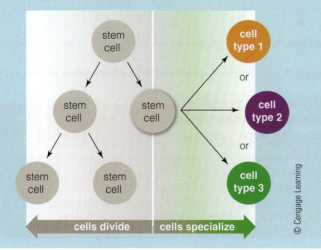
© Cengage Learning

cells divide   cells specialize

Figure 4.6 Skin substitutes are grown in the laboratory. A cultured skin substitute called Apligraf can be sutured over a wound to help prevent infection and speed up the healing process. (Courtesy of © Organogensis, Inc., www.organogenesis.com)

4.6 Cell Junctions: Holding Tissues Together

- Junctions between the cells in a tissue knit the cells firmly together, stop leaks, and serve as communication channels.

- Links to Plasma membrane 3.4, Cytoskeleton 3.9

Our tissues and organs would fall into disarray if there were not some way for individual cells to "stick together" and to communicate. Cell junctions meet these needs. Junctions are most common where substances must not leak from one body compartment to another.

adhering junction A weldlike junction between cells that keeps cells tightly attached to one another.

gap junction A channel that connects the cytoplasm of neighboring cells.

tight junction A strand of protein that helps stop leaks between cells in a tissue by forming a gasket-like seal.

Figure 4.7 shows some examples of cell junctions. **Tight junctions** (Figure 4.7A) are strands of protein that help stop leaks across a tissue. The strands form gasket-like seals that prevent molecules from moving easily across the junction. In epithelium tight junctions allow the epithelial cells to control what enters the body. For instance, while food is being digested, various types of nutrient molecules can diffuse into epithelial cells or enter them selectively by active transport, but tight junctions keep those needed molecules from slipping *between* cells. Tight junctions also prevent the highly acidic gastric fluid in your stomach from leaking out and digesting proteins of your own body instead of those you consume in food.

Adhering junctions (Figure 4.7B) cement cells together. One type, sometimes called desmosomes, is like a spot weld at the plasma membranes of two adjacent cells. They are anchored to the cytoskeleton in each cell and help hold cells together in tissues that often stretch, such as epithelium of the skin, the lungs, beating heart muscle, and the stomach.

Gap junctions (Figure 4.7C) are channels that connect the cytoplasm of neighboring cells. They help cells communicate because ions and small molecules can pass through them from cell to cell. Smooth muscle and cardiac muscle have the most gap junctions. As you will read in Chapter 6, ions moving through them from muscle cell to muscle cell play an important role in the contraction of whole muscles.

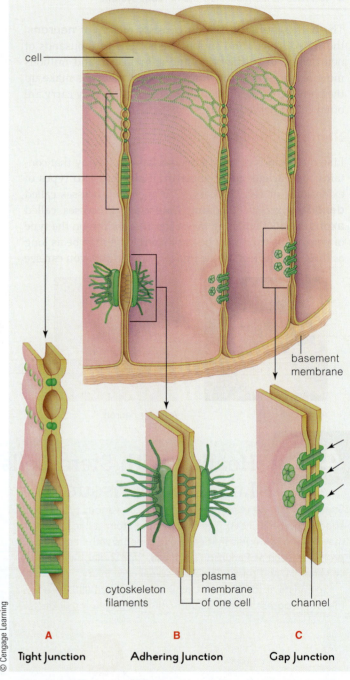

© Cengage Learning

A Tight Junction **B** Adhering Junction **C** Gap Junction

Figure 4.7 Animated! Junctions knit cells together in tissues.

THINK OUTSIDE THE BOOK

Stem cell therapy is still in its infancy. To date, we have seen genuine progress with only a few diseases and disorders. Still, some hopeful patients ill with a crippling or possibly fatal disease travel to countries where clinics hold out the promise of a "stem cell cure." Using the Web, check out what the American Medical Association (AMA) says about such "stem cell tourism." Is it risky? What advice does the AMA provide for those who may be considering it?

TAKE-HOME MESSAGE

WHAT DO CELL JUNCTIONS DO?

- Cell junctions hold cells together in tissues or allow communication between cells.
- Tight junctions help stop leaks in a tissue.
- Adhering junctions cement cells together in a tissue.
- Gap junctions are channels that allow ions and small molecules to cross between cells.

■ Thin, sheetlike membranes cover many body surfaces and cavities. Some provide protection. Others both protect and lubricate organs.

A *membrane* is a thin, sheetlike tissue covering. In the body we find two basic types—epithelial membranes and connective tissue membranes.

In epithelial membranes, epithelium pairs with connective tissue

Epithelial membranes consist of a sheet of epithelium atop connective tissue. Examples are **mucous membranes**, also called mucosae (singular: mucosa). These are the pink, moist membranes lining the tubes and cavities of your digestive, respiratory, urinary, and reproductive systems (Figure 4.8A). Most mucous membranes, like the lining of the stomach, contain glands and are specialized to secrete substances, absorb them, or both. Some of the glands are single cells. For example, goblet cells—so named because their shape resembles a stemmed glass—secrete mucous. Other mucous membranes have no glands. The mucous membranes lining the urinary tract (including the tubes that carry urine out) are like this.

Serous membranes are epithelial membranes that occur in paired sheets. Imagine one paper sack inside another, with a narrow space between them, and you'll get the idea. Serous membranes don't have glands, but the layers do secrete a fluid that fills the space between them. Examples include the membranes that line the chest cavity and enclose the heart and lungs (Figure 4.8B). Among other functions, serous membranes help anchor internal organs in place and provide lubricated smooth surfaces that prevent chafing between adjacent organs or between organs and the body wall.

You know a third type of epithelial membrane, the **cutaneous membrane**, as your skin (Figure 4.8C). Its tissues are part of one of the body's major organ systems, the integumentary system—the topic of Section 4.9.

Membranes in joints consist only of connective tissue

A few membranes consist only of connective tissue. These **synovial membranes** (Figure 4.8D) line cavities of the body's movable joints. They contain cells that secrete fluid that lubricates the ends of moving bones or prevents friction between a bone and a moving tendon.

cutaneous membrane
The skin, a type of epithelial membrane.

mucous membrane
Epithelial membrane that lines tubes and cavities of the digestive, respiratory, urinary, and reproductive systems.

serous membrane Type of epithelial membrane that occurs in paired sheets. Serous membranes lack glands but secrete fluid that fills the space between the sheets.

synovial membrane
Connective tissue membrane that lines joint cavities.

TAKE-HOME MESSAGE

WHAT ARE THE FUNCTIONS OF MEMBRANES?

- Membranes protect and sometimes lubricate many body surfaces and cavities.
- The various epithelial membranes (mucous, serous, and cutaneous) consist of epithelium overlying connective tissue. Most contain glands.
- Synovial membranes consist only of connective tissue. They line joint cavities and produce fluid that lubricates the joint.

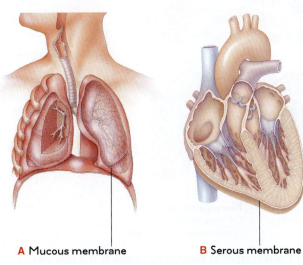

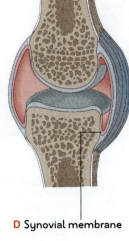

A Mucous membrane   **B** Serous membrane   **C** Cutaneous membrane (skin)   **D** Synovial membrane

A, B, D: © Cengage Learning; C: © Fabian Cevallos/Sygma/Corbis

Figure 4.8 Membranes cover many body surfaces and line body cavities.

4.8 Organs and Organ Systems

abdominal cavity Body cavity that contains the stomach, liver, intestines, kidneys, and some other organs.

cranial cavity Body cavity that encloses the brain.

organ Structure, such as the heart, built of two or more kinds of tissue that together perform one or more functions.

organ system Combination of two or more organs that work in a coordinated way to carry out a specific function.

pelvic cavity Body cavity that encloses reproductive organs, the bladder, and rectum.

spinal cavity Body cavity that encloses the spinal cord.

thoracic cavity Body cavity that contains the heart and lungs.

- The human body's organs are organized into eleven organ systems.

- Link to Levels of biological organization 1.3

An **organ** is a combination of two or more kinds of tissue that together perform one or more functions. As an example, the stomach contains all four of the tissue types you have read about in previous sections (Figure 4.9A). Its wall is mainly muscle, and nerves help regulate muscle contractions that mix and move food. Connective tissue provides support, while the stomach lining is epithelium.

The stomach and many other major organs are located inside body cavities shown in Figure 4.9B. The **cranial cavity** and **spinal cavity** house your brain and spinal cord—the central nervous system.

Your heart and lungs reside in the **thoracic cavity**—essentially, inside your chest. Below the thoracic cavity is the **abdominal cavity**, which holds your stomach, liver, pancreas, most of the intestine, and other organs. Reproductive organs, the bladder, and the rectum are located in the lower abdominal cavity in a region often called the **pelvic cavity**.

In an **organ system**, two or more organs "cooperate" to carry out a major body function (Figure 4.10). For example, interactions between your skeletal and muscular systems allow you to move about. Blood in the cardiovascular system rapidly carries nutrients and other substances to cells and transports products and wastes away from them. Your respiratory system delivers oxygen from air to your cardiovascular system and takes up carbon dioxide wastes from it—and so it goes, throughout the entire body.

TAKE-HOME MESSAGE

WHAT ARE ORGANS AND ORGAN SYSTEMS?

- An organ is a combination of two or more kinds of tissues that together perform one or more functions.
- Organ systems consist of two or more organs that together perform a specific function.
- The five body cavities contain many major organs.

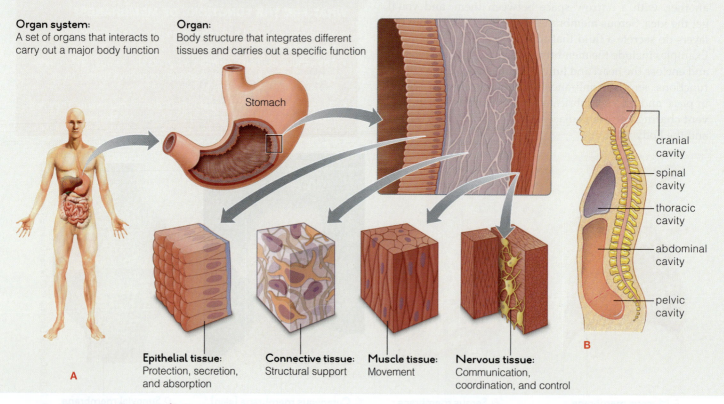

Organ system:
A set of organs that interacts to carry out a major body function

Organ:
Body structure that integrates different tissues and carries out a specific function

Stomach

Epithelial tissue: Protection, secretion, and absorption

Connective tissue: Structural support

Muscle tissue: Movement

Nervous tissue: Communication, coordination, and control

A

B

cranial cavity

spinal cavity

thoracic cavity

abdominal cavity

pelvic cavity

Figure 4.9 Animated! An organ consists of two or more tissues. **A** The four types of tissue in the stomach. **B** A side view of major body cavities where many organs are located. (© Cengage Learning)

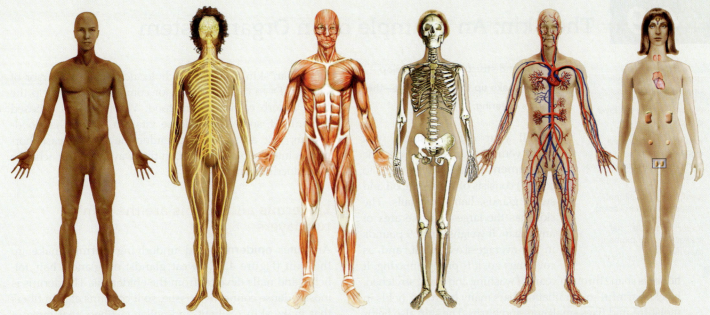

Integumentary System

Protects body from injury, dehydration, and some microbes; helps control body temperature; excretes wastes; receives sensory information.

Nervous System

Detects external and internal stimuli; controls and coordinates the responses to stimuli; integrates all organ system activities.

Muscular System

Moves body and its parts; maintains posture; generates heat by increasing metabolic activity.

Skeletal System

Supports and protects body parts; provides muscle attachment sites; produces red blood cells; stores calcium, phosphorus.

Cardiovascular System

Rapidly transports many materials to and from cells; helps stabilize pH and temperature.

Endocrine System

Hormonally controls body functioning; works with nervous system to integrate body functions.

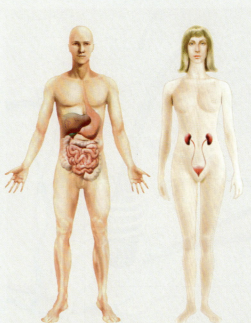
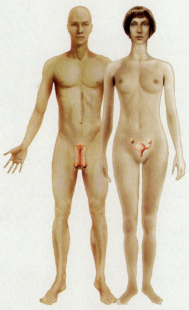

Lymphatic System

Collects and returns tissue fluid to the blood; defends the body against infection as part of the immune system.

Respiratory System

Delivers oxygen to all living cells; removes carbon dioxide wastes of cells; helps regulate the pH of blood.

Digestive System

Ingests food and water; mechanically, chemically breaks down food and absorbs small molecules into internal environment; eliminates food residues.

Urinary System

Maintains the volume and chemical composition of blood and tissue fluid; excretes unneeded fluid and blood-borne wastes.

Reproductive System

Female: Produces eggs; after fertilization, provides a protected environment for the development of a fetus. *Male*: Produces and transfers sperm to the female. Hormones of both systems also influence other organ systems.

Figure 4.10 Animated! The body has eleven organ systems. Not shown is the immune system, which consists mainly of cells called lymphocytes. (© Cengage Learning)

The Skin: An Example of an Organ System

■ **Skin and structures that develop from it make up the integument—the body's covering.**

Of all your organ systems, you know your integument the best. The **integument** (from Latin *integere*, "to cover") consists of your skin, oil and sweat glands, hair, and nails. The skin has the largest surface area of any organ. It weighs about 9 pounds in an average-sized adult, and as coverings go, it is pretty amazing. It holds its shape through years of washing and being stretched, blocks harmful solar radiation, bars many microbes, holds in moisture, and fixes small cuts and burns. The skin also helps regulate body temperature, and signals from its sensory receptors help the brain assess what's going on in the outside world. Yet except for places subjected to regular abrasion (such as your palms and the soles of your feet), your skin is generally not much thicker than a sheet of construction paper. It is even thinner in some places, such as the eyelids.

Human skin also makes cholecalciferol, a precursor of vitamin D—a catchall name for compounds that help the body absorb calcium from food. When skin is exposed to sunlight, some cells release vitamin D into the bloodstream, just as specialized endocrine cells release hormones into the bloodstream. In this way your skin acts like an endocrine gland.

Epidermis and dermis are the skin's two layers

An outer **epidermis** and underlying **dermis** make up the skin (Figure 4.11). Sweat glands, oil glands, hair follicles, and nails develop from the epidermis. The dermis is mostly dense connective tissue, so it contains elastin fibers that make skin resilient and collagen fibers that make it strong. The epidermis and dermis form the cutaneous membrane you read about in Section 4.7. Under the dermis is a subcutaneous ("under the skin") layer called the hypodermis. This loose connective tissue anchors the skin while allowing it to move a bit. It also contains fat that helps insulate the body and cushions some of its parts.

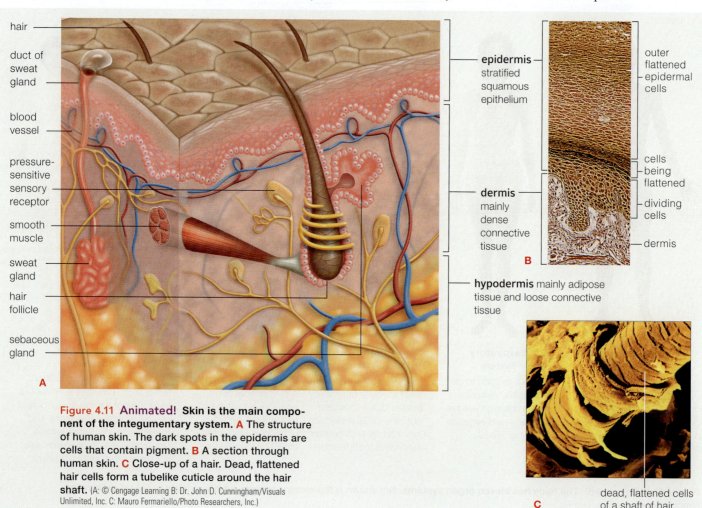

hair

duct of sweat gland

blood vessel

pressure-sensitive sensory receptor

smooth muscle

sweat gland

hair follicle

sebaceous gland

A

epidermis
stratified squamous epithelium

dermis
mainly dense connective tissue

B

hypodermis mainly adipose tissue and loose connective tissue

outer flattened epidermal cells

cells being flattened

dividing cells

dermis

dead, flattened cells of a shaft of hair

C

Figure 4.11 Animated! Skin is the main component of the integumentary system. A The structure of human skin. The dark spots in the epidermis are cells that contain pigment. B A section through human skin. C Close-up of a hair. Dead, flattened hair cells form a tubelike cuticle around the hair shaft. (A: © Cengage Learning B: Dr. John D. Cunningham/Visuals Unlimited, Inc. C: Mauro Fermariello/Photo Researchers, Inc.)

The epidermis is stratified squamous epithelium. Its cells arise in deeper layers and are pushed toward the surface as new cells arise beneath them. (This efficient replacement is one reason why the skin can mend minor damage so quickly.) As cells move upward, they become flattened, lose their nucleus, and die. Eventually they rub off or flake away.

Most cells of the epidermis are **keratinocytes**. These cells make keratin, a tough, water-insoluble protein. By the time they reach the skin surface and have died, all that remains are the keratin fibers inside plasma membranes. This helps make the skin's outermost layer—the stratum corneum—tough and waterproof.

In the deepest layer of epidermis, cells called **melanocytes** produce a brown-black pigment called melanin. The pigment is transferred to keratinocytes and helps give skin its color. Human skin color varies due to differences in the distribution and activity of those cells. A yellow-orange pigment in the dermis, called carotene, also contributes some color. Pale Caucasian skin has only a little melanin, so the pigment hemoglobin inside red blood cells shows through thin-walled blood vessels and the epidermis itself, both of which are transparent. Naturally brown or black skin contains more melanin.

The epidermis also contains some defensive cells. *Langerhans cells* are phagocytes ("cell eaters"). They consume bacteria or viruses, mobilizing the immune system in the process. *Granstein cells* may help control immune responses in the skin.

Small blood vessels and sensitive nerve endings lace through the dermis, and hair follicles, sweat glands, and oil glands are embedded in it. On the palms and soles of the feet it also has ridges that push up corresponding ridges on the epidermis. These ridges loop and curve in the patterns we call fingerprints. Determined mainly by genes, the pattern is different for each of us, even identical twins.

Sweat glands and other structures develop from epidermis

The body has about 2.5 million sweat glands. Sweat is 99 percent water; it also contains dissolved salts, traces of ammonia and other wastes, vitamin C, and other substances. A subset of sweat glands that are in the palms, soles of the feet, forehead, and armpits is important for cooling the body when it becomes overheated. Another type of sweat gland is abundant in the skin around the genitals. Stress, pain, and sexual foreplay all can increase the amount of sweat they secrete.

Oil glands (or *sebaceous glands*) are everywhere except on the palms and the soles of the feet. The oily substance they release, called sebum, softens and lubricates the hair and skin. Other secretions kill harmful bacteria.

A hair is mostly keratinized cells, rooted in skin with a shaft above its surface. As cells divide near the root's base, older cells are pushed upward, then flatten and die. Flattened cells of the shaft's outer layer overlap (Figure 4.11C) and may frizz out as "split ends." On average the scalp has

Figure 4.12 Vitiligo is a disorder caused by the death of melanocytes. Lee Thomas is an African American television reporter who has vitiligo. The disorder has turned his hands white and produced white blotches on his face and arms. (© Michael Shore Photography)

about 100,000 hairs. However, genes, nutrition, hormones, and stress affect hair growth and density.

Skin disorders are common

The dense connective tissue of the dermis makes it quite tough, but this protection has limits. For example, steady abrasion—as might happen if you wear a too-tight shoe—separates the epidermis from the dermis, the gap fills with a watery fluid, and you get a **blister**.

Acne is a skin inflammation that develops when bacteria infect oil glands. **Cold sores** are caused by a type of herpesvirus. In the disorder **vitiligo** (Figure 4.12), melanocytes die and white patches form on the skin. The cause is not known, but people of all races are affected.

Ultraviolet (UV) radiation stimulates the melanin-producing cells of the epidermis. Prolonged sun exposure increases melanin levels and light-skinned people become tanned. Tanning gives some protection against UV radiation, but over the years elastin fibers in the dermis clump. The skin loses its resiliency and begins to look leathery and wrinkled.

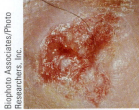

Squamous cell carcinoma

UV radiation, including from tanning lamps, also can trigger cancer. The **squamous cell carcinoma** shown above is a common and easily treatable form of skin cancer. Much more serious is **malignant melanoma**, which forms a dark, uneven, raised lesion (*right*). It is a grave threat because in its later stages it spreads quickly to other parts of the body.

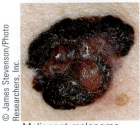

Malignant melanoma

Biophoto Associates/Photo Researchers, Inc.

© James Stevenson/Photo Researchers, Inc.

TAKE-HOME MESSAGE

WHAT IS THE INTEGUMENTARY SYSTEM?

- The integumentary system is the body's covering. It consists of skin and structures that develop from it.
- With its layers of keratinized and melanin-shielded epidermal cells, skin helps the body conserve water, limit damage from ultraviolet radiation, and resist mechanical stress.
- Hair, oil glands, sweat glands, and nails are derived from the skin's epidermis.

Homeostasis: The Body in Balance

- Cells and more complex body parts function properly only when conditions inside the body are stable.
- Links to Life's characteristics 1.1, Acid–base balance 2.7

The internal environment is a pool of extracellular fluid

The trillions of cells in your body all are bathed in fluid—about 15 liters, or a little less than 4 gallons. This fluid, called **extracellular** ("outside the cell") **fluid**, is what we mean by the "internal environment." Much of the extracellular fluid is *interstitial*, meaning that it fills spaces between cells and tissues. The rest is blood plasma, the fluid portion of blood. Substances constantly enter and leave interstitial fluid as cells draw nutrients from it and expel metabolic waste products into it. Those substances can include ions, compounds such as water, and other materials.

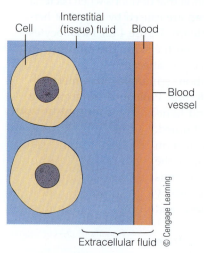

Cell — Interstitial (tissue) fluid — Blood — Blood vessel — Extracellular fluid — © Cengage Learning

All this chemical traffic means that the chemical makeup and volume of extracellular fluid change from moment to moment. If the changes are drastic, they can have drastic effects on cell activities. The number and type of ions in extracellular fluid (such as H^+) are especially crucial, because they must be kept at levels that allow metabolism to continue normally.

As you read in Chapter 1, homeostasis means "staying the same." We use this term because the mechanisms of homeostasis maintain stability in the chemical makeup and volume of extracellular fluid.

In maintaining homeostasis, all components of the body work together in the following general way:

- Each cell engages in metabolic activities that ensure its own survival.
- Tissues, which consist of cells, perform one or more activities that contribute to the survival of the whole body.
- Together, the operations of individual cells, tissues, organs, and organ systems help keep the extracellular fluid in a stable state—a state of homeostasis that allows cells to survive.

Homeostasis requires the interaction of sensors, integrators, and effectors

Three "partners" must interact to maintain homeostasis. They are sensory receptors, integrators, and effectors (Figure 4.13). **Sensory receptors** are cells or cell parts that can detect a stimulus—a specific change in the environment. For a simple example, if someone taps you on the shoulder, there is a change in pressure on your skin. Receptors in the skin translate the stimulus into a signal, which can be sent to the brain. Your brain is an **integrator**, a control center where different bits of information are pulled together in the selection of a response. It can send signals to muscles, glands, or both. Your muscles and glands are **effectors**—they carry out the response, which in this case might include turning your head to see if someone is there. Of course, you cannot keep your head turned indefinitely, because eventually you must eat, use the bathroom, and perform other tasks that maintain body operating conditions.

How does the brain deal with physiological change? Receptors inform it about how things *are* operating, but the brain also maintains information about how things *should be* operating—that is, information from "set points." When some condition in the body shifts sharply from a set point, the brain brings it back within proper range. It does this by sending signals that cause specific muscles and glands to step up or reduce their activity. Set points are important in a great many physiological mechanisms, including those that influence eating, breathing, thirst, and urination, to name a few.

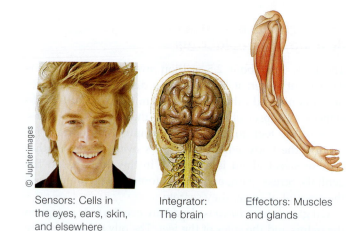

Sensors: Cells in the eyes, ears, skin, and elsewhere

Integrator: The brain

Effectors: Muscles and glands

Figure 4.13 Different body structures function as sensors, integrators, and effectors. In humans the brain and spinal cord are integrators. Organs of the endocrine system and muscles are effectors. (© Cengage Learning)

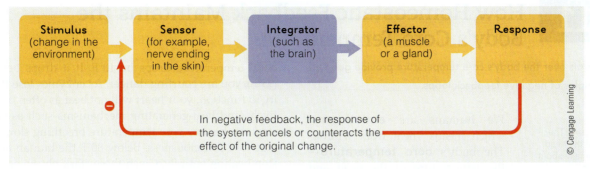

In negative feedback, the response of the system cancels or counteracts the effect of the original change.

© Cengage Learning

Figure 4.14 Animated! Negative feedback is the main mechanism for maintaining homeostasis.

Negative feedback is the main control mechanism of homeostasis

Mechanisms for feedback help keep physical and chemical aspects of the body within tolerable ranges. In **negative feedback**, an activity alters a condition in the external or internal environment—that is, extracellular fluid. Sensory receptors detect the change and send this information to an integrator. The integrator (often, the brain) triggers a response that reverses the altered condition (Figure 4.14). Negative feedback is the main mechanism of control for maintaining homeostasis.

As an analogy, a thermostat-controlled heating system works by negative feedback. The thermostat senses the air temperature and mechanically compares it to a preset point on a thermometer built into the furnace controls. When the temperature falls below the preset point, the thermostat signals a switch that turns on the heating unit. When the air warms enough to match the preset level, the thermostat signals the switch to shut off the heating unit. In Section 4.11 you will learn how negative feedback helps regulate body temperature.

Positive feedback plays a role outside of homeostasis

In a few situations *positive feedback* operates. In this type of mechanism, a chain of events intensify a change from an original condition. Eventually, though, the intensifying feedback reverses the change. There are not many examples of positive feedback in body functions that affect the makeup of extracellular fluid, so positive feedback does not have a major role in homeostasis. Positive feedback does occur in the body, however. A familiar example is childbirth. During labor a fetus exerts pressure on the walls of its mother's uterus. The pressure stimulates the production and secretion of a hormone (oxytocin) that causes the mother's uterine muscles to contract and exert pressure on the fetus, which exerts more pressure on the uterine wall, and so on until the fetus is expelled.

As the body monitors and responds to information about the external world and the internal environment, its organ systems must operate in a coordinated way. In upcoming chapters we will be asking four important questions about how organ systems function:

1. What physical or chemical aspect of the internal environment is each organ system working to maintain as conditions change?

2. How is each organ system kept informed of changes?

3. How does each system process incoming information?

4. What are the responses?

As you will see, all organ systems operate under precise controls of the nervous system and the endocrine system.

effector Tissue, gland, or other body part that carries out a response ordered by an integrator.

extracellular fluid Blood plasma and tissue fluid.

integrator Control center, such as the brain, that compares a detected environmental change with a set point and activates a response.

negative feedback Control mechanism of homeostasis that reverses change to body system if it exceeds a set point.

sensory receptor Cell or cell part that can detect some type of stimulus—a change in the environment.

TAKE-HOME MESSAGE

WHAT ARE HOMEOSTATIC CONTROLS?

- Homeostatic controls are mechanisms that maintain the characteristics of the internal environment within ranges that allow cells to function properly.
- Negative feedback is the main control operating to maintain homeostasis.
- Body parts that serve as sensory receptors, integrators, and effectors monitor conditions in the body and carry out negative feedback.

How Homeostatic Feedback Maintains the Body's Core Temperature

■ Controls over the body's core temperature provide good examples of negative feedback loops.

core temperature The temperature of the head and torso, normally about 37°C, or 98.6°F.

endotherm An animal whose body heat is generated "from within," by the metabolic processes of its cells.

hyperthermia Condition in which the body core temperature rises above the normal range.

hypothermia Condition in which the body core temperature falls below the normal range.

nonshivering heat production When brown adipose tissue releases energy as heat rather than storing it as ATP.

We humans are **endotherms**, which means "heat from within." The body's **core temperature**—the temperature of the head and torso—is about 37°C, or 98.6°F. It is controlled mainly by metabolic activity, which produces heat, and by negative feedback loops. These homeostatic controls adjust physiological responses for conserving or getting rid of heat (Figure 4.15). We can assist the physiological controls by altering our behavior—for example, by changing clothes or switching on a furnace or an air-conditioner.

Metabolism produces heat. If that heat were to build up internally, your core temperature would steadily rise. Above 41°C (105.8°F), some enzymes become denatured and virtually shut down. By the same token, the rate of enzyme activity generally *decreases* by at least half when body temperature drops by 10°F. If it drops below 35°C (95°F), you are courting danger. As enzymes lose their ability to function, your heart will not beat as often or as effectively, and heat-generating mechanisms such as shivering stop. At this low core temperature breathing slows, so you may lose consciousness. Below 80°F the human heart may stop beating entirely. Given these stark physiological facts, humans require mechanisms that help maintain the core body temperature within narrow limits.

Excess heat must be eliminated

Table 4.3 summarizes the main responses to heat stress. They are governed by the hypothalamus, a structure in the brain in which there are both neurons and endocrine cells. When core temperature rises above a set point, the hypothalamus orders blood vessels in the skin to dilate. This widening, called *vasodilation*, allows more blood to flow through the vessels, where the excess heat that blood carries is dissipated.

The hypothalamus also can activate sweat glands and increase the amount of body heat lost via evaporation. With roughly 2.5 million sweat glands in skin, lots of heat is dissipated when the water in sweat evaporates. With prolonged heavy sweating the body also loses key salts, especially sodium chloride. Losing too many of these electrolytes can make you feel woozy. "Sports drinks" replenish electrolytes.

Sometimes peripheral blood flow and evaporative heat loss can't adequately counter heat stress. The result is **hyperthermia**, in which the core temperature rises above normal. A realtively modest increase causes *heat exhaustion*, in which blood pressure drops due to vasodilation and water losses from heavy sweating. The skin feels cold and clammy, and the person may collapse.

When heat stress is severe enough to completely break down the body's temperature controls, *heat stroke* occurs. Sweating stops, the skin

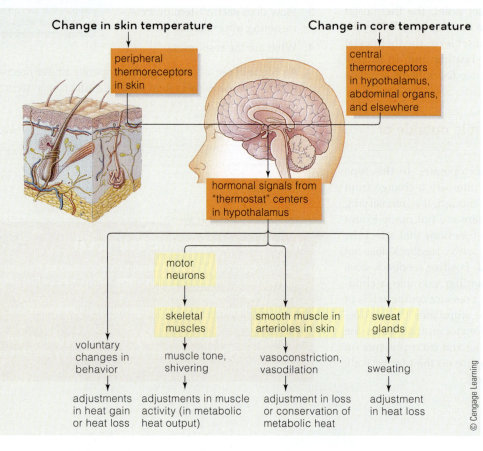

Change in skin temperature

peripheral thermoreceptors in skin

Change in core temperature

central thermoreceptors in hypothalamus, abdominal organs, and elsewhere

hormonal signals from "thermostat" centers in hypothalamus

motor neurons

skeletal muscles

smooth muscle in arterioles in skin

sweat glands

voluntary changes in behavior

muscle tone, shivering

vasoconstriction, vasodilation

sweating

adjustments in heat gain or heat loss

adjustments in muscle activity (in metabolic heat output)

adjustment in loss or conservation of metabolic heat

adjustment in heat loss

© Cengage Learning

Figure 4.15 Animated! Homeostatic controls regulate internal body temperature.

TABLE 4.3 Summary of Human Responses to Cold Stress and to Heat Stress		
Environmental Stimulus	Main Responses	Outcome
DROP IN TEMPERATURE	Vasoconstriction of blood vessels in skin; pilomotor response; behavior changes (e.g., putting on a sweater)	Heat is conserved
	Increased muscle activity; shivering; nonshivering heat production	More heat is produced
RISE IN TEMPERATURE	Vasodilation of blood vessels in skin; sweating; changes in behavior; heavy breathing	Heat is dissipated from body
	Reduced muscle activity	Less heat is produced

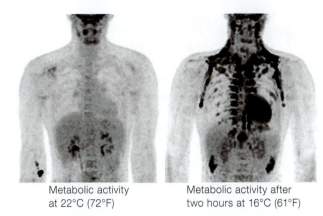

Metabolic activity at 22°C (72°F)   Metabolic activity after two hours at 16°C (61°F)

Figure 4.16 Nonshivering heat production uses the body's reserves of brown fat. People who are cold-adapted generally have more stored brown fat. (From "Cold-Activated Brown Adipose Tissue in Healthy Men." van Marken Lichtenbelt, Wouter D. et al. Article DOI: 10.1056/NEJMoa0808718. © 2009 The New England Journal of Medicine. Reprinted with permission.)

becomes dry, and the core body temperature rapidly rises to a level that can be lethal.

When someone has a fever, the hypothalamus has reset the "thermostat" that dictates what the body's core temperature will be. The normal response mechanisms occur but they maintain a higher temperature. When a fever "breaks," peripheral vasodilation and sweating increase as the body works to restore the normal core temperature. The controlled increase in core temperature during a fever seems to boost immune responses, so using fever-reducing drugs may actually interfere with fever's beneficial effects. A severe fever always requires medical attention because of the dangers it poses.

Several responses counteract cold

Table 4.3 also summarizes the major responses to cold stress, which the hypothalamus also regulates. When the outside temperature drops, thermoreceptors (*thermo-* means "heat") at the body surface detect the decrease. When their signals reach the hypothalamus, neurons signal smooth muscle in the walls of certain skin blood vessels to contract, and the blood vessels narrow. This narrowing, called *vasoconstriction*, reduces blood flow to capillaries near the body surface, so your body retains heat. When your hands or feet get cold, as much as 99 percent of the blood that would otherwise flow to your skin is diverted.

In the pilomotor response to a drop in outside temperature, your body hair can "stand on end." This happens because smooth muscle controlling the erection of body hair is stimulated to contract. This creates a layer of still air close to the skin that reduces heat loss. (This response is most effective in mammals with more body hair than humans!) Heat loss can be restricted even more by behaviors that reduce the amount of body surface exposed for heat exchange, as when you put on a sweater or hold your arms tightly against your body.

When other responses can't counteract cold stress, signals from the hypothalamus step up skeletal muscle contractions, similar to the low-level contractions that produce muscle tone. The result is shivering—skeletal muscles contract ten to twenty times per second, boosting heat production throughout the body.

Severe exposure to cold can lead to a hormone-driven response that speeds up cell metabolism. This **nonshivering heat production** occurs in a type of brownish adipose tissue called "brown fat" (Figure 4.16). Heat is generated as the lipid molecules are broken down. In babies (who can't shiver) brown fat makes up about 5 percent of body weight; most adults have some in their neck and upper chest.

In **hypothermia**, body core temperature falls below the normal range. A drop of only a few degrees leads to mental confusion. Further cooling can cause coma and death. Some victims of extreme hypothermia, mainly children, have survived prolonged immersion in ice-cold water. One reason is that mammals, including humans, have a dive reflex. When the body is submerged, the heart rate slows and blood is shunted to the brain and other vital organs.

Freezing often destroys tissues, a condition we call *frostbite*. Frozen cells may be saved if thawing is precisely controlled. This sometimes can be done in a hospital.

TAKE-HOME MESSAGE

HOW IS BODY TEMPERATURE REGULATED?

- The hypothalamus (in the brain) regulates physiological mechanisms that adjust the body's core temperature.
- Responses to heat stress include dilation of blood vessels near the body surface and evaporative heat loss.
- Responses to cold stress include constriction of blood vessels near the body surface, the pilomotor response, shivering, and nonshivering heat production.

As epithelium, your skin contains fibers of collagen and elastin. These structural proteins have different properties that you can see in action when you pull on a patch of skin. Notice that even if you pull firmly, the skin doesn't tear. Which type of protein fiber gives the skin that tensile strength? Which type returns the skin to its original shape when you let go?

SUMMARY

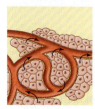

Section 4.1 A tissue is a group of similar cells that perform the same function (Table 4.4). Epithelial tissue covers body surfaces and lines internal cavities. Each kind of epithelium has one surface exposed to body fluids or the outside environment; the opposite surface rests on a basement membrane between it and underlying tissue.

Glands are derived from epithelium. Exocrine glands release substances (such as saliva and tears) onto the surface of an epithelium through ducts or tubes. Endocrine glands secrete hormones directly into extracellular fluid.

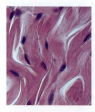

Section 4.2 Connective tissues bind, support, strengthen, and protect other tissues. Most have fibers of structural proteins (especially collagen), fibroblasts, and other cells within a matrix. They include fibrous connective tissue and specialized connective tissues such as cartilage, bone, adipose tissue, and blood.

Section 4.3 Muscle tissue contracts. It helps move the body or its parts. The three types of muscle tissue are skeletal muscle, smooth muscle, and cardiac muscle.

Section 4.4 Nervous tissue receives and integrates information from inside and outside the body and sends signals for responses. Neurons and the support cells called neuroglia are the main cells in nervous tissue.

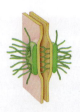

Section 4.6 Tight junctions help prevent substances from leaking across a tissue. Adhering junctions bind cells together in tissues. Gap junctions link the cytoplasm of neighboring cells.

Section 4.7 Membranes cover all body surfaces and cavities. Those made of epithelium include mucous and serous membranes. Connective tissue membranes include the synovial membranes of certain joints. The skin is a cutaneous membrane.

Section 4.8 Different tissues combine to form an organ. Body organs are located in five major cavities: the cranial cavity (brain); spinal cavity (spinal cord); thoracic cavity (heart and lungs); abdominal cavity (stomach, liver, most of the intestine, other organs); and pelvic cavity (reproductive organs, bladder, rectum). The various organs in the body are arranged into eleven organ systems. In an organ system, two or more organs interact in ways that contribute to the body's survival. Each system performs a specific function, such as transporting blood (cardiovascular system) or reproduction.

Section 4.9 An example of an organ system is the integument, or skin. Skin has an outer epidermis and an underlying dermis. Most epidermal cells are keratinocytes, which make the protein keratin. Keratin makes the skin's outer layer tough and waterproof. Melanocytes in the epidermis produce pigment that gives skin its color. Hair, nails, sweat glands, and oil glands are derived from the epidermis.

Skin protects the rest of the body from abrasion, invading bacteria, ultraviolet radiation, and dehydration. It helps control internal temperature, contains cells that synthesize vitamin D, and serves as a blood reservoir for the rest of the body. Receptors in skin are essential for detecting environmental stimuli.

Section 4.10 Extracellular fluid (blood and tissue fluid) is the body's internal environment. Tissues, organs, and organ systems work together to maintain the stable state of homeostasis in this environment. Maintaining homeostasis requires sensory receptors, which can detect a stimulus,

TABLE 4.4 Summary of Basic Tissue Types in the Human Body

Tissue	Function	Characteristics
EPITHELIUM	Covers body surface; lines internal cavities and tubes	One free surface; opposite surface rests on basement membrane supported by connective tissue
CONNECTIVE TISSUE	Binds, supports, adds strength; some provide protection or insulation	Cells surrounded by a matrix (ground substance) containing structural proteins except in blood
FIBROUS CONNECTIVE TISSUES		
Loose	Elasticity, diffusion	Cells and fibers loosely arranged
Dense	Support, elasticity	Several forms. One has collagen fibers in various orientations in the matrix; it occurs in skin and as capsules around some organs. Another form has collagen fibers in parallel bundles; it occurs in ligaments, tendons
Elastic	Elasticity	Mainly elastin fibers; occurs in organs that must stretch
SPECIALIZED CONNECTIVE TISSUES		
Cartilage	Support, flexibility, low-friction surface	Matrix solid but pliable; no blood supply
Bone	Support, protection, movement	Matrix hardened by minerals
Adipose tissue	Insulation, padding, energy storage	Soft matrix around large, fat-filled cells
Blood	Transport	Liquid matrix (plasma) containing blood cells, many other substances
MUSCLE TISSUE	Movement of the body and its parts	Made up of arrays of contractile cells
NERVOUS TISSUE	Communication between body parts; coordination, regulation of cell activity	Made up of neurons and support cells (neuroglia)

integrators that interpret it, and effectors that carry out a response. In negative feedback, a change in a condition triggers a response that reverses the change. Negative feedback is the main control mechanism of homeostasis.

Section 4.11 Physiological responses that govern temperature rely on negative feedback controls that respond to heat stress and cold stress.

REVIEW QUESTIONS

1. List the general characteristics of epithelium, and then describe the basic types of epithelial tissues in terms of specific characteristics and functions.

2. List the major types of connective tissues; add the names and characteristics of their specific types.

3. Identify and describe the tissues shown below.

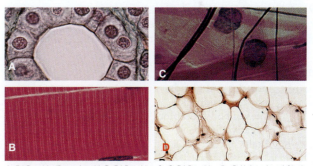

A: Ed Reschke/Peter Arnold; B: Ed Reschke; C: © Ed Reschke; D: © University of Cincinnati, Raymond Walters College, Biology

4. List the types of cell junctions and their functions.

5. List the basic types of membranes in the body.

6. Define the terms *tissue*, *organ*, and *organ system*. List the body's eleven major organ systems.

7. What are some functions of skin?

8. Define homeostasis.

9. What is extracellular fluid, and how does the concept of homeostasis pertain to it?

1. _____ tissues have closely linked cells and one free surface.
   - a. Muscle
   - b. Nerve
   - c. Connective
   - d. Epithelial

2. Most _____ has collagen and elastin fibers.
   - a. muscle tissue
   - b. nervous tissue
   - c. connective tissue
   - d. epithelial tissue

3. _____ , a specialized connective tissue, is mostly plasma with cellular components and various dissolved substances.
   - a. Irregular connective tissue
   - b. Blood
   - c. Cartilage
   - d. Bone

4. _____ tissue detects and integrates information about changes and controls responses to changes.
   - a. Muscle
   - b. Nervous
   - c. Connective
   - d. Epithelial

5. _____ can shorten (contract).
   - a. Muscle tissue
   - b. Nervous tissue
   - c. Connective tissue
   - d. Epithelial tissue

6. After you eat too many carbohydrates and proteins, your body converts the excess to storage fats, which accumulate in _____ .
   - a. loose connective tissue
   - b. dense connective tissue
   - c. adipose tissue
   - d. both b and c

7. The body's internal environment consists of _____ .
   - a. blood plasma
   - b. interstitial fluid
   - c. the five body cavities
   - d. both a and b

8. In _____ , physical and chemical aspects of the body are being kept within tolerable ranges by controlling mechanisms.
   - a. positive feedback
   - b. negative feedback
   - c. homeostasis
   - d. metastasis

9. Fill in the blanks: _____ detect specific environmental changes, an _____ pulls different bits of information together in the selection of a response, and _____ carry out the response.

10. Match the concepts:
    - ____ muscles and glands
    - ____ positive feedback
    - ____ sites of body receptors
    - ____ negative feedback
    - ____ brain
    - a. integrating center
    - b. reverses an altered condition
    - c. eyes and ears
    - d. effectors
    - e. intensifies the original condition

CRITICAL THINKING

1. In people who have the genetic disorder anhidrotic ectodermal dysplasia, patches of tissue have no sweat glands. What kind of tissue does it affect?

2. The disease called scurvy results from a deficiency of vitamin C, which the body uses to synthesize collagen. Explain why scurvy sufferers tend to lose teeth, and why any wounds heal much more slowly than normal, if at all.

3. The man pictured in Figure 4.17 wears several dozen ornaments in his skin, most of them applied by piercing.

Your Future

The blood remaining in the umbilical cord after a birth contains small numbers of stem cells that can give rise to new blood cells. Although such stem cells are in short supply, they already are used to help reestablish a supply of healthy blood cells in patients who suffer from leukemia and some other blood cancers. They are a popular option because unlike donated blood from other sources, cord blood does not have to genetically match the patient's blood. Scientists have been looking for ways to get cord blood cells to multiply more rapidly, so that more will be available for suffering patients. They are making progress, so it may not be long before cord blood stem cells are much more widely available.

© Juergen Berger, Max-Planck Institute/Photo Researchers, Inc.

Among the skin's many functions, it serves as a barrier to potentially dangerous bacteria, and some people object to extensive body piercing on the grounds that it opens the door to infections. Explain why you do or don't agree with this objection.

© Sean Sprague/Stock Boston

Figure 4.17 This man has chosen to undergo heavy body piercing.

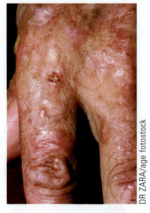

DR ZARA/age fotostock

Figure 4.18 Sun exposure causes ulcers and blisters to form on the skin of a person affected by porphyria.

4. Various forms of porphyria, a genetic disorder, affect humans. In one form, affected people lack enzymes of a metabolic pathway that forms heme, the iron-containing group in hemoglobin. Intermediate chemicals called porphyrins accumulate and cause terrible symptoms, especially if the person is exposed to sunlight. Sores and scars form on the skin (Figure 4.18). Thick hair grows on the face and hands. The gums shrink away from the teeth and the canine teeth can begin to look like fangs. Symptoms get worse if the person drinks alcohol or eats garlic. People with porphyria can avoid sunlight and aggravating substances. They also can get injections of heme from normal red blood cells. If you are familiar with vampire stories, which date from centuries ago, can you think of a reason why they may have arisen among people who knew nothing about the cause of porphyria?

THE SKELETAL SYSTEM

LINKS TO EARLIER CONCEPTS

As you study the skeletal system, you will learn more about the structure and functions of bone tissue, cartilage, and some other connective tissues (4.2) that are major components of the system.

KEY CONCEPTS

The Structure and Functions of Bones

Bones are built of bone tissue. They store minerals, protect and support soft organs, and function in body movement. Some bones contain marrow where blood cells develop. Section 5.1

The Skeleton

The skeleton's key function is to serve as the body's internal framework. Its 206 bones are organized into two parts, the axial skeleton and the appendicular skeleton. Sections 5.2–5.4

Joints

At joints, bones touch or are in close contact with one another. Some of these connections permit adjoining bones to move in ways that in turn move body parts, such as the limbs. Section 5.5

Disorders of the Skeleton

Disorders that affect our bones usually prevent them from functioning normally. In addition to breaks and arthritis, the skeleton may be impaired by cancer, infections, and other conditions. Section 5.6

CONNECTIONS:
The Skeletal System in Homeostasis Section 5.7

Top: Ed Reschke/Peter Arnold; Middle: both © Cengage Learning; Bottom: © Prof. P. Motta, Dept. of Anatomy, Univ. of La Sapienza, Rome/SPL/Photo Researchers, Inc.

At eighteen, Susanna was a college freshman with a problem that sometimes made her feel more like eighty. After three years of playing on her high school's girl's soccer team, she had so much pain in her knees that each day she downed multiple doses of an NSAID (a nonsteroidal anti-inflammatory drug) just to make it through her hectic day of classes, a part-time job, and commuting to and from campus on her bike. Susanna had become another statistic in the records of the U.S. Consumer Product Safety Commission—one of the roughly 140,000 Americans under the age of twenty-four who annually suffer some type of soccer injury. Of that total, 65 percent of the injuries affect players' knees and ankles, with knee damage more common in girls. By the time she finished college, Susanna's knees were seriously affected by osteoarthritis, a disorder in which joints become painfully stiff because the cartilage lining is breaking down or bone spurs have formed there.

Susanna's story introduces the skeletal system, our topic in this chapter. Its parts help provide a sturdy framework for the body's soft flesh, and bones partner with muscles to bring about the movements we take for granted as we work, play, and go about daily life.

Mike Powell/Getty Images

Homeostasis Preview
The skeleton and bones have many roles in homeostasis. In addition to providing physical support, attachment points for skeletal muscles, and protection for soft body parts, the skeleton helps maintain proper calcium balance in the blood. Stem cells in bone marrow are the source of our billions of blood cells.

- Bone tissue is a form of connective tissue hardened by the mineral calcium.

- Link to Connective tissues 4.2

Bone is a connective tissue, so it is a blend of living cells and a matrix that contains fibers. Bones are covered by a sturdy two-layer membrane called the periosteum (meaning "around the bone"). The membrane's outer layer is dense connective tissue, and the inner layer contains bone cells called **osteoblasts** ("bone formers"). As bone develops, the osteoblasts secrete collagen and some elastin, as well as carbohydrates and other proteins. With time, this matrix around osteoblasts hardens when salts of the mineral calcium are deposited in it. Each osteoblast is trapped in a space, or lacuna, in the matrix (*lacuna* means hole). At this point their bone-forming function ends and they are called **osteocytes** (*osteo* = bone; *cyte* = cell).

The minerals and collagen in bone tissue make it hard, but it is the collagen that gives our bones the strength to withstand the mechanical stresses associated with activities such as standing, lifting, and tugging.

There are two kinds of bone tissue

Bones contain two kinds of tissue, compact bone and spongy bone. Figure 5.1 shows where these tissues are in a long bone such as the femur (thighbone). As its name suggests, **compact bone** is a dense tissue that looks solid and smooth. In a long bone, it forms the bone's shaft and the outer part of its two ends. A cavity inside the shaft contains bone marrow.

Compact bone tissue forms in thin, circular layers around small central canals. Each set of layers is called an **osteon** (os-tee-ahn; sometimes called a *Haversian system*). The canals connect with each other and serve as channels for blood vessels and nerves that transport substances to and from osteocytes (Figure 5.1A, right, and Figure 5.1B). Osteocytes also extend slender cell processes into narrow channels called canaliculi that run between lacunae. These "little canals" allow nutrients to move through the hard matrix from osteocyte to osteocyte. Wastes can be removed the same way.

The bone tissue *inside* a long bone's shaft and at its ends looks like a sponge. Tiny, flattened struts are fused together to make up this **spongy bone** tissue, which looks lacy and delicate but actually is quite firm and strong.

A bone develops on a cartilage model

An early embryo has a rubbery skeleton that consists of cartilage and membranes. Yet, after only about two months of life in the womb, this flexible framework is transformed into a bony skeleton. Once again, we can look at the development of a long bone as an example.

As you can see at the top of Figure 5.2, a cartilage "model" provides the pattern for each long bone. Once the outer membrane is in place on the model, the bone-forming osteoblasts become active and a bony "collar" forms around the cartilage shaft. Then the cartilage inside the shaft calcifies, and blood vessels, nerves, and elements including osteoblasts begin to infiltrate the forming bone. Soon, the marrow cavity forms and osteoblasts produce the matrix that will become mineralized with calcium.

Each end of a long bone is called an **epiphysis** (e-PIF-uh-sis). As long as a person is growing, each epiphysis is separated from the bone shaft by an *epiphyseal plate* of cartilage.

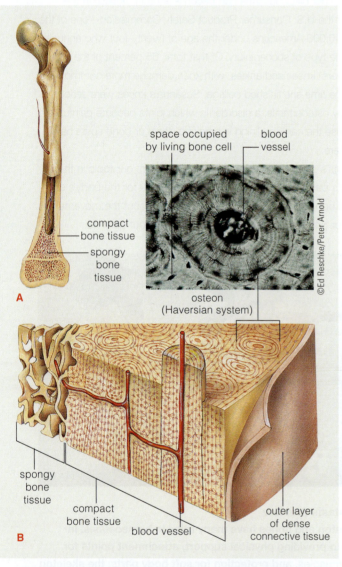

space occupied by living bone cell

blood vessel

compact bone tissue

spongy bone tissue

©Ed Reschke/Peter Arnold

A

osteon (Haversian system)

spongy bone tissue

compact bone tissue

blood vessel

outer layer of dense connective tissue

B

Figure 5.1 Animated! Bones contain both compact and spongy bone tissue. **A** Spongy and compact bone tissue in a femur (thighbone). **B** The canal in the center of each osteon contains blood vessels and nerves. The blood vessel carries substances to and from osteocytes, the living bone cells in small spaces in the bone tissue. Narrow tunnels called canaliculi connect neighboring spaces. (© Cengage Learning)

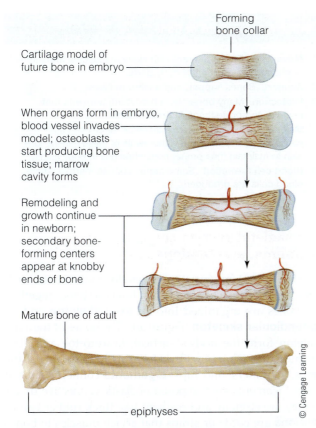

Forming bone collar

Cartilage model of future bone in embryo

When organs form in embryo, blood vessel invades model; osteoblasts start producing bone tissue; marrow cavity forms

Remodeling and growth continue in newborn; secondary bone-forming centers appear at knobby ends of bone

Mature bone of adult

epiphyses

© Cengage Learning

Figure 5.2 A long bone forms on a cartilage model. First, osteoblasts begin to function in a cartilage model in the embryo. The bone-forming cells are active first in the shaft, then at the knobby ends. In time, cartilage is left only in the epiphyses at the ends of the shaft.

Human growth hormone (GH) prevents the plates from calcifying, so the bone can lengthen. When growth stops, usually when people reach their late teens or early twenties, bone replaces the cartilage plates.

Bone tissue is constantly "remodeled"

Calcium is constantly entering and leaving a person's bones. Calcium is deposited when osteoblasts form bone, and it is withdrawn when "bone breaker" cells called **osteoclasts** break down the matrix of bone tissue. This ongoing calcium recycling is called **bone remodeling**, and it has several important functions.

Regularly breaking down "old" bone and replacing it with fresh tissue helps keep bone resilient, so it is less likely to become brittle and break. When a bone is subjected to mechanical stress, such as load-bearing exercise, the remodeling process is adjusted so that more bone is deposited than removed. That is why the bones of regular exercisers are denser and stronger than the bones of couch potatoes. On the other hand, when the body must heal

a broken bone, osteoclasts release more calcium than usual from bone matrix. Osteoblasts then use the calcium to repair the injured bone tissue.

A child's body requires lots of calcium to meet the combined demands of bone growth and other needs for the calcium stored in bones. Along with dietary calcium, remodeling helps meet the demand. For example, the diameter of a growing child's thighbones increases as osteoblasts form bone at the surface of each shaft. At the same time, however, osteoclasts break down a small amount of bone tissue *inside* the shaft. Thus the child's thighbones become thicker and stronger to support the increasing body weight, but they don't get too heavy.

Bone remodeling also plays a key role in maintaining homeostasis of the blood level of calcium. Neither our nervous system nor our muscles can function properly unless the blood level of calcium stays within a narrow range. When the level falls below this range, a hormone called PTH stimulates osteoclasts to break down bone and release calcium to the blood. If the level rises too high, another hormone, calcitonin, stimulates osteoblasts to *deposit* calcium in bone tissue. Notice that this control mechanism is an example of negative feedback. You will read more about it in Chapter 15, when we take a closer look at hormones.

bone Connective tissue that functions in movement and locomotion, protection of other organs, mineral storage, and (in some bones) blood cell production.

bone remodeling The recycling of calcium as it is deposited and withdrawn from bones.

compact bone The denser, solid-looking type of bone tissue, which forms in slender, circular layers.

epiphysis The end of a long bone.

osteoblast Type of bone cell that forms the matrix of bone tissue, which eventually becomes mineralized.

osteoclast Type of bone cell that breaks down the matrix of bone tissue.

osteocyte Name for a bone-forming cell (osteoblast) after the matrix around it has become mineralized and the cell stops forming bone matrix.

osteon Each set of thin, circular layers that forms in compact bone. The layers encircle a channel for blood vessels and nerves.

spongy bone The lacy, more open type of bone tissue. In long bones it occurs inside the bone shaft.

TAKE-HOME MESSAGE

WHAT IS BONE?

- Bone is a form of connective tissue that consists of living cells and a nonliving matrix that is hardened by the mineral calcium.
- Bones contain two types of bone tissue—dense compact bone and lacy but strong spongy bone.
- Bones grow, become strong, and are repaired through the process called bone remodeling.
- Bone remodeling is important in homeostasis because it plays a major role in maintaining the proper balance of calcium in the blood.

The Skeletal System: The Body's Bony Framework

- Bones are the major organs of the skeletal system. Their functions include providing a hard surface against which muscles can exert force to move body parts.

- Link to Muscle tissue 4.3

appendicular skeleton Portion of the skeleton that includes bones of the limbs, shoulders, and hips.

axial skeleton Portion of the skeleton that forms the body's vertical axis—the skull bones, vertebral column, and rib cage.

bone marrow Connective tissue in some bones, where blood cells form.

ligament Connective tissue that connects bones at joints.

tendon Straplike connective tissue that attaches muscles to bones or to other muscles.

An adult's skeleton contains 206 bones. These bones vary in size and shape, from bones in the ear that are the size of a watch battery to massive thighbones (Figure 5.3). Some, like the thighbone, are long and slender. Others, like the ankle bones, are quite short. Still others, such as the sternum (breastbone), are flat, and still others, such as spinal vertebrae, are "irregular." All bones contain bone tissue, however, and other connective tissue lines their surfaces and internal cavities. At joints there is cartilage where one bone meets or "articulates" with another. Other tissues associated with bones include nervous tissue and epithelium, which occurs in the walls of blood vessels that carry substances to and from bones. Clearly, bones are complex organs!

Some bones, such as long bones, have cavities that contain the connective tissue called **bone marrow**. In children most marrow-containing bones have red bone marrow, in which blood cells form. With time, however, much of this red marrow is replaced by fat-rich yellow marrow, where no blood cells form. For this reason, most of an adult's blood cells form in red bone marrow in irregular bones, such as the hip bone, and in flat bones, such as the sternum. If you lose a lot of blood, yellow marrow in your long bones can convert to red marrow, which makes red blood cells.

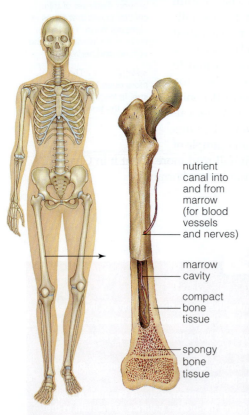

nutrient canal into and from marrow (for blood vessels and nerves)

marrow cavity

compact bone tissue

spongy bone tissue

Figure 5.3 The femur (thighbone) is a typical long bone. (© Cengage Learning)

TABLE 5.1 Functions of Bone

1. **Movement.** Bones interact with skeletal muscles to maintain or change the position of body parts.
2. **Support.** Bones support and anchor muscles.
3. **Protection.** Many bones form hard compartments that enclose and protect soft internal organs.
4. **Mineral storage.** Bones are a reservoir for calcium and phosphorus. Deposits and withdrawals of these mineral ions help to maintain their proper concentrations in body fluids.
5. **Blood cell formation.** Some bones contain marrow where blood cells are produced.

The skeletal system consists of bones, ligaments, and tendons

The skeletal system consists of bones along with joints, cartilages, and straplike ligaments that hold our bones together. The bones are organized into an **axial skeleton** and an **appendicular skeleton** (Figure 5.4). The bones of the axial skeleton form the body's vertical, head-to-toe axis. The appendicular ("hanging") skeleton includes bones of the limbs, shoulders, and hips. **Ligaments** connect bones at joints. Ligaments are composed of elastic connective tissue, so they are stretchy and resilient like thick rubber bands. **Tendons** are cords or straps that attach muscles to bones or to other muscles. They are built of connective tissue packed with collagen fibers, which make tendons strong.

Bones have several important functions

Bones contribute to homeostasis in many ways (Table 5.1). For instance, bones that support and anchor skeletal muscles help maintain or change the positions of our body parts. Some form hard compartments that enclose and protect other organs; for example, the skull encloses and protects the brain, and the rib cage protects the lungs. As noted in Section 5.1, bones also serve as a "pantry" where the body can store calcium. Because the calcium in bone is in the form of the compound calcium phosphate, bone also is a storage depot for phosphorus.

TAKE-HOME MESSAGE

WHAT ARE THE PARTS OF THE SKELETAL SYSTEM?

- In the skeletal system, bones are organized into two parts—the axial skeleton and the appendicular skeleton.
- Ligaments connect bones at joints, and tendons attach bones to muscles (or to other muscles).
- Bones contribute to homeostasis by providing body support, enabling movement, and storing minerals.
- Some bones contain red marrow where blood cells form. In adults, some other bones contain fatty yellow marrow where no blood cells form.

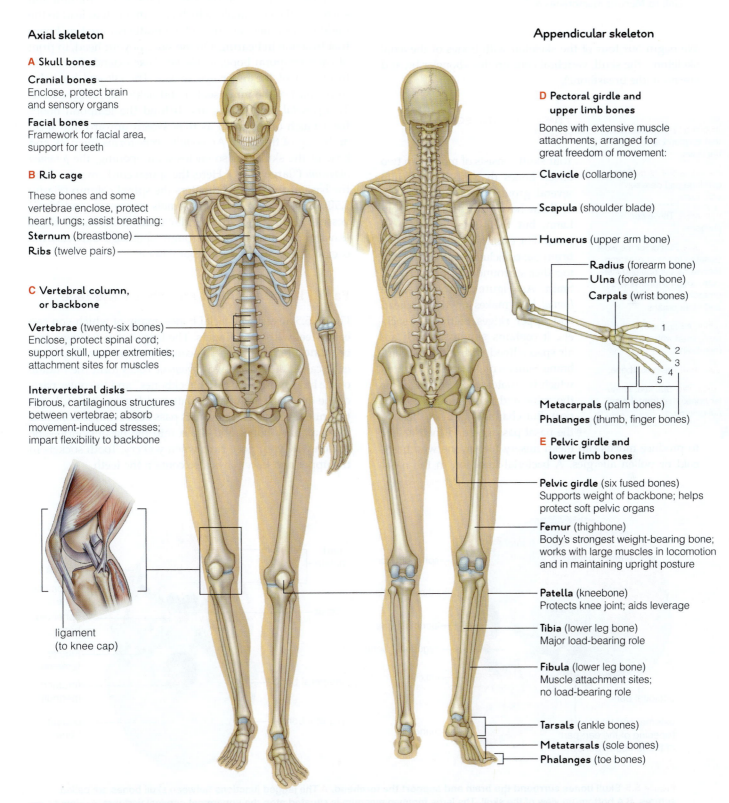

Axial skeleton

A Skull bones

Cranial bones —
Enclose, protect brain
and sensory organs

Facial bones —
Framework for facial area,
support for teeth

B Rib cage

These bones and some
vertebrae enclose, protect
heart, lungs; assist breathing:

Sternum (breastbone) —
Ribs (twelve pairs) —

**C Vertebral column,
or backbone**

Vertebrae (twenty-six bones) —
Enclose, protect spinal cord;
support skull, upper extremities;
attachment sites for muscles

Intervertebral disks —
Fibrous, cartilaginous structures
between vertebrae; absorb
movement-induced stresses;
impart flexibility to backbone

ligament
(to knee cap)

Appendicular skeleton

**D Pectoral girdle and
upper limb bones**

Bones with extensive muscle
attachments, arranged for
great freedom of movement:

Clavicle (collarbone)

Scapula (shoulder blade)

Humerus (upper arm bone)

Radius (forearm bone)
Ulna (forearm bone)
Carpals (wrist bones)

1
2
3
4
5

Metacarpals (palm bones)
Phalanges (thumb, finger bones)

**E Pelvic girdle and
lower limb bones**

Pelvic girdle (six fused bones)
Supports weight of backbone; helps
protect soft pelvic organs

Femur (thighbone)
Body's strongest weight-bearing bone;
works with large muscles in locomotion
and in maintaining upright posture

Patella (kneebone)
Protects knee joint; aids leverage

Tibia (lower leg bone)
Major load-bearing role

Fibula (lower leg bone)
Muscle attachment sites;
no load-bearing role

Tarsals (ankle bones)

Metatarsals (sole bones)

Phalanges (toe bones)

Figure 5.4 **The skeletal system is divided into the axial skeleton and the appendicular skeleton. The blue areas are cartilage.**
(© Cengage Learning)

- The axial skeleton includes bones that make up the body's vertical axis and protect many internal organs.

- Link to Mucous membranes 4.7

We begin our tour of the skeleton with bones of the axial skeleton—the skull, vertebral column (backbone), ribs, and sternum (the breastbone).

brain case Skull bones that surround and protect the brain.

intervertebral disk Fibro-cartilage pad between vertebrae.

mandible The lower jawbone.

rib cage Portion of the axial skeleton in the upper torso, formed by the ribs and sternum, which supports and protects the heart, lungs, and other organs.

sinus Air space in the skull, lined with mucous membrane.

sternum The breastbone.

vertebrae Stacked, irregular bones that form the spinal column.

The skull protects the brain

Your skull consists of more than two dozen bones that are divided into several groups. By tradition many of them have names derived from Latin, but their roles are easy to grasp. For example, the cranium, or **brain case**, includes eight bones that together surround and protect your brain. As Figure 5.5A shows, the *frontal bone* makes up the forehead and upper ridges of the eye sockets. It contains **sinuses**, which are air spaces lined with mucous membrane. Sinuses make the skull lighter, which translates into less weight for the spine and neck muscles to support. But channels connect them to the nasal passages, and their ability to produce mucus can mean misery for anyone who has a cold or pollen allergies. A bacterial infection in the nasal passages can spread to the sinuses, causing *sinusitis*. Figure 5.5C shows sinuses in the cranial and facial bones.

Temporal bones form the lower sides of the cranium and surround the ear canals, which are tunnels that lead to the middle and inner ear. Inside the middle ear are tiny bones that function in hearing. On the sides of your head, in front of each temporal bone, a *sphenoid bone* extends inward to form part of the inner eye socket. The *ethmoid bone* also forms part of the inner socket and helps support the nose. Two *parietal bones* above and behind the temporal bones form much of the skull as they sweep upward and meet at the top of the head. An *occipital bone* forms the back and base of the skull. It also encloses an opening, the *foramen magnum* ("large hole"). Here, the spinal cord emerges from the base of the brain and enters the spinal column (Figure 5.5B). Other openings are channels for nerves and blood vessels. For instance, the jugular veins, which carry blood leaving the brain, pass through openings between the occipital bone and each temporal bone.

Facial bones support and shape the face

Figure 5.5 also shows facial bones, many of which you can easily feel with your fingers. The largest is your lower jaw, or **mandible**. The upper jaw consists of two *maxillary bones,* each called a *maxilla*. Two *zygomatic bones* form the middle of the hard bumps we call "cheekbones" and the outer parts of the eye sockets. A small, flattened *lacrimal bone* fills out the inner eye socket. Tear ducts pass between this bone and the maxillary bones and drain into the nasal cavity—one reason why your nose runs when you cry. Tooth sockets in the upper and lower jaws also contain the teeth.

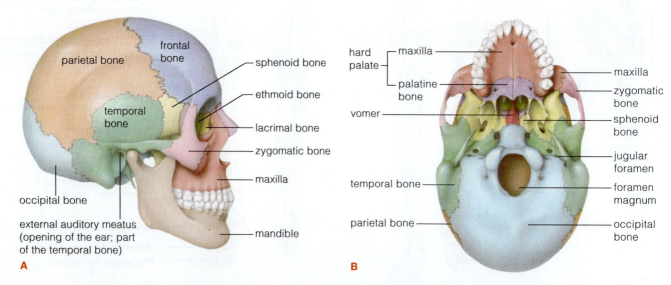

Figure 5.5 **Skull bones surround the brain and support the forehead. A** The jagged junctions between skull bones are called sutures. **B** A bottom-up view of the skull. The large foramen magnum is situated atop the uppermost cervical vertebra. **C** Sinuses in bones in the skull and face. (© Cengage Learning)

Palatine bones make up part of the floor and side wall of the nasal cavity. (Extensions of these bones, together with the maxillary bones, form the back of the hard palate, the "roof" of your mouth.) A *vomer bone* forms part of the nasal septum, a thin "wall" that divides the nasal cavity into two sections.

The vertebral column is the backbone

The flexible, curved vertebral column—your backbone or spine—runs from the base of the skull to the hip bones (pelvic girdle). This arrangement transmits the weight of a person's torso to the lower limbs. As a result, people who gain too much weight may develop problems with their knees and ankles because those joints are not designed to bear such a heavy load. The **vertebrae** are stacked and have bony projections that form a protected channel for the delicate spinal cord. As sketched in Figure 5.6, humans have seven *cervical* vertebrae in the neck, twelve *thoracic* vertebrae in the chest area, and five *lumbar* vertebrae in the lower back. During the course of human evolution, five other vertebrae have become fused to form the sacrum, and several more have become fused to form the coccyx, or "tailbone." Counting these, there are thirty-three vertebrae in all.

Roughly a quarter of your spine's length consists of **intervertebral disks**—compressible pads of fibrocartilage sandwiched between vertebrae. The disks serve as shock absorbers and flex points. They are thickest between cervical vertebrae and between lumbar vertebrae. Severe or rapid shocks, as well as changes due to aging, can cause a disk to *herniate* or "slip." If the slipped disk ruptures, its jellylike core may squeeze out, making matters worse. And if the changes compress neighboring nerves or the spinal cord, the result can be excruciating pain and the

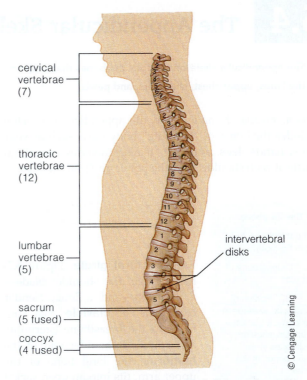

Figure 5.6 **Vertebrae and interverterbral disks make up the vertebral column (backbone).** The cranium balances on the column's top vertebra.

loss of mobility that often comes with pain. Depending on the situation, treatment can range from bed rest and use of painkilling drugs to surgery.

The ribs and sternum support and help protect internal organs

In addition to protecting the spinal cord, absorbing shocks, and providing flexibility, the vertebral column also serves as an attachment point for twelve pairs of ribs, which in turn serve as a scaffolding for the thoracic cavity, the body cavity of the upper torso. The upper ribs also attach to the paddle-shaped **sternum** (see Figure 5.4B). As you will read in later chapters, this **rib cage** helps protect the lungs, heart, and other internal organs and is vitally important in breathing.

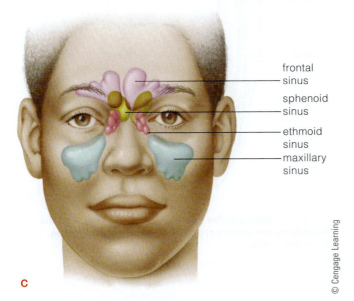

c

TAKE-HOME MESSAGE

WHAT ARE THE PARTS OF THE AXIAL SKELETON?

- The axial skeleton consists of bones that make up the body's vertical axis.
- The axial skeleton includes the skull and facial bones, the vertebral column (backbone), and the ribs and sternum.
- Portions of the axial skeleton, such as the skull and rib cage, protect key soft internal organs such as the brain and heart.
- Intervertebral disks absorb shocks and serve as flex points.

THE SKELETAL SYSTEM **93**

■ The appendicular skeleton includes the bones that support the limbs, upper chest, shoulders, and pelvis.

Append means "to hang," and the appendicular skeleton includes the bones of "hanging" body parts such as your arms, hands, legs, and feet. It also includes a pectoral girdle at each shoulder and the pelvic girdle at the hips.

clavicle The collarbone.

femur The thighbone, largest bone in the body.

humerus The long bone of the upper arm.

pectoral girdle Portion of the appendicular skeleton in the upper body; it consists of the scapulas (shoulder blades), clavicles (collarbones), and bones of the upper limbs.

pelvic girdle Portion of the appendicular skeleton in the lower body; it consists of bones of the pelvis and lower limbs.

radius Long bone on the inner (thumb) side of the forearm.

scapula The shoulder blade.

ulna Long bone on the outer (pinky finger) side of the forearm.

The pectoral girdle and upper limbs provide flexibility

Each **pectoral girdle** (Figure 5.7) has a large, flat shoulder blade—a **scapula**—and a long, slender collarbone, or **clavicle**, that connects to the breastbone (sternum). The rounded shoulder end of the **humerus**, the long bone of the upper arm, fits into an open socket in the scapula. Your arms can move in a great many ways; they can swing in wide circles and back and forth, lift objects, or tug on a rope. Such freedom of movement is possible because muscles only loosely attach the pectoral girdles and upper limbs to the rest of the body. Although the arrangement is sturdy enough under normal conditions, it is vulnerable to strong blows. Fall on an outstretched arm and you might fracture your clavicle or dislocate your shoulder. In all but the elderly the collarbone is the bone most frequently broken.

Each of your upper limbs includes thirty separate bones. The humerus connects with two bones of the forearm—the **radius** (on the thumb side) and the **ulna** (on the "pinky finger" side). The upper end of the ulna joins the lower end of the humerus to form the elbow joint. The bony bump sometimes (mistakenly) called the "wrist bone" is the lower end of the ulna.

The radius and ulna join the hand at the wrist joint, where they meet eight small, curved *carpal* bones. Ligaments attach these bones to the long bones. Blood vessels, nerves, and tendons pass in sheaths over the wrist; when a blow, constant pressure, or repetitive movement (such as typing) damages these tendons, the result can be a painful disorder called carpal tunnel syndrome (Section 5.6). The bones of the hand, the five *metacarpals*, end at the knuckles. *Phalanges* are the bones of the fingers.

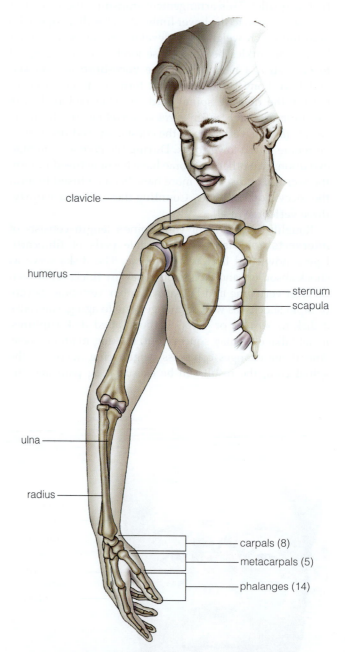

Figure 5.7 Bones of the pectoral girdle, the arm, and the hand form the upper part of the appendicular skeleton.
(© Cengage Learning)

The pelvic girdle and lower limbs support body weight

For most of us, our shoulders and arms are much more flexible than our hips and legs. Why? Although there are similarities in the basic "design" of both girdles, this lower part of the appendicular skeleton is adapted to bear the body's entire weight when we are standing. The **pelvic girdle** (Figure 5.8) is much more massive than the combined pectoral girdles, and it is attached to the axial skeleton by extremely strong ligaments. It forms an open basin: A pair of *coxal bones* attach to the lower spine (sacrum) in back, then curve forward and meet at the *pubic arch*. ("Hip bones" are actually the upper *iliac* regions of the coxal bones.) This combined structure is the *pelvis*. In females the pelvis is broader than in males, and it shows other structural differences that are evolutionary adaptations for childbearing. A forensic scientist or paleontologist examining skeletal remains can easily establish the sex of the deceased if a pelvis is present.

The legs contain the body's largest bones. In terms of length, the thighbone, or **femur**, ranks number one. It is also extremely strong. When you run or jump, your femurs routinely withstand stresses of several tons per square inch (aided by contracting leg muscles). The femur's ball-like upper end fits snugly into a deep socket in the coxal (hip) bone. The other end connects with one of the bones of the lower leg, the thick, load-bearing tibia on the inner (big toe) side. A slender fibula parallels the tibia on the outer (little toe) side. The tibia is your shinbone. A triangular kneecap, the patella, helps protect the knee joint. As Susanna's story in the chapter introduction noted, however, athletes often damage their knees.

The ankle and foot bones correspond closely to those of the wrist and hand. *Tarsal* bones make up the ankle and heel, and the foot contains five long bones, the *metatarsals*. The largest metatarsal, leading to the big toe, is thicker and stronger than the others to support a great deal of body weight. Like fingers, the toes contain phalanges.

TAKE-HOME MESSAGE

WHAT ARE THE PARTS OF THE APPENDICULAR SKELETON?

- The appendicular skeleton includes bones of the limbs, a pectoral girdle at the shoulders, and a pelvic girdle at the hips.
- The thighbone (femur) is the largest bone in the body and also one of the strongest.
- The wrists and hands and the ankles and feet have corresponding sets of bones known respectively as carpals and metacarpals and tarsals and metatarsals.

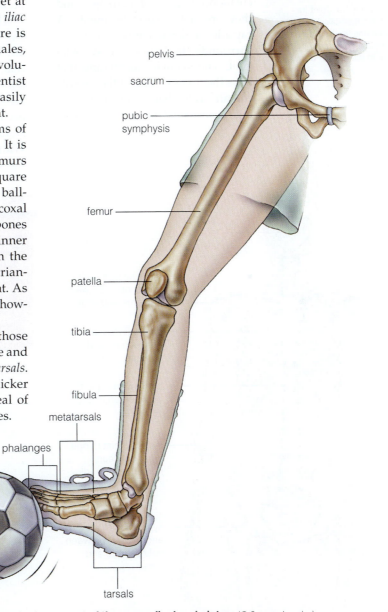

Figure 5.8 The pelvic girdle, the leg, and the foot form the lower part of the appendicular skeleton. (© Cengage Learning)

5.5 | Joints: Connections between Bones

■ Joints are areas of contact or near contact between bones. All joints have some form of connective tissue that bridges the gap between bones.

■ Link to Synovial membranes 4.7

cartilaginous joint Joint in which cartilage fills the space between bones.

fibrous joint Joint consisting of fibrous connective tissue that connects the bones. The joint has no cavity.

synovial joint Joint in which a fluid-filled cavity separates the linked bones; the most common type of joint in the human body.

There are three main types of joints in the skeletal system (Figure 5.9A). In the most common type of joint, called a **synovial joint**, adjoining bones are separated by a cavity (Figure 5.9B). The articulating ends of the bones are covered with a cushioning layer of cartilage, and they are stabilized by ligaments. A capsule of dense connective tissue surrounds the bones of a synovial joint. The synovial membrane that lines the inner surface of the capsule contains cells that secrete a lubricating *synovial fluid* into the joint cavity.

Synovial joints are built to allow movement. In hinge-like synovial joints such as the knee and elbow, the motion is limited to simple flexing and extending (straightening). The ball-and-socket joints at the hips are known as freely movable joints because they are capable of a wider range of movements: They can rotate and move in different planes—for instance, up-down or side-to-side. Figure 5.10 shows these and some other ways body parts can move at joints.

In a **cartilaginous joint**, cartilage fills the space between bones, so only slight movement is possible. The intervertebral disks between vertebrae are examples. Similar joints occur between the breastbone and some of the ribs.

There is no cavity in a **fibrous joint**, and fibrous connective tissue unites the bones. An adult's fibrous joints generally don't allow movement. Examples are the fibrous joints that hold your teeth in their sockets. In a fetus, fibrous joints loosely connect the flat skull bones. During childbirth, these loose connections allow the bones to slide over each other, preventing skull fractures. A newborn baby's skull still has fibrous joints and soft areas called fontanels. With time the joints harden into *sutures*. Much later in life the skull bones may fuse completely.

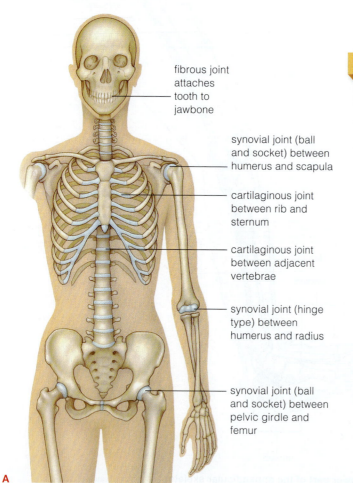

fibrous joint attaches tooth to jawbone

synovial joint (ball and socket) between humerus and scapula

cartilaginous joint between rib and sternum

cartilaginous joint between adjacent vertebrae

synovial joint (hinge type) between humerus and radius

synovial joint (ball and socket) between pelvic girdle and femur

A

> **TAKE-HOME MESSAGE**
>
> ### WHAT IS A JOINT?
>
> • A joint connects one bone to another.
> • In all joints, connective tissue bridges the gap between bones.
> • Synovial (freely movable) joints include the hinge-like knee joint and the ball-and-socket joints at the hips.
> • Cartilaginous joints have cartilage in the space between bones. They allow only slight movement. In fibrous joints, fibrous connective tissue joins the bones.

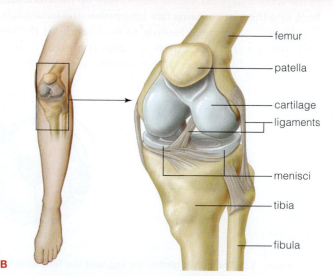

femur

patella

cartilage

ligaments

menisci

tibia

fibula

B

Figure 5.9 The three main types of joints between bones are synovial joints, cartilaginous joints, and fibrous joints. A shows locations of the different types of joints. **B** shows the knee joint, a synovial joint that is the largest and most complex joint in the body. (© Cengage Learning)

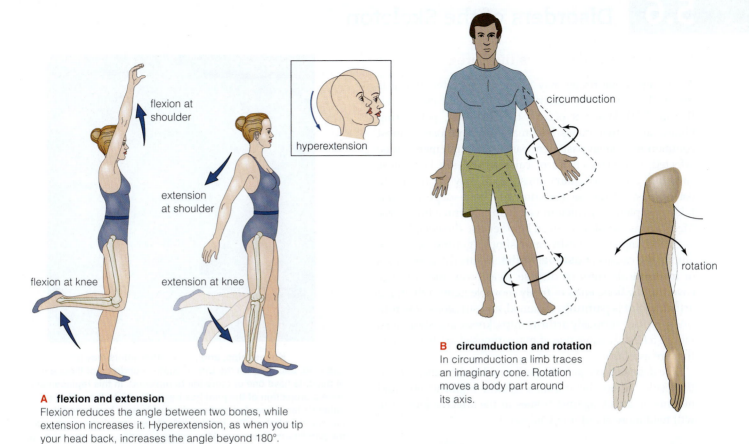

A flexion and extension
Flexion reduces the angle between two bones, while extension increases it. Hyperextension, as when you tip your head back, increases the angle beyond 180°.

B circumduction and rotation
In circumduction a limb traces an imaginary cone. Rotation moves a body part around its axis.

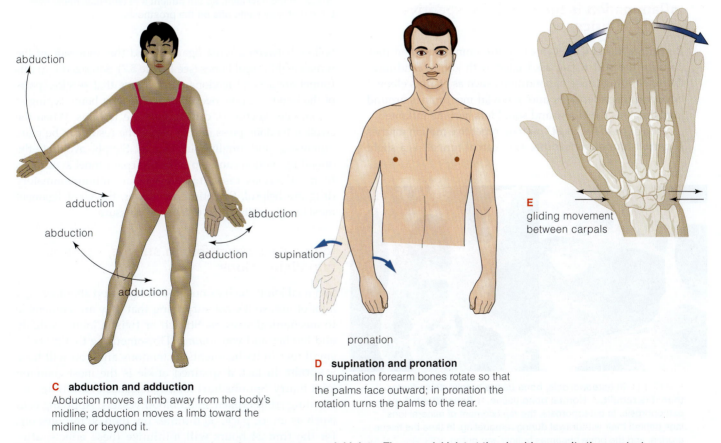

C abduction and adduction
Abduction moves a limb away from the body's midline; adduction moves a limb toward the midline or beyond it.

D supination and pronation
In supination forearm bones rotate so that the palms face outward; in pronation the rotation turns the palms to the rear.

E
gliding movement between carpals

Figure 5.10 A–E Body parts can move in various ways at synovial joints. The synovial joint at the shoulder permits the greatest range of movement. (© Cengage Learning)

Tissue in bones or joints may break down

As we age, bone tissue may break down faster than it is renewed. This steady deterioration is called *osteoporosis* (Figure 5.11). When it occurs, the backbone, pelvis (hip bones), and other bones lose mass. Osteoporosis is most common in women past menopause, although men can be affected, too. Deficiencies of calcium and sex hormones, smoking, and a sedentary lifestyle all may contribute to osteoporosis. Exercise (to stimulate bone deposits) and taking in plenty of calcium can help minimize bone loss. Medications can slow or even help reverse the bone loss.

Sports injuries, obesity, and simply getting older are among the causes of **osteoarthritis**. In this disorder, years of mechanical stress or disease wear away the cartilage covering the bone ends of freely movable joints. Often, the arthritic joint is painfully inflamed, and surgeons now routinely replace seriously arthritic hips, knees, and shoulders. Figure 5.12 shows just one of many types of artificial joints that are available.

The degenerative joint condition called *rheumatoid arthritis* results when the immune system malfunctions and mounts an attack against tissues in the affected joint. You will read more about it in Chapter 9.

Inflammation is the culprit in repetitive motion injuries

Repetitive movements can cause inflammation when they damage the soft tissue associated with joints. Tendinitis, the underlying cause of conditions such as "tennis elbow," develops when tendons and synovial membranes around joints such as the elbow and shoulders become inflamed.

Today one of the most common repetitive motion injuries is **carpal tunnel syndrome**. The "carpal tunnel" is a slight

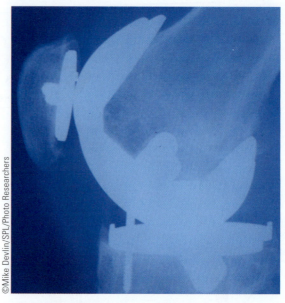

Figure 5.12 **Knees, hips, and some other joints** may be replaced. Each year in the United States hundreds of thousands of patients have one or more joints replaced. In this replacement knee a projection of the joint has been fitted into the end of the patient's femur (*center*) and another projection has been fitted into the tibia below. The hatlike disk at the upper left attaches to the patella—the kneecap. After surgery, walking and standing put stress on the new joint, so the patient's osteoblasts make new bone that grows into pits on the prosthesis.

hollow between a wrist ligament and the underside of the wrist's eight carpal bones (see Figure 5.7). Squeezed into this tunnel are several tendons and a nerve that services parts of the hand. Chronic overuse, such as long hours typing at a computer keyboard, can inflame the tendons. When the swollen tendons press on the nerve, the result can be pain, numbness, and tingling in fingers. Simply avoiding the offending motion can help relieve carpal tunnel syndrome. In more serious cases injections of an anti-inflammatory drug are helpful. Sometimes, however, the wrist ligament must be surgically cut to relieve the pressure.

Joints are susceptible to strains, sprains, and dislocations

Synovial joints such as our knees, hips, and shoulders get a lot of use, so it's not surprising that they are vulnerable to mechanical stresses. Stretch or twist a joint suddenly and too far, and you *strain* it. Do something that makes a small tear in its ligaments or tendons and you will have a **sprain**. In fact, a sprained ankle is the most common joint injury. Sprains hurt mainly because of swelling and bleeding from broken small blood vessels. Applying cold (such as an ice pack, 30 minutes on, then 30 minutes off) for the first 24 hours will minimize these effects; after that, doctors usually advise applying heat, such as a

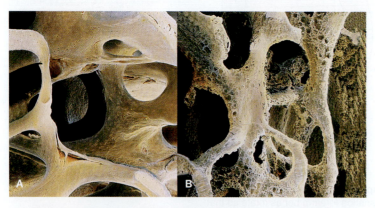

Figure 5.11 **In osteoporosis, bone tissue breaks down faster than it is rebuilt. A** Normal bone tissue. **B** Bone affected by osteoporosis. In osteoporosis, the replacement of mineral ions lags behind their withdrawal during remodeling. In time the tissue erodes, and the bone becomes hollow and brittle. (BOTH: ©Prof. P. Motta, Dept. of Anatomy, Univ. of La Sapienza, Rome/SPL/Photo Researchers, Inc).

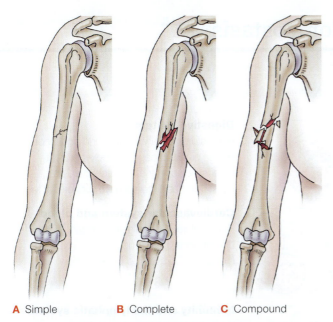

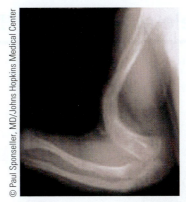

Figure 5.14 *Osteogenesis imperfecta* is a genetic bone disorder. *Left*: An X-ray of an arm bone deformed by OI. *Right*: Tiffany, who was born with OI.

A Simple   **B** Complete   **C** Compound

Figure 5.13 **Bone fractures range from simple to much more serious compound fractures.** (© Cengage Learning)

heating pad. The warmth speeds healing by increasing blood circulation to the injured tissue.

A blow can *dislocate* a joint—that is, the two bones will no longer be in contact. During collision sports such as football, a blow to a knee often tears a ligament. If the torn part is not reattached within ten days, phagocytic cells in the knee joint's synovial fluid will attack and destroy the damaged tissue.

Bones break in various ways

Most bone breaks can be classed as either a simple or closed fracture, a complete fracture, or a compound fracture. As you can probably tell from the drawings in Figure 5.13, a **simple fracture** is the least serious injury because the bone ends don't do much damage to the surrounding soft tissue. A **complete fracture,** in which the bone separates into two pieces and soft tissue is damaged, is more serious. Worse is a **compound fracture**, in part because broken ends or shards of bone puncture the skin, creating an open wound and the chance of infection. A surgeon may have difficulty reattaching all the pieces of a bone that has been shattered in this way.

When a bone breaks into pieces, the pieces must quickly be reset into their normal alignment. Otherwise it's unlikely that the bone will heal properly. Its functioning may be impaired for the rest of a person's life. In addition to the pins and casts that may be used to hold healing bones in place, the injured area may be stimulated with electricity, which speeds healing.

Joint and bone injuries tend to heal faster when we're younger. Changes that come with aging and bad habits such as smoking cigarettes slow the body's ability to repair itself.

Genetic diseases, infections, and cancer all may affect the skeleton

Some skeletal disorders are inherited, and a few cause lifelong difficulties. An example is *osteogenesis imperfecta* or OI (Figure 5.14). In this incurable disease, the collagen in bone tissue is defective, so the bones are extremely brittle and break easily. Children with OI must have surgeries to set the fractures and often have stunted growth.

Bones (and bone marrow) can become infected when a bacterial infection elsewhere spreads (via the bloodstream) or when the microbe enters an open wound. Antibiotics usually can cure the problem, although severe cases may require surgery to clean out the affected bone tissue.

The bone cancer called *osteosarcoma* strikes people young and old. It often develops in a long bone in a limb, or in a joint such as the hip or knee. The most common treatment of a primary bone cancer is amputation of the limb involved. Like many other types of cancer, bone cancer often is curable if caught early. Unfortunately, many bone cancer cases involve cancer that has spread from another site in the body. The image at right is of a bone scan that shows red "hot spots" where cancer has spread to many sites in the patient's skeleton.

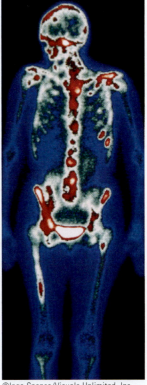

©Inga Spence/Visuals Unlimited, Inc.

CONNECTIONS

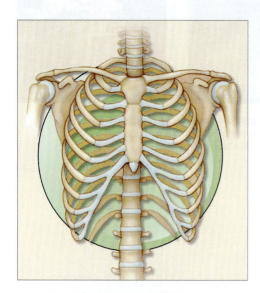

The Skeletal System

The skeleton supports and helps protect soft body parts. Bones, joints, tendons, and ligaments all have essential roles in moving the body and its parts. Bone is a reservoir for calcium, which is vital for many body functions including muscle contractions, the transmission of nerve impulses, and blood clotting. Calcium also is required for the proper functioning of some enzymes and of proteins in the cell plasma membrane.

Integumentary system
The skeleton provides support for skin and the muscles below it.

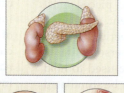

Muscular system
Skeletal muscles attach to bones, which serve as levers for body movements. Bone calcium may be released as needed to maintain blood levels required for muscle contractions.

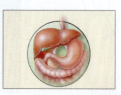

Digestive system
Bone stores dietary calcium and phosphorus. Bones of the rib cage and pelvis protect organs including the stomach, liver, and intestines. Facial bones have sockets for teeth.

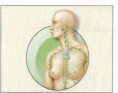

Cardiovascular system and blood
Bone calcium is available for heart contractions that pump blood. All types of blood cells form in red bone marrow.

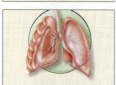

Immunity and the lymphatic system
White blood cells that function in body defenses form in bone marrow.

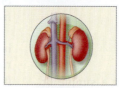

Respiratory system
The rib cage and sternum protect the lungs. Muscles used in breathing attach to ribs and associated cartilages.

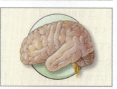

Urinary system
The rib cage partially protects the kidneys. The pelvis helps protect the bladder.

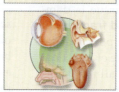

Nervous system
The skull protects the brain. Vertebrae protect the spinal cord. Bone calcium stores may be released as needed to maintain blood levels required for transmission of nerve impulses.

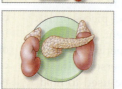

Sensory systems
Skull and facial bones surround and protect sensory organs in the head. Calcium in bones helps maintain blood levels required for transmission of sensory nerve impulses.

Endocrine system
Calcium may be released as needed to maintain blood levels required for the formation and secretion of many hormones.

Reproductive system
Pelvic bones protect female reproductive organs and associated glands in males. Calcium is available to help nourish a fetus and for milk production in a nursing mother.

EXPLORE
ON YOUR OWN

When it comes to the skeleton and joints, your own body can be a great learning tool.

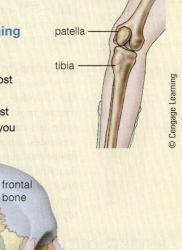

- Feel along the back of your neck beginning at your hairline. Can you feel any lumps made by the bony processes of your spinal vertebrae (Figure 5.6)? Locate the C7 vertebra, which in most people is the most prominent. Can you feel it at the base of your neck?
- While seated, feel your kneecap—the patella—move as you flex and extend your lower leg. Just below the patella you should also be able to feel a ligament that attaches it to your tibia. Can you find the upper protuberance of your tibia? Moving your fingers around to outside of the joint, can you feel the knobby upper part of the fibula?
- Using the diagram at right as a guide, see if you can locate the ridges of your frontal bone above your eyebrows; the arching part of your zygomatic bone, which forms your "cheekbones"; and the joint where your lower jaw articulates with the temporal bone.

SUMMARY

Section 5.1 Bones are organs that contain bone tissue and other connective tissues, nerves, and blood vessels.

A bone develops as osteoblasts secrete collagen fibers and a matrix of protein and carbohydrate. Calcium salts are deposited and harden the matrix. Mature living bone cells, osteocytes, are located inside spaces (lacunae) in the bone tissue.

Bones have both compact and spongy bone tissue. Denser compact bone is organized as thin, circular layers called osteons. In spongy bone, needlelike struts are fused in a lattice.

A cartilage model provides the pattern for a developing bone. Long bones lengthen at their ends (epiphyses) until early adulthood when bone growth ends.

Bones grow, gain strength, and are repaired by remodeling. In this process, osteoblasts deposit bone and osteoclasts break it down.

Section 5.2 As the main elements of the skeleton, bones interact with skeletal muscles to move body parts. Bones also store minerals and help protect and support other body parts. Ligaments connect bones at joints; tendons attach muscles to bones or to other muscles. Some bones, including the sternum, hip bones, and femur, contain bone marrow. Blood cells are produced in red bone marrow.

Section 5.3 The skeleton has an axial portion and an appendicular portion (Table 5.2). The axial skeleton forms the body's vertical axis and is a central support structure. In the spine, intervertebral disks of fibrocartilage are shock pads and flex points.

Skull bones form the brain case, which protects the brain. Sinuses in the frontal bone reduce the skull's weight.

TABLE 5.2 Review of the Skeleton's Parts

APPENDICULAR SKELETON

Pectoral girdle: clavicle and scapula
Arm: humerus, radius, ulna
Wrist and hand: carpals, metacarpals, phalanges (of fingers)
Pelvic girdle (6 fused bones at the hip)
Leg: femur (thighbone), patella, tibia, fibula
Ankle and foot: tarsals, metatarsals, phalanges (of toes)

AXIAL SKELETON

Skull: cranial bones and facial bones
Rib cage: sternum (breastbone) and ribs (12 pairs)
Vertebral column: vertebrae (26)

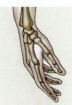

Section 5.4 The appendicular skeleton (Table 5.2) provides support for upright posture and interacts with skeletal muscles in most movements.

Sections 5.5, 5.6 In its partnership with skeletal muscles, the skeleton works like a system of levers in which rigid rods (bones) move about at fixed points (joints).

In a synovial joint, a fluid-filled cavity separates adjoining bones. Such joints are freely movable. In cartilaginous joints, cartilage fills the space between bones and allows only slight movements. In fibrous joints, fibrous connective tissue knits the bones together.

REVIEW QUESTIONS

1. Describe the basic elements of bone tissue.

2. What are the two types of bone tissue, and how are they different?

3. Describe how bone first develops.

4. Explain why bone remodeling is important, and give its steps.

5. Name the two main divisions of the skeleton.

6. How does a tendon differ from a ligament?

7. What are intervertebral disks made of and what is their function?

8. What is a joint?

9. What is the defining feature of a synovial joint?

SELF-QUIZ *Answers in Appendix V*

1. Fill in the blanks: The _____ and _____ systems work together to move the body and specific body parts.

2. Bone tissue contains _____.
   a. living cells   d. all of these
   b. collagen fibers   e. only a and b
   c. calcium and phosphorus

3. _____ are shock pads and flex points.
   a. Vertebrae   c. Lumbar bones
   b. Cervical bones   d. Intervertebral disks

4. The hollow center of an osteon (Haversian system) provides space for what vital part of compact bone tissue?
   a. marrow   c. a blood vessel
   b. collagen fibers   d. osteocytes

5. _____ is a type of connective tissue; _____ form(s) in it.
   a. An osteon; collagen
   b. Bone marrow; blood cells
   c. Bone; an osteocyte
   d. A sinus; bone marrow

6. Mineralization of bone tissue requires _____.
   a. calcium ions   c. elastin
   b. osteoclasts   d. all of the above

7. Bone remodeling has all of the following functions except _____.
   a. helps maintain homeostasis in blood level of calcium
   b. replaces "old" bone with fresh bone tissue
   c. exchanges collagen with elastin fibers for flexibility
   d. strengthens bones subjected to mechanical stress

8. Fill in the blanks: The axial skeleton consists of the _____, while the appendicular skeleton consists of the _____.

9. Fill in the blank: Ligaments are important components of the skeletal system because they _____.

10. Match the terms and definitions.
    _____ bone   a. spaces in certain skull bones
    _____ collagen   b. all in the hands
    _____ synovial fluid   c. blood cell production
    _____ osteocyte   d. a fibrous protein
    _____ marrow   e. mature bone cell
    _____ metacarpals   f. lubrication
    _____ mandible   g. mineralized connective
    _____ sinuses   tissue
                             h. the lower jaw

CRITICAL THINKING

1. Growth hormone, or GH, is used medically to spur growth in children who are unusually short because they have a GH deficiency. However, it is useless for a short but otherwise normal 25-year-old to request GH treatment in order to grow taller. Why?

2. If bleached human bones found lying in the desert were carefully examined, would osteons be present? How about osteocytes and a marrow cavity?

3. For young women, the recommended daily allowance (RDA) of calcium is 1,000 milligrams. For a 60-year-old woman, however, the RDA is 1,200 milligrams a day. What might happen to an older woman's bones without the larger amount?

4. The anterior cruciate ligament (ACL) helps stabilize the knee joint. It is easily injured by hyperextension of the knee. How would you have to move your lower leg to cause a hyperextension injury?

THE MUSCULAR SYSTEM

LINKS TO EARLIER CONCEPTS

Building on Chapter 5's discussion of the skeletal system, in this chapter you will discover how skeletal muscles partner with bones to move the body and its parts.

You will learn how the proteins actin and myosin work together in muscle contraction (3.9). Our discussion also will draw on your knowledge of how ATP fuels cell activities (3.8) and how active transport moves substances into and out of cells (3.11).

KEY CONCEPTS

Types of Muscle Tissue

The body contains skeletal muscle, smooth muscle, and cardiac muscle. Muscle cells produce force by contracting. Section 6.1

What Skeletal Muscles Do

Skeletal muscles pull on bones to move body parts. They are arranged as pairs or groups. Often, the action of one muscle opposes or reverses the action of another. Section 6.2

How Muscles Work

In a muscle cell, the action of units called sarcomeres is the basis for muscle contraction, which is controlled by motor neurons. Sections 6.3–6.6, 6.8

Disorders of the Muscular System Section 6.7

CONNECTIONS:
Homeostasis and the Muscular System Section 6.9

Top: © Don Fawcett/Visuals Unlimited; Middle: both © Cengage Learning; Bottom: Maxisport/Shutterstock.com

Many athletes want to boost their performance

by increasing the size and strength of their muscles. Matt Musick, who recently graduated from a university where he played varsity football and studied business, got the strong skeletal muscles he wants by spending long hours in the weight room. Not every athlete takes this approach. Numerous sports superstars have admitted to using banned anabolic (tissue-building) steroids, human growth hormone (HGH), or taking other unethical shortcuts to bulking up muscles. Some competitive amateurs, including college and high school athletes, do doping as well, despite possible health side effects that include liver tumors, shrinking testicles in males, severe acne, and infertility. Several years ago, however, drug testing of athletes at all levels began to increase. Today the National Institute on Drug Abuse (NIDA) reports a downward trend in doping in the United States. That said, reliable statistics on the extent of doping are hard to come by, and so far there hasn't been much research on ways to treat the physical damage doping can cause.

In this chapter we look at the natural roles of body muscles. Our main focus will be on skeletal muscles—the muscles that partner with the skeleton to bring about body movements.

Matthew Musick

Homeostasis Preview

Muscle contractions move body parts and substances such as blood and food to be digested. They also produce much of the body's heat.

The Body's Three Kinds of Muscle

- In all three kinds of muscle in the body, groups of cells contract to produce movement.

- Links to Muscle tissue 4.3, Nervous tissue 4.4, Tissue membranes 4.7

The three kinds of muscle have different structures and functions

In Chapter 4 we introduced the three basic kinds of muscle tissue—skeletal muscle, smooth muscle, and cardiac muscle. Together they make up about 50 percent of the body. In all of them, cells specialized to contract bring about some type of movement.

Most of the body's muscle tissue is **skeletal muscle**, which interacts with the skeleton to move body parts. Its long, thin cells are often called muscle "fibers" (Figure 6.1A). And unlike other body cells, skeletal muscle fibers have more than one nucleus. As you may remember from Section 4.3, the internal structure of muscle fibers gives them a striated, or striped, appearance, and bundles of them form skeletal muscles.

Smooth muscle is found in the walls of hollow organs and of tubes, such as blood vessels (Figure 6.1B). Its cells are smaller than skeletal muscle cells, and they do not look striped—hence the "smooth" name for this muscle tissue. Junctions link smooth muscle cells, which often are organized into sheets.

Cardiac muscle is found only in the heart (Figure 6.1C). It looks striated, like skeletal muscle. Unlike skeletal and smooth muscle, however, cardiac muscle can contract without

stimulation by signals from the nervous system. Special junctions between its cells allow the contraction signals to pass between them so fast that for all intents and purposes the cells contract as a single unit.

We do not have conscious control over contractions of cardiac muscle and smooth muscle, so they are said to be "involuntary" muscles. We *can* control many of our skeletal muscles, so they are "voluntary" muscles. Figure 6.2 shows the major skeletal muscles in the body. Some are close to the surface, others deep in the body wall. Some, such as facial muscles, attach to the skin. The trunk has muscles of the thorax (chest), spine, abdominal wall, and pelvic cavity. And of course, other muscle groups attach to limb bones.

When we speak of the body's **muscular system**, we're talking about skeletal muscle—the focus of the rest of this chapter. Skeletal muscle interacts with the skeleton to move the body, its limbs, or other parts. Those movements range from delicate adjustments that help you keep your balance to the cool moves you might execute on a dance floor. Our skeletal muscles also help stabilize joints and generate body heat.

cardiac muscle Muscle tissue of the heart.

muscular system Organ system that consists of skeletal muscles, which partner with bones to move body parts.

skeletal muscle Muscle tissue of the skeletal system.

smooth muscle Muscle tissue in the walls of tubes and hollow organs.

TAKE-HOME MESSAGE

WHAT TYPES OF MUSCLE TISSUE OCCUR IN THE BODY?

- The body's muscle tissue includes skeletal, smooth, and cardiac muscle tissue.
- Skeletal muscle makes up the muscular system, which partners with the skleleton to move body parts.
- Smooth muscle is found in the walls of organs and of tubes, such as blood vessels. Cardiac muscle is the muscle of the heart.

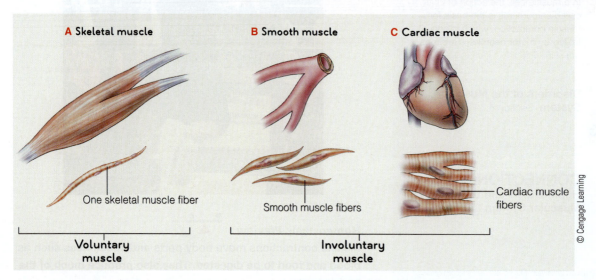

A Skeletal muscle   B Smooth muscle   C Cardiac muscle

One skeletal muscle fiber

Smooth muscle fibers

Cardiac muscle fibers

Voluntary muscle

Involuntary muscle

© Cengage Learning

Figure 6.1 Animated! Muscle tissue in the human body includes skeletal muscle, smooth muscle, and cardiac (heart) muscle.

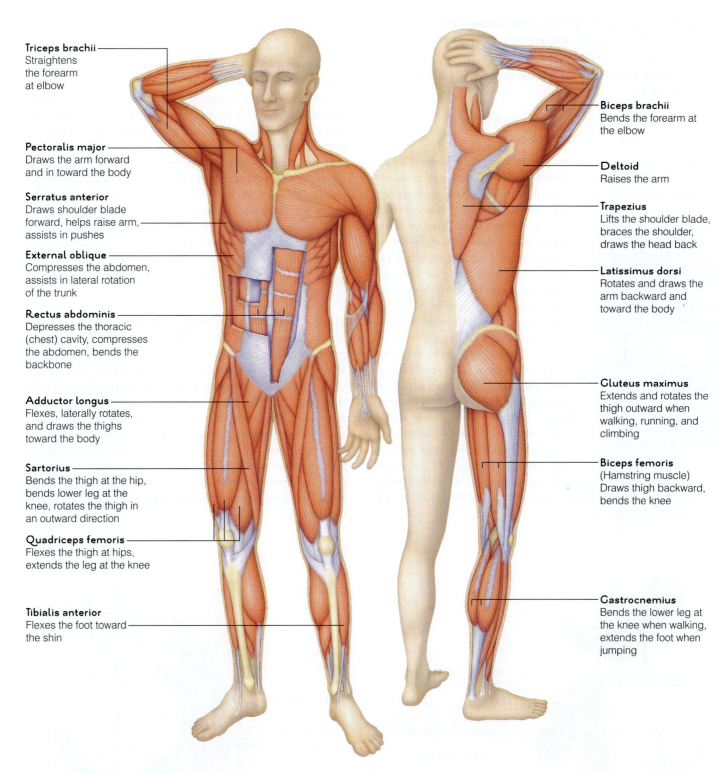

Triceps brachii
Straightens
the forearm
at elbow

Pectoralis major
Draws the arm forward
and in toward the body

Serratus anterior
Draws shoulder blade
forward, helps raise arm,
assists in pushes

External oblique
Compresses the abdomen,
assists in lateral rotation
of the trunk

Rectus abdominis
Depresses the thoracic
(chest) cavity, compresses
the abdomen, bends the
backbone

Adductor longus
Flexes, laterally rotates,
and draws the thighs
toward the body

Sartorius
Bends the thigh at the hip,
bends lower leg at the
knee, rotates the thigh in
an outward direction

Quadriceps femoris
Flexes the thigh at hips,
extends the leg at the knee

Tibialis anterior
Flexes the foot toward
the shin

Biceps brachii
Bends the forearm at
the elbow

Deltoid
Raises the arm

Trapezius
Lifts the shoulder blade,
braces the shoulder,
draws the head back

Latissimus dorsi
Rotates and draws the
arm backward and
toward the body

Gluteus maximus
Extends and rotates the
thigh outward when
walking, running, and
climbing

Biceps femoris
(Hamstring muscle)
Draws thigh backward,
bends the knee

Gastrocnemius
Bends the lower leg at
the knee when walking,
extends the foot when
jumping

Figure 6.2 Some of the major muscles of the muscular system. (© Cengage Learning)

6.2 | The Structure and Function of Skeletal Muscles

- Muscle cells produce force by contracting. After a muscle contracts, it can relax and lengthen. Skeletal muscles attach to and interact with bones.
- Links to Metabolism 3.13, Connective tissue 4.2, Muscle tissue 4.3

A whole skeletal muscle consists of bundled muscle cells

A skeletal muscle contains bundles of muscle fibers (Figure 6.3). Each fiber contain threadlike **myofibrils** (*myo-* refers to skeletal muscle). As you will read in Section 6.4, these structures contain the units that contract a muscle fiber. There may be hundreds, even thousands, of fibers in a muscle, all bundled together by connective tissue that extends past them to form tendons. You may remember from Chapter 5 that a *tendon* is a strap of dense connective tissue that attaches a muscle to bone or to another muscle. Tendons make joints more stable by helping keep the adjoining bones properly aligned. Tendons often rub against bones, but they slide inside fluid-filled sacs called **bursae** (BURR-see; singular: bursa) that help reduce the friction (Figure 6.4). In some cases a bursa is elongated into a tendon sheath that folds around a tendon. Your knees, wrists, and finger joints all have tendon sheaths.

> **bursae** Fluid-filled sacs around tendons that help reduce friction as the tendons move.
>
> **insertion** The end of a muscle attached to the bone that moves the most when the muscle contracts.
>
> **myofibrils** Threadlike structures in a muscle fiber that contain the units of contraction.
>
> **origin** The end of a muscle that is attached to a bone that stays relatively motionless during a movement.

Bones and skeletal muscles work like a system of levers

You have more than 600 skeletal muscles, and each one helps produce some kind of body movement. In general, one end of a muscle, called the **origin**, is attached to a bone that stays relatively motionless during a movement. The other end of the muscle, called the **insertion**, is attached to the bone that moves the most (Figure 6.5). In effect, the skeleton and the muscles attached to it are like a system of levers in which bones (rigid rods) move near joints (fixed points). When a skeletal muscle contracts, it pulls on the bones it attaches to. Because muscles attach very close to most joints, a muscle only has to contract a short distance to produce a major movement.

Many muscles are arranged as pairs or in groups

Many muscles are arranged as pairs or groups. Some work in opposition (that is, antagonistically) so that the action of one opposes or reverses the action of the other. Figure 6.5 shows an antagonistic muscle pair, the biceps and triceps of the arm. Try extending your right arm in front of you, then place your left hand over the biceps in the upper arm and slowly "bend the elbow." Can you feel the biceps contract? When the biceps relaxes and its partner (the triceps) contracts, your arm straightens. This kind of coordinated action comes partly from *reciprocal innervation* by nerves from the spinal cord. When one muscle group is stimulated, no signals are sent to the opposing group, so it does not contract.

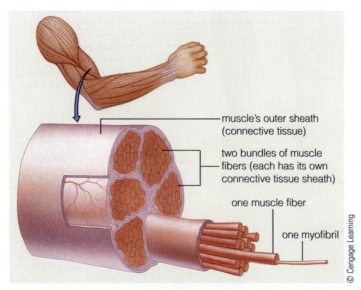

muscle's outer sheath (connective tissue)

two bundles of muscle fibers (each has its own connective tissue sheath)

one muscle fiber

one myofibril

© Cengage Learning

Figure 6.3 Animated! In skeletal muscle, the muscle fibers are bundled together inside a wrapping of connective tissue.

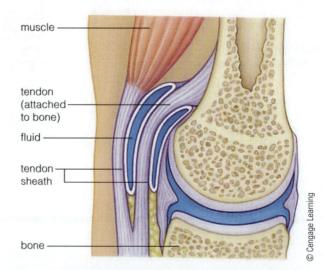

muscle

tendon (attached to bone)

fluid

tendon sheath

bone

© Cengage Learning

Figure 6.4 A bursa encloses lubricating fluid that prevents friction when the attached bone moves. The bursa is called a tendon sheath when it wraps all the way around a tendon.

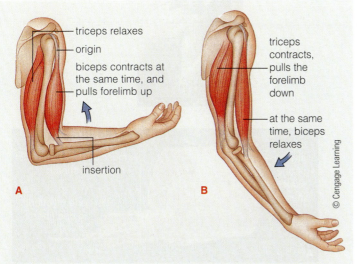

Figure 6.5 Animated! **Arm movements demonstrate the action of opposing muscle groups. A** When the triceps relaxes and its opposing partner (biceps) contracts, the elbow joint flexes and the forearm bends up. **B** When the triceps contracts and the biceps relaxes, the forearm is extended down.

Other muscles work in a synergistic, or support, role. Their contraction adds force or helps stabilize another contracting muscle. If you make a fist while keeping your wrist straight, synergist muscles are stabilizing your wrist joint while muscles in your hand are doing the "heavy lifting" of closing your fingers.

Skeletal muscle includes "fast" and "slow" types

Your body has two basic types of skeletal muscle (Figure 6.6A). "Slow" or "red" muscle appears crimson because its fibers are packed with myoglobin, a reddish protein that binds oxygen for the cell's use in making ATP. Red muscle also is served by larger numbers of the tiny blood vessels called capillaries. (Red muscle is the dark meat in chicken and turkey.) Red muscle contracts fairly slowly, but because its fibers are so well equipped to make lots of ATP, the contractions can be sustained for a long time. For example, some muscles of the back and legs—called postural muscles because they aid body support—must contract for long periods when a person is standing. They have a high proportion of red muscle fibers. By contrast, the muscles of your hand have fewer capillaries and relatively more "fast" or "white" muscle fibers, in which there are fewer mitochondria and less myoglobin. Fast muscle can contract rapidly and powerfully for short periods, but it can't sustain contractions for long periods. This is why you get writer's cramp if you write longhand for an extended period.

When an athlete trains rigorously, one goal is to increase the relative size and contractile strength of fast or slow

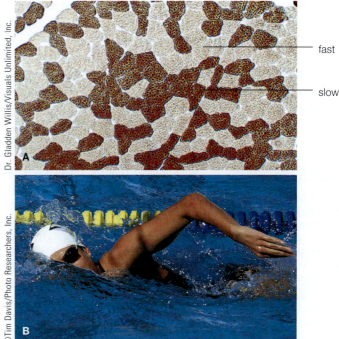

Figure 6.6 **A skeletal muscle has "fast" and "slow" fibers. A** This micrograph shows a cross section of the different kinds of fibers in a skeletal muscle. The lighter "white fibers" are fast muscle. They have little myoglobin and fewer mitochondria than the dark red fibers, which are slow muscle. **B** The shoulder muscles of a distance swimmer can work for an extended time because they contain many well-developed slow muscle fibers.

fibers in muscles. The type of sport determines which type of fiber is targeted. A sprinter will benefit from larger, stronger fast muscle fibers in the thighs, while a distance swimmer (Figure 6.6B) will train to increase the number of mitochondria in the shoulder muscle fibers.

TAKE-HOME MESSAGE

WHAT IS THE GENERAL STRUCTURE AND FUNCTION OF A SKELETAL MUSCLE?

• A skeletal muscle consists of muscle cells bundled together by connective tissue. Tendons strap skeletal muscles to bone.

• When a skeletal muscle contracts, it pulls on a bone to produce movement.

• In many movements, the action of one muscle opposes or reverses the action of another.

• Red or "slow" skeletal muscle fibers have features that support slow, long-lasting contractions.

• White or "fast" skeletal muscle fibers are specialized for rapid, strong bursts of contraction.

6.3 How Muscles Contract

- Bones move when the skeletal muscles attached to them contract and pull them.

A muscle contracts when its cells shorten

A skeletal muscle contracts when the individual muscle fibers in it shorten. In turn, each muscle fiber shortens when units of contraction inside its myofibrils shorten. Each of these basic units of contraction is a **sarcomere**.

Bundles of fibers in a skeletal muscle run parallel along the muscle's length (Figure 6.7A). Looking a bit deeper, each of the myofibrils in a muscle fiber is divided into bands (Figure 6.7B). The bands appear as an alternating light–dark pattern when they are stained and viewed under a microscope. Bands in neighboring myofibrils line up closely, which is why a skeletal muscle fiber looks striped. The dark bands are called Z bands. They mark the ends of each sarcomere (Figure 6.7C).

Inside a sarcomere are many filaments, some thick, others thin. These thick and thin filaments are arranged in an overlapping array shown in the image in Figure 6.7C. Each thin filament is like two strands of beads, twisted together, with one end attached to a Z band. The "beads" are molecules of **actin** (Figure 6.7D), a globular protein that can contract.

Each thick filament is made of molecules of the protein **myosin**. A myosin molecule has a tail and a double head. In a thick filament many of them are bundled together so that all the heads stick out (Figure 6.7E), away from the sarcomere's center.

As you can see in Figure 6.7, muscle bundles, muscle fibers, myofibrils, and their filaments all run in the same direction. This alignment focuses the force of a contracting muscle. All sarcomeres in all fibers of a muscle work together and pull a bone in the same direction.

actin Beadlike contractile protein in muscle fibers.

myosin Contractile protein in muscle fibers that has a tail and a double head.

rigor mortis Stiffening of muscles after death due to the lack of ATP energy to release muscle contraction.

sarcomere The basic unit of contraction in skeletal muscles.

sliding filament mechanism The mechanism by which skeletal muscles contract; sarcomeres contract (shorten) when myosin filaments slide along and pull actin filaments toward the center of the sarcomere.

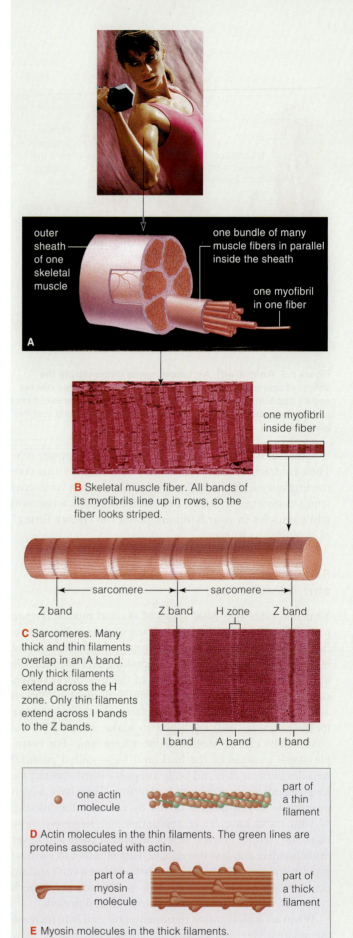

A

outer sheath of one skeletal muscle

one bundle of many muscle fibers in parallel inside the sheath

one myofibril in one fiber

B Skeletal muscle fiber. All bands of its myofibrils line up in rows, so the fiber looks striped.

one myofibril inside fiber

sarcomere — sarcomere

Z band   Z band   H zone   Z band

C Sarcomeres. Many thick and thin filaments overlap in an A band. Only thick filaments extend across the H zone. Only thin filaments extend across I bands to the Z bands.

I band   A band   I band

one actin molecule

part of a thin filament

D Actin molecules in the thin filaments. The green lines are proteins associated with actin.

part of a myosin molecule

part of a thick filament

E Myosin molecules in the thick filaments.

Figure 6.7 Animated! This diagram zooms down through skeletal muscle from a biceps to filaments of the proteins actin and myosin.
(Top: ©Steve Cole/Photodisc/Getty Images; B, C: © Don Fawcett/Visuals Unlimited; A, D, E: © Cengage Learning)

Muscle cells shorten when actin filaments slide over myosin

A **sliding filament mechanism** explains how interactions between thick and thin filaments allow muscle fibers to contract. In a contraction, all the myosin filaments stay in place. They use short "power strokes" to slide the sets of actin filaments over them, toward the sarcomere's center. Pulling both sets of filaments shrinks the length of the sarcomere (Figures 6.8A and 6.8B). Each power stroke is driven by energy from ATP.

Each myosin head repeatedly "grabs" binding sites on a nearby actin filament (Figure 6.8C). The head is an ATPase, a type of enzyme. It binds ATP and catalyzes a phosphate-group transfer that powers the reaction.

A rise in the concentration of calcium ions causes the myosin head to attach to the actin (Figure 6.8D). This link tilts the myosin head and pulls the actin filament toward the sarcomere's center (Figures 6.8E–F). Next, with the help of energy from ATP, the myosin head's grip on actin is broken and the head returns to its starting position (Figure 6.8G). Each time a sarcomere contracts, hundreds of myosin heads make a series of short strokes down the length of actin filaments.

When someone dies, her or his body cells stop making ATP. In muscles this means that the myosin cross-bridges with actin can't break apart after a power stroke. As a result skeletal muscles "lock up," a stiffening called **rigor mortis** ("stiffness of death"). Rigor mortis lasts for 24 to 60 hours, or until the natural decomposition of dead tissues gets under way. Understanding this sequence helps crime investigators pinpoint when a suspicious death occurred.

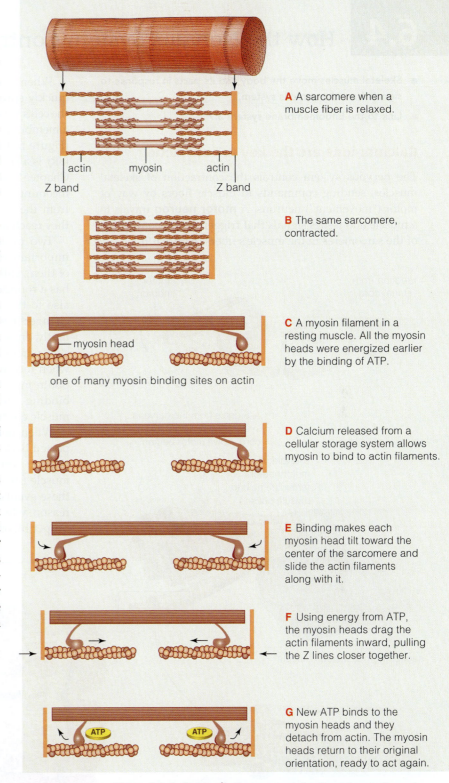

A A sarcomere when a muscle fiber is relaxed.

actin   myosin   actin

Z band   Z band

B The same sarcomere, contracted.

C A myosin filament in a resting muscle. All the myosin heads were energized earlier by the binding of ATP.

myosin head

one of many myosin binding sites on actin

D Calcium released from a cellular storage system allows myosin to bind to actin filaments.

E Binding makes each myosin head tilt toward the center of the sarcomere and slide the actin filaments along with it.

F Using energy from ATP, the myosin heads drag the actin filaments inward, pulling the Z lines closer together.

G New ATP binds to the myosin heads and they detach from actin. The myosin heads return to their original orientation, ready to act again.

ATP

Figure 6.8 Animated! Sarcomeres shorten when actin and myosin filaments interact. This interaction is the sliding filament mechanism of muscle contraction. (© Cengage Learning)

6.4 How the Nervous System Controls Muscle Contraction

- Skeletal muscles move the body and its parts in response to signals from the nervous system.
- Link to the Endomembrane system 3.7

Calcium ions are the key to contraction

The nervous system controls the contraction of skeletal muscles, sending commands to muscle fibers by way of motor ("movement") neurons. A **motor neuron** travels to a muscle and issues signals that trigger or halt contraction of the sarcomeres in the muscle's fibers (Figure 6.9).

When nerve impulses arrive at a muscle fiber, they quickly spread. Eventually they reach small extensions of the cell's plasma membrane. These "T tubules" connect with a membrane system that laces around the fiber's myofibrils (Figure 6.9D). The system, called the *sarcoplasmic reticulum* (SR), is a version of the endoplasmic reticulum described in Chapter 3. SR takes up and releases calcium ions (Ca^{++}). An incoming nerve impulse triggers the release of calcium ions from the SR. The ions diffuse into myofibrils, and when they reach actin filaments the stage is set for contraction.

Two proteins on the surface of actin filaments have important roles in muscle contraction (Figure 6.10). One of them, called troponin ("tropo-" means turn or change), has a rounded shape. It attaches to the actin filament and also to the second protein, called tropomyosin ("myosin changer"), which winds along the actin filament. Importantly, in a resting muscle fiber troponin covers up the sites where myosin can link up with actin. This changes when incoming calcium binds to troponin. Then the troponin moves, twisting tropomyosin away from the actin binding sites. Myosin now can attach to the sites, and muscle contraction can occur.

When nerve impulses stop, calcium is actively transported back into the SR. Tropomyosin covers the binding sites on actin again, myosin can't bind to actin, and the muscle fiber relaxes. Notice the importance of calcium in these events. Its central role in muscle contraction is one reason why mechanisms of homeostasis that maintain proper blood levels of calcium are so important.

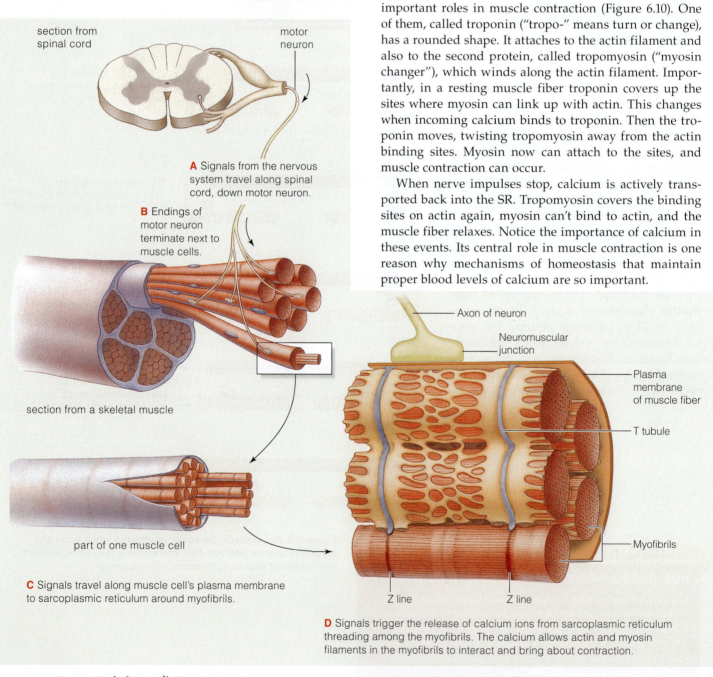

section from spinal cord

motor neuron

A Signals from the nervous system travel along spinal cord, down motor neuron.

B Endings of motor neuron terminate next to muscle cells.

section from a skeletal muscle

part of one muscle cell

C Signals travel along muscle cell's plasma membrane to sarcoplasmic reticulum around myofibrils.

Axon of neuron

Neuromuscular junction

Plasma membrane of muscle fiber

T tubule

Myofibrils

Z line   Z line

D Signals trigger the release of calcium ions from sarcoplasmic reticulum threading among the myofibrils. The calcium allows actin and myosin filaments in the myofibrils to interact and bring about contraction.

Figure 6.9 Animated! Signals from the nervous system stimulate contraction of skeletal muscle. (© Cengage Learning)

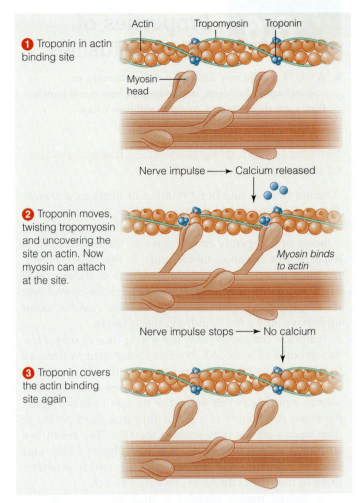

① Troponin in actin binding site

Actin   Tropomyosin   Troponin

Myosin head

Nerve impulse → Calcium released

② Troponin moves, twisting tropomyosin and uncovering the site on actin. Now myosin can attach at the site.

Myosin binds to actin

Nerve impulse stops → No calcium

③ Troponin covers the actin binding site again

Figure 6.10 Animated! Proteins uncover and then cover sites on actin where myosin can attach. Calcium ions must bind to the protein troponin and then release it in order for the sliding filament mechanism of muscle contraction to operate. (© Cengage Learning)

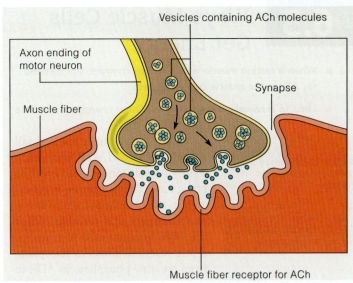

Vesicles containing ACh molecules

Axon ending of motor neuron

Muscle fiber

Synapse

Muscle fiber receptor for ACh

Figure 6.11 Animated! A chemical messenger called a neurotransmitter carries a signal across a neuromuscular junction. (© Cengage Learning)

Each year in the United States about 2 million people have injections of Botox to smooth out facial wrinkles. Made by the bacterium *Clostridium botulinum*, Botox blocks the release of ACh, so the muscle contractions that produce wrinkles stop for a while. The muscle-relaxing effect lasts four to six months and can have side effects, such as droopy eyelids. Botox also is used to treat disorders. For example, it may relieve abnormal muscle contractions that trouble stroke patients. Only a physician can legally prescribe Botox.

motor neuron The type of neuron that carries nervous system signals to skeletal muscles.

neuromuscular junction Sites where endings of neuron axons come close to the cell membranes of muscle fibers.

neurotransmitter A chemical messenger that carries signals from a neuron to a receiving cell across a synapse.

Neurons signal muscle cells at neuromuscular junctions

A motor neuron has long extensions called *axons* that carry nerve impulses. The nerve impulses that stimulate a skeletal muscle fiber arrive at **neuromuscular junctions**. These are places where the branched endings of axons come close to muscle fiber membranes, as you can see in Figure 6.9B and D, and in Figure 6.11. Between the neuron endings and each muscle cell is a gap called a *synapse*. A type of chemical messenger, a **neurotransmitter** called ACh (for acetylcholine), carries the signals from a motor neuron across the gap.

The signaling between a neuron and a muscle cell takes place in steps. As these steps take place, calcium ions from the extracellular fluid flow inside the axon endings, and vesicles in each ending release ACh. If enough ACh binds to receptors on the muscle cell membrane, the events that cause the muscle cell to contract may get under way. ACh can excite or inhibit muscle and gland cells, as well as some cells in the brain and spinal cord.

THINK OUTSIDE THE BOOK

Botox is potentially quite dangerous, since it blocks the release of ACh. Yet each year millions of people around the world receive Botox injections for cosmetic purposes. Research this topic on the website of the National Institutes of Health (www.nih.gov) to learn more about both the potential health dangers of Botox and how it is used in medical treatments.

TAKE-HOME MESSAGE

HOW DOES THE NERVOUS SYSTEM CONTROL THE CONTRACTION OF MUSCLE CELLS?

- Nerve impulses spark the release of calcium ions from a membrane system around a muscle cell's myofibrils.
- Nerve impulses pass from a neuron to a muscle cell across neuromuscular junctions.

6.5 Ways Muscle Cells Get Energy

- When a resting muscle is ordered to contract, the demand for ATP in the muscle cell skyrockets.

- Links to How cells make ATP 3.14, Other energy sources 3.16

muscle fatigue A physiological state in which a skeletal muscle cannot contract.

oxygen debt A state in which muscles require more ATP than aerobic cellular respiration can provide.

A resting muscle fiber has a small amount of stored ATP and much more of a substance called creatine phosphate. This substance is generated in the fiber from natural stores of the amino acid creatine. When the fiber is stimulated to contract, a fast reaction transfers phosphate from creatine phosphate to ADP, to form more ATP. This reaction can fuel contractions until a slower ATP-forming pathway, such as aerobic cellular respiration, can start up (Figure 6.12).

Normally, most of the ATP for muscle contraction comes from the oxygen-using reactions of cellular respiration. If you exercise hard, however, your respiratory and circulatory systems may not be able to deliver enough oxygen for aerobic cellular respiration in some muscles. Then, glycolysis (which does not use oxygen) will contribute more of the ATP being formed. Muscle cells rely on glycolysis until there is too little stored glycogen to provide glucose or until **muscle fatigue** sets in. This is a state in which a muscle can no longer contract. One cause of fatigue may be an **oxygen debt** that results when muscles need more ATP than aerobic cellular respiration can deliver. They then switch to glycolysis, which produces lactic acid. Along with the already low ATP supply, the rising acidity hampers the contraction of muscle cells. It also causes the "burn" you might feel while working out. Deep, rapid breathing helps repay the oxygen debt.

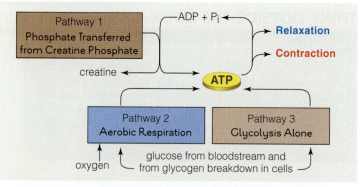

Figure 6.12 Animated! Three metabolic pathways can form ATP in active muscle cells. (© Cengage Learning)

TAKE-HOME MESSAGE

WHAT ARE ENERGY SOURCES FOR MUSCLE FIBERS?

- Muscle fibers may sometimes use creatine phosphate, glucose, and/or glycolysis alone to form ATP.

6.6 Properties of Whole Muscles

- A whole muscle may contract weakly, strongly, or somewhere in between, depending on how many muscle fibers are stimulated to contract and how often.

Several factors determine the characteristics of a muscle contraction

A motor neuron supplies a number of fibers in a muscle. The motor neuron and the muscle fibers it synapses with form a **motor unit** (Figure 6.13). The number of fibers in a motor unit depends on how precise the muscle control must be. For instance, motor units in the bulky, powerful thigh muscles may include hundreds of thousands of fibers. In contrast, we need much more precise control over the tiny muscles that move the eye. In these muscles, motor units have only a few hundred muscle fibers.

A muscle contraction may last a long time or only a few thousandths of a second. When a motor neuron fires, all the fibers in its motor unit contract briefly. This response is a **muscle twitch** (Figure 6.14A). If a new nerve impulse arrives before a twitch ends, the muscle twitches again. Repeated stimulation of a motor unit in a short period of time makes all the twitches run together. The result is a sustained contraction called **tetanus** (Figure 6.14B). Our muscles normally contract in this way, which generates three or four times the force of a single twitch.

A skeletal muscle contains a large number of muscle fibers, but not all of them contract at the same time. If a muscle is contracting only weakly—say, as your forearm muscles do when you pick up a pencil—it is because the nervous system is activating only a few of the muscle's motor units. In stronger contractions (when you heft a stack of books) more motor units are stimulated. Even when a muscle is relaxed, however, some of its motor units are contracted. This steady, low-level contracted state is called **muscle tone**. It helps maintain muscles in general good health and is important in stabilizing the skeleton's movable joints.

Muscle tension is the force that a contracting muscle exerts on an object, such as a bone. Opposing this force is a load, either the weight of an object or gravity's pull on the muscle. A stimulated muscle shortens only when muscle tension exceeds the opposing forces.

You have probably heard of isotonic and isometric exercise. "Isotonic" means same or steady tension. When a muscle contracts isotonically, it shortens as it moves a load (Figure 6.15A). "Isometric" means same length. When a muscle contracts isometrically it contracts and develops tension but doesn't shorten. This happens when you attempt to lift an object that is too heavy (Figure 6.15B).

Tired muscles can't generate much force

When steady, strong stimulation keeps a muscle in a state of tetanus, the muscle eventually becomes fatigued. Then

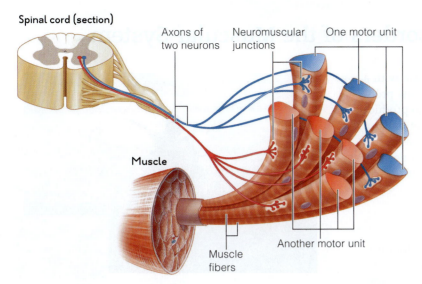

Figure 6.13 **Muscle cells are organized into motor units.** (© Cengage Learning)

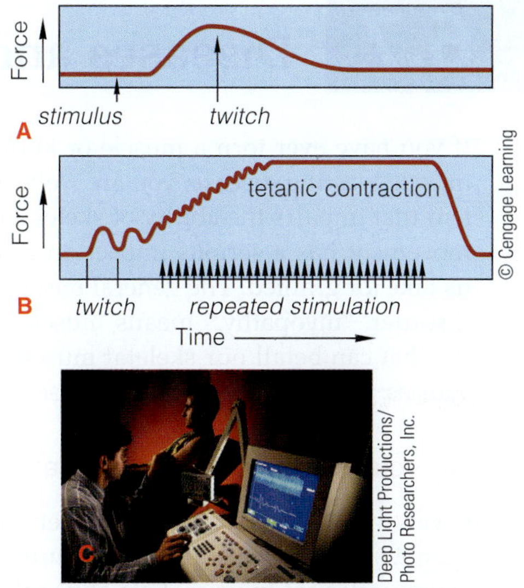

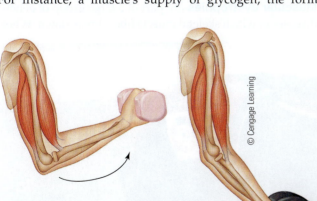

A
stimulus   *twitch*

B
twitch   *repeated stimulation*
tetanic contraction
Time →

© Cengage Learning

Deep Light Productions/
Photo Researchers, Inc.

Figure 6.14 **Each contraction of a motor unit is a muscle twitch. This figure shows recordings of twitches in muscles artificially stimulated in different ways. A A single twitch. B About 20 twitches per second cause tetanic contraction. C A technician monitoring activity in a patient's muscle in real time.**

the muscle's ability to generate force—that is, to develop tension—plummets. After a few minutes of rest, however, a fatigued muscle will be able to contract again.

How long does this recovery take? That depends in part on how long and how often the muscle was stimulated before. Muscles trained by a pattern of brief, intense exercise fatigue recover rapidly. Regular weight lifting provides this sort of muscle training. Muscles used in prolonged, moderate exercise fatigue slowly but take longer to recover, often up to a day. In addition to oxygen debt, other factors may contribute to muscle fatigue. For instance, a muscle's supply of glycogen, the form

in which muscles hold glucose in reserve for energy, may run out. In addition, research suggests that in tiring muscles, calcium starts to leak out of the SR (sarcoplasmic reticulum, Section 6.4). As a result there is less calcium available to spark muscle contractions. It's possible that such mechanisms are "safety valves" that help prevent serious muscle damage from overexercising.

motor unit Unit consisting of a motor neuron and the muscle fibers it controls.

muscle tension The force that a contracting muscle exerts on an object.

muscle tone A steady, low-level state of contraction of a skeletal muscle.

muscle twitch One contraction of a motor unit.

tetanus Sustained muscle contraction that develops when motor units are repeatedly stimulated in a short period of time, so that individual twitches are combined.

© Cengage Learning

A Isotonic contraction. Muscle tension is greater than the opposing force and the muscle shortens, as when you lift a light weight.

B Isometric contraction. Muscle tension is less than the opposing force and the muscle remains at the same length, rather than shortening.

Figure 6.15 **Muscle contractions may be isotonic or isometric. A In an isotonic contraction, the load is less than a muscle's peak capacity to contract, so the muscle can contract, shorten, and lift the load. B In an isometric contraction, the load exceeds the muscle's peak capacity. The muscle contracts but can't shorten.**

TAKE-HOME MESSAGE

WHAT ARE THE PROPERTIES OF WHOLE MUSCLES?

- In a whole muscle, the fibers are organized into motor units. This arrangement permits variations in how whole muscles contract.
- A motor unit consists of a motor neuron and the muscle cells it serves. In a muscle twitch, the cells contract simultaneously.
- The number of motor units in a muscle correlates with how precisely the nervous system must control a muscle's activity.
- Skeletal muscles normally contract in a sustained manner called tetanus. Healthy muscles maintain good muscle tone even when they are relaxed.

6.7 Diseases and Disorders of the Muscular System

If you have ever torn a muscle or known someone with a muscle-wasting disease, you are well aware that any problem that impairs the ability of skeletal muscles to produce movement has a serious impact on activities that most of us take for granted. The general medical term for a muscle disorder, "myopathy," means muscle disease. In general, ills that can befall our skeletal muscles fall into three categories: injuries, disease, and disuse.

Muscle injuries include strains and tears

Given that our muscular system gets almost constant use, it's not surprising that the most common disorders of skeletal muscles are injuries. Lots of people, and athletes especially, strain a muscle at some point in their lives (Figure 6.16). The injury happens when a movement stretches or tears muscle fibers. Usually, there is some bleeding into the damaged area, which causes swelling and a painful muscle spasm. The usual first aid is an ice pack, followed by resting the affected muscle and using anti-inflammatory drugs such as ibuprofen.

When a whole muscle is torn, the aftereffects can last a lifetime. If scar tissue develops while the tear mends, the healed muscle may be shorter than before. As a result, it may not function as effectively.

Cramps and spasms are abnormal contractions

In a **muscle spasm**, a muscle suddenly and involuntarily contracts. A **muscle cramp** is a painful muscle spasm that doesn't immediately release. Any skeletal muscle can cramp, but the usual "victims" are calf and thigh muscles. In some cases the real culprit is a deficiency of potassium,

healthy muscle

DMD muscle

Figure 6.17 Muscular dystrophies are inherited disorders. Duchenne muscular dystrophy is most common in boys. Above, the top image shows healthy muscle fibers. The lower image shows muscle from a DMD patient. The fibers are misshapen and have other characteristics used to diagnose the disease.

which is needed for the proper transmission of nerve impulses to muscles and other tissues. Gentle stretching and massage may coax a cramped muscle to release.

Most people experience occasional muscle "tics." These minor, involuntary twitches are common in muscles of the face and eyelids and may be triggered by anxiety or some other psycho-emotional cause.

Muscular dystrophies destroy muscle fibers

Muscular dystrophies are part of a large group of genetic diseases in which skeletal muscle fibers break down. Whole

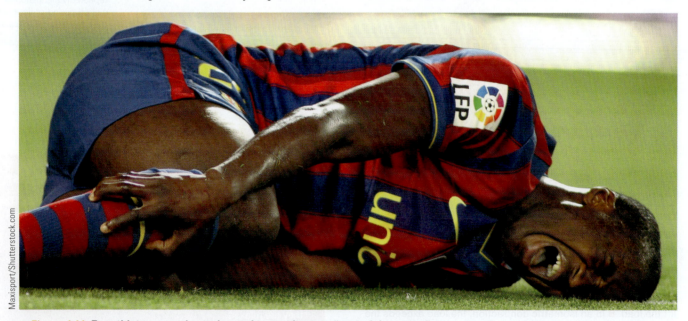

Figure 6.16 For athletes, muscle strains and tears often are "part of the game."

muscles in turn weaken and shrivel. **Duchenne muscular dystrophy** (DMD) is the most common form in children (Figure 6.17). It is caused by a single mutant gene that interferes with the ability of sarcomeres in muscle cells to contract. Affected youngsters usually are confined to a wheelchair by their teens, and most die by their early twenties.

Myotonic muscular dystrophy is usually seen in adults. It generally affects only the hands and feet and is not life-threatening. "Myo-" means muscle, and the name of this disorder indicates that affected muscles contract strongly but don't relax in the normal way.

Scientists have recently made strides in efforts to develop effective treatments for muscular dystrophies. As yet there is no cure, but that sad fact is simply spurring the pace of research.

Bacterial infections can interfere with nervous system signals to muscles

Section 6.4 mentioned the use of *Clostridium botulinum* toxin for Botox injections. This microorganism normally lives in soil. When it contaminates food in unsterilized cans or jars, it produces the botulinum toxin, which causes the deadly food poisoning called **botulism**. The toxin stops motor neurons from releasing ACh, the neurotransmitter that triggers muscle contractions. As a result, muscles become paralyzed. Swift treatment with an antitoxin is the only way to prevent death due to paralysis of the heart muscle and the skeletal muscles involved in breathing.

A similar microbe, *Clostridium tetani*, lives in the gastrointestinal (GI) tract of animals such as cattle and horses. (It may also inhabit the human GI tract.) *C. tetani* spores, a resting stage of the microbe, may be in soil that contains manure. If they enter a wound, the microbe becomes active and produces a toxin that causes the disease **tetanus**.

Unlike the healthy state of steady, low-level muscle contraction of the same name (Section 6.6), the disease tetanus is life-threatening. The bacterial toxin travels to the spinal cord, where it blocks nervous system signals that release skeletal muscles from contraction. The muscles go into continuing spasms called spastic paralysis. A patient's fists and jaw may stay clenched (which is why the disease sometimes is called "lockjaw") and the spine may arch in a stiff curve. Death comes when paralysis reaches the heart and muscles used in breathing.

Today a tetanus vaccine can confer immunity to the disease, and in developed countries such as the United States nearly all people are immunized as children, with periodic "booster" shots recommended for adults. Vaccines were not available for soldiers who sustained battlefield wounds in early wars, and many suffered an agonizing death due to tetanus (Figure 6.18). Globally, the disease kills about 200,000 people each year, mostly women who must give birth in unsanitary conditions.

Cancer may develop in muscle tissue

Cancers that affect the body's soft tissues are a form of **sarcoma** (the prefix "sarc-" means tissue). Luckily, cancer that begins in muscle tissue is relatively rare—only about 1 percent of each year's new cancer cases. It is most common in children and young adults, and about two-thirds of cases involve malignancies that develop in skeletal muscle. This form of cancer is known as rhabdomyosarcoma.

The exact cause of rhabdomyosarcoma is not known, although, as with all cancers, genetic changes are the direct triggers. Having certain rare connective tissue disorders increases the risk. Experience shows that patients must be treated with a three-pronged therapy: surgery to remove as much of the tumor as possible, then chemotherapy and radiation to kill any remaining cancerous cells. When patients undergo this demanding treatment regimen, the chances of a cure are excellent.

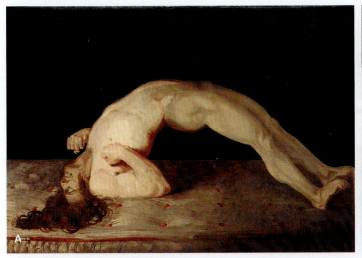
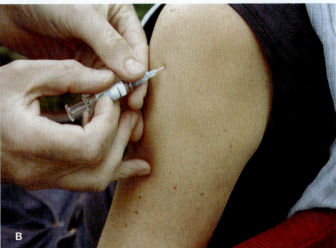

Figure 6.18 **The disease tetanus "freezes" muscles in a contracted state. A** This painting depicts a soldier dying of the disease tetanus in a military hospital in the 1800s after the bacterium *Clostridium tetani* infected a battlefield wound. **B** The tetanus vaccine has saved countless lives in countries where it is readily available. (A: Royal College of Surgeons, Edinburgh; B: © Photopat med/Alamy)

Muscle cells adapt to the activity demanded of them. When severe nerve damage or prolonged bed rest prevents a muscle from being used, the muscle will rapidly begin to waste away, or *atrophy* (AT-row-fy). Over time, affected muscles can lose up to three-fourths of their mass, with a corresponding loss of strength. It is more common for the skeletal muscles of a sedentary person to stay basically healthy, but to be less able to respond to physical demands in the same way that well-worked muscles can.

The best way to maintain or improve the work capacity of your muscles is to exercise them—that is, to increase the demands on muscle fibers to contract. To increase muscle endurance (how long a muscle can sustain contractions), nothing beats regular **aerobic exercise**—activities such as brisk walking, biking, jogging, swimming, and aerobics classes (Figure 6.19). Aerobic exercise works muscles at a rate at which the body can keep them supplied with oxygen. It affects muscle fibers in several ways:

aerobic exercise Exercise that works muscles at a rate that does not exceed the body's ability to keep them supplied with oxygen (in blood).

strength training Intense, short-duration exercise that produces larger, stronger skeletal muscles.

1. There is an increase in the number and the size of mitochondria, the organelles that make ATP.

2. The number of blood capillaries supplying muscle tissue increases. This increased blood supply brings more oxygen and nutrients to the muscle tissue and removes metabolic wastes more efficiently.

3. Muscle tissues contain more of the oxygen-binding pigment myoglobin.

In general, exercise affects a muscle enzyme called LPL (lipoprotein lipase). LPL allows a muscle to take up fatty acids and triglycerides from the blood (see Section 11.12). A low level of LPL is associated with increased risk of cardiovascular disease. Recent studies show that people who sit for long periods tend to have low LPL, even if they exercise at other times. Experts recommend that people who must sit for long periods take frequent short breaks to stand up or walk, because activity that puts weight on leg muscles encourages the steady production of LPL.

Together, these changes produce skeletal muscles that are more efficient metabolically, that maintain their tone, and that can work longer without becoming fatigued. In addition, research suggests that the more skeletal muscles are exercised, the more they produce the so-called "exercise hormone" called irisin, which apparently converts stored white fat to the "energy-burning" brown fat mentioned in Section 4.11. Higher irisin levels may help explain why regular exercise helps keep off excess weight.

Strength training involves intense, short-duration exercise, such as weight lifting. It affects fast muscle fibers, which form more myofibrils and make more of the enzymes used in glycolysis (which forms some ATP). These changes translate into whole muscles that are larger and stronger, but such bulging muscles fatigue rapidly so they don't have much endurance. Fitness experts generally recommend a workout plan that combines strength training and aerobic workouts.

Starting at about age 30, the tension, or physical force, a person's muscles can muster begins to decrease. This means that, once you enter your fourth decade of life, you may exercise just as long and intensely as a younger person but your muscles cannot adapt to the workouts to the same extent. Even so, being physically active is extremely beneficial. Aerobic exercise improves your endurance and blood circulation, and even modest strength training slows the loss of skeletal muscle tissue that is an inevitable part of aging.

Figure 6.19 Physical activity is important for muscle health throughout life. **A** Aerobic exercise builds endurance and improves overall muscle function. **B** Regularly sitting for long periods is associated with an increased risk of cardiovascular disease.

CONNECTIONS

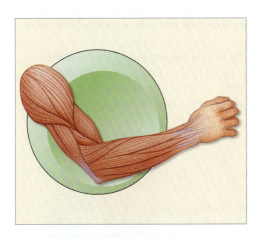

Muscle Tissue and the Muscular System

The muscular system works with the skeleton to bring about body movements. Contractions of skeletal muscles also stabilize joints and body positions. Muscle tissue produces much of the body's metabolic heat.

Smooth muscle forms the walls of hollow organs, blood vessels, ducts, and tubes. Its contractions move substances including blood and food that is being digested. Internal sphincters that control the passage of food, feces, and urine also consist of smooth muscle.

Cardiac muscle forms the wall of the heart. Its contractions move blood throughout the body via the cardiovascular system.

Integumentary system
Skeletal muscle provides support for skin. Many facial muscles, especially those used for making facial expressions such as smiling, attach to skin instead of to bones.

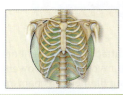

Skeletal system
Skeletal muscles attach to bones, which serve as levers for body movements. The muscles also stabilize movable joints.

Digestive system
Abdominal muscles support many digestive organs. Other skeletal muscles operate in chewing and swallowing. Contractions of smooth muscle move material through the system.

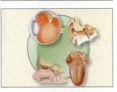

Cardiovascular system and blood
Contractions of cardiac (heart) muscle pump blood. Smooth muscle in blood vessels allows adjustments in blood flow in different body regions. Contraction of leg muscles helps return blood to the heart.

Immunity and the lymphatic system
Smooth muscle forms the walls of lymphatic system vessels. Skeletal muscle helps support lymph nodes in various parts of the body.

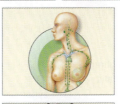

Respiratory system
The diaphragm and skeletal muscles attached to the ribs function in breathing and help clear airways by coughing. Smooth muscle in airways allows changes in air flow to and from the lungs.

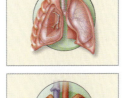

Urinary system
Abdominal muscles help support the kidneys and bladder. Smooth muscle in the bladder is strong and stretchable enough to store urine; its contractions move urine out of the body.

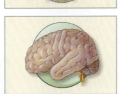

Nervous system
All types of muscle tissue respond to nerve impulses to carry out a wide variety of body functions. Skeletal muscles help support the spine and head.

Sensory systems
Skeletal muscles move the eyes and contain sensory receptors that provide information about changes in body position.

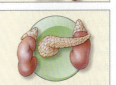

Endocrine system
Skeletal muscles help support endocrine organs such as the pancreas and thyroid gland and produce a hormone that indirectly influences the metabolism of stored body fat.

Reproductive system
Muscle contractions move eggs and sperm. Contraction of smooth muscle in the uterus expels a fetus during childbirth and assists with shedding of the uterine lining (menstruation).

EXPLORE
ON YOUR OWN

A good way to improve your understanding of your muscular system is to explore the movements of your own muscles. Try the following quick exercises.

Human hands don't contain many of the muscles that control hand movements. Instead, as you can see in Figure 6.20A, most of those muscles are in the forearm. Tendons extending from one muscle, the flexor digitorum superficialis (the superficial finger flexor), bend your fingers. Place one hand on the top of the opposite forearm, and then wiggle your fingers on that side or make a fist several times. Can you feel the "finger flexor" in action?

Place your fingers on the skin above your nose, between your eyebrows. Now frown. The muscle you feel pulling your eyebrows together is the corrugator supercilii. One effect of its contraction is to "corrugate" the skin of your forehead into vertical wrinkles.

A grin calls into action other facial muscles, including the zygomaticus major (Figure 6.20B). On either side of the skull, this muscle originates on the cheekbones and inserts at the corners of the mouth. To feel it contract, place the tips of your index fingers at the corners of your mouth, and then smile.

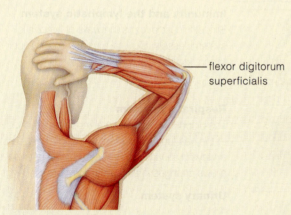

flexor digitorum superficialis

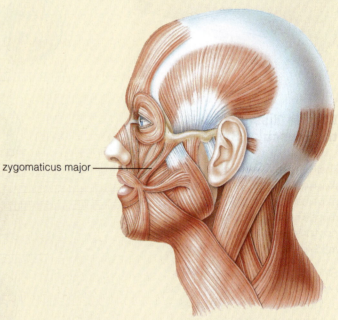

zygomaticus major

A The flexor digitorum superficialis, a forearm muscle that helps move the fingers.

Figure 6.20 Explore these muscles! (© Cengage Learning)

B The zygomaticus major, which helps you smile.

SUMMARY

Section 6.1 The body's muscle tissue includes skeletal, smooth, and cardiac muscle. Despite having different functions, cells in all three types of muscle generate force by contracting.

Section 6.2 The muscular system consists of more than 600 skeletal muscles, which transmit force to bones and move body limbs or other parts (Table 6.1). Skeletal muscles also help to stabilize joints and generate body heat. Each one contains bundles of muscle fibers (muscle cells) wrapped in connective tissue.

Tendons connect skeletal muscle to bones. The origin end of a muscle attaches to the bone that moves least during a movement. The insertion end attaches to the bone that moves most. Some muscles work antagonistically—the action of one opposes or reverses the action of the other. Synergist muscles assist each other's movements.

Section 6.3 Bones move when they are pulled by the shortening, or contraction, of skeletal muscles. This shortening occurs because individual muscle fibers are shortening. Skeletal muscle fibers contain threadlike myofibrils, which are divided lengthwise into sarcomeres, the basic units of contraction. Each sarcomere consists of an array of filaments of the proteins actin (thin) and myosin (thick):

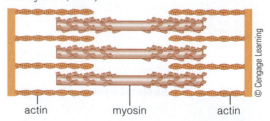

actin   myosin   actin

To shorten a sarcomere, the myosin attaches to a neighboring actin, and the actin slides over the myosin. ATP powers this interaction, which is called the sliding filament mechanism of muscle contraction.

TABLE 6.1 Review of Skeletal Muscle
FUNCTION OF SKELETAL MUSCLE: Contraction (shortening) that moves the body and its parts. **MAJOR COMPONENTS OF SKELETAL MUSCLE CELLS:** **Myofibrils:** Strands containing filaments of the contractile-proteins actin and myosin. **Sarcomeres:** The basic units of muscle contraction. ***Other:*** **Motor unit:** A motor neuron and the muscle fibers it controls. **Neuromuscular junction:** Synapse between a motor neuron and muscle fibers.

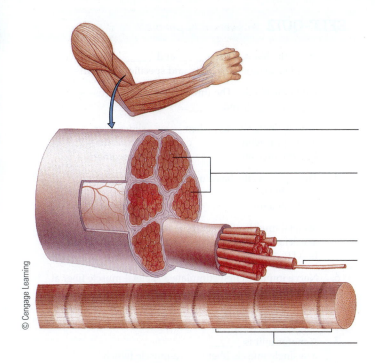

© Cengage Learning

 **Section 6.4** Nerve impulses make skeletal muscle fibers contract. They do this by triggering the release of calcium ions from sarcoplasmic reticulum, a membrane system that wraps around myofibrils in the muscle fiber. The calcium alters proteins on actin filaments so that the heads of myosin molecules can bind to actin.

A neuromuscular junction is a synapse between a motor neuron and a muscle fiber. A nerve impulse triggers the release of a neurotransmitter called ACh into the synapse. This starts the events that cause the fiber to contract.

 Section 6.5 The ATP required for muscle contraction can come from cellular respiration, from glycolysis alone, or from the generation of ATP from creatine phosphate. When muscles use more ATP than aerobic respiration can provide, an oxygen debt may develop in muscle tissue.

 Section 6.6 A motor neuron and the muscle fibers it controls form a motor unit. When a stimulus activates enough motor units, it produces a muscle twitch. If a series of twitches occur close together, a sustained contraction called tetanus develops. Skeletal muscles normally operate near or at tetanus. Important functional properties of whole muscles include the force they exert (tension), muscle tone, and fatigue.

REVIEW QUESTIONS

1. In a general sense, how do skeletal muscles produce movement?

2. In the diagram above, label the fine structure of a muscle, down to one of its myofibrils. Identify the basic unit of contraction in a myofibril.

3. How do actin and myosin interact in a sarcomere to bring about muscle contraction? What roles do ATP and calcium play?

4. How does a muscle fiber incur an oxygen debt?

5. What is the function of the sarcoplasmic reticulum in muscle cell contraction?

6. Explain why (*a*) calcium ions and (*b*) ACh are vital for muscle contraction.

7. What is a motor unit? Why does a rapid series of muscle twitches yield a stronger overall contraction than a single twitch?

8. What are the structural and functional differences between "slow" and "fast" muscle?

1. Fill in the blanks: The _____ and _____ systems work together to move the body and specific body parts.

2. Fill in the blanks: The three types of muscle tissue are _____, _____, and _____.

3. Muscle fibers shorten when _____ slides over _____.
   a. myosin, actin
   b. actin, myosin
   c. myoglobin, actin
   d. myosin, sarcomeres

4. The _____ is the basic unit of muscle contraction.
   a. myofibril
   b. sarcomere
   c. muscle fiber
   d. myosin filament

5. Skeletal muscle contraction requires _____.
   a. calcium ions
   b. ATP
   c. arrival of a nerve impulse
   d. all of the above

6. Nerve impulses first stimulate a skeletal muscle fiber at _____.
   a. T tubules
   b. sarcomeres
   c. neuromuscular junctions
   d. actin binding sites

7. A motor unit is _____.
   a. a single muscle fiber
   b. a single sarcomere
   c. a muscle twitch
   d. a motor neuron and the fibers it synapses with

8. Muscle fatigue _____.
   a. occurs when ATP runs out
   b. may be caused by oxygen debt
   c. is a state in which a muscle can no longer contract
   d. all of the above

9. Muscle tone is_____.
   a. the same as muscle tension
   b. a steady, low-level state of contraction
   c. not present when a muscle is relaxed
   d. all of the above

10. Match the M words with their defining feature.
    _____ muscle
    _____ muscle twitch
    _____ muscle tension
    _____ myosin
    _____ motor neuron
    _____ myofibrils
    _____ muscle fatigue
    a. actin's partner
    b. delivers contraction signal
    c. a muscle cannot contract
    d. motor unit response
    e. force exerted by cross-bridges
    f. muscle cells bundled in connective tissue
    g. threadlike parts in a muscle fiber

Your Future

Surgeons at the University of California, Davis Medical Center recently announced that they had succeeded in developing artificial skeletal muscle that can allow patients with facial paralysis to blink. The "muscle tissue" is made from a rubbery silicon polymer. Electrodes embedded in it send electrical signals that stimulate it to contract and relax. The doctors involved in developing this new technology believe that it is a major first step in the development of artificial muscles that can help patients with other types of paralysis.

CRITICAL THINKING

1. You are training athletes for the 100-meter dash. They need muscles specialized for speed and strength, *not* endurance. What muscle characteristics would your training regimen aim to develop? How would you alter it to train a long-distance swimmer?

2. Jay thinks he has torn a muscle in his calf while doing yardwork. His best friend tells him that the tear will likely heal itself over time, and because his yard is still a mess he decides to "work through the pain." Do you think this plan is OK? Explain why or why not.

3. Curare, a poison extracted from a South American shrub, blocks the binding of ACh by muscle cells. What do you suppose would happen to your muscles, including the ones involved in breathing, if a toxic dose of curare entered your bloodstream?

4. At the gym Sean gets on a stair-climbing machine and "climbs" as fast as he can for fifteen minutes. At the end of that time he is breathing hard and his quadriceps and other leg muscles are aching. What is the physiological explanation for these symptoms?

5. In training for a marathon, Maria plans to secretly take a performance-enhancing drug because she believes it will help her place in the top five finishers and she desperately wants to build a reputation as a world-class competitive marathoner. What is your opinion on this plan?

CIRCULATION: THE HEART AND BLOOD VESSELS

LINKS TO EARLIER CONCEPTS

This chapter discusses cardiac muscle (4.3) and the junctions between cardiac muscle cells (4.6).

Blood vessels are built from layers of epithelium, connective tissue, and smooth muscle (4.1–4.3).

Lipoproteins and cholesterol (2.10, 2.12) are two important factors in our discussion of cardiovascular health.

KEY CONCEPTS

Circulating Blood

The cardiovascular system consists of a pump—the heart—and blood vessels that carry pumped blood and substances in it throughout the body. Section 7.1

Pumping Blood

The heart's pumping action drives blood under pressure through two circuits—one to the lungs and one to other body regions. Sections 7.2–7.5

Blood Vessels

Blood vessels, including arteries, arterioles, capillaries, venules, and veins, are specialized for different blood transport functions. Sections 7.6–7.7

Disorders of the Cardiovascular System

Sections 7.8–7.9

CONNECTIONS: The Cardiovascular System and Blood in Homeostasis

Section 7.10

Top and second from top: © Cengage Learning; Second from bottom: ©Biophoto Associates/Photo Researchers, Inc.; Bottom: ©Biophoto Associates/Photo Researchers, Inc.

Matt Nader, the healthy-looking young man shown in the photo, had a close brush with death while playing in a high school football game. With little warning, Matt collapsed after his heart abruptly stopped beating—an event called sudden cardiac arrest, or SCA. Like many other young people who suffer SCA, Matt had an unsuspected genetic condition that caused the problem. Matt's luck hadn't run out, however. His parents, both trained medical professionals, were at the game. They immediately started CPR (cardiopulmonary resuscitation), which includes sets of chest compressions that keep blood flowing to the brain, lungs, and other organs. If CPR is started within four to six minutes, a cardiac arrest victim's chances of surviving rise by 50 percent.

While CPR keeps blood moving, restarting a stopped heart requires a defibrillator. This device delivers a jolt of electricity to the chest. With luck the shock reactivates the heart's "pacemaker," which stimulates the heartbeat. Luckily for Matt, his school had an automated external defibrillator (AED). Matt was the first person it had ever been used on, and it saved his life. Many schools, senior centers, shopping malls, hotels, and airports now keep an AED on hand. According to the American Heart Association, such emergency measures save more than 300,000 lives each year.

Matt Nader's story introduces our topic in this chapter—the structure and functioning of the heart and blood vessels. Together these organs make up the cardiovascular system.

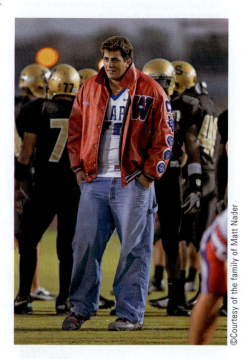

©Courtesy of the family of Matt Nader

Homeostasis Preview

The cardiovascular system delivers oxygen, nutrients, hormones, and other substances to body cells. It carries away wastes and substances cells produce. Flowing blood also carries excess heat to the body surface.

- The cardiovascular system is built to rapidly transport blood to every living cell in the body.

- Links to Diffusion 3.10, Metabolism 3.13

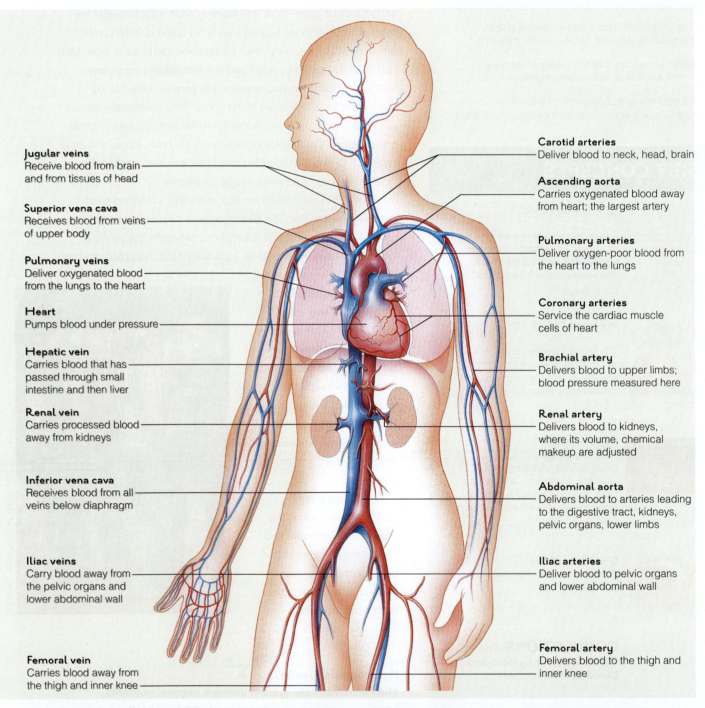

Jugular veins
Receive blood from brain and from tissues of head

Superior vena cava
Receives blood from veins of upper body

Pulmonary veins
Deliver oxygenated blood from the lungs to the heart

Heart
Pumps blood under pressure

Hepatic vein
Carries blood that has passed through small intestine and then liver

Renal vein
Carries processed blood away from kidneys

Inferior vena cava
Receives blood from all veins below diaphragm

Iliac veins
Carry blood away from the pelvic organs and lower abdominal wall

Femoral vein
Carries blood away from the thigh and inner knee

Carotid arteries
Deliver blood to neck, head, brain

Ascending aorta
Carries oxygenated blood away from heart; the largest artery

Pulmonary arteries
Deliver oxygen-poor blood from the heart to the lungs

Coronary arteries
Service the cardiac muscle cells of heart

Brachial artery
Delivers blood to upper limbs; blood pressure measured here

Renal artery
Delivers blood to kidneys, where its volume, chemical makeup are adjusted

Abdominal aorta
Delivers blood to arteries leading to the digestive tract, kidneys, pelvic organs, lower limbs

Iliac arteries
Deliver blood to pelvic organs and lower abdominal wall

Femoral artery
Delivers blood to the thigh and inner knee

Figure 7.1 Animated! The heart and blood vessels make up the cardiovascular system. Arteries, which carry oxygenated blood to tissues, are shaded red. Veins, which carry deoxygenated blood away from tissues, are shaded blue. Notice, however, that for the pulmonary arteries and veins the roles are reversed. (© Cengage Learning)

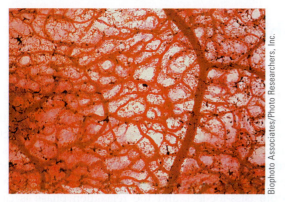

Figure 7.2 The cardiovascular system includes networks of branching blood vessels.

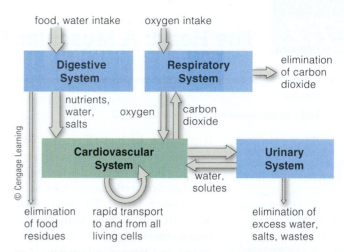

Figure 7.3 Together with the other organ systems shown here, the cardiovascular system helps maintain stable chemical and physical conditions in the extracellular fluid.

The heart and blood vessels make up the cardiovascular system

"Cardiovascular" comes from the Greek *kardia* (heart) and the Latin *vasculum* (vessel). As you can see in Figure 7.1 the **cardiovascular system** has two main elements, the heart and blood vessels.

- The **heart** is a muscular pump that generates the pressure required to move blood throughout the body.
- Blood vessels are tubes of different diameters that transport blood (Figure 7.2).

The heart pumps blood at high pressure into **arteries**, which have a large diameter. From there blood flows into smaller and narrower vessels called **arterioles**, which branch into even narrower **capillaries**. Fluids and solutes diffuse out of the capillaries into the extracellular fluid and eventually enter body cells. In the reverse route, fluid and solutes move from body cells into the extracellular fluid and from there into the blood in capillaries. Blood flows from capillaries into small **venules**, then into large-diameter **veins** that return blood to the heart.

As you will read later on, the volume of blood flowing to a particular part of the body is adjustable. So is the rate at which it flows. This flexibility permits the cardiovascular system to deliver blood in ways that suit conditions in different parts of the body. For example, blood flows rapidly through arteries, but in capillaries it must flow slowly so that there is time for substances moving to and from cells to diffuse into and out of the extracellular fluid.

The cardiovascular system is linked to the lymphatic system

The heart's pumping action puts pressure on blood flowing through the cardiovascular system. Partly because of this pressure, small amounts of water and some proteins dissolved in blood are forced out and become part of extracellular fluid (the fluid outside cells). A network of drainage vessels picks up excess extracellular fluid and usable substances in it and returns them to the cardiovascular system. This vessel network is part of the lymphatic system, which includes

organs with major roles in body defenses. We consider it more fully in Chapter 9.

Blood circulation is essential for maintaining homeostasis

The heart and blood vessels are sometimes referred to as the "circulatory system." This name is apt because blood circulates through the system, bringing body cells such essentials as oxygen and nutrients from food. Circulating blood also takes away the wastes produced by our metabolism, along with excess heat. In fact, cells depend on blood to constantly pick up and deliver an extremely diverse range of substances, including those that move into or out of the digestive system, the urinary system, and the respiratory system (Figure 7.3).

arteriole A blood vessel that connects arteries with capillaries.

artery A large-diameter blood vessel that carries blood away from the heart at high pressure.

capillary A narrow blood vessel that functions in the exchange of substances between blood and the extracellular fluid.

cardiovascular system The heart and blood vessels, which together move blood throughout the body.

heart The muscular, blood-pumping organ of the cardiovascular system. It generates pressure required to move blood through the system.

vein Large-diameter vessel that returns blood to the heart.

venule Vessel that receives blood from capillaries; the smallest-diameter vessel in the venous system.

TAKE-HOME MESSAGE

WHAT IS THE CARDIOVASCULAR SYSTEM?

- The cardiovascular system consists of the heart and blood vessels. The system transports blood to and from all living cells in the body.
- Some fluid and dissolved substances in blood move from blood vessels to the extracellular fluid.
- The lymphatic system collects excess extracellular fluid and returns it—along with useful substances it contains—to the cardiovascular system.

The Heart: A Muscular Double Pump

- The heart is a durable pump that consists mainly of cardiac muscle.
- Links to Epithelium 4.1, Muscle tissue 4.3

aorta Artery carrying blood pumped by the left ventricle (to the systemic circulation).

aortic valve Valve that opens from the left ventricle into the aorta.

atrioventricular valve Valve through which blood flows from an atrium to a ventricle.

atrium Either of the upper heart chambers, one above each ventricle.

cardiac cycle The sequence of contraction and relaxation of the heart chambers.

coronary circulation Arteries and veins that service the heart.

diastole The relaxation phase of the cardiac cycle.

myocardium Cardiac muscle tissue of the heart wall.

pulmonary valve Valve that opens from the right ventricle into the pulmonary artery.

systole The contraction phase of the cardiac cycle.

ventricle Large heart chamber located below the atrium in each side of the heart.

If you look back at Figure 7.1, you can see that your heart is located roughly in the center of your chest. Its structure reflects its role as a long-lasting pump. The heart wall is mostly cardiac muscle tissue, the **myocardium** (Figure 7.4A). A tough, fibrous sac, the *pericardium* ("peri" = around), surrounds, protects, and lubricates it. The heart's chambers have a smooth lining (*endocardium*) composed of connective tissue and a layer of epithelial cells. The epithelial cell layer, known as *endothelium*, also lines blood vessels.

The heart has two halves and four chambers

A thick wall, the *septum*, divides the heart into two halves, right and left. Each half has two chambers: an **atrium** (plural: atria) located above a larger **ventricle**. Flaps of membrane separate the two chambers and serve as a one-way **atrioventricular valve** (AV valve) between them. The AV valve in the right half of the heart is called a *tricuspid valve* because its three flaps come together in pointed cusps (Figure 7.4B). In the heart's left half, the AV valve consists of just two flaps; it is called the *bicuspid valve* or *mitral valve*. Tough, collagen-reinforced strands (*chordae tendineae*, or "heartstrings") connect the AV valve flaps to cone-shaped muscles that extend out from the ventricle wall. When a blood-filled ventricle contracts, this arrangement prevents the flaps from opening backward into the atrium. Each half of the heart also has a valve between the ventricle and the arteries leading away from it. The **pulmonary valve** controls blood flow to the pulmonary artery, and the **aortic valve** controls blood flow to the aorta. Because both these valves are shaped like a half-moon, they are also known as "semilunar" valves. During a heartbeat, the valves open and close in ways that keep blood moving in one direction, out of the heart.

Arteries and veins that serve only the heart provide what is called the **coronary circulation**. Two coronary arteries service most of the cardiac muscle (Figure 7.5). They branch off the **aorta,** the major artery carrying blood away from the heart. Coronary veins empty blood into the right atrium.

In a "heartbeat," the heart's chambers contract, then relax

Blood is pumped each time the heart beats. It takes less than a second for a "heartbeat"—one sequence of contraction and relaxation of the heart chambers. This sequence is the **cardiac cycle** (Figure 7.6). It occurs almost simultaneously in both sides of the heart. The contraction phase is called **systole** (SISS-toe-lee), and the relaxation phase is called **diastole** (dye-ASS-toe-lee).

During the cardiac cycle, the ventricles relax before the atria contract, and the ventricles contract when the atria relax. When the relaxed atria are filling with blood, the fluid pressure inside them rises and the AV valves open. Blood flows into the ventricles, which are 80 percent filled by the time the atria contract. As the filled ventricles begin to contract, fluid pressure inside *them* increases, forcing the AV valves shut. The rising pressure forces the aortic and pulmonary valves open—and blood flows out of the heart and into the aorta and pulmonary artery. Now the ventricles relax, and the valves close. For about half a second the atria and ventricles are all in diastole. Then the blood-filled atria contract, and the cycle repeats.

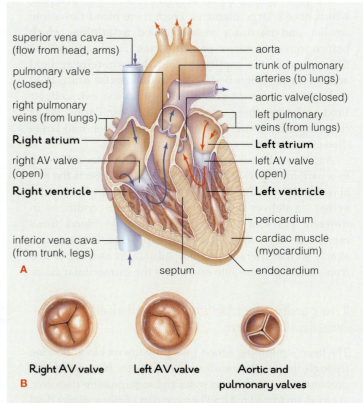

superior vena cava (flow from head, arms)

aorta

pulmonary valve (closed)

trunk of pulmonary arteries (to lungs)

aortic valve(closed)

right pulmonary veins (from lungs)

left pulmonary veins (from lungs)

Right atrium

Left atrium

right AV valve (open)

left AV valve (open)

Right ventricle

Left ventricle

pericardium

cardiac muscle (myocardium)

inferior vena cava (from trunk, legs)

A

septum

endocardium

Right AV valve   **Left AV valve**   Aortic and pulmonary valves

B

Figure 7.4 Animated! **The heart is divided into right and left halves. A** The heart's internal anatomy. Blue arrows represent blood flow into and out of the right ventricle. Red arrows represent blood flow into and out of the left ventricle. **B** The shapes of heart valves. (© Cengage Learning)

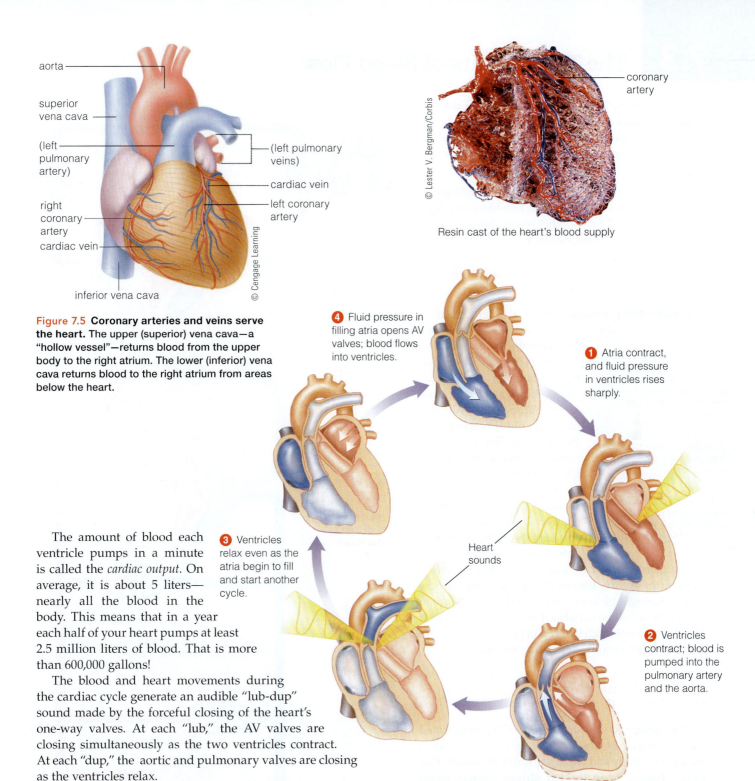

aorta

superior
vena cava

(left
pulmonary
artery)

(left pulmonary
veins)

cardiac vein

left coronary
artery

right
coronary
artery

cardiac vein

inferior vena cava

© Cengage Learning

Figure 7.5 Coronary arteries and veins serve the heart. The upper (superior) vena cava—a "hollow vessel"—returns blood from the upper body to the right atrium. The lower (inferior) vena cava returns blood to the right atrium from areas below the heart.

coronary artery

© Lester V. Bergman/Corbis

Resin cast of the heart's blood supply

4 Fluid pressure in filling atria opens AV valves; blood flows into ventricles.

1 Atria contract, and fluid pressure in ventricles rises sharply.

Heart sounds

3 Ventricles relax even as the atria begin to fill and start another cycle.

2 Ventricles contract; blood is pumped into the pulmonary artery and the aorta.

The amount of blood each ventricle pumps in a minute is called the *cardiac output*. On average, it is about 5 liters—nearly all the blood in the body. This means that in a year each half of your heart pumps at least 2.5 million liters of blood. That is more than 600,000 gallons!

The blood and heart movements during the cardiac cycle generate an audible "lub-dup" sound made by the forceful closing of the heart's one-way valves. At each "lub," the AV valves are closing simultaneously as the two ventricles contract. At each "dup," the aortic and pulmonary valves are closing as the ventricles relax.

Figure 7.6 Animated! The heart beats in a sequence called the **cardiac cycle.** (© Cengage Learning)

■ Each half of the heart pumps blood. The two side-by-side pumps are the basis of two cardiovascular circuits through the body, each with its own set of arteries, arterioles, capillaries, venules, and veins.

Every day, your blood travels roughly 12,000 miles, making the equivalent of four coast-to-coast trips across the United States. This blood flow occurs in the two circuits we now consider.

In the pulmonary circuit, blood picks up oxygen in the lungs

The **pulmonary circuit**, which is diagrammed in Figure 7.7A at right, receives blood from tissues and circulates it through the lungs for gas exchange. The circuit begins as blood from tissues enters the right atrium, then moves through the AV valve into the right ventricle. As the ventricle fills, the atrium contracts. Blood arriving in the right ventricle is fairly low in oxygen and high in carbon dioxide. When the ventricle contracts, the blood moves through the right semilunar valve into the *main* pulmonary artery, then into the *right* and *left* pulmonary arteries. These arteries carry the blood to the two lungs, where (in capillaries) it picks up oxygen and gives up carbon dioxide that will be exhaled. The freshly oxygenated blood returns through two sets of pulmonary veins to the heart's left atrium, completing the circuit.

hepatic portal system System of blood vessels that transport blood from the digestive tract to and from the liver.

pulmonary circuit The short path in which blood flows through the lungs for gas exchange.

systemic circuit The long path in which blood flows from the heart to tissues and back to the heart.

In the systemic circuit, blood travels to and from tissues

In the **systemic circuit** (Figure 7.7B), oxygenated blood pumped by the left half of the heart moves through the body and returns to the right atrium. This circuit begins when the left atrium receives blood from pulmonary veins, and this blood moves through an AV (bicuspid) valve to the left ventricle. This chamber contracts with great force, sending blood coursing through a semilunar valve into the aorta.

As the aorta arches over the heart and descends into the torso (as the abdominal aorta) (see Figure 7.1), major arteries branch off it, funneling blood to organs and tissues where O$_2$ is used and CO$_2$ is produced. For example, in a resting person, each minute a fifth of the blood pumped into the systemic circulation enters the kidneys (Figure 7.7C) via *renal arteries*. Deoxygenated

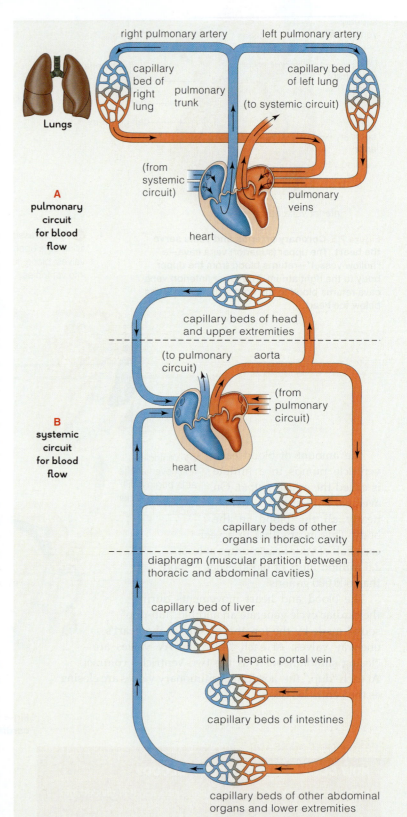

A pulmonary circuit for blood flow

right pulmonary artery — left pulmonary artery — capillary bed of right lung — pulmonary trunk — capillary bed of left lung — (to systemic circuit) — Lungs — (from systemic circuit) — pulmonary veins — heart

B systemic circuit for blood flow

capillary beds of head and upper extremities — (to pulmonary circuit) — aorta — (from pulmonary circuit) — heart — capillary beds of other organs in thoracic cavity — diaphragm (muscular partition between thoracic and abdominal cavities) — capillary bed of liver — hepatic portal vein — capillary beds of intestines — capillary beds of other abdominal organs and lower extremities

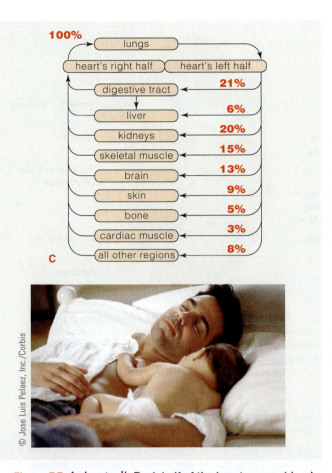

100%

lungs

heart's right half | heart's left half

digestive tract — 21%

liver — 6%

kidneys — 20%

skeletal muscle — 15%

brain — 13%

skin — 9%

bone — 5%

cardiac muscle — 3%

all other regions — 8%

C

Figure 7.7 Animated! Each half of the heart pumps blood in a different circuit. The **A** pulmonary and **B** systemic circuits for blood flow in the cardiovascular system. **C** How the heart's output is distributed in people napping. (© Cengage Learning)

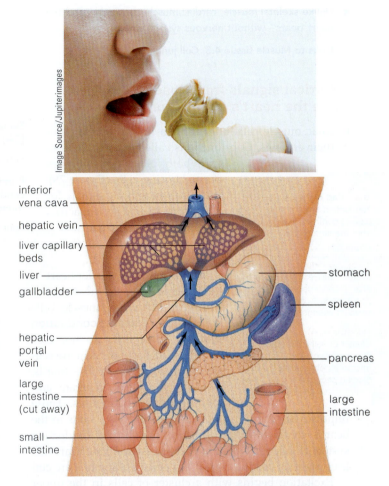

inferior vena cava

hepatic vein

liver capillary beds

liver

gallbladder

hepatic portal vein

large intestine (cut away)

small intestine

stomach

spleen

pancreas

large intestine

Figure 7.8 Blood from the digestive tract detours to the liver. Arrows show the direction in which blood flows. (© Cengage Learning)

blood returns to the right half of the heart, where it enters the pulmonary circuit. Notice that in both the pulmonary and the systemic circuits, blood travels through arteries, arterioles, capillaries, and venules, finally returning to the heart in veins. Blood from the head, arms, and chest arrives through the *superior vena cava*. The *inferior vena cava* collects blood from the lower part of the body.

Because the heart pumps constantly, the volume of flow through the entire system each minute is equal to the volume of blood returned to the heart each minute.

Blood from the digestive tract is shunted through the liver for processing

As you can see in Figure 7.8, blood passing through capillary beds in the digestive tract travels to another capillary bed in the liver. After a meal, the *hepatic portal vein* brings nutrient-laden blood to this capillary bed. As blood seeps through it, the liver can remove impurities and

process absorbed substances. The vessels involved in this detour collectively are called the **hepatic portal system**. You will read more about this topic in Chapter 11.

Blood leaving the liver's capillary bed enters the general circulation through a *hepatic vein*. The liver receives oxygenated blood via the *hepatic artery*.

TAKE-HOME MESSAGE

WHAT ARE THE TWO CIRCUITS IN WHICH BLOOD FLOWS THROUGH THE BODY?

- Blood flows through a pulmonary circuit and a systemic circuit.
- The pulmonary circuit carries blood through the lungs for gas exchange. The systemic circuit transports blood to and from tissues.
- After meals, the blood in capillary beds in the digestive tract is diverted to the liver for processing. Blood then returns to the general circulation.

7.4 How Cardiac Muscle Contracts

- Unlike skeletal muscle, cardiac muscle contracts—and the heart beats—without nervous system orders.

- Links to Muscle tissue 4.3, Cell junctions 4.6

Electrical signals from "pacemaker" cells drive the heart's contractions

Cardiac muscle cells branch, then link to one another at their endings. Gap junctions called *intercalated discs* span both plasma membranes of neighboring cells (Figure 7.9). With each heartbeat, signals for contraction spread so fast across the junctions that cardiac muscle cells contract together, almost as if they were a single unit.

atrioventricular (AV) node Part of the cardiac conduction system that passes contraction signals from the atria to the ventricles.

cardiac conduction system Self-exciting heart muscle cells that spontaneously generate and conduct electrical signals.

sinoatrial (SA) node Cluster of self-exciting cells that establish a regular heartbeat; also called the cardiac pacemaker.

Where do the signals for heart contractions come from? About 1 percent of cardiac muscle cells function as the **cardiac conduction system**. These cells do not contract. Instead, some of them are self-exciting "pacemaker" cells—that is, they spontaneously generate and conduct electrical impulses. Those impulses are the signals that stimulate contractions in the heart's contractile cells. Because the cardiac conduction system is independent of the nervous system, the heart will keep right on beating even if all nerves leading to it are cut!

Excitation begins with a cluster of cells in the upper wall of the right atrium (Figure 7.10). About seventy times a minute, this **sinoatrial (SA) node** generates signals that stimulate waves of excitation. Each wave spreads swiftly over both atria and causes them to contract. It then reaches the **atrioventricular (AV) node** in the septum dividing the two atria.

When a stimulus reaches the AV node, it slows but keeps moving along bundles of conducting fibers that extend to the ventricles. At places along each bundle, cells called

— cardiac muscle cell

intercalated disc where cardiac muscle cells meet

— cardiac muscle cell

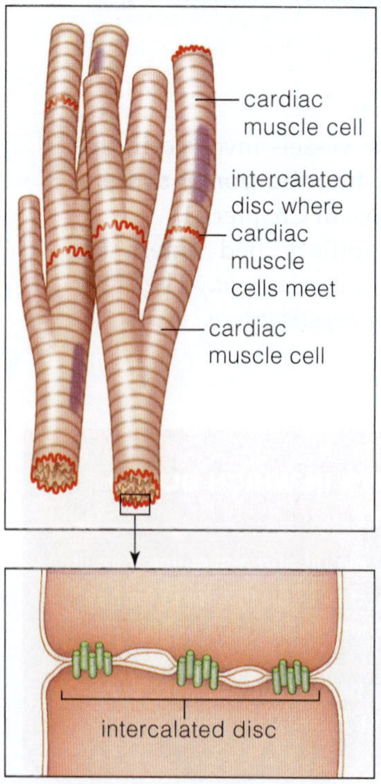

Figure 7.9 Intercalated discs form communication junctions between cardiac muscle cells. Signals travel rapidly across the junctions and cause cells to contract nearly in unison. (© Cengage Learning)

intercalated disc

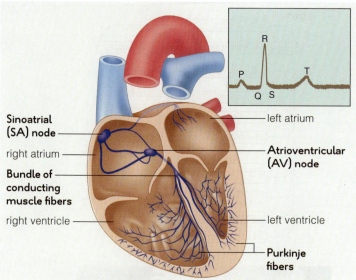

Figure 7.10 Animated! In the cardiac conduction system, pacemaker cells produce electrical signals. In the recording of a heartbeat above, letters indicate three waves of electrical activity that were caused by the spread of impulses across cardiac muscle. (© Cengage Learning)

Sinoatrial (SA) node

right atrium

Bundle of conducting muscle fibers

right ventricle

left atrium

Atrioventricular (AV) node

left ventricle

Purkinje fibers

Purkinje fibers pass the signal on to contractile muscle cells in each ventricle. The slow conduction in the AV node is an important part of this sequence. It gives the atria time to finish contracting before the wave of excitation spreads to the ventricles.

Of all cells of the cardiac conduction system, the SA node fires off impulses at the fastest rate and is the first region to respond in each cardiac cycle. It is called the "intrinsic cardiac pacemaker" because its self-generated rhythmic firing is the basis for the normal rate of heartbeat. People whose SA node chronically malfunctions may have an artificial pacemaker implanted to provide a regular stimulus for their heart contractions.

The nervous system adjusts heart activity

The nervous system can adjust the rate and strength of cardiac muscle contraction. Stimulation by one set of nerves can increase heart activity, while stimulation by another set of nerves can slow it. The control centers for these adjustments are in the spinal cord and parts of the brain. They are discussed more fully in Chapter 13.

TAKE-HOME MESSAGE

WHAT IS THE CARDIAC CONDUCTION SYSTEM?

- The cardiac conduction system consists of specialized cardiac muscle cells that stimulate heart contractions. The system's signals stimulate a rhythmic cycle of contraction, first in the heart atria, then in the ventricles.

7.5 Blood Pressure

- Heart contractions generate blood pressure, which changes as blood moves through the systemic circuit.

Blood exerts pressure against the walls of blood vessels

Blood pressure is the fluid pressure that blood exerts against vessel walls. Blood pressure is highest in the aorta; then it drops along the systemic circuit. The pressure typically is measured when a person is at rest (Figure 7.11). For an adult, the National Heart, Lung, and Blood Institute has established blood pressure values under 120/80 as

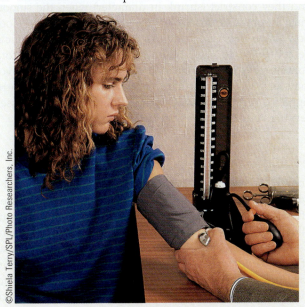

Figure 7.11 Animated! Measuring blood pressure is one way to monitor cardiovascular health. A hollow cuff attached to a pressure gauge is wrapped around the upper arm. The cuff is inflated to a pressure above the highest pressure of the cardiac cycle—at systole, when ventricles contract. As air in the cuff is slowly released, some blood flows into the artery in the arm. The turbulent flow causes soft tapping sounds. When the tapping starts, the gauge's value is the systolic pressure, measured in millimeters of mercury (Hg). This value measures how far the pressure would force mercury to move upward in a narrow glass column. Now more air is released from the cuff. Just after the sounds grow dull and muffled, blood is flowing steadily, so the turbulence and tapping end. The silence corresponds to diastolic pressure at the end of a cardiac cycle, before the heart pumps out blood.

TABLE 7.1 Blood Pressure Values (mm of Hg)		
	Systolic	**Diastolic**
NORMAL	100–119	60–79
HYPOTENSION	Less than 100	Less than 60
PREHYPERTENSION	120–139	80–139
HYPERTENSION	140 and up	90 and up

Risk Factors for Hypertension

1. Smoking
2. Obesity
3. Sedentary lifestyle
4. Chronic stress
5. A diet low in fruits, vegetables, dairy foods, and other sources of potassium and calcium
6. Excessive salt intake (in some individuals)
7. Poor salt management by the kidneys, usually due to disease
8. Factors not related to lifestyle including advancing age, male gender, and being of African descent

Figure 7.12 A variety of factors may cause hypertension. (© Cengage Learning)

the healthiest (Table 7.1). The first number, *systolic pressure*, is the peak of pressure in the aorta while the left ventricle contracts and pushes blood into the aorta. The second number, *diastolic pressure*, measures the lowest blood pressure in the aorta, when blood is flowing out of it and the heart is relaxed.

> **blood pressure** The fluid pressure that blood exerts against the walls of blood vessels. It is measured in millimeters of mercury.
>
> **hypertension** Chronically elevated blood pressure.

Values for systolic and diastolic pressure provide important health information. Chronically elevated blood pressure, or **hypertension**, can be associated with various ills, such as atherosclerosis (Section 7.8). The chart in Figure 7.12 lists some major causes and risk factors. Hypertension can lead to a stroke or heart attack. Each year it indirectly kills about 180,000 Americans, many of whom may not have had any outward symptoms. Roughly 40 million people in the United States are unaware that they have hypertension. Personal blood pressure monitors are marketed as tools for keeping tabs on blood pressure.

Abnormally *low* blood pressure is called *hypotension*. This condition can develop when for some reason there is not enough water in blood plasma—for instance, if there are too few proteins in the blood to "pull" water in by osmosis. A large blood loss also can cause blood pressure to plummet. Such a drastic decrease is one sign of a dangerous condition called *circulatory shock*.

TAKE-HOME MESSAGE

WHAT IS BLOOD PRESURE?

- Blood pressure is the fluid pressure blood exerts against the walls of blood vessels. Heart contractions generate blood pressure.
- Systolic pressure is the peak of pressure in the aorta while blood pumped by the left ventricle is flowing into it.
- Diastolic pressure measures the lowest blood pressure in the aorta, when blood is flowing out of it.

■ There are differences in how different kinds of blood vessels manage blood flow and blood pressure.

■ Links to Epithelium 4.1, Connective tissues 4.2

Arteries are large, strong blood pipelines

The wall of an artery has several tissue layers (Figure 7.13A). The outer layer is mainly collagen, which anchors the vessel to the tissue it runs through. A thick middle layer of smooth muscle is sandwiched between thinner layers containing elastin. The innermost layer is a thin sheet of endothelium. Together these layers form a thick, muscular, and elastic wall. In a large artery the wall bulges slightly under the pressure surge caused when a ventricle contracts. In arteries near the body surface, as in the wrist, you can feel the surges as your **pulse**.

The bulging of artery walls helps keep blood flowing on through the system. How? For a moment, some of the blood pumped during the systole phase of each cardiac cycle is stored in the "bulge"; the elastic recoil of the artery then forces that stored blood onward during diastole, when heart chambers are relaxed. In addition to having stretchable walls, arteries also have large diameters. For this reason, they present little resistance to blood flow, so blood pressure in large arteries is quite stable (Figure 7.14).

Arterioles are control points for blood flow

Arteries branch into narrower arterioles, which have a wall built of rings of smooth muscle over a single layer of elastic fibers (Figure 7.13B). Being built this way, arterioles dilate (enlarge in diameter) when the smooth muscle relaxes, and they constrict (shrink in diameter) when the smooth muscle contracts. Arterioles offer more resistance to blood flow than other vessels do. As the blood flow slows, it can be controlled in ways that adjust how much of the total volume goes to different body regions. For example, you may feel sleepy after a large meal in part because control signals divert blood away from your brain and into vessels serving your digestive system.

Capillaries are specialized for diffusion

Your body has about 2 miles of arteries and veins but a whopping 62,000 miles of capillaries. These tiny vessels often interlace in **capillary beds**, and their structure allows substances to readily diffuse between blood and tissue fluid. Specifically, a capillary has the thinnest wall of any blood vessel—a single layer of flat endothelium (Figure 7.13C). As you might guess, the body's capacity for maintaining homeostasis depends heavily on the diffusion of gases (oxygen and carbon dioxide), nutrients, and wastes that occurs across the walls of capillaries.

Blood can't move fast in capillaries. However, because there are so many capillaries and capillary beds, they present less total resistance to flow than do the arterioles leading into them, so overall blood pressure drops more slowly in them.

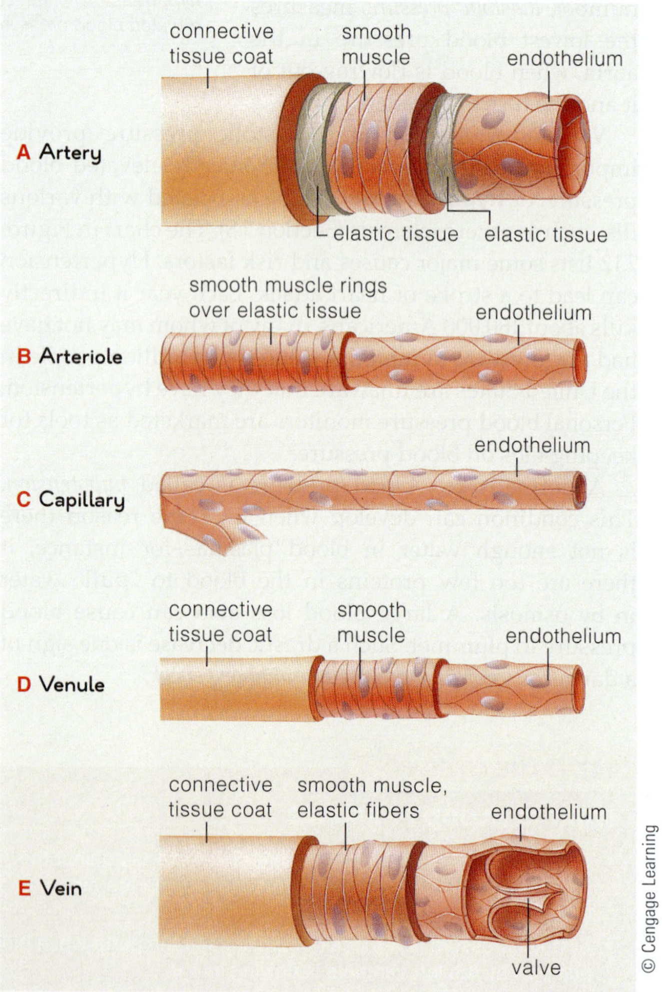

A Artery

connective tissue coat — smooth muscle — endothelium — elastic tissue — elastic tissue

B Arteriole

smooth muscle rings over elastic tissue — endothelium

C Capillary

endothelium

D Venule

connective tissue coat — smooth muscle — endothelium

E Vein

connective tissue coat — smooth muscle, elastic fibers — endothelium — valve

© Cengage Learning

Figure 7.13 Animated! The structure of a blood vessel matches its function.

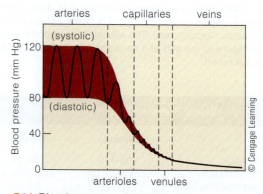

© Cengage Learning

Figure 7.14 Blood pressure changes as blood flows through different parts of the cardiovascular system.

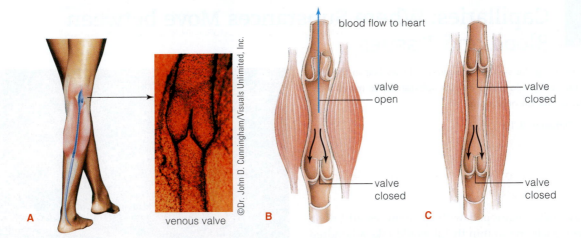

blood flow to heart

valve open

valve closed

valve closed

valve closed

Figure 7.15 **Contracting skeletal muscles help keep blood in veins flowing toward the heart. A** Skeletal muscles nestle against veins. **B** Contracting muscles provide a "push" that keeps blood flowing forward. **C** When skeletal muscles relax, valves in the vein shut—preventing backflow. (© Cengage Learning)

venous valve

©Dr. John D. Cunningham/Visuals Unlimited, Inc.

A   B   C

Venules and veins return blood to the heart

Capillaries merge into venules, or "little veins," which in turn merge into large-diameter veins. Venules function a little like capillaries, in that some solutes diffuse across their relatively thin walls (Figure 7.13D).

Veins are large-diameter, low-resistance transport tubes to the heart (Figure 7.13E). Their valves prevent backflow. When blood starts moving backward due to gravity, it pushes the valves closed. Unlike an arterial wall, a vein wall can bulge quite a bit under pressure. Thus veins are reservoirs for variable volumes of blood. Together, the veins of an adult can hold up to 50 to 60 percent of the total blood volume.

When a person's blood must circulate faster (for instance, during exercise), the smooth muscle in veins contracts. The wall stiffens, the vein bulges less, and venous pressure rises—so more blood flows to the heart. Venous pressure also rises when contracting skeletal muscle—especially in the legs and abdomen—bulges against adjacent veins. This muscle activity helps return blood through the venous system (Figure 7.15).

Obesity, pregnancy, and other factors can weaken venous valves. The walls of a *varicose vein* have become overstretched because, over time, weak valves have allowed blood to pool there.

Many vessels have roles in homeostatic mechanisms that help control blood pressure

Some arteries, all arterioles, and even veins have roles in homeostatic mechanisms that help maintain adequate blood pressure over time. Centers in the brain monitor resting blood pressure. When the pressure rises abnormally, they order slower, less forceful heart contractions. They also order smooth muscle in arterioles to relax. The result is **vasodilation**—an enlargement (dilation) of the vessel diameter. On the other hand, when the centers detect an abnormal *decrease* in blood pressure, they command the heart to beat faster and contract more forcefully. Neural signals also cause the smooth muscle

of arterioles to contract. The result is **vasoconstriction**, a narrowing of the vessel diameter. In some parts of the body arterioles have receptors for hormones that trigger vasoconstriction or vasodilation, thus helping to maintain blood pressure.

Carotid arteries in the neck, in the arch of the aorta, and elsewhere contain pressure sensors called baroreceptors. In what is called a **baroreceptor reflex**, the sensors monitor changes in mean arterial pressure ("mean" = the midpoint) and send signals to centers in the brain. The brain centers use this information to coordinate the rate and strength of heartbeats with changes in the diameter of arterioles and veins. The baroreceptor reflex thus helps keep blood pressure within normal limits in the face of sudden changes, such as when you leap up from a chair.

baroreceptor reflex Automatic response by sensors (in the carotid arteries) sensitive to changes in arterial blood pressure. Brain centers that receive the signals order the response.

capillary bed An interlacing network of capillaries.

carotid arteries Arteries that service the head and neck and have blood pressure sensors associated with them.

pulse Pressure surge that may be felt in arteries near the body surface when a heart ventricle contracts.

vasoconstriction Narrowing of a blood vessel's diameter.

vasodilation Enlargement of a blood vessel's diameter.

TAKE-HOME MESSAGE

HOW DOES A BLOOD VESSEL'S STRUCTURE SUIT ITS PARTICULAR FUNCTION IN THE CARDIOVASCULAR SYSTEM?

- Arteries have thick, elastic walls and are the main pipelines for oxygenated blood. Smooth muscle in arterioles allows them to dilate and constrict. Arterioles function as control points for blood flow and blood pressure.
- Capillaries are where substances diffuse between the blood and extracellular fluid in tissues.
- Blood moves back to the heart through venules and veins. Valves in veins prevent the backflow of blood due to gravity.

7.7 Capillaries: Where Substances Move between Blood and Tissues

- Blood entering the systemic circulation flows fast in the aorta, but has to slow in order for substances to move into and out of the bloodstream.

- Link to Diffusion 3.10

A vast network of capillaries brings blood close to nearly all body cells

Your body comes equipped with as many as 40 billion capillaries—each one so thin that it would take a hundred of them to equal the thickness of a human hair. And at least one of these tiny vessels is next to living cells in nearly all body tissues.

In addition to forming a vast network of vessels (Figure 7.16A), this branching system also affects the speed at which blood flows through it. Recall from Sections 7.5–7.6 that the flow is fastest in the aorta, quickly "loses steam" in the more numerous arterioles, and slows to a relative crawl in the narrow capillaries (Figure 7.16B). The flow of blood speeds up again as blood moves into veins for the return trip to the heart.

Why have such an extensive system of capillaries in which blood slows to a snail's pace? As you have read, capillaries are where all the substances that enter and leave cells are exchanged with the blood. Many of these exchanges occur by diffusion—but diffusion is a slow process that is not efficient over long distances. In a large, multicellular organism such as a human, having billions of narrow capillaries solves both these problems. There is a capillary close to nearly every cell, and in each one the blood is barely moving. As blood "creeps" along in capillaries, there is time for the necessary exchanges of fluid and solutes to take place. In fact, most solutes that enter and leave the bloodstream diffuse across capillary walls.

Some substances pass through "pores" in capillary walls

Some substances enter and leave capillaries by way of slitlike areas between the cells of capillary walls (Figure 7.16C). These "pores" are filled with water. They are passages for substances that cannot diffuse through the lipid bilayer of the cells that make up the capillary wall, but that *can* dissolve in water.

When the blood pressure inside a capillary is greater than pressure from the extracellular fluid outside, water and solutes may be forced out of the vessel—a type of fluid movement called "bulk flow" (Figure 7.17). Various factors affect this process, but on balance, a little more water leaves capillaries than enters them. You may remember that Section 7.1 mentioned a close association between the cardiovascular system and the lymphatic system. The lymphatic system, which consists of lymph vessels, lymph

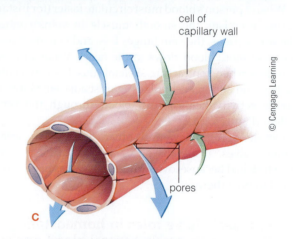

cell of capillary wall

pores

Figure 7.16 Capillaries deliver blood close to cells. A A resin cast showing a dense network of capillaries. **B** Red blood cells moving single file in capillaries. **C** How substances pass through slitlike pores in the wall of a capillary.

THINK OUTSIDE THE BOOK

In the condition called telangiectasia, capillaries, arterioles, or veins show up on the body surface as "spider veins." Most people have at least a few by the time they reach age 30. Research this topic to find out the most common causes of spider veins and recommended treatments for them.

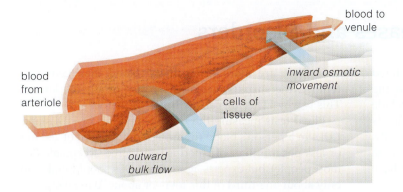

blood to venule

inward osmotic movement

cells of tissue

blood from arteriole

outward bulk flow

Figure 7.17 Fluid may move by bulk flow out of a capillary bed.
(© Cengage Learning)

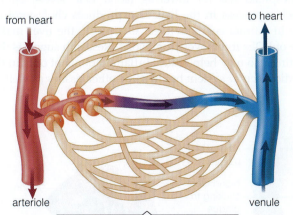

A Fully relaxed

precapillary sphincters

thoroughfare channel

capillaries

from heart

to heart

arteriole

venule

Fully relaxed arteriole and sphincter muscles allow maximum blood flow through the arterioles and capillary networks.

B Fully contracted

from heart

to heart

arteriole

venule

Fully contracted arteriole and sphincter muscles allow only a minimal amount of blood to flow through a vessel called a thoroughfare channel.

Figure 7.18 This diagram shows the general direction of blood flow through a capillary bed when vessel walls are fully relaxed (A) and fully contracted (B). A precapillary sphincter wraps around the base of each capillary. (© Cengage Learning)

nodes, and some other organs, receives fluid that leaves capillaries and returns it to the blood. This system also plays a major role in body defense, the subject of Chapter 9.

Overall, the movements of fluid and solutes into and out of capillaries help maintain blood pressure by adding water to, or subtracting it from, blood plasma. The fluid traffic also helps maintain the proper fluid balance between blood and surrounding tissues.

Blood in capillary beds flows onward to venules

Capillary beds are the "turnaround points" for blood in the cardiovascular system. They receive blood from arterioles, and after the blood flows through the bed it enters channels that converge into venules—the beginning of its return trip to the heart.

At the point where a capillary branches into the capillary bed, a wispy ring of smooth muscle wraps around it. This structure, a **precapillary sphincter**, regulates the flow of blood into the capillary. The smooth muscle is sensitive to chemical changes in the capillary bed. It can contract and prevent blood from entering the capillary, or it can relax and let blood flow in (Figure 7.18).

For example, if you sit quietly and listen to music, only about one-tenth of the capillaries in your skeletal muscles are open. But if you decide to get up and dance, precapillary sphincters will sense the demand for more blood flow to your muscles to deliver oxygen and carry away carbon dioxide. Many more of the sphincters will relax, allowing a rush of blood into the muscle tissue. The same mechanism brings blood to the surface of your skin when you blush or become flushed with heat.

> **precapillary sphincter**
> A ring of smooth muscle that regulates the flow of blood into a capillary.

TAKE-HOME MESSAGE

HOW DO SUBSTANCES MOVE BETWEEN BLOOD IN CAPILLARIES AND THE SURROUNDING EXTRACELLULAR FLUID?

- Substances enter and leave the blood in capillaries—and into and out of the extracellular fluid—by diffusion, through capillary pores, or by bulk flow.
- Movements of water and other substances into and out of capillaries help maintain blood pressure and the proper fluid balance between blood and tissues.

Heart disease is the leading cause of death in the United States. As we age, all of us are at increased risk of developing a cardiovascular disorder. Other major risk factors include a family history of heart trouble, high levels of blood lipids such as cholesterol and trans fats, hypertension, obesity, smoking, and lack of exercise. Interestingly, however, more than half of people who suffer heart attacks do not have any of these risk factors.

Scientists studying this puzzle have focused on inflammation, a defense response discussed in Chapter 9. Infections can trigger inflammation, which in turn causes the liver to make *C-reactive protein*, which also is implicated in heart disease. This link is why infection-related inflammation and C-reactive protein are listed in Table 7.2. Homocysteine, an amino acid, is released as certain proteins are broken down. Too much of it in the blood also may cause damage that is a first step in a major cardiovascular disorder, atherosclerosis.

Arteries can clog or weaken

In *arteriosclerosis*—hardening of the arteries—arteries become thicker and stiffen. In **atherosclerosis**, however, the condition gets worse as cholesterol and other lipids build up in the artery wall.

Having too many lipids in the blood is a major risk factor for atherosclerosis. This lipid overload may be due to a variety of factors, including personal genetics (family history) and a diet high in cholesterol and trans fat. In the blood, proteins called LDLs (*low-density lipoproteins*) bind cholesterol and other fats and carry them to body cells. Proteins called HDLs (*high-density lipoproteins*) also pick up cholesterol in the blood, but they carry it to the liver where it is processed. Eventually it moves into the intestine and is excreted in feces. Because HDLs help remove excess cholesterol from the body, they are often termed "good cholesterol."

If there are more LDLs in the blood than cells can remove, the surplus increases the risk of atherosclerosis. This is why LDLs are called "bad cholesterol."

Blood tests measure the relative amounts of HDLs and LDLs in a person's blood (in milligrams). A total of 200 mg or less per milliliter of blood is considered acceptable (for most people), but experts agree that LDLs should make up only about one-third of this total, or about 70 to 80 mg.

When LDLs infiltrate artery walls, cholesterol builds up there. Defensive cells called macrophages ("big eaters") also move into the wall and begin to remove LDLs by the process of phagocytosis mentioned in Section 3.11. Next, inflammation sets in and a fibrous net forms over a growing, artery-clogging mass called an atherosclerotic **plaque** (Figure 7.19B). Calcium deposits may enter the plaque and harden both it and the vessel wall.

Sometimes abnormal blood clots form at the site of a plaque. If a clot sticks to the plaque, it is called a *thrombus*. If it floats off into the bloodstream it becomes an *embolus*. A thrombus can grow big enough to completely block an artery. An embolus may dangerously clog a smaller vessel in the heart, lungs, or elsewhere.

Surgery may be the only answer for a severely blocked coronary artery. In a *coronary bypass*, a section of a large vessel taken from the chest is stitched to the aorta and to the coronary artery below the affected region (Figure 7.19C). In *laser angioplasty*, laser beams vaporize the plaques. In *balloon angioplasty*, a small balloon is inflated inside a blocked artery to flatten a plaque so there is more room in the artery. A small wire cylinder called a stent may be inserted to help keep the artery open.

"Plaque-busting" drugs called statins have come into widespread use for lowering blood LDL levels. Statins block a process in the liver that produces cholesterol for use in normal cell activities. As a result, the liver makes use of LDLs it removes from blood. Research suggests that statins also may help reduce inflammation, not only in the cardiovascular system but in some other tissues as well.

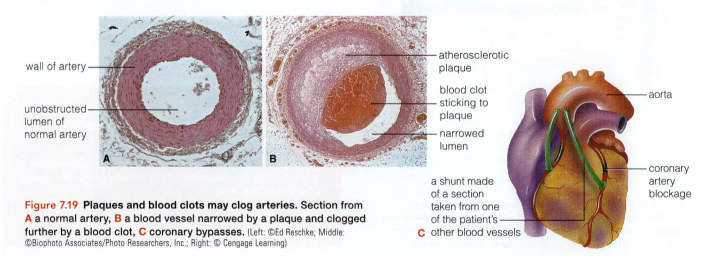

wall of artery

unobstructed lumen of normal artery

A

B

atherosclerotic plaque

blood clot sticking to plaque

narrowed lumen

aorta

coronary artery blockage

a shunt made of a section taken from one of the patient's

C other blood vessels

Figure 7.19 Plaques and blood clots may clog arteries. Section from A a normal artery, B a blood vessel narrowed by a plaque and clogged further by a blood clot, C coronary bypasses. (Left: ©Ed Reschke; Middle: ©Biophoto Associates/Photo Researchers, Inc.; Right: © Cengage Learning)

TABLE 7.2 Major Risk Factors for Cardiovascular Disease

1. Inherited predisposition
2. Elevated blood lipids (cholesterol, trans fats)
3. Hypertension
4. Obesity
5. Smoking
6. Lack of exercise
7. Age 50+
8. Inflammation due to infections
9. High blood levels of C-reactive protein
10. Elevated blood levels of homocysteine

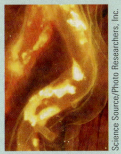

Plaques in heart arteries

Science Source/Photo Researchers, Inc.

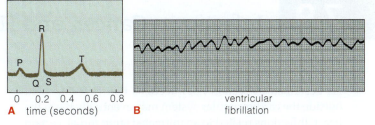

A time (seconds)

B ventricular fibrillation

Figure 7.20 An ECG tracing can reveal abnormal heart activity. A ECG of a normal heartbeat. The P wave is generated by electrical signals from the SA node that stimulate contraction of the atria. As the stimulus moves over the ventricles, it is recorded as the QRS wave complex. The T wave marks the brief period when the ventricles are resting. **B** A recording of ventricular fibrillation.
(© Cengage Learning)

Disease, an injury, or an inborn defect can weaken an artery so that part of its wall balloons outward. This pouch-like weak spot is called an **aneurysm** (ANN-yoo-rizm). Aneurysms can develop in various parts of the cardiovascular system, including the aorta and vessels in the brain and abdomen. If an aneurysm bursts, it can cause serious and even fatal blood loss. A minor aneurysm may not present any immediate worry, but in the brain, especially, an aneurysm is potentially so dangerous that it requires immediate medical treatment.

Heart damage can lead to heart attack and heart failure

A **heart attack**—medically, a *myocardial infarction* (MI)—is damage to or death of heart muscle due to reduced blood flow to the affected region. Usually the attack happens when blood-starved heart muscle no longer receives enough oxygen. Warning signs include pain or a sensation of squeezing behind the breastbone, pain or numbness radiating down the left arm, sweating, and nausea. Women more often experience neck and back pain, fatigue, vague indigestion, a fast heartbeat, shortness of breath, and low blood pressure. Risk factors include atherosclerosis, a circulating embolus, and high blood pressure. In **heart failure** (HF), the heart is weakened and cannot pump enough blood to meet the body's needs. Even basic exertion such as walking can become difficult.

Arrhythmias are abnormal heart rhythms

An electrocardiogram, or ECG, is a recording of the electrical activity of the cardiac cycle (Figure 7.20A). ECGs reveal **arrhythmias**, or irregular heart rhythms. Some arrhythmias are abnormal, others are not. For example, endurance athletes may have a below-average resting cardiac rate, or *bradycardia*, which is an adaptation to regular strenuous exercise. A cardiac rate above 100 beats per minute, called *tachycardia*, occurs normally during exercise or stressful situations. Serious tachycardia can be triggered by drugs (including caffeine, nicotine, alcohol, and cocaine) and excessive thyroid hormones, among other factors.

In the arrhythmia called **atrial fibrillation (AFib)**, rapid, disorganized electrical impulses hamper contractions of the heart atria, reducing normal blood flow into the ventricles. **Ventricular fibrillation** is by far the most dangerous arrythmia. In parts of the ventricles, the cardiac muscle contracts haphazardly, so blood isn't pumped normally. This is what happens in sudden cardiac arrest (SCA), as described in the chapter introduction. Ventricular fibrillation is a major medical emergency. With luck, a strong electrical jolt to the patient's heart from an AED, or the use of defibrillating drugs, can restore a normal rhythm before the damage is too serious.

A heart-healthy lifestyle may help prevent cardiovascular disease

Given an unhealthy diet and sedentary lifestyle, the early signs of atherosclerosis begin to show up in the arteries of children as young as age 10 and steadily worsen as the years pass. The best strategy for avoiding cardiovascular disease is to limit your intake of foods rich in cholesterol and trans fats, get plenty of exercise, and avoid smoking. Regular exercise and moderate fat intake also help keep weight under control. Exercise is a great stress reliever and helps keep muscles and bones fit and strong. Tobacco smoke harms just about every body system, so not smoking helps your whole body stay as healthy as possible.

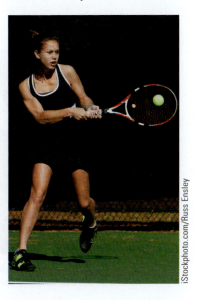

iStockphoto.com/Russ Ensley

Infections may seriously damage the heart

As described in Section 7.8, infections that first take hold outside the cardiovascular system may eventually harm the heart. Infections related to an untreated strep throat, certain dental procedures, or IV drug abuse are in this category.

Strep infections are caused by strains of *Streptococcus* bacteria (Figure 7.21). If the illness isn't treated with an antibiotic, it may lead to **rheumatic fever**. In this disorder, the body produces defensive antibodies that attack the invading bacteria—but they also mistakenly attack heart valves. Although in affluent countries most people who develop a strep infection get treatment, rheumatic fever still is the most common cause of heart valve disease. It is an example of an autoimmune disorder, a topic we will discuss in Chapter 9.

Microbes that enter the bloodstream during a dental procedure or on a contaminated IV needle may attack heart valves directly. This condition is called **endocarditis** ("inside the heart"). People who have an existing valve problem due to aging or some other heart disorder often are advised to take an antibiotic before having dental work. Endocarditis is a major hazard for IV drug users. It can rapidly destroy infected valves and cause sudden heart failure.

Heart problems also can be a complication of **Lyme disease**, which is caused by the bacterium *Borrelia burgdorferi* and spread by ticks. At first the body responds to a Lyme infection with a "bull's-eye" rash (Figure 7.22). Later the joints may become inflamed, as may the heart muscle (the myocardium). Heart inflammation, called **myocarditis**, produces an irregular heart rhythm that manifests as dizzy spells and other other symptoms. Measles caused by the rubella virus in unvaccinated people can also damage the heart muscle.

Alcohol abuse and recreational drugs also may cause heart inflammation. When someone dies of a cocaine overdose, an autopsy often reveals myocarditis. Cocaine, amphetamines, and habitual, heavy alcohol use all can cause cardiomyopathy, or weakness of the heart muscle, that in turn may lead to heart failure.

Is there such a thing as heart cancer?

Although the reason is a mystery, cancer almost never starts in the heart muscle or blood vessels. More often, a cancer that begins elsewhere in the body, such as the skin cancer malignant melanoma, spreads to the heart. Even more often, the heart or vessels are damaged by cancer treatments such as radiation or chemotherapy.

Inborn heart defects are fairly common

You may have heard of "blue babies," infants born with a hole in some part of the heart wall, so that the heart doesn't pump blood efficiently. In fact, thousands of babies enter the world each year with some type of heart defect. Depending on the problem, one or more surgeries may be required to repair it.

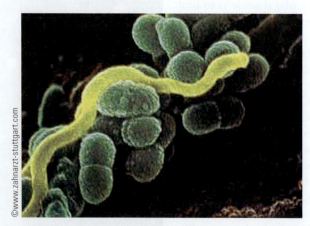

Figure 7.21 *Streptococcus* bacteria cause strep infections. In this image the bacteria are colored *green*.

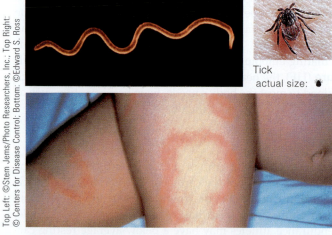

Tick actual size:

Top Left: ©Stem Jems/Photo Researchers, Inc.; Top Right: © Centers for Disease Control; Bottom: ©Edward S. Ross

Figure 7.22 **Heart damage may be a complication of Lyme disease.** *Borrelia burgdorferi (above left)* is the Lyme bacterium. The lower photograph shows the bull's-eye rash that is a key symptom of Lyme disease, now the most common tick-borne disease in the United States.

CONNECTIONS

 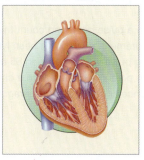

The Cardiovascular System and Blood

The heart pumps blood into blood vessels that transport blood throughout the body. In this way the system delivers blood's cargoes to body cells and carries away potentially toxic wastes and other unneeded materials.

Blood pressure generated by heart contractions helps keep blood flowing through the cardiovascular system.

Mechanisms that widen or narrow the diameter of arterioles and capillaries allow adjustments in blood flow to different body regions as conditions warrant.

As described in Chapter 8, blood is the medium that transports nutrients, oxygen, hormones, cell wastes, and other substances. It also carries and distributes a great deal of body heat.

Blood's ability to clot allows the body to sustain minor wounds without a serious loss of blood.

Integumentary system
Adjustments to blood flow at the skin's surface help regulate body temperature. Blood-clotting mechanisms help repair skin injuries.

Skeletal system
Stem cells in bone marrow produce blood cells. Circulating blood delivers calcium and phosphate used to form bone tissue.

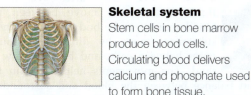

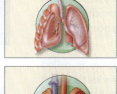

Digestive system
The bloodstream circulates nutrients from food digestion to cells. The liver receives and processes certain nutrients via the hepatic portal system.

Muscular system
Circulating blood distributes heat produced by active skeletal muscles. Contraction of leg muscles helps return venous blood to the heart.

Immunity and the lymphatic system
The lymphatic system picks up fluid lost from capillaries and returns it to the bloodstream, helping to maintain normal blood pressure. Circulating blood also carries many types of white blood cells, which function in defense.

Respiratory system
Blood pumped by the heart picks up inhaled oxygen from the lungs and delivers carbon dioxide to the lungs to be exhaled.

Urinary system
The kidneys filter impurities and other unneeded substances from blood and form urine that removes them from the body. The kidney hormone erythropoietin stimulates the formation of red blood cells.

Nervous system
Centers in the brain and spinal cord adjust the rate and strength of heart contractions and help maintain proper blood pressure by adjusting the diameter of arterioles.

Sensory systems
Sensors in the carotid arteries help monitor blood pressure. Sensory perceptions related to mental or physiological states may trigger changes in local blood flow (as in blushing, sexual arousal).

Endocrine system
Nearly all hormones reach their targets via the bloodstream. Certain cells in the heart atria release a hormone (ANP) that helps regulate blood pressure.

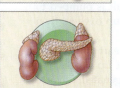 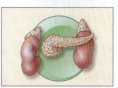

Reproductive system
Reproductive hormones, including estrogens and testosterone, travel in the bloodstream. Arterioles in organs of sexual intercourse dilate at times of arousal. Blood vessels of the placenta help maintain homeostasis in a developing fetus.

EXPLORE
ON YOUR OWN

©Lori Adamski Peek/Stone/Getty Images

As described in Section 7.6, a pulse is the pressure wave created during each cardiac cycle as the body's elastic arteries expand and then recoil. Common pulse points—places where an artery lies close to the body surface—include the radial artery at the inside of the wrist and the carotid artery at the front of the neck. Monitoring your pulse is an easy way to observe how a change in your posture or activity affects your heart rate.

To take your pulse, simply press your fingers on a pulse point and count the number of "beats" during one minute. For this exercise, take your first measurement after you've been lying down for a few minutes. If you are a healthy adult, it's likely that your resting pulse will be between 65 and 70 beats per minute. Now sit up and take your pulse again. Did the change in posture correlate with a change in your pulse? Now run in place for 30 seconds and take your pulse rate once again. In a short paragraph, describe what changes in your heart's activity led to the pulse differences.

SUMMARY

Section 7.1 The cardiovascular system consists of the heart and blood vessels including arteries, arterioles, capillaries, venules, and veins. The system helps maintain homeostasis by providing rapid internal transport of substances to and from cells.

Section 7.2 The heart muscle is called the myocardium. A septum divides the heart into two halves, each with two chambers, an atrium and a ventricle. Valves in each half help control the direction of blood flow. These include aortic, pulmonary, and atrioventricular valves. Coronary arteries provide much of the heart's blood supply. They branch off the aorta, which carries oxygenated blood away from the heart.

Blood is pumped each time the heart beats, in a cardiac cycle of contraction and relaxation. Systole, the contraction phase, alternates with the relaxation phase, called diastole.

Section 7.3 The partition between the heart's two halves separates the blood flow into two circuits, one pulmonary and the other systemic.

In the pulmonary circuit, deoxygenated blood in the heart's right half is pumped to capillary beds in the lungs. The blood picks up oxygen, then flows to the heart's left atrium.

In the systemic circuit, the left half of the heart pumps oxygenated blood to body tissues. There, cells take up oxygen and release carbon dioxide. The blood, now deoxygenated, flows to the heart's right atrium.

Section 7.4 Electrical impulses stimulate heart contractions via the heart's cardiac conduction system. In the right atrium, a sinoatrial node—the cardiac pacemaker—generates the impulses and establishes a regular heartbeat. Signals from the SA node

pass to the atrioventricular node, a way station for stimulation that triggers contraction of the ventricles. The nervous system can adjust the rate and strength of heart contractions.

Section 7.5 Blood pressure is the fluid pressure blood exerts against vessel walls. It is highest in the aorta, which receives blood pumped by the left ventricle, and drops along the systemic circuit.

Section 7.6 Arteries are strong, elastic pressure reservoirs. They smooth out pressure changes resulting from heartbeats and so smooth out blood flow. When a ventricle contracts, it causes a pressure surge, or pulse, in large arteries.

Arterioles are control points for distributing different volumes of blood to different regions.

Capillary beds are diffusion zones where blood and extracellular fluid exchange substances.

Venules overlap capillaries and veins somewhat in function. Some solutes diffuse across their walls.

Veins are blood reservoirs that can be tapped to adjust the volume of flow back to the heart. Valves in some veins, in the limbs, prevent blood returning to the heart from flowing backward due to gravity.

Blood vessels help control blood pressure. Arterioles dilate when centers in the brain detect an abnormal rise in blood pressure. If blood pressure falls below a set point, the centers trigger vasoconstriction of arterioles. Baroreceptors in carotid arteries provide short-term blood pressure control by way of signals that adjust the pressure when sudden changes occur.

Section 7.7 Capillaries are where fluids and solutes move between the bloodstream and body cells. These substances move by diffusion, through pores between cells, and by bulk flow of fluid. The movements help maintain the proper fluid balance between the blood and surrounding tissues, and also help maintain proper blood volume.

REVIEW QUESTIONS

1. List the functions of the cardiovascular system.

2. Define a "heartbeat," giving the sequence of events that make it up.

3. What is the difference between the systemic and pulmonary circuits?

4. Explain the function of (a) the sinoatrial node and (b) the atrioventricular node.

5. State the main function of blood capillaries. Name the main ways substances cross the walls of capillaries.

6. State the main functions of venules and veins. What forces work together in returning venous blood to the heart?

7. Label the heart's main parts in the diagram below.

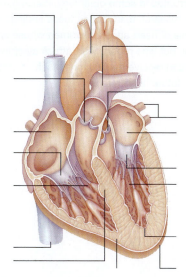

© Cengage Learning

SELF-QUIZ *Answers in Appendix V*

1. Cells obtain nutrients from and deposit waste into _____.
   a. blood   c. each other
   b. lymph vessels   d. both a and b

2. Fill in the blanks: The contraction phase of the heartbeat is _____; the relaxation phase is _____.

3. In the pulmonary circuit, the heart's _____ half pumps _____ blood to capillary beds inside the lungs; then _____ blood flows to the heart.
   a. left; deoxygenated; oxygenated
   b. right; deoxygenated; oxygenated
   c. left; oxygenated; deoxygenated
   d. right; oxygenated; deoxygenated

4. In the systemic circuit, the heart's _____ half pumps _____ blood to all body regions; then _____ blood flows to the heart.
   a. left; deoxygenated; oxygenated
   b. right; deoxygenated; oxygenated
   c. left; oxygenated; deoxygenated
   d. right; oxygenated; deoxygenated

5. After you eat, blood passing through the GI tract travels through the _____ to a capillary bed in the _____.
   a. aorta; liver
   b. hepatic portal vein; liver
   c. hepatic vein; spleen
   d. renal arteries; kidneys

6. The cardiac pacemaker _____.
   a. sets the normal rate of heartbeat
   b. is the same as the AV node
   c. establishes resting blood pressure
   d. all of these are correct

7. Blood pressure is _____.
   a. generated by heart contractions
   b. exerted against vessel walls
   c. healthiest when it is under 120/80
   d. all of these are correct

8. Blood pressure is highest in _____ and lowest in _____.
   a. arteries; veins   c. arteries; ventricles
   b. arteries; relaxed atria d. arterioles; veins

9. _____ contraction drives blood through the systemic and pulmonary circuits; outside the heart, blood pressure is highest in the _____.
   a. Atrial; ventricles   c. Ventricular; arteries
   b. Atrial; atria   d. Ventricular; aorta

10. Match the type of blood vessel with its major function.
    _____ arteries   a. diffusion
    _____ arterioles   b. control of blood distribution
    _____ capillaries   c. transport, blood volume
    _____ veins   reservoirs
                          d. blood transport and pressure
                             regulators

11. Match these three circulation components with their descriptions.

    _____ capillary beds   a. two atria, two ventricles
    _____ heart chambers   b. driving force for blood
    _____ heart contractions c. zones of diffusion

CRITICAL THINKING

1. A patient suffering from hypertension may receive drugs that decrease the heart's output, dilate arterioles, or increase urine production. In each case, how would the drug treatment help relieve hypertension?

2. Heavy smokers often develop abnormally high blood pressure. The nicotine in tobacco is a potent vasoconstrictor. Explain the connection between these two facts, including what kind of blood vessels are likely affected.

3. Before antibiotics were available, it wasn't uncommon for people in the United States (and elsewhere) to develop rheumatic fever. The infection can trigger an inflammation that ultimately damages valves in the heart. How must this disease affect the heart's functioning? What kinds of symptoms would arise as a result?

4. Several years ago the deaths of several airline travelers led to warnings about "economy-class syndrome." The idea is that economy-class passengers don't have as much leg room as passengers in more expensive seats, so they are more likely to sit essentially motionless for long periods on flights—conditions that may allow blood to pool and clots to form in the legs. This condition is called deep-vein thrombosis, or DVT. Given what you know about blood flow in the veins, explain why periodically getting up and moving around in the plane's cabin during a long flight may lower the risk that a clot will form.

Your Future

©Biophoto Associates/ Photo Researchers, Inc.

Section 7.8 noted that the signs of heart attack differ in males and females. New research suggests that gender differences many also apply to diagnosing heart attacks in the first place. Researchers at the Cardiac and Vascular Institute at New York University/Langone studied a group of fifty women who all had suffered heart attacks, even though 38 percent of the patients did not have coronary arteries seriously clogged by atherosclerotic plaques. Because imaging methods (called angiography) didn't reveal obvious clogs, affected women were initially told they hadn't had a heart attack at all. The NYU study discovered that the heart attacks were real and had been triggered when a relatively small plaque had suddenly ruptured or become disrupted in some other way, causing a major blockage. The research team hopes these findings will spur earlier diagnosis of heart attack in women whose angiograms appear normal, but who report heart attack symptoms. Then such patients can receive medications, such as statins, that can help avert future heart attacks.

BLOOD

This chapter expands on your study of the cardiovascular system in Chapter 7. Blood cells arise in red bone marrow (5.2). Red blood cells contain the oxygen-carrying protein hemoglobin (2.12).

This chapter's discussion of blood typing shows a key function of recognition proteins embedded in cell plasma membranes (3.4).

Section 8.7 on blood clotting provides good examples of how enzymes catalyze chemical reactions that are vital to life (2.8).

KEY CONCEPTS

Components and Functions of Blood

Blood consists of plasma, red blood cells, white blood cells, and platelets. Red blood cells carry O_2 and CO_2, white blood cells function in defense, and platelets help clot blood. Circulating blood helps maintain proper pH and body temperature. Sections 8.1–8.3

Blood Types

Surface markers on red blood cells establish each person's blood type. Sections 8.4–8.5

Blood Clotting

Mechanisms that clot blood help prevent blood loss. Section 8.7

Blood Disorders Section 8.8

From top: National Cancer Institute/Photo Researchers, Inc.; © Lester V. Bergman & Associates, Inc./Project Masters, Inc.; © Professor Pietro M. Motta/Photo Researchers, Inc.; Dr. Stanley Flegler/Visuals Unlimited

Every two seconds, someone in the United States needs a blood transfusion. Worldwide, hospital patients require almost 80 million units of donated blood. Most recipients are surgery patients and people who have lost blood due to a serious injury.

A blood bank has to ensure that donated blood doesn't carry disease-causing agents such as hepatitis viruses and HIV, the human immunodeficiency virus that causes AIDS. Donated blood also must be chemically analyzed to determine its "type"—a gene-based characteristic that varies from person to person. The match between donated blood and a recipient's own blood must be close enough to keep the recipient's immune system from attacking the replacement blood cells.

Researchers have been trying to develop a blood substitute that can be used in emergencies when it's not feasible to match blood types. As you will learn in this chapter, however, blood is extremely complex. The more scientists try to develop blood substitutes, the more we understand just how remarkable a substance courses through our arteries and veins.

Gustoimages/Photo Researchers, Inc.

Homeostasis Preview

Blood transports oxygen, nutrients, and other materials to and from cells. It also carries ions with roles in maintaining an appropriate pH in extracellular fluid. Blood-borne proteins have many physiological roles. Some function in blood clotting, which prevents a dangerous loss of blood when minor injuries occur.

Blood: Plasma, Blood Cells, and Platelets

- Human blood is a sticky fluid that consists of water, blood cells, and other substances.

- Links to Properties of water 2.5, Proteins 2.11, Osmosis 3.10, Skeleton 5.2

The old saying is true—**blood** really is thicker than water. This unusual fluid consists of plasma, blood cells, and cell fragments called platelets. If you are an adult woman of average size, your body has about 4 to 5 liters of blood. Males have slightly more. In all, blood amounts to about 6 to 8 percent of your body weight.

Plasma is the fluid part of blood

If you whirl a prepared blood sample in a centrifuge, the test tube's contents should look like what you see in Figure 8.1. About 55 percent of whole blood is **plasma**. Plasma is mostly water. It transports blood cells and fragments called platelets, and over a hundred other substances. Most of these "substances" are different plasma proteins, which have a variety of functions.

Plasma proteins determine blood's fluid volume—how much of it is water. Two-thirds of plasma proteins are albumin molecules made in the liver. Because there is so much of it—that is, because its concentration is so high—albumin has a major influence on the osmotic movement of water into and out of blood. Albumin also carries many substances in blood, from wastes to therapeutic drugs.

Other plasma proteins include certain hormones and proteins involved in immunity and blood clotting. Lipoproteins carry lipids, and still other plasma proteins transport fat-soluble vitamins. Plasma proteins are in high demand for various therapeutic uses, and in many countries people who wish to donate plasma can do so at government-regulated collection centers.

Plasma also contains ions, glucose and other simple sugars, amino acids, various communication molecules, and dissolved gases—mostly oxygen, carbon dioxide, and nitrogen. The ions (such as Na^+, Cl^-, H^+, and K^+) help maintain the volume and pH of extracellular fluid.

The blood cells in your test tube arose from stem cells in red bone marrow. Remember from Chapter 4 that a stem cell is like a "blank slate"—it stays unspecialized and retains the ability to divide. Some of the daughter cells, however, differentiate—they become specialized to carry out particular functions. As Figure 8.2 shows, the formation of specialized blood cells begins with pluripotent ("many powers") stem cells from which two lines of precursor cells arise. The descendants of *lymphoid* stem cells circulate mainly in the lymphatic system. *Myeloid* ("from bone marrow") stem cells give rise to the other types of circulating blood cells.

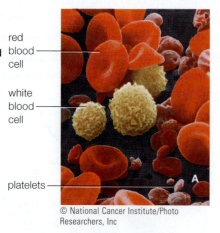

Figure 8.1 Blood consists of cells, platelets, and plasma. In the micrograph (A) the *red* cells are red blood cells. Platelets are *pink*. The wrinkled *gold balls* are white blood cells. The plasma portion (B) makes up a little over half of the volume of blood.

red blood cell

white blood cell

platelets

© National Cancer Institute/Photo Researchers, Inc

Components	Relative Amounts	Functions
Plasma Portion (*50%–60% of total volume*):		
1. Water	91%–92% of plasma volume	Solvent
2. Plasma proteins (albumin, globulins, fibrinogen, etc.)	7%–8%	Defense, clotting, lipid transport, roles in extracellular fluid volume, etc.
3. Ions, sugars, lipids, amino acids, hormones, vitamins, dissolved gases	1%–2%	Roles in extracellular fluid volume, pH, etc.
Cellular Portion (*40%–50% of total volume*):		
1. White blood cells:		
Neutrophils	3,000–6,750	Phagocytosis during inflammation
Lymphocytes	1,000–2,700	Immune responses
Monocytes (macrophages)	150–720	Phagocytosis in all defense responses
Eosinophils	100–360	Defense against parasitic worms
Basophils	25–90	Secrete substances for inflammatory response and for fat removal from blood
2. Platelets	250,000–300,000	Roles in clotting
3. Red blood cells	4,800,000–5,400,000 per microliter	Oxygen, carbon dioxide transport

B

© Cengage Learning

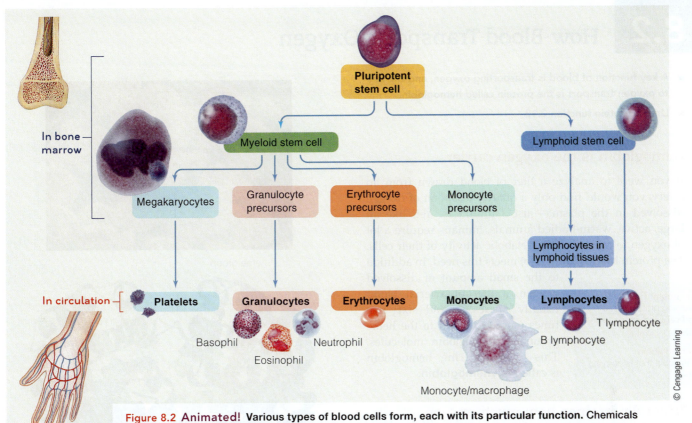

Figure 8.2 Animated! Various types of blood cells form, each with its particular function. Chemicals called growth factors stimulate the growth and specialization of the different subgroups of blood cells.

Red blood cells carry oxygen and CO$_2$

About 45 percent of whole blood—the bottom portion in the test tube in Figure 8.1—consists of **erythrocytes**, or **red blood cells**. Each red blood cell is a biconcave disk, like a thick pancake with a dimple on each side. The cell's red color comes from the iron-containing protein hemoglobin. Hemoglobin transports oxygen that body cells need for aerobic respiration. Red blood cells also carry away some carbon dioxide wastes.

White blood cells defend and clean up

Leukocytes, or **white blood cells**, make up a tiny fraction of whole blood. Even so, they play crucial roles in body defense. Some remove dead or worn-out cells, or material identified as foreign to the body. Others target or destroy disease agents such as bacteria or viruses. Most white blood cells go to work after they squeeze out of blood vessels and enter tissues. How many are in the body at any one time varies, depending on your activity level and whether you are healthy or fighting an infection.

In different types of white blood cells, the nucleus varies in its size and shape, and there are other differences as well. For example, some contain granules that become visible when the cells are stained for viewing under a microscope. This group (called granulocytes) includes neutrophils, eosinophils, and basophils. All have roles in body defenses that you will read more about in Chapter 9.

Other leukocytes don't have visible granules in their cytoplasm (and so are agranulocytes). Those known as

monocytes develop into macrophages, "big eaters" that enter tissues and engulf and destroy invading microbes and debris. Another type, lymphocytes, operates in immune responses discussed in Chapter 9. Some types of white blood cells may live for years, but most types live for only a few days or, during a major infection, perhaps a few hours.

Platelets help clot blood

Some stem cells in bone marrow develop into "giant" cells called megakaryocytes ("mega" = large). These cells shed bits of cytoplasm enclosed in a plasma membrane. The fragments, known as **platelets**, last only about a week, but millions are always circulating in our blood. Platelets release substances that begin the process of blood clotting described in Section 8.7.

blood Fluid connective tissue consisting of plasma, blood cells, and platelets.

erythrocyte A red blood cell.

leukocyte A white blood cell. Two subgroups of white blood cells, the granulocytes and agranulocytes, give rise to various types of white blood cells involved in body defenses.

plasma The fluid portion of whole blood.

platelet A cell fragment that produces some of the substances required for blood clotting.

8.2 How Blood Transports Oxygen

- A key function of blood is transporting oxygen, and the key to oxygen transport is the protein called hemoglobin.
- Link to Protein function 2.12

Hemoglobin is the oxygen carrier

If you were to analyze a liter of blood drawn from an artery, you would find only a quarter teaspoon of oxygen dissolved in the plasma—just 3 milliliters. Yet, like all large, active, warm-bodied animals, humans require a lot of oxygen to maintain the metabolic activity of their cells. The protein **hemoglobin** (Hb) meets this need. In addition to the small amount of dissolved oxygen, a liter of arterial blood usually carries around sixty-five times more O_2 bound to the heme groups of hemoglobin molecules. This oxygen-bearing hemoglobin is called **oxyhemoglobin**.

> **hemoglobin** The iron-containing protein in red blood cells that binds to oxygen.
>
> **oxyhemoglobin** Hemoglobin that is carrying oxygen.

What determines how much oxygen hemoglobin can carry?

As conditions change in different tissues and organs, so does the tendency of hemoglobin to bind with and hold on to oxygen. Several factors influence this process. The most important factor is how much oxygen is present relative to the amount of carbon dioxide. Other factors are the temperature and acidity of tissues. Hemoglobin is most likely to bind oxygen in places where blood plasma contains a relatively large amount of oxygen, where the temperature is relatively cool, and where the pH is roughly neutral. This is exactly the environment in our lungs, where the blood must take on oxygen. By contrast, metabolic activity in cells *uses* oxygen. It also increases both the temperature and the acidity (lowers the pH) of tissues. Under those conditions, the oxyhemoglobin of red blood cells arriving in tissue capillaries tends to release oxygen, which then can enter cells. We can summarize these events this way:

The protein portion of hemoglobin also carries some of the carbon dioxide wastes that cells produce, along with hydrogen ions (H^+) that affect the pH of body fluids. You'll read more about hemoglobin in Chapter 10, where we consider the many interacting elements that enable the respiratory system to transport gases efficiently to and from body cells.

You can see the structure of a hemoglobin molecule in Figure 8.3. Notice that it has two parts: the protein globin and heme groups that contain iron. Globin is built of four linked polypeptide chains, and each chain is associated

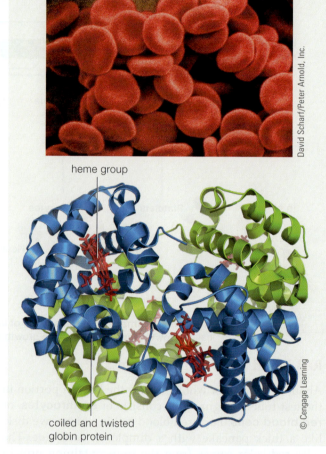

Figure 8.3 **Animated!** The iron in hemoglobin binds oxygen. This diagram represents hemoglobin, which is a globular protein that has four iron-containing heme groups. Oxygen binds to the iron in heme groups, which is one reason why humans require iron as a mineral nutrient.

with a heme group. It is the iron molecule at the center of each heme group that binds oxygen. Therefore, each hemoglobin molecule can carry four oxygen atoms.

Oxygen in the lungs diffuses into the blood plasma and then into individual red blood cells. There it binds with the iron in hemoglobin. This oxyhemoglobin is deep red. Hemoglobin that is depleted of oxygen looks purplish, especially when it is observed through skin and the walls of blood vessels.

> **TAKE-HOME MESSAGE**
>
> **HOW DOES BLOOD TRANSPORT OXYGEN?**
>
> - Hemoglobin in red blood cells carries oxygen.
> - Oxygen binds to iron in heme groups in each hemoglobin molecule.
> - The relative amounts of oxygen and carbon dioxide present in blood and the temperature and acidity of tissues affect how much oxygen hemoglobin binds—and therefore the amount of oxygen available to tissues.

8.3 Making New Red Blood Cells

■ Red blood cells do not live long. In response to hormones, stem cells in bone marrow constantly produce new ones.

Each second, about 3 million new red blood cells enter your bloodstream. Each one gradually loses its nucleus and other organelles. Such structures are unnecessary because red blood cells do not divide or make new proteins.

Red blood cells have enough enzymes and other proteins to function for about 120 days. As they near the end of their life, die, or become damaged or abnormal, phagocytes called macrophages remove them from the blood. Much of this cleanup occurs in the spleen, which is located in the upper left abdomen. As a macrophage dismantles a hemoglobin molecule, amino acids from its proteins return to the bloodstream and the iron in its heme groups returns to red bone marrow, where it may be recycled in new red blood cells. The rest of the heme group is converted to the orangish pigment bilirubin. Liver cells take up this pigment, which is mixed with bile that is released into the small intestine during digestion.

Steady replacements from stem cells in bone marrow keep a person's red blood cell count fairly constant over time. A **Complete Blood Count (CBC)** is a tally of the number of red blood cells, white blood cells, and platelets in a microliter of blood. On average, an adult male's red blood cell count is around 5.4 million. In an adult female the count averages about 4.8 million red blood cells.

Having a stable red blood cell count is important for homeostasis, because body cells need a reliable supply of oxygen. Your kidneys make erythropoietin (EPO).

This hormone stimulates the production of new red blood cells when they are needed.

The process relies on a negative feedback loop (Figure 8.4). In this loop, the kidneys monitor the level of oxygen in your blood. When it falls below a set point, kidney cells detect the change and soon release EPO. It stimulates stem cells in bone marrow to produce more red blood cells. As new red blood cells enter your bloodstream, the blood can carry more oxygen and the oxygen level rises in your blood and tissues. This information feeds back to the kidneys. They make less erythropoietin, and production of red blood cells in bone marrow drops.

In "blood doping," some of an athlete's blood is withdrawn and stored. Erythropoietin then stimulates the production of replacement red blood cells. The stored blood is reinjected several days prior to an athletic event, so that the athlete has more than the normal number of red blood cells to carry oxygen to body muscles—and an unethical competitive advantage. Some cyclists, runners, and other "distance" athletes have used lab-made EPO, even though it is a banned performance-enhancing drug. Better drug testing is helping to curb this practice.

> **Complete Blood Count (CBC)** The number of red blood cells, white blood cells, and platelets in a microliter of blood.

TAKE-HOME MESSAGE

HOW DOES THE BODY MAKE NEW RED BLOOD CELLS?

• As needed, the kidneys release erythropoietin, a hormone that stimulates the production of new red blood cells by stem cells in bone marrow.

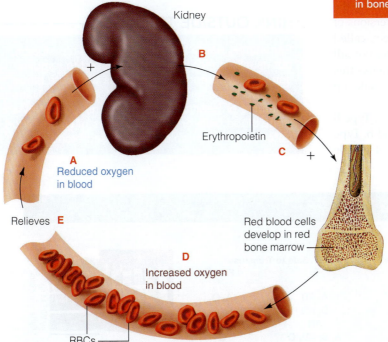

A The kidneys detect reduced O_2 in the blood.

B When less O_2 is delivered to the kidneys, they secrete the hormone erythropoietin into the blood.

C Erythropoietin stimulates production of red blood cells in bone marrow.

D The additional circulating RBCs increase O_2 carried in blood.

E The increased O_2 relieves the initial stimulus that triggered erythropoietin secretion.

Figure 8.4 **A negative feedback loop helps maintain a normal red blood cell count.** (© Cengage Learning)

8.4 | Different Red Blood Cells

- The different human blood types are due to variations in the surface markers on red blood cells.
- Link to Plasma membrane 3.4

Each of your body cells has markers on its surface that mark the cell as "self." Your genes have determined the chemical characteristics of these self markers, which vary from person to person. The variations are medically important because the markers on cells and substances that are *not* part of an individual's own body are antigens. An *antigen* is a chemical characteristic of a cell, particle, or substance that causes the immune system to mount an immune response. Defensive proteins called *antibodies* identify and attach to antigens in a process that is a major topic of Chapter 9. For now, it's important to know that for each type of antigen, the body makes a specific type of antibody that can bind to it.

To date, biologists have identified at least thirty common self markers on human red blood cells, and many more rare ones. Because each kind of marker can have several forms, the different forms are often called "blood groups." Two of them, the Rh blood group and the ABO blood group, are extremely important in situations where the blood of two people mixes. We will consider the Rh blood group in Section 8.5. For now, let's look more closely at the ABO blood group, which is a vital consideration in blood transfusions.

Self markers on red blood cells include the ABO group of blood types

One of our genes carries the instructions for building the ABO self markers on red blood cells. Different versions of this gene carry instructions for different markers, called type A and type B. A third version of the gene does not call for a marker, and red blood cells of someone who has this gene are dubbed type O. Together, these markers make up the ABO blood group (Table 8.1).

In type A blood, red blood cells bear A markers. Type B blood has B markers, and type AB has both A and B. Type AB blood is quite rare, but a large percentage of people have type O red blood cells—they have neither A nor B markers. Depending on your ABO blood type, your blood plasma also will contain antibodies to other blood types, even if you have never been exposed to them. As you will read shortly, a severe immune response takes place when incompatible blood types are mixed. This is why donated blood must undergo the chemical analysis called **ABO blood typing**.

Mixing incompatible blood types can cause the clumping called agglutination

As you can see in Table 8.1, if you are type A, your body does not have antibodies against A markers but does have them against B markers. If you are type B, you don't have antibodies against B markers, but you do have antibodies against A markers. If you are type AB, you do not have antibodies against either form of the marker. If you are type O, however, you have antibodies against *both* forms of the marker, so you can only receive blood from another type O individual.

In theory, type O people are "universal donors," because they have neither A nor B antigens, and—again, only in theory—type AB people are "universal recipients." In fact, however, as already noted, there are *many* markers associated with our red blood cells, and any of them can trigger a defense response in which the recipient's antibodies attack the donor's blood cells. This defense response is called **agglutination** (Figure 8.5). When the mixing of incompatible blood causes agglutination, antibodies act against the "foreign" cells and cause them to clump. The clumps

THINK OUTSIDE THE BOOK

Sometimes whole blood is used for a transfusion, but usually donor blood is processed to remove some of its components, such as the white blood cells. Do some research on this topic at the National Institutes of Health website (www.nih.gov). Why are white blood cells or other elements of whole blood potentially undesirable in transfused blood?

TABLE 8.1 Animated! ABO Blood Types

Blood Type	Antigens on Plasma Membranes of RBCs	Antibodies in Blood	Safe to Transfuse To	Safe to Transfuse From
A	A	Anti-B	A, AB	A, O
B	B	Anti-A	B, AB	B, O
AB	A + B	none	AB	A, B, AB, O
O	—	Anti-A, Anti-B	A, B, AB, O	O

Tan Wei Ming/Shutterstock.com

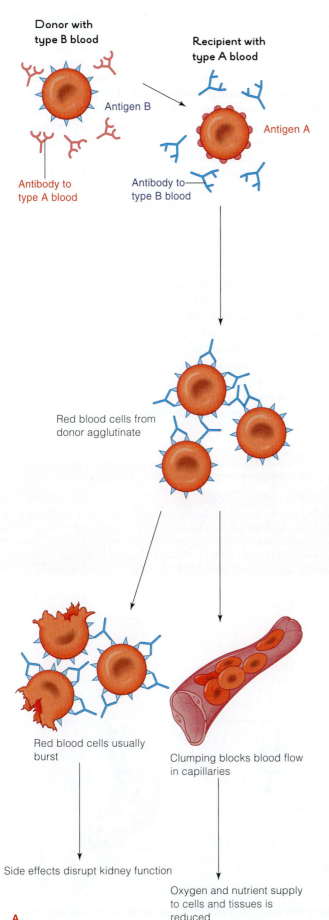

Donor with type B blood

Recipient with type A blood

Antigen B

Antigen A

Antibody to type A blood

Antibody to type B blood

Red blood cells from donor agglutinate

Red blood cells usually burst

Clumping blocks blood flow in capillaries

Side effects disrupt kidney function

Oxygen and nutrient supply to cells and tissues is reduced

A

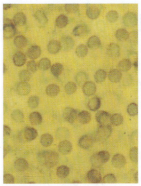

B

Compatible blood cells

Incompatible blood cells

Figure 8.5 Mixing incompatible blood types causes agglutination, or clumping. A Example of an agglutination reaction. This diagram shows what happens when type B blood is transfused into a person who has type A blood. **B** What an agglutination reaction looks like. In the micrograph on the left, commingled red blood cells are compatible and have not clumped. The cells on the right are a mix of incompatible ABO types, and they have clumped together. Donated blood is typed in order to avoid an agglutination response when the blood is transfused into another person. (© Cengage Learning)

can clog small blood vessels, severely damaging tissues throughout the body and sometimes even causing death.

For safety's sake, some people bank their own blood

As the chapter introduction noted, a safe blood transfusion requires accurately matched blood types. Donor blood must also be free of viruses or other disease-causing agents. Although hospital blood supplies are carefully screened, some people who are slated for elective surgery take the extra precaution of pre-donating some of their own blood for an autologous ("from one's self") transfusion. This means they have some of their own blood removed and stored before the procedure so it can be used during the surgery if a transfusion is necessary.

We turn next to the Rh blood group. Agglutination is also a danger when mismatched Rh blood types mix.

ABO blood typing Chemical analysis to determine which self marker or markers from the ABO blood group occur on a person's red blood cells.

agglutination Clumping of red blood cells when incompatible blood types mix.

TAKE-HOME MESSAGE

WHAT IS A BLOOD TYPE?

- Like all cells, red blood cells bear genetically determined "self" markers on their surface. Some of these markers determine a person's blood type.
- Each type of marker may have several forms, which are collectively called a blood group.
- Major blood groups include the ABO group and the Rh group.
- When incompatible blood types mix, an agglutination response occurs in which antibodies cause potentially fatal clumping of red blood cells.

- Another surface marker on red blood cells that can cause agglutination is the Rh factor, which was first identified in the blood of rhesus monkeys.

Rh blood typing looks for an Rh marker

> **Rh blood typing** Chemical test that determines whether red blood cells bear an Rh marker.

Rh blood typing determines the presence or absence of an Rh marker. If your blood cells bear this marker, you are Rh$^+$ (positive). If they don't have the marker, you are Rh$^-$ (negative). When a person's blood type is determined, the ABO blood type and Rh type are usually combined. For instance, if your blood is type A and Rh negative, your blood type will be given as type A$^-$.

Most people don't have antibodies against the Rh marker. But an Rh$^-$ person who receives a transfusion of Rh$^+$ blood will make antibodies against the marker, and these will continue circulating in the person's bloodstream.

If an Rh$^-$ woman becomes pregnant by an Rh$^+$ man, there is a chance the fetus will be Rh$^+$. During pregnancy or childbirth, some of the fetal red blood cells may leak into the mother's bloodstream. If they do, her body will produce antibodies against Rh (Figure 8.6). If she gets pregnant *again*, Rh antibodies will enter the bloodstream of this new fetus. If its blood is Rh$^+$, its mother's antibodies will cause its red blood cells to swell and burst.

In extreme cases, called **hemolytic disease of the newborn**, so many red blood cells are destroyed that the fetus dies. If the condition is diagnosed before or during a live birth, the baby can survive by having its blood replaced with transfusions free of Rh antibodies.

Currently, a known Rh$^-$ woman can be treated after her first pregnancy with an anti-Rh gamma globulin (RhoGam) that will protect her next fetus. The drug will inactivate Rh$^+$ fetal blood cells circulating in the mother's bloodstream before she can become sensitized and begin producing anti-Rh antibodies. In non-maternity cases, an Rh$^-$ person who receives a transfusion of Rh$^+$ blood also can have a severe negative reaction if he or she has previously been exposed to the Rh marker.

There are also many other markers on red blood cells

Besides the Rh and AB blood marker proteins, hundreds of others are now known to exist. These markers are a bit like needles in a haystack—they are widely scattered within the human population and usually don't cause problems in transfusions. Reactions do occur, though, and except in extreme emergencies, hospitals use a method called *cross-matching* to exclude the possibility that blood to be transfused and that of a patient might be incompatible due to the presence of a rare blood cell marker outside the ABO and Rh groups.

> **TAKE-HOME MESSAGE**
>
> **WHAT IS THE PURPOSE OF RH BLOOD TYPING?**
>
> - In some people, red blood cells are marked with an Rh protein. If this Rh$^+$ blood mixes with the Rh$^-$ blood of someone else, the Rh$^-$ individual will develop antibodies against it. The antibodies will trigger an immune response against Rh$^+$ red blood cells if the person is exposed to them again.

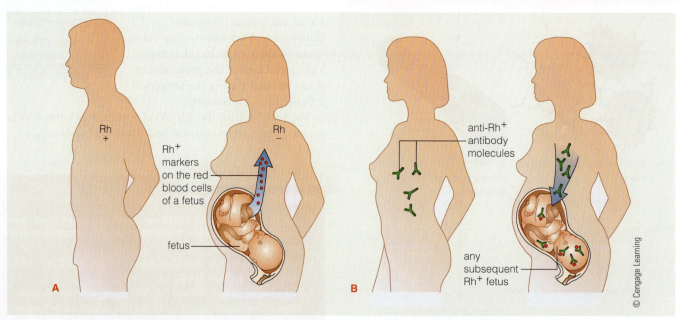

Figure 8.6 Animated! In some cases antibodies develop in response to Rh1 blood. **A** Blood cells from an Rh$^+$ fetus leak into the Rh$^-$ mother's bloodstream. **B** The mother now develops antibodies against a subsequent Rh$^+$ fetus.

What Can Your Blood Say about You?

Laboratory blood tests can reveal a great deal about the functioning of your kidneys and liver, including levels of electrolytes and other important metabolites. Your physician can build a "risk profile" for heart disease by measuring blood levels of HDL and LDL cholesterol and blood lipids called triglycerides. A total blood count would help determine whether you are anemic or battling an infection.

Some tests, like the "stick" test for monitoring blood glucose, work with only a few drops of blood. Other blood tests, such as the Complete Blood Count (CBC) listed in Table 8.2, may require several vials of blood. It usually doesn't matter whether blood is drawn from an artery or vein, although veins just inside the elbow or on the top of the hand are common drawing sites because they are close to the body surface and easily accessible.

The blood's chemical makeup is literally a "fluid situation." Even when a person is completely at rest, the composition of his or her blood constantly changes as cells, tissues, and organs carry out basic functions such as taking up oxygen or nutrients and releasing wastes. To get as clear a picture as possible of the internal situation that's being

CBC Test	Normal Range Results*
Red blood cell (varies with altitude)	Male: 5 to 6 million cells/mcL
	Female: 4 to 5 million cells/mcL
White blood cell	4,500 to 10,000 cells/mcL
Platelets	140,000 to 450,000 cells/mcL
Hemoglobin (varies with altitude)	Male: 14 to 17 gm/dL
	Female: 12 to 15 gm/dL
Hematocrit (varies with altitude)	Male: 41% to 50%
	Female: 36% to 44%

© Cengage Learning

* Cells/mcL 5 cells per microliter; gm/dL 5 grams per deciliter. Hematocrit measures the percentage of whole blood made up by red blood cells.

TABLE 8.2 Some Common Blood Tests

Test	Measures	Useful for Monitoring/ Diagnosing
ALB	Blood level of albumin	Liver/kidney disease, malnutrition
A1C	Blood sugar level	Diabetes
BMP	Basic Metabolic Panel: Tests include blood sugar, kidney enzymes, electrolytes, blood pH	Diabetes, kidney disease, hypertension
BUN	Blood urea nitrogen	Kidney function, many related disorders
CBC	Complete Blood Count: Number of red blood cells/amount of hemoglobin, type and number of white blood cells, platelet count	Anemias, infections
Allergen-specific IgE test/RAST	Circulating antibodies to allergens	Allergies
Creatinine	Blood level of creatinine, a substance released from metabollically active muscle	Kidney function, many related disorders
HIV antibody	Circulating antibodies to the human immunodeficiency virus	HIV infection
Lipid profile	Blood levels of LDL/HDL cholesterol, triglycerides	Cardiovascular disease risk
PT, AMT	PT: blood clotting factors made in the liver; AMT: liver enzyme	Hepatitis
Quantitative hCG	Circulating human chorionic hormone (hCG)	Pregnancy
TSH	Blood level of thyroid-stimulating hormone (TSH)	Thyroid disorders

Photo: David Scharf/Peter Arnold, Inc.

checked, the physician ordering a blood test may ask you to fast, avoid caffeine or alcohol, delay taking medications or vitamins, or temporarily adjust your usual routine in some other way. Even altitude can affect the results of testing for the presence of red blood cells and hemoglobin. In people who live at high altitiude—where the air contains less oxygen than at sea level—the kidneys stimulate the production of more red blood cells, so more cells and the hemoglobin they contain circulate in the bloodstream.

When test results are evaluated, other factors that can affect test values, such as a person's age, weight, and gender, must be taken into account. For example, the blood of a female typically contains less creatinine than is present in a male of the same age and weight. Creatinine is formed when muscle cells break down proteins, and females typically have less muscle mass than males do. Laboratories interpret test results using ranges they have established for normal and abnormal values. Standard ranges may vary from laboratory to laboratory, so it's important to fully discuss test results with your physician. The chart above shows some ranges for a Complete Blood Count established by the National Heart Lung and Blood Institute.

Hemostasis and Blood Clotting

■ Small blood vessels can easily tear or be damaged by a cut or blow. To maintain homeostasis, it is essential for small tears to be quickly repaired.

Hemostasis prevents blood loss

Hemostasis means "stopping bleeding." It involves responses that stop bleeding when a blood vessel is torn or punctured and so helps prevent the excessive loss of blood. The damaged vessel constricts, platelets plug up the tear, and blood coagulates, or clots (Figure 8.7). Although hemostasis can seal tears or punctures in only relatively small blood vessels, most cuts and punctures fall into this category.

hemostasis The name for mechanisms that slow or stop the loss of blood from a ruptured blood vessel.

When a blood vessel is ruptured (Figure 8.7), smooth muscle in the damaged vessel wall contracts in an automatic response called a spasm. The muscle contraction constricts the blood vessel, so blood flow through it slows or stops. This response can last for up to half an hour, and it is vital in stemming the immediate loss of blood. Then, while the flow of blood slows, platelets arrive and clump together, creating a temporary plug in the damaged wall. They also release the hormone serotonin and other chemicals that help prolong the spasm and attract more platelets. Lastly, blood coagulates—that is, it converts to a gel—and forms a clot.

Factors in blood are one trigger for blood clotting

Two different mechanisms can cause a blood clot to form. The first is called an "intrinsic" clotting mechanism because it involves substances that are in the blood itself. Figure 8.7 diagrams this process. It gets under way when a protein in the blood plasma, called "factor X," is activated. This triggers reactions that produce thrombin. This is an enzyme that acts on a rod-shaped protein called fibrinogen. The fibrinogen rods stick together, forming long threads of fibrin. The fibrin threads also stick to one another. The result is a net that entangles blood cells and platelets, as you can see in the micrograph in Figure 8.7. The entire mass is a blood clot. With time, the clot becomes more compact, drawing the torn walls of the vessel back together.

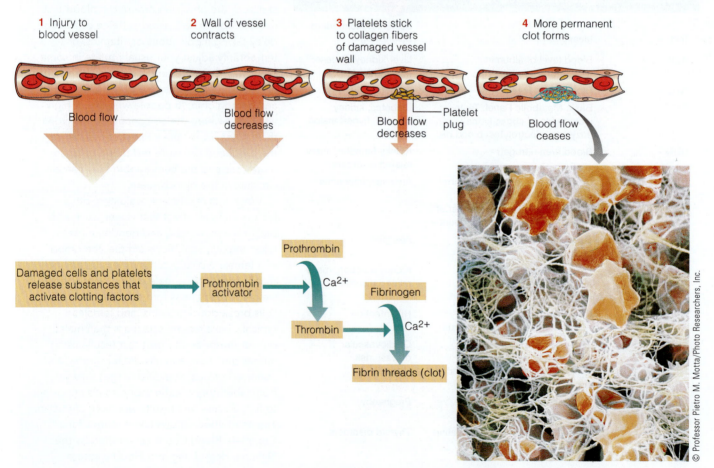

1 Injury to blood vessel

2 Wall of vessel contracts

3 Platelets stick to collagen fibers of damaged vessel wall

4 More permanent clot forms

Blood flow

Blood flow decreases

Blood flow decreases

Platelet plug

Blood flow ceases

Damaged cells and platelets release substances that activate clotting factors → Prothrombin activator → Prothrombin → Ca^{2+} → Thrombin → Fibrinogen → Ca^{2+} → Fibrin threads (clot)

© Professor Pietro M. Motta/Photo Researchers, Inc.

Figure 8.7 A blood clot forms in four steps. The micrograph shows red blood cells trapped in a fibrin net. (© Cengage Learning)

Factors from damaged tissue also can cause a clot to form

Blood also can coagulate through an extrinsic clotting mechanism. "Extrinsic" means that the reactions leading to clotting are triggered by the release of enzymes and other substances *outside* the blood. These chemicals come from damaged blood vessels or from tissue around the damaged area. The substances lead to the formation of thrombin, and the remaining steps are like the steps of the intrinsic pathway.

Because aspirin reduces the aggregation of platelets, it is sometimes prescribed in small doses to help prevent blood clots. A clot that forms in an unbroken blood vessel can be a serious threat because it can block the flow of blood. You may remember from Section 7.8 that a clot that stays where it forms is called a thrombus. The condition is called a **thrombosis**.

Even worse is an embolus, a clot that breaks free and circulates in the bloodstream. Someone who suffers an **embolism** in the heart, lungs, brain, or some other organ may suddenly die when the roving clot shuts down the organ's blood supply. This is what usually happens with a **stroke**. A blood clot blocks the flow of blood to some part of the brain and the affected tissue dies. Strokes can be mild to severe. In serious cases the person may be paralyzed on one side of the body and have trouble speaking. Physical therapy and speech therapy may help minimize the long-term effects.

The disease **hemophilia** is a genetic disorder in which the blood does not contain the usual clotting factors and so does not clot properly. You will read more about this disorder in Chapter 20.

The formation of a blood clot is a first step in healing wounds

When the skin is punctured or torn, blood clotting gets under way immediately to help seal the breach (Figure 8.8). With minor cuts, it usually takes less than 30 minutes for a clot to seal off injured vessels. In a few more hours, phagocytes are at work cleaning up debris and a scab has begun to form. This quick action is vital to minimize blood loss and the chances of infection.

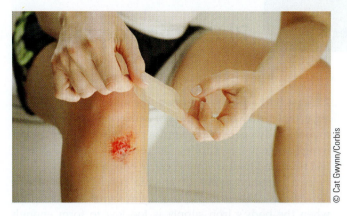

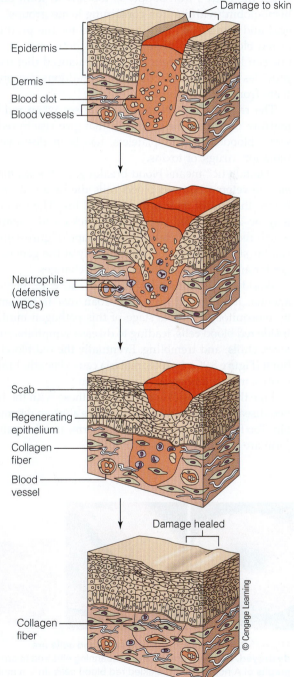

Figure 8.8 **Blood clotting helps heal a wound in the skin.**

8.8 | Blood Disorders

Anemias are red blood cell disorders

At least half a dozen **anemias** (meaning "no blood") are signs that red blood cells are not delivering enough oxygen to meet body needs. All anemias result from other underlying problems. To varying degrees they make a person feel tired and listless, among other symptoms.

Two common types of anemia result from nutrient deficiencies. For example, *iron-deficiency anemia* develops when the body's iron supply is too low to form enough hemoglobin (with its iron-containing heme groups). Folic acid and vitamin B_{12} both are needed for the production of red blood cells in bone marrow. A deficiency of either one can lead to *pernicious anemia*. A balanced diet usually provides both nutrients, but other conditions can prevent them from being absorbed.

The rare malady *aplastic anemia* arises when red bone marrow, including the stem cells that give rise to red and white blood cells and platelets, has been destroyed by radiation, drugs, or toxins.

"Hemolytic" means blood breaking, and *hemolytic anemias* develop when red blood cells die or are destroyed before the end of their normal useful life. The root cause may be an inherited defect, as in sickle-cell anemia, in which red blood cells take a sickle shape (Figure 8.9B) and can burst. Chapter 20 looks more fully at the genetic trigger for and bodywide effects of these changes.

Worldwide, **malaria** is a major cause of hemolytic anemia. It is caused by a protozoan that is transmitted by mosquitoes. One life stage of this pathogen multiplies inside red blood cells, leading to disease symptoms such as fever, chills, and trembling. Eventually the red blood cells burst (Figure 8.9C). In 2008, malaria caused nearly 1 million deaths, mostly among African children.

Like those with sickle cell anemia, those with the inherited disease **thalassemia** also produce abnormal hemoglobin. Too few healthy red blood cells form, and those that do form are thin and extremely fragile.

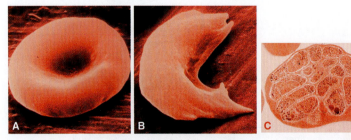

Figure 8.9 In hemolytic anemias red blood cells are destroyed. Here you are looking at scanning electron micrographs of **A** normal and **B** sickled red blood cells. In **C** a multiplying life stage of the microorganism that causes malaria is about to burst open a red blood cell. (A, B: Dr. Stanley Flegler/Visuals Unlimited, Inc.; C: © Moredun Scientific Ltd/Photo Researchers, Inc.)

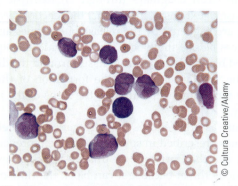

Figure 8.10 This image shows blood from a person with chronic myelogenous leukemia. Abnormal white blood cells (*purple*) are starting to crowd out normal cells.

Viruses and leukemias affect white blood cells

Our white blood cells also can be affected by disease. For example, **infectious mononucleosis** is caused by the Epstein-Barr virus. The infection triggers the overproduction of lymphocytes. The patient feels achy and tired and runs a low-grade fever for several weeks as the highly contagious disease runs its course.

The most notorious virus that attacks white blood cells is HIV, the human immunodeficiency virus, which causes AIDS. Its ability to kill lymphocytes of the immune system is a major topic in Chapter 9.

Leukemias are often called "blood cancers." In fact they are the result of cancer in bone marrow. The word "leukemia" means white blood, and the hallmark of leukemias (like other cancers) is runaway multiplication of the abnormal cells. This unchecked growth of white blood cells destroys healthy bone marrow.

In the most serious forms of leukemia, which tend to strike children, the marrow cavities in bones become choked with cancerous white blood cells. As other types of blood cells (and stem cells) are excluded, typical symptoms of leukemia develop—fever, weight loss, anemia, internal bleeding, pain, and susceptibility to infections. Modern treatments now save thousands of lives, and there is hope that experimental gene therapies may provide more help. Figure 8.10 shows cells of one type of leukemia, called **chronic myelogenous leukemia**.

Carbon monoxide poisoning prevents hemoglobin from binding oxygen

Carbon monoxide, or CO, is a colorless, odorless gas. It is present in auto exhaust fumes and in smoke from burning wood, coal, charcoal, and tobacco. It binds to hemoglobin at least 200 times more tightly than oxygen does. As a result, breathing even tiny amounts of it can tie up half of the body's hemoglobin and prevent tissues from receiving the oxygen they need. CO poisoning is especially dangerous

What is your "Blood IQ"? To find out how much you know about blood and public blood supplies, visit www.nhlbi.nih.gov, the website of the National Heart Lung and Blood Institute, and take the Blood IQ test. The website offers information about blood, blood donation, and even current research on blood substitutes and other topics.

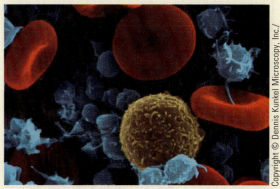

Eye of Science/Photo Researchers, Inc.

Figure 8.11 *Staphylococcus aureus* **bacteria destroy red blood cells and prevent blood from clotting.**

because an affected person may not realize that the symptoms—headache and feeling "woozy"—are signs of life-threatening distress.

Toxins can poison the blood

Some bacteria release toxins into the blood, a condition called **septicemia**. One of our scariest bacterial foes is *Staphylococcus aureus*, or simply "staph A" (Figure 8.11). Although this bacterium lives harmlessly in some people, in others it produces enzymes that destroy red blood cells and prevent blood clotting. Some strains have become highly resistant to antibiotics. One of them, MRSA (methicillin-resistant staph A), can kill, and it is most common where you might least expect it—in health care facilities, including hospitals.

Metabolic poisons in the body cause **toxemia**. For example, the kidneys normally remove many toxic wastes from blood. In a person whose kidneys don't function well due to disease or some other cause, the buildup of certain wastes prevents the normal replacement of red blood cells. It also prevents platelets from functioning. Thus the person becomes anemic and blood doesn't clot properly.

SUMMARY

Section 8.1 Blood consists of watery plasma, red and white blood cells, and cell fragments called platelets. In addition to blood cells and platelets, plasma transports proteins, simple sugars, amino acids, mineral ions, vitamins, hormones, and oxygen and carbon dioxide gases that are dissolved in plasma water.

Stem cells in bone marrow give rise to red blood cells (erythrocytes), white blood cells (leukocytes), and platelets. Red blood cells carry oxygen and some carbon dioxide. White blood cells are involved in body defenses and debris removal. One subgroup (granulocytes) includes neutrophils, basophils, and eosinophils. Another group (agranulocytes) includes lymphocytes and monocytes, which develop into macrophages that scavenge debris and cleanse tissues of foreign material. Lymphocytes destroy specific microbes and other agents of disease. Platelets produce substances that initiate blood clotting.

Section 8.2 Red blood cells contain hemoglobin, an iron-containing protein that binds reversibly with oxygen, forming oxyhemoglobin. Hemoglobin in red blood cells also carries some carbon dioxide to the lungs to be exhaled.

Section 8.3 Red blood cells live for about 120 days. A cell count measures the number of them in a microliter of blood. Macrophages remove dead or damaged red blood cells while stem cells provide replacements.

Sections 8.4, 8.5 Blood type is determined by proteins on the surface of red blood cells. The four main human blood types are A, B, AB, and O. Agglutination is a defense response activated when a person's blood mixes with an incompatible type. Rh blood typing determines the presence or absence of Rh factors (+ or −) on red blood cells.

Section 8.7 Mechanisms of hemostasis slow or stop bleeding. These events include spasms that constrict blood vessels, the formation of platelet plugs, and blood clotting.

REVIEW QUESTIONS

1. What is blood plasma, and what is its function?

2. What are the cellular components of blood? Where do the various kinds come from?

3. Add the missing labels to this diagram of hemoglobin. Then, on a separate sheet of paper, list the factors that affect the tendency of hemoglobin to bind with oxygen.

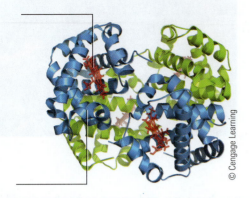

© Cengage Learning

4. What is an agglutination response? How it can be avoided when blood is transfused?

5. What is the function of hemostasis? What are the two ways a blood clot can form?

SELF-QUIZ *Answers in Appendix V*

1. Which of the following statements about blood plasma is/are accurate?
   a. Blood plasma is mostly water.
   b. Blood plasma transports only a few substances, including blood cells, several proteins, and some hormones.
   c. Blood plasma transports blood cells, as well as proteins and many other substances.
   d. Molecules of albumin make up the majority of blood plasma proteins.

2. The _____ produces red blood cells, which transport _____ and some _____.
   a. liver; oxygen; mineral ions
   b. liver; oxygen; carbon dioxide
   c. bone marrow; oxygen; hormones
   d. bone marrow; oxygen; carbon dioxide

3. The _____ produces white blood cells, which function in _____ and _____.
   a. liver; oxygen transport; defense
   b. lymph glands; oxygen transport; stabilizing pH
   c. bone marrow; day-to-day housekeeping; defense
   d. bone marrow; stabilizing pH; defense

4. In the lungs, the main factor in boosting the tendency of hemoglobin to bind with and hold oxygen is _____.
   a. temperature
   b. the amount of O_2 relative to the amount of CO_2 in plasma
   c. acidity (pH)
   d. all are equally important

5. Which of the following statements about red blood cells is false?
   a. They live about 120 days before being replaced.
   b. They lack a nucleus and other organelles when they are mature.
   c. They are replaced when the hormone EPO stimulates stem cells in bone marrow.
   d. They are replaced as part of a negative feedback loop in which kidney cells monitor blood oxygen levels.
   e. All the above statements are true except b, because all cells require a nucleus to in order to function properly.

Your Future

© Professor Pietro M. Motta/ Photo Researchers, Inc.

Dangerous blood clots are more common in people who have cancer or blood poisoning (septicemia). Although the reason wasn't clear, medical researchers once suspected it might have something to do with inflammation, which is part of the body's response to infection or the onset of cancer. Now scientists in Germany have identified genetic changes triggered by inflammation that short-circuit feedback mechanisms which normally shut down the activity of thrombin. As described in Section 8.7, this enzyme promotes the formation of a clot's fibrin net. When normal controls on thrombin go awry, blood clots are much more likely to form. Thrombin also promotes the formation of new blood vessels that cancerous tumors need to grow. In the not-too-distant future, drugs and other therapies that slow "runaway" thrombin production may be powerful new tools in treating clotting associated with cancer and other diseases as well.

6. Blood typing determines the particular _____ on a person's _____.
   a. antibodies; white blood cells
   b. self markers; red blood cells
   c. antibodies; red blood cells
   d. self markers; white blood cells

7. The formation of a blood clot _____.
   a. is an alternative to hemostasis
   b. can occur only through an intrinsic mechanism that involves substances in the blood itself
   c. is just one of several mechanisms of hemostasis
   d. relies on the action of a single clotting factor

8. Match the blood terms with the best description.
   _____ red blood cell   a. plug leaks
   _____ platelets   b. blood markers
   _____ stem cell   c. blood cell source
   _____ plasma   d. erythrocyte
   _____ A, B, O   e. more than half of whole blood

CRITICAL THINKING

1. Thrombocytopenia (throm-bo-sye-tow-PEE-ne-ah) is a disorder that develops when certain drugs, bone marrow cancer, or radiation destroys red bone marrow, including stem cells that give rise to platelets. Predict a likely symptom of this disorder.

2. As the text described, when a person's red blood cell count drops, the kidneys receive less oxygen. In response they release erythropoietin, which prompts the bone marrow to make more red blood cells. As the rising number of red blood cells carry more oxygen to the kidneys, they stop releasing the hormone. What type of homeostatic control mechanism are we talking about here?

3. Blood-feeding insects such as mosquitos inject an anticoagulant into the victim's bloodstream. How does this benefit the insect?

9 IMMUNITY AND DISEASE

LINKS TO EARLIER CONCEPTS

The skin (4.9) and white blood cells (8.1) have major roles in defense. This chapter also draws on what you have learned about proteins (2.11) and how cells take up substances by the mechanisms of endocytosis and phagocytosis (3.11).

You will see how circulating blood (7.3) serves as a highway for defensive cells and substances. And you will now learn more about how the cardiovascular system interacts with the body-wide network of vessels and organs that form the lymphatic system (7.1).

KEY CONCEPTS

Defenses against Disease

The body's defenses against disease threats include physical barriers and two interacting sets of cells and proteins. The lymphatic system has key roles in defense. Sections 9.1–9.3

Innate and Adaptive Immunity

Inborn defenses provide innate immunity, while encounters with agents of disease provide adaptive immunity. White blood cells and defensive proteins act in both kinds of immune responses. Sections 9.4–9.8

Immune System Disorders

Flawed or failed immune mechanisms result in allergies, cancer, autoimmune disorders, and immune deficiencies such as AIDS. Sections 9.9–9.10

Patterns of Infectious Disease

Infectious diseases spread in predictable ways. Understanding these patterns is helpful in avoiding infections. Section 9.11

Hope, grit, and a donated heart have allowed Kelly Perkins to conquer some of the highest mountains in the world. Now in her forties, Kelly had a heart transplant in 1995 after a viral infection triggered inflammation that seriously damaged her own heart. After her surgery, she took up climbing, and trekked up Mt. Fuji in Japan, Mt. Kilimanjaro in Tanzania, and more than a dozen other peaks, including 14,505-foot Mt. Whitney in California—all with her husband, Craig, by her side. Her memoir about her experiences became a best-selling book that continues to inspire transplant patients and their families.

Kelly's story launches our study of the body's responses to disease-causing viruses, bacteria, and other potential threats. In her case, an invading virus unleashed powerful defenses that cleared the threat from her body but left in their wake a battle-scarred heart. Much more often, however, the body's defensive weapons work amazingly well to preserve and restore our health.

- Every day we encounter a vast number of health threats. Body defenses include physical barriers and two interacting sets of cells and proteins.
- Links to Skin 4.9, Blood cells 8.1

adaptive immunity A set of immune defenses that are tailored to the particular pathogens that enter the body.

antigen A molecule or particle that the immune system recognizes as nonself.

B cell B lymphocyte. Cells derived from them make antibodies.

basophil Circulating white blood cell; factor in inflammation.

cytokines Signaling chemicals released by cells of the immune system.

dendritic cell Phagocytic white blood cell; dendritic cells can mobilize adaptive immunity.

eosinophil White blood cell that targets large parasites, such as worms.

immune system A system of interacting white blood cells that defend the body.

immunity The body's ability to resist and fight infections.

innate immunity The body's inborn, general defenses against infection.

lymphocytes B cells, T cells, and other white blood cells that are active mainly in tissues and organs of the lymphatic system.

macrophage Phagocytic white blood cell in tissue fluid.

mast cell White blood cell in many tissues; role in inflammation.

neutrophil Phagocyte that follows chemical trails to infected, inflamed, or damaged tissues.

T cell T lymphocyte. T cells target abnormal body cells, among other roles in adaptive immunity.

Three lines of defense protect the body

We can't really avoid the viruses, bacteria, fungi, protozoa, and parasitic worms that cause disease. These pathogens are in the air we breathe, the food we eat, and on everything we touch. This means that our survival depends on having effective defenses against them. Biologists sometimes portray the body's anti-infection mechanisms as three "lines of defense." This approach can make it easier to remember what each "line" does, but it is important to remember that all the defenses we will discuss in this chapter function as parts of a whole.

The first defensive line is the body's array of physical and chemical barriers to infection, such as intact skin and the linings of body cavities and tubes. These barriers, which we discuss further in Section 9.3, are not part of the immune system. Instead, the **immune system** is a "cellular system" because white blood cells perform most of its core functions (Figure 9.1). As you'll soon see, the immune system's two interacting arms respond differently to threats. One is a bit like a community's "first responder" squad that rushes to the scene of a mishap. In addition to providing immediate "first aid," this rapid response mobilizes reinforcements—appropriate "specialists" that can deal with particular types of threats.

You may remember from Section 8.4 that an **antigen** is something that the body identifies as nonself and that triggers an immune response. Virus particles, foreign cells, toxins, and cancer cells all have antigens on their surface. Most

TABLE 9.1 Comparison of Innate and Adaptive Immunity

	Innate Immunity	Adaptive Immunity
RESPONSE TIME	Immediate	7–10 days
ANTIGEN DETECTION	About 1,000 preset receptors	Vast number of receptors for specific antigens
TRIGGERS	Nonself chemical cues on or in pathogens	Antigens of pathogens, toxins, proteins on altered body cells
MEMORY	None	Long-term

antigens are proteins, lipids, or the large sugar molecules called oligosaccharides. **Immunity** is the body's overall ability to resist and combat something that is nonself.

We are born with some general immune defenses and acquire other, specific ones

There are two categories of immune responses. Each of us is born with some preset responses to infection that are carried out by certain white blood cells and proteins in blood. These activities are the body's second line of defense. They are triggered by chemical cues (such as certain proteins) that are present on or in a variety of pathogens. The responses begin within minutes and they provide **innate immunity**.

Innate immune responses are general, rather like a disinfectant that can kill many different species of bacteria in a bathroom. Even so, innate responses can wipe out many invaders before an infection sets in.

When an innate immune response starts, it also unleashes the third line of defense, or **adaptive immunity**. Our adaptive immunity changes as we go through life. Its responses are tailored to target the particular attackers that chance to enter the body—a given species of bacteria, a particular virus, a toxin, or an abnormal cell such as a cancer cell.

In adaptive immunity, huge numbers of white blood cells mount a counterattack against the invasion. The cells in this "army" all have receptors for a particular antigen and destroy anything that bears it. Biologists estimate that adaptive immunity can produce white blood cell armies with receptors for billions of different antigens. This means that adaptive responses can be mounted against a vast array of potential pathogens. Adaptive immune responses take a week or so to develop, but they leave behind cells that "remember" an antigen and protect against it for a long time, perhaps even for life. Table 9.1 compares basic aspects of innate and adaptive immunity.

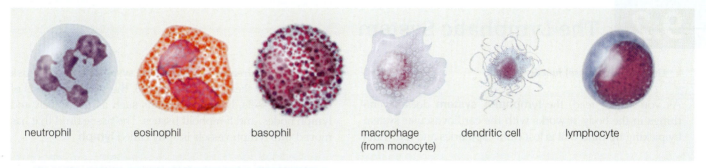

neutrophil   eosinophil   basophil   macrophage (from monocyte)   dendritic cell   lymphocyte

Figure 9.1 Animated! An array of white blood cells carry out immune responses. These sketches show some of the major types. (© Cengage Learning)

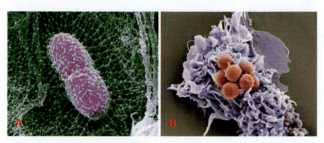

Figure 9.2 White blood cells include phagocytes and cells with other roles. **A** Neutrophils are phagocytes. Here they have surrounded bacteria (*purple*) in lung tissue. **B** Here a fluffy-looking dendritic cell (*lavender*) is engulfing cells of a fungus (*orange*). (A: © 2010, Papayannopoulos et al. Originally published in J. Cell Biol. 191:677–691. doi: 10.1083/jcb.201006052 (Image by Volker Brinkman and Abdul Hakkim); B: Prof. Matthias Gunzer/Photo Researchers, Inc.)

White blood cells and their chemicals are the defenders in immune responses

As you've just read, white blood cells operate in both the innate and adaptive immune responses. Many circulate in blood, while some enter tissues. All release substances that help muster or strengthen defense responses. These chemicals include several types of **cytokines**, "cell movers" that promote and regulate many aspects of immunity. Some white blood cells also secrete enzymes and toxins that kill microbes.

Section 8.1 introduced the main white blood cells that circulate in blood. Of the trillions of these cells in the body, about two-thirds are **neutrophils**, which are phagocytes. **Eosinophils** target parasites, such as worms, that are too big for phagocytosis. They also play a part in allergies. **Basophils** release substances from their granules that cause inflammation.

Three other types of white blood cells perform their functions while in tissues. Like basophils, **mast cells** have granules containing chemicals that cause inflammation (Figure 9.2). **Macrophages** are large phagocytes that arise from circulating monocytes. Each one can engulf as many as one hundred bacteria! **Dendritic cells** alert the adaptive immune system when they detect antigens.

Central roles in adaptive immunity are filled by the cells known as **lymphocytes**. Most of their activities occur in tissues and organs of the lymphatic system (Section 9.2). Those called **B cells** and **T cells** are the white blood cells that can recognize specific antigens. As you'll read later in this chapter, cells derived from B cells make the defensive proteins called antibodies. Some T cells kill abnormal body cells while others help activate B cells.

Many white blood cells circulate in lymph, watery fluid carried in vessels of the lymphatic system. As the next section describes, this system, which has major roles in defense, also works with the cardiovascular system in moving substances throughout the body.

Gender and stress influence the immune system

Other things being equal, studies show that in females immune responses to invading pathogens are stronger than in males. This "turbocharged" immunity has a down side, however. Females are statistically more likely to develop common autoimmune disorders, such as multiple sclerosis, lupus, and Hashimoto's disease (which affects the thyroid gland). As you will read in Section 9.9, in those and other autoimmune disorders, an overvigilant immune system attacks normal cells or proteins.

Prolonged fatigue, emotional upsets, and other forms of ongoing stress can result in less effective immune responses. This is because hormones that are produced in response to stress suppress the chemical triggers that help launch immune responses. The folk wisdom that you risk getting sick when your "resistance is down" comes from everyday observations of such biological effects of stress.

TAKE-HOME MESSAGE

WHAT ARE THE BODY'S THREE LINES OF DEFENSE?

- The three lines of defense are physical barriers, innate immunity, and adaptive immunity.
- Immune responses are executed by white blood cells and the chemicals they release.
- Innate immunity is inborn. Its responses occur quickly but do not provide long-term protection against specific pathogens.
- Adaptive immunity develops over a week or so, and it provides long-term protection against particular pathogens.

9.2 The Lymphatic System

■ Link to Blood vessel function 7.6

As you've just read, the **lymphatic system** does several things in the body. It works with the cardiovascular system by picking up fluid that is lost from capillaries and returning it to the bloodstream. The lymphatic system's other key task is defense. As sketched in Figure 9.3, the system consists of drainage vessels, lymphoid organs such as the spleen and lymph nodes, and lymphoid tissues. The tissue fluid that has moved into lymph vessels is aptly called **lymph**.

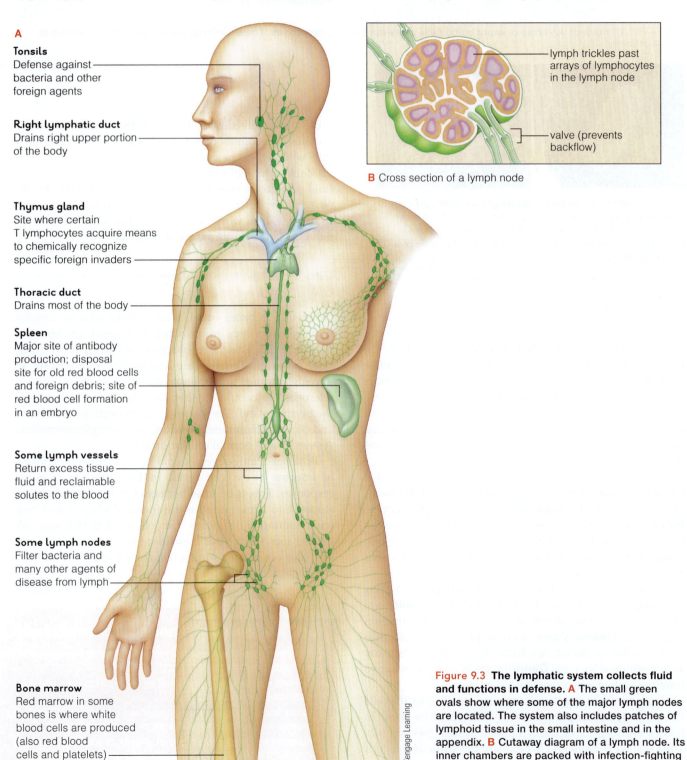

A

Tonsils
Defense against bacteria and other foreign agents

Right lymphatic duct
Drains right upper portion of the body

Thymus gland
Site where certain T lymphocytes acquire means to chemically recognize specific foreign invaders

Thoracic duct
Drains most of the body

Spleen
Major site of antibody production; disposal site for old red blood cells and foreign debris; site of red blood cell formation in an embryo

Some lymph vessels
Return excess tissue fluid and reclaimable solutes to the blood

Some lymph nodes
Filter bacteria and many other agents of disease from lymph

Bone marrow
Red marrow in some bones is where white blood cells are produced (also red blood cells and platelets)

lymph trickles past arrays of lymphocytes in the lymph node

valve (prevents backflow)

B Cross section of a lymph node

Figure 9.3 The lymphatic system collects fluid and functions in defense. A The small green ovals show where some of the major lymph nodes are located. The system also includes patches of lymphoid tissue in the small intestine and in the appendix. **B** Cutaway diagram of a lymph node. Its inner chambers are packed with infection-fighting white blood cells.

© Cengage Learning

The lymph vascular system functions in drainage, delivery, and disposal

The **lymph vascular system** consists of lymph capillaries and other vessels that collect water and dissolved substances from tissue fluid and transport them to ducts of the cardiovascular system. The lymph vascular system has three functions, which we could call the "three Ds"—drainage, delivery, and disposal.

To begin with, the system's vessels are drainage channels. They collect water and solutes that have leaked out of the blood in capillary beds (due to fluid pressure there) and return those substances to the bloodstream. The system also picks up fats the body has absorbed from the small intestine and delivers them to the bloodstream. Finally, lymphatic vessels transport foreign material and cellular debris from body tissues to the lymph vascular system's disposal centers, the lymph nodes.

The lymph vascular system starts at capillary beds (Figure 9.4A), where fluid enters the lymph capillaries. These capillaries don't have an obvious entrance. Instead, water and solutes move into their tips at flaplike "valves." These are areas where endothelial cells overlap (see Figure 9.4A).

Lymph capillaries merge into larger lymph vessels. Like veins, these vessels have smooth muscle in their walls and valves that prevent backflow. They converge into collecting ducts that drain into veins in the lower neck. This is how the lymph fluid is returned to circulating blood. Movements of skeletal muscles and of the rib cage (during breathing) help move fluid through the lymph vessels, just as they do for veins.

Lymphoid organs and lymphatic tissues are specialized for body defense

Several elements of the lymphatic system operate in body defenses. These parts include the lymph nodes, the spleen, and the thymus. They also include the tonsils and patches of tissue in the small intestine, in the appendix, and in airways leading to the lungs.

The **lymph nodes** are located at intervals along lymph vessels (Figure 9.3B). Before lymph enters the bloodstream, it trickles through at least one of these nodes. A lymph node has several chambers where white blood cells accumulate after they have been produced in bone marrow. During an infection, lymph nodes become battlegrounds where armies of lymphocytes form and where foreign agents are destroyed. Macrophages in the nodes help clear the lymph of bacteria and other unwanted substances (Figure 9.4B).

The **spleen** is the lymphatic system's largest organ. It filters blood and also serves as a holding station for lymphocytes. The spleen has inner chambers filled with soft red and white tissue called "pulp." The red pulp is a storage reservoir of red blood cells and macrophages. (In a developing embryo, the spleen produces red blood cells.) In the white pulp, masses of lymphocytes are arrayed close to blood vessels. If an invader reaches the spleen during

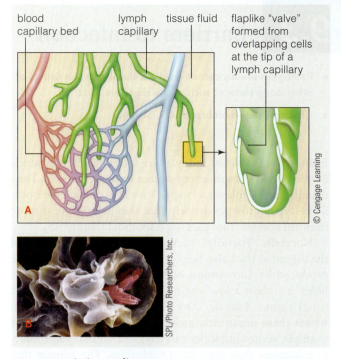

Figure 9.4 Animated! Lymph capillaries collect fluid and direct it through lymph nodes. A This diagram shows lymph capillaries at the start of the lymph vascular system. **B** The macrophage shown here has intercepted a bacterium (*pink*) that can cause tuberculosis.

an infection, the lymphocytes are mobilized to destroy it, just as they are in lymph nodes.

The **thymus** is where T cells multiply and become specialized to combat specific foreign antigens. You will soon be learning more about how these cells function.

lymph The tissue fluid that moves in lymph vessels.

lymph nodes Organs of the lymphatic system that filter lymph; they are located at intervals along lymph vessels.

lymph vascular system Lymph capillaries and other vessels of the lymphatic system.

lymphatic system Organs and tissues that return tissue fluid to the cardiovascular system and have roles in defense.

spleen Lymphoid organ that filters blood and serves as a reservoir for lymphocytes.

thymus Lymphoid organ where T cells multiply and mature.

TAKE-HOME MESSAGE

WHAT IS THE LYMPHATIC SYSTEM?

- The lymphatic system consists of lymph nodes, lymph vessels, and other organs and tissues that collect and filter tissue fluid.
- Lymph vessels return tissue fluid lost from capillaries to the blood, transport fats, and carry debris and foreign material to lymph nodes.
- Lymph nodes, the spleen, and the thymus all function in body defense.

9.3 Barriers to Infection

- Pathogens usually cannot get past the skin or the linings of other body surfaces such as the digestive tract.
- Links to Tissue membranes 4.7, Skin 4.9

Even if you showered today, there are probably thousands of microorganisms on every square inch of your skin. They usually are harmless as long as they stay outside the body. Some types grow so densely that they help prevent more harmful species from gaining a foothold (Figure 9.5).

Normally "friendly" bacteria in the mucosal lining of the digestive tract also help protect you. In females, lactate produced by *Lactobacillus* bacteria in the vaginal mucosa helps maintain a low vaginal pH that most bacteria and fungi cannot tolerate. Any change in the conditions in which these organisms grow can cause an infection. For example, some antibiotics can trigger a vaginal yeast infection because they also kill *Lactobacillus*. The fungus that causes *athlete's foot* may begin to grow between your toes if the skin there is often moist and warm.

The inner walls of the respiratory airways leading to your lungs are coated with sticky mucus. That mucus contains protective substances such as **lysozyme**, an enzyme that chemically attacks and helps destroy many bacteria. Broomlike cilia in the airways sweep out the pathogens.

Lysozyme and some other chemicals in tears, saliva, and gastric fluid offer more protection. Urine's low pH and flushing action help bar pathogens from the urinary tract. In adults, mild diarrhea can rid the lower GI tract of pathogens. In children, diarrhea serves the same function but must be controlled to prevent dangerous dehydration.

lysozyme An enzyme that destroys many bacteria.

Figure 9.5 Many types of bacteria live on body surfaces. This image shows *Staphylococcus epidermis*, the most common species of bacterium on human skin. (© David Scharf/Peter Arnold)

TAKE-HOME MESSAGE

WHAT PHYSICAL AND CHEMICAL BARRIERS HELP PREVENT AN INFECTION?

- Intact skin and mucous membranes are physical barriers to infection.
- Chemicals such as lysozyme in mucus help destroy some microbes.

9.4 Innate Immunity

- Phagocytosis, inflammation, and fever are the body's "first responder" mechanisms that act at once to counter threats in general and prevent infection.
- Links to Cells 3.1, Blood cells 8.1, Blood clotting 8.7

White blood cells and proteins carry out innate immune responses. Extremely important in these responses is a set of proteins called the **complement system**, or simply *complement*. These proteins circulate in the blood and tissue fluid. They are inactive until one makes contact with a pathogen. The interaction unleashes a cascade of reactions that activate ever more complement proteins, until the molecules flood a damaged area.

Activated complement attracts phagocytic white blood cells to invaded tissues. It also blankets pathogens with a "complement coat." The coating helps phagocytes attach to the invader, making it easier to engulf and kill. Some complement proteins form **membrane attack complexes** that create a pore—that is, a hole—in the cell wall or plasma membrane of invading bacteria (Figure 9.6). The punctured cell then disintegrates.

When a pathogen enters the body, macrophages in tissue fluid are usually the first defenders on the scene. They engulf and destroy virtually anything other than healthy body cells. If a macrophage detects an antigen, it releases cytokines that attract more macrophages as well as other white blood cells.

Activated complement and cytokines released by macrophages both trigger the phenomenon we call **inflammation**. This is a fast, general response to tissue damage or infection (Figure 9.7A). In inflammation, mast cells (Figure 9.7B) and basophils respond to an antigen or to the cascade of complement proteins by releasing histamine and other substances. **Histamine** is a chemical messenger that makes arterioles

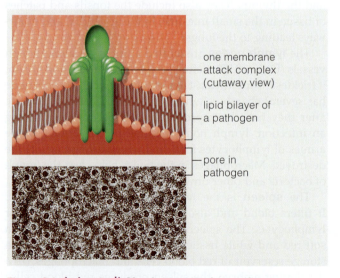

one membrane attack complex (cutaway view)

lipid bilayer of a pathogen

pore in pathogen

Figure 9.6 Animated! Membrane attack complexes can form holes in the plasma membrane of bacteria. The damaged cell then dies. (Top: © Cengage Learning; bottom: © Robert R. Dourmashkin, Courtesy of Clinical Research Centre, Harrow, England)

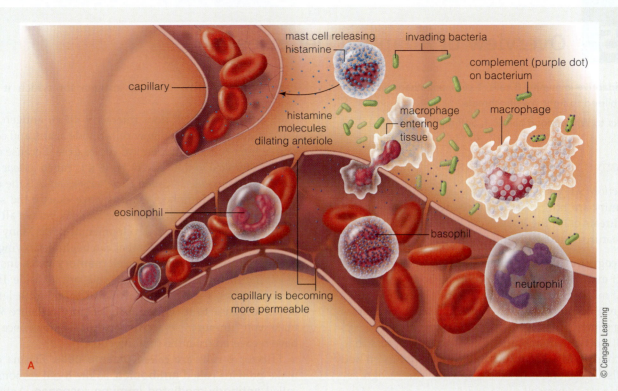

© Antonio Zamora, www.scientificpsychic.com

blood vessel

mast cell

Figure 9.7 Animated! Inflammation is a general response to tissue damage. A This diagram illustrates how invading bacteria might trigger inflammation. In addition to combating the attack, the process helps prepare the damaged tissue for repair. **B** Mast cells are located close to blood vessels, nerves, and mucous membranes near the body surface.

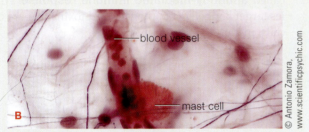

Macrophage engulfing an invading cell

in the tissue dilate, so more blood flows through them. As a result, the tissue reddens and warms with blood-borne metabolic heat. Histamine also makes capillaries leak. The narrow gaps between the cells of the capillary wall become a bit wider, so plasma proteins and phagocytes slip out through them. Water flows out as well. Due to these and other changes, the tissue balloons with fluid. This swelling is called *edema*. The pain that comes with inflammation is due to edema and the effects of inflammatory chemicals.

The plasma proteins leaking into tissue fluid include blood clotting factors. Clots can wall off inflamed areas and delay the spread of microbes into nearby tissues.

A **fever** is a core body temperature above the normal 37°C (98.6°F). Fever develops when macrophages release cytokines called interleukins, which stimulate the brain to release prostaglandins. These signaling molecules in turn can raise the set point on the hypothalamic thermostat, which controls core temperature.

Fevers are not usually harmful. In fact, a fever of about 39°C (100°F) is actually helpful. Among other benefits, it increases body temperature to a level that is too hot for many pathogens to function normally. However, a fever that rises above 42°C (107.6°F) is a medical emergency because it can result in organ damage or death.

Phagocytosis, inflammation, and fever rid the body of most pathogens before they do major harm. If an infection does take hold, the adaptive immune system takes over. We turn to this topic next.

complement system A set of inactive proteins in blood and tissue fluid; the proteins are activated as part of innate immunity.

fever A core body temperature above 37°C (98.6°F).

histamine Chemical released by mast cells and basophils that helps cause inflammation; it dilates arterioles and makes them leak.

inflammation A general response to tissue damage. Symptoms are warmth, redness, swelling, and pain.

membrane attack complex Structure formed by complement proteins that punctures bacteria, which then die.

TAKE-HOME MESSAGE

WHAT IS INNATE IMMUNITY?

- Innate immunity is the body's general defense that includes the complement system, inflammation, and fever.
- Complement proteins kill pathogens or chemically attract phagocytes that can destroy them.
- Mast cells and basophils release histamine and other substances that cause inflammation.

9.5 Overview of Adaptive Defenses

■ In adaptive immunity, B cells, T cells, and phagocytes are mobilized to fight specific threats.

■ Links to White blood cells 8.1, Self markers 8.4

Adaptive immunity has four key features

In Section 9.1 you read that the responses of adaptive immunity are not preset. They change throughout life as new threats enter the body. Now we are ready to consider four features that are central to adaptive immunity.

1. **Recognition of self versus nonself:** Like the ABO self markers on red blood cells, all body cells have self markers called MHC markers. (They are named after the genes that code for them.) MHC markers are some of the proteins that stick out above a cell's plasma membrane. T cells have receptors that recognize **MHC markers** and other self tags on body cells. These and other receptors also can recognize antigens as nonself.

2. **Specificity:** Each B or T cell makes receptors for only one kind of antigen. A receptor and its antigen fit together, something like a lock and key. B cells make **B cell receptors**, and T cells make **T cell receptors**.

3. **Diversity:** B and T cells collectively may have receptors for more than 2 billion different antigens.

4. **Memory:** Some of the B and T cells formed during a first response to an invader are held in reserve for future battles with it.

Let's now learn a little more about these features, which are the defining characteristics of adaptive defenses.

Lymphocytes become specialized for different roles in adaptive immunity

As you know, lymphocytes arise from stem cells in red bone marrow. Forming B cells continue developing in bone marrow, but cells that will specialize as T cells travel via the blood to the thymus gland, where they complete their development. When B and T cells are mature, most move into lymph nodes, the spleen, and other lymphatic system tissues. Each cell is studded with its unique receptors and is capable of becoming active in an adaptive immune response. As you will see, however, activation requires that the B or T cell

meet and recognize the one antigen for which it has receptors.

When a B or T cell does interact with an antigen, the cell divides, the resulting new cells divide again, and so on until a huge number of identical copies exist. Because each copy is identical to the "parent" cell, it is called a *clone* and the copying process is called *clonal expansion*. All the cells in the resulting lymphocyte army have exactly the same receptors, so all recognize the same antigen, but not all will continue down the same developmental path. Many differentiate into **effector cells** that can begin destroying the enemy right away. Others become **memory cells**. Instead of joining the first battle, memory cells are set aside. If the threat returns, they will be available to mount a larger, faster response to it (Figure 9.8). Memory cells are what make you "immune" to a given cold or flu virus once you have recovered from the first infection.

Effector B cells are called **plasma cells**. Plasma cells make proteins called **antibodies**, so the B cell response is called **antibody-mediated immune response**. Antibodies target antigens of pathogens and toxins that are outside cells, in blood or tissue fluid.

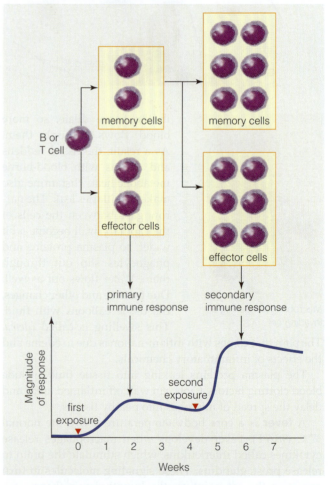

antibodies Proteins made by activated B cells that are the cells' antigen receptors.

antibody-mediated immune response An adaptive immune response in which antibodies are produced against an antigen.

antigen-presenting cell A cell that processes an antigen in ways that allow a T cell to detect it.

B cell receptor Antigen receptor on a B cell.

cell-mediated immune response An adaptive immune response mounted by T cells.

cytotoxic T cell Effector T cell that kills body cells that have been infected by a virus or altered by cancer.

Figure 9.8 Animated! Activated lymphocytes produce effector cells and memory cells. (© Cengage Learning)

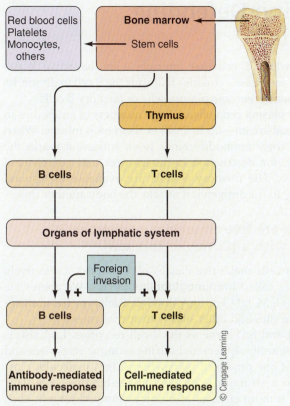

Figure 9.9 **Animated!** The two arms of adaptive immunity are the antibody-mediated response and the cell-mediated response.

Two types of effector T cells form. **Cytotoxic T cells** kill target cells. Hence the T cell response is called **cell-mediated immune response** (Figure 9.9). **Helper T cells** release cytokines that help boost adaptive immune responses. Some of these chemicals are "interleukins" (meaning between leukocytes) that promote B-cell activity. Others are growth factors that spur the activity of cytotoxic T cells. Still others serve as a call to arms for neutrophils and other white blood cells that operate in innate responses.

Even if a T cell encounters an antigen that is a match for its receptors, the antigen will be "invisible" until it is processed by an **antigen-presenting cell**. Although macrophages and B cells may provide this service, the presenter often is a dendritic cell. First, by endocytosis, the presenter engulfs a cell or material that includes an antigen. Then enzymes (from the presenter cell's lysosomes) cut the antigen into pieces. Some of the pieces then are joined with MHC markers to form an antigen–MHC complex. This structure is part self (the MHC marker) and part nonself (the antigen).

The presenter cell displays the complex at its surface (Figure 9.10). A T cell that has the receptor for the antigen part of the complex can bind to it—and a cell-mediated immune response can get under way.

effector cell A B or T cell that has been sensitized to an antigen as an adaptive immune response gets under way. Effectors begin to act as soon as they are sensitized.

helper T cell Effector T cell that stimulates adaptive immune responses.

memory cell A B or T cell that has been sensitized to an antigen but remains in reserve and acts in a secondary response.

MHC marker A self-recognition protein on a body cell.

plasma cell Cell derived from an activated B cell; it produces antibodies that operate in antibody-mediated immunity.

T cell receptor Antigen receptor on a T cell.

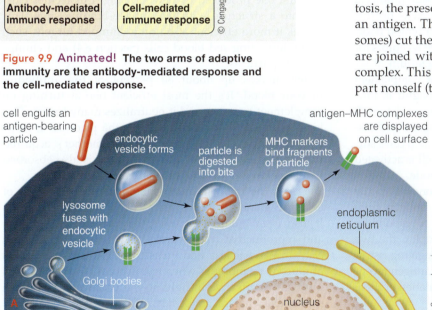

Figure 9.10 **Antigen-presenting cells process antigens.** **A** This diagram shows how an antigen-presenting cell forms an antigen–MHC complex—the chemical flag that can launch an immune response by lymphocytes. **B** At right you see a dendritic cell interacting with a T cell. The interaction activates the T cell.

TAKE-HOME MESSAGE

WHAT ARE THE BODY'S ADAPTIVE IMMUNE DEFENSES?

- An adaptive immune response is one that counteracts specific threats, such as a particular virus or species of bacteria.
- Adaptive responses are amazingly diverse. They also produce memory cells that can mount a faster, stronger response to an antigen that enters the body again.
- Antigen-presenting cells expose T cells to processed antigens, which they can recognize.
- Helper T cells produce cytokines, substances that stimulate immune responses of B cells, cytotoxic T cells, and other white blood cells.

Antibody-Mediated Immunity: Defending against Threats outside Cells

- Different kinds of antibodies form during antibody-mediated defenses.

Antibody-mediated immune responses produce a flood of antibodies

When a B cell forms, the genetic mechanisms involved ensure that it has receptors for only one antigen. If a B cell is activated, the antibodies that the resulting plasma cells make will target the same antigen. Figure 9.11 shows the typical Y shape of a simple antibody. The place where an antibody can bind an antigen usually is near the tip of the two "arms." As antibodies form, they are embedded in a B cell's plasma membrane so that the two arms stick out.

Antibody-mediated immune responses unfold in various parts of the lymphatic system, especially the spleen and lymph nodes. Figure 9.12 provides an example of the basic events that occur when a harmful bacterium enters a tissue.

To begin with, complement proteins in tissue fluid coat the bacterial cell. When a lymphatic vessel picks up this fluid and carries it—and the bacterium—to a lymph node, it's likely that an inactive B cell there will have receptors for an antigen on the bacterium. If so, the microbe's coat of complement signals the B cell to engulf it. Like dendritic cells and macrophages, B cells can process antigens and display them (in an antigen–MHC complex) at the B cell's surface. When this step is accomplished, the B cell is activated and the antigen is exposed to T cells in the node.

immunoglobulin Ig; an antibody.

Meanwhile a dendritic cell also is responding to the invasion. It, too, engulfs and processes an antigen from the invading bacterium. When receptors of a responding helper T cell bind to this antigen–MHC complex, the two cells trade signals. The T cell begins to divide, giving rise to effector and memory cells that have its same antigen receptors. When you have an infection, the accumulation of these T cells makes your lymph nodes swell.

A helper T cell that interacts with the active B cell begins to release cytokines that spur the B cell to divide. Its descendants become plasma cells or memory B cells.

The plasma cells release huge numbers of antibodies in the bloodstream—up to 2,000 of them each minute. When any of these antibodies binds to an antigen, it marks the invader for destruction by phagocytes and complement proteins. The memory B cells are available to respond quickly to the antigen if it attacks the body another time.

There are five classes of antibodies, each with a particular function

Plasma cells make five classes of antibodies. Collectively they are called **immunoglobulins**, or Igs. We abbreviate them as IgM, IgD, IgG, IgA, and IgE. Each type has antigen-binding sites and other sites with special roles.

IgM and IgD serve as the B cell receptors. IgM also is the first antibody secreted during immune responses and the first one produced by newborns. IgM molecules cluster into a structure with ten antigen-binding sites. This makes it more efficient at binding clumped targets, such as agglutinating red blood cells (Section 8.4) and clumps of virus particles.

IgG makes up about 80 percent of the antibodies in your blood. It's the most efficient one at turning on complement proteins, and it neutralizes many toxins. This long-lasting antibody easily crosses the placenta. It helps protect the developing fetus with the mother's acquired immunities. IgG secreted into early milk is also absorbed into a suckling newborn's bloodstream.

IgA is the main immunoglobulin in the secretions of exocrine glands, such as tears, saliva, and breast milk. It also is in mucus that coats the respiratory, digestive, and reproductive tracts—areas to which microbes have easy access. Bacteria and viruses can't attach to the cells of mucous membranes when IgA is bound to them. In this way, IgA is effective in fighting the pathogens that cause salmonella, cholera, gonorrhea, influenza, and polio.

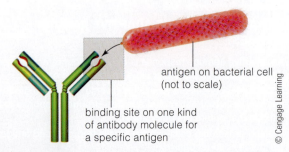

antigen on bacterial cell (not to scale)

binding site on one kind of antibody molecule for a specific antigen

© Cengage Learning

Figure 9.11 Antibodies can bind to antigens. Each kind of antibody can bind only one kind of antigen. The antigen fits into grooves and bumps on the antibody molecules. In this example the antibody has bound to a species of bacteria.

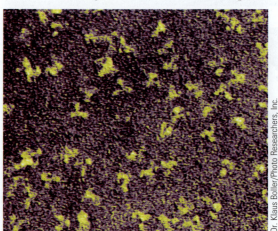

Human IgG antibodies (*yellow*)

Dr. Klaus Boller/Photo Researchers, Inc.

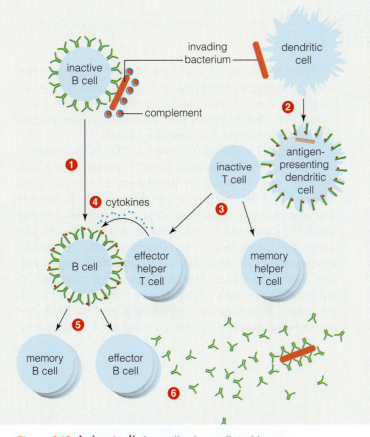

1 The B cell receptors on an inactive (naive) B cell bind to the antigen on a bacterium. Then the B cell engulfs it. Fragments of the bacterium bind MHC markers, and the complexes are displayed at the surface of the now-activated B cell.

2 A dendritic cell engulfs the same kind of bacterium that the B cell encountered. Digested fragments of the bacterium bind to MHC markers, and the complexes are displayed at the dendritic cell's surface. The dendritic cell is now an antigen-presenting cell.

3 The antigen–MHC complexes on the antigen-presenting cell are recognized by antigen receptors on an inactive T cell. The T cell now divides and gives rise to effector and memory helper T cells.

4 Antigen receptors of one of the effector helper T cells bind antigen–MHC complexes on the B cell. Binding makes the T cell secrete cytokines.

5 The cytokines stimulate the B cell to divide, giving rise to many identical B cells. The cells give rise to effector B cells and memory B cells.

6 The effector B cells begin making and secreting huge numbers of IgA, IgG, or IgE, all of which recognize the same antigen as the original B cell receptor. The new antibodies circulate throughout the body and bind to any remaining bacteria.

Figure 9.12 Animated! An antibody-mediated immune response occurs when B cells make antibodies to an antigen. In this example, the invader is a bacterium. (© Cengage Learning)

IgE is involved in allergic reactions, including asthma, hay fever, and hives. IgE also triggers inflammation after attacks by parasitic worms and other pathogens. When it binds to an antigen, basophils and mast cells release histamine that causes the inflammation response. Table 9.2 summarizes the roles of antibodies.

TAKE-HOME MESSAGE

WHAT IS AN ANTIBODY-MEDIATED IMMUNE RESPONSE?

- In an antibody-mediated response, antibodies form and bind to antigens of pathogens and toxins that are outside cells.
- The steps that result in an antibody-mediated response produce plasma cells, which make antibodies, and memory B cells.
- Antibodies do not directly kill pathogens. They flag them for destruction by other defenders.
- Plasma cells secrete five classes of antibodies (immunoglobulins) that help protect the body against diverse threats.

TABLE 9.2 Antibodies in the Human Body

Secreted antibodies

IgG — Main antibody in blood; activates complement, neutralizes toxins; protects fetus and is secreted in early milk.

IgA — Abundant in exocrine gland secretions (e.g., tears, saliva, milk, mucus), where it occurs in the form shown here. Interferes with binding of pathogens to body cells.

Membrane-bound antibodies

IgE — Anchored to surface of basophils, mast cells, eosinophils, and some dendritic cells. IgE binding to antigen induces anchoring cell to release histamines and cytokines. Factor in allergies and asthma.

IgD — B cell receptor.

IgM 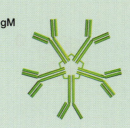 — B cell receptor, as a monomer. Also is secreted as pentamer (group of five), as shown here.

9.7 Cell-Mediated Responses: Combating Threats inside Cells

- Responses by antibodies can't reach threats inside cells. Accordingly, when cells become infected or altered in harmful ways, other "warrior" cells must come to the defense.

- Link to Blood types 8.4

apoptosis Programmed cell death, which is governed by genes.

NK cell Natural killer cell; a lymphocyte that kills virus-infected cells and some types of cancer cells.

Many pathogens evade antibodies. They hide in body cells, kill them, and often reproduce inside them. They are exposed only briefly after they slip out of one cell and before they infect others. Viruses, bacteria, and some fungi and protozoans all can enter cells. Cell-mediated immune responses are the body's weapons against these dangers as well as against abnormal body cells such as cancer cells.

Figure 9.13 gives an overview of how a cell-mediated immune response takes place. It gets under way when a cell such as a dendritic cell presents an antigen to T cells that have receptors that can recognize the antigen. The response also produces memory T cells.

Cytotoxic T cells release various "killer" substances. *Perforins* are proteins that literally perforate the target cell, making holes in its plasma membrane. This "direct hit" kills the target cell in much the same way that complement membrane attack complexes do (Figure 9.14). Cytotoxic T cells also secrete chemicals that cause the target cell to self-destruct. This genetically programmed cell death is called **apoptosis** (a-poh-TOE-sys). The term comes from a Greek word meaning "to fall apart," and that's what happens to the cell. As it disintegrates, its cytoplasm dribbles out, and its DNA and organelles are broken up. After a cytotoxic T cell has done

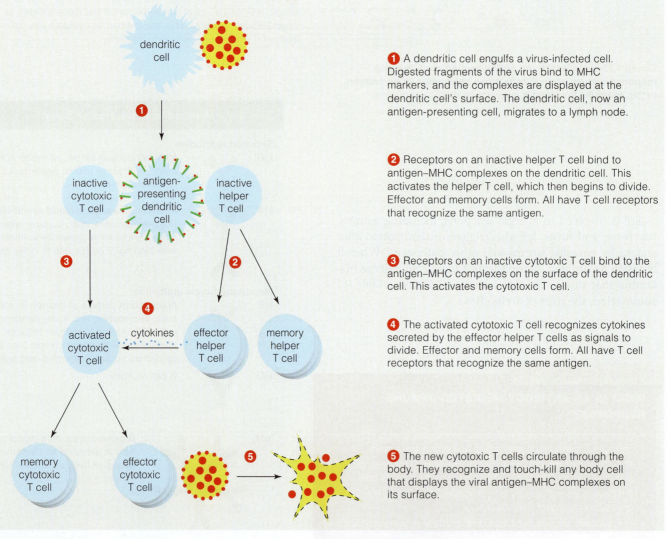

1 A dendritic cell engulfs a virus-infected cell. Digested fragments of the virus bind to MHC markers, and the complexes are displayed at the dendritic cell's surface. The dendritic cell, now an antigen-presenting cell, migrates to a lymph node.

2 Receptors on an inactive helper T cell bind to antigen–MHC complexes on the dendritic cell. This activates the helper T cell, which then begins to divide. Effector and memory cells form. All have T cell receptors that recognize the same antigen.

3 Receptors on an inactive cytotoxic T cell bind to the antigen–MHC complexes on the surface of the dendritic cell. This activates the cytotoxic T cell.

4 The activated cytotoxic T cell recognizes cytokines secreted by the effector helper T cells as signals to divide. Effector and memory cells form. All have T cell receptors that recognize the same antigen.

5 The new cytotoxic T cells circulate through the body. They recognize and touch-kill any body cell that displays the viral antigen–MHC complexes on its surface.

Figure 9.13 Animated! T cells are the warriors in cell-mediated immune responses. (© Cengage Learning)

its defensive work, it disengages from the doomed cell and moves on.

Other kinds of cells make more general responses. These cells include macrophages as well as lymphocytes called **NK cells** (natural killers). NK cells are present in tissues and organs of the lymphatic system. They can detect and kill virus-infected body cells and some cancer cells.

Helper T cell cytokines stimulate NK cells, but NK cells don't need to have an antigen presented to them. Instead, they simply attack any body cell that has too few or altered MHC markers, or that antibodies have tagged for destruction. They also kill body cells flagged with chemical "stress markers" that develop when a cell is infected or has become cancerous.

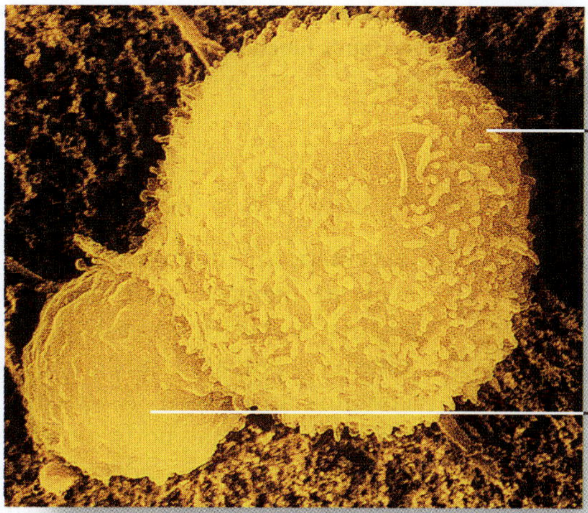

Figure 9.14 **This image shows a cytotoxic T cell killing a tumor cell.**

Cytotoxic T cells cause the body to reject transplanted tissue

Cytotoxic T cells cause the rejection of tissue and organ transplants. This is partly because features of the MHC markers on donor cells differ enough from the recipient's to be recognized as antigens.

To help prevent rejection, before an organ is transplanted the MHC markers of a potential donor are analyzed to determine how closely they match those of the patient. Because such tissue grafts generally succeed only when the donor and recipient share at least 75 percent of their MHC markers, the best donor is a close relative of the recipient, such as a parent or sibling, who is likely to have a similar genetic makeup.

More commonly, however, the donated organ comes from a fresh cadaver. In addition to having well-matched MHC markers, the donor and recipient also must have compatible blood types (Section 8.4).

After surgery, the organ recipient receives drugs that suppress the immune system. The treatment also may include other therapies designed to fend off an attack by B and T cells. As with Kelly Perkins, the heart transplant patient described in the chapter introduction, suppression

of the immune system means that the patient must take large doses of antibiotics to control infections. In spite of the difficulties, many organ recipients survive for years beyond the surgery and lead highly active lives.

Interestingly, not all transplanted tissues provoke a recipient's immune defenses. Two examples are tissues of the eye and the testicles. In simple terms, the plasma membrane of cells of these organs is thought to bear receptors that can detect activated lymphocytes. Before such a defender can launch an attack, the protein signals the soon-to-be-besieged cell to secrete a chemical that triggers apoptosis in the approaching lymphocytes—so the attack is usually averted. Our ability to readily transplant the cornea—the outermost layer of the eye that is vital to clear vision—depends on this mechanism. The Think Outside the Book feature on this page discusses a medical condition for which a cornea transplant is the only cure.

THINK OUTSIDE THE BOOK

A cornea transplant is the only cure for advanced cases of *keratoconus*, a condition in which the cornea thins and becomes cone-shaped. As the disease progresses, an affected person's vision becomes seriously distorted. Usually both eyes are affected to some degree. Patients often are diagnosed in their teens or early twenties. The National Keratoconus Foundation (www.nkcf.org) reports that about 1 in 2,000 people develop the disorder, although its cause is unknown. What else can you learn about keratoconus? What does a cornea transplant entail?

TAKE-HOME MESSAGE

WHAT IS A CELL-MEDIATED IMMUNE RESPONSE?

- A cell-mediated immune response occurs when certain types of defensive cells attack infected or chemically altered body cells.
- Cytotoxic T cells target specific antigens.
- NK cells, macrophages, and various other white blood cells make other, nonspecific responses.

■ Modern science has developed powerful weapons that can enhance the immune system's functioning or harness it in new ways to treat disease.

active immunity Immunity that develops after a person receives a vaccine, which stimulates the immune system to produce antibodies against a particular pathogen.

immunotherapy A clinical therapy that uses immune system cells or substances to treat disease.

interferon Type of cytokine released by cells that are infected by a virus. Uninfected cells may respond by producing substances that prevent the virus from multiplying.

monoclonal antibody Antibodies made in the laboratory by cells cloned from a single plasma (B) cell.

passive immunity Immunity conferred by injected antibodies to a pathogen. It does not stimulate the recipient's immune system to produce antibodies.

vaccine A prepared substance that contains an antigen. Most vaccines are made using dead or weakened antigens.

Vaccination stimulates immunity

Vaccination ("immunization") is a way to increase your immunity against a specific disease. A **vaccine** is a prepared substance that contains an antigen. A vaccine is injected into the body or taken orally, sometimes according to a schedule (Table 9.3). The first injection elicits a primary immune response that confers **active immunity**. A later booster shot elicits a secondary response, in which more effector cells and memory cells form. The booster can provide long-lasting protection.

Many vaccines are made from killed or extremely weakened pathogens. For example, weakened poliovirus particles are used for the Sabin polio vaccine. Worldwide vaccinations with a weakened relative of the smallpox virus allowed a successful global effort to eradicate the disease (Figure 9.15). Other vaccines are made using inactivated forms of natural toxins, such as the bacterial toxin that causes tetanus.

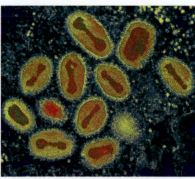

Figure 9.15 Smallpox was eradicated by a global vaccination effort. This girl survived smallpox, which can leave behind heavily pitted scars. Vaccinations stopped in 1972 and the last known naturally occuring case was recorded in 1977, in Somalia. The image on the right shows smallpox virus particles.
(Left: CDC/James Hicks; Right: Eye of Science/Photo Researchers, Inc.)

Today many vaccines are made with genetically engineered viruses (Chapter 21). These harmless "transgenic" viruses incorporate genes from three or more different viruses in their genetic material. After a person is vaccinated with an engineered virus, body cells use the new genes to produce antigens, and immunity is established.

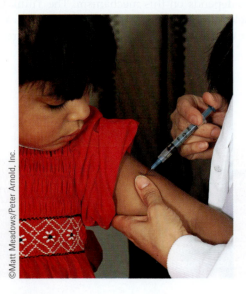

TABLE 9.3 Recommended Immunization Schedule for Children

Vaccine	Age of Vaccination
Hepatitis B	Birth
Hepatitis B boosters	1–2 months and 6–18 months
Rotavirus	2, 4, and 6 months
DTP: diphtheria, tetanus, and pertussis (whooping cough)	2, 4, and 6 months
DTP boosters	15–18 months, 4–6 years, and 11–12 years
HiB (*Haemophilus influenzae*)	2, 4, and 6 months
HiB booster	12–15 months
Pneumococcal	2, 4, and 6 months
Pneumococcal booster	12–15 months
Inactivated poliovirus	6–18 months
Inactivated poliovirus boosters	4–6 years
Influenza	Yearly, 6 months to 18 years
MMR (measles, mumps, rubella)	12–18 months
MMR booster	4–6 years
Varicella (chicken pox)	12–18 months
Varicella booster	4–6 years
Hepatitis A series	1–4 years
HPV series	11–12 years
Meningococcal	11–12 years

Source: Centers for Disease Control (CDC), 2012

Antibodies provide "borrowed" immunity

People who are already infected with pathogens, such as those that cause tetanus, measles, hepatitis B, and rabies, may be helped by injections of antibodies that confer **passive immunity**. A patient receives antibodies that have been purified from another source, preferably someone whose adaptive immune system already has produced a large amount of the antibody. The result is "passive" immunity because the recipient's own B cells are not producing antibodies or memory B cells. While the helpful effect doesn't last long, the injected antibodies may counter the immediate attack.

Vaccines are powerful weapons, but they can fail or have adverse effects. In rare cases, a vaccine can damage the nervous system or result in chronic immunological problems. A physician can explain the risks and benefits.

Monoclonal antibodies are used in research and medicine

Commercially prepared **monoclonal antibodies** harness antibodies for medical and research uses (Figure 9.16A). The word "monoclonal" (meaning one clone) refers to the fact that the antibodies are made by cells cloned from just a single antibody-producing plasma cell.

At one time laboratory mice were the "factories" for making monoclonal antibodies. Today most monoclonal antibodies are produced using genetically altered bacteria. Genetically engineered plants such as corn also are being used to make antibodies that may be both cost-effective and safe (few plant pathogens can infect people). The first "plantibody" to be used on human volunteers prevented infection by a bacterium that causes tooth decay.

Monoclonal antibodies have become useful tools in diagnosing health conditions. Because they can recognize and bind to specific antigens, they can detect substances in the body—a bacterial cell, another antibody, or a chemical—even if only a tiny amount is present. Uses include home pregnancy tests (Figure 9.16B). and screening for prostate cancer and some sexually transmitted diseases. As you'll read next, monoclonal antibodies also have potential uses as "magic bullets" to deliver drugs used to treat certain forms of cancer.

Immunotherapies reinforce defenses

Immunotherapy bolsters defenses against infections and cancer cells by manipulating the body's own immune mechanisms. Cytokines that activate B and T cells are being used to treat some cancers. Monoclonal antibodies are another weapon. For example, some aggressive breast cancers have telltale HER2 proteins at their surface. The drug Herceptin is a monoclonal antibody that binds to the proteins and draws a response from NK cells. The drug can be a double-edged sword, however, because some healthy body cells also have HER2 proteins, and they are attacked as well.

Monoclonal antibodies also can be bound to poisons to make *immunotoxins*. When these substances bind to an antigen on a cancer cell, they enter the cell and block processes that allow it to survive and multiply. Some experimental immunotoxins are being tested against HIV, the virus that causes AIDS.

Most body cells that become infected by a virus can make and secrete antiviral cytokines called **interferons**. When an interferon reaches each an uninfected cell, it triggers a chemical attack that prevents the virus from multiplying. Genetically engineered gamma interferon is used to treat hepatitis C, a chronic, potentially lethal viral disease that impairs liver functioning. Some kinds of cells produce beta interferon. This protein has been approved for the treatment of a type of **multiple sclerosis**, a disease in which the immune system mounts an attack on parts of the nervous system.

TAKE-HOME MESSAGE

WHAT ARE SOME APPLICATIONS OF IMMUNOLOGY?

- Vaccination, antibodies, and natural antiviral chemicals have been harnessed to enhance immunity to specific diseases.
- Vaccination (immunization) can stimulate the production of both effector and memory lymphocytes.
- Monoclonal antibodies and cytokines such as interferons are important tools in medical research, testing, and the treatment of various diseases.

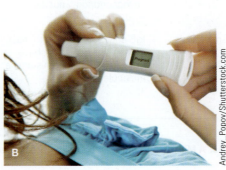

©Simon Fraser/Photo Researchers, Inc.

Andrey_Popov/Shutterstock.com

Figure 9.16 Monoclonal antibodies have a variety of uses in research and medicine. A In the laboratory monoclonal antibodies are stored frozen in liquid nitrogen. **B** The test strips in home pregnancy tests contain monoclonal antibodies that will bind to molecules of the hormone HCG (human chorionic gonadotropin), which is present in urine soon after a woman becomes pregnant.

In allergies, harmless substances provoke an immune attack

Most allergies won't kill you, but they sure can make you miserable. In at least 15 percent of the people in the United States, normally harmless substances can provoke immune responses. These substances are **allergens**, and the response to them is an **allergy**. Common allergens are pollen (Figure 9.17A), a variety of foods and drugs, dust mites, fungal spores, insect venom, and ingredients in cosmetics. Some responses start within minutes; others are delayed. Either way, the allergens trigger mild to severe inflammation of mucous membranes and in some cases other tissues as well.

Some people are genetically predisposed to allergies. Infections, emotional stress, or changes in air temperature also may cause the reactions. When an allergic person first is exposed to certain antigens, IgE antibodies are secreted and bind to mast cells (Figure 9.17B). When the IgE binds an allergen, mast cells secrete prostaglandins, histamine, and other substances that fan inflammation. They also cause an affected person's airways to constrict, and sometime trigger the itchy raised welts called hives (Figure 9.18A). In **hay fever**, the allergic response produces stuffed sinuses, a drippy nose, and sneezing.

In food allergies a particular food is interpreted as an "invader." The most common culprits are shellfish, eggs, and wheat. Depending on the person and the food involved, symptoms typically include diarrhea, vomiting, and sometimes swelling or tingling of mucous membranes. Some food allergies can be lethal. For example, in people who are allergic to peanuts even a tiny amount can trigger **anaphylactic shock**—a whole-body allergic response that produces frightening symptoms. Within moments air passages to the lungs close almost completely. Fluid gushes from dilated blood vessels all through the body. Blood pressure plummets, which can cause the person's cardiovascular system to collapse (Figure 9.18B).

Anaphylactic shock is also a concern for people who are allergic to wasp or bee venom, for a single sting can kill them. One emergency treatment is an injection of the hormone epinephrine. People who know they are at risk (usually because they've already had a bad reaction to an allergen) can carry injectable epinephrine with them, just in case.

Allergen (antigen) enters the body

A

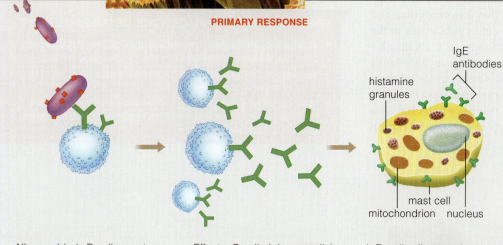
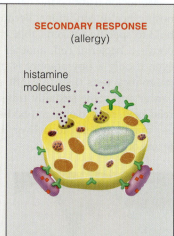

PRIMARY RESPONSE

SECONDARY RESPONSE
(allergy)

IgE antibodies

histamine granules

histamine molecules

mast cell

mitochondrion   nucleus

Allergen binds B cell receptors; the sensitized B cell now processes the antigen and, with the help of T cells (not shown), proceeds through the steps leading to cell proliferation.

Effector B cells (plasma cells) produce and secrete IgE antibodies to the allergen.

IgE antibodies attach to mast cells in tissues, which have granules containing histamine molecules.

After the first exposure, when the allergen enters the body it binds with IgE antibodies on mast cells; binding stimulates the mast cell to release histamine and other substances.

B

Figure 9.17 Allergies are misguided immune responses. A Micrograph of ragweed pollen. **B** The basic steps leading to an allergic response. (© Cengage Learning)

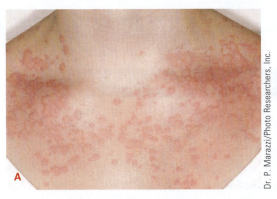

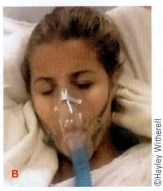

A
B

Dr. P. Marazzi/Photo Researchers, Inc.

©Hayley Witherell

Figure 9.18 **Hives and anaphylactic shock are other types of allergic reactions.** **A** The welts called hives may itch intensely. **B** An emergency treatment for anaphylactic shock includes receiving oxygen.

©James Stevenson/Photo Researchers, Inc.

Figure 9.19 **This person's hand is crippled by rheumatoid arthritis.**

As their name suggests, antihistamines are anti-inflammatory drugs that counteract the histamine released by basophils and mast cells. Many people use them to relieve short-term allergy symptoms. In some cases a sufferer may undergo a desensitization program. Following skin tests that identify offending allergens, inflammatory responses to some of them can be blocked if the patient's body can be stimulated to make IgG instead of IgE. Gradually, larger and larger doses of specific allergens are administered. Each time, the person's body produces more IgG molecules and memory cells. The IgG will bind with an allergen and block its attachment to IgE. Inflammation is blocked, too.

Autoimmune disorders attack "self"

Normally, a B or T cell does not have receptors that can "see" a body cell's MHC self tags as antigens of an invader. This **immunological tolerance** is what protects the body's own cells from attack by the immune system. In fact, a newly forming B or T cell that recognizes healthy body cells as foreign undergoes apoptosis and dies. This weeding out of self-reactive B and T cells goes on throughout a person's life. When it goes awry, the result is **autoimmunity**, in which the immune system's weapons are unleashed against normal body cells or proteins. An example is **rheumatoid arthritis (RA)**. People with RA are genetically predisposed to it. Their macrophages and T and B cells become activated by antigens associated with the joints. Immune responses are mounted against their body's collagen molecules and also apparently against antibodies that have bound to an (as yet unknown) antigen. Joint tissues suffer more damage from inflammation and the complement system (Figure 9.19). Malfunctioning repair mechanisms make the problem worse. Eventually the affected joints become immobile.

Another autoimmune disease is **type 1 diabetes**. This is a type of diabetes mellitus in which the pancreas does not secrete enough of the hormone insulin for proper absorption of glucose from the blood. In type 1 diabetes, the immune system attacks and destroys the insulin-secreting cells. A viral infection may trigger the response. Chapter 15 looks at the various forms of diabetes in more detail.

Systemic lupus erythematosus (SLE) mainly affects younger women, but other people develop it also. A common symptom is a "butterfly" rash on the face that extends from cheek to cheek across the nose. The rash is one sign that the affected person has developed antibodies to her or his own DNA and other "self" components. Antigen–antibody complexes accumulate in joints, blood vessel walls, the skin, and the kidneys. Other symptoms include fatigue, painful arthritis, and in some cases a near-total breakdown of kidney function. Medicines can help relieve many SLE symptoms, but there is no cure.

As you read in Section 9.1, autoimmunity is far more common in women. We know that the receptor for estrogen is involved in certain genetic controls. Is the receptor also implicated in autoimmune responses? Researchers are exploring the question.

Immune responses can be deficient

The term **immunodeficiency** applies when a person's immune system is weakened or lacking altogether. When the body has too few properly functioning lymphocytes, its immune responses are not effective. Both T and B cells are in short supply in the disorder known as **severe combined immune deficiency (SCID)**. SCID usually is inherited, and infants born with it may die early in life. Lacking adequate immune responses, they are extremely vulnerable to infections that are not life-threatening to other people. One type of SCID is now being treated by gene therapy (Chapter 21).

The human immune deficiency virus (HIV) disables several kinds of white blood cells. This is how it causes AIDS, or acquired immune deficiency syndrome, which we consider in the next section.

allergen A substance that causes an allergic reaction.

allergy An immune response caused by a normally harmless substance.

anaphylactic shock A whole-body allergic reaction.

autoimmunity An immune response unleashed against normal (self) body cells or proteins.

immunodeficiency Weakened of absent immune responses.

immunological tolerance The lack of an immune response against normal body cells.

9.10 HIV and AIDS

FOCUS ON HEALTH

AIDS (acquired immune deficiency syndrome) is a group of diseases caused by infection with **HIV**, the human immune deficiency virus. HIV infects cells that have a certain type of surface receptor. Macrophages, dendritic cells, and helper T cells have this receptor. Because HIV kills lymphocytes, it leaves the body vulnerable to infections and rare forms of cancer.

There is no way to rid the body of the known forms of the virus, HIV-I and HIV-II. Sooner or later, people who are infected begin to develop symptoms of illness. Diagnostic signs of AIDS include a severely depressed immune system, a positive HIV test, and an "indicator disease," including types of pneumonia, recurrent yeast infections, cancer, and drug-resistant tuberculosis. According to the World Health Organization, in 2010 nearly 34 million people were living with HIV/AIDS (Table 9.4).

HIV is transmitted in body fluids

HIV is transmitted when body fluids, especially blood and semen, of an infected person enter another person's tissues. The virus can enter through any kind of cut or abrasion, anywhere on or in the body. HIV-infected blood also can be present on toothbrushes and razors; on needles used to inject drugs, pierce ears, or do acupuncture; and on contaminated medical equipment.

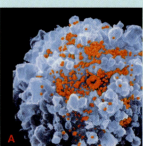

The most common mode of transmission is sex with an infected partner. HIV in semen and vaginal secretions enters a partner's body through epithelium lining the penis, vagina, rectum, or (rarely) mouth. Anything that damages the linings, such as

TABLE 9.4 Global HIV and AIDS Cases

Region	AIDS Cases	New HIV cases
Sub-Saharan Africa	22,900,000	1,900,000
Asia	4,800,000	350,000
Central Asia/East Europe	1,500,000	160,000
Latin America	1,500,000	110,000
North America	1,300,000	58,000
Western/Central Europe	1,100,000	30,000
Middle East/North Africa	470,000	59,000
Caribbean Islands	200,000	12,000
Oceania	59,000	13,000
Approx. worldwide total	33,829,000	2,700,000

Source: WHO (World Health Organization) HIV/AIDS Progress Report, 2010 data

other sexually transmitted diseases, anal intercourse, or rough sex, increases the odds that the virus will be transmitted.

HIV is not effectively transmitted by food, air, water, casual contact, or insect bites. However, infected mothers can transmit HIV to their babies during pregnancy, birth, and breast-feeding (Figure 9.20B).

About half of HIV-infected adults worldwide are women. Some of those infections are due to intravenous drug abuse, but most are the result of sexual contact with infected men. In recent years, more young adults in the United States have died from AIDS than from any other single cause.

HIV infection begins a fateful struggle

HIV is a retrovirus, which means that its primary genetic instructions are in the form of RNA, not DNA. Each virus particle has a lipid envelope, a bit of plasma membrane that enclosed it as it budded from an infected cell. Proteins spike from the envelope, extend across it, or line its inner surface. Inside the envelope, so-called "viral coat" proteins enclose RNA and an enzyme called reverse transcriptase. This enzyme uses RNA as a template for making DNA (the reverse of a more common process in which DNA is the template for making RNA). The newly formed DNA makes up genes that are then inserted into one of the host cell's chromosomes. Eventually the genetic message in the DNA is "rewritten" back into RNA, and these RNA instructions are then translated into protein (Figure 9.21). Chapter 21 explains this process more fully.

After HIV infects a person, virus particles enter the bloodstream. At this stage, many people have flulike symptoms as the adaptive immune response begins. B cells make antibodies that can be detected by an HIV test. Armies of helper T cells and killer T cells also form. With

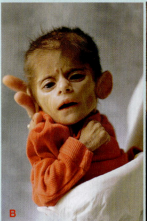

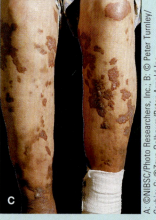

A: ©NIBSC/Photo Researchers, Inc.; B: © Peter Turnley/Corbis; C: ©Zeva Delbaum/Peter Arnold Inc.

Figure 9.20 HIV disables the immune system. A A human T cell (*blue*) infected with HIV (*red*). **B** This Romanian baby contracted AIDS from his mother's breast milk and later died. **C** Lesions of Kaposi's sarcoma, a cancer that is common in adult AIDS patients.

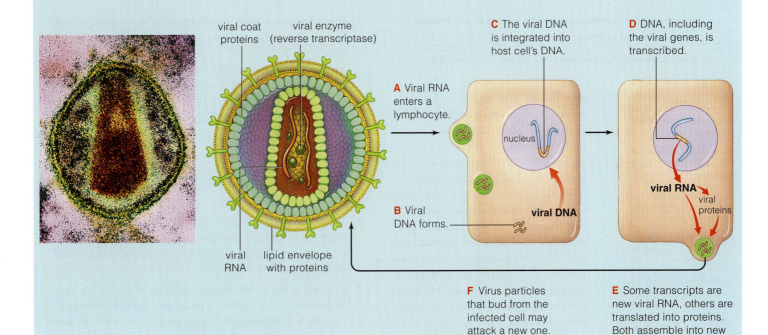

viral coat proteins

viral enzyme (reverse transcriptase)

C The viral DNA is integrated into host cell's DNA.

D DNA, including the viral genes, is transcribed.

A Viral RNA enters a lymphocyte.

nucleus

B Viral DNA forms.

viral DNA

viral RNA

viral proteins

viral RNA

lipid envelope with proteins

F Virus particles that bud from the infected cell may attack a new one.

E Some transcripts are new viral RNA, others are translated into proteins. Both assemble into new virus particles.

Figure 9.21 Animated! This diagram summarizes the steps by which HIV replicates inside a cell. (©NIBSC/Photo Researchers, Inc.; © Cengage Learning)

time, however, the adaptive immune response begins to slow as up to 1 billion new virus particles are built every day. They bud from the plasma membrane of an infected helper T cell or are released when the membrane ruptures.

Over time, billions of HIV particles and masses of infected T cells accumulate in lymph nodes. The number of circulating virus particles also increases and the body produces fewer and fewer helper T cells to replace those it has lost. As the number of healthy helper T cells drops, the person may lose weight and experience symptoms such as fatigue, nausea, heavy night sweats, enlarged lymph nodes, and a series of minor infections. With time, one or more of the typical AIDS indicator diseases appear. These are the diseases that eventually kill the individual.

Can drugs and vaccines be used to help fight HIV?

Drugs can't cure infected people, because there is no way to remove HIV genes that are already inserted into someone's DNA. Also, HIV mutates rapidly, so it can rapidly develop resistance to drugs. Even so, researchers have developed a fairly effective arsenal of anti-HIV drugs. Protease inhibitors block the action of HIV protease, an enzyme required for the assembly of new virus particles. Other drugs inhibit an enzyme that allows the virus to replicate itself.

At present the preferred treatment is antiretroviral therapy, or ART. With this therapy, patients receive a drug "cocktail" that often consists of a protease inhibitor and at least two anti-HIV drugs. This regimen can sometimes suppress HIV, at least for a time. The World Health Organization credits the increasing availability of ART for much of the decline in AIDS deaths in recent years. The drug cocktails may have serious side effects, however, and because they are expensive, patients in wealthier countries have easiest access to them.

The search also is on for compounds that might prevent HIV from entering human cells. Such "entry inhibitors" are now being tested.

Making an effective HIV vaccine has proven to be a tall order. Because HIV mutates rapidly, there may be many different genetic forms in a single person, and each presents the immune system with a different antigen. No single vaccine has yet been devised that can keep up with this challenge. Despite the obstacles, researchers are hopeful. At present, several HIV vaccines are undergoing clinical trials in various parts of the world.

Most HIV infections result from a personal choice to have unprotected sex or to use a shared needle for intravenous drugs. Education programs around the world are having an effect on the spread of the virus. In many—but not all—countries, the incidence of new cases of HIV each year is slowing. Even so, many battles remain before we come close to winning the global battle against AIDS.

9.11 | Patterns of Infectious Disease

- Infections that can threaten health spread in predictable ways and occur in predictable patterns.

- Links to Antibiotic resistance 1.7, Blood disorders 8.8

endemic disease Disease that occurs more or less continously.

epidemic A disease rate that exceeds what would be expected.

nosocomial infection An infection acquired in a hospital.

pandemic Widespread occurrence of an epidemic disease.

sporadic disease A disease that breaks out irregularly.

virulence A measure of the likelihood that a pathogen will make its host seriously ill.

Our innate and adaptive defenses evolved to protect the body from pathogens and dangerous abnormalities such as cancerous cells. This is how defense responses make a vital contribution to homeostasis.

Pathogens are ranked according to their **virulence**—how likely it is that the pathogen will make its host seriously ill. Virulence depends on how fast the pathogen can invade tissues, the degree of damage it causes, and which tissues it targets. For example, a virus that can cause pneumonia is more virulent than one that causes the sniffles. Rabies viruses are highly virulent because they target the brain. Antibiotic resistance in some bacteria has made them highly virulent. Infectious-disease experts have instituted a worldwide surveillance system to flag new resistant strains before they become established.

Pathogens spread in four ways

By definition, an infectious disease can be transmitted from person to person. There are four common modes of transmission:

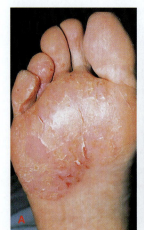

Figure 9.22 Many infections are spread by contact or when a pathogen is inhaled. A Athlete's foot, caused by a fungus that lives in warm, damp places such as shower stalls. **B** Sneezes, used tissues, and contaminated hands all may spread a virus. (Left: ©Dr. P. Marazzi/Photo Researchers, Inc.; Right: © Yoav Levy/Phototake—All rights reserved.)

1. **Direct contact** with a pathogen, as by touching open sores or body fluids from an infected person. (This is where "contagious" comes from; the Latin *contagio* means touch or contact.) Infected people can transfer pathogens from their hands, mouth, or genitals.

2. **Indirect contact**, as by touching doorknobs, tissues, diapers, or other objects previously in contact with an infected person. As already noted, food, water and surfaces can be contaminated by pathogens. Some common infections are caused by organisms that are nearly always present in our surroundings (Figure 9.22A).

3. **Inhaling pathogens**, such as cold and flu viruses, that have been spewed into the air by uncovered coughs and sneezes (Figure 9.22B). This is the most common mode of transmission.

4. **Contact with a vector**, such as flies, fleas, and ticks. A disease vector carries a pathogen from an infected person or contaminated material to new hosts. In some cases, part of the pathogen's life cycle must take place inside the vector, which is an intermediate host. For example, mosquitoes are the intermediate hosts for the West Nile virus and parasites that cause malaria.

Every year 5 to 10 percent of hospital patients come down with a **nosocomial infection**—one that is acquired in a hospital. The MRSA infection mentioned in Section 8.8 is an example. Nosocomial infections are so common because anyone sick enough to be hospitalized may have a compromised immune system, and invasive medical procedures give bacteria easy access to tissues. Also, the intensive use of antibiotics in hospitals increases the chances that antibiotic-resistant pathogens will be present there. Hospitals usually are careful to monitor patients vulnerable to nosocomial infection.

Diseases occur in several patterns

Infectious diseases may occur in different patterns. In an **epidemic**, a disease rate increases to a level above what we would predict, based on experience. When cholera spread throughout Haiti in 2011, that was an epidemic. When epidemics break out in several countries around the world in a given time span, they collectively are called a **pandemic**. HIV/AIDS is pandemic. A **sporadic disease**, such as whooping cough, breaks out irregularly. Usually, relatively few people are affected. An **endemic disease**, such as the common cold, occurs more or less continuously. Many of the diseases listed in Table 9.5 are endemic in various parts of the world.

You may recall from Chapter 1 that emerging diseases are caused by pathogens that once were present only in a limited geographical area or that only recently have begun

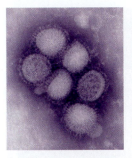

H1N1 virus particles

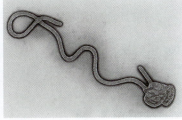

An Ebola virus

Left: CDC/C. S. Goldsmith and A. Balish
Right: ©CAMR/A. Barry Dowsett/Photo
Researchers, Inc.

Figure 9.23 Emerging disease pathogens include the H1N1 virus (*left*) and the Ebola virus (*right*). Another pathogen of increasing concern is the virus that causes avian influenza ("bird flu") that you will learn more about in Chapter 10.

to infect humans. One is the "swine flu" pandemic that spread around the world in 2009. It is caused by the H1N1 virus (Figure 9.23), which is genetically similar to a virus that causes flu in pigs. Most people infected with H1N1 have relatively mild symptoms. That's not the case with the Ebola virus, which is carried by bats. Although still limited to a few parts of Africa, Ebola causes massive bleeding and kills 90 percent of the humans it infects.

There are many public and personal strategies for preventing infection

The best way to combat any disease is to prevent it in the first place (Figure 9.24). With infectious diseases, prevention depends on knowing how a disease is transmitted and what the pathogen reservoir is. Preventive measures recognize that the human body, soil, water, and other animals all are reservoirs for a range of pathogens.

TABLE 9.5 Infectious Diseases: Some Common Global Health Threats*		
Disease	**Type of Pathogen**	**Estimated Deaths per Year**
Diarrheas (includes amoebic dysentery, cryptosporidiosis)	Protozoa, virus, and bacteria	31 million
Various respiratory infections (pneumonia, viral influenza, diphtheria, strep infections)	Virus, bacteria	71 million
Malaria	Protozoa	2.7 million
Tuberculosis	Bacteria	2.4 million
Hepatitis (includes A, B, C, D, E)	Virus	1–2 million
Measles	Virus	220,000
Schistosomiasis	Worm	200,000
Whooping cough	Bacteria	100,000
Hookworm	Worm	50,000+

*Does not include AIDS-related deaths.

The World Health Report 2010

Respiratory tract
Preventative measures:
- Hand washing
- Cover mouth when coughing or sneezing
- Proper disposal of used tissues
- Vaccination programs

GI tract
Preventative measures:
- Hand washing
- Proper food storage, handling, and cooking
- Good public sanitation (sewage, drinking water)

Blood
Preventative measures:
- Avoid/prevent needle sharing/ IV drug abuse
- Maintain pure public blood supplies
- Vaccination programs against blood-borne pathogens (e.g., hepatitis B)

Skin
Preventative measures:
- Hand washing
- Limit contact with items used by an infected person

© Cengage Learning

A

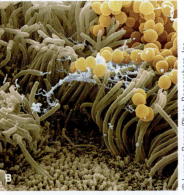

B

©Juergen Berger/Photo Researchers, Inc.

Figure 9.24 It's helpful to know how pathogens spread and what their reservoirs are. A Some recommended strategies for preventing the spread of infectious disease. **B** *Staphylococcus aureus* bacteria (*yellow balls*) sticking to cilia of a person's nasal epithelium. This strain is common on the skin and on the epithelial lining of the nose, throat, and intestines. It is a leading cause of bacterial disease in humans.

TAKE-HOME MESSAGE

HOW DO DISEASE PATHOGENS SPREAD?

- Disease pathogens are spread by direct or indirect contact, by being inhaled, or by vectors (such as insects).
- Some disease organisms are extremely virulent—they can cause severe illness.
- Simple hand washing is a good strategy for avoiding many common infectious organisms.

EXPLORE
ON YOUR OWN

The photograph in Figure 9.25 shows a reaction to a skin test for tuberculosis. For this test, a health care worker scratches a bit of TB antigen into a small patch of a patient's skin. In people who have a positive reaction to the test, a red swelling develops at the scratch site, usually within a day or two. Even in a person with no medical history of the disease, this response is visible evidence of immunological memory. It shows that there has been an immune response against the tuberculosis bacterium, which the person's immune system must have encountered at some time in the past. Tests for allergies work the same way.

In many communities, a TB test is required for people who are applying for jobs that involve public contact, such as teaching in the public schools. To learn more about this public health measure, find out if the test is required in your community, where it is available, and why public health authorities believe it is important.

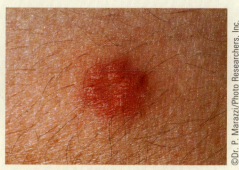

Figure 9.25 This skin eruption indicates a positive reaction to a tuberculosis skin test.

©Dr. P. Marazzi/Photo Researchers, Inc.

SUMMARY

Section 9.1 The body protects itself from pathogens with general and specific responses of white blood cells and chemicals they release. Inborn responses provide innate immunity. Responses after the body detects antigens of specific pathogens provide adaptive immunity (Table 9.6). An antigen is a molecule that triggers an immune response against itself. Chemicals called cytokines help organize or strengthen immune responses.

Section 9.2 T and B cells are stationed in lymph nodes, the spleen, and other parts of the lymphatic system. Lymph vessels also recover water and dissolved substances that have escaped from the bloodstream and return them to the general circulation.

Section 9.3 The body's first line of defense against pathogens includes physical barriers such as intact skin and mucous membranes and chemical barriers such as tears, saliva, gastric juice, and lysozyme in mucus. Urine and diarrhea help flush pathogens from the urinary tract and GI tract.

Section 9.4 General, innate immune responses may stop an infection from setting in. Macrophages are "first responders" that engulf and digest foreign agents and clean up damaged tissue. Complement proteins bind to pathogens and kill them by inserting membrane attack complexes into the invader's plasma membrane. They also attract phagocytes.

Activated complement and cytokines from macrophages trigger inflammation, a fast, local response to tissue damage (Table 9.7). Signs of inflammation include redness, warmth, and pain. Chemical signals triggered by infection can cause a fever.

Section 9.5 Adaptive defenses attack specific pathogens. They combat a great diversity of antigens, and generate memory T and B cells that provide extended immunity. When B cells and T cells recognize an antigen, they are activated and multiply to form large populations of identical cells.

Plasma cells are effector B cells that make antibodies which bind specific antigens and flag them to be destroyed by phagocytes or other defender cells.

TABLE 9.6 The Body's Three Lines of Defense

NONSPECIFIC

Barriers

Intact skin; mucous membranes at other body surfaces

Infection-fighting chemicals in tears, saliva, gastric fluid

Resident bacteria that outcompete pathogens

Flushing effect of tears, saliva, urination, diarrhea, sneezing, and coughing

Innate Immune Responses

Complement system

Inflammation
    Fast-acting white blood cells (neutrophils, eosinophils, and basophils), macrophages
    Also blood-clotting proteins, infection-fighting cytokines, NK cells

SPECIFIC

Adaptive Immune Responses

T cells, B cells/plasma cells

Cytokines such as interleukins, other chemical weapons (such as antibodies, perforins)

TABLE 9.7 Some Chemical Weapons of Immunity

COMPLEMENT	Proteins directly kill cells; stimulate lymphocytes
CYTOKINES	Communication chemicals
Interleukins	Cause inflammation and fever; cause T cells and B cells to divide and specialize; stimulate bone marrow stem cells; attract phagocytes; activate NK cells
Interferons	Confer resistance to viruses; activate NK cells

T cells provide cell-mediated immunity. T cells recognize combinations of antigen fragments and MHC self markers. These complexes are produced by antigen-presenting cells. Cytotoxic T cells are effectors that attack intruders directly. Helper T cells release cytokines that mobilize and strengthen defense responses.

 Section 9.6 Antibodies target pathogens outside cells. In the antibody-mediated response, plasma cells secrete large numbers of antibodies that circulate in the bloodstream. Antibodies are proteins called immunoglobulins. Each binds one kind of antigen. The five classes of antibodies are IgG, IgD, IgE, IgA, and IgM.

 Section 9.7 Cell-mediated responses destroy infected cells, cancer cells, and cells of tissue or organ transplants. Cytotoxic T cells secrete chemicals that can trigger apoptosis (programmed cell death) in an invading cell.

 **Section 9.8** Vaccination provokes an active immune response, including the production of memory cells. Injections of antibodies help fight infection by conferring short-term passive immunity. Monoclonal antibodies are used in medical research, testing, and the treatment of various diseases.

 Section 9.11 Disease pathogens spread by contact, by being inhaled, or by vectors. Highly virulent pathogens are the most serious threats to health.

REVIEW QUESTIONS

1. While you're jogging in the surf, your toes land on a jellyfish. Soon the bottoms of your toes are swollen, red, and warm to the touch. Using the diagram below as a guide, describe how these signs of inflammation came about.

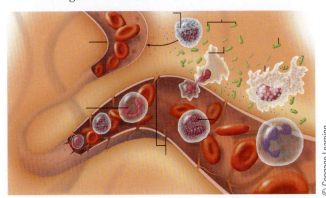
© Cengage Learning

2. Distinguish between
   a. neutrophil and macrophage
   b. cytotoxic T cell and natural killer cell
   c. effector cell and memory cell
   d. antigen and antibody

3. What is the difference between innate immunity and adaptive immunity?

4. What is the difference between an allergy and an autoimmune response?

SELF-QUIZ *Answers in Appendix V*

1. _____ are barriers to pathogens at body surfaces.
   a. Intact skin and mucous membranes
   b. Tears, saliva, and gastric fluid
   c. Resident bacteria
   d. All are correct

2. Complement proteins function in defense by _____.
   a. neutralizing toxins
   b. enhancing resident bacteria
   c. promoting inflammation
   d. forming pores that cause pathogens to disintegrate
   e. both a and b are correct
   f. both c and d are correct

3. _____ are molecules that lymphocytes recognize as foreign and that elicit an immune response.
   a. Interleukins   d. Antigens
   b. Antibodies   e. Histamines
   c. Immunoglobulins

4. Another term for *antibodies* is _____; there are _____ classes of these molecules.
   a. B cells; three
   b. immunoglobulins; three
   c. B cells; five
   d. immunoglobulins; five

5. Antibody-mediated responses work best against _____.
   a. pathogens inside cells   d. both b and c
   b. pathogens outside cells   e. all are correct
   c. toxins

6. Cell-mediated responses work best against _____.
   a. pathogens inside cells   d. both b and c
   b. pathogens outside cells   e. all are correct
   c. toxins

7. The most common antigens are _____.
   a. nucleotides   c. steroids
   b. triglycerides   d. proteins

8. The ability to develop a secondary immune response is based on _____.
   a. memory cells   d. effector cytotoxic
   b. circulating antibodies   T cells
   c. plasma cells   e. mast cells

9. Tears are part of the body's defensive arsenal. What defense category do they fall into, and why?

10. Match the immunity concepts:
    ___ inflammation   a. neutrophil
    ___ antibody secretion   b. plasma cell
    ___ phagocyte   c. nonspecific response
    ___ immunological memory   d. purposely causing
    ___ vaccination   memory cell
    ___ allergy   production
                            e. basis of secondary
                               immune response
                            f. nonprotective
                               immune response

1. New research suggests a link between some microbes that normally live in the body and seemingly unrelated major illnesses. The gum disease called periodontitis itself is not life-threatening, for instance, but it is a fairly good predictor for heart attacks. Bacteria that cause gum disease can trigger inflammation. Thinking back to your reading in Chapter 7, how do you suppose that this response also may be harmful to the heart?

2. Given what you now know about how foreign invaders trigger immune responses, explain why mutated forms of viruses, which have altered surface proteins, pose a monitoring problem for a person's memory cells.

3. Researchers have been trying to develop a way to get the immune system to accept foreign tissue as "self." Can you think of some clinical applications for such a development?

4. Elena developed chicken pox when she was in kindergarten. Later in life, when her children developed chicken pox, she stayed healthy even though she was exposed to countless virus particles each day. Explain why.

5. By the 1790s when English physician Edward Jenner (*right*) was treating patients, people all over the world had been trying to protect themselves against the scourge of smallpox for centuries. Jenner observed that people who caught cowpox, a similar but less virulent disease, never got smallpox, and this led him to wonder if something about cowpox was protective. To test his hypothesis, he injected a young boy with material from cowpox scabs. After the boy's bout of cowpox was over, Jenner injected him with pus from a smallpox sore. The boy stayed healthy, and the episode led to the discovery of vaccination—a term that literally means "encowment." What do you think would happen if a physician tried this experiment today?

Your Future

As noted in Section 9.10, there are different genetic subtypes of the human immune deficiency virus (HIV), which causes AIDS. A successful HIV vaccine will have to combat these various viral strains. Recently cancer researchers at Harvard University's Dana-Farber Cancer Institute have made major progress toward that goal. They discovered that anti-HIV monoclonal antibodies can attach to a certain region of the HIV protein envelope (called the V3 loop) in more than one HIV strain—including one that is responsible for nearly two-thirds of HIV infections worldwide. Next steps include figuring out how to make a vaccine that will stimulate the body to make its own antibodies capable of attaching to the V3 loop—allowing the immune system to ward off the virus before a serious infection sets in.

©The Granger Collection, New York

THE RESPIRATORY SYSTEM

In this chapter your knowledge of concentration gradients and diffusion (3.10) will help you understand the mechanisms that move oxygen into and carbon dioxide out of the body.

You will see how the respiratory system works together with the cardiovascular system (7.1) to supply oxygen and remove carbon dioxide.

You will also learn how hemoglobin and red blood cells function in gas exchange (8.2).

KEY CONCEPTS

The Respiratory System

Respiration provides the body with the oxygen for aerobic respiration in cells. It also removes waste carbon dioxide. These gases enter and leave the body by way of the respiratory system. Sections 10.1–10.4

Gas Exchange

Oxygen and carbon dioxide are exchanged across the thin walls of microscopic sacs in the lungs called alveoli. Circulating blood carries gases to and from the lungs. Section 10.5

Breathing Controls

The nervous system controls the rate, depth, and rhythmic pattern of breathing. Other controls match air flow to blood flow. Section 10.6

Disorders of the Respiratory System Sections 10.7–10.8

CONNECTIONS:
The Respiratory System in Homeostasis Section 10.9

Top three: © Cengage Learning; Bottom: O. Auerbach/Visuals Unlimited, Inc.

Each day in the United States several thousand teens try their first cigarette. It's often a nasty experience: The first time a person lights up, the irritants in smoke typically cause coughing, burning in the throat, and often nausea. These responses occur because tobacco smoke is toxic to human tissues. These days most people know that smoking is a major risk factor for lung cancer. It also is linked with cancers of the tongue, throat, and other tissues and organs. For example, females who start smoking in their teens are about 70 percent more likely to develop breast cancer than those who don't smoke. Tobacco smoke also elevates blood pressure and blood levels of LDL ("bad") cholesterol, and it lowers levels of "good" cholesterol (HDL). In fact, studies show that smoking doubles the risk of heart disease. Even with all this evidence of smoking's ill effects, many smokers struggle mightily to quit, because the nicotine in cigarette smoke is highly addictive.

Anything we inhale that enters the bloodstream gets there by way of the respiratory system, our focus in this chapter. Its parts collectively have one basic job—to bring in oxygen cells need for their metabolism, and to dispose of cells' carbon dioxide wastes.

Timothy Large/Shutterstock.com

Homeostasis Preview

Cells require oxygen for making ATP in mitochondria and also must get rid of potentially toxic carbon dioxide the reactions produce. The respiratory system exchanges these gases with the bloodstream.

■ Getting oxygen from air and releasing carbon dioxide wastes are the basic functions of the respiratory system.

Airways are pathways for moving air

The lungs and airways are the centerpieces of the human **respiratory system** (Figure 10.1). When a person breathes quietly, air typically enters and leaves the system by way of the nose. Hairs at the entrance to the nasal cavity and in its ciliated epithelial lining filter out large particles, such as dust, from incoming air. The air also is warmed in the nose and picks up moisture from mucus. A septum (wall) of bone and cartilage separates the nasal cavity's two chambers. Channels link the cavity with paranasal sinuses above and behind it (which is why nasal sprays can relieve mucus-clogged sinuses). Tear glands produce moisture that drains into the nasal cavity. Crying increases the flow, which is why your nose runs when you cry.

From the nasal cavity, air moves into the **pharynx**. This is the entrance to both the **larynx** (an airway) and the esophagus (which leads to the stomach). Nine pieces of cartilage form the larynx. One of these, the thyroid cartilage, is the Adam's apple.

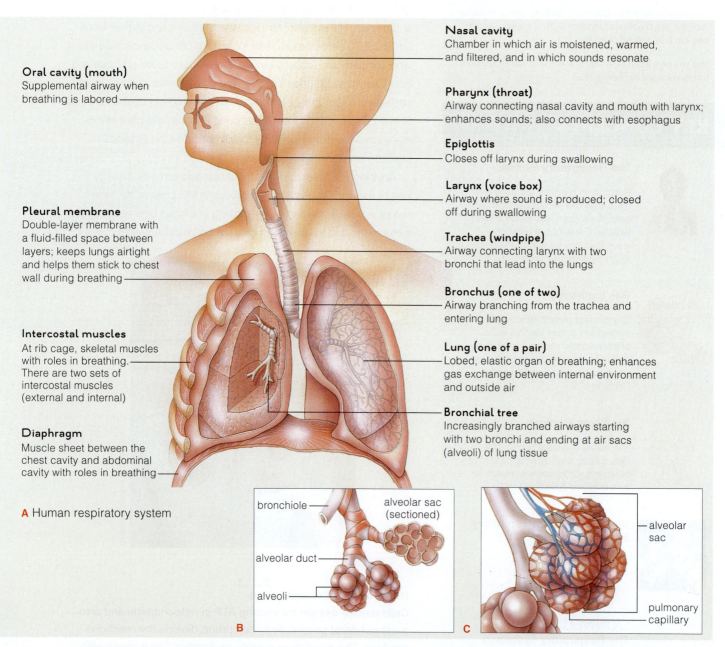

Oral cavity (mouth)
Supplemental airway when breathing is labored

Pleural membrane
Double-layer membrane with a fluid-filled space between layers; keeps lungs airtight and helps them stick to chest wall during breathing

Intercostal muscles
At rib cage, skeletal muscles with roles in breathing. There are two sets of intercostal muscles (external and internal)

Diaphragm
Muscle sheet between the chest cavity and abdominal cavity with roles in breathing

A Human respiratory system

Nasal cavity
Chamber in which air is moistened, warmed, and filtered, and in which sounds resonate

Pharynx (throat)
Airway connecting nasal cavity and mouth with larynx; enhances sounds; also connects with esophagus

Epiglottis
Closes off larynx during swallowing

Larynx (voice box)
Airway where sound is produced; closed off during swallowing

Trachea (windpipe)
Airway connecting larynx with two bronchi that lead into the lungs

Bronchus (one of two)
Airway branching from the trachea and entering lung

Lung (one of a pair)
Lobed, elastic organ of breathing; enhances gas exchange between internal environment and outside air

Bronchial tree
Increasingly branched airways starting with two bronchi and ending at air sacs (alveoli) of lung tissue

B
bronchiole
alveolar sac (sectioned)
alveolar duct
alveoli

C
alveolar sac
pulmonary capillary

Figure 10.1 Animated! The respiratory system includes the lungs and airways. Also shown are the diaphragm and other structures with secondary roles in respiration. (© Cengage Learning)

Figure 10.2 The lining of the airways includes mucus-secreting cells (*orange*) and tufts of hairlike cilia.

CNRI/Photo Researchers, Inc.

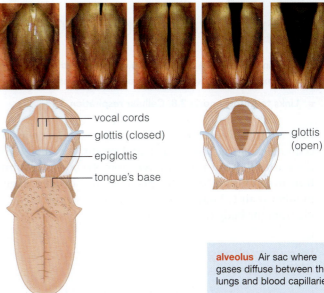

Figure 10.3 The vocal cords are at the upper opening of the larynx. Contraction of skeletal muscle in the cords changes the width of the glottis, the gap between them. The sketches show what the glottis looks like when it is closed and opened. (Courtesy of Kay Elemetrics Corporation; © Cengage Learning)

The flaplike **epiglottis**, attached to the larynx, points up during breathing. When you swallow, the larynx moves up so that the epiglottis partly covers the opening of the larynx. This helps prevent food from entering the respiratory tract and causing choking.

From the larynx, air moves into the "windpipe" or **trachea** (TRAY-key-uh). Press gently at the lower front of your neck, and you can feel some of the bands of cartilage that ring the tube, adding strength and helping to keep it open. The trachea branches into two airways, one leading to each lung. Each airway is a **bronchus** (BRONG-cuss; plural: BRONG-kee). The epithelial lining of bronchi includes mucus-secreting cells and cilia. Figure 10.2 shows a close-up of these cilia. The mucus traps bacteria and airborne particles, then the upward-beating cilia sweep the debris-laden mucus toward the mouth.

Just above the larynx, horizontal folds of an elastic mucous membrane form the **vocal cords** (Figure 10.3). When you exhale, air rushes through the *glottis*, a gap between the cords that opens to the larynx. Air moving through it makes the cords vibrate. By controlling the vibrations we can make sounds. Using our lips, teeth, tongue, and the soft roof of the mouth (the soft palate), we can form these sounds into vocalizations such as speech.

Gases are exchanged in the lungs

Your **lungs** are cone-shaped organs separated from each other by the heart. The left lung has two lobes, the right lung three. The lungs are located inside the rib cage above the **diaphragm**, a sheet of muscle between the thoracic (chest) and abdominal cavities. The lungs are soft, spongy, and elastic and don't attach directly to the chest wall. Instead, each lung is enclosed by a pair of thin membranes called **pleurae** (singular: pleura). This arrangement is not unlike your fist pushed into an inflated balloon. A lung occupies the same sort of position as your fist, and the pleural membrane folds back on itself (as the balloon does) to form a closed pleural sac. A narrow intrapleural space ("intra-" means between) separates the membrane's two facing surfaces. A thin film of lubricating fluid in the space reduces chafing between the membranes.

Inside each lung, the bronchi narrow as they branch and form "bronchial trees." These narrowing airways are **bronchioles**. Their narrowest portions deep in the lungs are *respiratory bronchioles*. In each lung, about 150 million tiny air sacs bulge out from their walls. Each sac is an **alveolus** (plural: alveoli). Alveoli are where gases diffuse between the lungs and blood capillaries (Figures 10.1B and 10.1C). Together the millions of alveoli provide a huge surface area for this exchange of gases. If they were stretched out as a single layer, they would cover the body several times over—or the floor of a racquetball court!

alveolus Air sac where gases diffuse between the lungs and blood capillaries.

bronchiole Narrow passageway that ends in clusters of alveoli.

bronchus Airway that leads directly to a lung.

diaphragm Sheet of muscle separating the thoracic and abdominal cavities.

epiglottis Flap that closes off the larynx during swallowing.

larynx The voice box, the airway where sound is produced.

lungs Lobed organs where gas exchange occurs.

pharynx The throat; it connects to the larynx and esophagus.

pleurae Paired membranes that enclose each lung.

respiratory system Organ system that consists of the lungs and airways.

trachea The windpipe; it branches into the two bronchi.

vocal cords Horizontal folds above the larynx that vibrate as air passes upward between them.

TAKE-HOME MESSAGE

WHAT ARE THE MAIN STRUCTURES AND FUNCTIONS OF THE RESPIRATORY SYSTEM?

- The lungs and airways are the main components of the respiratory system.
- The respiratory system's basic functions are taking oxygen into the body and removing carbon dioxide.
- In alveoli inside the lungs, oxygen enters lung capillaries, and carbon dioxide leaves them to be exhaled.

10.2 Respiration = Gas Exchange

- All living cells in the body rely on respiration to supply them with oxygen and dispose of carbon dioxide wastes.

- Links to Mitochondria 3.8, Cellular respiration 3.15

Chapter 3 discussed how aerobic cellular respiration inside cell mitochondria uses oxygen and produces carbon dioxide wastes that must be removed from the body. **Respiration**, in contrast, refers to the processes that deliver oxygen in inhaled air to body cells and remove waste carbon dioxide from the body (Figure 10.4).

In gas exchange, oxygen and carbon dioxide diffuse down a pressure gradient

Gas exchange in the body relies on the tendency of oxygen and carbon dioxide to diffuse down their respective concentration gradients—or, as we say for gases, their *pressure gradients*. When molecules of either gas are more concentrated outside the body, they tend to move into the body, and vice versa.

> **respiration** The processes that together deliver oxygen from the air to body cells and remove waste carbon dioxide to the outside.
>
> **respiratory surface** The thin, moist surface across which oxygen and carbon dioxide diffuse during respiration; the thin walls of alveoli in the lungs provide this surface.

At sea level the air is about 78 percent nitrogen, 21 percent oxygen, 0.04 percent carbon dioxide, and 0.96 percent other gases. Atmospheric pressure at sea level is about 760 mm Hg, as measured by a mercury barometer (Figure 10.5). Each gas accounts for only *part* of the total pressure exerted by the whole mix of gases. Oxygen's partial pressure is 21 percent of 760, about 160 mm Hg. Carbon dioxide's partial pressure is about 0.3 mm Hg.

Gases are exchanged across a thin, moist respiratory surface

Meeting the metabolic needs of a large, active animal such as a human requires extremely efficient gas exchange. Various factors influence the process. To start with, gases enter and leave the body by crossing a **respiratory surface** of thin, moist epithelium (Figure 10.6A). The surface must be thin—at most, one or two cells thick—because gases

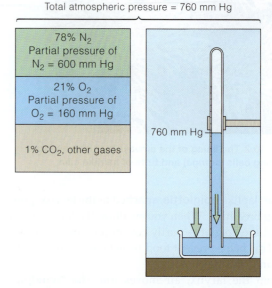

Figure 10.5 **Each gas in air exerts part of the total air pressure. This is the meaning of "partial pressure." Hg is the chemical symbol for the element mercury.** (© Cengage Learning)

only diffuse rapidly over short distances. The respiratory surface must be moist because gases can't diffuse across it unless they are dissolved in fluid. The thin walls of the millions of alveoli in the lungs meet these requirements.

Oxygen and carbon dioxide also move between body cells and tissue fluid. Blood circulated by the cardiovascular system carries these gases to and from the tissue fluid that bathes cells (Figure 10.6B).

Two factors affect how many gas molecules can move into and out of lung alveoli in a given period of time. The first is surface area, and the second is the partial pressure gradient across it. Diffusion occurs faster when the surface area is large and the gradient is steep. The millions of alveoli in your lungs provide a huge surface area for gas exchange. As we see next, the interaction between hemoglobin and oxygen helps maintain a steep gradient that in turn helps bring oxygen into the lungs.

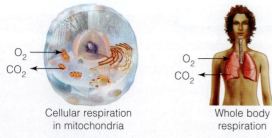

Cellular respiration in mitochondria

Whole body respiration

Figure 10.4 **Respiration is the exchange of inhaled oxygen for waste carbon dioxide, which is exhaled.** (© Cengage Learning)

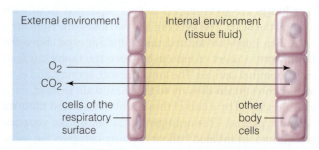

A Cells of the respiratory surface exchange gases with both the external and internal environment.

B Other body cells exchange gases with the internal environment.

Figure 10.6 **Gases are exchanged in the lungs and in tissues.** (© Cengage Learning)

When hemoglobin binds oxygen, it helps maintain the steep pressure gradient

Gas exchange also gets a boost from the hemoglobin in red blood cells. Each hemoglobin molecule binds oxygen molecules in the lungs, where the oxygen concentration is

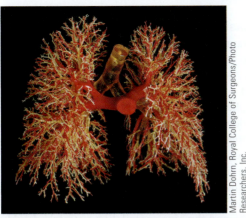

Figure 10.7 Dense networks of airways and blood vessels in the lungs support gas exchange there. In the resin cast shown in this photograph, airways are white and blood vessels are red.

Martin Dohrn, Royal College of Surgeons/Photo Researchers, Inc.

high. When blood carries red blood cells into tissues where the oxygen concentration is low, hemoglobin *releases* oxygen. Thus, by carrying oxygen away from the respiratory surface, hemoglobin helps maintain the pressure gradient that helps draw oxygen into the lungs—and into the blood in lung capillaries. Figure 10.7 gives you an idea of the intricate networks of airways and blood vessels in the lungs. Later in this chapter you will learn more about the way oxygen binds to and is released from hemoglobin.

TAKE-HOME MESSAGE

WHAT IS RESPIRATION?

- Respiration is the overall exchange of oxygen (O_2) and carbon dioxide (CO_2) between the external environment and body cells.
- Gas exchange in the respiratory system depends on steep partial pressure gradients between the outside and inside of the body.
- The larger the respiratory surface and the larger the partial pressure gradient, the faster gases diffuse.
- The cardiovascular system transports O_2 and CO_2 between the lungs and tissues.
- When hemoglobin in red blood cells binds oxygen, it helps maintain the pressure gradient that draws air into the lungs.

10.3 | Breathing at High Altitude and Underwater

SCIENCE COMES TO LIFE

In environments where there is less oxygen than normal, such as at high altitude or underwater, the rules of gas exchange change. For instance, the partial pressure of oxygen falls the higher you go (Figure 10.8). A person who isn't acclimatized to the thinner air at high altitude can become *hypoxic*—meaning that tissues are chronically short of oxygen. Above 2,400 meters (about 8,000 feet), the brain's respiratory centers trigger *hyperventilation*—faster, deeper breathing—to compensate for the oxygen deficiency. People with heart disease (which impairs blood pumping) or respiratory problems such as asthma may experience severe symptoms at high altitides, such as the heart pain called angina. Such pain indicates that the heart muscle is receiving too little oxygen.

When you swim or dive, there may be plenty of oxygen dissolved in the water but the human body has no way to extract it. (Gills do this for a fish.) People trained to dive without oxygen tanks can stay submerged only for about three minutes.

Deep divers risk nitrogen narcosis or "raptures of the deep." This condition develops because water pressure increases the deeper you go, and at about 45 meters (150 feet) dangerous amounts of nitrogen gas (N_2) start to become dissolved in tissue fluid and move into cells. In brain cells the nitrogen interferes with nerve impulses, and the diver becomes euphoric and drowsy. If a diver ascends from depth too quickly, the falling pressure causes N_2 to enter

the blood faster than it can be exhaled, so nitrogen bubbles may form in blood and tissues. The resulting pain (especially in joints) is called "the bends" or decompression sickness.

Galen Rowell/Peter Arnold, Inc.

Figure 10.8 High altitudes and underwater environments present major challenges to breathing. This photograph shows a climber approaching the summit of Mt. Everest, where the air contains much less oxygen than air at sea level.

- You will take about 500 million breaths by age 75—and probably more because young children breathe faster than adults do.
- Links to the Axial skeleton 5.3, Skeletal muscles 6.2

When you breathe, air pressure gradients reverse in a cycle

Breathing ventilates the lungs in a continuous in/out pattern called a **respiratory cycle**. Ventilation has two phases. First, **inspiration**—or inhalation—draws a breath of air into the airways. Then, in the phase of **expiration**, or exhalation, a breath moves out.

In each respiratory cycle, the volume of the chest cavity increases, then decreases (Figure 10.9). At the same time, pressure gradients between the lungs and the air outside the body are *reversed*. To understand how this shift affects breathing, it helps to remember that air in your airways (oxygen, carbon dioxide, and the other atmospheric gases) is at the same pressure as the outside atmosphere.

Before you inhale, the pressure inside all your alveoli (called *intrapulmonary pressure*) is also the same as that of outside air.

The basic respiratory cycle   As you start to inhale, the diaphragm contracts and flattens, and external intercostal muscle movements lift the rib cage up and out (Figure 10.9A). As the chest cavity expands, the lungs expand too. At that time, the air pressure in alveoli is lower than the atmospheric pressure. Fresh air follows this gradient and flows down the airways, then into the alveoli. If you take a deep breath, the volume of the chest cavity increases even more because contracting neck muscles raise the sternum and the first two ribs.

During normal, quiet breathing, expiration is passive. The muscles that contracted to bring about inspiration simply relax and the lungs recoil, like a stretched rubber band. As the lung volume shrinks, the air in the alveoli is compressed. Because pressure in the sacs now is greater than the outside atmospheric pressure, air follows the gradient and moves out of the lungs (Figure 10.9B).

If your lungs must rapidly expel more air—for instance, when you huff and puff while working out—expiration becomes active. Muscles in the wall of the abdomen contract, pushing your diaphragm upward, and other muscle movements reduce the volume of the chest cavity even more. Add to these changes the natural recoil of the lungs, and a great deal of air in the lungs is pushed outward.

Another pressure gradient aids the process   A negative pressure gradient *outside* the lungs contributes to the respiratory cycle. Atmospheric pressure is a little bit higher than the pressure in the pleural sac that wraps around the lungs. The pressure difference is enough to make the lungs stretch and fill the expanded chest cavity. It keeps the lungs snug against the chest wall even when air is being exhaled, when the lung volume is much smaller than the space inside the chest cavity. As a result, when the chest cavity expands with the next breath, so do the lungs.

You may recall from Chapter 2 that the hydrogen bonds between water molecules prevent them from being easily pulled apart. This cohesiveness of water molecules in the

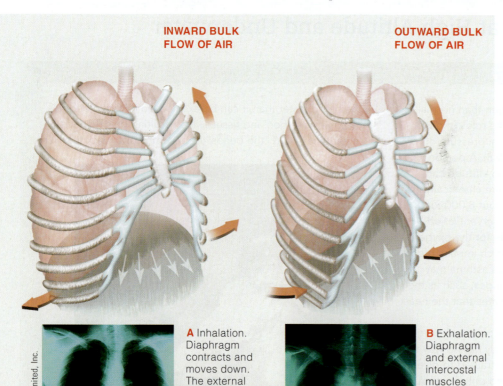

INWARD BULK FLOW OF AIR

OUTWARD BULK FLOW OF AIR

Both: SIU/Visuals Unlimited, Inc.

A Inhalation. Diaphragm contracts and moves down. The external intercostal muscles contract and lift the rib cage upward and outward. The lung volume expands.

B Exhalation. Diaphragm and external intercostal muscles return to the resting positions. Rib cage moves down. Lungs recoil passively.

Figure 10.9 Animated! The volume of the chest cavity increases, then decreases during a respiratory cycle. The X-ray image in **A** shows how taking a deep breath changes the volume of the chest (thoracic) cavity. Part **B** shows how the volume shrinks after exhalation. (© Cengage Learning)

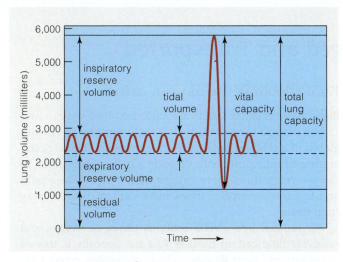

Figure 10.10 **Animated!** Lung volume changes during quiet breathing and during forced inspiration and expiration. In this graph you can see "spikes" above and below the normal tidal volume. (© Cengage Learning)

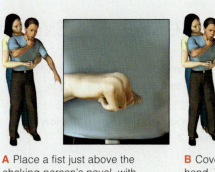

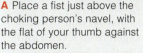

A Place a fist just above the choking person's navel, with the flat of your thumb against the abdomen.

B Cover the fist with your other hand. Thrust both fists up and in with enough force to lift the person off his or her feet.

Figure 10.11 **Animated!** The Heimlich maneuver is an emergency procedure designed to save the life of an adult who is choking. (© Cengage Learning)

fluid in the pleural sac also helps your lungs hug the chest wall, in much the same way that two wet panes of glass resist being pulled apart.

Pneumothorax, or a "collapsed lung," is caused by an injury or illness that allows air to enter the pleural cavity. The lungs can't expand normally and breathing becomes difficult and painful.

How much air is in a "breath"?

About 500 milliliters (two cupfuls) of air enters or leaves your lungs in a normal breath. This volume of air is called **tidal volume**. You can increase the amount of air you inhale or exhale, however. In addition to air taken in as part of the tidal volume, a person can forcibly inhale roughly 3,100 milliliters of air, called the *inspiratory reserve volume*. By forcibly exhaling, you can expel an additional *expiratory reserve volume* of about 1,200 milliliters of air. **Vital capacity** is the maximum volume of air that can move out of the lungs after you inhale as deeply as possible. It is about 4,800 milliliters for a healthy young man and about 3,800 milliliters for a healthy young woman. As a practical matter, people rarely take in more than half their vital capacity, even when they breathe deeply during strenuous exercise. At the end of your deepest exhalation, your lungs still are not completely emptied of air; another roughly 1,200 milliliters of *residual volume* remains (Figure 10.10).

How much of the 500 milliliters of inspired air is actually available for gas exchange? Between breaths, about 150 milliliters of exhaled "dead" air remains in the airways and never reaches the alveoli. Thus only about 350 (500 − 150) milliliters of fresh air reaches the alveoli each time you inhale. An adult typically breathes at least twelve times per minute. This rate of ventilation supplies the alveoli with 4,200 (350 × 12) milliliters of fresh air every 60 seconds. This is about the volume of soda pop in four 1-liter bottles.

When food "goes down the wrong way" and enters the trachea (instead of the esophagus), it's impossible to inhale or exhale normally. A choking person can suffocate in just a few minutes. The Heimlich maneuver can dislodge an object from the trachea by elevating the diaphragm muscle (Figure 10.11). This reduces the chest volume, forcing air up the trachea. With luck, the air will rush out with enough force to eject the item.

expiration Exhalation of a breath.

inspiration Inhalation of a breath.

respiratory cycle The in/out pattern of breathing.

tidal volume The volume of air entering and leaving the lungs during a normal breath.

vital capacity The maximum volume of air that can move out of the lungs after the deepest possible inhalation.

THINK OUTSIDE THE BOOK

By some estimates the Heimlich maneuver saves more than 100,000 lives each year. Use online resources to learn more about the procedure. Can it be used for emergencies other than choking? Can it be used on anyone? Can you use it on yourself?

TAKE-HOME MESSAGE

WHAT IS THE RESPIRATORY CYCLE?

- The in/out air movements of breathing make up the respiratory cycle.
- Breathing air movements occur when lung volume changes as the chest cavity expands and shrinks.
- These changes alter the pressure gradients between the lungs and outside air.

10.5 How Gases Are Exchanged and Transported

- Gas exchange during respiration provides body cells with oxygen for cellular respiration and picks up the carbon dioxide cells produce as a waste product.
- Links to Acid and base balance 2.7, Diffusion 3.10, How blood transports oxygen 8.2

Physiologists divide respiration into external and internal phases. *External* respiration moves oxygen from alveoli into the blood and moves carbon dioxide in the opposite direction. During *internal* respiration, oxygen moves from the blood into tissues, and carbon dioxide moves from tissues into the blood.

Alveoli are built for gas exchange

The alveoli in your lungs are ideally constructed for their function of gas exchange. The wall of each alveolus is a single layer of epithelial cells, supported by a gossamer-thin basement membrane. Hugging the alveoli are lung capillaries (Figure 10.12A). They, too, have an extremely thin basement membrane around their wall. In between the two basement membranes is a film of fluid. It may seem like a lot of layers, but the **respiratory membrane** they form is far narrower than even a fine baby hair (Figures 10.12B and 10.12C). This is why oxygen and carbon dioxide can diffuse rapidly across it—oxygen moving in and carbon dioxide moving out.

Some cells in the epithelium of alveoli secrete *pulmonary surfactant*. This substance reduces the surface tension of the watery film between alveoli. Without it, the force of surface tension can collapse the delicate alveoli. This can happen to premature babies whose underdeveloped lungs do not yet have working surfactant-secreting cells. The result is a dangerous disorder called **infant respiratory distress syndrome**.

Figure 10.12 Animated! Gases are exchanged between blood in pulmonary capillaries and air in alveoli. (A: Dr. Richard Kessel & Dr. Randy Kardon/Tissues & Organs/Visuals Unlimited, Inc.; B, C: © Cengage Learning)

Hemoglobin is the oxygen carrier

Blood plasma cannot carry enough dissolved oxygen and carbon dioxide to meet the body's requirements. The hemoglobin in red blood cells solves this problem. Hemoglobin binds and transports both O_2 and CO_2. It enables blood to carry seventy times more oxygen than it otherwise would and to carry seventeen times more carbon dioxide away from tissues.

Air inhaled into alveoli contains plenty of oxygen and relatively little carbon dioxide. Just the opposite is true of blood arriving from tissues—which, remember, enters lung capillaries at the "end" of the pulmonary circuit (Section 7.3). Thus, in the lungs, oxygen diffuses down its pressure gradient into the blood plasma and then into red blood cells, where up to four oxygen molecules rapidly form a weak, reversible bond with each molecule of hemoglobin. Hemoglobin with oxygen bound to it is called **oxyhemoglobin**, or HbO_2.

The amount of HbO_2 that forms depends on several factors. One is the partial pressure of oxygen—that is, the relative amount of oxygen in blood plasma. In general, the higher its partial pressure, the more oxygen hemoglobin will pick up, until oxygen is attached to all four hemes in

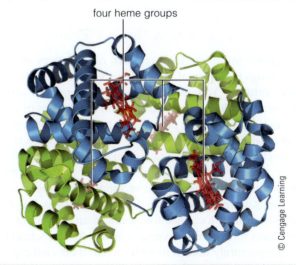

four heme groups

© Cengage Learning

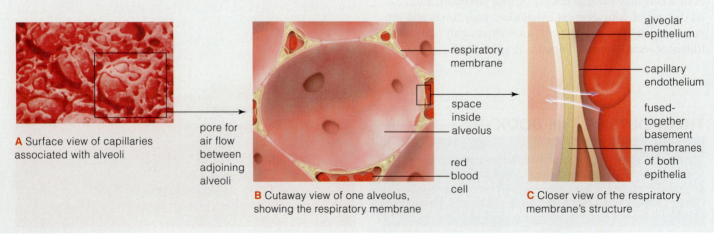

A Surface view of capillaries associated with alveoli

pore for air flow between adjoining alveoli

B Cutaway view of one alveolus, showing the respiratory membrane

respiratory membrane

space inside alveolus

red blood cell

alveolar epithelium

capillary endothelium

fused-together basement membranes of both epithelia

C Closer view of the respiratory membrane's structure

the hemoglobin molecule. HbO_2 will give up its oxygen in tissues where the partial pressure of oxygen is lower than in the blood. Figure 10.13 gives an idea of the pressure gradients in different body regions.

In highly active tissues—which have a greater demand for oxygen—the chemical conditions loosen hemoglobin's grip on oxygen. For example, the binding of oxygen weakens as temperature rises or as increasing acidity lowers the pH. Several events contribute to a falling pH. The reaction that forms HbO_2 releases hydrogen ions (H^+), making the blood more acidic. Blood pH also falls as the level of CO_2 given off by active cells increases.

When tissues are chronically short of oxygen, red blood cells increase their production of a compound called 2,3-diphosphoglycerate, or DPG for short. DPG reversibly binds hemoglobin. As more of it binds to hemoglobin, the more easily hemoglobin binds oxygen. This makes more oxygen available to tissues.

Hemoglobin and blood plasma both carry carbon dioxide

As you know, aerobic respiration in cells produces carbon dioxide as a waste. For this reason, there is more carbon dioxide in metabolically active tissues than in the blood in the nearby capillaries. So, following its pressure gradient, carbon dioxide diffuses into these capillaries. It will be carried toward the lungs in three ways. About 7 percent stays dissolved in plasma. About another 23 percent binds with hemoglobin in red blood cells, forming the compound **carbaminohemoglobin** ($HbCO_2$). Most of the carbon dioxide, about 70 percent, combines with water to form bicarbonate (HCO_3^-). The reaction has two steps. First carbonic acid (H_2CO_3) forms; then it dissociates (that is, it separates) into bicarbonate ions and hydrogen ions:

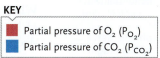

$$CO_2 + H_2O \rightleftharpoons \underset{\text{carbonic acid}}{H_2CO_3} \rightleftharpoons \underset{\text{bicarbonate}}{HCO_3^-} + H^+$$

This reaction occurs in blood plasma and red blood cells. However, it is faster in red blood cells, which contain carbonic anhydrase. This enzyme increases the reaction rate by at least 250 times. Newly formed bicarbonate in red blood cells diffuses into the plasma, which will carry it to the lungs. The reactions rapidly "sop up" carbon dioxide in the blood and help maintain the gradient that keeps CO_2 diffusing from tissue fluid into the bloodstream.

The reactions that make bicarbonate are reversed in alveoli, where the partial pressure of carbon dioxide is *lower* than it is in surrounding capillaries. The CO_2 that forms as the reactions go in reverse diffuses into the alveoli and is exhaled.

If you look again at the chemical reactions outlined in the pink shaded area above, you can see that the steps that

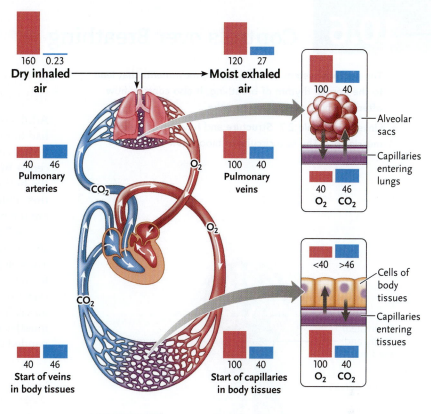

Figure 10.13 **Partial pressure gradients for oxygen and carbon dioxide change as blood travels through the cardiovascular system.** Remember that each gas moves from regions of higher to lower partial pressure. (© Cengage Learning)

form bicarbonate also produce some H^+, which makes blood more acid. What happens to these hydrogen ions? Hemoglobin binds some of them and thus acts as a buffer (Section 2.7). Certain proteins in blood plasma also bind H^+. These buffering mechanisms are extremely important in homeostasis, because they help prevent an abnormal decline in blood pH.

carbaminohemoglobin
Hemoglobin with carbon dioxide bound to it.

oxyhemoglobin
Hemoglobin with oxygen bound to it.

respiratory membrane
Two-layer membrane between the walls of lung capillaries and alveoli; blood gases diffuse across it.

TAKE-HOME MESSAGE

HOW ARE BLOOD GASES EXCHANGED AND TRANSPORTED?

- Oxygen and carbon dioxide diffuse into and out of capillaries following their partial pressure gradients.
- Hemoglobin in red blood cells greatly increases the oxygen-carrying capacity of the blood.
- Hemoglobin and blood plasma also carry carbon dioxide.
- In plasma, most carbon dioxide is transported in the form of bicarbonate.
- Buffers help prevent the blood from becoming too acid due to H^+ being released when bicarbonate forms.

10.6   Controls over Breathing

- The nervous system controls muscle movements that lead to the normal rhythm of breathing. It also controls how often and how deeply you breathe.

- Links to pH scale 2.7, Structure and function of skeletal muscles 6.2, The two circuits of blood flow 7.3

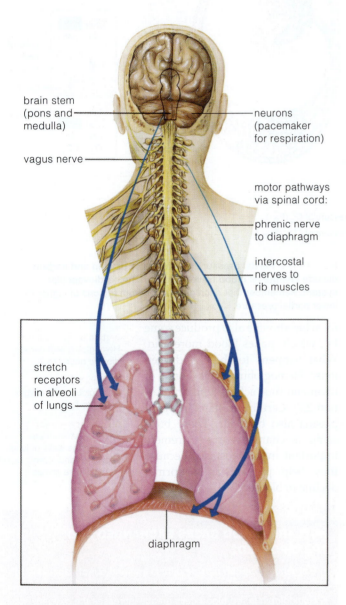

brain stem
(pons and
medulla)

vagus nerve

neurons
(pacemaker
for respiration)

motor pathways
via spinal cord:

phrenic nerve
to diaphragm

intercostal
nerves to
rib muscles

stretch
receptors
in alveoli
of lungs

diaphragm

Figure 10.14 Respiration centers in the brain control the basic operations of breathing. In quiet breathing, centers in the brain stem coordinate signals to the diaphragm and muscles that move the rib cage, triggering inhalation. When a person breathes deeply or rapidly, another center receives signals from stretch receptors in the lungs and coordinates signals for exhalation. (© Cengage Learning)

A respiratory pacemaker in the brain sets the basic rhythm of breathing

Adults usually take about 12 to 15 breaths a minute. If you had to remember to inhale and exhale each time, could you do it, even when you sleep? Luckily for us all, a respiratory center in the medulla in the brain stem, at the lower rear of the brain, provides this service. Like the heart's SA node, this center contains neurons that fire spontaneously. They are the pacemaker for respiration.

As Figure 10.14 suggests, signals from the respiratory center travel nerve pathways to the diaphragm and chest. These signals stimulate the rib cage muscles and diaphragm to contract. As you read in Section 10.4, this causes the rib cage to expand, and you inhale a breath as air moves into the lungs. In between nerve impulses, the diaphragm and chest muscles relax. Elastic recoil returns the rib cage to its unexpanded state, and you exhale as air in the lungs moves out.

Carbon dioxide is the main trigger for controls over the rate and depth of breathing

While the respiratory center governs the basic operations of breathing, other controls determine how rapidly and deeply the lungs are ventilated. Overall, these controls monitor three aspects blood chemistry: the levels of carbon dioxide and oxygen in the bloodstream and the acidity, or pH, of blood. Sensory receptors that respond to chemicals are called *chemoreceptors*. Some of the sensors are in the brain stem, and others monitor the blood flowing through arteries.

You might guess that the amount of oxygen in blood is the most important factor in respiratory control systems, but actually brain stem chemoreceptors are more sensitive to levels of carbon dioxide. The receptors also detect hydrogen ions that are produced when dissolved CO_2 leaves the bloodstream and enters fluid that bathes the medulla. In this fluid (called *cerebrospinal fluid*) the drop in pH that goes along with increasing H^+ indicates that the blood is becoming more acidic. The brain's respiratory centers respond to this signal (Figure 10.15). In short order breathing becomes more rapid and deeper. Soon the blood level of CO_2 falls—and so does the blood's acidity. Notice that this is another example of a negative feedback loop helping to maintain homeostasis.

The brain also receives information about blood gases and pH from chemoreceptors in arteries. These receptors include **carotid bodies**, where the carotid arteries branch to the brain, and **aortic bodies** in artery walls near the heart. Both types of receptors detect changes in levels of carbon dioxide and oxygen in the blood. They also detect changes in blood pH. When there is too little oxygen in the blood relative to carbon dioxide and hydrogen ions, the brain responds by increasing the ventilation rate, so more oxygen can be delivered to tissues.

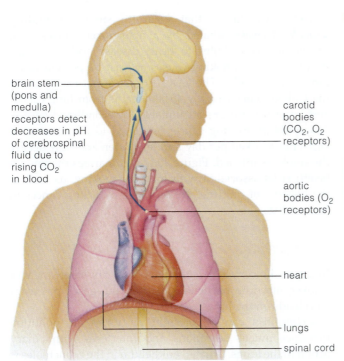

brain stem (pons and medulla) receptors detect decreases in pH of cerebrospinal fluid due to rising CO_2 in blood

carotid bodies (CO_2, O_2 receptors)

aortic bodies (O_2 receptors)

heart

lungs

spinal cord

Figure 10.15 Sensors in arteries and the brain monitor carbon dioxide, oxygen, and blood pH. (© Cengage Learning)

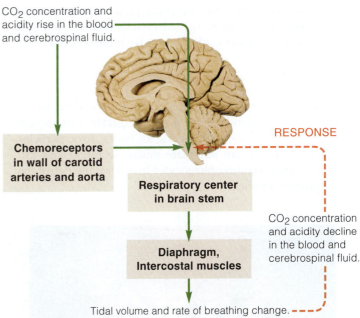

STIMULUS

CO_2 concentration and acidity rise in the blood and cerebrospinal fluid.

RESPONSE

Chemoreceptors in wall of carotid arteries and aorta

Respiratory center in brain stem

CO_2 concentration and acidity decline in the blood and cerebrospinal fluid.

Diaphragm, Intercostal muscles

Tidal volume and rate of breathing change.

Figure 10.16 Breathing patterns change with a person's activity level. (© Cengage Learning; photo: Ralph Hutchings/Visuals Unlimited, Inc.)

Overall, the mechanisms that control our breathing allow gas exchange to match the body's activity level. For example, if you start exercising, your skeletal muscles immediately require more oxygen and begin producing more CO_2. As you have just read, these changes prompt the brain's respiratory center to step up its signals to the breathing muscles (Figure 10.16).

Other controls help match air flow to blood flow

Controls over air flow also operate in the millions of lung alveoli. For example, if you get nervous, your heart may start pumping hard and fast but your lungs may not be ventilating at a corresponding pace. If too little carbon dioxide is moving out of the lungs, the rising blood level of CO_2 makes smooth muscle in the walls of bronchioles relax and widen, so more air flows through them. On the other hand, an abnormal *decrease* in the level of carbon dioxide in the lungs causes the bronchiole walls to constrict, so less air flows through them. Shifting oxygen levels have a similar effect. If you breathe in oxygen faster than it can enter blood capillaries, the oxygen level rises in parts of the lungs, capillaries dilate. As more blood flows through them, it can pick up more oxygen. When less oxygen is available in the lungs, the vessels constrict and less blood moves through them.

Only minor aspects of breathing are under conscious control

Reflexes such as swallowing or coughing briefly halt breathing. You also can deliberately alter your breathing pattern, as when you change your normal breathing rhythm to talk, laugh, or sing, or when you hold your breath underwater. At most, however, you can only hold your breath for two or three minutes. As CO_2 builds up and your blood's chemistry shifts, "orders" from the nervous system force you to take a breath.

aortic bodies Receptors in the aorta that detect shifts in blood levels of carbon dioxide and oxygen.

carotid bodies Receptors in carotid arteries that detect shifts in blood levels of carbon dioxide and oxygen.

TAKE-HOME MESSAGE

HOW IS BREATHING CONTROLLED?

- Respiratory centers in the brain stem control the rhythmic pattern of breathing.
- Brain centers that adjust the rate and depth of breathing receive information mainly from sensors that monitor blood levels of carbon dioxide.
- These controls contribute to homeostasis by helping to maintain proper levels of carbon dioxide, oxygen, and hydrogen ions in arterial blood.

10.7 Respiratory System Disorders: Tobacco, Irritants, and Apnea

A variety of infections and other disorders can prevent the respiratory system from functioning properly. Some of these problems develop when we inadvertently inhale a pathogen or noxious substances, while others we bring on ourselves.

Tobacco is an avoidable threat

People who start smoking tobacco begin wreaking havoc on their lungs. Smoke from a single cigarette can prevent cilia in bronchioles from beating for hours. Toxic particles smoke contains can stimulate mucus secretion and kill the infection-fighting phagocytes that normally patrol the respiratory epithelium.

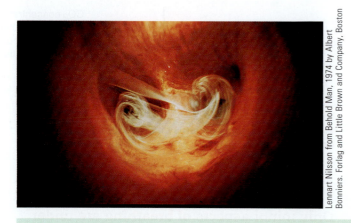

Lennart Nilsson from Behold Man, 1974 by Albert Bonniers. Forlag and Little Brown and Company, Boston

Today we know that cigarette smoke, including *secondhand smoke* inhaled by a nonsmoker, causes lung cancer and contributes to heart disease and other ills. In all, cigarette smoking causes at least 80 percent of all lung cancer deaths. The CDC estimates that each year in the United States roughly 45,000 deaths from lung cancer and heart disease are attributable to secondhand smoke. Susceptibility to lung cancer is related to the number of cigarettes smoked per day and how often and how deeply the smoke is inhaled. Figure 10.17 summarizes the known health risks associated with tobacco smoking, as well as the benefits of quitting. We look again at lung cancer in Section 10.8.

Irritants cause other disorders

In cities, in certain occupations, and anywhere near a smoker, airborne particles and irritating gases put extra workloads on the lungs.

Bronchitis can be brought on when air pollution increases mucus secretions and interferes with ciliary action in the lungs. Ciliated epithelium in the bronchioles is especially sensitive to cigarette smoke. Mucus and the particles it traps—including bacteria—accumulate in airways, coughing starts, and the bronchial walls become inflamed. Bacteria or chemical agents start destroying the wall tissue.

Effects of Smoking	Benefits of Quitting
Shortened life expectancy Nonsmokers live about 8.3 years longer than those who smoke two packs a day from their midtwenties on.	Cumulative risk reduction; after 10–15 years, the life expectancy of ex-smokers approaches that of nonsmokers.
Chronic bronchitis, emphysema Smokers have 4–25 times higher risk of dying from these diseases than do nonsmokers.	Greater chance of improving lung function and slowing down rate of deterioration.
Cancer of lungs Cigarette smoking is the major cause.	After 10–15 years, risk approaches that of nonsmokers.
Cancer of mouth 3–10 times greater risk among smokers.	After 10–15 years, risk is reduced to that of nonsmokers.
Cancer of larynx 2.9–17.7 times more frequent among smokers.	After 10 years, risk is reduced to that of nonsmokers.
Cancer of esophagus 2–9 times greater risk of dying from this.	Risk proportional to amount smoked; quitting should reduce it.
Cancer of pancreas 2–5 times greater risk of dying from this.	Risk proportional to amount smoked; quitting should reduce it.
Cancer of bladder 7–10 times greater risk for smokers.	Risk decreases gradually over 7 years to that of nonsmokers.
Cardiovascular disease Cigarette smoking is a major contributing factor in heart attacks, strokes, and atherosclerosis.	Risk for heart attack declines rapidly, for stroke declines more gradually, and for atherosclerosis it levels off.
Impact on offspring Women who smoke during pregnancy have more stillbirths, and the weight of liveborns is lower than the average (which makes babies more vulnerable to disease and death).	When smoking stops before fourth month of pregnancy, risk of stillbirth and lower birth weight eliminated.
Impaired immunity More allergic responses, destruction of white blood cells (macrophages) in respiratory tract.	Avoidable by not smoking.
Slow bone healing Surgically cut or broken bones may take 30 percent longer to heal in smokers, perhaps because smoking depletes the body of vitamin C and reduces the amount of oxygen delivered to tissues. Reduced vitamin C and reduced oxygen interfere with formation of collagen fibers in bone (and many other tissues).	Avoidable by not smoking.

Figure 10.17 From the American Cancer Society, a list of the risks incurred by smoking and the benefits of quitting. The photograph shows swirls of cigarette smoke at the entrance to the two bronchi that lead into the lungs.

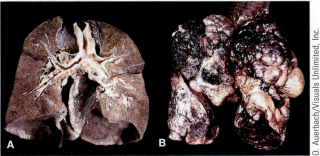

Figure 10.18 **Emphysema ravages the lungs. A** Normal human lungs that have been chemically preserved. **B** Lungs from a person with emphysema.

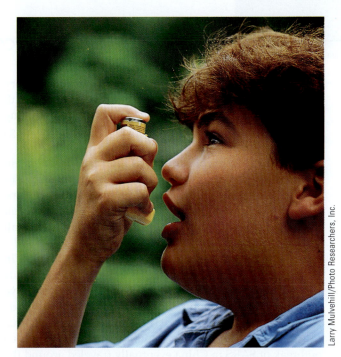

Figure 10.19 **Many asthma sufferers must use an aerosol inhaler.**

Cilia in the lining die, and mucus-secreting cells multiply as the body attempts to get rid of the accumulating debris. Eventually scar tissue forms and can block parts of the respiratory tract.

In an otherwise healthy person, even acute bronchitis is treatable with antibiotics. When inflammation continues, however, scar tissue builds up and the bronchi become chronically clogged with mucus. Also, the walls of some alveoli break down and become surrounded by stiffer fibrous tissue. The result is **emphysema**, in which the lungs are so distended and inelastic that gases cannot be exchanged efficiently (Figure 10.18). Running, walking, even exhaling can be difficult. About 1.3 million people in the United States have emphysema.

Smoking, frequent colds, and other respiratory ailments sometimes make a person susceptible to emphysema. Many emphysema sufferers lack a normal gene coding for a protein that inhibits tissue-destroying enzymes made by bacteria. Emphysema can develop over 20 or 30 years. By the time the disease is detected, however, the lungs are permanently damaged.

Millions of people suffer from **asthma**, a disorder in which the bronchioles suddenly narrow when the smooth muscle in their walls contracts in strong spasms. At the same time, mucus gushes from the bronchial epithelium, clogging the constricted passages even more. Breathing can become extremely difficult so quickly that the victim may feel in imminent danger of suffocating. Triggers include allergens such as pollen, dairy products, shellfish, pet hairs, flavorings, or even the dung of tiny mites in house dust. In susceptible people, attacks also can be triggered by noxious fumes, cold air or water, stress, strenuous exercise, or a respiratory infection. African Americans have a greater risk of developing the disease. While the reasons aren't fully understood, the incidence of asthma in the United States has grown rapidly in the last several decades. Some health experts believe that increased air pollution is at least partly to blame.

Many asthma sufferers rely on aerosol inhalers, which squirt a fine mist into the airways (Figure 10.19). A drug in the mist dilates bronchial passages and helps restore free breathing. Some devices contain powerful steroids that can harm the immune system, so inhalers should be used only with medical supervision.

Apnea is a condition in which breathing controls malfunction

As described in Section 10.6, respiration usually is on "autopilot," controlled by the brain's respiratory center. In some situations, however, a person can fail to breathe in the usual pattern. Breathing that stops briefly and then resumes spontaneously is called *apnea*. During certain times in the normal sleep cycle, breathing may stop for one or two seconds or even minutes—in extreme cases, as often as 500 times a night. This *sleep apnea* can be a contributing factor in heavy snoring.

Aging also takes a toll on the respiratory system. Sleep apnea is a common problem in the elderly, because the mechanisms for sensing a change in oxygen and carbon dioxide levels gradually become less effective over the years. Also, as we age, our lungs lose some of their elasticity, so ventilation of the lungs less efficient. Obese people often have a problem with sleep apnea, because fat deposits in the neck obstruct the airways.

Inhaled viruses, bacteria, or fungi all can infect respiratory organs. A dry cough, chest pain, and shortness of breath are symptoms of **pneumonia**. The infection inflames lung tissue, and then fluid (from edema) builds up in the lungs and makes breathing difficult.

Strains of *Streptococcus pneumoniae* can cause pneumonia and other infections. This bacterium often causes outbreaks of illness among children at day-care centers. Penicillin or some other antibiotic is the usual treatment for bacterial pneumonia. Unfortunately, today half of all strains of *S. pneumoniae* are antibiotic-resistant.

Sometimes the trigger for pneumonia is **influenza**, in which an infection that began in the nose or throat spreads to the lungs. There are many flu viruses, but several have made headlines recently. One is the H1N1 "swine flu" virus mentioned in Section 9.11. Another is the virus that causes SARS—severe acute respiratory syndrome. A 2003 outbreak of SARS in China eventually traveled around the globe (Figure 10.20A). So-called bird flu, or avian influenza, is caused by the H5N1 virus. To date it has killed about 330 people, nearly all of whom had close contact with infected wild birds. Health authorities worry that the virus may mutate in a way that allows human-to-human transmission.

Mycobacterium tuberculosis

CAMR/A. Barry Dowsett/Photo Researchers, Inc.

Tuberculosis (TB) is a lung infection caused by the bacterium *Mycobacterium tuberculosis* (*left*). It starts with flulike symptoms but eventually can destroy patches of lung tissue and can spread to other parts of the body. Antibiotics usually can cure TB, but newer drug-resistant strains of *M. tuberculosis* have made treatment much more challenging. Untreated TB can be fatal.

Lung cancer (Figure 10.21) kills more people than any other cancer. Long-term tobacco smoking is the overwhelming risk factor. In the body, some compounds in tobacco smoke and coal tar are converted to carcinogens (cancer-causing substances). They trigger genetic damage leading to lung cancer. Other risks are exposure to asbestos, radiation, and industrial chemicals such as arsenic.

Recently the incidence of lung cancer has fallen among men but risen among women. This shift is thought to be due to a rise in the relative number of female smokers several decades ago. Warning signs include cough, shortness of breath, chest pain, bloody phlegm, unexplained weight loss, and frequent respiratory infections or pneumonia.

Four types of lung cancer account for 90 percent of cases. About one-third of lung cancers are **squamous cell carcinomas** in squamous epithelium in the bronchi. About 48 percent are either **adenocarcinomas** or **large-cell carcinomas**. The most aggressive type, **small-cell carcinoma**, kills most patients within five years.

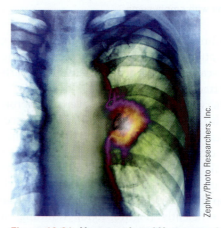

Zephyr/Photo Researchers, Inc.

Figure 10.21 Here a colored X-ray reveals a malignant lung tumor (*purple and orange*).

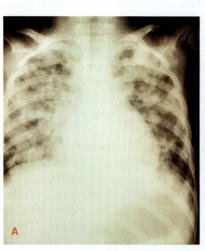

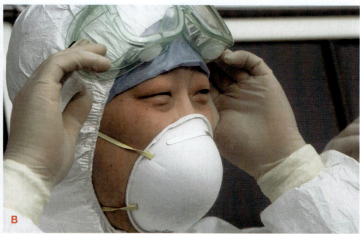

A. CDC/Dr. Joel D. Meyers B: © WHO, Pierre-Michel Virot, photographer

Figure 10.20 Many pathogens can infect the lungs. **A** X-ray showing pneumonia. Fluid has filled the lungs. **B** A health care worker in China wears protection against the SARS virus, which causes a form of influenza.

CONNECTIONS

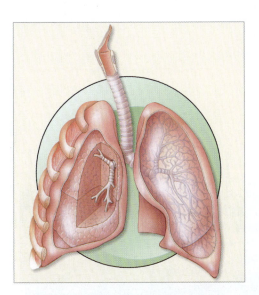

The Respiratory System

The airways and lungs bring in air and deliver the oxygen it contains into the bloodstream for transport to all living body cells. Exhaled air eliminates waste carbon dioxide that is produced as cells carry out aerobic respiration.

Muscular system
Respiratory controls over the rate and depth of breathing adjust oxygen intake and removal of carbon dioxide to service the changing demands of muscle tissue.

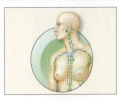

Cardiovascular system and blood
Adjustments in elimination of CO_2 help manage hydrogen ions (H^+) in blood and so help maintain blood pH (acid-base balance).

Immunity and the lymphatic system
Cilia and mucus in airways trap foreign material, functioning as physical barriers to infection.

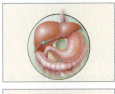

Digestive system
Voluntary contraction of the diaphragm muscle may aid voiding of feces from the large intestine/rectum.

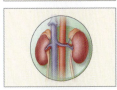

Urinary system
Binding H^+ by hemoglobin complements kidney functions that help maintain pH of blood and tissue fluid.

Nervous system
Air vibrating vocal cords allows an individual to produce spoken language.

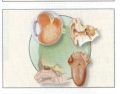

Sensory systems
Epithelium in the nose contains sensory receptors for smell (olfaction).

Endocrine system
Cells in the lungs form an enzyme (angiotensin-converting enzyme) that acts in the formation of the hormone angiotensin II, which influences formation of urine in the kidneys.

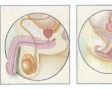

Reproductive system
Via the mother's cardiovascular system and the placenta, the respiratory system supplies oxygen to and removes carbon dioxide from the blood of a developing fetus.

EXPLORE
ON YOUR OWN

Air pollution is a serious problem in many parts of the world.
Even if you don't live near a large urban area, you may be breathing the kinds of air pollutants shown in the chart in Figure 10.22. The ultrafine particulates can stay in the air for weeks or months before they settle to Earth or are washed down by rain, and all of them are known to cause respiratory problems, especially in people who have asthma or emphysema.

Explore this health issue by finding out if your community monitors its air quality. If so, what do authorities consider to be the greatest threats to the health of you and your fellow citizens? Where do these pollutants come from?

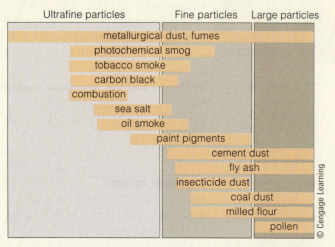

Ultrafine particles	Fine particles	Large particles
metallurgical dust, fumes		
photochemical smog		
tobacco smoke		
carbon black		
combustion		
sea salt		
oil smoke		
paint pigments		
	cement dust	
	fly ash	
	insecticide dust	
	coal dust	
	milled flour	
	pollen	

© Cengage Learning

Figure 10.22 Particles from numerous sources may be present in the air you breathe.

SUMMARY

Section 10.1 The respiratory system brings air, which contains oxygen, into the body and disposes of carbon dioxide by way of exhaled air.

Airways include the nasal cavity, pharynx, larynx, trachea, bronchi, and bronchioles. Gas exchange occurs in millions of saclike alveoli located at the end of the terminal respiratory bronchioles. Airways lead to the lungs, which are elastic organs located in the rib cage above the diaphragm. They are separated by the heart.

Section 10.2 Respiration brings oxygen from air into the blood and removes carbon dioxide from blood. Both these processes occur in the lungs. The cardiovascular system partners with the respiratory system as it circulates blood throughout the body.

Air is a mixture of oxygen, carbon dioxide, and other gases. Each gas exerts a partial pressure, and each tends to move (diffuse) from areas of higher to lower partial pressure. Following pressure gradients, oxygen diffuses into deoxygenated blood in the lungs, and carbon dioxide diffuses from the blood into the lungs to be exhaled.

In respiration, oxygen and carbon dioxide diffuse across a respiratory surface—a moist, thin layer of epithelium in the alveoli of the lungs. Airways carry gases to and from one side of the respiratory surface, and blood vessels carry gases to and away from the other side.

Section 10.4 Breathing ventilates the lungs in a respiratory cycle. During inspiration (inhalation), the chest cavity expands, pressure in the lungs falls below atmospheric pressure, and air flows into the lungs. During normal expiration (exhalation), these steps are reversed.

The volume of air in a normal breath, called the tidal volume, is about 500 milliliters. Vital capacity is the maximum volume of air that can move out of the lungs after you inhale as deeply as possible.

Section 10.5 Driven by its partial pressure gradient, oxygen in the lungs diffuses from alveoli into pulmonary capillaries. Then it diffuses into red blood cells and binds with hemoglobin, forming oxyhemoglobin. In tissues where cells are metabolically active, hemoglobin gives up oxygen, which diffuses out of the capillaries, across tissue fluid, and into cells.

Hemoglobin binds with or releases oxygen in response to shifts in oxygen levels, carbon dioxide levels, pH, and temperature.

Driven by its partial pressure gradient, carbon dioxide diffuses from cells across tissue fluid and into the bloodstream. Most CO_2 reacts with water to form bicarbonate; the reactions are speeded by the enzyme carbonic anhydrase. They are reversed in the lungs, where carbon dioxide diffuses from lung capillaries into the air spaces of the alveoli, then is exhaled.

Section 10.6 Gas exchange is regulated by the nervous system and by chemical controls in the lungs. A respiratory pacemaker in the medulla (part of the brain stem) sets the normal, automatic rhythm of breathing in and out (ventilation).

The nervous system monitors the levels of oxygen and carbon dioxide in arterial blood by way of sensory receptors. These include carotid bodies (at branches of carotid arteries leading to the brain), aortic bodies (in an arterial wall near the heart), and receptors in the medulla of the brain. Blood levels of carbon dioxide are most important in triggering nervous system commands that adjust the rate and depth of breathing.

REVIEW QUESTIONS

1. In the diagram below, label the parts of the respiratory system and the structures that enclose some of its parts.

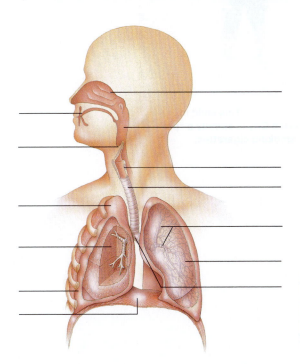

© Cengage Learning

2. What is the difference between respiration and aerobic cellular respiration?

3. Explain what a partial pressure gradient is and how such gradients figure in gas exchange.

4. What is oxyhemoglobin? Where does it form?

5. What drives oxygen from the air spaces in alveoli, through tissue fluid, and across capillary epithelium? What drives carbon dioxide in the opposite direction?

6. How does hemoglobin help maintain the oxygen partial pressure gradient during gas transport in the body?

7. What reactions enhance the transport of carbon dioxide throughout the body? How is carbon dioxide moved out of the body?

8. How do nerve impulses from the brain regulate ventilation of the lungs? How are the rate and depth of breathing controlled?

9. Why does your breathing rate increase when you exercise? What happens to your heart rate at the same time—and why?

SELF-QUIZ *Answers in Appendix V*

1. A partial pressure gradient of oxygen exists between _____.
   a. air and lungs
   b. lungs and metabolically active tissues
   c. air at sea level and air at high altitudes
   d. all of the above

2. The _____ is an airway that connects the nose and mouth with the _____.
   a. oral cavity; larynx
   b. pharynx; trachea
   c. trachea; pharynx
   d. pharynx; larynx

3. Oxygen in air must diffuse across _____ to enter the blood.
   a. pleural sacs   c. a moist respiratory surface
   b. alveolar sacs   d. both b and c

4. Each lung encloses a _____.
   a. diaphragm   c. pleural sac
   b. bronchial tree   d. both b and c

5. Gas exchange occurs at the _____.
   a. two bronchi   c. alveoli
   b. pleural sacs   d. both b and c

6. Breathing _____.
   a. ventilates the lungs
   b. draws air into airways
   c. expels air from airways
   d. causes reversals in pressure gradients
   e. all of the above

7. After oxygen diffuses into lung capillaries, it also diffuses into _____ and binds with _____.
   a. tissue fluid; red blood cells
   b. tissue fluid; carbon dioxide
   c. red blood cells; hemoglobin
   d. red blood cells; carbon dioxide

8. Due to its partial pressure gradient, carbon dioxide diffuses from cells into tissue fluid and into the _____; in the lungs, carbon dioxide diffuses into the _____.
   a. alveoli; bronchioles
   b. bloodstream; bronchioles
   c. alveoli; bloodstream
   d. bloodstream; alveoli

9. Hemoglobin performs which of the following respiratory functions?
   a. transports oxygen
   b. transports some carbon dioxide
   c. acts as a buffer to help maintain blood pH
   d. all of the above

10. Most carbon dioxide in the blood is in the form of _____.
    a. carbon dioxide
    b. carbon monoxide
    c. carbonic acid
    d. bicarbonate

CRITICAL THINKING

1. Cases of accidental carbon monoxide poisoning occur when someone builds a charcoal fire in an enclosed area. Assuming help arrives in time, what would be the most effective treatment: placing the victim outdoors in fresh air or administering pure oxygen? Explain your answer.

2. Skin divers sometimes purposely hyperventilate. Doing so doesn't increase the oxygen available to tissues. It does raise blood pH (making it more alkaline), and it decreases the blood level of carbon dioxide. Based on your reading in this chapter, how is hyperventilation likely to affect the neural controls over breathing?

3. When you sneeze, abdominal muscles abruptly contract, pushing your diaphragm upward. Given the discussion of the respiratory cycle in Section 10.4, why does this change expel air out your nose?

4. Underwater, we humans can't compete with whales and other air-breathing marine mammals, which can stay submerged for extended periods. At the beach one day you meet a surfer who tells you that special training could allow her to swim underwater without breathing for an entire hour. From what you know of respiratory physiology, explain why she is mistaken.

5. Physiologists have discovered that the nicotine in tobacco is as addictive as heroin. The cigarette-smoking child in Figure 10.23 probably is already addicted, and certainly has already begun to endanger her health. Based on the discussion in Section 10.7, what negative health effects might she develop in the coming years?

Your Future

Researchers at Duke University are working on a new therapy aimed at helping smokers kick the habit. They've developed an inhaler that delivers a blend of vaporized nicotine and pyruvate (the same molecule that forms during glycolysis). A chemical reaction between the two produces microscopic particles that a smoker can inhale just like cigarette smoke. The smoker receives enough nicotine to reduce cravings during the withdrawal period, but with far less lung irritation and less nictoine overall. Next steps aim to show that the new system is safe and effective for long-term use.

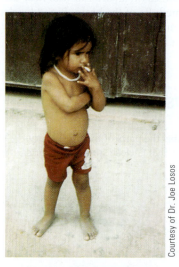

Courtesy of Dr. Joe Losos

Figure 10.23 This child in Mexico City is already a "pro" at smoking cigarettes.

This chapter explains how digestion breaks down carbohydrates (2.9), proteins (2.11), and lipids (2.10) in food.

Nutrient molecules enter the bloodstream by way of transport mechanisms that include diffusion, osmosis, and active transport (3.10–3.11).

Also relevant here is the ability of many types of body cells to extract energy from various types of biological molecules (3.16).

KEY CONCEPTS

The Digestive System

The digestive system mechanically and chemically breaks down food, absorbs nutrients, and eliminates the residues. Sections 11.1–11.8

Disorders of the Digestive System Sections 11.9–11.10

Nutrition and Body Weight

Food should supply the nutrients, vitamins, and minerals body cells require. Body weight depends on the balance between energy from food and energy used for bodily functions. Sections 11.12–11.15

CONNECTIONS:
The Digestive System in Homeostasis Section 11.11

Top: © Cengage Learning; Middle: Peter Hawtin, University of Southampton/Photo Researchers, Inc.; Bottom: © Cengage Learning/Gary Head

Since the 1990s, the numbers of overweight and obese individuals have risen dramatically all over the developed world. In North America roughly two of every three people are overweight, about half of them heavy enough to be considered obese. Adults and children, males and females, and all ethnic groups are included in these statistics. The excess weight is a risk factor for diabetes, heart disease, osteoarthritis, and some forms of cancer.

Storing fat comes naturally to our species. As with other mammals, our adipose tissues are packed with fat-storing cells. This energy warehouse evolved among our early ancestors, who unlike ourselves didn't have reliable sources of food. Stored body fat helped them through lean times. Once these cells form, they are in the body to stay. When you consume more calories than you burn, the cells fill with fat droplets.

Eating and body weight are all part of the bigger picture of food digestion and nutrition. In this chapter we begin by looking at how the digestive system brings nutrients into the body. Then we consider the body's nutritional needs and health issues related to maintaining a healthy body weight, poor eating habits, and eating disorders.

Simon Law, flickr.com/people/sfllaw

Homeostasis Preview

Except for oxygen, food digestion and the absorption of nutrients provide all the raw materials cells require to survive.

11.1 Overview of the Digestive System

- The digestive system is basically a tube with two openings—the mouth, where food enters, and the anus, where solid wastes exit.

- Link to Exocrine glands 4.1

The **digestive system** is a long tube in which food is broken down and from which the nutrients food contains are absorbed. It extends from the mouth to the anus and is often called the gastrointestinal (GI) tract (Figure 11.1). Stretched out, the GI tract would be 6.5 to 9 meters (21 to 30 feet) long in an adult.

An interesting fact about the GI tract is that while food or leftover residues are in it, technically the material is still outside the body. Nutrients don't "officially" enter the body until they move from the *lumen*—the space inside the digestive tube—into the bloodstream. Blood delivers nutrients to cells throughout the body.

From beginning to end, epithelium lines the surfaces facing the lumen. The lining is coated with thick, moist mucus that protects the wall of the tube and enhances the diffusion of substances across it.

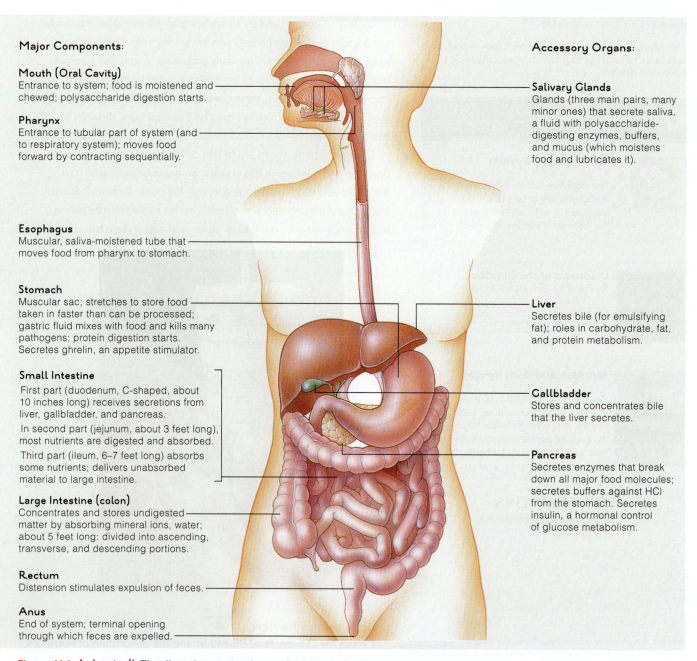

Major Components:

Mouth (Oral Cavity)
Entrance to system; food is moistened and chewed; polysaccharide digestion starts.

Pharynx
Entrance to tubular part of system (and to respiratory system); moves food forward by contracting sequentially.

Esophagus
Muscular, saliva-moistened tube that moves food from pharynx to stomach.

Stomach
Muscular sac; stretches to store food taken in faster than can be processed; gastric fluid mixes with food and kills many pathogens; protein digestion starts. Secretes ghrelin, an appetite stimulator.

Small Intestine
First part (duodenum, C-shaped, about 10 inches long) receives secretions from liver, gallbladder, and pancreas.

In second part (jejunum, about 3 feet long), most nutrients are digested and absorbed.

Third part (ileum, 6–7 feet long) absorbs some nutrients; delivers unabsorbed material to large intestine.

Large Intestine (colon)
Concentrates and stores undigested matter by absorbing mineral ions, water; about 5 feet long: divided into ascending, transverse, and descending portions.

Rectum
Distension stimulates expulsion of feces.

Anus
End of system; terminal opening through which feces are expelled.

Accessory Organs:

Salivary Glands
Glands (three main pairs, many minor ones) that secrete saliva, a fluid with polysaccharide-digesting enzymes, buffers, and mucus (which moistens food and lubricates it).

Liver
Secretes bile (for emulsifying fat); roles in carbohydrate, fat, and protein metabolism.

Gallbladder
Stores and concentrates bile that the liver secretes.

Pancreas
Secretes enzymes that break down all major food molecules; secretes buffers against HCl from the stomach. Secretes insulin, a hormonal control of glucose metabolism.

Figure 11.1 Animated! **The digestive system has major and accessory organs.** (© Cengage Learning)

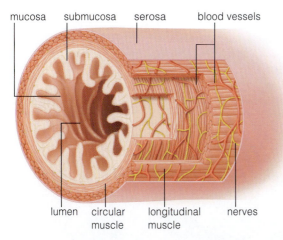

mucosa   submucosa   serosa   blood vessels

lumen   circular   longitudinal   nerves
         muscle   muscle

Figure 11.2 Four layers make up the wall of the digestive tract. This diagram shows the layers in the small intestine. (© Cengage Learning)

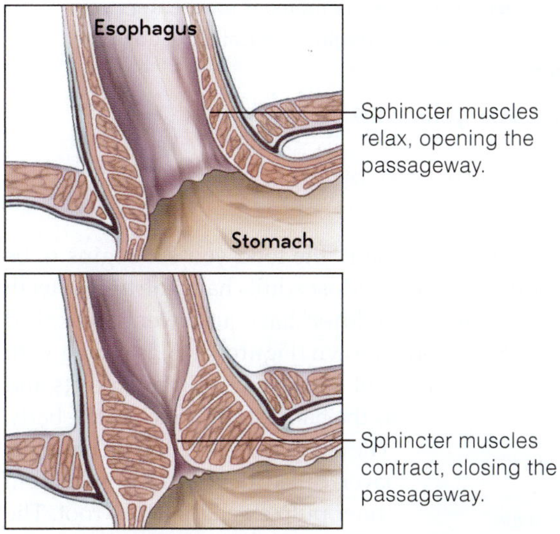

Esophagus

Sphincter muscles relax, opening the passageway.

Stomach

Sphincter muscles contract, closing the passageway.

Figure 11.3 Sphincters help regulate the passage of food through parts of the GI tract. This diagram shows the sphincter where the esophagus opens into the stomach. (© Cengage Learning)

When we eat, food advances in one direction, from the mouth (the oral cavity) through the pharynx, the esophagus, stomach, small intestine, and large intestine. The large intestine ends in the rectum, anal canal, and anus.

The digestive tube has four layers

From the esophagus onward, the digestive tube wall has four layers (Figure 11.2). The innermost layer is a mucosa of epithelial cells. It lines the lumen, the space through which food passes. The mucosa is surrounded by the submucosa, a layer of connective tissue with blood and lymph vessels and nerve cells. The next layer is smooth muscle—usually two sublayers, one circling the tube and the other oriented lengthwise. An outer layer, the serosa, is a very thin serous membrane (Section 4.7). Circular arrays of smooth muscle called **sphincters** are located at the junctions between sections of the GI tract. As sphincters contract and relax, they control the movement of material in the tube. For example, a gastroesophageal sphincter controls the passage of food from the esophagus into the stomach (Figure 11.3).

The different parts of the digestive system perform five basic functions

To extract nutrients from food, the digestive system's parts carry out five types of tasks:

1. **Mechanical processing** and **motility.** Movements of various parts, such as the teeth, tongue, and muscle layers, break up, mix, and propel food along.

2. **Secretion.** Glands and accessory organs release enzymes or other chemicals used in digestion and absorption.

3. **Digestion.** Food is chemically broken down to nutrient molecules small enough to be absorbed.

4. **Absorption.** Digested nutrients and fluid pass across the tube wall and into blood or lymph.

5. **Elimination.** Undigested and unabsorbed residues are excreted from the end of the GI tract.

Once nutrients from food have entered the bloodstream, cells throughout the body can take them up for use in all aspects of metabolism.

digestive system The tube where food is digested and absorbed and undigested residues are expelled. Also called the GI tract.

sphincter An array of circular muscles that regulates the passage of material between neighboring sections of the digestive tube.

TAKE-HOME MESSAGE

WHAT DOES THE DIGESTIVE SYSTEM DO?

- The digestive system breaks down food mechanically and chemically, absorbs nutrients, and eliminates residues.
- The digestive tube, or gastrointestinal (GI) tract, extends from the mouth to the anus.
- For most of its length, the tube wall consists of four layers, including smooth muscle.
- Substances in the GI tract don't enter the body until they pass across the tube wall into the bloodstream or lymph.

Chewing and Swallowing: Food Processing Begins

- Food processing begins the moment food enters your mouth, where enzymes begin chemical digestion of starches.
- Link to Carbohydrates 2.9

The teeth tear and grind bulk food into smaller chunks

In the oral cavity, or mouth, the food you eat begins to be broken apart by chewing. Most adults have thirty-two teeth (Figure 11.4A). Young children have just twenty so-called primary teeth. A tooth's crown (Figure 11.4B) is coated with tooth enamel. It consists of hardened calcium deposits and is the hardest substance in the body. The enamel covers a living, bonelike layer called dentin. Dentin and an inner pulp extend into the root. The pulp cavity contains blood vessels and nerves.

The shape of a tooth fits its function. Chisel-shaped incisors bite off chunks of food, and cone-shaped canines (cuspids) tear it. Premolars and molars, with broad crowns and rounded cusps, grind it.

> **bolus** A ball of chewed and swallowed food.
>
> **palate** The roof of the mouth.
>
> **salivary amylase** Enzyme in saliva that begins the chemical digestion of starch.
>
> **salivary glands** Glands that produce saliva, a mix of water, enzymes, and other substances.

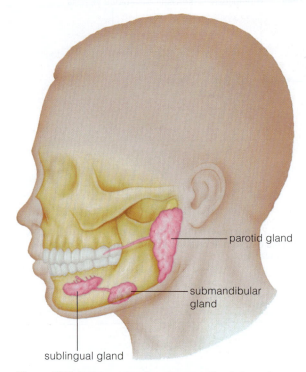

Figure 11.5 Salivary glands release saliva into various regions of the mouth. (© Cengage Learning)

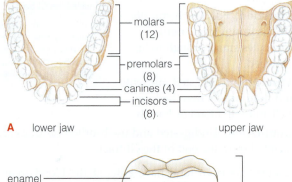

A lower jaw   upper jaw

molars (12)
premolars (8)
canines (4)
incisors (8)

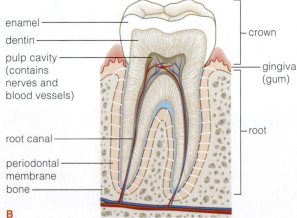

B

enamel
dentin
pulp cavity (contains nerves and blood vessels)
root canal
periodontal membrane
bone
crown
gingiva (gum)
root

Figure 11.4 The structure of a tooth, including its shape, fits its function. (© Cengage Learning)

Enzymes in saliva begin the chemical digestion of food

Chewing mixes food with saliva from several **salivary glands** (Figure 11.5). A large parotid gland nestles just in front of each ear. Submandibular glands lie just below the lower jaw in the floor of the mouth, and sublingual glands are under your tongue. The tongue itself is skeletal muscle covered by a membrane. As described in Chapter 14, its taste receptors respond to dissolved chemicals.

Saliva is mostly water, but it includes other substances. An important one is the enzyme **salivary amylase**, which breaks down starch; chew on a soda cracker and you can feel it becoming mushy as salivary amylase goes to work. A buffer, bicarbonate (HCO_3^-), keeps the pH of your mouth between 6.5 and 11.5, a range within which salivary amylase can function. Saliva also contains mucins, proteins that help bind food bits into a lubricated ball. Once it is swallowed, this ball of chewed food is called a **bolus** (ʙow-lus). Starch digestion continues in the stomach until acids there inactivate salivary amylase.

Behind the upper teeth is a bone-reinforced section of the **palate**—the roof of the mouth. It provides a hard surface against which the tongue can press food it is mixing

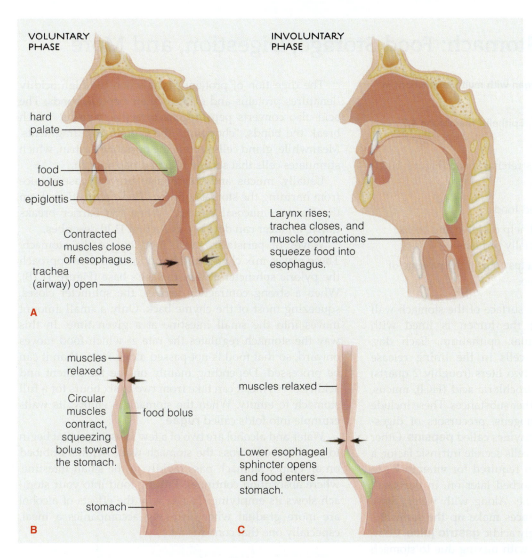

VOLUNTARY PHASE

- hard palate
- food bolus
- epiglottis
- Contracted muscles close off esophagus.
- trachea (airway) open

A

INVOLUNTARY PHASE

- Larynx rises; trachea closes, and muscle contractions squeeze food into esophagus.

- muscles relaxed
- Circular muscles contract, squeezing bolus toward the stomach.
- food bolus
- stomach

B

- muscles relaxed
- Lower esophageal sphincter opens and food enters stomach.

C

Figure 11.6 Animated!
Swallowing and peristalsis move food toward the stomach. **A** Contractions of the tongue push the food bolus into the pharynx. Next, the vocal cords seal off the larynx, and the epiglottis bends downward, helping to keep the trachea closed. Contractions of throat muscles then squeeze the food bolus into the esophagus. **B, C** Finally, peristalsis in the esophagus moves the bolus through a sphincter, and food enters the stomach.
(© Cengage Learning)

with saliva. Tongue muscle contractions force the bolus into the **pharynx** (FARE-inks), the throat. This passageway connects with the windpipe, or *trachea* (Figure 11.6), which leads to the lungs. It also connects with the **esophagus**, which leads to the stomach. Mucus secreted by the membrane lining the pharynx and esophagus lubricates the bolus, helping move food on its way.

Swallowing has voluntary and involuntary phases

Swallowing food might seem simple, but it involves a sequence of events (Figure 11.6). Swallowing begins when voluntary skeletal muscle contractions push a bolus into the pharynx, stimulating sensory receptors in the pharynx wall. The receptors trigger a reflex in which involuntary muscle contractions keep food from moving up into your nose and down into the trachea. As this reflex occurs, the vocal cords are stretched tight across the entrance to the larynx (your "voice box"). Then, the flaplike epiglottis is pressed down over the vocal cords as a secondary seal. For a moment, breathing stops as food moves into the esophagus,

so you normally don't choke when you swallow. When swallowed food reaches the lower esophagus, it passes through a sphincter into the stomach (Figures 11.6B and 11.6C). Waves of muscle contractions called **peristalsis** (pare-ih-STAL-sis) help push the food bolus along.

esophagus The passageway leading from the pharynx to the stomach.

peristalsis Rhythmic smooth muscle contractions that propel food through the GI tract.

pharynx The throat.

The Stomach: Food Storage, Digestion, and More

- **The stomach is a complex organ with multiple functions in processing food.**

- **Links to Exocrine glands 4.1, Epithelial membrane 4.7**

The stomach is a muscular, stretchable sac (Figure 11.7A) with three functions:

1. It mixes and stores ingested food.

2. It produces secretions that help dissolve and break down food particles, especially proteins.

3. By way of a sphincter, it helps control the passage of food into the small intestine.

chyme The pasty stomach contents formed from the mixing of food with gastric juice.

gastric juice The fluid formed as glands in the stomach lining release HCl, mucus, enzymes, gastrin, bicarbonate, and other substances.

pepsin Enzyme in gastric juice that helps digest proteins.

rugae Folds in the inner wall (mucosa) of the stomach.

The surface of the stomach wall facing the lumen is lined with glandular epithelium. Each day, gland cells in the lining release about two liters (roughly 2 quarts) of hydrochloric acid (HCl), mucus, and other substances. These include pepsinogens, precursors of digestive enzymes called **pepsins**. Other gland cells secrete intrinsic factor, a protein required for vitamin B_{12} to be absorbed later on, in the small intestine. Along with water, these substances make up the stomach's strongly acidic **gastric juice**. Combined with mixing due to stomach contractions, the acidity converts swallowed boluses into thick, pasty **chyme** (KIME). The acidity also kills most microbes in food.

The digestion of proteins starts when the high acidity denatures proteins and exposes their peptide bonds. The acid also converts pepsinogens to active pepsins, which break the bonds, "chopping" the protein into fragments. Meanwhile, gland cells secrete the hormone gastrin, which stimulates cells that secrete HCl and pepsinogen.

Usually, mucus and bicarbonate prevent gastric juice from harming the stomach lining. These protections form the "gastric mucosal barrier." When the barrier breaks down, an ulcer can develop, as Section 11.10 describes.

Waves of peristalsis move food out of the stomach. These waves mix chyme and build force as they approach the pyloric sphincter at the stomach's base (Figure 11.7B). When a strong contraction arrives, the sphincter closes, squeezing most of the chyme back. Only a small amount moves into the small intestine at a given time. In this way the stomach regulates the rate at which food moves onward, so that food is not passed along faster than it can be processed. Depending mainly on the fat content and acidity of chyme, it can take from two to six hours for a full stomach to empty. When the stomach is empty, its walls crumple into folds called **rugae**.

Water and alcohol are two of a few substances that begin to be absorbed across the stomach wall. Liquids imbibed on an empty stomach pass rapidly to the small intestine, where absorption continues. Putting food into your stomach slows its emptying. This is why the effects of alcohol are more gradual when drinking accompanies a meal, especially one that contains fat.

TAKE-HOME MESSAGE

WHAT IS THE STOMACH'S ROLE IN DIGESTION?

- The stomach receives and stores swallowed food. It also produces gastric juice that begins chemical digestion, especially of proteins.

- A sphincter helps control the pace at which the stomach contents empty into the small intestine.

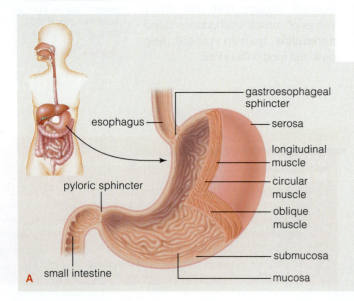

esophagus — gastroesophageal sphincter
serosa
longitudinal muscle
circular muscle
oblique muscle
pyloric sphincter
submucosa
small intestine
mucosa

A

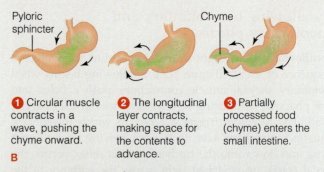

Pyloric sphincter   Chyme

1 Circular muscle contracts in a wave, pushing the chyme onward.

2 The longitudinal layer contracts, making space for the contents to advance.

3 Partially processed food (chyme) enters the small intestine.

B

Figure 11.7 Animated! The stomach's structure allows it to store and mix food and move it onward. **A** Structure of the stomach. **B** How a peristaltic wave moves down the stomach.
(© Cengage Learning)

The Small Intestine: A Huge Surface for Digestion and Absorption

■ **The structure of the small intestine wall is the key to its ability to absorb nutrients.**

Your small intestine is about an inch and a half in diameter and 6 meters (20 feet) long. It absorbs most nutrients in the food you eat. Figures 11.8A and 11.8B show how densely folded the mucosa is, and how the folds all stick out like ruffles into the lumen. Each fold has even smaller, hairlike projections (Figure 11.8C). Each "finger" is a **villus** (plural: villi). Small blood vessels (an arteriole and a vein) and a lymph vessel in each villus move substances to and from the bloodstream. Gland cells in the mucosal lining release digestive enzymes.

Most cells in the epithelium covering a villus have a threadlike projection of their plasma membrane. This projection is called a **microvillus** (plural: microvilli). Each epithelial cell has about 1,700 microvilli—a dense array that gives the epithelium of villi its common name, the **brush border** (Figures 11.8D and 11.8E).

What is the benefit of so many folds and projections from the intestinal mucosa? Together, they greatly increase the surface area for absorbing nutrients from chyme. Without that huge surface area, absorption would take place too slowly to sustain life.

brush border The collective array of microvilli on epithelial cells lining the intestinal mucosa.

microvillus Any of the hundreds of microscopic projections of the plasma membranes of epithelial cells that cover villi.

villus A fingerlike projection from the mucosa (inner surface) of the small intestine.

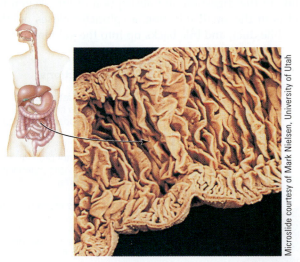

Microslide courtesy of Mark Nielsen, University of Utah

A Longitudinal cross section through the small intestine showing its highly folded lining.

TAKE-HOME MESSAGE

WHAT IS THE ROLE OF THE SMALL INTESTINE?

- The small intestine is where most nutrients in food are absorbed.
- The small intestine provides a vast surface area for nutrient absorption because it has a densely folded mucosa that in turn has millions of villi and hundreds of millions of microvilli.

camilla$$/shutterstock.com

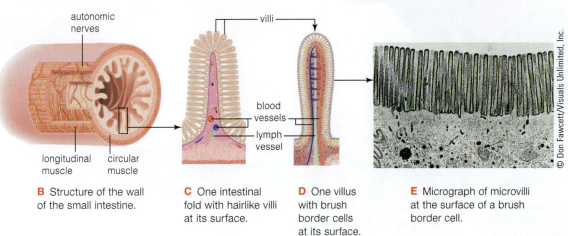

© Don Fawcett/Visuals Unlimited, Inc.

autonomic nerves

villi

blood vessels

lymph vessel

longitudinal muscle

circular muscle

B Structure of the wall of the small intestine.

C One intestinal fold with hairlike villi at its surface.

D One villus with brush border cells at its surface.

E Micrograph of microvilli at the surface of a brush border cell.

Figure 11.8 The small intestine has a very large surface area. (© Cengage Learning)

- The pancreas, gallbladder, and liver assist digestion but are outside the digestive tube.

- Links to pH and buffers 2.7, Exocrine glands 4.1

The pancreas produces key digestive enzymes

The **pancreas** nestles behind and below the stomach (Figure 11.9). It contains exocrine cells that release digestive enzymes into the duodenum, the first section of the small intestine. The pancreas also contains endocrine cells that release hormones into the blood. The hormones help regulate blood sugar, a topic of Chapter 15.

There are four types of pancreatic enzymes, which can chemically dismantle the four major categories of food—complex carbohydrates, proteins, lipids, and nucleic acids. These enzymes work best when the pH is neutral or slightly alkaline, so the "pancreatic juice" also contains bicarbonate (HCO_3^-), which neutralizes the acid in chyme moving into the duodenum from the stomach. Depending on how often and what type of food you eat, your pancreas may make two quarts of this fluid each day.

The liver makes bile and the gallbladder stores it

When the digestive system is processing food, a yellowish fluid called bile is released into the upper small intestine. Making bile is a digestive role of the **liver**. This large organ secretes as much as 1,500 mL, or almost 1.6 quarts, of bile every day. **Bile** is a blend of substances including water and bile salts synthesized from cholesterol. Bile salts aid in the digestion and absorption of fats, as you will read in Section 11.6. Bile is stored in the **gallbladder**, a small sac tucked behind the liver. As needed, the gallbladder contracts and empties bile into the small intestine where it aids in the digestion and absorption of fats. When no food is in the small intestine, a sphincter closes off the main bile duct, and bile backs up into the gallbladder.

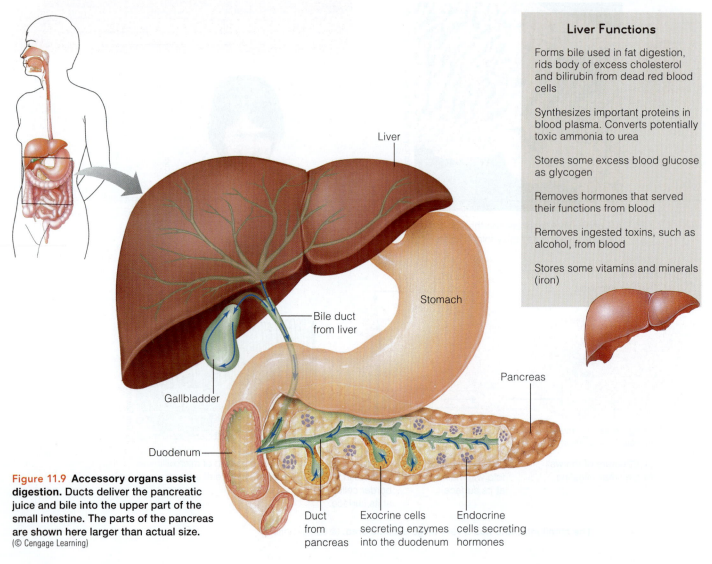

Liver Functions

Forms bile used in fat digestion, rids body of excess cholesterol and bilirubin from dead red blood cells

Synthesizes important proteins in blood plasma. Converts potentially toxic ammonia to urea

Stores some excess blood glucose as glycogen

Removes hormones that served their functions from blood

Removes ingested toxins, such as alcohol, from blood

Stores some vitamins and minerals (iron)

Liver

Stomach

Pancreas

Bile duct from liver

Gallbladder

Duodenum

Duct from pancreas

Exocrine cells secreting enzymes into the duodenum

Endocrine cells secreting hormones

Figure 11.9 Accessory organs assist digestion. Ducts deliver the pancreatic juice and bile into the upper part of the small intestine. The parts of the pancreas are shown here larger than actual size.
(© Cengage Learning)

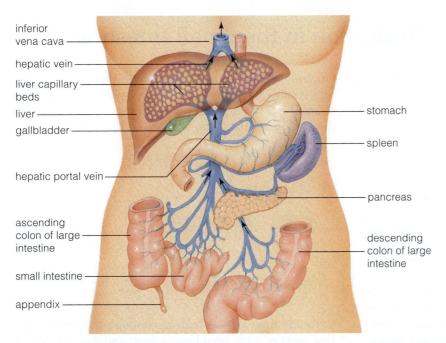

inferior vena cava

hepatic vein

liver capillary beds

liver

gallbladder

hepatic portal vein

ascending colon of large intestine

small intestine

appendix

stomach

spleen

pancreas

descending colon of large intestine

A gallstone

© C. James Webb/Phototake

Figure 11.10 **The hepatic portal system diverts nutrient-rich blood to the liver.** Arrows show the direction of blood flow. The photograph at right shows a large gallstone that has been sliced in half. Cholesterol gives it a pale color. (© Cengage Learning)

The liver is a multipurpose organ

Besides its digestive functions, the liver processes incoming nutrients into substances the body requires. A system of blood vessels called the **hepatic portal system** diverts blood from the small intestine to the liver. As you can see in Figure 11.10, blood entering the liver in this system arrives in the hepatic portal vein and returns to the general circulation via the hepatic vein. After a meal, when blood from the small intestine enters the system loaded with nutrients, liver cells manage this bonanza in various ways. For example, if the blood contains more of the sugar glucose than body cells can take up at the time, the liver removes some of the excess and stores it as glycogen. The liver also stores several vitamins and minerals and forms the active form of Vitamin D, which is essential for the uptake of calcium from digested food. Liver cells use arriving amino acids to synthesize proteins such as the albumin in blood plasma, or process and reship them in a form cells throughout the body can use to make ATP.

As you will read in Section 11.6, digested lipids don't enter the hepatic portal system, but they do reach the liver in the general circulation. Liver cells may use some of these lipids to make lipoproteins, including the HDLs and LDLs that carry cholesterol.

The chart in Figure 11.9 lists some other major liver functions. For instance, the liver removes alcohol and other potential toxins, such as ammonia produced by the breakdown of amino acids. The ammonia is converted to urea, a much less toxic waste product that is excreted in urine. Liver cells also take up bilirubin, a pigment that forms as aging or damaged red blood cells are broken down and the hemoglobin in them is recycled. Bilirubin is added to bile and eventually is excreted in feces. In addition, the liver also inactivates many hormones, which move via the blood to the kidneys and are excreted in urine.

Bile often contains cholesterol apart from that used to synthesize bile salts. This excess may form a gallstone in the gallbladder (Figure 11.10 *right*). If the gallbladder is surgically removed—usually due to the painful presence of gallstones—the duct that connects it to the small intestine enlarges and takes on the role of bile storage. This is why millions of people are walking around minus their gallbladder, with few or no ill effects.

bile Fluid that contains bile salts; it forms in the liver.

gallbladder Organ that stores bile from the liver.

hepatic portal system A system of blood vessels that divert blood from the small intestine to the liver for processing, then return it to the bloodstream.

liver Organ that produces bile salts used in fat digestion. Its other roles include storing excess glucose in blood and detoxifying waste ammonia from protein digestion.

pancreas The source of enzymes that dismantle complex food molecules; it also produces hormones that regulate blood sugar.

TAKE-HOME MESSAGE

WHAT ARE THE ACCESSORY ORGANS OF DIGESTION, AND WHAT DO THEY DO?

- Accessory organs of digestion are the pancreas, gallbladder, and liver.
- The pancreas produces digestive enzymes that are released into the small intestine.
- The liver produces bile salts that are stored in the gallbladder (as bile).
- Beyond digestion, the liver's many functions include removing toxins from the blood and helping to manage blood levels of glucose and other substances.

- Absorption moves nutrients into the internal environment—tissue fluid and the bloodstream.

- Links to Buffers 2.7, Osmosis 3.10

Each day about 9 liters (10 quarts) of fluid enters the first section of the small intestine, the **duodenum** (doo-oh-DEE-num). This fluid includes chyme along with enzymes and other substances from the pancreas, liver, and gallbladder. Most digestion and nutrient absorption occurs in the next section, the 3-foot-long **jejunum**. Some nutrients are absorbed while the remaining material is moving through the **ileum**, the last section of the small intestine, on its way to the large intestine.

Chyme entering the duodenum triggers hormone signals that stimulate a brief flood of digestive enzymes from the pancreas. As part of pancreatic juice, these enzymes act on carbohydrates, fats, proteins, and nucleic acids (Table 11.1 and Figure 11.11). For example, like pepsin in the stomach, the pancreatic enzymes trypsin and chymotrypsin digest the polypeptide chains of proteins into peptide fragments. The fragments are then broken down to amino acids by different peptidases (which are on the surface of the intestinal mucosa). Recall from Section 11.5 that the pancreas also secretes bicarbonate that buffers stomach acid, maintaining a chemical environment in which pancreatic enzymes can function.

Fat digestion requires enzymes called lipases. Bile salts in bile secreted by the liver (and delivered via the gallbladder) make fat digestion more efficient. Bile salts are like a detergent—they emulsify, or break up, large units of fat into smaller ones. How does this process work? Most fats in the average diet are triglycerides, which tend to clump into big fat globules in chyme. When peristalsis mixes chyme, the globules break up into droplets that become coated with bile salts (see step 5 in Figure 11.11). These droplets, called micelles (my-CELLS), give fat-digesting enzymes a much greater surface area to act on. So, because triglycerides are emulsified, they can be broken down much faster to monoglycerides and fatty acids, molecules that are small enough to be absorbed. Micelles also may contain fat-soluble vitamins.

When a substance is absorbed, it crosses the intestine lining into the bloodstream. Due partly to the vast absorptive surface area of the small intestine, this process is very

duodenum The first section of the small intestine, where chyme and digestive enzymes enter.

ileum Final section of the small intestine, where absorption is completed and residues move toward the large intestine.

jejunum Middle section of the small intestine, where most nutrients are digested and absorbed.

lacteals Lymph vessels that take up triglycerides from digested fat and deliver them to the bloodstream.

segmentation Mechanical mixing of digested food moving through the small intestine.

efficient. **Segmentation** helps, too. In this process, rings of smooth muscle in the wall repeatedly contract and relax. The result is a back-and-forth movement that mixes digested material and forces it against the wall:

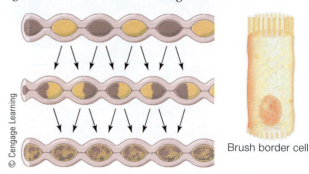

© Cengage Learning

Brush border cell

By the time food is halfway through the small intestine, most of it has been broken apart and digested. Water crosses the intestine lining by osmosis, and cells in the lining also selectively absorb minerals. Transport proteins in the plasma membrane of brush border cells actively move some nutrients, such as the monosaccharide glucose and amino acids, across the lining. After glucose and amino acids are absorbed, they move into tissue fluid and then directly into blood vessels (steps 1–4 in Figure 11.11).

Additional steps occur before digested lipids move into the bloodstream. After lipases digest micelles, the fatty acids and monoglycerides enter brush border cells, just as glucose and amino acids do. (The bile salts that formed the droplets are recycled.) There, fatty acids and monoglycerides quickly reunite into triglycerides. Then triglycerides combine with proteins into particles that leave the cells by exocytosis and enter tissue fluid. They don't directly enter blood vessels, however. Instead they cross into lymph vessels known as **lacteals**, which drain into the general circulation.

TAKE-HOME MESSAGE

HOW ARE DIFFERENT KINDS OF NUTRIENTS ABSORBED IN THE SMALL INTESTINE?

- In the small intestine, chemical and mechanical processes break down large organic molecules to smaller molecules that can be absorbed.

- Enzymes from the pancreas act on carbohydrates, fats, proteins, and nucleic acids in chyme. Bile salts emulsify large fat globules, allowing fats to be more easily digested.

- Simple sugars and amino acids pass through brush border cells that line the surface of intestinal villi, then move into the blood.

- Digested lipids pass through brush border cells, then into lacteals, then into the bloodstream.

TABLE 11.1 Major Enzymes of Digestion and What They Do

Enzyme	Released by:	Active in:	Breaks down:	Resulting Products
DIGESTING CARBOHYDRATES				
Salivary amylase	Salivary glands	Mouth, stomach	Polysaccharides	Disaccharides, oligosaccharides
Pancreatic amylase	Pancreas	Small intestine	Polysaccharides	Disaccharides, monosaccharides
Disaccharidases	Intestinal lining	Small intestine	Disaccharides	MONOSACCHARIDES* (e.g., glucose)
DIGESTING PROTEINS				
Pepsins	Stomach lining	Stomach	Proteins	Protein fragments
Trypsin and chymotrypsin	Pancreas	Small intestine	Proteins	Protein fragments
Carboxypeptidase	Pancreas	Small intestine	Peptides	AMINO ACIDS*
Aminopeptidase	Intestinal lining	Small intestine	Peptides	AMINO ACIDS*
DIGESTING FATS				
Lipases	Pancreas	Small intestine	Triglycerides	FATTY ACIDS, MONOGLYCERIDES*
DIGESTING NUCLEIC ACIDS				
Pancreatic nucleases	Pancreas	Small intestine	DNA, RNA	NUCLEOTIDES*
Intestinal nucleases	Intestinal lining	Small intestine	Nucleotides	NUCLEOTIDE BASES, MONOSACCHARIDES*

*Products small enough to be absorbed into the bloodstream.
The four basic kinds of biological molecules are shown in blue type.

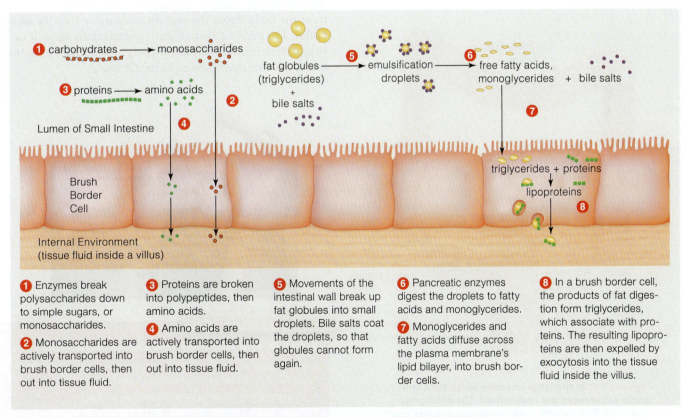

1 Enzymes break polysaccharides down to simple sugars, or monosaccharides.

2 Monosaccharides are actively transported into brush border cells, then out into tissue fluid.

3 Proteins are broken into polypeptides, then amino acids.

4 Amino acids are actively transported into brush border cells, then out into tissue fluid.

5 Movements of the intestinal wall break up fat globules into small droplets. Bile salts coat the droplets, so that globules cannot form again.

6 Pancreatic enzymes digest the droplets to fatty acids and monoglycerides.

7 Monoglycerides and fatty acids diffuse across the plasma membrane's lipid bilayer, into brush border cells.

8 In a brush border cell, the products of fat digestion form triglycerides, which associate with proteins. The resulting lipoproteins are then expelled by exocytosis into the tissue fluid inside the villus.

Figure 11.11 Animated! Different types of nutrients are absorbed by different mechanisms. (© Cengage Learning)

- Anything not absorbed in the small intestine moves into the large intestine.

- Link to Osmosis 3.10

anal canal The short passageway through which feces move from the rectum to the anus.

anus The terminal opening of the GI tract.

appendix A small, slender pouch off the cecum that contains lymphocytes. It doesn't function in digestion.

colon The portion of the large intestine that connects at its upper end to the small intestine and at its lower end to the rectum.

rectum The region of the large intestine that stores uneliminated feces.

The large intestine is about 1.2 meters (5 feet) long. It begins as a blind pouch called the cecum (SEE-cum). The cecum merges with the **colon**, which is divided into four regions in an inverted U-shape. The *ascending* colon travels up the right side of the abdomen, the *transverse* colon continues across to the left side, and the *descending* colon then turns downward. The *sigmoid* colon makes an S-curve and connects with the **rectum** (Figure 11.12A).

Cells in the colon's lining actively transport sodium ions out of the tube. When the ion concentration there falls, water moves out by osmosis and returns to the bloodstream. As water leaves, material left in the colon is gradually concentrated into feces, a mixture of the remaining water, undigested and unabsorbed matter, and bacteria. It is stored and then eliminated. The typical brown color of feces comes mainly from bile pigments.

Bacteria make up almost a third of the dry weight of feces. In fact, at least 57 species of bacteria, including *Escherichia coli*, normally inhabit our intestines and are nourished by the food residues there. Their metabolism produces useful fatty acids and some vitamins (such as vitamins K and B_{12}). These substances are absorbed across the colon lining as waves of peristalsis push material against its absorptive surface. Feces of humans and other animals can contain a variety of disease-causing organisms, too. Health officials use evidence of "coliform bacteria," including *E. coli*, in water and food supplies as a measure of fecal contamination in general (Figure 11.12B).

Your **appendix** projects from the cecum like the little finger of a glove. It doesn't function in digestion but does contain patches of lymphoid tissue where B and T cells are present. These lymphocytes may attack parasites or harmful bacteria consumed in food.

Shortly after you eat, signals from the nervous system and hormones direct large portions of the ascending and transverse colon to contract at the same time. Within a few seconds, residues in the colon may move as much as three-fourths of the colon's length and make way for incoming food. When feces distend the wall of the rectum, the stretching triggers defecation—elimination of feces from the body. From the rectum feces move into the **anal canal**. The nervous system also controls defecation. It can stimulate or inhibit contractions of sphincter muscles at the **anus**, the terminal opening of the GI tract.

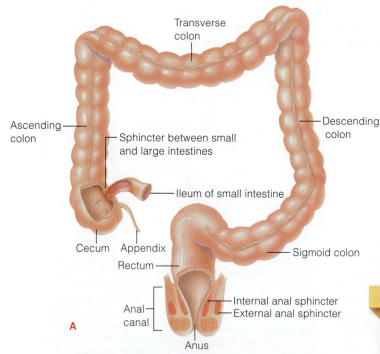

Transverse colon

Ascending colon

Sphincter between small and large intestines

Descending colon

Ileum of small intestine

Cecum Appendix

Rectum

Sigmoid colon

Anal canal

Internal anal sphincter
External anal sphincter

A

Anus

Figure 11.12 In the large intestine feces form and some substances are reabsorbed. The photograph shows Dr. Rita Colwell, a microbiologist who has spent many years working to improve the quality of drinking water in countries such as Bangladesh. (© *Cengage Learning*)

TAKE-HOME MESSAGE

WHAT IS THE ROLE OF THE LARGE INTESTINE?

- In the large intestine, water, salts, certain vitamins, and other useful substances are reabsorbed from food residues.
- The remaining material is eliminated as feces.
- Water, salts, certain vitamins, and other useful substances are reabsorbed from food residues. The remaining material is eliminated as feces.

11.8 Controls over Digestion

■ Nerves and hormones regulate food digestion.

Signals from the nervous system and hormones from endocrine cells jointly regulate digestion. These controls are sensitive to the presence of food in the GI tract and the food's chemical makeup (Figure 11.13).

When you take food in your mouth—and sometimes when you merely think about eating—sensory receptors in your mouth stimulate the salivary glands to release saliva. Food entering the stomach stretches the stomach walls, and then those of the small intestine. This stretching also triggers signals from sensory receptors. Some of the signals give you (by way of processing in your brain) that "full" feeling after you eat. Others can lead to the muscle contractions of peristalsis or the release of digestive enzymes and other substances. Centers in the brain coordinate these activities with factors such as how much blood is flowing to the small intestine, where nutrients are being absorbed.

There are several types of endocrine cells in the GI tract (Table 11.2). For example, one type secretes the hormone gastrin into the bloodstream when the stomach contains protein. Gastrin mainly stimulates the release of hydrochloric acid (HCl), which you may recall is a key ingredient in gastric juice. After the stomach has emptied out, the increased acidity there causes another type of endocrine cell to release somatostatin, which shuts down HCl secretion so that conditions in the stomach are less acidic. Notice that this is an example of negative feedback.

Hormones also come from endocrine cells in the small intestine. One of them, secretin, signals the pancreas to release bicarbonate when acid enters the duodenum. When fat enters the small intestine, a hormone called CCK (for cholecystokinin) is released. CCK spurs the pancreas to release enzymes and triggers gallbladder contractions that deliver bile into the small intestine. Secretin and CCK also slow the rate at which the stomach empties—the mechanism mentioned in Section 11.3 that prevents food from entering the small intestine faster than it can be processed there. Yet another hormone, GIP (for glucose-dependent insulinotropic peptide) is released when fat and glucose are in the small intestine. GIP stimulates the release of insulin from the pancreas, which is required for cells to take up glucose.

Rolf Bruderer/age fotostock

TABLE 11.2 Hormonal Controls of Digestion

Hormone	Source	Effects on Digestive System
Gastrin	Stomach	Increases acid secretion by stomach
Somatostatin	Pancreas	Halts HCl secretion in the empty stomach
Cholecystokinin (CCK)	Small intestine	Increases enzyme secretion by pancreas and causes contraction of gallbladder
Secretin	Small intestine	Increases bicarbonate secretion by pancreas and slows contractions in the small intestine
GIP	Small intestine	Stimulates pancreas to release insulin

Hormone controls

Acidic chyme stimulates release of the hormone secretin in the small intestine. Secretin inhibits motility in the small intestine and stimulates HCO_3^- secretion into the duodenum.

Fat in chyme stimulates release of the hormone CCK (cholecystokinin). CCK inhibits stomach emptying and stimulates secretion of pancreatic enzymes.

Food entering the GI tract stimulates GIP secretion, which triggers release of the hormone insulin. Insulin stimulates the uptake of glucose from blood.

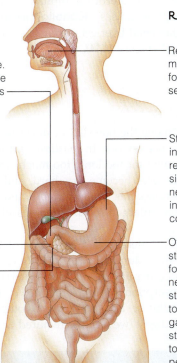

Receptor controls

Receptors in the mouth respond to food by increasing secretion of saliva.

Stretch receptors in the stomach respond to food, signaling the nervous system to increase stomach contractions.

Other receptors in the stomach respond to food, signaling the nervous system to stimulate the stomach to secrete the hormone gastrin. It in turn stimulates the stomach to secrete HCl and pepsinogen.

Figure 11.13 Hormones and the nervous system act to control various aspects of digestion. (© Cengage Learning)

TAKE-HOME MESSAGE

HOW DOES THE BODY REGULATE FOOD DIGESTION?

- Signals from nerves and hormones control activity in the digestive system.
- Sensory receptors in the mouth and in the wall of the stomach and small intestine trigger nervous system controls.

"Heartburn" is an upper GI tract disorder

The main symptom of **gastroesophageal reflux disease**, or GERD, is often called "heartburn," but it has nothing to do with the heart. With this common disorder acidic chyme backs up into the esophagus when the lower esophageal sphincter doesn't close properly. The irritation causes burning in the upper chest and throat. Mild cases often can be controlled by over-the-counter drugs that reduce stomach acid and by limiting intake of acidic foods such as tomatoes, orange juice, coffee, and alcoholic beverages.

Hepatitis and cirrhosis strike the liver

Hepatitis is inflammation of the liver. Obesity, certain drugs, and environmental toxins may trigger it. Some types are caused by viruses that are transmitted in body fluids such as blood and semen. The inflammation may subside with treatment, although some patients suffer major, irreversible damage for which the only option is a liver transplant. Long-term inflammation due to heavy alcohol consumption causes **alcoholic cirrhosis** (sir-OH-sis), in which damaged liver cells are replaced by connective tissue "scars" (Figure 11.14).

Colon problems range from constipation to cancer

It's normal to "move the bowels," or defecate, from three times a day to once a week. In *constipation*, food residues remain in the colon for too long, too much water is reabsorbed, and the feces become dry, hard, and difficult to eliminate. Constipation is uncomfortable, and it is a common cause of the enlarged rectal blood vessels known as hemorrhoids.

Constipation is often caused by a lack of bulk in the diet. "Bulk" is the volume of fiber (mainly cellulose from plant foods) and other undigested food material that is not decreased by absorption in the colon. Much of it is *insoluble fiber* such as cellulose and other plant compounds that humans cannot digest (we lack the required enzymes) and

that does not easily dissolve in water. Wheat bran and the edible skins of fruits are just two examples. (Plant carbohydrates such as fruit pectins that swell or dissolve in water are *soluble fiber*.)

If you eat too little fiber, you are much more likely to be in the 50 percent of the U.S. population in whom the colon has formed *diverticula*—knoblike sacs where the inner colon lining protrudes through the wall of the large intestine. Inflammation of a diverticulum is called **diverticulitis**, and it can have quite serious complications, including peritonitis, if an inflamed diverticulum ruptures. Much more common is **diverticulosis** (Figure 11.15), in which diverticula are there but have not (yet) become inflamed.

Have you ever heard of someone having a "spastic colon"? This problematical condition also is known as **IBS**, or **irritable bowel syndrome**. IBS is the most common intestinal disorder. It often begins in early to mid adulthood, and it affects twice as many women as men. The direct trigger of IBS symptoms—abdominal pain and alternating diarrhea and constipation—is a disturbance in the smooth muscle contractions that move material through the colon. New research implicates a prior bacterial infection as the root cause in some cases. Reports that consuming probiotics ("gut-friendly" bacteria such as *Bifidobacterium infantis*) helps tame post-infectious IBS support this hypothesis.

Crohn's disease is an inflammatory disorder that affects various organs including the eyes, liver, skin, and intestines. In some patients the intestinal lining is so severely damaged that much of the intestine must be removed (Figure 11.16). Although Crohn's isn't curable, new treatment options are helping patients live with the disease more comfortably than ever before.

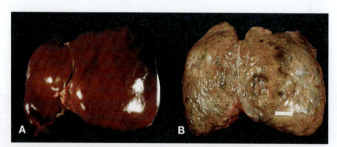

Figure 11.14 Long-term heavy alcohol use can cause cirrhosis of the liver. **A** A healthy liver. **B** A cirrhotic liver, in which a great deal of scarring has occurred. (A: Southern Illinois University/Photo Researchers, Inc.; B: Martin M. Rotker/Photo Researchers, Inc.)

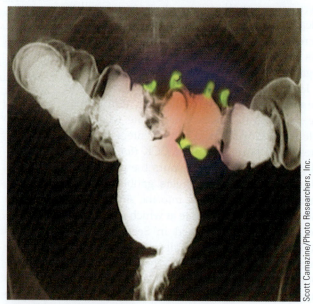

Scott Camazine/Photo Researchers, Inc.

Figure 11.15 Green areas in this X-ray are knoblike diverticula.

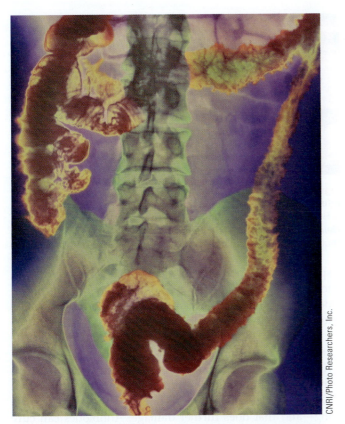

Figure 11.16 **Crohn's disease can do severe intestinal harm.** The blotchy areas in this X-ray image are ulcers in the wall of the intestine.

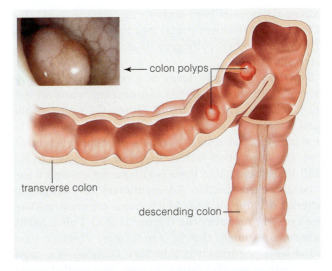

Figure 11.17 **A polyp on the colon wall may be a precursor to colorectal cancer.** The sketch shows one site where a colon polyp might be found—here, in the transverse colon. The photo shows a polyp found during a colonoscopy. Suspicious polyps can be removed and examined for the presence of cancer. (© Cengage Learning; inset photo: Science Photo Library/Custom Medical Stock Photo, Inc.)

Cancer of the colon or rectum—or **colorectal cancer**— is the number-two cancer diagnosis in the United States. It is second only to lung cancer and accounts for about 20 percent of all cancer deaths. The first internal sign of colorectal cancer may be a round, depressed area of abnormal cells. Another common early warning sign is a growth called a polyp that develops on the colon wall and becomes malignant (Figure 11.17). Fortunately, many precancerous growths and early cases of colon cancer can be detected by colonoscopy. After the patient is mildly sedated, a physician inserts a viewing tube into the colon and can examine it for polyps and other signs of disease.

Outward signs of colorectal cancer include a change in bowel habits, blood in feces, or rectal bleeding. People over age 50 have the highest risk. The tendency to develop polyps, and colorectal cancer, can run in families, but usually there is no obvious genetic link. Because colorectal cancer is much more common in Western societies, some experts have proposed that the typical high-fat, low-fiber Western diet may be a factor, and there is a lot of active research on the issue. Studies suggest that low doses of aspirin or NSAIDs (nonsteroidal anti-inflammatory drugs such as ibuprofen) may reduce the risk of developing precancerous polyps. Chapter 22 looks in more detail at the causes of cancer.

Malabsorption disorders prevent nutrients from being properly absorbed

Anything that interferes with the small intestine's ability to take up nutrients can lead to a **malabsorption disorder**. As many as 50 million adults in the U.S. develop **lactose intolerance**, a disorder that results from a deficiency of the enzyme lactase. It prevents normal digestion and absorption of lactose, the sugar found in milk and many milk products. Nausea, cramps, bloating, and diarrhea are common symptoms. People who have **celiac disease**, or **gluten intolerance**, are hypersensitive to gluten, a form of protein in wheat, rye, and barley. The disorder involves an autoimmune response in which lymphocytes attack the villi of brush border cells. Symptoms can range from lethargy and rashes to joint pain, mouth sores, and osteoporosis. People with gluten intolerance can control their symptoms by eating a gluten-free diet.

Other malabsorption disorders are associated with diseases that affect the pancreas, including the genetic condition **cystic fibrosis (CF)**. Patients with CF don't make the necessary pancreatic enzymes for normal digestion and absorption of fats and other nutrients. CF also affects the lungs, as you will read in later chapters.

11.10 Infections in the Digestive System

Because it opens at both ends to the outside world, the GI tract is a convenient portal into the body for bacteria, viruses, and other pathogens that contaminate foods, water, and hands.

Diarrhea, or watery feces, is a common effect of an intestinal infection. Diarrhea can develop when an irritant (such as a bacterial toxin) causes the lining of the small intestine to secrete more water and salts than the large intestine can absorb. It can also develop when infections, stress, or other factors speed up peristalsis in the small intestine, so that there isn't time for enough water to be absorbed. Section 3.5 mentioned the dangerous bacterial disease cholera. Much more common is diarrhea caused by a **rotavirus** (Figure 11.18A). Public health authorities estimate that by age 2, most children have had at least one rotavirus infection. Diarrhea is a cause for concern in small children because they easily become dehydrated, losing water and salts that nerve and muscle cells need to function properly. Figure 11.18B shows the protozoan *Giardia intestinalis*, which causes **giardiasis**. It forms cysts that enter water or food in contaminated feces. Symptoms include explosive diarrhea and "rotten egg" belches.

Several harmful strains of *E. coli* bacteria infect the GI tract (Figure 11.18C). One of them, called O157:H7, normally lives in the intestines of cattle. If a person eats ground beef or some other food that is contaminated with this microbe, it can cause a dangerous form of diarrhea that is complicated by anemia. A few cases have led to kidney failure and death.

Bacteria that cause **dental caries**—tooth decay—flourish on food residues in the mouth, especially sugars (Figure 11.19A). Daily brushing and flossing are the best way to avoid a bacterial infection of the gums, which can lead to **gingivitis** (jin-juh-vy-tus). This inflammation can

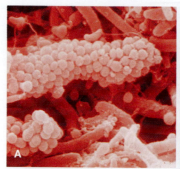
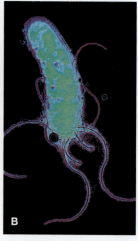

Figure 11.19 Various bacteria infect the mouth and stomach. A Bacteria on a human tooth. **B** *Helicobacter pylori*, the bacterium that causes most peptic ulcers.

A: Stanley Flegler/Visuals Unlimited, Inc. B: Peter Hawtin, University of Southampton/Photo Researchers, Inc.

spread to the periodontal membrane that helps anchor each tooth in the jaw. Untreated periodontal disease can slowly destroy a tooth's bony socket, which can lead to loss of the tooth and other complications.

A **peptic ulcer** is an open sore in the lining of the stomach or small intestine. Most are caused by the bacterium *Helicobacter pylori* (Figure 11.19B). It produces a toxin that inflames the stomach lining and causes damage that allows hydrogen ions and pepsins to diffuse into the lining—and that does further damage. Antibiotics can cure peptic ulcers caused by *H. pylori*. They don't help with the 20 percent of ulcers related to factors such as chronic stress, smoking, and overuse of aspirin and alcohol. *H. pylori* also is responsible for some cases of **gastritis** (an inflammation of the GI tract) and **stomach cancer**.

If you have ever had a case of "food poisoning," your stomach or intestines have been colonized by bacteria such as *Salmonella*, which can contaminate meat, poultry, and eggs (Figure 11.20).

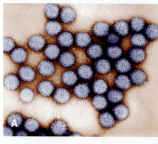

Figure 11.18 Intestinal infections can cause diarrhea. A Rotavirus particles isolated from a diarrhea sample. **B** *Giardia intestinalis*. **C** *E. coli* O157:H7. (A: © CDC/Bryon Skinner B: Dr. Stan Erlandsen, University of Minnesota C: Courtesy of Dr. Michael S. Donnenberg)

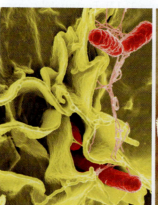

Figure 11.20 Food and water may also harbor protozoa. Shown at left is a colored image of a *Salmonella* bacterium (*pink*) invading a human cell. (Left: © Rocky Mountain Laboratories Right: © Cengage Learning)

CONNECTIONS

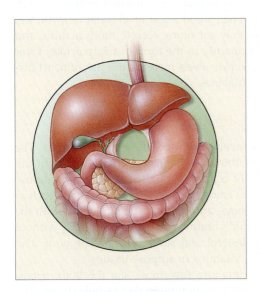

The Digestive System

Body cells require nutrients for energy and for the processes that build new cells and cell parts.

The digestive system contributes to homeostasis by breaking down bulk food to nutrients, vitamins, and minerals that can be absorbed into the bloodstream. It also absorbs water and stores and eliminates solid wastes as feces.

Integumentary system

Excess calories may be stored as insulating fat in the hypodermis; absorbed copper is used in making melanin.

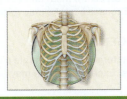

Skeletal system

Absorbed calcium and phosphorus are major components of bone tissue.

Muscular system

Absorbed calcium, potassium, and sodium are needed for muscle contraction; the lactic acid produced by working muscles is converted to glucose in the liver.

Cardiovascular system and blood

Absorbed water maintains blood volume; iron is used in red blood cells to make hemoglobin; vitamin K made in the colon is used in clotting; many plasma proteins are made in the liver.

Immunity and the lymphatic system

Gastric juice and stomach enzymes help destroy microorganisms; diarrrhea helps flush microbes from the intestines.

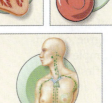

Respiratory system

Absorbed nutrients nourish lungs and other organs; water provides moisture needed for gas exchange; the stomach and liver provide physical support for the diaphragm.

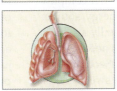

Urinary system

The kidneys use absorbed water in forming urine; absorbed sodium is essential in adjustments to the body's acid–base balance and the water content of urine.

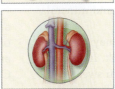

Nervous system

Glucose from digested carbohydrates or formed in the liver is the basic source of energy for brain cells; absorbed sodium and potassium are requried to generate nerve impulses.

Sensory systems

Absorbed nutrients nourish all sensory organs; vitamin A makes pigments used in vision.

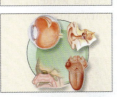

Endocrine system

Pancreas, stomach, and intestinal hormones help regulate hunger and digestion; insulin and glucagon from the pancreas regulate blood sugar; the liver deactivates several hormones.

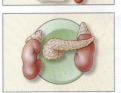

Reproductive system

Absorbed nutrients support development of sperm and eggs and sustain growing offspring during pregnancy.

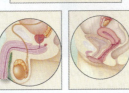

- Diet has a major effect on body functions because it supplies major nutrients as well as vitamins and minerals.

- Links to Carbohydrates, lipids, and proteins 2.9–2.11

The nutrients we absorb are burned as fuel to provide energy and used as building blocks to build and replace tissues. In this section we focus on the three main classes of nutrients—carbohydrates, lipids, and proteins—and we take a look at guidelines for what makes up a healthy diet.

Complex carbohydrates are best

There are many views on the definition of a "proper" diet, but just about all nutritionists can agree on this point: The healthiest carbohydrates are "complex" ones such as starch—the type of carbohydrate in fleshy fruits, cereal grains, and legumes, including peas and beans (see Figure 11.21).

Human digestive enzymes easily break down complex carbohydrates to glucose, the body's chief energy source. Foods rich in complex carbohydrates also usually are high in fiber, including the insoluble fiber that adds needed bulk to feces and helps prevent constipation (see Section 11.9). By contrast, simple sugars such as those in sweets don't have much fiber, and they lack the vitamins and minerals of whole foods.

A person who eats lots of packaged food may consume up to two pounds of refined sugars per week. Ingredient labels may list these sugars as corn syrup, corn sweeteners, and dextrose. They represent "empty calories" because they add to our caloric intake but meet no other nutritional needs. Highly refined carbohydrates also have a high **glycemic index** (GI). This index ranks foods by their effect on blood glucose during the first two hours after a meal. For example, white rice and breads or crackers made with refined white flour have a high GI. They are digested quickly and cause a surge in the blood levels of sugar and insulin.

Figure 11.21 Fruits, vegetables, and whole grains are good sources of both soluble and insoluble fiber. (Left: Brian Chase/Shutterstock.com Right: © Paul Poplis Photography, Inc./Stockfood.com)

Circulating insulin makes cells take up glucose quickly, and it also prevents cells from using stored fat as fuel. At the same time, glucose that is not needed as fuel for cells is stored as fat. When blood sugar levels later fall, you feel hungry. So you may eat more, secrete more insulin, and keep storing fat, mainly in the form of triglycerides. Over time, high triglyceride levels increase the risk of heart disease and type 2 diabetes.

Some fats are healthier than others

The body can't survive without fats and other lipids. The phospholipid lecithin and the sterol cholesterol both are building blocks of cell membranes. Fat stored in adipose tissue serves as an energy reserve, cushions organs such as the eyes and kidneys, and provides insulation beneath the skin. A young child's brain won't develop properly without a supply of cholesterol and saturated fat. The body also stores fat-soluble vitamins in adipose tissues.

The liver can manufacture most fats the body needs, including cholesterol, from protein and carbohydrates. The ones it cannot make are **essential fatty acids**, but whole foods and vegetable oils provide plenty of them. Linoleic acid is an example. You can get enough of it by consuming just one teaspoon a day of corn oil, olive oil, or some other polyunsaturated fat.

Animal fats—the fat in butter, cheese, and fatty meat—are rich in saturated fats and cholesterol. Eating too much of these kinds of foods increases the risk for heart disease and stroke, as well as for certain cancers. As described in Chapter 2, trans fatty acids, or "trans fats," are also bad for

TABLE 11.3   Main Types of Dietary Lipids
POLYUNSATURATED FATTY ACIDS:

Liquid at room temperature; essential for health.
  Omega-3 fatty acids
    Alpha-linolenic acid and its derivatives
    Sources: Nut oils, vegetable oils, oily fish
  Omega-6 fatty acids
    Linoleic acid and its derivatives
    Sources: Nut oils, vegetable oils, meat

MONOUNSATURATED FATTY ACIDS:
Liquid at room temperature. Main dietary source is olive oil. Beneficial in moderation.
SATURATED FATTY ACIDS:
Solid at room temperature. Main sources are meat and dairy products, palm and coconut oils. Excessive intake may raise risk of heart disease.
TRANS FATTY ACIDS (HYDROGENATED FATS):
Solid at room temperature. Manufactured from vegetable oils and used in many processed foods. Excessive intake may raise risk of heart disease.

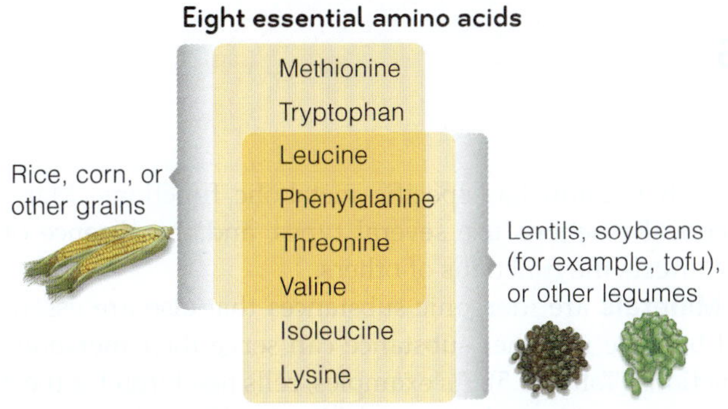

Eight essential amino acids

Rice, corn, or other grains

Methionine
Tryptophan
Leucine
Phenylalanine
Threonine
Valine
Isoleucine
Lysine

Lentils, soybeans (for example, tofu), or other legumes

Figure 11.22 Numerous foods can supply the eight essential amino acids. (© Cengage Learning)

ChooseMyPlate.gov/USDA

Figure 11.23 The USDA formulates nutritional guidelines. Recommended proportions to add up to a daily 2,000 kilocalorie intake for sedentary females aged 10 to 30. The recommended intake and serving sizes are larger for males and highly active females and less for older females.

the cardiovascular system. Food labels are now required to show the amounts of trans fats, saturated fats, and cholesterol per serving. Table 11.3 lists the main types of lipids in food.

Proteins are body-building nutrients

When the digestive system breaks down and absorbs proteins, their amino acids become available for protein synthesis in cells. Of the twenty common amino acids, eight are **essential amino acids**. Our cells cannot make them, so we must obtain them from food. The eight are isoleucine, leucine, lysine, methionine, phenylalanine, threonine, tryptophan, and valine (Figure 11.22).

Most animal proteins are complete, meaning their ratios of amino acids match human nutritional needs. Nearly all plant proteins are incomplete, meaning they lack one or more of the essential amino acids. (The proteins of quinoa, pronounced KEEN-wah, are an exception.) To get required amino acids from a vegetarian diet, one must combine plant foods so the amino acids missing from one are present in others. Examples are combining beans with rice, cornbread with chili, tofu with rice, and lentils with wheat bread.

There are several guidelines for healthy eating

Scientists at the Department of Agriculture (USDA) and other U.S. government agencies study diets that may help prevent health problems such as heart disease, type 2 diabetes, and certain cancers. It periodically updates its nutritional guidelines. Figure 11.23 shows the most recent USDA guidelines, presented in the form of a plate with recommended portions of various types of foods. The guidelines suggest eating less meat and more dark green and orange vegetables, fruits, and milk products. They call for eating less of foods containing refined grains (such as white flour and white rice), trans fats and saturated fats, and refined sugars.

Respected alternative diets include the Mediterranean diet, which is associated with a lower risk of heart disease, among other chronic ills. It emphasizes grains, fruits, and vegetables. Its main fat is olive oil, an excellent antioxidant.

The diet also limits weekly intakes of animal protein, eggs, and refined sugars.

Also popular are "low-carb" diets: fewer carbohydrates and more proteins and fats. People often lose weight rapidly on such diets. However, their long-term effects on organs such as the kidneys are not yet known. High-fat, high-protein diets make the kidneys work harder, raising the risk of kidney stones and other kidney problems. Studies show that following a low-carb diet for six months does not increase LDLs, the "bad" form of cholesterol. Still, given all the evidence that a diet high in saturated fat increases the risk of heart disease, low-carb dieters are advised to obtain their protein from fish, lean meat, or vegetable sources.

essential amino acid An amino acid that cannot be made in the body and so must be obtained from the diet.

essential fatty acid A fatty acid that cannot be made in the body and so must be obtained from the diet.

glycemic index Ranking of foods by their effect on blood glucose during the two hours following a meal.

TAKE-HOME MESSAGE

WHAT IS A "HEALTHY DIET"?

- A healthy diet must provide essential nutrients in the proper proportions and amounts.
- Complex carbohydrates provide nutrients and fiber without adding "empty" calories.
- Fats and other lipids are used for building cell membranes, energy stores, and other needs.
- Various foods, including vegetable oils, provide essential fatty acids.
- Proteins provide essential amino acids.

11.13 Vitamins and Minerals

■ Vitamins and minerals are essential for normal body functioning.

Vitamins are organic substances with essential functions in metabolism (Table 11.4). No other substances can play their roles. In the course of evolution, animal cells have lost the ability to synthesize these substances, so we must obtain vitamins from food.

Each vitamin has specific metabolic functions. Many chemical reactions use several types, and the absence of one affects the functions of others.

Minerals are inorganic substances that also are essential because no other substance can serve their metabolic functions (Table 11.5). As examples, cells need iron for their electron transport chains, red blood cells can't function

TABLE 11.4 Major Vitamins: Sources, Functions, and Effects of Deficiencies or Excesses*

Vitamin	Common Sources	Main Functions	Effects of Chronic Deficiency	Effects of Extreme Excess
FAT-SOLUBLE VITAMINS				
A	Its precursor comes from beta-carotene in yellow fruits, yellow or green leafy vegetables; also in fortified milk, egg yolk, fish, liver	Used in synthesis of visual pigments, bone, teeth; maintains epithelia	Dry, scaly skin; lowered resistance to infections; night blindness; permanent blindness	Malformed fetuses; hair loss; changes in skin; liver and bone damage; bone pain
D	Inactive form made in skin, activated in liver, kidneys; in fatty fish, egg yolk, fortified milk products	Promotes bone growth and mineralization; enhances calcium absorption; possible role in immunity	Bone deformities (rickets) in children; bone softening in adults	Retarded growth; kidney damage; calcium deposits in soft tissues
E	Whole grains, dark green vegetables, vegetable oils	Counters effects of free radicals; helps maintain cell membranes; blocks breakdown of vitamins A and C in gut	Lysis of red blood cells; nerve damage	Muscle weakness, fatigue, headaches, nausea
K	Enterobacteria form most of it; also in green leafy vegetables, cabbage	Blood clotting; ATP formation via electron transport	Abnormal blood clotting; severe bleeding (hemorrhaging)	Anemia; liver damage and jaundice
WATER-SOLUBLE VITAMINS				
B_1 (thiamin)	Whole grains, green leafy vegetables, legumes, lean meats, eggs	Connective tissue formation; folate utilization; coenzyme action	Water retention in tissues; tingling sensations; heart changes; poor coordination	None reported from food; possible shock reaction from repeated injections
B_2 (riboflavin)	Whole grains, poultry, fish, egg white, milk	Coenzyme action	Skin lesions	None reported
B_3 (niacin)	Green leafy vegetables, potatoes, peanuts, poultry, fish, pork, beef	Coenzyme action	Contributes to pellagra (damage to skin, gut, nervous system, etc.)	Skin flushing; possible liver damage
B_6	Spinach, tomatoes, potatoes, meats	Coenzyme in amino acid metabolism	Skin, muscle, and nerve damage; anemia	Impaired coordination; numbness in feet
Pantothenic acid	In many foods (meats, yeast, egg yolk especially)	Coenzyme in glucose metabolism, fatty acid and steroid synthesis	Fatigue; tingling in hands; headaches; nausea	None reported; may cause diarrhea occasionally
Folate (folic acid)	Dark green vegetables, whole grains, yeast, lean meats; enterobacteria produce some folate	Coenzyme in nucleic acid and amino acid metabolism	A type of anemia; inflamed tongue; diarrhea; impaired growth; mental disorders	Masks vitamin B_{12} deficiency
B_{12}	Poultry, fish, red meat, dairy foods (not butter)	Coenzyme in nucleic acid metabolism	A type of anemia; impaired nerve function	None reported
Biotin	Legumes, egg yolk; colon bacteria produce some	Coenzyme in fat, glycogen formation and in amino acid metabolism	Scaly skin (dermatitis); sore tongue; depression; anemia	None reported
C (ascorbic acid)	Fruits and vegetables, especially citrus, berries, cantaloupe, cabbage, broccoli, green pepper	Collagen synthesis; possibly inhibits effects of free radicals; structural role in bone, cartilage, and teeth; used in carbohydrate metabolism	Scurvy; poor wound healing; impaired immunity	Diarrhea, other digestive upsets; may alter results of some diagnostic tests

* Guidelines for appropriate daily intakes are being worked out by the Food and Drug Administration.

without iron in hemoglobin (the oxygen-carrying pigment in blood), and neurons require sodium and potassium.

People who are in good health and who eat a balanced diet of whole foods probably get the vitamins and minerals they need. Yet nearly half of Americans regularly "self-prescribe" vitamin and mineral supplements, including several touted as antioxidants effective in slowing normal aging and preventing cancer. Is this a sensible strategy? A growing body of scientific evidence suggests that the decision to take supplemental vitamins or minerals should be tailored to personal circumstances and planned with the help of a physician. One reason for caution is that excessive amounts of many vitamins and minerals can be useless or even harmful. For example, very large doses of vitamin E appear not to protect against cancer, nor do they extend lifespan generally. Large doses of the fat-soluble vitamins D and A can build up in tissues, especially in the liver, and interfere with normal metabolism. High doses of vitamin C can produce digestive upsets.

Recent research suggests that a healthy diet should provide various **phytochemicals**, also called phytonutrients. These compounds, which occur naturally in various plants, are thought to reduce the risk of certain cancers and other disorders. Examples include the lycopene in tomatoes, flavonoids in yellow/orange fruits such as cantaloupe and citrus fruits, lutein in leafy greens, and the thousands of phytochemicals in broccoli, kale, and other members of the cabbage family.

minerals Inorganic substances with an essential role in metabolism.

phytochemicals Plant molecules that are not an essential part of the human diet but that may reduce the risk of certain disorders.

vitamins Organic substances with an essential role in metabolism.

TAKE-HOME MESSAGE

WHAT ARE VITAMINS AND MINERALS?

- Vitamins and minerals are substances that have specific metabolic functions no other nutrients can serve.

TABLE 11.5 Major Minerals: Sources, Functions, and Effects of Deficiencies or Excesses*

Mineral	Common Sources	Main Functions	Chronic Deficiency	Extreme Excess
Calcium	Dairy products, dark green vegetables, dried legumes	Bone, tooth formation; blood clotting; neural and muscle action	Stunted growth; possibly diminished bone mass (osteoporosis)	Impaired absorption of other minerals; kidney stones in susceptible people
Chloride	Table salt (usually too much in diet)	HCl formation in stomach; contributes to body's acid–base balance; neural action	Muscle cramps; impaired growth; poor appetite	Contributes to high blood pressure in certain people
Copper	Nuts, legumes, seafood, drinking water	Used in synthesis of melanin, hemoglobin, and some transport chain components	Anemia; changes in bone and blood vessels	Nausea; liver damage
Fluorine	Fluoridated water, tea, seafood	Bone, tooth maintenance	Tooth decay	Digestive upsets; mottled teeth and deformed skeleton in chronic cases
Iodine	Marine fish, shellfish, iodized salt, dairy products	Thyroid hormone formation	Enlarged thyroid (goiter), with metabolic disorders	Toxic goiter
Iron	Whole grains, green leafy vegetables, legumes, nuts, eggs, lean meat, molasses, dried fruit, shellfish	Formation of hemoglobin and cytochrome (transport chain component)	Iron-deficiency anemia; impaired immune function	Liver damage; shock; heart failure
Magnesium	Whole grains, legumes, nuts, dairy products	Coenzyme role in ATP–ADP cycle; roles in muscle, nerve function	Weak, sore muscles; impaired neural function	Impaired neural function
Phosphorus	Whole grains, poultry, red meat	Component of bone, teeth, nucleic acids, ATP, phospholipids	Muscular weakness; loss of minerals from bone	Impaired absorption of minerals into bone
Potassium	Diet alone provides ample amounts	Muscle and neural function; roles in protein synthesis and body's acid–base balance	Muscular weakness	Muscular weakness; paralysis; heart failure
Sodium	Table salt; diet provides ample to excessive amounts	Key role in body's salt–water balance; roles in muscle and neural function	Muscle cramps	High blood pressure in susceptible people
Sulfur	Proteins in diet	Component of body proteins	None reported	None likely
Zinc	Whole grains, legumes, nuts, meats, seafood	Component of digestive enzymes; roles in normal growth, wound healing, sperm formation, and taste and smell	Impaired growth; scaly skin; impaired immune function	Nausea, vomiting, diarrhea; impaired immune function and anemia

* Guidelines for appropriate daily intakes are being worked out by the Food and Drug Administration.

11.14 Food Energy and Body Weight

- **Attitudes about body weight often are cultural, but excess weight also raises important health issues.**

The "fat epidemic" noted in this chapter's introduction is spreading around the world. Lifestyles are becoming more sedentary, and many people simply are eating more: Studies show that since the 1970s portion sizes in most restaurants have doubled. This is one reason why the FDA guidelines noted in Section 11.12 don't say "servings" of food but focus on relative amounts instead.

The scientific standard for body weight is based on the ratio of weight to height. Today many authorities use body mass index, or **BMI**, as a general standard (Table 11.6). BMI is determined by the formula

$$\text{BMI} = \frac{\text{weight (pounds)} \times 703}{\text{height (inches)}^2}$$

In general, a person who is *overweight* has a BMI of 25 to 29.9. When BMI enters this range, the health risk begins to rise. Smoking, a family history of heart disease, and having an "apple shape" (fat stored above the waist) increase the risk. **Obesity** is an excess of body fat and it corresponds to a BMI of 30 or more. The World Health Organization has declared obesity a major global health concern, in part because its harmful effects on health are so serious—increasing the risk of not only type 2 diabetes and heart disease but also osteoarthritis, high blood pressure, kidney stones, and many other ailments.

BMI is only a general standard for determining weight status. Why? One reason is that muscle weighs more than fat, and BMI doesn't factor in the ratio of muscle to fat in an individual's body. The ratio is higher in males, whose bodies tend to be more muscular, and it also varies depending on a person's race, age, and general fitness. For instance, a trained athlete's well-developed muscles may push him or her beyond the "healthy weight" BMI category even though the person is not overfat and in fact is in good physical condition. Taking into account the correlation between

TABLE 11.6 Animated! Body Mass Index (BMI)

Height	Under-weight (<18.5)		Healthy Weight (18.5–24.9)						Overweight (25–29.9)						Obese (≥30)								
	18	19	20	21	22	23	24	25	26	27	28	29	30	31	32	33	34	35	36	37	38	39	40
											Body weight (in pounds)												
4'10"	86	91	96	100	105	110	115	119	124	129	134	138	143	148	153	158	162	167	172	177	181	186	191
4'11"	89	94	99	104	109	114	119	124	128	133	138	143	148	153	158	163	168	173	178	183	188	193	198
5'0"	92	97	102	107	112	118	123	128	133	138	143	148	153	158	163	168	174	179	184	189	194	199	204
5'1"	95	100	106	111	116	122	127	132	137	143	148	153	158	164	169	174	180	185	190	195	201	206	211
5'2"	98	104	109	115	120	126	131	136	142	147	153	158	164	169	175	180	186	191	196	202	207	213	218
5'3"	102	107	113	118	124	130	135	141	146	152	158	163	169	175	180	186	191	197	203	208	214	220	225
5'4"	105	110	116	122	128	134	140	145	151	157	163	169	174	180	186	192	197	204	209	215	221	227	232
5'5"	108	114	120	126	132	138	144	150	156	162	168	174	180	186	192	198	204	210	216	222	228	234	240
5'6"	112	118	124	130	136	142	148	155	161	167	173	179	186	192	198	204	210	216	223	229	235	241	247
5'7"	115	121	127	134	140	146	153	159	166	172	178	185	191	198	204	211	217	223	230	236	242	249	255
5'8"	118	125	131	138	144	151	158	164	171	177	184	190	197	203	210	216	223	230	236	243	249	256	262
5'9"	122	128	135	142	149	155	162	169	176	182	189	196	203	209	216	223	230	236	243	250	257	263	270
5'10"	126	132	139	146	153	160	167	174	181	188	195	202	209	216	222	229	236	243	250	257	264	271	278
5'11"	129	136	143	150	157	165	172	179	186	193	200	208	215	222	229	236	243	250	257	265	272	279	286
6'0"	132	140	147	154	162	169	177	184	191	199	206	213	221	228	235	242	250	258	265	272	279	287	294
6'1"	136	144	151	159	166	174	182	189	197	204	212	219	227	235	242	250	257	265	272	280	288	295	302
6'2"	141	148	155	163	171	179	186	194	202	210	218	225	233	241	249	256	264	272	280	287	295	303	311
6'3"	144	152	160	168	176	184	192	200	208	216	224	232	240	248	256	264	272	279	287	295	303	311	319
6'4"	148	156	164	172	180	189	197	205	213	221	230	238	246	254	263	271	279	287	295	304	312	320	328
6'5"	151	160	168	176	185	193	202	210	218	227	235	244	252	261	269	277	286	294	303	311	319	328	336
6'6"	155	164	172	181	190	198	207	216	224	233	241	250	259	267	276	284	293	302	310	319	328	336	345

Based on values established by the National Heart, Lung and Blood Institute and published by the Weight-control Information Network (WIN) of the National Institute of Diabetes and Digestive and Kidney Diseases (NIDDK). win.niddk.nih.gov.

mid-body fat stores and obesity, many authorities believe that the best overall indicator of "weight fitness" is a combination of BMI and a measurement of an individual's waist circumference.

When someone is overweight, the usual culprit is an unbalanced "energy equation" in which too many food calories are taken in while too few calories are burned. We measure food energy in **kilocalories** (kcal). A kilocalorie is 1,000 calories of heat energy. (Calorie, with a capital *C*, is shorthand for a kilocalorie.) A value called **basal metabolic rate (BMR)** measures the amount of energy needed to sustain basic body functions. As a general rule, the younger you are, the higher your BMR. But like BMI, BMR also varies from person to person, and it is influenced by the amount of muscle tissue in the body, emotions, hormones, and differences in physical activity. Adding BMR to the kcal needed for other demands (such as body movements) gives the total amount of food energy you need to fuel your daily life.

To figure out how many kcal you should take in daily to maintain a desired weight, multiply that weight (in pounds) by 10 if you are sedentary, by 15 if you are fairly active, and by 20 if highly active. From the value you get this way, subtract the following amount:

Age	20–34	Subtract	0
	35–44		100
	45–54		200
	55–64		300
	Over 65		400

For instance, if you want to weigh 120 pounds and are very active, 120 × 20 = 2,400 kilocalories. If you are 35 years old and moderately active, then you should take in a total of 1,800 − 100, or 1,700 kcal a day. Along with this rough estimate, factors such as height and gender also must be considered. Because males tend to have more muscle, they tend to burn more calories (they have a higher BMR); hence an active woman needs fewer kilocalories than an active man of the same height and weight. Nor does she need as many as another active woman who weighs the same but is several inches taller.

Genes, hormones, and activity affect weight

Unlike hunger, which is the physiological drive to take in food, **appetite** is a desire to eat apart from any physical need. Although various factors influence both appetite and body weight, scientists have identified several genes that code for hormones that influence appetite. One of them, ghrelin (GRELL-in), is made in the stomach. It stimulates a person's appetite. When your stomach is empty, more ghrelin is released. The level falls again after a meal.

The fat-filled cells in adipose tissue produce a second hormone, leptin, which acts on certain cells in the brain. It suppresses appetite and so may help prevent overeating. When leptin was first discovered, some physiologists wondered if obese people might have a leptin deficiency, but further study showed that cases of leptin deficiency are

Figure 11.24 Physical activity helps maintain a healthy body weight.

© Cengage Learning/Gary Head

rare. Some researchers have speculated that some people with weight problems have developed "leptin resistance," in which the hormone does not have its usual appetite-suppressing effects in the brain.

For most people, maintaining a healthy weight over the years requires balancing their "energy budget" so that energy in—calories in food—equals energy used by body cells. Losing a pound of fat requires expending about 3,500 kcal. Weight-loss diets may accomplish this deficit temporarily, but over the long haul keeping off excess weight means pairing a moderate reduction in caloric intake with an increase in physical activity (Figure 11.24). Exercise also increases the mass of skeletal muscles, and even at rest muscle burns more calories than other types of tissues.

appetite The desire to eat, apart from the physical need for food.

BMI Body mass index, a measure of the ratio of weight to height.

BMR Basal metabolic rate, the amount of energy needed to sustain body functions.

kilocalories The standard measure of food energy. One kcal equals 1,000 calories of heat energy.

obesity An excess of body fat with a BMI of 30 or higher.

Decades of scientific studies support a link between having a healthy weight and overall physical well-being. Although some people with a BMI above 30 exercise enough to maintain overall fitness, research shows that most overweight people are too sedentary. On balance, carrying significant excess weight over a period of years correlates strongly with increased risk of chronic diseases and disorders, including type 2 diabetes, atherosclerosis, hypertension, and sleep apnea.

A condition called **metabolic syndrome** factors into this risk picture. Medically, a syndrome is a cluster of symptoms. Symptoms of metabolic syndrome include chronic high blood pressure, elevated triglycerides in the blood, lower "good" cholesterol (HDL), and a reduced capacity of body cells to respond to the hormone insulin, which promotes the uptake of blood sugar. Another symptom is a marked increase in abdominal fat—a waist measurement of 35 inches or more in women and 40 inches or more in men.

A build-up of excess fat triggers genetic changes that result in inflammation. This fat-related inflammation may be a factor in the strong correlation between obesity and heart disease.

metabolic syndrome
A cluster of symptoms that increase the risk of disorders such as diabetes and atherosclerosis.

Surgery can be an option for treating extreme obesity

Some extremely obese people have surgery that reduces the amount of food they can eat. Typical candidates are people with a BMI of 40 or higher, or who have an obesity-related disease. One option is gastric bypass surgery, which closes off most of the stomach and bypasses several feet of the small intestine (Figure 11.25A). A less drastic choice is gastric banding. There are several types of banding procedures, but in each a plastic band cinches off part of the stomach so that only a small pouch can hold food. A device placed under the skin of the abdomen allows the patient's doctor to adjust the band (Figure 11.25B). With either type of surgery, the person can eat only a small amount before feeling full, which reduces the amounts of nutrients that can be absorbed.

Some obese people who have weight-loss surgery lose as much as 70 percent of their excess fat. The surgery isn't an easy way out, however. Minor to serious complications can arise, and the patient must be willing to commit to long-term lifestyle changes that include healthy diet and regular exercise.

Courtesy of Lisa Hyche

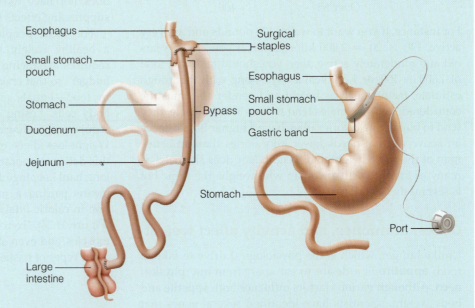

Esophagus — Surgical staples
Small stomach pouch
Stomach
Duodenum — Bypass
Jejunum

Esophagus
Small stomach pouch
Gastric band
Stomach

Large intestine

Port

A gastric bypass
In gastric bypass, surgical staples restrict the stomach to a small pouch that the surgeon connects to the lower jejunum, bypassing the rest of the stomach, the entire duodenum, and the upper jejunum. The stomach pouch holds only a few tablespoons of food at a time.

B gastric banding
A gastric band is placed slightly beyond the gastroesophageal sphincter at the entrance to the stomach, creating a small pouch that can receive food. The opening is adjustable by way of a port placed under the skin of the abdomen.

Figure 11.25 Gastric surgery is an option for treating severe obesity. The photographs show a woman before and after she underwent gastric bypass surgery. (© Cengage Learning)

Eating disorders can be life threatening

It's probably safe to say that most of us overeat from time to time. Some people, though, develop a habit of **binge eating**—they eat an abnormally large amount of food in a few hours and do so at least twice a week for 6 months or more. Emotional factors are at the root of this and other forms of compulsive eating. Psychological counseling may help a binge eater regain a healthier perspective on food.

Emotions sometimes can influence weight gain and loss to a dangerously extreme degree. People who suffer from **anorexia nervosa** see themselves as fat no matter how thin they become. An anorexic purposely starves despite feeling hungry and may overexercise as well. Common side effects include heart arrythmias, osteoporosis, and cessation of menstruation (*amenorrhea*). Although anorexia nervosa is most common among younger women, more cases are being reported among older women and young men.

Anorexia nervosa typically is rooted in a complex blend of psychological and emotional issues and social factors. Female dancers or athletes may become anorexic as they try to achieve an ultra-thin body they see as crucial to their success. Untreated anorexia can kill, but people who receive treatment—usually, a long-term process—can return to a healthy weight (Figure 11.26).

Another eating extreme is the binge–purge disorder called **bulimia nervosa**. *Bulimic* means "having an oxlike appetite." A bulimic person might consume as much as 50,000 calories at one sitting and then purposely vomit, take a laxative, or both. Unlike anorexia, bulimia is easier to hide from the notice of others because the individual doesn't become seriously emaciated. As a result, some bulimics live with their disorder for many years. Even so, the cumulative effects of bulimia can be extremely destructive. Chronic vomiting can erode the enamel from a person's teeth (due to stomach acid) and rupture the stomach. In severe cases it also can cause chemical imbalances that lead to heart and kidney failure. Options for help include psychological counseling, including supportive group therapy. Because depression is a factor in a fair number of cases, antidepressants can help as well. Table 11.7 lists some major indicators for anorexia and bulimia.

anorexia nervosa Eating disorder in which a person purposely starves and may become dangerously thin.

binge eating Routinely eating an abnormally large quantity of food within a few hours.

bulimia nervosa Eating disorder in which a person alternately binges and purges (as by forced vomiting or use of laxatives).

Eric Gaillard/REUTERS

Figure 11.26 Anorexia nervosa is a serious but curable eating disorder. The photograph shows French cyclist Leontien Zijlaard, who was treated for anorexia nervosa. Four years later she won three Olympic gold medals.

TABLE 11.7 Major Indicators for Anorexia and Bulimia

ANOREXIA

- Refusal to maintain body weight at or above a minimal normal weight for age and height (defined as weight loss resulting in body weight less than 85 percent of that expected for age and height)
- Intense fear of gaining weight or becoming fat, despite being underweight
- Distorted body image or denial that the current body weight is below a minimum healthy standard
- In women, no menstrual periods for at least three months

BULIMIA

- Binge eating that includes a sense of lack of control over the behavior
- Inappropriate compensating behavior to prevent weight gain, such as self-induced vomiting, misuse of laxatives or enemas, and excessive exercise
- Binging/inappropriate compensating behavior (purging or overexercising) occurs at least twice a week for 3 months
- Overconcern with body weight and shape

Adapted from guidelines of the American Psychiatric Association, *Diagnostic and Statistical Manual of Mental Disorders*, 2000.

shabaneiro/Shutterstock.com

EXPLORE
ON YOUR OWN

This is an exercise you can eat when you're done. All you need is a food item like a slice of pizza or a sandwich and paper for jotting notes.

To begin, analyze your meal, noting the various kinds of biological molecules it includes. (For this exercise, ignore nucleic acids.) Then, beginning with your mouth and teeth, write what happens to your meal as it moves through your digestive system. Consider the following questions: What kinds of enzymes act on the different components of the meal (such as lettuce or meat), and where do they act, as it is digested? What mechanical processes aid digestion? Which ones can you consciously control? Using the tables in Section 11.13, list the vitamins and minerals that your meal likely contains. Finally, analyze your meal in terms of its contribution (or lack of one) to a balanced diet.

SUMMARY

Section 11.1 The digestive system breaks food down into molecules that are small enough to be absorbed into the bloodstream. It also stores and eliminates unabsorbed materials and promotes homeostasis by its interactions with other organ systems.

The gastrointestinal tract includes the mouth, pharynx, esophagus, stomach, small intestine, and large intestine. Its associated accessory organs include salivary glands, the liver, the gallbladder, and the pancreas (Table 11.8).

The GI tract is lined with mucous membrane. From the esophagus onward its wall consists of four layers: an innermost mucosa, then the submucosa, then smooth muscle, then the serosa. Sphincters at either end of the stomach and at other locations within the GI tract control the forward movement of ingested material.

Sections 11.2, 11.3 Starch digestion begins in the mouth or oral cavity, where the salivary glands secrete saliva, which contains salivary amylase. Chewed food mixes with saliva to form a bolus that is swallowed. Waves of peristalsis move each bolus down the esophagus to the stomach.

Protein digestion begins in the stomach, where gastric fluid containing pepsins and other substances is secreted. The stomach contents are reduced to a watery chyme that passes through a sphincter into the small intestine.

Section 11.4 Digestion is completed and most nutrients are absorbed in the small intestine, which has a large surface area for absorption due to its many villi and microvilli.

Section 11.5 Enzymes and some other substances secreted by the pancreas, the liver, and the gallbladder aid digestion. Bile (secreted by the liver and then stored and released into the small intestine by the gallbladder) contains bile salts that speed up the digestion of fats. Micelles aid the absorption of fatty acids and triglycerides. The hepatic portal system diverts nutrient-laden blood to the liver for processing.

Section 11.6 In the small intestine, segmentation mixes material and forces it close to the absorptive surface. Absorbed glucose and amino acids move into blood vessels in intestinal villi. Triglycerides enter lacteals, then move into blood vessels.

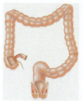

Section 11.7 Peristalsis moves wastes into the large intestine. Water is reabsorbed in the colon; wastes (feces) move on to the rectum and into the anal canal and are eliminated via the anus. The appendix projects from the upper part of the large intestine. It may have a role in immunity.

Section 11.8 The nervous and endocrine systems govern the digestive system. Many controls operate in response to the volume and composition of food in the gut. They cause changes in muscle activity and in the secretion rates of hormones or enzymes.

Sections 11.12, 11.13 Complex carbohydrates are the body's preferred energy source. The diet also must provide eight essential amino acids, some essential fatty acids, vitamins, and minerals. Vitamins are organic substances; minerals are inorganic. Both have metabolic functions no other nutrients can serve. Phytochemicals are useful nutrients obtained from plants.

Section 11.14 Food energy is measured in kilocalories. The basal metabolic rate is the amount of kilocalories needed to sustain the body when a person is awake and resting. To maintain a healthy weight a person's total energy output must balance caloric intake. Obesity is a health-threatening condition that increases the risk of type 2 diabetes, heart trouble, and other diseases and disorders.

TABLE 11.8 Summary of the Digestive System

MAJOR COMPONENTS

Mouth (oral cavity)	Start of digestive system, where food is chewed moistened, polysaccharide digestion begins
Pharynx	Entrance to tubular parts of digestive and respiratory systems
Esophagus	Muscular tube, moistened by saliva, that moves food from pharynx to stomach
Stomach	Sac where food mixes with gastric fluid and protein digestion begins; stretches to store food taken in faster than can be processed; gastric fluid destroys many microbes
Small intestine	The first part (duodenum) receives secretions from the liver, gallbladder, and pancreas
	Most nutrients are digested, absorbed in second part (jejunum)
	Some nutrients absorbed in last part (ileum), which delivers unabsorbed material to colon
Colon (large intestine)	Concentrates and stores undigested matter (by absorbing mineral ions and water)
Rectum	Distension triggers expulsion of feces
Anus	Terminal opening of digestive system

ACCESSORY ORGANS

Salivary glands	Glands (three main pairs, many minor ones) that secrete saliva, a fluid with polysaccharide-digesting enzymes, buffers, and mucus (which moistens and lubricates ingested food)
Pancreas	Secretes enzymes that digest all major food molecules and buffers against HCl from stomach
Liver	Secretes bile (used in fat emulsification); role in carbohydrate, fat, and protein metabolism
Gallbladder	Stores and concentrates bile from the liver

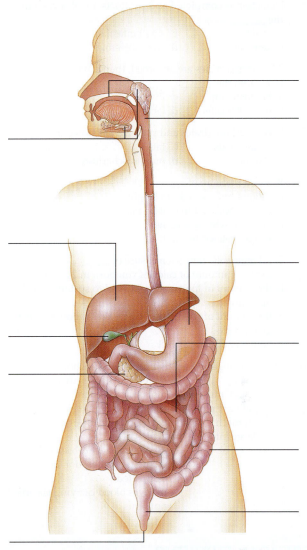

Figure 11.27 Fill in the blanks for substances that cross the lining of the small intestine. (© Cengage Learning)

REVIEW QUESTIONS

1. What are the main functions of the stomach? What roles do enzymes and hormones play?

2. Explain the differences between the digestion roles of the small and large intestines. Does the appendix also have a digestive function?

3. List the organs and accessory organs of the digestive system. On a separate piece of paper, list the main functions of each organ.

4. Define peristalsis, and list the regions of the GI tract where it occurs. Be sure to mention segmentation in your answer.

5. Using the black lines shown in Figure 11.27, name the types of nutrient molecules present at each site that are small enough to be absorbed across the small intestine's lining.

SELF-QUIZ *Answers in Appendix V*

1. Fill in the blanks: Different regions of the digestive system specialize in _____ and _____ food and in _____ unabsorbed food residues.

2. Fill in the blanks: Maintaining normal body weight requires that _____ intake be balanced by _____ output.

3. Fill in the blank: The preferred energy sources for the body are _____.

4. Fill in the blanks: The human body cannot produce its own vitamins or minerals, nor can it produce certain _____ and _____.

5. Digestion is completed and products are absorbed in the _____.
   a. mouth
   b. stomach
   c. small intestine
   d. large intestine

6. After triglycerides are absorbed, they leave the cell and move into the _____.
   a. bloodstream
   b. intestinal cells
   c. liver
   d. lacteals

7. Excess carbohydrates and proteins are stored as _____.
   a. amino acids
   b. starches
   c. fats
   d. monosaccharides

8. BMI is a measure of _____.
   a. ratio of body weight to height
   b. ratio of body fat to muscle mass
   c. body muscle mass alone
   d. weight-related health risk

9. Basal metabolic rate is a measure of _____.
   a. the total amount of calories you burn in 24 hours
   b. the amount of food energy needed to sustain basic body operations
   c. the amount of energy burned by skeletal muscle in a given period
   d. both a and b are correct

10. Match the digestive system parts and functions.
    ____ liver
    ____ small intestine
    ____ salivary glands
    ____ stomach
    ____ large intestine

    a. secrete substances that moisten food, start polysaccharide breakdown
    b. where protein digestion begins
    c. where water is reabsorbed
    d. where most digestion is completed
    e. receives blood carrying absorbed nutrients

CRITICAL THINKING

1. A glass of whole milk contains lactose, protein, triglycerides (in butterfat), vitamins, and minerals. Explain what happens to each component when it passes through your digestive tract.

2. Some nutritionists claim that the secret to long life is to be slightly underweight as an adult. If a person's weight is related partly to diet, partly to activity level, and partly to genetics, what underlying factors could be at work to generate statistics that support this claim?

3. As a person ages, the number of body cells steadily decreases and energy needs decline. If you were planning an older person's diet, what kind(s) of nutrients would you emphasize, and why? Which ones would you recommend an aging person eat less of?

4. Along the lines of question 3, formulate a healthy diet for an actively growing 7-year-old.

5. The food label in Figure 11.28 lists the nutrients and other substances in a package of ready-to-eat macaroni and cheese. Based on your reading in this chapter, how would you rate this product's "healthiness" in terms of fats and carbohydrates?

Your Future

Not everyone who inherits a predisposition for a given disease will actually develop it. Researchers in the emerging field of nutritional genomics are working to discover if diet influences which susceptible people eventually fall ill with a disease that is "in their genes."

Brian Chase/Shutterstock.com

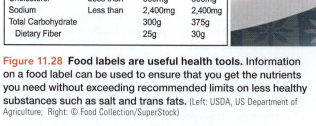

Nutrition Facts
Serving Size 1 cup (228g)
Servings Per Container 2

Amount Per Serving

Calories 250	Calories from Fat 110

	% Daily Value*
Total Fat 12g	18%
Saturated Fat 3g	15%
Trans Fat 1.5g	
Cholesterol 30mg	10%
Sodium 470mg	20%
Total Carbohydrate 31g	10%
Dietary Fiber 0g	0%
Sugars 5g	
Protein 5g	

Vitamin A	4%
Vitamin C	2%
Calcium	20%
Iron	4%

* Percent Daily Values are based on a 2,000 calorie diet. Your Daily Values may be higher or lower depending on your calorie needs:

	Calories:	2,000	2,500
Total Fat	Less than	65g	80g
Sat Fat	Less than	20g	25g
Cholesterol	Less than	300mg	300mg
Sodium	Less than	2,400mg	2,400mg
Total Carbohydrate		300g	375g
Dietary Fiber		25g	30g

Figure 11.28 Food labels are useful health tools. Information on a food label can be used to ensure that you get the nutrients you need without exceeding recommended limits on less healthy substances such as salt and trans fats. (Left: USDA, US Department of Agriculture; Right: © Food Collection/SuperStock)

LINKS TO EARLIER CONCEPTS

Studying the urinary system will tap your knowledge of pH and buffer systems (2.7) and of osmosis and transport mechanisms (3.10, 3.11).

You will also use what you have learned about blood circulation by the cardiovascular system (7.1) and the movement of substances into and out of blood capillaries (7.7).

KEY CONCEPTS

Maintaining the Extracellular Fluid

The body must eliminate chemical wastes from extracellular fluid, including the blood, and manage the levels of water and solutes in it. The urinary system performs this task. Section 12.1

The Urinary System

The urinary system consists of the kidneys, ureters, bladder, and urethra. In the kidneys, structures called nephrons filter substances from the blood, eliminating unneeded ones in urine. Section 12.2

How the Kidneys Form Urine

The kidneys form urine in steps called filtration, reabsorption, and secretion. Hormones and a thirst mechanism adjust the chemical makeup of urine. Sections 12.3–12.5

Disorders of the Urinary System Sections 12.6–12.7

CONNECTIONS:
The Urinary System in Homeostasis Section 12.8

Light or dark? Clear or cloudy? A lot or a little?

Like blood, urine can tell a lot about a person's health. Acidic urine can signal metabolic problems, while alkaline urine may be a sign of a bacterial infection. Too much protein in urine might mean the kidneys are not functioning properly. Specialized urine tests can detect chemicals produced by cancers of the kidney, bladder, and prostate gland. Do-it-yourself urine tests are popular for monitoring a woman's fertile period or early signs she may be pregnant. A test for older women may reveal declining hormone levels that signal the onset of menopause.

Not everyone is eager to have their urine tested. Competitive athletes can be stripped of honors or medals when mandatory urine tests reveal they use prohibited drugs. If you use marijuana, cocaine, or other kinds of illegal drugs, urine also can tell the tale.

That urine can be such a trusty indicator of health, the presence of hormones, and drug use is a tribute to the urinary system. As you will discover in this chapter, the kidneys are the urinary system's all-important blood filters.

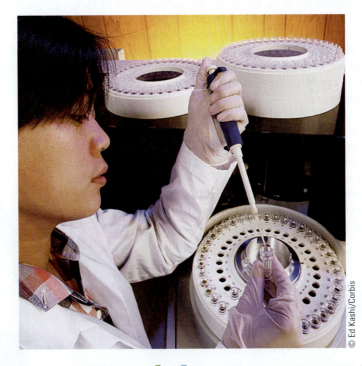

© Ed Kashi/Corbis

Homeostasis Preview
Blood filtering by the kidneys rids the body of excess water and excess or harmful substances in the blood and other body fluids.

The Challenge: Shifts in Extracellular Fluid

- The chemical makeup of body fluid changes constantly as water and solutes enter and leave it.

- Links to Water and life 2.5, Condensation reactions 2.8, Metabolism 3.13, Homeostasis 4.10, Blood 8.1, Nutrient absorption 11.4

If you are an adult female in good health, by weight your body is about 50 percent fluid. If you are an adult male, the ratio is about 60 percent. This fluid is extremely important both in the composition of body structures and in nearly all body functions. Chapter 4 introduced the concept of two "fluid compartments" in the body—one that is inside cells, and a second that is outside cells. This concept, summarized in Figure 12.1, is the starting point for understanding why a "fluid management" system is so important in body functioning.

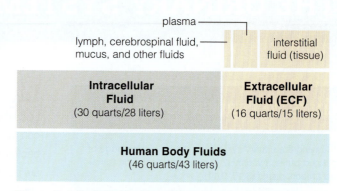

Figure 12.1 **Body fluid occurs in two compartments—one inside cells and the other outside cells. Extracellular fluid consists of blood and tissue (interstitial) fluid.** (© Cengage Learning)

The urinary system adjusts fluid that is outside cells

Section 4.10 explained that tissue fluid fills the spaces between cells and other components of tissues. Blood, which is mostly watery plasma, circulates in blood vessels. As you may remember, tissue fluid, blood plasma, and the relatively small amounts of other fluids (such as in lymph) outside cells together make up the body's *extracellular fluid*, or ECF.

The fluid *inside* cells is *intracellular fluid*. A variety of gases and other substances move constantly between intracellular and extracellular fluid. Those exchanges are crucial for keeping cells functioning smoothly. They can't occur properly unless the volume and composition of the ECF are stable.

Yet the ECF is always changing, because gases, cell products, ions, and other materials enter or leave it. To maintain stable conditions in the ECF, especially the concentrations of water and vital ions such as sodium (Na^+) and potassium (K^+), there must be mechanisms that remove substances as they enter the extracellular fluid or add needed ones as they leave it. The *urinary system* performs this task. Before examining how it operates, we'll

now take a general look at the traffic of substances into and out of extracellular fluid.

The body gains water from food and metabolic processes

Ordinarily, each day you take in about as much water as your body loses (Table 12.1). Some of the water is absorbed from foods and liquids you consume. The rest is produced during metabolic reactions, including cellular respiration and condensation reactions.

Thirst influences how much water we take in. When there is a water deficit in body tissues, the brain "urges" us to seek out water—for example, from a water fountain or a cold drink from the refrigerator. We'll discuss this thirst mechanism later in the chapter.

TABLE 12.1 Normal Daily Balance between Water Gain and Water Loss in Adult Humans			
Water Gain (milliliters)		**Water Loss (milliliters)**	
Ingested in solids:	850	Urine:	1,500
Ingested as liquids:	1,400	Feces:	200
Metabolically derived:	350	Evaporation:	900
	2,600		2,600

The body loses water in urine, sweat, feces, and by evaporation

Water leaves the body in four ways: excretion in urine, evaporation from the lungs and skin, sweating, and in feces. Of these four routes, **urinary excretion** is the form of water loss over which the body has the most control. Urinary excretion eliminates excess water, as well as excess or harmful solutes, in the form of **urine**. Some water also evaporates from our skin and from the respiratory surfaces of the lungs. These are sometimes called "insensible" water losses, because a person is not always aware they are taking place. As noted in Chapter 11, normally very little water that enters the GI tract is lost; most is absorbed and only a little is eliminated in feces.

Solutes enter extracellular fluid from food, respiration, and metabolism

Three main sources add solutes to the body's extracellular fluid. Food supplies nutrients (including glucose) and mineral ions (such as potassium and sodium ions) that are absorbed from the GI tract. Many of us also consume many drugs and food additives. The respiratory system brings oxygen into the blood. Last but not least, living cells continually secrete substances, including carbon dioxide, into tissue fluid and circulating blood. Figure 12.2 gives a snapshot of major interactions between the urinary system and other organ systems.

Solutes leave the ECF by urinary excretion, in sweat, and during breathing

Metabolic wastes, mineral ions, and other solutes leave extracellular fluid in several ways. Metabolism produces more than 200 waste substances. Carbon dioxide is the most abundant one, and we get rid of it by exhaling it from our lungs. All other major wastes leave in urine.

Important metabolic wastes include by-products of processes that break down nucleic acids and proteins. Dismantling nucleic acids produces one of these wastes, uric acid. Another one, ammonia, forms in "deamination" reactions, which remove the nitrogen-containing amino groups from amino acids. Ammonia is highly toxic if it accumulates in the body. Reactions in the liver combine ammonia with carbon dioxide, producing the much less toxic **urea**. Accordingly, urea is the main waste product when cells break down proteins. About half of the urea filtered from blood in the kidneys is reabsorbed. The rest is excreted. Protein breakdown also produces creatine, phosphoric acid, sulfuric acid, and small amounts of other nitrogen-containing compounds, some of which are toxic. These also are excreted.

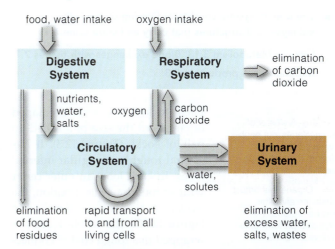

Figure 12.2 The activities of the urinary system coordinate with those of other organ systems. (© Cengage Learning)

Sweat carries away a small percentage of urea, but most nitrogen-containing wastes are removed by the kidneys while they filter other wastes and excess water from the blood. The kidneys also help maintain the balance of important ions such as sodium, potassium, and calcium. These ions are sometimes called **electrolytes** because a solution in which they are dissolved will carry an electric current. Chapter 13 describes the crucial roles electrolytes have in the nervous system.

Normally only a little of the water and solutes that enter the kidneys leaves as urine. In fact, except when you drink lots of fluid (without exercise), all but about 1 percent of the water is returned to the blood. However, the chemical composition of the fluid that is returned has been adjusted in vital ways. Just how this happens will be our focus in the next few sections.

electrolytes Ions that carry an electric current when dissolved in fluid.

urea The main waste product from protein breakdown.

urinary excretion Removal of excess or unwanted water and solutes from the body, in urine.

urine Fluid produced and eliminated from the body during the process of urinary excretion.

The Urinary System: Built for Filtering and Waste Disposal

- The urinary system consists of filtering organs—the kidneys—and structures that carry and store urine.

- Links to Metabolism 3.13, Blood exchanges with tissues 7.7, Red blood cell production 8.3

glomerulus A cluster of blood capillaries in a nephron, where substances move from the blood into the nephron.

kidneys Organs that adjust fluid balance and filter wastes from blood. The kidneys also perform other physiological functions.

nephrons Blood-filtering units in the kidneys.

ureter Tube that carries urine from kidneys to the bladder.

urethra Tube that carries urine to the outside of the body.

urinary bladder Hollow organ that stores urine.

Each **kidney** is a bean-shaped organ about the size of a rolled-up pair of socks (Figure 12.3). It has several roughly triangular internal lobes. In each lobe, an outer *cortex* wraps around a central region, the *medulla*, as you can see sketched in Figure 12.3C. The whole kidney is wrapped in a tough coat of connective tissue, the *renal capsule* (from the Latin *renes*, meaning kidneys). A kidney's central cavity is called the *renal pelvis*.

Our kidneys have several functions. They produce the hormone erythropoietin, which stimulates the production of red blood cells (Section 8.3). They also convert vitamin D to a form that stimulates the small intestine to absorb calcium in food. In addition, kidneys make the enzyme renin, which helps regulate blood pressure, as you will read later in this chapter. The main function of kidneys, however, is to remove metabolic wastes from the blood and adjust fluid balance in the body.

In addition to the two kidneys, the urinary system includes "plumbing" that transports or stores urine. Once urine has formed in a kidney, it flows into a tubelike **ureter**, then on into the **urinary bladder**, where it is stored until you urinate. Urine leaves the bladder through the **urethra**, a tube that opens at the body surface.

Nephrons are the kidney filters

Each kidney lobe contains blood vessels and more than a million slender tubes called **nephrons**. Nephrons are the structures that filter water and solutes from blood.

A nephron is shaped a bit like the piping under a sink (Figure 12.4A). Its wall is a single layer of epithelial cells, but the cells and junctions between them vary in different parts of the tube. Water and solutes pass easily through some parts, but other parts block solutes unless they are moved across by active transport (Section 3.11).

As shown in Figure 12.4B, the nephron wall balloons around a cluster of blood capillaries called the **glomerulus** (glo-MARE-yoo-luss; plural: glomeruli). The cuplike wall

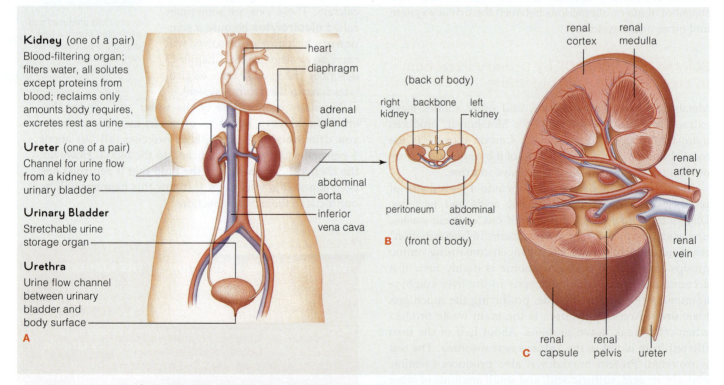

Kidney (one of a pair)
Blood-filtering organ; filters water, all solutes except proteins from blood; reclaims only amounts body requires, excretes rest as urine

Ureter (one of a pair)
Channel for urine flow from a kidney to urinary bladder

Urinary Bladder
Stretchable urine storage organ

Urethra
Urine flow channel between urinary bladder and body surface

A

heart
diaphragm
adrenal gland
abdominal aorta
inferior vena cava

(back of body)
right kidney
backbone
left kidney
peritoneum
abdominal cavity
B (front of body)

renal cortex
renal medulla
renal artery
renal vein
C capsule
renal pelvis
ureter

Figure 12.3 Animated! **The urinary system consists of the kidneys and several other parts. A** Structure and function of the urinary system. **B** The two kidneys, ureters, and urinary bladder are located between the abdominal cavity's wall and its lining, the peritoneum. **C** Internal structure of a kidney. (© Cengage Learning)

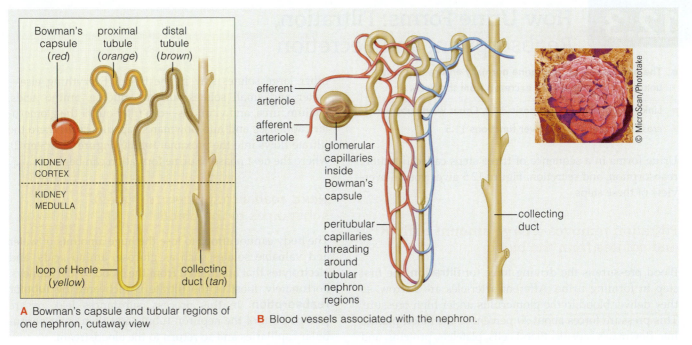

A Bowman's capsule and tubular regions of one nephron, cutaway view

B Blood vessels associated with the nephron.

Figure 12.4 Animated! Interacting with two sets of capillaries, nephrons are a kidney's blood-filtering units.
A Diagram of a nephron. **B** The arterioles and capillaries associated with a nephron. (© Cengage Learning)

region, called the **Bowman's** (glomerular) **capsule**, receives the substances filtered from blood. The rest of the nephron is a winding tubule ("little tube"). Filtrate flows from the cup into the **proximal tubule** (*proximal* means "next to"), then through a hairpin-shaped **loop of Henle** and into the **distal tubule** ("most distant" from Bowman's capsule). This part of the nephron tubule empties into a collecting duct.

Special vessels transport blood to, in, and away from nephrons

Each hour, about 75 gallons of blood course through your kidneys, delivered by the renal arteries. An *afferent arteriole* brings blood to each nephron (*afferent* means "carrying toward"). The blood flows into the glomerulus inside Bowman's capsule. These capillaries are not like capillaries in other parts of the body. Specialized pores between the cells of their walls make them much more permeable than other capillaries. Thus it is much easier for water and solutes to move across the wall.

Glomerular capillaries don't channel blood to venules, as other capillaries do (Section 7.7). Instead, they merge to form an *efferent* ("carrying away from") *arteriole*.

This arteriole branches into **peritubular** ("around the tubule") **capillaries**. As you can see in Figure 12.4B, the peritubular capillaries weave around a nephron's tubules. They merge into venules, which carry filtered blood out of the kidneys.

Bowman's capsule The cuplike region of a nephron that receives water and solutes filtered from blood.

distal tubule The part of the nephron tubule farthest from the Bowman's capsule.

loop of Henle The hairpin-shaped midsection of a nephron's tubule.

peritubular capillaries The set of capillaries that weave around a nephron's tubules.

proximal tubule The portion of the nephron tubule closest to the Bowman's capsule.

TAKE-HOME MESSAGE

WHAT DO KIDNEY NEPHRONS DO?

- Kidney nephrons filter water and solutes from blood. A cluster of capillaries called a glomerulus is the nephron's blood-filtering unit.
- The capillaries in a glomerulus have pores in their walls that make the vessels unusually permeable.
- Arterioles transport blood to and from nephrons. Peritubular capillaries weave around nephron tubules and deliver filtered blood back to the general circulation.

How Urine Forms: Filtration, Reabsorption, and Secretion

- The processes that form urine normally ensure that only unneeded substances are excreted from the body.

- Links to Diffusion and osmosis 3.10, Other ways substances cross membranes 3.11, Liver functions 11.5

Urine forms in a sequence of three steps called filtration, reabsorption, and secretion. Figure 12.5 gives you an overview of these steps.

Filtration removes a large amount of fluid and solutes from the blood

Blood pressure is the driving force for **filtration**, the first step in forming urine. Afferent arterioles are narrow, so they deliver blood to the glomerulus under high pressure. This pressure forces about 20 percent of the blood plasma into Bowman's capsule. Blood cells, platelets, proteins, and other large solutes stay in the blood. Everything else—water and small solutes such as glucose, amino acids, sodium, urea, and vitamins—can filter out of the glomerular capillaries and into Bowman's capsule. From there the filtrate flows into the proximal tubule (Figure 12.5, Step 2), where the next phase of urine formation can begin.

Next, reabsorption returns useful substances to the blood

The body cannot afford to lose the huge amounts of water and valuable solutes such as glucose, amino acids, and electrolytes that are filtered from the blood by the kidneys. Fortunately, most of the filtrate is recovered by **tubular reabsorption**. In this process, substances leak or are pumped out of the nephron tubule and then enter peritubular capillaries and so return to the bloodstream.

❶ **Glomerular filtration**
Filtered plasma forced out of glomerular capillaries by blood pressure enters Bowman's capsule.

❷ **Tubular reabsorption**
Essential ions, nutrients, water, and some urea in the filtrate return to the blood. *Green* arrows indicate reabsorption.

❸ **Tubular secretion**
Wastes and excess ions are moved from the blood into the filtrate for elimination in urine. *Blue* arrows indicate secretion.

❹ Hormones that alter permeability of distal tubules and collecting ducts adjust urine concentration.

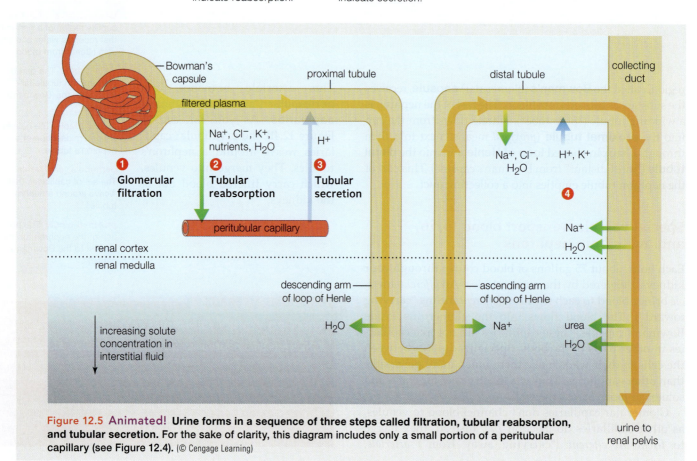

Figure 12.5 Animated! Urine forms in a sequence of three steps called filtration, tubular reabsorption, and tubular secretion. For the sake of clarity, this diagram includes only a small portion of a peritubular capillary (see Figure 12.4). (© Cengage Learning)

Most reabsorption takes place across the walls of proximal tubules. As in all parts of the tubule, the walls in this area are only one cell thick. Step 2 in Figure 12.5 shows what happens with water, glucose, and salt (ions of sodium, Na$^+$, and chloride, Cl$^-$). All these substances can diffuse from the filtrate in a tubule into and through the cells of the tubule wall. On the outer side of the cells, active transport (through proteins in the cells' plasma membranes) moves glucose and Na$^+$ into the tissue fluid. Sodium ions (Na$^+$) are positively charged, and negatively charged ions, including chloride (Cl$^-$), follow the sodium.

As the concentration of solutes rises in the fluid, water moves out of the tubule cells by osmosis. In a final step, solutes are actively transported into peritubular capillaries and water again follows by osmosis. These substances now have been reabsorbed. The solutes and water that remain in the tubule become part of urine.

Reabsorption usually returns almost 99 percent of the filtrate's water, all of the glucose and most amino acids, all but about 0.5 percent of the salt (sodium and chloride ions), and 50 percent of the urea to the blood (Table 12.2).

Secretion rids the body of excess hydrogen ions and some other substances

Tubular secretion takes up unwanted substances that have been transported out of peritubular capillaries and adds them to the urine that is forming in nephron tubules. It is summarized in Step 3 of Figure 12.5. Among other functions, this highly controlled process rids the body of urea and of excess hydrogen ions (H$^+$) and potassium ions (K$^+$).

Secretion is crucial to maintaining the body's acid–base balance, which you will read about in a later section. It also helps ensure that some wastes (such as uric acid and some breakdown products of hemoglobin) and foreign substances (such as antibiotics and some pesticides) do not build up in the blood. The drug testing noted in the chapter introduction relies on the use of urinalysis to detect drug residues that have been secreted into urine.

Homeostasis requires that the total volume of fluid in the blood and tissues stay fairly stable. Blood and tissue fluid are mostly water, and while your kidneys are removing impurities from your blood they are also adjusting the amount of water that is excreted in urine or returned to the bloodstream. These adjustments, which we will look at in more detail in Section 12.4, are represented by Step 4 in Figure 12.5.

Urination is a controllable reflex

You probably don't need to be told that *urination* is urine flow from the body. Urination is a reflex response. As the bladder fills, tension increases in the smooth muscle of its strong walls. Where the bladder joins the urethra, an *internal urethral sphincter* built of smooth muscle helps prevent urine from flowing into the urethra (Figure 12.6). As tension in the bladder wall increases, though, the sphincter relaxes; at the same time, the bladder walls contract and force urine through the urethra.

Skeletal muscle forms an *external urethral sphincter* closer to the urethral opening. Learning to control it is the basis of urinary "toilet training" in young children.

TABLE 12.2 Average Daily Reabsorption Values for a Few Substances			
	Amount Filtered	Percentage Excreted	Percentage Reabsorbed
Water	180 liters	1	99
Glucose	180 grams	0	100
Amino acids	2 grams	5	95
Sodium ions	630 grams	0.5	99.5
Urea	54 grams	50	50

filtration In nephrons, blood pressure forces water and small solutes in blood plasma out of glomerular capillaries and into the Bowman's capsule.

tubular reabsorption Substances move from the filtrate inside a kidney tubule into the peritubular capillaries.

tubular secretion Substances move out of peritubular capilaries and into the filtrate in kidney tubules.

Urinary bladder

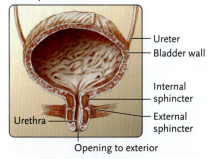

Figure 12.6 **Sphincter muscles control the passage of urine into the urethra. The external sphincter is skeletal muscle and allows voluntary control of urination.** (© Cengage Learning)

How Kidneys Help Manage Fluid Balance and Blood Pressure

- The kidneys concentrate urine before it flows to the bladder. These concentration mechanisms help regulate blood volume and blood pressure.

- Links to Chemical bonds 2.3, Liver functions 11.5

Overall, the total volume of your body fluids, including blood plasma, doesn't vary much. This is because during reabsorption, the kidneys adjust how much water and salt (sodium + chloride ions) the body conserves or excretes in urine. As you know, blood and tissue fluid are mostly water. In general, when the volume of blood increases or decreases, so does blood pressure. The kidneys help ensure that the volume of extracellular fluid, and blood in particular, stays within a normal range.

Water follows salt as urine forms

Although about two-thirds of filtered salt and water is reabsorbed in the proximal tubule, the filtrate usually still contains more of both than the body can afford to lose in urine. This situation is addressed as the filtrate enters the loop of Henle, which descends into the kidney medulla (Figure 12.7). There the loop is surrounded by extremely salty tissue fluid. Water can pass through the thin wall of the loop's descending limb, so more water moves out by osmosis and is reabsorbed. As the water leaves, the salt concentration in the fluid still inside the descending limb increases until it matches that in the fluid outside.

ADH Antidiuretic hormone; it stimulates nephrons to reabsorb water.

aldosterone A hormone that concentrates urine by reducing the amount of water leaving distal tubules.

juxtaglomerular apparatus The region where efferent arterioles lie close to a nephron's distal tubules.

Now the filtrate "rounds the turn" of the loop and enters the ascending limb. The wall of this part of the nephron tubule doesn't allow water to pass through. This is an important variation in the tubule's structure, because here sodium is actively transported out of the ascending limb—but water can't move with it.

The filtrate now moves into the distal tubule. Its cells continue to remove salt but don't also let water escape. Hence, a dilute urine moves on into the collecting duct.

Naturally, as salt leaves the filtrate moving through a nephron tubule, the concentration of solutes rises outside the tubule and falls inside it. This steep gradient helps drive the reabsorption of valuable solutes, which move into peritubular capillaries. It also draws water out of the descending limb by osmosis.

Urea boosts the gradient. As water is reabsorbed, urea left in the filtrate becomes concentrated. Some of it will be excreted in urine, but when filtrate enters the final portion of the collecting duct, some urea also will diffuse out—so the concentration of solutes in the inner medulla rises even more.

Drink a large glass of water and the next time you "go," your urine may be pale and dilute. If you sleep eight hours without a break, your urine will be concentrated and darker yellow. As described next, hormones control how much water the kidneys add to urine. These controls also adjust blood pressure.

Hormones control whether kidneys make urine that is concentrated or dilute

When you don't take in as much water as your body loses, the salt concentration in your blood rises. In the brain, receptors sense this change and trigger the release of antidiuretic hormone, or **ADH**. It acts on cells in distal tubules and collecting ducts so that more water moves out of them and is reabsorbed into the blood (Figure 12.8). As a result, the urine becomes more concentrated. Gradually the additional water in blood reduces the salt concentration there. It also increases the blood volume and blood pressure. Then a negative feedback loop inhibits the release of ADH (Figure 12.9).

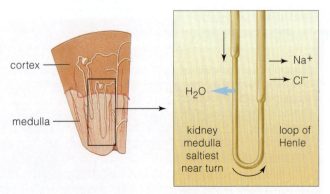

Figure 12.7 Water and salt are reabsorbed in the loop of Henle. (© Cengage Learning)

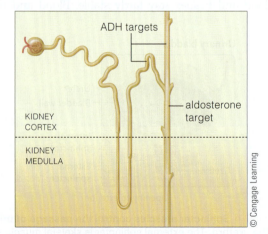

Figure 12.8 ADH and aldosterone act in different parts of kidney nephrons.

Reduced blood volume also affects cells in the afferent arterioles that bring blood to nephrons. These cells release the enzyme renin. They are part of the **juxtaglomerular apparatus** (Figure 12.10A). *Juxta-* means "next to," and this "apparatus" is an area where arterioles of the glomerulus come into contact with a nephron's distal tubule.

Renin triggers reactions that produce a protein called angiotensin I and then convert it to angiotensin II. Among other effects, angiotensin II stimulates cells of the adrenal cortex, the outer portion of a gland perched on top of each kidney, to secrete the hormone **aldosterone** (Figures 12.3 and 12.10B). Aldosterone causes cells of the distal tubules and collecting ducts to reabsorb sodium faster, so less of it and less water are excreted. By limiting the loss of water, this process also influences blood pressure.

What must the kidneys do to make dilute urine? Not much. Urine is automatically dilute as long as ADH levels are low, so little of the hormone acts on the distal tubules and collecting ducts.

A *diuretic* is a substance that promotes the loss of water in urine. For example, caffeine reduces the reabsorption of sodium along nephron tubules, so more water is excreted.

A thirst center monitors sodium

What makes you thirsty when you don't drink enough? The concentration of salt in your blood has risen, and this change reduces the amount of saliva your salivary glands produce. A drier mouth stimulates nerve endings that signal a *thirst center* in the brain. The center also receives signals from the same sensors that stimulate the release of ADH. In this case the signals are relayed to a part of the brain that "tells" you to find and drink fluid.

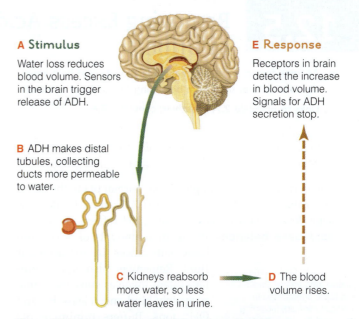

A Stimulus
Water loss reduces blood volume. Sensors in the brain trigger release of ADH.

B ADH makes distal tubules, collecting ducts more permeable to water.

C Kidneys reabsorb more water, so less water leaves in urine.

D The blood volume rises.

E Response
Receptors in brain detect the increase in blood volume. Signals for ADH secretion stop.

Figure 12.9 A negative feedback loop from the kidneys to the brain helps adjust the fluid volume of the blood.
(© Cengage Learning)

TAKE-HOME MESSAGE

HOW DO THE KIDNEYS HELP MANAGE THE BODY'S FLUID BALANCE?

- To maintain the volume of extracellular fluid, hormones adjust the amount of water urine contains.
- ADH stimulates the kidneys to conserve water. It acts on distal tubules and collecting ducts.
- Aldosterone promotes the reabsorption of sodium, which indirectly increases the amount of water the body retains.

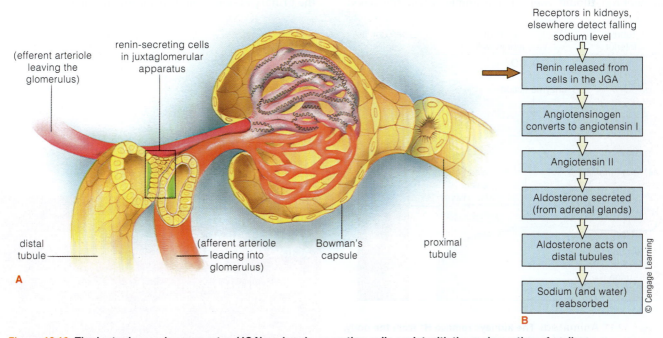

(efferent arteriole leaving the glomerulus)

renin-secreting cells in juxtaglomerular apparatus

distal tubule

(afferent arteriole leading into glomerulus)

Bowman's capsule

proximal tubule

A

Receptors in kidneys, elsewhere detect falling sodium level

Renin released from cells in the JGA

Angiotensinogen converts to angiotensin I

Angiotensin II

Aldosterone secreted (from adrenal glands)

Aldosterone acts on distal tubules

Sodium (and water) reabsorbed

B

© Cengage Learning

Figure 12.10 The juxtaglomerular apparatus (JGA) and renin-secreting cells assist with the reabsorption of sodium.

Removing Excess Acids and Other Substances in Urine

- As urine forms, nephrons make adjustments that help keep the extracellular fluid from becoming too acidic or too basic.
- Links to pH scale 2.7, Breathing controls 10.6

The kidneys play a key role in maintaining the balance of acids and bases in the blood

You may recall from Chapter 2 that normal pH in the blood and other body fluids is between 7.37 and 7.43. Because acids lower pH and bases raise it, pH reflects the body's **acid–base balance**—the relative amounts of acidic and basic substances in extracellular fluid. Remember also that a buffer system involves substances that reversibly bind and release H^+ and OH^- ions. Buffers minimize pH changes as acidic or basic molecules enter or leave body fluids.

Chapter 10 described how bicarbonate (HCO_3^-) serves as a buffer in the lungs. It forms when carbon dioxide combines with water. The bicarbonate then reacts with H^+ to form carbonic acid, and enzyme action converts carbonic acid into water and carbon dioxide. The CO_2 is exhaled, while the hydrogen ions are now a part of water molecules. H^+ is not eliminated permanently, however. Only the kidneys can do that. They also restore the buffer bicarbonate.

Depending on changes in the acid–base balance of the blood that enters nephrons, the kidneys can either excrete bicarbonate or form new bicarbonate and add it to the blood. The necessary chemical reactions go on in the cells of nephron tubule walls. For example, when the blood

acid-base balance The relative amounts of acidic and basic substances in extracellular fluid, including the blood.

metabolic acidosis Higher than normal acidity in blood and other body fluids.

metbolic alkalosis Higher than normal alkalinity in blood and other body fluids.

is too acidic (a too high concentration of H^+), water and carbon dioxide combine with the help of an enzyme. They form carbonic acid that then can be broken into bicarbonate and H^+. Figure 12.11 summarizes these steps.

As you can see, bicarbonate produced in the reactions moves into peritubular capillaries. It ends up circulating in the blood, where it buffers excess H^+. When the blood is too basic (alkaline), chemical adjustments in the kidneys normally ensure that less bicarbonate is reabsorbed into the bloodstream.

The H^+ that is formed in the tubule cells is secreted into the filtrate in the tubule. There the excess H^+ may combine with phosphate ions, ammonia (NH_3), or even bicarbonate. In this way the excess H^+ is excreted.

Various factors may cause serious acid–base imbalances

Maintaining proper blood pH is crucial to homeostasis. If the pH of blood falls outside the normal range for long, the most serious impact occurs in the central nervous system (brain and spinal cord). When severe diarrhea, kidney disease, or some other problem prevents kidneys from excreting enough acid, the result is **metabolic acidosis.** Then nerve cells cannot communicate properly and an affected person may fall into a fatal coma.

Severe vomiting or dehydration, hormonal disorders, and overuse of antacids are common causes of **metabolic alkalosis**, or blood that is too basic. Then nerve cells are overstimulated, so a person may suffer muscle spasms, nervousness, or convulsions. In the next two sections you will find information about other major disorders that prevent the urinary system from functioning normally.

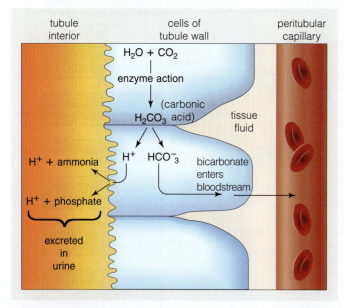

Figure 12.11 Animated! The kidneys remove H^+ from the body, preventing the blood from becoming too acidic. (© Cengage Learning)

TAKE-HOME MESSAGE

HOW DO THE KIDNEYS HELP KEEP THE EXTRACELLULAR FLUID FROM BECOMING TOO ACIDIC OR TOO BASIC?

- The urinary system eliminates excess hydrogen ions and also replenishes bicarbonate used in buffering reactions.
- Serious acid-base imbalances include metabolic acidosis (body fluids that are too acidic) and metabolic alkalosis (body fluids that are too basic).

Kidney Disorders: When Kidneys Fail

Many people don't realize how much good health depends on normal kidney function. Disorders or injuries that interfere with it may cause only mild distress, but often the impact is quite serious.

Kidney stones are deposits of uric acid, calcium salts, and other substances that have settled out of urine. Smaller kidney stones usually are eliminated naturally during urination. Larger ones can become lodged in the renal pelvis or ureter or even in the bladder or urethra. The blockage can partially dam urine flow and cause intense pain and kidney damage. Large kidney stones must be removed medically or surgically. A procedure called *lithotripsy* uses high-energy sound waves to break up the stone into fragments that are small enough to pass out in the urine.

Glomerulonephritis is an umbrella term for several disorders that can lead to kidney failure. Two major ones are chronic high blood pressure and diabetes, both of which damage kidney capillaries. Sometimes the flow of blood through the glomeruli all but stops.

At any given time, roughly 1 million people in the United States have kidneys so impaired that they can only minimally filter the blood and form urine. This loss of kidney function means that toxic by-products of protein breakdown can accumulate in the bloodstream. Patients can suffer nausea, fatigue, and memory loss. In advanced cases, death may result. A kidney dialysis machine can restore the proper solute balances (Figure 12.12A). Like the kidneys, the machine helps maintain healthy volume and composition of extracellular fluid by selectively removing and adding solutes to the patient's bloodstream.

In a dialysis process, substances in one solution can be exchanged with those in a chemically different solution by crossing a permeable membrane. In *hemodialysis*, a dialysis machine is connected to an artery or a vein, and then blood is pumped through tubes made of a material similar to cellophane. The tubes are submerged in warm water that contains a precise mix of salts, glucose, and other substances. As blood flows through the tubes, the wastes dissolved in it diffuse out, so solute concentrates return to a normal range. The cleansed blood then returns to the patient's body.

Patients usually receive hemodialysis three times a week, although for some it is a daily need. Patients with reversible kidney disorders may receive dialysis until they recover. In chronic cases, the procedure must be used for the rest of the patient's life or until a healthy kidney can be transplanted.

Polycystic kidney disease is an inherited disorder in which cysts (semisolid masses) form in the kidneys and in many cases gradually destroy normal kidney tissue (*right*). Frequent urinary tract infections are a common early symptom; in severe cases, dialysis or a kidney transplant are the only real options for treatment. With treatment and the proper diet, many people with chronic kidney disease are able to pursue a surprisingly active, close-to-normal lifestyle (Figure 12.12B).

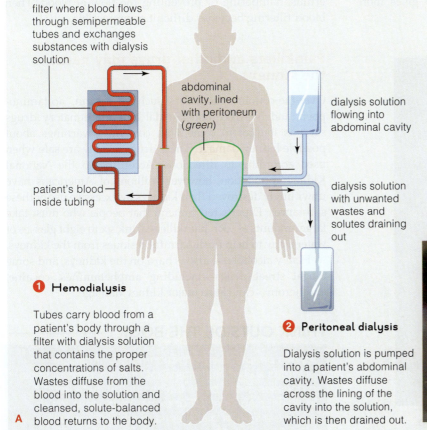

filter where blood flows through semipermeable tubes and exchanges substances with dialysis solution

patient's blood inside tubing

abdominal cavity, lined with peritoneum (*green*)

dialysis solution flowing into abdominal cavity

dialysis solution with unwanted wastes and solutes draining out

❶ Hemodialysis

Tubes carry blood from a patient's body through a filter with dialysis solution that contains the proper concentrations of salts. Wastes diffuse from the blood into the solution and cleansed, solute-balanced blood returns to the body.

❷ Peritoneal dialysis

Dialysis solution is pumped into a patient's abdominal cavity. Wastes diffuse across the lining of the cavity into the solution, which is then drained out.

A

B

Polycystic kidney

Arthur Glauberman/Photo Researchers/Getty Images

© Air Force News/Photo by Tech. Sgt. Timothy Beardsley

Figure 12.12 Animated! Kidney dialysis cleanses the blood of patients with kidney failure. **A** Two options for dialysis: hemodialysis and peritoneal dialysis. **B** Karole Hurtley, who lives with kidney failure. Despite severe kidney disease requiring daily peritoneal dialysis, Karole became a national champion in karate at age 13. (© Cengage Learning)

Urinary system cancer is on the rise

Carcinomas of the bladder and kidney (Figure 12.13) account for about 100,000 new cancer cases each year, a number that is increasing. The incidence is higher in males, and smoking and exposure to certain industrial chemicals are major risk factors. Kidney cancer easily metastasizes via the bloodstream to the lungs, bone, and liver. An inherited type, called Wilms tumor, is one of the most common of all childhood cancers.

Urinary tract infections are common

Urinary tract infections plague millions of people. Women especially are susceptible to bladder infections because of their urinary anatomy: The female urethra is short, just a little over 1 inch long. (An adult male's urethra is about 9 inches long.) The outer opening of a female's urethra also is close to the anus, so it is fairly easy for bacteria from outside the body to make their way to a female's bladder and trigger an inflammation called **cystitis**—or even all the way to the kidneys to cause **pyelonephritis**.

In both sexes, urinary tract infections sometimes result from sexually transmitted microbes, including the microorganisms that cause *chlamydia*. Chapter 16 gives more information on this topic.

Nephritis is an inflammation of the kidneys. It can be caused by various factors, including bacterial infections. As you may remember from Chapter 9, inflamed tissue tends to swell as fluid accumulates in it. However, because a kidney is "trapped" inside the tough renal capsule, it can't increase in size. As a result, pressure builds up in or around the capillaries that service the glomerulus, hampering or preventing the flow of blood. Then blood filtering becomes difficult or impossible.

Painkillers and other drugs may harm the kidneys

Over-the-counter painkillers such as aspirin, acetaminophen, and NSAIDs (nonsteroidal anti-inflammatory drugs such as ibuprofen) come with consumer warnings about possible kidney damage. These drugs usually are safe when used according to directions. According to the National Kidney Foundation, however, millions of Americans have unwittingly destroyed their kidneys by excessive use of these substances. Experts recommend that people who must take large amounts of such painkillers drink six to eight glasses of water a day to help flush harmful residues from the kidneys.

Heavy alcohol use also is hard on the kidneys, and some illegal street drugs—including amphetamines, cocaine, and heroin—can cause major kidney damage.

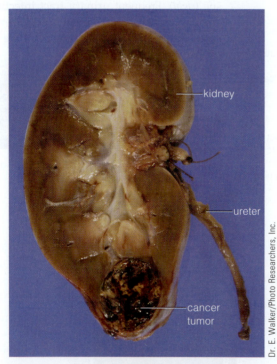

Figure 12.13 More than 50,000 cases of kidney cancer are diagnosed each year in the United States. This photograph shows a tumor in a kidney, which has been cut open to reveal the cancer's location (the circular area on the *bottom*).

THINK OUTSIDE THE BOOK

As this chapter's introduction noted, urinalysis provides a chemical snapshot of many physiological processes in the body. It also can be quite helpful in diagnosing illness. For example, glucose in urine may be a sign of diabetes, and white blood cells (pus) frequently indicate a urinary tract infection. Research this topic to find out some other common medical uses of urinalysis.

CONNECTIONS

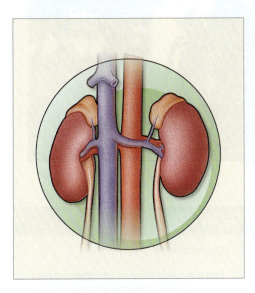

The Urinary System

The kidneys adjust the chemical composition of the extracellular fluid, including blood, in ways that are essential to the survival of the body as a whole. They remove toxic nitrogenous wastes from the breakdown of proteins. They also maintain the balance of water, electrolytes, and acids and bases in the blood. The bladder, ureters, and urethra provide for the storage and elimination of wastes in urine.

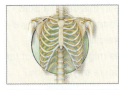

Skeletal system
The kidneys adjust blood levels of calcium and phosphate used in building bone.

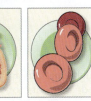

Muscular system
Adjustments in elimination of CO_2 help manage hydrogen ions (H^+) in blood and so help maintain blood pH (acid–base balance).

Cardiovascular system and blood
Kidney adjustments to water content of urine help maintain blood volume. The kidney hormone erythropoietin stimulates the production of red blood cells.

Immunity and the lymphatic system
Urine washes pathogens out of the urethra. Adjustments to water and solutes in blood plasma help assure adequate volume and chemical composition of lymph.

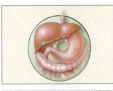

Respiratory system
Kidney adjustments in acid–base balance complement shifts in the depth and rate of breathing that help maintain pH of blood and tissue fluid.

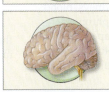

Digestive system
Kidneys convert vitamin D to a form that aids absorption of calcium from food.

Nervous system
Kidney management of acid–base balance helps ensure that nerve cells can function properly.

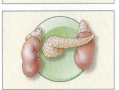

Endocrine system
Kidney hormone erythropoietin is part of overall hormonal controls in the body. Adjustments that help maintain blood volume assist with hormone transport in the bloodstream.

Reproductive system
The male urethra also serves as a channel for sperm ejaculated during intercourse.

EXPLORE
ON YOUR OWN

The rider and horse shown at the right are living examples of the mammalian body's ability to cool itself by producing sweat. Since sweat is mostly water, how is heavy sweating likely to affect the concentration of urine, especially if the athlete—in this case, a polo player—doesn't remember to drink fluid during the match? (You may well have observed this effect in your own body after exercise.)

Drink 1 quart of water in one hour. What changes might you expect (and can you observe) in your kidney function and the nature of your urine?

© David Jennings/The Image Works

SUMMARY

Section 12.1 Extracellular fluid (ECF) contains various types and amounts of substances dissolved in water. The ECF includes tissue fluid and blood plasma. The following processes maintain a healthy balance in the volume and chemical composition of ECF:

The body absorbs water from the GI tract and gains it from condensation reactions of metabolism. Water is lost by urinary excretion, evaporation from the lungs and skin, sweating, and elimination in feces.

Solutes are gained by absorption from the GI tract, secretion by cells, respiration, and metabolism. They are lost by excretion, respiration, and sweating. These solutes include electrolytes (sodium, potassium, calcium ions).

The kidneys control losses of water and solutes by adjusting the volume and chemical makeup of urine.

Section 12.2 The urinary system consists of two kidneys, two ureters, a urinary bladder, and a urethra. In the kidneys, blood is filtered and urine forms in nephrons.

A nephron starts as a cup-shaped capsule that is followed by three tubelike regions: the proximal tubule, loop of Henle, and distal tubule, which empties into a collecting duct.

The Bowman's (glomerular) capsule surrounds a set of highly permeable capillaries. Together, they are a blood-filtering unit, the glomerulus.

Section 12.3 Urine forms through a sequence of steps: filtration, reabsorption, and secretion (Table 12.3).

Filtration of blood at the glomerulus of a nephron transfers water and small solutes into the nephron.

In reabsorption, needed water and solutes leave the nephron tubule and enter the peritubular capillaries that thread around the tubule. Many solutes are reabsorbed when they diffuse down their concentration gradients back into the

TABLE 12.3 Processes of Urine Formation	
Process	**Characteristics**
FILTRATION	Pressure generated by heartbeats drives water and small solutes (not proteins) out of glomerulus capillaries and into Bowman's capsule, the entrance to the nephron.
REABSORPTION	Most water and solutes in the filtrate move from a nephron's tubule into interstitial fluid around the nephron, then into blood inside the peritubular capillaries.
SECRETION	Urea, H$^+$, and some other solutes move out of peritubular capillaries, into interstitial fluid, then into the filtrate inside the nephron for excretion in urine.

bloodstream. Others, such as sodium, are reabsorbed by active transport. Water is reabsorbed by osmosis. A small amount of water and solutes remains in the nephron.

In secretion some ions and a few other substances leave the peritubular capillaries and enter the nephron for disposal in urine.

Section 12.4 During reabsorption in kidney nephrons, water and salt are reabsorbed or excreted as required to conserve or eliminate water. The mechanisms that concentrate urine also help regulate blood volume and blood pressure.

Urine becomes more or less concentrated by the action of two hormones, ADH and aldosterone, on cells of distal tubules and collecting ducts.

ADH is secreted when the body must conserve water; it increases reabsorption from the distal nephron tubule and collecting ducts. Inhibition of ADH allows more water to be excreted.

Aldosterone conserves sodium by increasing its reabsorption in the distal tubule. It is secreted when cells in the juxtaglomerular apparatus (next to the distal tubule) secrete renin, an enzyme that triggers reactions that lead to aldosterone secretion. More sodium is excreted when aldosterone is inhibited. Because "water follows salt," aldosterone influences how much water is reabsorbed into the bloodstream.

 Section 12.5 Together with the respiratory system and other mechanisms, the kidneys also help maintain the body's overall acid–base balance. They help regulate pH by eliminating excess hydrogen ions and replenishing the supply of bicarbonate, which acts as a buffer elsewhere in the body.

REVIEW QUESTIONS

1. Label the parts of this kidney and nephron.

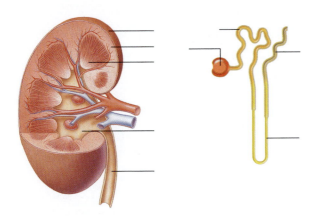

© Cengage Learning

2. How does the formation of urine help maintain the body's internal environment?

3. Explain what is meant when we talk about filtration, reabsorption, and secretion in the kidneys.

4. Which hormone or hormones promote (*a*) water conservation, (*b*) sodium conservation, and (*c*) thirst behavior?

5. Explain how the kidneys help to maintain the balance of acids and bases in extracellular fluid.

SELF-QUIZ *Answers in Appendix V*

1. The body gains water by _____.
   a. absorption in the gut   c. responding to thirst
   b. metabolism   d. all of the above

2. The body loses water by way of the _____.
   a. skin   d. urinary system
   b. lungs   e. c and d
   c. digestive system   f. a through d

3. Water and small solutes enter nephrons during _____.
   a. filtration   c. secretion
   b. reabsorption   d. both a and b

4. Kidneys return water and small solutes to blood by _____.
   a. filtration   c. secretion
   b. reabsorption   d. both a and c

5. Some substances move out of the peritubular capillaries and are moved into the nephron during _____.
   a. filtration   c. secretion
   b. reabsorption   d. both a and c

6. Reabsorption depends on _____.
   a. osmosis across the nephron wall
   b. active transport of sodium across the nephron wall
   c. a steep solute concentration gradient
   d. all of the above

7. _____ directly promotes water conservation.
   a. ADH   c. Aldosterone
   b. Renin   d. both b and c

8. _____ enhances sodium reabsorption.
   a. ADH   c. Aldosterone
   b. Renin   d. both b and c

9. Changes in acid-base balance may cause the kidneys to _____.
   a. reabsorb and excrete bicarbonate
   b. form and release bicarbonate to the blood
   c. release ADH
   d. either a or b, depending on circumstances

10. Match the following salt–water balance concepts:
    _____ aldosterone   a. blood filter of a nephron
    _____ nephron   b. controls sodium reabsorption
    _____ thirst mechanism   c. occurs at nephron tubules
    _____ reabsorption   d. site of urine formation
    _____ glomerulus   e. controls water gain

CRITICAL THINKING

1. A urinalysis reveals that the patient's urine contains glucose, hemoglobin, and white blood cells (pus). Are any of these substances abnormal in urine? Explain.

2. As a person ages, nephron tubules lose some of their ability to concentrate urine. What is the effect of this change?

3. Fatty tissue holds the kidneys in place. Extremely rapid weight loss may cause this tissue to shrink so that the kidneys slip from their normal position. On rare occasions, the slippage can put a kink in one or both ureters and block urine flow. Suggest what might then happen to the kidneys.

4. Licorice is used as a remedy in Chinese traditional medicine and also is a flavoring for candy. When licorice is eaten, one of its components triggers the formation of a compound that mimics aldosterone and binds to receptors for it. Based on this information, explain why people who have high blood pressure are advised to avoid eating much licorice.

5. Drinking too much water can be a bad thing. If someone sweats heavily and drinks lots of water, their sodium levels drop. The resulting "water intoxication" can be fatal. Why is the sodium balance so important?

6. As the text noted, two-thirds of the water and solutes that the body reclaims by reabsorption in nephrons occurs in the proximal tubule. Proximal tubule cells have large numbers of mitochondria and demand a great deal of oxygen. Explain why.

Your Future

On average, roughly 75,000 Americans are on the waiting list for a kidney transplant. The wait time can vary from a few months to several years, during which time the patient's own kidneys are increasingly likely to fail. About 4,000 die each year before they can be matched to a willing donor.

Many factors are involved—for example, self markers of potential donors and recipients may not match closely, or perhaps someone who has signed on to donate a kidney is no longer a good candidate for health-related or other reasons. Until recently, compiling and sorting all the relevant biological and personal data was a logistical nightmare. Now researchers at several universities have created donor registry computer software that can quickly sift through all the information and generate a series of options. All it takes is one ready-and-willing donor to start the process. Then, the software can quickly crank out a plan that matches that donor to the best potential recipient. Sometimes the first match has a domino effect, freeing up another "ready" kidney to go to another, more appropriate patient, and so on. As more potential donors and recipients are added to the system, the hope is that the waiting time for kidney transplants will become much shorter and many more lives will be saved.

LINKS TO EARLIER CONCEPTS

This chapter expands on Chapter 4's introduction to neurons and other cells that make up the body's nervous tissue (4.4).

Remembering the structure of plasma membranes (3.4) and how substances move across them (3.10–3.11) will help you understand how neurons produce nerve impulses.

You'll also get a fuller picture of how nervous system signals make skeletal muscles contract (6.4).

KEY CONCEPTS

How Neurons Work

The operation of the nervous system depends on the capacity of neurons to produce electrical signals and transmit them to other cells. Sections 13.1–13.4

The Nervous System

Different parts of the nervous system detect information, process it, and then select or control muscles and glands that carry out responses. Sections 13.5–13.6

The Brain

The brain is a master controller that receives, processes, stores, and retrieves information. It also coordinates responses by adjusting body activities. Sections 13.7–13.10

Disorders of the Nervous System Sections 13.11–13.12

CONNECTIONS:
The Nervous System and Homeostasis Section 13.13

"Ecstasy" is an illegal but popular "rave" drug that relieves anxiety and produces a mild high. Ecstasy's active ingredient, MDMA, causes brain neurons to release abnormally large amounts of serotonin, a signaling molecule that helps regulate functions ranging from memory and learning to appetite and sleep. An overdose can kill. That is how Lorna Spinks died at the age of nineteen. The lower photograph at right was taken minutes after her death.

Some people who use Ecstasy to get high may think it is safe used "in moderation" because of research reports that MDMA may help some patients suffering from clinically diagnosed anxiety. What they may not know, however, is that studies also document how repeated "recreational" MDMA use may disrupt the normal flow of serotonin in a way that results in lasting damage in the brain. The damage contributes to loss of concentration, depression, and memory problems. The more often you use Ecstasy, the worse the damage becomes.

In this chapter we look at how the healthy nervous system manages a wide range of body functions. We will start by considering how neurons are built and operate. Then we'll examine how neurons interact in the nervous system and how the brain serves as the body's master control center.

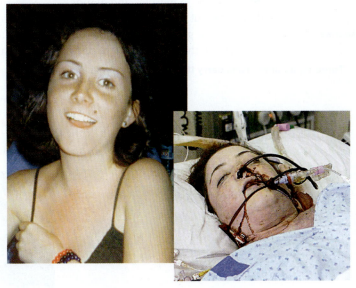

Manni Mason's Pictures, Masons News Ltd./South West News Service

Homeostasis Preview

Along with chemical signals from the endocrine system, the nervous system provides the communication required to monitor, adjust, and regulate body functions.

Neurons: The Communication Specialists

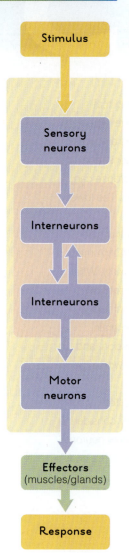

Figure 13.1 **Three types of neurons carry the nervous system's messages.** (© Cengage Learning)

- Three types of neurons are the nervous system's communication cells.

- Links to Nervous tissue 4.4, The plasma membrane 3.4

The **nervous system** detects information about external and internal conditions. It then integrates those inputs and selects or controls muscles and glands that carry out responses (Figure 13.1).

Neurons are the communication lines of the nervous system

Three types of neurons carry out the functions of the nervous system:

- **Sensory neurons** collect information about stimuli (such as light or touch) and relay it to the spinal cord and brain.

- **Interneurons** in the spinal cord and brain receive and process sensory input and send signals to other neurons.

- **Motor neurons** relay signals from interneurons to muscles and glands that carry out responses. Because muscles and glands produce the ultimate effect, they are called effectors.

A neuron has a large *cell body* that contains its nucleus and most organelles. The cell body and branched extensions called **dendrites** are "input zones" for arriving information. Near the cell body is a patch of the neuron's plasma membrane that serves as a "trigger zone." In motor neurons and interneurons the trigger zone is called the axon hillock ("little hill"). In this area information travels toward a slender and often long extension called an **axon**, the neuron's "conducting zone." As you can see in the diagram of a motor neuron in Figure 13.2, dendrites tend to be shorter than axons. Their number and length vary, depending on the type of neuron. The axon's endings are "output zones" where messages are sent to other cells.

Recall from Chapter 4 that roughly 90 percent of your nervous system consists of cells called glia (neuroglia). Glia help maintain the proper concentrations of vital ions in the fluid around neurons and assist in the formation of connections between brain neurons. Some physically support and protect neurons. Others provide insulation that allows signals to move along sensory and motor neurons with lightning speed.

Properties of a neuron's plasma membrane allow it to carry signals

Neurons are suited for communication partly because they are excitable—that is, a neuron can respond to certain stimuli by producing an electrical signal.

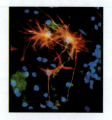

Star-shaped glia, called astrocytes (Nancy Kedersha/Harvard Medical School/Peter Arnold)

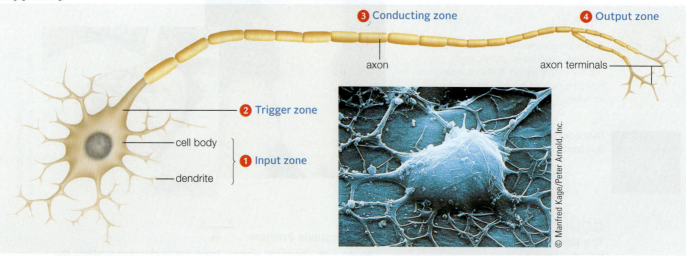

3 Conducting zone   **4** Output zone

axon   axon terminals

2 Trigger zone

cell body

1 Input zone

dendrite

© Manfred Kage/Peter Arnold, Inc.

Figure 13.2 **Animated! A motor neuron has four main function zones. 1** Dendrites and the cell body are an input zone where incoming signals arrive. **2** The signal spreads to a trigger zone at the start of an axon. **3** An axon is a conducting zone that passes signals toward the neuron's terminal endings. **4** Axon terminals a signal output zone. (© Cengage Learning)

You may remember from Chapter 3 that the plasma membrane's lipid bilayer prevents charged substances—such as ions of potassium (K⁺) and sodium (Na⁺)—from freely crossing it. Even so, ions can cross the membrane through channel proteins that span the bilayer (Figure 13.3). Some channels are always open, so that ions can steadily "leak"—by diffusion—in or out. Other channels open like gates under the proper circumstances. These controls mean that the concentrations of an ion can be different on either side of the plasma membrane.

For example, in a resting neuron, the gated sodium channels are closed and the plasma membrane allows only a little sodium to leak inward. The membrane is more permeable to K⁺. As a result, each ion has its own concentration gradient across the membrane (Figure 13.3A). Following the rules of diffusion, sodium tends to move in and potassium tends to move out.

For several reasons, on balance the cytoplasm next to the membrane is more negative than the fluid just outside the membrane. Electrical charges may be measured in millivolts, and for many neurons, the steady charge difference across the plasma membrane is about −70 millivolts. The minus indicates that the cytoplasm side of the membrane is more negative than than the outer side of the membrane. This difference is called **resting membrane potential**. The term means that the charge difference has the potential to do physiological work in the body. That "work" is the launching of a nerve impulse.

Various kinds of signals occur in the nervous system, but not all of them spark nerve impulses. Only a signal that is strong enough when it reaches a resting neuron's input zone may spread to a trigger zone. When a strong enough signal does arrive, however, it can cause the voltage difference across the plasma membrane to reverse, just for an instant. In the following section, we see how these reversals produce nervous system signals.

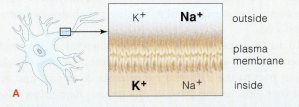

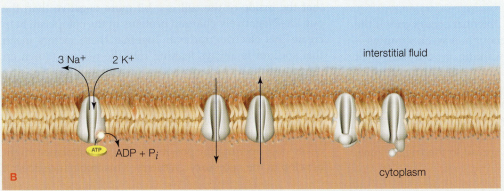

Sodium–potassium cotransporters actively transport three Na⁺ out of a neuron for every two K⁺ they pump in.

Passive transporters allow K⁺ ions to move across the plasma membrane, down their concentration gradient.

Voltage-gated channels for Na⁺ or K⁺ are closed in a neuron at rest (*left*), but open when it is excited (*right*).

Figure 13.3 Animated! Ions produce an electrical gradient across a neuron's plasma membrane. A Gradients of sodium (Na⁺) and potassium (K⁺) ions across a neuron's plasma membrane. **B** How ions cross the plasma membrane of a neuron. They are selectively allowed to cross at protein channels and pumps that span the membrane. (© Cengage Learning)

TAKE-HOME MESSAGE

WHAT KINDS OF CELLS CARRY NERVOUS SYSTEM SIGNALS?

- Sensory neurons, interneurons, and motor neurons all carry nervous sysstem signals, or nerve impulses.
- In a resting neuron, differences in the concentrations of Na⁺ and K⁺ across the plasma membrane produce a difference in electrical charge across the plasma membrane.
- This charge difference is called the resting membrane potential. It sets the stage for a neuron to do its physiological work of firing a nerve impulse.

Nerve Impulses = Action Potentials

- A nerve impulse fires when a signal causes a neuron's resting membrane potential to reverse.

- Link to Concentration and electric gradients 3.10

action potential A nerve impulse.

sodium–potassium pump A carrier protein through which active transport moves potassium ions into a neuron and sodium ions outward.

threshold The minimum change in the voltage difference across a neuron's plasma membrane that will trigger a nerve impulse.

When an adequate signal reaches a resting neuron's input zone, a change occurs in the membrane. Sodium gates in it open, and Na$^+$ rushes into the neuron. Sodium ions have a positive charge, so as they flow in, the cytoplasm next to the plasma membrane becomes less negative (Figure 13.4A and 13.4B). Then, more gates open, more sodium enters, and so on—an example of positive feedback. When the voltage difference across the neuron plasma membrane shifts by a minimum amount called the **threshold**, the result is a nerve impulse or **action potential**.

The threshold for an action potential can be reached where a neuron's plasma membrane has voltage-sensitive gated channels for sodium ions. When the threshold level is reached, the opening of more sodium gates doesn't depend any longer on the strength of the stimulus. The gates open on their own.

Keep in mind that an action potential occurs only if the stimulus to a neuron is strong enough. A weak stimulus—say, pressure from a tiny insect walking on your skin—that arrives at an input zone may not upset the ion balance enough to cause an action potential. This is because input zones don't have gated sodium channels, so sodium can't flood in there. On the other hand, a neuron's trigger zone is packed with sodium channels. If a stimulus that reaches an input zone is strong enough to spread to the trigger zone, an action potential may "fire."

Action potentials travel away from their starting point

To transmit messages within the body, action potentials must spread to other neurons or to cells in muscles or glands. Each action potential propagates itself, moving away from its starting point. This self-propagation occurs in part because the changes in membrane potential leading to an action potential don't lose strength. When the change spreads from one patch of a neuron's plasma membrane to another patch, about the same number of gated channels open (Figure 13.4C and 13.4D).

A neuron can't "fire" again until ion pumps restore its resting potential

When a signal causes an action potential in a neuron's trigger zone, that area of the cell's plasma membrane can't receive another signal until its resting membrane potential is restored.

To understand how the resting potential is restored, remember that a neuron's resting membrane potential is due in part to the different concentrations of Na$^+$ and K$^+$ on either side of the plasma membrane. Remember also that the inside of the cell is a bit more negative than the outside. Negatively charged proteins in the cytoplasm help create this electric gradient. Together these factors mean that sodium is always leaking into the neuron (down an electrochemical gradient), and potassium is always leaking out (down its concentration gradient).

A neuron can't respond to an incoming signal unless the proper concentration and electric gradients across its plasma membrane are in place. Yet the Na$^+$ and K$^+$ leaks never stop, opening the possibility that an imbalance might develop in the necessary gradients. This imbalance doesn't

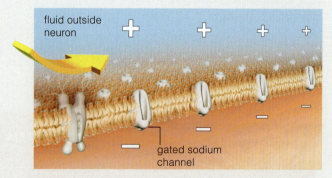

1 In a membrane at rest, the inside of the neuron is negative relative to the outside. An electrical disturbance (*yellow arrow*) spreads from an input zone to an adjacent trigger zone of the membrane, which has a large number of gated sodium channels.

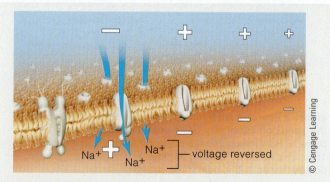

2 A strong disturbance initiates an action potential. Sodium gates open. Sodium flows in, reducing the negativity inside the neuron. The change causes more gates to open, and so on until threshold is reached and the voltage difference across the membrane reverses.

Figure 13.4 Animated! **An inward flood of sodium ions triggers an action potential. 1, 2** Steps leading to an action potential. **3, 4** How an action potential propagates, or travels, along a neuron.

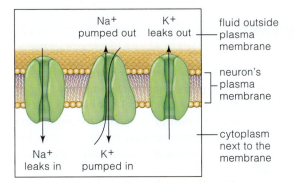

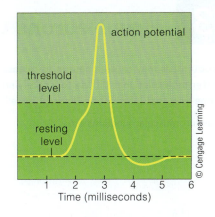

Figure 13.5 Sodium–potassium pumps maintain ion gradients across a neuron's plasma membrane. This pumping, and additional leaking of ions, maintain the proper balance of ions across a resting neuron's plasma membrane. (© Cengage Learning)

Figure 13.6 The action potential spikes when threshold is reached.

develop, however, because a resting neuron uses energy to power a pumping mechanism that maintains the gradients. Carrier proteins called **sodium–potassium pumps** span the neuron's membrane (Figure 13.5). With energy from ATP, they actively transport potassium *into* the neuron and transport sodium *out*.

Action potentials are "all-or-nothing"

There is no such thing as a "weak" or "strong" action potential. Every action potential in a neuron spikes to the same level above threshold as an all-or-nothing event. That is, once the positive-feedback cycle of opening sodium gates starts, nothing will stop the full spiking. If threshold is not reached, the disturbance to the plasma membrane will fade away as soon as the stimulus is removed. Figure 13.6 shows a recording of the voltage difference across a neuron's plasma membrane before, during, and after an action potential.

Each spike lasts for about a millisecond. At the place on the membrane where the charge reversed, the gated sodium channels close and the influx of sodium stops.

About halfway through the action potential, potassium channels open, so potassium ions flow out and restore the original voltage difference across the membrane. Sodium–potassium pumps restore the ion gradients. After the resting membrane potential has been restored, most potassium gates are closed and sodium gates are in their initial state, ready to be opened again when a suitable stimulus arrives.

TAKE-HOME MESSAGE

HOW DOES AN ACTION POTENTIAL, OR NERVE IMPULSE, TAKE PLACE?

- An action potential, or nerve impulse, occurs when a neuron's resting membrane potential briefly reverses.
- Action potentials self-propagate and always move away from the trigger zone.
- After an action potential, sodium–potassium pumps restore the neuron's resting potential.
- An action potential is all-or-nothing. Once the spiking starts, nothing can stop it.

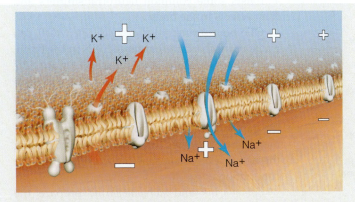

3 At the next patch of membrane, another group of gated sodium channels open. In the previous patch, some K⁺ moves out through other gated channels. That region becomes negative again.

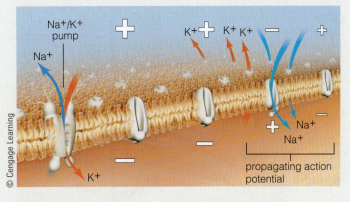

4 After each action potential, the sodium and potassium concentration gradients in a patch of membrane are not yet fully restored. Active transport at sodium–potassium pumps restores them.

How Neurons Communicate

- Action potentials may cause a neuron to release neurotransmitter molecules that diffuse to a receiving cell. This is one way that information flows from cell to cell.

- Link to Neuromuscular junctions 6.6

Action potentials can stimulate neurons to release the chemical signals called **neurotransmitters**. These molecules diffuse across a **chemical synapse**, a narrow gap between a neuron's output zone and the input zone of a neighboring cell. Some chemical synapses occur between neurons, others between a neuron and a muscle cell or gland cell.

At a chemical synapse, one of the two cells stores neurotransmitter molecules in synaptic vesicles in its cytoplasm. This is the *presynaptic* cell. The cell's plasma membrane has gated channels for calcium ions, and they open when an action potential arrives. There are more calcium ions outside the cell, and when they flow in (down their gradient), synaptic vesicles fuse with the plasma membrane, discharging their content. Neurotransmitter molecules now pour into the synapse, diffuse across it, and bind with receptor proteins on the plasma membrane of the *post*synaptic, or receiving, cell. Binding changes the shape of these proteins, so that a channel opens up through them. Ions then diffuse through the channels and enter the receiving cell.

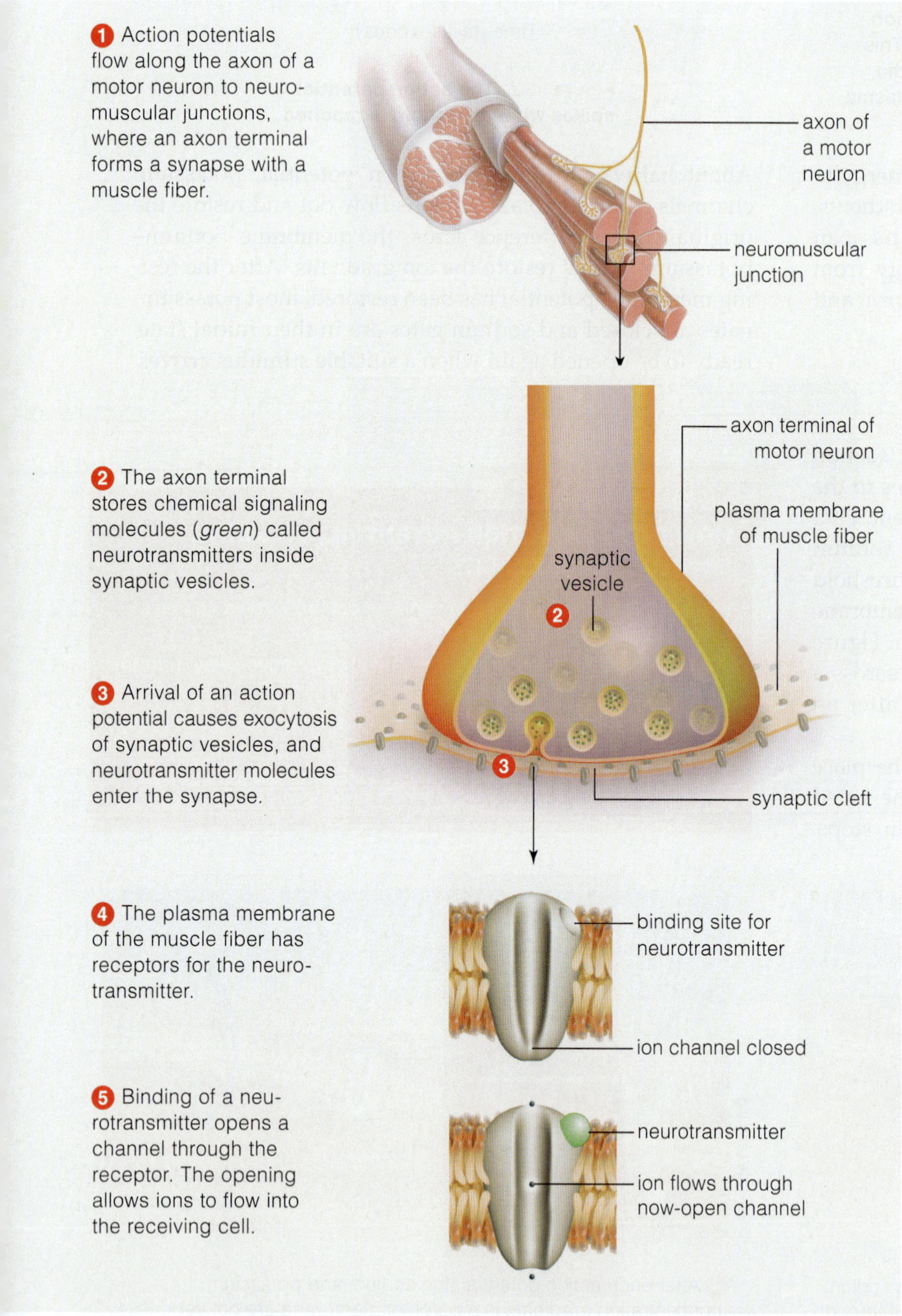

❶ Action potentials flow along the axon of a motor neuron to neuromuscular junctions, where an axon terminal forms a synapse with a muscle fiber.

— axon of a motor neuron

— neuromuscular junction

❷ The axon terminal stores chemical signaling molecules (*green*) called neurotransmitters inside synaptic vesicles.

— axon terminal of motor neuron

— plasma membrane of muscle fiber

synaptic vesicle

❸ Arrival of an action potential causes exocytosis of synaptic vesicles, and neurotransmitter molecules enter the synapse.

— synaptic cleft

❹ The plasma membrane of the muscle fiber has receptors for the neurotransmitter.

— binding site for neurotransmitter

— ion channel closed

❺ Binding of a neurotransmitter opens a channel through the receptor. The opening allows ions to flow into the receiving cell.

— neurotransmitter

— ion flows through now-open channel

Figure 13.7 Animated! A neuromuscular junction forms between axon endings of motor neurons and skeletal muscle fibers. (© Cengage Learning)

Neurotransmitters can excite or inhibit a receiving cell

How a receiving cell responds to a neurotransmitter depends on the type and amount of a neurotransmitter, the kinds of receptors the cell has, and some other factors. *Exciting* signals help drive the membrane toward an action potential. *Inhibiting* signals have the opposite effect. Table 13.1 lists some common neurotransmitters and their effects in the body.

One neurotransmitter, *acetylcholine (ACh)*, can excite *or* inhibit different target cells in the brain, spinal cord, glands, and muscles. Figure 13.7 shows a neuromuscular junction chemical synapse between a motor neuron and a muscle cell. ACh released from the neuron diffuses across the gap and binds to receptors on the muscle cell membrane. It excites this kind of cell, triggering the action potentials that cause skeletal muscle contractions.

Epinephrine and *norepinephrine* prepare the body to respond to stress or excitement. *Dopamine* acts in fine motor control and influences some type of learning. *GABA* inhibits the release of other neurotransmitters.

TABLE 13.1 Major Neurotransmitters and Their Effects

Neurotransmitter	Examples of Effects
Acetylcholine (ACh)	Causes skeletal muscle contraction; affects mood and memory
Epinephrine and norepinephrine	Speed heart rate; dilate the pupils and airways to lungs; slow GI tract contractions; increase anxiety
Dopamine	Reduces excitatory effects of other neurotransmitters; roles in memory, learning, fine motor coordination
Serotonin	Elevates mood; has a role in memory and learning
GABA	Inhibits the release of other neurotransmitters

Serotonin acts on brain cells that govern emotional states, sleeping, sensory perception, and regulation of body temperature. Some neurons secrete *nitric oxide* (NO), a gas that controls blood vessel dilation. It is not stored in synaptic vesicles but instead is manufactured as needed. As an example, a sexually aroused male has an erection when NO calls on blood vessels in his penis to dilate, allowing blood to rush in.

Neuromodulators can magnify or dampen the effects of a neurotransmitter. These substances include natural painkillers called *endorphins*. Endorphins inhibit nerves from releasing substance P, which conveys information about pain. In athletes who exercise beyond normal fatigue, endorphins can produce a euphoric high.

Competing signals are "summed up"

At any moment, many signals are washing over the input zones of a receiving neuron. All of them are graded potentials (their magnitude can be large or small), and they compete for control of the membrane potential at the trigger zone. The ones called EPSPs (for excitatory postsynaptic potentials) *depolarize* the membrane—they bring it closer to threshold. On the other hand, IPSPs (inhibitory postsynaptic potentials) may *hyperpolarize* the membrane (drive it away from threshold) or help keep the membrane at its resting level.

Synaptic integration tallies up the competing signals that reach an input zone of a neuron at the same time—a little like adding up the pros and cons of a certain course of action. This process, called *summation*, is how signals arriving at a neuron are suppressed, reinforced, or sent onward to other cells in the body.

Integration occurs when neurotransmitter molecules from more than one presynaptic cell reach a neuron's input zone at the same time. Signals also are integrated after a neurotransmitter is released repeatedly, over a short time period, from a neuron that is responding to a rapid series of action potentials.

Neurotransmitter molecules must be removed from the synapse

The flow of signals through the nervous system depends on the rapid, controlled removal of neurotransmitter molecules from synapses. Some of the neurotransmitter molecules diffuse out of the gap. Enzymes cleave others in the synapse, as when acetylcholinesterase breaks down ACh. Also, membrane transport proteins actively pump the neurotransmitter molecules back into presynaptic cells or into neighboring neuroglia.

Certain drugs can block the reuptake of particular neurotransmitters. For example, some antidepressant drugs elevate a depressed person's mood by blocking the reuptake of serotonin. Others shift the balance of a combination of neurotransmitters, such as serotonin and norepinephrine.

chemical synapse A gap between two neurons or between a neuron and muscle cell or gland cell.

neuromodulator A substance that can modify the effects of a neurotransmitter.

neurotransmitter A chemical that carries neural messages across a chemical synapse.

synaptic integration Process in which the competing signals arriving at a neuron are summed up before the neuron responds.

THINK OUTSIDE THE BOOK

Antidepressants have side effects, and some older studies have suggested that in a small number of patients, especially children and teens, the use of an antidepressant may increase the risk of suicide. More recent research casts doubt on this supposed suicide risk. Learn more about the uses, side effects, and concerns about antidepressants at the website of the National Institute of Mental Health (www.nimh.nih.gov).

TAKE-HOME MESSAGE

WHAT IS A NEUROTRANSMITTER?

- Neurotransmitters are chemicals that carry signals between the cells at a chemical synapse.
- A neurotransmitter may excite or inhibit the the activity of a target cell.
- By way of synaptic integration, nervous system messages can be reinforced or downplayed, sent onward or suppressed.

■ Once a message is sent in the nervous system, where it goes depends on how neurons are organized in the body.

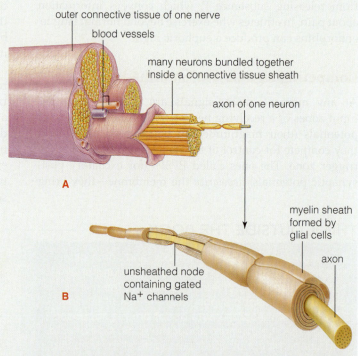

TABLE 13.2 Summary of the Components of Nerves	
NEURON	Nervous system cell specialized for communication
NERVE FIBER	Long axon of one neuron
NERVE	Long axons of several neurons enclosed by connective tissue

Nerves are long-distance lines

Nerves are communication lines between the brain or spinal cord and the rest of the body. A **nerve** consists of nerve fibers, which are the long axons of sensory neurons, motor neurons, or both. Connective tissue encloses most of the axons like electrical cords inside a tube (Table 13.2 and Figure 13.8A). In the central nervous system (the brain and spinal cord) nerves are called **nerve tracts**.

Each axon has an insulating **myelin sheath**, which allows action potentials to propagate faster than they would otherwise. The sheath consists of glial cells that wrap around the long axons like jelly rolls. As you can see in Figure 13.8B, an exposed node, or gap, separates each cell from the next one. There, voltage-sensitive, gated sodium channels pepper the plasma membrane. In a manner of speaking, action potentials jump from node to node (a phenomenon that sometimes is called saltatory conduction, after a Latin word meaning "to jump"). The sheathed areas between nodes hamper the movement of ions across the plasma membrane, so stimulation tends to travel along the membrane until the next node in line. At each node, however, the flow of ions can produce a new action potential. In large sheathed axons, action potentials propagate at a remarkable 120 meters (nearly 400 feet) per second!

In the central nervous system, glial cells called oligodendrocytes form the myelin sheath. In the rest of the nervous system, glial cells called **Schwann cells** form the sheath.

Reflexes are the simplest nerve pathways

Sensory and motor neurons of certain nerves take part in automatic responses called reflexes. A **reflex** is a simple, programmed movement in response to a stimulus. It is always the same and takes place with conscious effort. In the simplest reflexes, sensory neurons synapse directly on motor neurons. In most reflex pathways, however, the sensory neurons also interact with several interneurons. These excite or inhibit motor neurons as needed for a coordinated response.

The stretch reflex contracts a muscle after gravity or some other load has stretched the muscle. Suppose you steadily hold out a bowl as someone loads peaches into it, adding weight to the bowl. When your hand starts to drop, the biceps muscle in your arm is stretched. This stretching activates receptors in *muscle spindles*. These are sensory organs in which specialized cells are enclosed in a sheath that runs parallel with the muscle. The receptor endings are the input zones of sensory neurons whose axons synapse with motor neurons in the spinal cord (Figure 13.9). Axons of the motor neurons lead back to the stretched muscle. Action potentials that reach the axon endings trigger the release of ACh, which triggers contraction. As long as receptors continue to send messages, the motor neurons are

Figure 13.8 Animated! **Nerves are bundled long axons of neurons. A** Structure of a nerve. **B** Structure of a sheathed axon. A myelin sheath formed by Schwann cells blocks the flow of ions except at nodes between Schwann cells.
(© Cengage Learning)

myelin sheath A wrapping of glial cells around the axon of a neuron. The sheath provides insulation that allows nerve impulses to propagate faster than they would otherwise.

nerve A bundle of neuron axons.

nerve tract Nerves in the central nervous system (the brain and spinal cord).

reflex A simple, stereotyped movement in response to a stimulus.

Schwann cells The type of glial cells that sheathe axons outside the central nervous system.

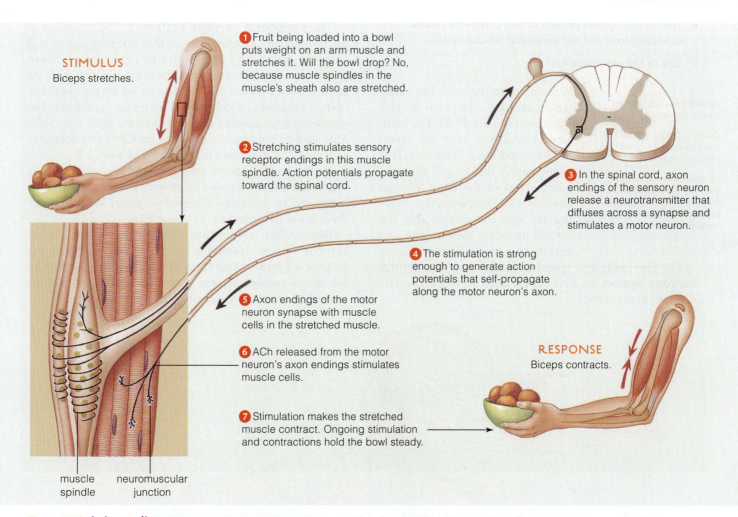

STIMULUS
Biceps stretches.

1 Fruit being loaded into a bowl puts weight on an arm muscle and stretches it. Will the bowl drop? No, because muscle spindles in the muscle's sheath also are stretched.

2 Stretching stimulates sensory receptor endings in this muscle spindle. Action potentials propagate toward the spinal cord.

3 In the spinal cord, axon endings of the sensory neuron release a neurotransmitter that diffuses across a synapse and stimulates a motor neuron.

4 The stimulation is strong enough to generate action potentials that self-propagate along the motor neuron's axon.

5 Axon endings of the motor neuron synapse with muscle cells in the stretched muscle.

6 ACh released from the motor neuron's axon endings stimulates muscle cells.

7 Stimulation makes the stretched muscle contract. Ongoing stimulation and contractions hold the bowl steady.

RESPONSE
Biceps contracts.

muscle spindle   neuromuscular junction

Figure 13.9 Animated! Reflexes are simple but important neural pathways. This diagram shows how nerves are organized in a reflex that operates when skeletal muscle stretches. (© Cengage Learning)

excited. This allows them to send signals to muscles that maintain your hand's position. This type of reflex is often called a *spinal reflex.*

In the brain and spinal cord, neurons interact in circuits

In your nervous system, sensory nerves relay information into the spinal cord, where they form chemical synapses with interneurons. The spinal cord and brain contain only interneurons, which integrate the signals. Many interneurons synapse with motor neurons, which carry signals away from the spinal cord and brain.

In the brain and spinal cord, blocks of hundreds or thousands of interneurons are parts of interacting circuits. Each block receives signals—some that excite, others that inhibit—and then integrates the messages and responds with new ones. For example, in some regions of the brain the circuits diverge—the processes of neurons in one block

fan out to form connections with other blocks. Elsewhere signals from many neurons are funneled to just a few. And in still other brain regions, neurons synapse back on themselves, repeating signals among themselves. These "reverberating" circuits include the ones that make your eye muscles twitch as you sleep.

13.5 Overview of the Nervous System

- The nervous system consists of two parts—the central nervous system and the peripheral nervous system.

- Link to Nervous tissue 4.4

central nervous system
The brain and spinal cord.

ganglia Clusters of neuron cell bodies in the peripheral nervous system.

peripheral nervous system Nerves and other nervous system structures outside the central nervous system.

Humans have the most intricately wired nervous system in the animal world (Figure 13.10). The brain alone contains at least 100 billion neurons, and many more form part of the nerves that branch throughout the rest of the body. We can simplify this complexity by dividing the nervous system into the central nervous system and the peripheral nervous system (Figure 13.12). The brain and spinal cord make up the **central nervous system** (in CNS). It contains all of the nervous system's interneurons.

The **peripheral nervous system** (in PNS) consists of thirty-one pairs of spinal nerves that carry signals to and from the spinal cord and twelve pairs of cranial nerves that carry signals to and from the brain (Figure 13.11). At some places in the PNS, cell bodies of several neurons occur in clusters called **ganglia** (singular: ganglion). The central and peripheral nervous systems both also have glia, such as the oligodendrocytes (in CNS) and Schwann cells (in PNS) described in Sections 13.1 and 13.4.

As Figure 13.12 shows, the peripheral nervous system is organized into *somatic* and *autonomic* subdivisions, and the autonomic nerves are subdivided yet again. We'll consider the roles of those nerves in Section 13.6.

Nerves that carry sensory information to the central nervous system sometimes are called *afferent* ("bringing to") nerves. Nerves that carry motor messages away from the central nervous system to muscles and glands may be termed *efferent* ("carrying outward") nerves.

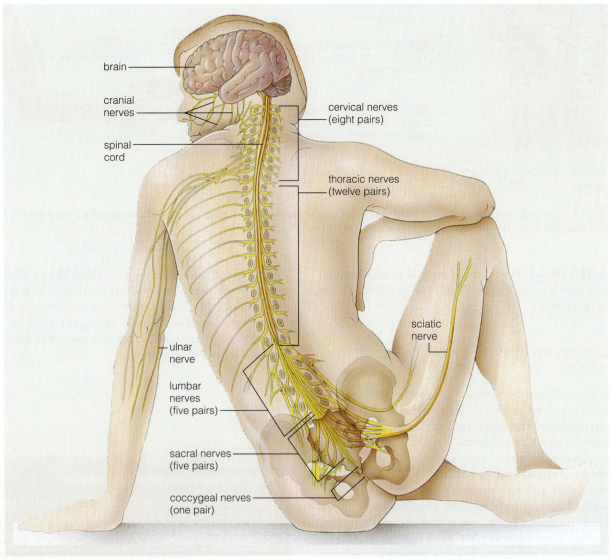

brain
cranial nerves
spinal cord
cervical nerves (eight pairs)
thoracic nerves (twelve pairs)
sciatic nerve
ulnar nerve
lumbar nerves (five pairs)
sacral nerves (five pairs)
coccygeal nerves (one pair)

Figure 13.10 Animated! Major parts of the nervous system include the brain, spinal cord, and nerves. (© Cengage Learning)

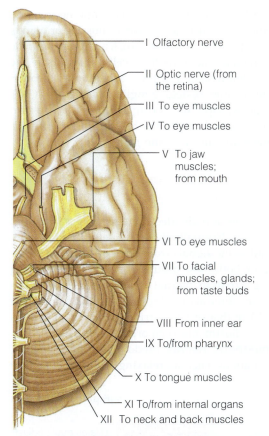

Figure 13.11 **Twelve pairs of cranial nerves extend from different regions of the brain stem.** Roman numerals are used to designate cranial nerves. (© Cengage Learning)

I Olfactory nerve

II Optic nerve (from the retina)

III To eye muscles

IV To eye muscles

V To jaw muscles; from mouth

VI To eye muscles

VII To facial muscles, glands; from taste buds

VIII From inner ear

IX To/from pharynx

X To tongue muscles

XI To/from internal organs

XII To neck and back muscles

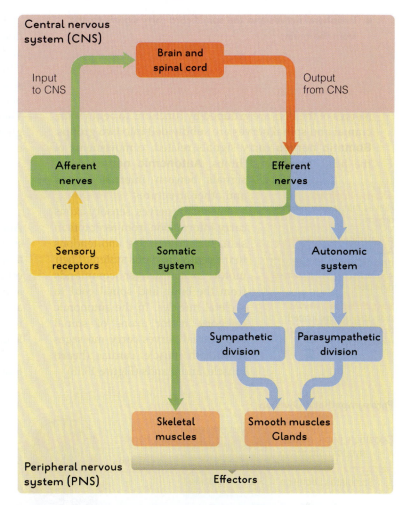

Figure 13.12 **The nervous system is subdivided into central and peripheral portions.** (© Cengage Learning)

Central nervous system (CNS)

Brain and spinal cord

Input to CNS

Output from CNS

Afferent nerves

Efferent nerves

Sensory receptors

Somatic system

Autonomic system

Sympathetic division

Parasympathetic division

Skeletal muscles

Smooth muscles Glands

Peripheral nervous system (PNS)

Effectors

Throughout our lives, our remarkable nervous system integrates the array of body functions in ways that help maintain homeostasis. Operations of the CNS also give us much of our "humanness," including our ability to reason and to appreciate the vast number of other living organisms with which we share the world. With this introduction, we now turn to a closer look at the two major parts of the nervous system.

© Bruce Beehler/Conservation International

TAKE-HOME MESSAGE

WHAT ARE THE TWO PARTS OF THE NERVOUS SYSTEM?

- The two parts of the nervous system are the central nervous system (CNS) and the peripheral nervous system (PNS).
- The central nervous system consists of the brain and spinal cord.
- The peripheral nervous system consists of branching spinal and cranial nerves that carry signals to and from the CNS.

Major Expressways: Peripheral Nerves and the Spinal Cord

■ Peripheral nerves and the spinal cord carry signals to and from the brain.

The peripheral nervous system consists of somatic and autonomic nerves

Nerves of the PNS are grouped by function. To begin with, cranial and spinal nerves are subdivided into two groups. **Somatic nerves** carry signals related to movements of the head, trunk, and limbs. **Autonomic nerves** carry signals beween internal organs and other structures.

autonomic nerves
Nerves that service internal organs.

parasympathetic nerves
Autonomic nerves that transmit signals for bodily housekeeping tasks such as digestion.

somatic nerves Nerves that carry signals related to head, trunk, and limb movements.

In somatic nerves, sensory axons carry information from receptors in skin, skeletal muscles, and tendons to the central nervous system. Their motor axons deliver commands from the brain and spinal cord to skeletal muscles. In the autonomic category, motor axons of spinal and cranial nerves carry messages to smooth muscle, cardiac (heart) muscle, and glands (Figure 13.13).

Unlike somatic neurons, single autonomic neurons do not extend the entire distance between muscles or glands and the central nervous system. Instead, preganglionic ("before a ganglion") neurons have cell bodies inside the spinal cord or brain stem, but their axons travel through nerves to autonomic system ganglia outside the CNS. There, the axons synapse with postganglionic ("after a ganglion") neurons, which make the actual connection with effectors—the body's muscles and glands.

Autonomic nerves are divided into parasympathetic and sympathetic groups

Autonomic nerves are divided into *parasympathetic* and *sympathetic* nerves. Normally these two sets of nerves work antagonistically—the signals from one oppose those of the other. However, both these groups of nerves carry exciting and inhibiting signals to internal organs. Often their signals arrive at the same time at muscle or gland cells and compete for control. When that situation arises, synaptic integration leads to minor adjustments in an organ's activity.

Parasympathetic nerves predominate during quiet, low-stress situations, such as relaxing. They tend to slow

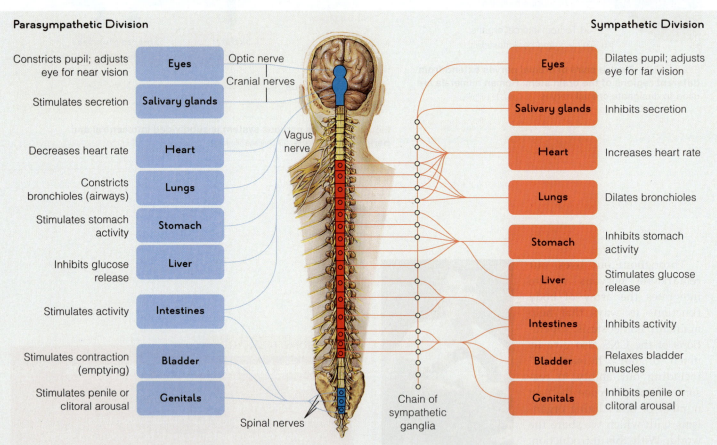

Parasympathetic Division

| | Sympathetic Division |

Constricts pupil; adjusts eye for near vision — **Eyes** — Optic nerve / Cranial nerves — **Eyes** — Dilates pupil; adjusts eye for far vision

Stimulates secretion — **Salivary glands** — **Salivary glands** — Inhibits secretion

Decreases heart rate — **Heart** — Vagus nerve — **Heart** — Increases heart rate

Constricts bronchioles (airways) — **Lungs** — **Lungs** — Dilates bronchioles

Stimulates stomach activity — **Stomach** — **Stomach** — Inhibits stomach activity

Inhibits glucose release — **Liver** — **Liver** — Stimulates glucose release

Stimulates activity — **Intestines** — **Intestines** — Inhibits activity

Stimulates contraction (emptying) — **Bladder** — **Bladder** — Relaxes bladder muscles

Stimulates penile or clitoral arousal — **Genitals** — **Genitals** — Inhibits penile or clitoral arousal

Spinal nerves

Chain of sympathetic ganglia

Figure 13.13 Animated! Autonomic nerves serve internal organs. This is a diagram of the major sympathetic and parasympathetic nerves leading out from the central nervous system to some major organs. There are *pairs* of both kinds of nerves, servicing the right and left halves of the body. The ganglia are clusters of cell bodies of the neurons that are bundled together in nerves. (© Cengage Learning)

Figure 13.14 **This young gymnast's sympathetic nervous system is highly active during his routine.**

down the body overall and divert energy to basic bodily housekeeping tasks, such as digestion.

Sympathetic nerves dominate at times of danger, stress, excitement, or strenuous physical activity (Figure 13.14). Among other effects, their signals increase the force and rate of the heartbeat, elevate blood pressure by constricting arterioles, increase the breathing rate, and dilate the pupils of the eyes so that more light can enter. This physiological shift is called the "fight–flight response" because it primes the body to respond to rapid-fire physical demands that might arise in an emergency (such as fighting hard or running away). The response suppresses activities that are less important during an emergency, such as digestion.

The spinal cord links the PNS and the brain

The **spinal cord** carries signals between the peripheral nervous system and the brain. It threads through a canal made of bones of the vertebral column (Figure 13.15). Most of the cord consists of nerve tracts (bundles of myelinated axons). Because the myelin sheaths of these axons are white, the tracts are called **white matter**. The cord also contains **gray matter** that consists of dendrites, cell bodies of neurons, interneurons, and glial cells. The cord lies inside a closed channel formed by the bones of the vertebral column. Those bones, and ligaments attached to them, protect the soft nervous tissue of the cord. So do the coverings called meninges discussed in Section 13.7.

Besides carrying signals between the brain and the peripheral nervous system, the spinal cord is a control center for reflexes that were described in Section 13.4. It also contributes to *autonomic reflexes* that deal with internal functions such as bladder emptying.

> **gray matter** Neuron cell bodies and dendrites, interneurons, and glia in the spinal cord.
>
> **spinal cord** Nervous tissue that links the brain with the peripheral nervous system; the cord also controls reflexes.
>
> **sympathetic nerves** Autonomic nerves that relay signals related to physiological arousal.
>
> **white matter** The nerve tracts of the spinal cord.

TAKE-HOME MESSAGE

WHAT IS THE PERIPHERAL NERVOUS SYSTEM?

- The peripheral nervous system (PNS) consists of the nerves traveling to and from the brain and spinal cord.
- PNS somatic nerves deal with skeletal muscle movements. Its autonomic nerves deal with internal organs and glands. Autonomic nerves are divided into parasympathetic nerves (for housekeeping functions) and sympathetic nerves (for aroused states).
- The spinal cord carries signals between peripheral nerves and the brain. It also is a control center for some reflexes.

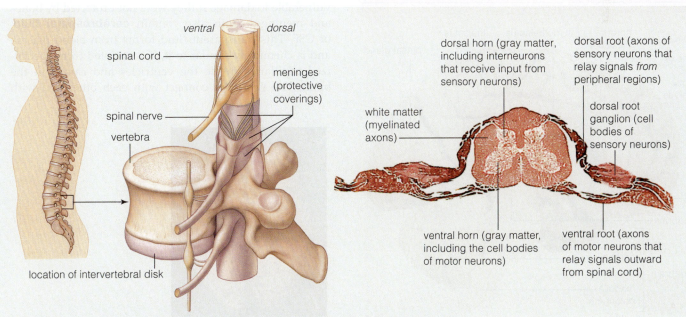

Figure 13.15 **Animated! The spinal cord connects the brain with the peripheral nervous system.** (*Right*) A cross section of the cord's gray matter resembles a butterfly. (Left: © Cengage Learning; Right: Manfred Kage/Peter Arnold, Inc.)

The Brain: Command Central

■ The brain is divided into three main regions, each one containing centers that manage specific biological tasks.

The spinal cord merges with the **brain**, which weighs about 3 pounds (1,300 grams) in an adult. Just as the bony vertebrae protect the spinal cord, the cranial bones of the skull, or cranium, protect the brain.

The brain's three main functional areas are the hindbrain, midbrain, and forebrain

The hindbrain sits atop the spinal cord (Figure 13.16). The portion just above the cord, the **medulla oblongata** (meh-DULL-uh ahb-lawn-GAH-tuh), helps govern breathing rhythm and the strength of heartbeats. It also controls reflexes such as swallowing, coughing, sneezing, and vomiting. Above the medulla is the **pons** (PAHNZ) which helps regulate breathing. *Pons* means "bridge," and nerve tracts extend through it to the midbrain. The **cerebellum** is the largest hindbrain region. It lies at the back of the brain and mainly coordinates voluntary movements.

The midbrain is the smallest of the three brain regions. It mainly relays information from sensory organs to the forebrain. Together, the pons, medulla, and midbrain form the **brain stem**. When the brain stem is damaged by a stroke, disease, or a head injury, the results can be severe or even fatal.

The forebrain is the most highly developed brain region. It includes the **cerebrum**, where information is processed, and sensory input and motor responses are integrated. The cerebrum consists of two **cerebral hemispheres**. A band of nerve tracts, the corpus callosum, carries signals between the hemispheres. The **thalamus** lies just below the corpus callosum. It is mainly a "switchboard" where incoming signals in sensory nerve tracts are relayed to clusters of neuron cell bodies called *basal nuclei* and then sent onward. The basal nuclei also process some outgoing motor information.

Located under the thalamus, the **hypothalamus** is the body's "supercenter" for controlling homeostatic adjustments in the activities of internal organs. It also helps to govern thirst, hunger, and sexual behavior.

Cerebrospinal fluid fills spaces in the brain and spinal cord

In addition to being shielded by its bony case, the brain is protected by three **meninges** (meh-NIN-jeez). These are membranes of connective tissue layered between the skull and the brain (Figure 13.17). Meninges cover the fragile CNS neurons and blood vessels that service the tissue. The leathery, outer membrane, the *dura mater,* is folded double around the brain. Its upper surface attaches to the skull. The lower surface is the brain's outer covering and separates its two hemispheres. A second membrane is called the *arachnoid,* and the even more delicate *pia mater* wraps the brain and spinal cord. The meninges also enclose spaces called ventricles.

Our brain and spinal cord would both be extremely vulnerable to damage if they were not protected by bones and meninges. Both also contain **cerebrospinal fluid**, or CSF. This transparent fluid forms from blood plasma and is chemically similar to it. It is secreted from specialized capillaries inside the ventricles and canals in the brain. The ventricles connect with each other and with

forebrain

corpus callosum

hypothalamus

thalamus

pineal gland

one of two optic nerves

midbrain

pons

cerebellum

hindbrain

medulla oblongata

A

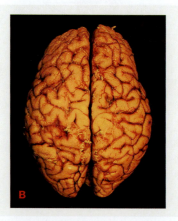

B

Figure 13.16 Animated! The brain has three major regions that are subdivided into functional areas. **A** Major brain areas in the right hemisphere. **B** The two brain hemispheres viewed from above. (A: Ralph Hutchings/Visuals Unlimited, Inc. B: Dr. Colin Chumbley/Photo Researchers, Inc.)

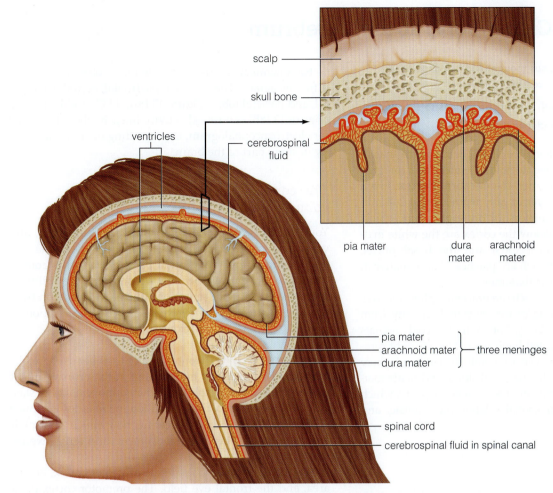

scalp

skull bone

ventricles

cerebrospinal fluid

pia mater   dura mater   arachnoid mater

pia mater
arachnoid mater   } three meninges
dura mater

spinal cord

cerebrospinal fluid in spinal canal

Figure 13.17 **The three meninges help protect the brain and spinal cord. Cerebrospinal fluid fills the space between the arachnoid and the pia mater.** (© Cengage Learning)

blood–brain barrier Features of brain capillaries that prevent many substances from passing from the blood into cerebrospinal fluid.

brain The master control center of the nervous system.

brain stem Brain region made up of the pons, medulla, and midbrain.

cerebellum The hindbrain region that coordinates voluntary movements.

cerebral hemispheres The two halves of the cerebrum.

cerebrospinal fluid Fluid that fills spaces in the brain and the spinal cord and cushions them against physical shocks; the CSF.

cerebrum The forebrain's main center for processing information.

hypothalamus Part of the brain that controls homeostatic adjustments in the functions of internal organs.

medulla oblongata The hindbrain center that controls reflexes such as swallowing, sneezing, and vomiting.

meninges The set of three membranes that cover and help protect the brain.

pons Hindbrain center that helps regulate breathing.

thalamus Forebrain region that relays sensory information.

the central canal of the spinal cord, and are filled with cerebrospinal fluid. The CSF also fills the space enclosed by the two inner meninges (the pia matter and arachnoid). Because this enclosed cerebrospinal fluid can't be compressed, it helps cushion the brain and spinal cord from jarring movements.

A **blood–brain barrier** helps control which bloodborne substances enter the CSF. The barrier is set up by the unusual structure of brain capillaries. Tight junctions between the cells of the capillary walls (Section 4.6) make the walls much less permeable than those of capillaries elsewhere in the body. Specialized transport proteins in the plasma membrane of wall cells allow glucose and a few other needed substances to move out of the bloodstream and into the CSF. Water crosses the barrier freely. So do lipid-soluble molecules, including oxygen and carbon dioxide, which diffuse through the membrane's lipid bilayer. This is one reason why lipid-soluble substances such as alcohol, nicotine, caffeine, and anesthetics can rapidly affect brain function.

The blood–brain barrier stops viruses, bacteria, many toxins, and hormones in blood from gaining access to most neurons in the brain and spinal cord. The barrier doesn't protect the hypothalamus, which has a central role in homeostasis. Instead the hypothalamus is directly exposed to the bloodstream and can monitor the chemical makeup and temperature of blood.

TAKE-HOME MESSAGE

WHAT ARE THE MAIN FUNCTIONS OF THE HINDBRAIN, MIDBRAIN, AND FOREBRAIN?

- In the hinddbrain and midbrain, various centers control reflexes related to body functions (such as coughing and vomiting). In the forebrain, the cerebrum handles overall processing and integration of sensory information and motor responses.
- Cerebrospinal fluid fills cavities and canals in the brain and spinal cord to provide a protective cushion.
- The blood–brain barrier prevents many potentially harmful substances in blood from entering the CSF.

- Our capacity for conscious thought and language arises from the activity of the cerebral cortex.

- The cortex interacts with other brain regions to shape our emotional responses and memories.

cerebral cortex The outer layer of gray matter of each cerebral hemisphere.

limbic system The brain region that governs emotions and influences related behavior. It includes parts of the thalamus, hypothalamus, the amygdala, and the hippocampus.

Each cerebral hemisphere has a deeply folded, outer layer of gray matter, the **cerebral cortex**. It is a layer of gray matter about 2–4 millimeters, or one-eighth inch, thick. Below the cortex are the white matter (axons) and the basal nuclei, which are patches of gray matter in the thalamus.

Each cerebral hemisphere receives and processes signals mainly from the opposite side of the body. For example, "cold" signals from an ice cube in your left hand travel to your right cerebral hemisphere, and vice versa. Overall, the left hemisphere deals mainly with speech, analytical skills, and mathematics. In most people it dominates the right hemisphere, which deals more with visual–spatial relationships, music, and other creative activities.

Each hemisphere also is divided into lobes that process different signals. The lobes are the frontal, occipital, temporal, and parietal lobes (Figure 13.18A). EEGs and PET scans (Figure 13.18B) can reveal activity in each lobe. EEG, short for electroencephalogram, is a recording of electrical activity in some part of the brain.

The cerebral cortex is the seat of consciousness

Your thoughts, memories, the ability to understand, and voluntary acts all begin in the cerebral cortex. The cortex is divided into three main parts. *Motor* areas control voluntary movements. *Sensory* areas govern the ability to grasp the meaning of sensations (that is, information from sensory organs). *Association* areas process information as needed to produce a conscious action.

Motor areas In the frontal lobe of each hemisphere, the whole body is spatially mapped out in the primary motor cortex. This area controls coordinated movements of skeletal muscles. Thumb, finger, and tongue muscles get much of the area's attention, indicating how much control is required for voluntary hand movements and verbal expression (Figure 13.19).

Also in the frontal lobe are the premotor cortex, Broca's area, and the frontal eye field. The premotor cortex deals with learned patterns or motor skills. Repetitive motor actions, such as bouncing a ball, are evidence that your motor cortex is coordinating the movements of several muscle groups. Broca's area (usually in the left hemisphere) and a corresponding area in the right hemisphere control the tongue, throat, and lip muscles used in speech. It kicks in when we are about to speak and even when we plan voluntary motor activities other than speaking (so you can talk on the phone and write down a message at the same time). Above Broca's area is the frontal eye field. It controls voluntary eye movements.

frontal lobe (planning of movements, aspects of memory, inhibition of unsuitable behaviors)

primary motor cortex

primary somatosensory cortex

parietal lobe (sensations from internal organs)

Wernicke's area

Broca's area

temporal lobe (hearing, advanced visual processing)

occipital lobe (vision)

© Cengage Learning

A

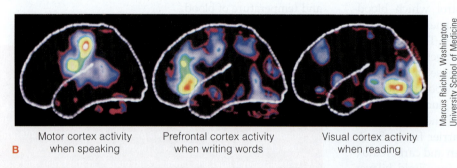

B　Motor cortex activity when speaking　　Prefrontal cortex activity when writing words　　Visual cortex activity when reading

Marcus Raichle, Washington University School of Medicine

Figure 13.18　The cerebrum is divided into hemispheres and lobes. A Lobes of the brain, showing the primary receiving and integrating centers of the cerebral cortex. **B** PET scans show brain regions that were active when a subject performed three specific language tasks: speaking, writing words, and reading.

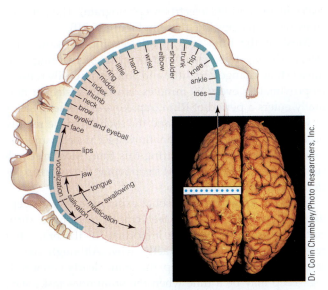

Figure 13.19 **The body is "mapped" in the primary motor cortex.** This diagram depicts a slice through the primary motor cortex of the left cerebral hemisphere. The distortions to the body draped over the diagram indicate which body parts are controlled with the greatest precision. (© Cengage Learning)

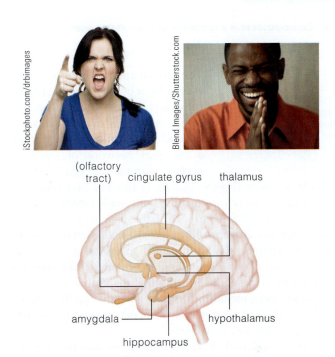

(olfactory tract) cingulate gyrus thalamus

amygdala

hippocampus

hypothalamus

Figure 13.20 **The limbic system operates in emotions and some other mental activities.** The amygdala and the cingulate gyrus are especially important in emotions. The hypothalamus is a clearinghouse for emotions and the activity of internal organs. (© Cengage Learning)

Sensory areas   Sensory areas occur in different parts of the cortex. In the parietal lobe, the body is spatially mapped out in the primary somatosensory cortex. This area is the main receiving center for sensory input from the skin and joints. The parietal lobe also has a primary cortical area dealing with perception of taste. At the back of the occipital lobe is the primary visual cortex, which receives sensory inputs from your eyes. Perception of sounds and of odors arises in primary cortical areas in each temporal lobe.

Association areas   Association areas occupy all parts of the cortex except the primary motor and sensory regions. Each integrates, analyzes, and responds to many inputs. For instance, the visual association area surrounds the primary visual cortex. It helps us recognize something we see by comparing it with visual memories. Neural activity in the most complex association area—the prefrontal cortex—is the basis for complex learning, intellect, and personality. Without it, we would be incapable of abstract thought, judgment, planning, and concern for others.

The limbic system governs emotions

The **limbic system** circles the upper brain stem. It includes parts of the thalamus along with the amygdala, the hypothalamus, and the hippocampus (Figure 13.20).

The limbic system influences the basic body functions regulated by the hypothalamus and brain stem and controls emotions. It is sometimes called the "emotional brain" because it produces emotional behaviors such as anger, pleasure, satisfaction, fear, and sexual arousal. In all these activities the limbic system interacts closely with the prefrontal cortex. Its connections with other brain regions bring about emotional responses such as smiling, surprise, blushing, or laughing.

TAKE-HOME MESSAGE

WHAT IS THE CEREBRAL CORTEX AND WHAT FUNCTIONS DOES IT CARRY OUT?

- The cerebral cortex forms the outer layer of gray matter of each hemisphere. It has motor, sensory, and association areas that collectively govern conscious behavior.
- Each cerebral hemisphere receives and processes responses to sensory input mainly from the opposite side of the body.
- The left hemisphere deals mainly with speech, analytical skills, and mathematics. It usually dominates the right hemisphere, which deals more with creative activity.
- The cerebral cortex interacts with the limbic system, which governs emotions.

- Consciousness is a spectrum of brain states such as alertness and stages of sleep.

reticular formation A network of brain stem neurons that processes sensory information and sends related signals to other parts of the CNS.

The spectrum of consciousness ranges from being wide awake and fully alert to drowsiness, sleep, and coma. All states of consciousness depend on the **reticular formation**, a two-part network of interconnected neurons that runs through the brain stem (Figure 13.21A). It receives and processes incoming sensory information, then sends signals to other parts of the CNS. One part of the formation, called the RAS (for *reticular activating system*) sends signals upward to the thalamus that stimulate it to arouse and activate the cerebral cortex. Depending on how much the cortex is stimulated, it responds in ways that determine the level of consciousness, including sleeping and waking. The central role of the RAS in this brain function is why brain stem damage often results in coma.

The second part of the reticular formation receives signals from the hypothalamus and relays them to spinal cord neurons. These signals govern skeletal muscle activity that helps maintain balance, posture, and muscle tone. The reticular system also filters incoming signals, helping the brain distinguish between important and unimportant ones. It is this filtering that enables you to sleep through many sounds but to waken to specific ones, such as a cat meowing to be let out or a baby crying.

The patterns for full alertness and other states of consciousness can be detected by electrodes placed on the scalp during an EEG (electroencephalogram). The patterns show up as tracings like those in Figure 13.21B.

Most of the time you spend sleeping is "slow-wave sleep." During this stage, your heart rate, breathing, and muscle tone change very little and you can be easily roused. Approximately every 90 minutes, however, a sleeper normally enters a period of REM (rapid-eye-movement) sleep, in which the eyelids flicker and the eyeballs move rapidly back and forth. Sleepers dream during REM sleep, and it is much harder to wake up during this time. Most research subjects awakened from REM sleep report they were experiencing vivid dreams.

You may know from personal experience that a sleep-deprived person tends to feel cranky and have difficulty concentrating. Sleep is important for the brain, but researchers don't know exactly why. Although neural activity changes during sleep, the brain clearly is not resting. Sleep may be a time when the brain does tasks such as consolidating memories and firming up connections involved in learning.

TAKE-HOME MESSAGE

WHAT IS CONSCIOUSNESS?

- Consciousness is a spectrum of mental activity that includes state of arousal and sleep.
- Signals from the reticular activitating system (RAS) of the reticular formation in the brain stem arouse and activate the cerebral cortex.

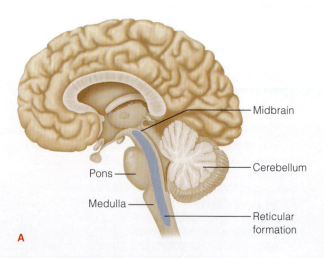

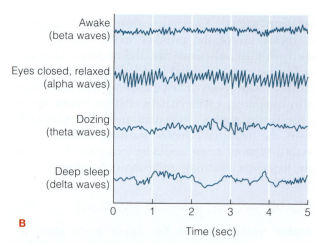

Figure 13.21 Several states of consciousness occur in the brain. A Location of the reticular formation in the brain stem. **B** EEG patterns for various stages of consciousness. (© Cengage Learning)

13.10 Memory

■ **Memory is how the brain stores and retrieves facts and other types of information.**

Learning and modifications of our behavior would be impossible without **memory**. The brain stores information in stages. The first is *short-term* storage of bits of sensory information—numbers, words of a sentence, and so on—for a few minutes or hours. In *long-term* storage, seemingly unlimited amounts of information get tucked away more or less permanently (Figure 13.22).

Only some of the sensory information reaching the cerebral cortex is transfered to short-term memory. Information is processed for relevance, so to speak. If irrelevant, it is forgotten; otherwise it is consolidated with the banks of information in long-term storage structures.

The brain processes facts separately from skills. Dates, names, faces, words, odors, and other bits of explicit information are facts that are stored together with the circumstance in which they were learned. Hence you might associate the smell of bread baking, say, with your grandmother's kitchen. This "fact" recall may be brief or long-term and is called *declarative* memory. By contrast, *skill memory* is gained by practicing specific motor activities. How to maneuver a snowboard or play a piano concerto is best recalled by actually performing it, rather than by remembering the circumstances in which the skill was first learned.

Separate memory circuits handle different kinds of input. A circuit leading to declarative memory (Figure 13.23A) starts with inputs at the sensory cortex that flow to the amygdala and hippocampus in the limbic system. The amygdala is the gatekeeper, connecting the sensory cortex with parts of the thalamus and with parts of the hippocampus that govern emotional states. Information

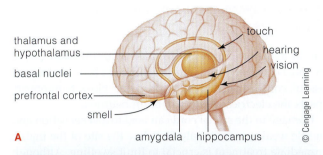

Figure 13.23 Memories of facts and skills are stored differently. A Possible circuits involved in declarative memory. **B** A snowboarder provides a dramatic demonstration of skill memory.

flows on to the prefrontal cortex, where multiple banks of fact memories are retrieved and used to stimulate or inhibit other parts of the brain. The new input also flows to basal nuclei, which send it back to the cortex in a feedback loop that reinforces the input until it can be consolidated in long-term storage.

Skill memory also starts at the sensory cortex, but this circuit routes sensory input to a region deeper in the brain that promotes motor responses (Figure 13.23B). Motor skills entail muscle conditioning. The skill memory circuit extends to the cerebellum, the brain region that coordinates motor activity.

Amnesia is a loss of fact memory. How severe the loss is depends on whether the hippocampus, amygdala, or both are damaged, as by a head blow. Amnesia does not affect a person's capacity to learn new skills.

amnesia The loss of fact memory.

memory Storage of information in the brain.

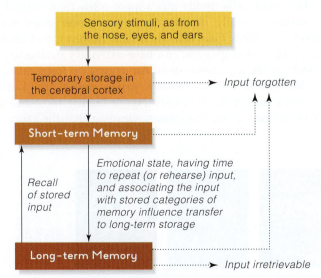

Figure 13.22 Memories are processed in two stages. Short-term memory is temporary. Long-term memories may be stored in the cerebral cortex for years. (© Cengage Learning)

TAKE-HOME MESSAGE

HOW DO MEMORIES FORM IN THE BRAIN?

- Memories form when sensory messages are procesed through short-term and long-term storage mechanisms.
- The storage mechanisms involve circuits between the cerebral cortex and parts of the limbic system, thalamus, and hypothalamus.

Physical injury is a common cause of nervous system damage

A blow to the head or neck can cause a **concussion**, one of the most common brain injuries. Blurred vision and a brief loss of consciousness result when the blow temporarily upsets the electrical activity of brain neurons.

Damage to the spinal cord can lead to lost sensation and muscle weakness or **paralysis** below the site of the injury. Immediate treatment is crucial to limit swelling. Although cord injuries usually have severe consequences, intensive therapy during the first year after an injury can improve the patient's long-term prognosis. Using nerve growth factors or stem cells to repair spinal cord injuries is a major area of medical research.

Brain injury, birth trauma, or other assaults can cause various forms of *epilepsy*, or **seizure disorders**. Some times the trigger is an inherited predisposition. A seizure results when the brain's normal electrical activity suddenly becomes chaotic. Worldwide, thousands of people develop recurrent seizures either as children or later in life. All but the most difficult cases usually respond well to drug therapy.

In some disorders, brain neurons break down

In 1817, physician James Parkinson observed troubling symptoms in certain people on the streets of London. They walked slowly, taking short, shuffling steps. And their limbs trembled, sometimes violently. Today we know that the culprit is a degenerative brain disorder that now is called **Parkinson's disease**, or PD (Figure 13.24A–C). In PD, neurons in parts of the thalamus (Section 13.7) begin to die. Those neurons make neurotransmitters (dopamine and norepinephrine) required for normal muscle function, so PD symptoms include muscle tremors and balance problems, among others. Multiple factors contribute to the development of PD. A head injury or exposure to pesticides in drinking water may increase the risk. Treatments include drugs that help replace absent neurotransmitters or surgical treatments that may relieve some symptoms. There is no cure.

Like PD, **Alzheimer's disease** involves the progressive degeneration of brain neurons. At the same time, there is an abnormal buildup of amyloid protein, leading to the loss of memory and intellectual functions. Alzheimer's disease is associated with advancing age, and we consider it again in our discussion of aging and the nervous system in Chapter 17.

Infections and cancer inflame or destroy brain tissue

Meningitis is an often fatal disease caused by a bacterial or viral infection. Symptoms include headache, a stiff neck, and vomiting. They develop when the meninges covering the brain and/or spinal cord become inflamed. **Encephalitis** is inflammation of the brain. It is usually caused by a viral infection, such as by the West Nile virus or a herpesvirus. Like meningitis, encephalitis can be extremely dangerous. Early symptoms include fever, confusion, and seizures. A form of **Creuzfeldt-Jakob disease** has occurred in people who ate beef from animals infected by a *prion*—a small infectious protein—that causes "mad cow disease," or *bovine spongiform encephalitis* (BSE). A BSE outbreak in Britain in the late 1990s raised public awareness of the potential danger of eating meat from infected animals. The infection causes holes in an affected person's brain tissue. The disease is rapidly debilitating and always fatal (Figure 13.25).

In cancer, cells divide much more often than normal. Neurons generally do not divide, so cancer does not develop in them. Glial cells do divide, however, and **glial cancers,** called gliomas, can have extremely destructive effects in the nervous system. An aggressive form called *glioblastoma multiforme* usually strikes males and kills within a year of the diagnosis (Figure 13.24D). Most cases of spinal cancer are metastases, meaning that the cancer

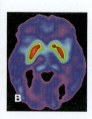
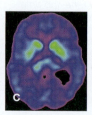
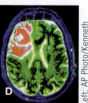

Left: AP Photo/Kenneth Lambert Center: From Neuro Via Clinical Research Program, Minneapolis Medical Center Right: © Collection CNRI/MedNet/Corbis

Figure 13.24 Many battle brain disease. A Parkinson's disease affects former heavyweight champion Muhammad Ali, actor Michael J. Fox., and about 500,000 others in the United States. PET scans from a healthy person **B** and an affected person **C.** The red area in **D** is a glioblastoma multiforme tumor in a patient's brain.

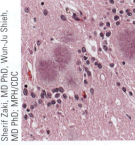

Figure 13.25 A prion causes variant Creuzfeldt-Jakob disease. Charlene Singh, here being cared for by her mother, died from the disease in 2004. She may have been exposed to BSE in Britain, where she grew up. The micrograph shows brain tissue of an affected person, with the characteristic holes and clusters of radiating prion protein fibers.

has spread to the spine from a primary cancer elsewhere in the body.

Some people are concerned that the radio waves emitted by cell phones could cause tumors. Although to date research hasn't shown increased brain tumor incidence related to cell phone use, studies have shown increased metabolic activity in cells near the phone. Some health officials recommend using a headset to keep the radiation-emitting part of the phone farther from the brain.

In young adults, the most common nervous system disease is **multiple sclerosis (MS)**. It is an autoimmune disease that may be triggered by a viral infection in susceptible people. MS involves progressive destruction of myelin sheaths of neurons in the central nervous system. The symptoms develop over time and include muscle weakness or stiffness, extreme fatigue, and slurred speech. A disorder called **Guillain-Barre syndrome** produces similar symptoms in the peripheral nervous system. It is caused by a viral or bacterial infection and usually is temporary.

Headaches only seem like brain "disorders"

One of the most common of all physical ailments is the pain we call **headache**. There are no sensory nerves in the brain, however, so it does not "feel pain." Instead, headache pain typically is due to tension (stretching) in muscles or blood vessels of the face, neck, and scalp.

Throbbing *migraine* headaches are infamous for being extremely painful and lasting for up to three days. In the United States alone, 28 million people, mainly female, suffer from migraines, which can be triggered by hormonal changes, fluorescent lights, certain foods (such as chocolate)—even changes in the weather.

Tension headaches and migraines are thought to be part of a continuum, and both are treated with drugs ranging from aspirin to prescription painkillers and drugs that act as neuromodulators (Section 13.3) to reduce the sensitivity of affected brain neurons to stimuli that trigger the headache.

Cluster headaches develop more often in men and are in a class by themselves. This type of headache produces a piercing pressure in one eye and may recur several times a day for weeks or months. Some sufferers have found the pain so unbearable that they have committed suicide.

Various neural disorders affect development, behavior, and mood

People affected by **ADHD**, or *attention deficit hyperactivity disorder*, have trouble concentrating, tend to fidget, and may be unusually impulsive. A lower than normal level of dopamine may be involved, and drugs used to treat ADHD increase brain dopamine levels. The well-known ADHD drug Ritalin prevents the reuptake of dopamine after it has been released at a synapse.

The mental state we call "mood" results at least in part from the interactions of several neurotransmitters, including serotonin and dopamine. Medications that adjust the levels of these substances in the brain are used to treat **mood disorders**. By some estimates, *depression* affects up to 17 percent of adults at some time in their lives. A clinically depressed person feels sad all the time and can't experience pleasure. Some depressed people lack energy; others may feel agitated or irritable. Several widely prescribed antidepressants, such as Paxil and Prozac, increase the amount of serotonin in the brain by preventing its reuptake. Antidepressants that prevent serotonin reuptake are also used to treat people who have anxiety disorders—ailments that can trigger extreme worry or panic in situations most people would consider normal.

Autism and related conditions including *Asperger's syndrome* are forms of persistent developmental disorders (PDDs) that usually show up in childhood. Affected youngsters experience mild to severe problems in thinking, language skills, and the capacity to relate to others. Research suggests that a family of "autism genes" may underlie PDDs. In some cases, affected children show major improvement with intensive behavioral therapy.

Disrupted thinking is the hallmark of **schizophrenia**. Patients experience paranoid delusions and often "hear voices" (auditory hallucinations). Holding a job or having normal social relationships often are impossible. Therapeutic drugs can help control symptoms. Multiple factors, including physical changes in the brain, may help trigger this devastating mental disorder.

Psychoactive drugs bind to neuron receptors in the brain. As a result, the neurons send or receive altered messages. The drugs typically affect parts of the brain that govern consciousness and behavior. Some also alter heart rate, respiration, sensory processing, and muscle coordination. Many affect a pleasure center in the hypothalamus and artificially fan the sense of pleasure we associate with eating, sex, or other activities.

Stimulants include caffeine, nicotine, cocaine, and amphetamines—including Ecstasy (MDMA). Nicotine mimics ACh, directly stimulating certain sensory receptors. It also increases the heart rate and blood pressure. At first amphetamines cause a flood of the neurotransmitters norepinephrine and dopamine, which stimulate the brain's pleasure center. Over time, however, the brain slows its production of those substances and depends more on the amphetamine. Chronic users may become psychotic, depressed, and malnourished. They may also develop heart problems. Cocaine stimulates the pleasure center by *blocking* the reabsorption of dopamine and other neurotransmitters. It also weakens the cardiovascular and immune systems.

Alcohol is a *depressant*, even though it produces a high at first. Drinking only an ounce or two diminishes judgment and can lead to disorientation and uncoordinated movements. *Blood alcohol concentration* (BAC) measures the percentage of alcohol in the blood. In most states, someone with a BAC of 0.08 per milliliter is considered legally drunk. When the BAC reaches 0.15 to 0.4, a drinker is visibly intoxicated and can't function normally. A BAC greater than 0.4 can kill.

Morphine, an *analgesic* (painkiller), is derived from the seed pods of the opium poppy. Like its cousin heroin, it blocks pain signals by binding with certain receptors on neurons in the central nervous system. Both morphine

TABLE 13.3 Warning Signs of Drug Addiction*
1. Tolerance—it takes increasing amounts of the drug to produce the same effect.
2. Habituation—it takes continued drug use over time to maintain self-perception of functioning normally.
3. Inability to stop or curtail use of the drug, even if there is persistent desire to do so.
4. Concealment—not wanting others to know of the drug use.
5. Extreme or dangerous behavior to get and use a drug, as by stealing, asking more than one doctor for prescriptions, or jeopardizing employment by drug use at work.
6. Deteriorating professional and personal relationships.
7. Anger and defensive behavior when someone suggests there may be a problem.
8. Preferring drug use over previous activities.

*Having three or more of these signs may be cause for concern.

and the synthetic version OxyContin produce euphoria. Thousands of people who obtained OxyContin illegally or by subterfuge have overdosed and died.

Marijuana is a *hallucinogen*. In low doses it slows but doesn't impair motor activity and causes mild euphoria. It can also cause visual hallucinations. Like alcohol, it skews the performance of complex tasks, such as driving.

The body eventually may develop drug *tolerance*, meaning that it takes larger or more frequent doses to produce the same effect. Tolerance reflects physical drug dependence. The liver produces enzymes that detoxify drugs in the blood. Tolerance develops when the level of those enzymes rises in response to the ongoing presence of the drug in the bloodstream. In effect, a drug user must increase his or her intake to stay ahead of the liver's growing ability (up to a point) to break down the drug.

In psychological drug dependence, or *habituation*, a user begins to crave the feelings associated with a particular drug. Without a steady supply of it the person can't "feel good" or function normally. Table 13.3 lists warning signs of potentially serious drug dependence. Habituation and tolerance both are evidence of addiction.

When different psychoactive drugs are used together, they can interact dangerously. For example, alcohol and barbiturates (such as Seconal and Nembutal) both depress the central nervous system. Used at the same time, they can depress respiratory centers in the brain enough to cause death.

Jamie Baker/Taxi/Getty Images

CONNECTIONS

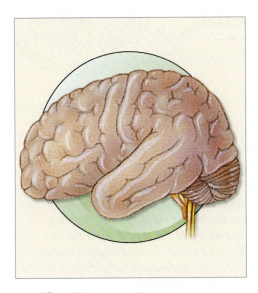

The Nervous System

The nervous system produces signals that flow between the brain and spinal cord and other parts of the body. Together with chemical signals from the endocrine system, these nerve impulses (action potentials) provide the communication required to monitor, adjust, and regulate all body functions.

Integumentary system
Sweat glands and skeletal muscles that move hair follicles receive signals from sympathetic nerves (autonomic division of the PNS). Sensory nerve endings detect pain, pressure, temperature.

Skeletal system
Sensory nerves that service bone tissue signal damage due to breaks or other physical harm.

Muscular system
Signals from motor areas stimulate skeletal muscle contractions required for movement; signals from the cerebellum coordinate motor activity and maintain posture; spinal cord governs reflex movements.

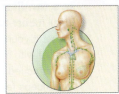

Cardiovascular system and blood
Centers in the brain stem help control the heart rate and help maintain proper blood pressure by adjusting the diameter of arterioles.

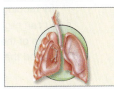

Immunity and the lymphatic system
Positive/negative mental states may strengthen/weaken some immune responses.

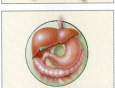

Respiratory system
Centers in the brain stem adjust the rate and depth of breathing.

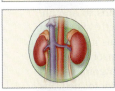

Digestive system
Parasympathetic nerves regulate various aspects of digestion. Sensory signals trigger release of saliva, feelings of hunger/fullness, and regulate peristalsis.

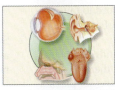

Urinary system
Reflexes govern urination (bladder emptying); parasympathetic nerves adjust blood flow to the kidneys.

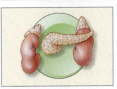

Sensory systems
Sensory association areas manage perception of sensory information, including injury to sense organs.

Endocrine system
Signals from the hypothalamus trigger the secretion of hormones from the pituitary gland. Signals from parasympathetic nerves regulate the release of hormones from the pancreas and adrenals.

Reproductive system
The hypothalamus governs the release of sex hormones and functioning of ovaries and testes. Sexual arousal and behavior depends on signals from the hypothalamus and the limbic system.

EXPLORE
ON YOUR OWN

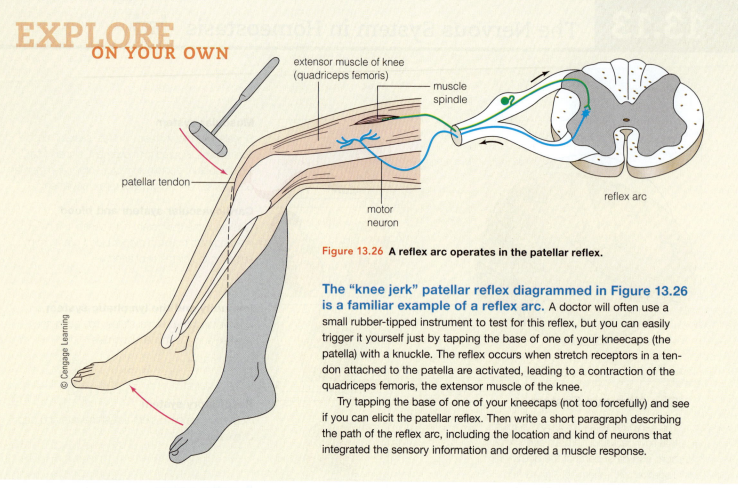

extensor muscle of knee
(quadriceps femoris)

muscle
spindle

patellar tendon

motor
neuron

reflex arc

© Cengage Learning

Figure 13.26 **A reflex arc operates in the patellar reflex.**

The "knee jerk" patellar reflex diagrammed in Figure 13.26 is a familiar example of a reflex arc. A doctor will often use a small rubber-tipped instrument to test for this reflex, but you can easily trigger it yourself just by tapping the base of one of your kneecaps (the patella) with a knuckle. The reflex occurs when stretch receptors in a tendon attached to the patella are activated, leading to a contraction of the quadriceps femoris, the extensor muscle of the knee.

Try tapping the base of one of your kneecaps (not too forcefully) and see if you can elicit the patellar reflex. Then write a short paragraph describing the path of the reflex arc, including the location and kind of neurons that integrated the sensory information and ordered a muscle response.

SUMMARY

Section 13.1 The nervous system detects, processes, and responds to stimuli. Sensory neurons respond directly to external or internal stimuli. Interneurons in the brain and spinal cord receive sensory signals, process them, and then send outgoing signals that influence other neurons. Motor neurons relay messages away from the brain and spinal cord to muscles or glands. Neuroglia provide various forms of physical or chemical support for neurons.

Neurons have extensions called axons and dendrites. Axons carry outgoing signals, and dendrites receive them. A resting neuron has a steady voltage difference across its plasma membrane. This difference is called the resting membrane potential.

A neuron maintains concentration gradients of various ions, notably sodium and potassium, across the membrane. Changes in this difference allow a neuron to send signals (nerve impulses).

Section 13.2 When the voltage difference across the membrane exceeds a threshold level, gated sodium channels in the membrane open and close rapidly and suddenly reverse the voltage difference. This reversal is a nerve impulse, or action potential. A sodium–potassium pump restores ion gradients after an action potential fires. Action potentials propagate away from the point of stimulation.

Section 13.3 Action potentials self-propagate along the neuron membrane until they reach a synapse with another neuron, muscle, or gland. The presynaptic cell releases a neurotransmitter into the synapse. The neurotransmitter excites or inhibits the receiving (postsynaptic) cell. Synaptic integration sums up the various signals acting on a neuron. Neuromodulators boost or reduce the effects of neurotransmitters.

Section 13.4 Nerves consist of the long axons of motor neurons, sensory neurons, or both. A myelin sheath formed by Schwann cells insulates each axon, so that action potentials propagate along it much more rapidly. Nerve pathways extend from neurons in one body region to neurons or effectors in different regions.

A reflex is a simple, stereotyped movement in response to a stimulus. In the simplest reflexes, sensory neurons directly signal motor neurons that act on muscle cells. In more complex reflexes, interneurons coordinate and refine the responses.

Sections 13.5, 13.6 The brain and spinal cord make up the central nervous system. The peripheral nervous system consists of nerves and ganglia in other body regions.

The peripheral nervous system's somatic nerves deal with skeletal muscles involved in voluntary body movements and sensations arising from skin, muscles, and joints. Its autonomic nerves deal with the functions of internal organs.

Autonomic nerves are subdivided into sympathetic and parasympathetic groups. Parasympathetic nerves govern basic tasks such as digestion and tend to slow the pace of other body functions. Signals from sympathetic nerves produce the fight–flight response, a state of intense arousal in situations that may demand increased activity.

Spinal cord nerve tracts carry signals between the brain and the PNS. The cord also is a center for many reflexes.

Section 13.7 The brain is divided into two cerebral hemispheres and has three main divisions (Table 13.4). It and the spinal cord are protected by bones (skull and vertebrae) and by the three meninges. Both are cushioned by cerebrospinal fluid. Specialized capillaries create a blood–brain barrier that prevents some blood-borne substances from reaching brain neurons.

In the forebrain the thalamus relays sensory information and helps coordinate motor responses. The hypothalamus monitors internal organs and influences behaviors related to their functions (such as thirst). The limbic system has roles in learning, memory, and emotional behavior.

Midbrain centers coordinate and relay some sensory information. The midbrain, medulla oblongata, and pons make up the brain stem.

The hindbrain includes the medulla oblongata, pons, and cerebellum. It contains reflex centers for vital functions and muscle coordination.

Section 13.8 The cerebral cortex is devoted to receiving and integrating information from sense organs and coordinating motor responses in muscles and glands.

Sections 13.9, 13.10 States of consciousness vary between total alertness and deep coma. The levels are governed by the RAS, the brain's reticular activating system. It is part of the reticular formation in the brain stem. Memory occurs in short-term and long-term stages. Long-term storage depends on chemical or structural changes in the brain.

TABLE 13.4 Summary of the Central Nervous System

FOREBRAIN	Cerebrum	Processes sensory inputs; initiates, controls skeletal muscle activity. Governs thought, memory, emotions
	Olfactory lobe	Relays sensory input from nose to olfactor centers of cerebrum
	Thalamus	Relays sensory signals to and from cerebral cortex; has role in memory
	Hypothalamus	With pituitary gland, a homeostatic control center; adjusts volume, composition, temperature of internal environment. Governs organ-related behaviors (e.g., sex, thirst, hunger) and expression of emotions
	Limbic system	Governs emotions; has roles in memory
	Pituitary gland	With hypothalamus, provides endocrine control of metabolism, growth, development
	Pineal gland	Helps control some circadian rhythms; also has role in reproductive physiology
MIDBRAIN (BRAIN STEM)	Roof of midbrain	In humans and other mammals, its reflex centers relay visual and auditory sensory input to the forebrain
HINDBRAIN (BRAIN STEM)	Pons	Some tracts bridge the cerebrum and cerebellum; others connect spinal cord with forebrain. With the medulla oblongata, controls rate and depth of respiration
	Medulla oblongata	Its tracts relay signals between spinal cord and pons; its reflex centers help control heart rate, adjustments in blood vessel diameter, respiratory rate, coughing, other vital functions
	Cerebellum	Coordinates motor activity for moving limbs and maintaining posture, and for spatial orientation
SPINAL CORD		Makes reflex connections for limb movements. Its tracts connect brain, peripheral nervous system

REVIEW QUESTIONS

1. Explain the difference between a sensory neuron, an interneuron, and a motor neuron.

2. What are the functional zones of a motor neuron?

3. Define an action potential.

4. What is a synapse? Explain the difference between an excitatory and an inhibitory synapse.

5. Explain what happens during synaptic integration.

6. What is a reflex? Describe what happens during a stretch reflex.

7. Distinguish between the following:
   a. neurons and nerves
   b. somatic system and autonomic system
   c. parasympathetic and sympathetic nerves

1. Fill in the blank: The nervous system senses, interprets, and issues commands for responses to _____.

2. Fill in the blank: A neuron responds to adequate stimulation with _____, a type of self-propagating signal.

3. Fill in the blank: When action potentials arrive at a synapse between a neuron and another cell, they stimulate the release of molecules of a _____ that diffuse over to that cell.

4. In the simplest kind of reflex, _____ directly signal _____, which act on muscle cells.
    a. sensory neurons; interneurons
    b. interneurons; motor neurons
    c. sensory neurons; motor neurons
    d. motor neurons; sensory neurons

5. The accelerating flow of _____ ions through gated channels across the membrane triggers an action potential.
    a. potassium
    b. sodium
    c. hydrogen
    d. a and b are correct

6. _____ nerves slow down the body overall and divert energy to housekeeping tasks; _____ nerves slow down housekeeping tasks and increase overall activity during times of heightened awareness, excitement, or danger.
    a. Autonomic; somatic
    b. Sympathetic; parasympathetic
    c. Parasympathetic; sympathetic

7. The three meninges are membranes of _____ tissue that _____.
    a. muscle; physically support the brain hemispheres
    b. connective; help protect the brain and spinal cord
    c. adipose; directly cushion the brain and spinal cord

8. Cerebrospinal fluid _____.
    a. fills cavities and canals in the brain and spinal cord
    b. is chemically similar to blood plasma
    c. cushions both the brain and spinal cord
    d. both a and c are correct, but not b
    e. a, b, and c are all correct

9. Match each of the following central nervous system regions with some of its functions.
    _____ spinal cord
    _____ medulla oblongata
    _____ hypothalamus
    _____ limbic system
    _____ cerebral cortex

    a. receives sensory input, integrates it with stored information, coordinates motor responses
    b. monitors internal organs and related behavior (e.g., hunger)
    c. governs emotions
    d. coordinates reflexes
    e. makes reflex connections for limb movements, internal organ activity

Your Future

Antidepressants typically take up to six weeks to provide patients with relief from their symptoms. Now researchers in the field of biological psychiatry are working to develop antidepressants that go to work within days. One candidate, a substance called scopolamine, has brought relief within 48 to 72 hours in early tests. More studies are under way to assess its long-term safety and effectiveness.

CRITICAL THINKING

1. Meningitis is an inflammation of the membranes that cover the brain and spinal cord. Diagnosis involves making a "spinal tap" (lumbar puncture) and analyzing a sample of cerebrospinal fluid for signs of infection. Why analyze this fluid and not blood?

2. In newborns and premature babies, the blood–brain barrier is not fully developed. Explain why this might be reason enough to pay careful attention to their diet.

3. In PET scans, red areas are brain regions that are most active, while blue, yellow, and green areas are least active. Figure 13.27 shows PET scans of normal brain activity (*left*) and of the brain of a person while using cocaine (*right*). The frontal lobes of the brain hemispheres are toward the top of the scans. Their neurons play major roles in reasoning and other intellectual functions. Looking at these scan images, how do you suppose cocaine may affect mental functioning?

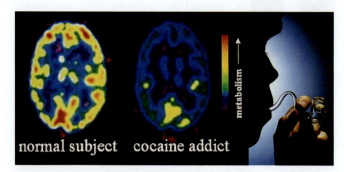

Figure 13.27 Crack cocaine has major effects on the brain.
(Left and center: Science VU/DOE/Visuals Unlimited, Inc.; Right: Ogden Gigli/Photo Researchers, Inc.)

SENSORY SYSTEMS

KEY CONCEPTS

Sensory Receptors and Pathways

Different kinds of sensory receptors detect different types of stimuli. When signals from sensory systems are decoded in the brain, we become aware of sights, sounds, odors, pain, and other sensations. Section 14.1

Somatic Senses

Receptors found at more than one location in the body produce somatic (body) sensations such as touch, pressure, temperature, and pain. Section 14.2

Special Senses

Receptors for the special senses detect chemicals (taste and smell), light (vision), sound waves (hearing), and changes in the body's position (balance). Sections 14.3–14.6, 14.8–14.9

Disorders of the Ears and Eyes Sections 14.7, 14.10

Cagey terrorists. Identity thieves hoping to get your social security number or a bank PIN. Is there any foolproof security against these threats? Biometric identification technologies may come close.

Biometrics measures data about some aspect of human physical makeup, such as digital fingerprints or individual patterns in the retina or iris of the eyes. One of the most reliable methods, iris scanning, relies on the spokelike arrangement of smooth muscle fibers in the iris, the colored surface part of the eyes. Like fingerprints, each person's iris pattern is different from that of every other person on Earth.

For iris scanning to be a solid identity check, each person's iris pattern must first be entered into an electronic database. The U.S. soldier shown at left is having an iris scan for this purpose. Someone who later wants to gain entry to a secure location looks into a scanning device that can instantly compare their eyes' iris pattern with the patterns stored in the database.

Several governments are considering requiring travelers to provide an "iris print" when applying for a passport or visa. Your bank could iris-print you when you open an account. Employers might require potential employees to allow an iris print as part of the job application process. Some already do.

Iris scanning takes advantage of a powerful natural device for gathering information, the human eye. In this chapter we look at the biological role of the eyes, ears, and other structures that make up our sensory systems. These systems are a major means by which the brain obtains information that it may use to help manage the body's biological affairs.

Sensory Receptors and Pathways

- Sensory systems notify the brain and spinal cord of specific changes inside and outside the body.

- Links to Action potentials 13.2, Information pathways 13.4, Sensory areas of the brain 13.8

In a sensory system, a stimulus activates receptors, which convert the stimulus to a nerve impulse—an action potential—that travels to the brain. There it may trigger a sensation or perception:

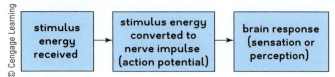

Technically, a **stimulus** (plural: stimuli) is a form of energy that activates receptor endings of a sensory neuron. That energy is converted to the electrochemical energy of action potentials—the nerve impulses by which the brain receives information and sends out commands in response. The brain's basic response is a **sensation**, which is conscious awareness of a stimulus. Higher-level processing results in a **perception**—an understanding of what the sensation means.

There are six main categories of sensory receptors. They reflect the type of stimulus that each kind of receptor detects. **Mechanoreceptors** detect changes in pressure, position, or acceleration. **Thermoreceptors** respond to heat or cold. **Nociceptors** (pain receptors) detect damage to tissues. **Chemoreceptors** detect chemicals dissolved in the fluid around them. **Osmoreceptors** detect changes in water volume (solute concentration) in a body fluid. **Photoreceptors** detect visible light (Table 14.1).

A

Regardless of their differences, all sensory receptors convert the stimulus to nerve impulses (action potentials).

Nerve impulses that move along sensory neurons are all the same. So how does the brain know what sort of sensory event has occurred? It assesses *which* nerves are carrying nerve impulses, the *frequency* of the nerve impulses on each axon in the nerve, and the *number* of axons that responded to the stimulus. Let's consider the steps involved in this processing.

TABLE 14.1 Animated! Major Categories of Sensory Receptors

Category	Examples	Stimulus
MECHANORECEPTORS		
Touch, pressure	Certain free nerve endings and Merkel discs in skin	Mechanical pressure against body surface
Baroreceptors	Carotid sinus (artery)	Pressure changes in blood
Stretch	Muscle spindle in skeletal muscle	Stretching of muscle
Auditory	Hair cells in organ inside ear	Vibrations (sound waves)
Balance	Hair cells in organ inside ear	Fluid movement
THERMORECEPTORS	Certain free nerve endings (heating, cooling)	Change in temperature
NOCICEPTORS (pain receptors)	Certain free nerve endings	Tissue damage (e.g., distortions, burns)
CHEMORECEPTORS		
Internal chemical sense	Carotid bodies in blood vessel wall	Substances (O_2, CO_2, etc.) dissolved in extracellular fluid
Taste	Taste receptors of tongue	Substances dissolved in saliva, etc.
Smell	Olfactory receptors of nose	Molecules in air
OSMORECEPTORS	Hypothalamic osmoreceptors	Change in water volume (solute concentration) of fluid around them
PHOTORECEPTORS		
Visual	Rods, cones of eye	Wavelengths of light

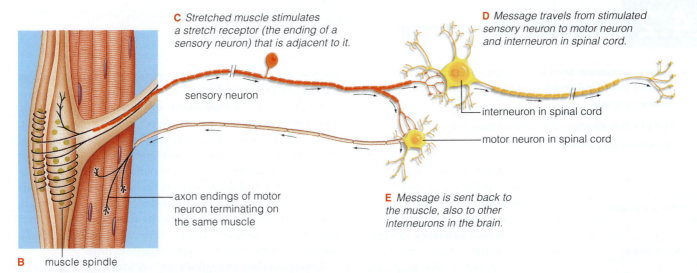

C *Stretched muscle stimulates a stretch receptor (the ending of a sensory neuron) that is adjacent to it.*

D *Message travels from stimulated sensory neuron to motor neuron and interneuron in spinal cord.*

sensory neuron

interneuron in spinal cord

motor neuron in spinal cord

axon endings of motor neuron terminating on the same muscle

E *Message is sent back to the muscle, also to other interneurons in the brain.*

B muscle spindle

Figure 14.1 Signals from stretch receptors in muscles provide an example of a sensory pathway.
A A dancer's complex moves depend on feedback from stretch receptors. B–E The diagram above depicts the path of impulses from receptors called muscle spindles to the spinal cord and brain.
(© Cengage Learning)

First, specific sensory areas of the brain can interpret action potentials only in certain ways. That is why you "see stars" when your eye is poked, even in the dark. The mechanical pressure on photoreceptors in the eye triggers signals that travel along the optic nerve. The brain always interprets signals from an optic nerve as "light." In fact, as you will read in Section 14.2, the brain has a detailed map of the sources of different sensory stimuli.

Second, a strong signal makes receptors fire nerve impulses more often and longer than a weak one does. So, while the same receptor in your ear can detect the sounds of a whisper and a screech, the brain senses the difference through variations in the signals each sound produces.

Third, the stronger a stimulus, the more sensory receptors respond. Gently tap a spot of skin on your arm and you activate only a few touch receptors. Press hard on the same spot and you activate more. The increase translates into nerve impulses in many sensory neurons at once. Your brain interprets the combined activity as an increase in the intensity of the stimulus.

In some cases the frequency of nerve impulses (how often they occur in a given period of time) slows or stops even when the stimulus continues at constant strength. For instance, after you put on a T-shirt, you quickly become only dimly aware of its pressure against your skin. This diminishing response to an ongoing stimulus is called **sensory adaptation**.

Some mechanoreceptors adapt rapidly to a sustained stimulus and only signal when it starts and stops. Other receptors adapt slowly or not at all; they help the brain monitor particular stimuli all the time.

The dancer in Figure 14.1A is holding his position in response to signals from his skin, skeletal muscles, joints, tendons, and ligaments. For example, how fast and how far a muscle stretches depends on activation of stretch receptors in muscle spindles (Figure 14.1B and Section 13.4). By responding to changes in the length of muscles, his brain helps him maintain his balance and posture.

In the rest of this chapter we explore examples of the body's sensory receptors. Receptors that are found at more than one location in the body contribute to somatic ("of the body") sensations. Other receptors are restricted to sense organs, such as the eyes or ears, and contribute to what are called the "special senses."

chemoreceptor Receptor that detects dissolved chemicals.

mechanoreceptor Receptor that detects changes in pressure, position, or acceleration.

nociceptor Receptor that detects tissue damage; a pain receptor.

osmoreceptor Receptor that detects change in water volume of a solution.

perception Understanding the meaning of a sensation.

photoreceptor Receptor that detects light.

sensation Awareness of a stimulus.

sensory adaptation Diminishing response to an ongoing stimulus.

stimulus Any form of energy that can activate a sensory neuron.

thermoreceptor Receptor that responds to heat or cold.

TAKE-HOME MESSAGE

WHAT ARE THE PARTS OF A SENSORY SYSTEM?

- A sensory system has receptors for specific stimuli, such as those in the eye that detect light. It also has nerve pathways that conduct information from receptors to the brain, and brain regions that receive and process the information.

- The brain senses a stimulus based on which nerves carry the incoming signals, the frequency of nerve impulses traveling along each axon in the nerve, and the number of axons that have been recruited.

14.2 Somatic Sensations

- Somatic sensations start with receptors near the body surface, in skeletal muscles, and in the walls of soft internal organs.

- Links to Skin structure 4.9, Inflammation 9.4, Somatosensory cortex 13.8

encapsulated receptor
A sensory receptor enclosed in epithelial or connective tissue.

free nerve ending
Dendrite of a sensory neuron.

somatic sensation
Sensory input from the various types of receptors that are scattered throughout the body, such as free nerve endings.

somatosensory cortex Region of the cerebrum that receives signals from somatic receptors.

Receptors for somatic senses are scattered in different parts of the body. **Somatic sensations** come about when signals from receptors reach the **somatosensory cortex** in the cerebrum. There, interneurons are organized like maps of individual parts of the body surface, just as they are for the motor cortex. The largest areas of the map correspond to body parts where sensory receptors are the most dense. These body parts, including the fingers, thumbs, and lips, have the sharpest sensory acuity and require the most intricate control (Figure 14.2).

Receptors near the body surface sense touch, pressure, and more

There are thousands of sensory receptors in your skin, providing information about touch, pressure, cold, warmth, and pain (Figure 14.3). Places with the most sensory receptors, such as the fingertips and the tip of the

tongue, are the most sensitive. Less sensitive areas, such as the back of the hand, have many fewer receptors.

Several types of **free nerve endings** in the epidermis and many connective tissues detect touch, pressure, heat, cold, or pain. These nerve endings are simple structures. Basically, they are thinly myelinated or unmyelinated ("naked") dendrites of sensory neurons. One type coils around hair follicles and detects the movement of the hair inside. That might be how, for instance, you become aware that a spider is gingerly making its way across your arm. Free nerve endings sensitive to chemicals such as histamine may be responsible for the sensation of itching.

Encapsulated receptors are enclosed in a capsule of epithelial or connective tissue and bear the names of their discoverers. One type, Merkel's discs, adapt slowly and are the most important receptors for steady touch. In the lips, fingertips, eyelids, nipples, and genitals there are many Meissner's corpuscles, which are sensitive to light touching. Deep in the dermis and in joint capsules are Ruffini endings, which respond to steady pressure.

The Pacinian corpuscles widely scattered in the skin's dermis are sensitive to deep pressure and vibrations. They also are located near freely movable joints (like shoulder and hip joints) and in some soft internal organs.

Sensing limb motions and changes in body position relies on mechanoreceptors in skin, skeletal muscles, joints, tendons, and ligaments. Examples include the stretch receptors of muscle spindles described in Section 14.1.

Pain is the perception of bodily injury

Pain is perceived injury to some body region. The body's most important pain receptors are called nociceptors (from the Latin word *nocere*, "to do harm"). Nociceptors are free nerve endings. Several million of them are distributed throughout the skin and in internal tissues, except for the brain.

Somatic pain starts with nociceptors in skin, skeletal muscles, joints, and tendons. One group is the source of prickling pain, like the jab of a pin when you stick your finger. Another contributes to itching or the feeling of warmth caused by chemicals such as histamine. Sensations of *visceral pain*, which is associated with internal organs, are related to muscle spasms, muscle fatigue, too little blood flow to organs, and other abnormal conditions.

When cells are damaged, they release chemicals that activate neighboring pain receptors. The most potent are bradykinins. They trigger the release of histamine, prostaglandins, and other substances associated with inflammation (Section 9.4).

When signals from pain receptors reach interneurons in the spinal cord, the interneurons release a chemical called substance P. One result is that the hypothalamus and midbrain send signals that call for the release of endorphins and enkephalins. These are natural opiates (morphinelike

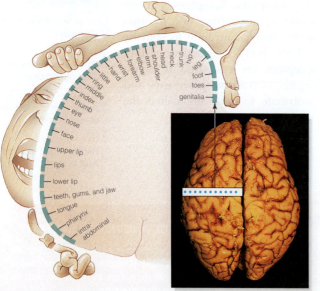

Figure 14.2 The somatosensory cortex "maps" body parts. This strip of cerebral cortex is a little wider than an inch (2.5 centimeters), from the top of the head to just above the ear.
(© Cengage Learning)

Dr. Colin Chumbley/Photo Researchers, Inc.

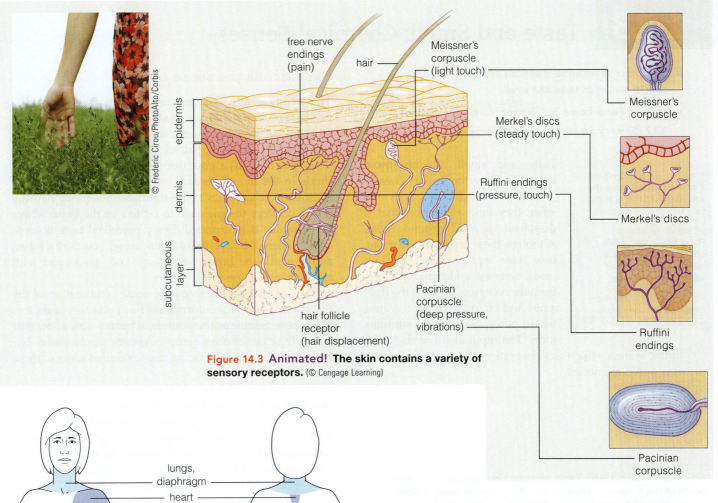

free nerve endings (pain)

hair

Meissner's corpuscle (light touch)

Meissner's corpuscle

Merkel's discs (steady touch)

Merkel's discs

Ruffini endings (pressure, touch)

Ruffini endings

Pacinian corpuscle (deep pressure, vibrations)

Pacinian corpuscle

epidermis

dermis

subcutaneous layer

hair follicle receptor (hair displacement)

© Frederic Cirou/PhotoAlto/Corbis

Figure 14.3 Animated! The skin contains a variety of sensory receptors. (© Cengage Learning)

lungs, diaphragm
heart
stomach
liver, gallbladder
small intestine
ovaries
colon
appendix
urinary bladder
kidney
ureter

Figure 14.4 In referred pain, the brain projects a sensation from an internal organ to an area of the skin. (© Cengage Learning)

substances) that, like morphine derived from opium poppies, reduce our ability to perceive pain. Morphine, hypnosis, and natural childbirth techniques may also stimulate the release of these natural opiates.

Referred pain is a matter of perception

A person's perception of pain often depends on the brain's ability to identify the affected tissue. Get hit in the face with a snowball and you "feel" the contact on facial skin. However, sensations of pain from some internal organs may be wrongly projected to part of the skin surface. This response, called *referred pain*, is related to the way the nervous system is built. Sensory information from the skin and from certain internal organs may enter the spinal cord along the same nerve pathways, so the brain can't accurately identify their source. For example, as shown in Figure 14.4, a heart attack can be felt as pain in skin above the heart and along the left shoulder and arm.

Referred pain is not the same as the *phantom pain* reported by amputees. Often they sense the presence of a missing body part, as if it were still there. In some undetermined way, sensory nerves that were cut during the amputation continue to respond to the trauma. The brain projects the pain back to the missing part, past the healed region.

TAKE-HOME MESSAGE

WHAT ARE SOMATIC SENSATIONS?

- Somatic sensations include touch, pressure, heat and cold, pain, limb motions, and changes in body positions. They are detected by free nerve endings and encapsulated receptors.

- Receptors sensitive to chemicals are responsible for the special senses of taste and smell.

- Links to Facial bones 5.3, Sensory centers in the forebrain 13.7, Sensory areas of the parietal lobe 13.8

chemical senses Senses that detect substances dissolved in fluid that is in contact with chemoreceptors.

olfactory receptors Receptors in the nose's olfactory epithelium that detect water-soluble or vaporized substances.

taste receptors The chemosensors in taste buds.

Taste and smell are **chemical senses**. They begin at chemoreceptors, which are activated when they bind a chemical that is dissolved in fluid around them. Although these receptors wear out, new ones replace them. In both cases, sensory information travels from the receptors through the thalamus and on to the cerebral cortex, where perceptions of the stimulus form. The input also travels to the limbic system, which can integrate it with emotional states and stored memories.

Gustation is the sense of taste

The technical term for taste is *gustation*. Sensory organs called taste buds hold the **taste receptors** (Figure 14.5). About 10,000 taste buds are scattered over your tongue, the roof of your mouth (the palate), and your throat.

A taste bud has a pore through which saliva and other fluids in the mouth contact the surface of receptors. The stimulated receptor in turn stimulates a sensory neuron, which conveys the message to centers in the brain where the stimulus is interpreted. Every perceived taste is some combination of five primary tastes: sweet, sour, salty, bitter, and umami (the brothy or savory taste associated with meats or aged cheese).

The flavors of most foods are some combination of the five basic tastes, plus information from olfactory receptors in the nose. Simple as this sounds, scientists now know that our taste sense involves complex genetic mechanisms. *Science Comes to Life* on the facing page examines some of these findings.

The olfactory element of taste is extremely important. In addition to odor molecules in inhaled air, molecules of volatile chemicals are released as you chew food. These waft up into the nasal passages. There, the "smell" inputs contribute to the perception of complex flavors. This is why anything that dulls your sense of smell—such as a head cold—also seems to diminish food's flavor.

Olfaction is the sense of smell

Olfactory receptors (Figure 14.6) detect water-soluble or easily vaporized substances. When odor molecules bind to receptors on olfactory neurons in cells of the nose's olfactory epithelium, the resulting nerve impulse travels directly to olfactory bulbs in the frontal area of the brain. There, other neurons forward the message to a center in the cerebral cortex, which interprets it as "fresh bread," "pine tree," or some other substance.

From an evolutionary perspective, olfaction is an ancient sense—and for good reason. Food, potential mates, and predators give off substances that can diffuse through air (or water) and so give clues or warnings of their whereabouts. Even with our rather insensitive sense of smell, we humans have about 10 million olfactory receptors in patches of olfactory epithelium in the upper nasal passages.

Just inside your nose, next to the vomer bone (Section 5.3), is a *vomeronasal organ*, or "sexual nose." (Some other mammals also have one.) Receptors in this tiny organ

Figure 14.5 Animated! Taste receptors are present in several areas of the tongue. A Taste buds. There may be several types of receptors in a taste bud; the diagram in B shows areas where receptors for a given "taste" dominate. (© Cengage Learning)

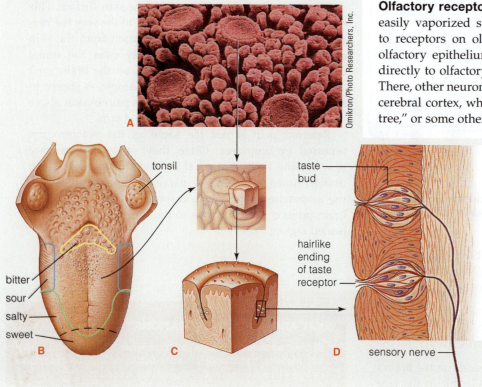

Omikron/Photo Researchers, Inc.

A

tonsil

bitter
sour
salty
sweet

B

taste bud

C

hairlike ending of taste receptor

D

sensory nerve

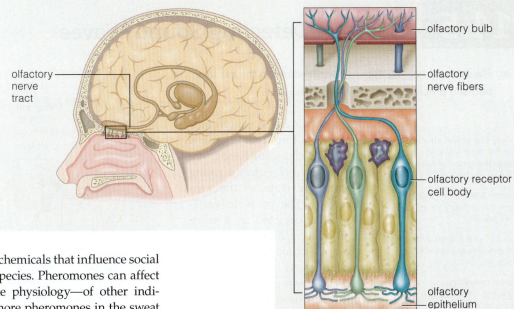

Figure 14.6 Animated!
A sensory pathway leads from olfactory receptors in the nose to primary receiving centers in the brain.
(© Cengage Learning)

olfactory nerve tract

olfactory bulb

olfactory nerve fibers

olfactory receptor cell body

olfactory epithelium

detect pheromones, which are chemicals that influence social interactions in many animal species. Pheromones can affect the behavior—and maybe the physiology—of other individuals. For instance, one or more pheromones in the sweat of females may account for the common observation that women of reproductive age who are in regular, close contact with one another often come to have their menstrual periods on a similar schedule. Many scientists are not convinced that pheromones operate in humans, however, and debate on the topic is always lively!

TAKE-HOME MESSAGE

WHAT ARE CHEMICAL SENSES?

- Taste and smell (olfaction) are the human chemical senses.
- Taste depends on receptors in taste buds in the tongue. The receptors bind molecules dissolved in fluid. The five primary tastes are sweet, sour, salty, bitter, and umami.
- Olfaction relies on receptors in patches of epithelium in the upper nasal passages. Olfactory neurons send signals directly to the olfactory bulbs in the brain.

14.4 Tasty Science

SCIENCE COMES TO LIFE

Taste buds help make eating one of life's pleasures. So how do the sensory receptors in our taste buds distinguish the tastes in different foods?

Each taste category such as sweet or sour is associated with particular "tastant" molecules. When you eat food, however, which taste category (or combination of them) you ultimately perceive depends on the nature of the triggering chemical and on how it is processed by the receptor. In each case, some event causes the receptor cell to release a neurotransmitter that triggers nerve impulses in a nearby sensory neuron.

For example, when you taste "salt," the receptor cell's response is due to the flow of Na^+ through sodium ion channels in its plasma membrane. Acidic tastant molecules release hydrogen ions that block certain ion channels. The blockage causes a receptor to respond with a "sour" message.

Cells that detect bitter substances may have receptors sensitive to as many as one hundred different trigger tastants. This diversity probably is a survival tool. Many toxic chemicals (including plant alkaloids such as nicotine and morphine) taste bitter, an adaptation that may help protect us from ingesting dangerous substances. Familiar bitter-tasting alkaloids are caffeine and quinine, the mouth-

puckering tastant in tonic water. And while many "sweet" tastants are sugars, others are amino acids or alcohols. Both bitter and sweet tastes are detected by specific proteins inside the receptor. The taste category called umami also is triggered by amino acids, notably glutamate. Its name was bestowed by the Japanese researcher who identified it.

Each taste bud has receptors that can respond to tastants in at least two—and in some cases all five—of the taste classes. Various tastants commingle (together with odors) into our perceptions of countless flavors.

Not all taste receptors are equally sensitive. "Bitter" ones tend to be extremely sensitive and so can detect tiny amounts of bitter tastants—and thus potential poisons. Sour tastants are needed in higher concentrations before the stimulus registers. Even higher levels of sweet and salty substances must be present for the stimulus to register. So why can relatively small amounts of artificial sweeteners so readily sweeten foods? Their molecular characteristics make them 150 times (aspartame) to more than 600 times (saccharin) as potent as plain sucrose.

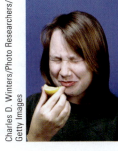

Charles D. Winters/Photo Researchers/Getty Images

Hearing: Detecting Sound Waves

- **The sense of hearing depends on structures in the ear that trap and process sounds traveling through air.**

cochlea The coiled structure in the inner ear that contains the organ of hearing (organ of Corti).

hair cells Mechanoreceptors that are the sensory receptors for sound.

organ of Corti Organ in the ear where sensory hair cells are located.

tectorial membrane The jellylike structure that bending hair cells press against in response to pressure waves in the cochlear fluid.

tympanic membrane The eardrum.

Sounds are waves of compressed air. They are a form of mechanical energy. If you clap your hands, you force out air molecules, creating a low-pressure state in the area they vacated. The pressure variations can be depicted as a wave form, and the *amplitude* of its peaks corresponds to loudness. The *frequency* of a sound is the number of wave cycles per second. Each cycle extends from the start of one wave to the start of the next (Figure 14.7).

The sense of hearing starts with vibration-sensitive mechanoreceptors deep in the ear. When sound waves travel down the ear's auditory canal, they reach a membrane and make it vibrate. The vibrations cause a fluid inside the ear to move, the way water in a waterbed sloshes. In your ear, the moving fluid bends the tips of hairs on mechanoreceptors. With enough bending, the result will be action potentials sent to the brain, where they are interpreted as sound.

The ear gathers "sound signals"

A human ear has three regions (Figure 14.8A), each with its own role in hearing. The *outer ear* is a pathway for sound waves to enter the ear, setting up vibrations. The vibrations are amplified in the *middle ear*. The *inner ear* contains the coiled **cochlea** (KAHK-lee-uh; Figure 14.8B), where vibrations of different sound frequencies are sorted out as they stimulate different patches of receptors. The inner ear also contains *semicircular canals*, which are involved in balance (Section 14.6).

Sensory hair cells are the key to hearing

Hearing begins when the outer ear's fleshy flaps collect and channel sound waves through the auditory canal to the

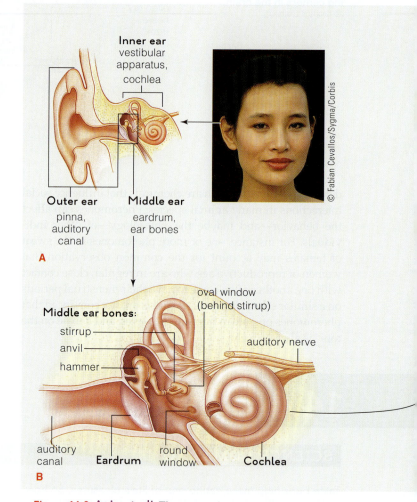

Figure 14.8 Animated! The ear gathers sound waves and converts them to nerve impulses. (© Cengage Learning)

tympanic membrane (the eardrum). Sound waves cause the membrane to vibrate, which in turn causes vibrations in a leverlike array of three tiny bones of the middle ear: the *malleus* ("hammer"), *incus* ("anvil"), and stirrup-shaped *stapes*. The vibrating bones transmit their motion to the *oval window*, an elastic membrane over the entrance to the cochlea. The oval window is much smaller than the tympanic membrane. So, as the middle-ear bones vibrate against its small surface with the full energy that struck the tympanic membrane, the force of the original vibrations is amplified.

Now the action shifts to the cochlea. If we could uncoil the cochlea, we would see that a fluid-filled chamber folds around an inner *cochlear duct* (Figure 14.8C). Each "arm" of the outer chamber functions as a separate compartment (the *scala vestibuli* and *scala tympani*, respectively). The amplified vibrations of the oval window create pressure waves in the fluid within the chambers. These waves are transmitted to the fluid in the cochlear duct. On the floor of the cochlear duct is a *basilar membrane*, and resting on the basilar membrane

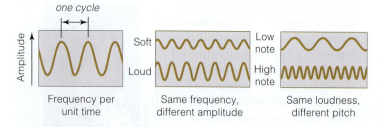

Figure 14.7 Sound travels in the form of a wave. (© Cengage Learning)

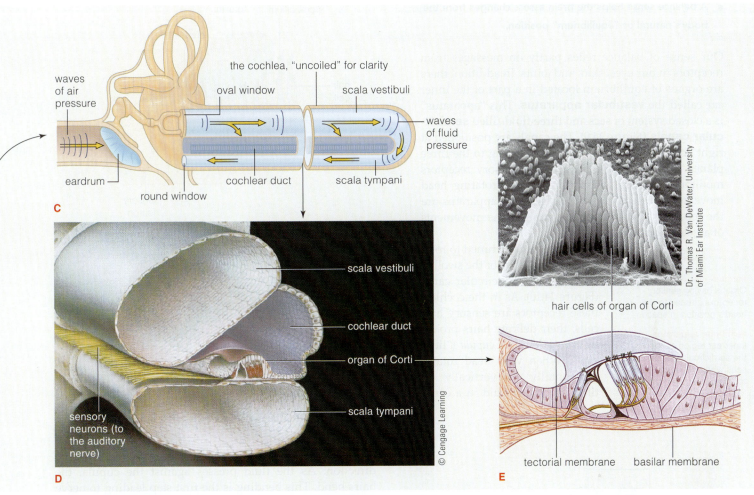

the cochlea, "uncoiled" for clarity

waves of air pressure

oval window

scala vestibuli

waves of fluid pressure

eardrum

round window

cochlear duct

scala tympani

C

scala vestibuli

cochlear duct

organ of Corti

sensory neurons (to the auditory nerve)

scala tympani

© Cengage Learning

D

Dr. Thomas R. Van DeWater, University of Miami Ear Institute

hair cells of organ of Corti

tectorial membrane

basilar membrane

E

is a specialized **organ of Corti**, which includes **hair cells**. These cells are the mechanoreceptors that serve as the sensory receptors for sound.

Slender projections at the tips of hair cells rest against an overhanging **tectorial** ("rooflike") **membrane**, which is not a membrane at all but a jellylike structure. When pressure waves in the cochlear fluid vibrate the basilar membrane, its movements can press hair cell projections against the tectorial membrane so that the projections bend like brush bristles. Affected hair cells release a neurotransmitter. It triggers action potentials in neurons of the auditory nerve, which carries them to the brain.

Different sound frequencies cause different parts of the basilar membrane to vibrate—and, accordingly, to bend different groups of hair cells. Apparently, the total number of hair cells stimulated in a given region determines the loudness of a sound. The perceived tone or "pitch" of a sound depends on the frequency of the vibrations that excite different groups of hair cells. The higher the frequency, the higher the pitch.

Eventually, pressure waves moving through the cochlea push against the *round window*, a membrane at the far end of the cochlea. As the round window bulges outward toward

the air-filled middle ear, it serves as a "release valve" for the force of the waves. Air also moves through an opening in the middle ear into the *eustachian tube*. This tube runs from the middle ear to the throat (pharynx), permitting air pressure in the middle ear to be equalized with the pressure of outside air. When you change altitude (say, during a plane trip), this equalizing process makes your ears pop.

Sounds such as amplified music and the thundering of jet engines are so intense that long-term exposure to them can permanently damage the inner ear (Section 14.7). Evolution has not equipped hair cells of the human ear to cope with such extremely loud, modern-day sounds.

TAKE-HOME MESSAGE

WHAT IS THE SENSE OF HEARING?

- Hearing is the perception of sound waves that are detected by mechanoreceptors in the ears.
- Hair cells in the inner ear are the receptors for sound.
- Hair cells are attached to membranes inside the cochlea.
- Pressure waves generated by sound cause membrane vibrations that bend hair cells. The bending produces nerve impulses in neurons of the auditory nerve.

Balance: Sensing the Body's Natural Position

■ A balance sense helps the brain assess changes from the body's natural or "equilibrium" position.

Our sense of balance relies partly on messages from receptors in our eyes, skin, and joints. In addition, there are organs of equilibrium located in a part of the inner ear called the **vestibular apparatus**. This "apparatus" is a closed system of sacs and three fluid-filled **semicircular canals** (Figure 14.9). The canals are positioned at right angles to one another, corresponding to the three planes of space. Inside them, some sensory receptors monitor dynamic equilibrium—that is, rotating head movements. Elsewhere in the vestibular apparatus are the receptors that monitor the straight-line movements of acceleration and deceleration.

semicircular canals
Organs having sensory receptors attuned to the head's position in space.

vestibular apparatus
Inner ear region containing the semicircular canals plus sacs where receptors monitor straight-line movements.

The receptors attuned to rotation are on a ridge of the swollen base of each semicircular canal (Figure 14.10). As in the cochlea, these receptors are sensory hair cells; their delicate hairs project up into a jellylike *cupula* ("little cap"). When your head rotates horizontally or vertically or tilts diagonally, fluid in a canal

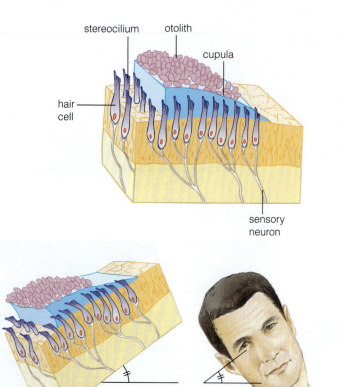

Figure 14.10 **Otoliths move when the head tilts.**
(© Cengage Learning)

corresponding to that direction moves in the opposite direction. As the fluid presses against the cupula, the hairs bend. This bending is the first step leading to nerve impulses that travel to the brain—in this case, along the vestibular nerve.

The receptors attuned to the head's position in space are located in two fluid-filled sacs in the vestibular apparatus, the utricle and saccule shown in Figure 14.9. Each sac contains an *otolith* organ, which has hair cells embedded in a jellylike "membrane." The material also contains hard bits of calcium carbonate called otoliths ("ear stones"). Movements of the membrane and otoliths signal changes in the head's orientation relative to gravity, as well as straight-line acceleration and deceleration. For example, if you tilt your head, the otoliths slide in that direction, the membrane mass shifts, and tips of the hair cells bend (Figure 14.10).

THINK OUTSIDE THE BOOK

A common disorder of the vestibular apparatus is called vertigo. It produces a sensation that the surroundings are spinning or whirling. A viral infection, head injury, or other conditions can cause vertigo. Research this topic on the Web or in the library. Is vertigo differerent from dizziness? How is the disorder treated?

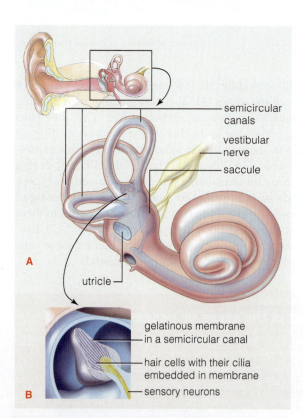

A

B

semicircular canals

vestibular nerve

saccule

utricle

gelatinous membrane in a semicircular canal

hair cells with their cilia embedded in membrane

sensory neurons

Figure 14.9 Animated! **The vestibular apparatus is an organ of equilibrium.** (© Cengage Learning)

Figure 14.11 Riding in a rollercoaster gives the vestibular apparatus a workout. It must process rapidly changing signals from the eyes, otoliths, and muscles. In some people a rollercoaster ride produces motion sickness leading to nausea or even vomiting.

The otoliths also press on hair cells if your head accelerates, as when you start running or are riding in an accelerating vehicle.

Nerve impulses from the vestibular apparatus travel to reflex centers in the brain stem. As the signals are processed along with information from your muscles and eyes, the brain orders compensating movements that help you keep your balance when you stand, walk, dance, or move your body in other ways. It also helps a you maintain "a sense of where you are" when you turn a cartwheel or ride in a rollercoaster (Figure 14.11).

Motion sickness can result when extreme or continuous motion overstimulates hair cells in the balance organs. It can also be caused by conflicting signals from the ears and eyes about motion or the head's position. If you are prone to motion sickness, you know all too well that nerve impulses triggered by the sensory input can reach a brain center that governs the vomiting reflex.

Although the hearing apparatus of our ears is remarkably sturdy, a variety of illnesses and injuries can damage it.

Children have short eustachian tubes, so they especially are susceptible to **otitis media**—a painful inflammation of the middle ear that usually is caused by the spread of a respiratory infection such as a cold. An antibiotic is the usual treatment, although resistant infections are now common. In some cases pus and fluid can build up and cause the eardrum to tear. The rupture usually will heal on its own.

Ear infections, taking lots of aspirin, and genetic factors can cause the ringing, whirring, or buzzing in the ears known as **tinnitus**. While the condition is not a serious health threat, it can be extremely annoying.

Deafness is the partial or complete inability to hear. Some people suffer from congenital (inborn) deafness, and in other cases aging, disease, or environmentally caused damage is the culprit. About one-third of adults in the United States will suffer significant hearing loss by the time they are 65. Researchers believe that most cases of this progressive deafness are due to the long-term effects of living in a noisy world.

The loudness of a sound is measured in decibels. A quiet conversation occurs at about 50 decibels. Rustling papers make noise at a mere 20 decibels. The delicate sensory hair cells in the inner ear (Figure 14.12) begin to be damaged when a person is exposed to sounds louder than about 75–85 decibels over long periods. Some MP3 players can crank out sound at well over 100 decibels. At 130 decibels—typical of a rock concert or shotgun blast—permanent damage can occur much more quickly. Protective earwear is a must for anyone who regularly operates noisy equipment or who works around noisy machinery such as aircraft.

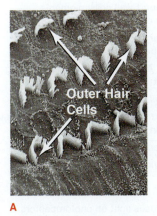
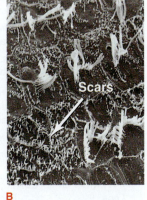

A   B

Figure 14.12 Noise is a danger to the ear's hair cells. A Healthy sensory hair cells of the inner ear. **B** Hair cells damaged by exposure to loud noise. (Robert E. Preston, courtesy Joseph E. Hawkins, Kresge Hearing Research Institute, University of Michigan Medical School)

14.8 Vision: An Overview

■ Vision requires a system of photoreceptors and brain centers that can receive and interpret the patterns of nerve impulses.

accommodation
Adjustments of the eye's lens that focus light precisely on the retina.

cornea The transparent structure that covers the eye's iris.

eyes Sensory organs that contain arrays of photoreceptors.

iris The pigmented eye region; light enters through the pupil in the iris.

lens Eye structure that focuses light on the retina.

retina A layer of tissue that contains the eye's photoreceptors.

vision Awareness of the characteristics of visual stimuli.

visual cortex Brain region that receives nerve impulses from the optic nerve.

The sense of **vision** is an awareness of the position, shape, brightness, distance, and movement of visual stimuli. Our **eyes** are sensory organs that contain tissue with a dense array of photoreceptors.

The eye is built to detect light

The eye has three layers (Table 14.2), sometimes called "tunics." The outer layer consists of a sclera and a transparent **cornea**. The middle layer consists mainly of a choroid, ciliary muscle, and iris. The key feature of the inner layer is the retina (Figure 14.13).

The *sclera* is the dense, fibrous "white" of the eye. It protects most of the eyeball, except for the region formed by the cornea. Moving inward, the thin, darkly pigmented

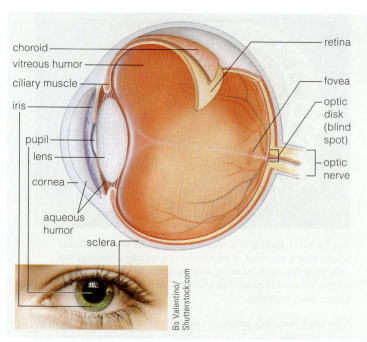

Figure 14.13 Animated! The eye is specialized to receive light and focus it on photoreceptors. (© Cengage Learning)

choroid lies under the sclera. It prevents light from scattering inside the eyeball and contains most of the eye's blood vessels.

Behind the transparent cornea is the round, pigmented **iris** (after *irid*, which means "colored circle"). The iris has more than 250 measurable features (such as pigments and fibrous tissues). This is why, as you read in the chapter introduction, the iris can be used for identification. Look closely at someone's eye, and you will see a "hole" in the center of the iris. This *pupil* is the entrance for light. When bright light hits the eye, circular muscles in the iris contract and shrink the pupil. In dim light, radial muscles contract and enlarge the pupil.

Behind the iris is a saucer-shaped **lens**, with onionlike layers of transparent proteins. Ligaments attach the lens to smooth muscle of the *ciliary body*; this muscle functions in focusing light, as we will see shortly. The lens focuses incoming light onto a dense layer of photoreceptor cells behind it, in the retina. A clear fluid, *aqueous humor* (body fluids were once called "humors"), bathes both sides of the lens. A jellylike substance (*vitreous humor*) fills the chamber behind the lens.

The **retina** is a thin layer of neural tissue at the back of the eyeball. It has a pigmented basement layer that covers the choroid. Resting on the basement layer are densely packed photoreceptors that are linked with a variety of neurons. Axons from some of these neurons converge to form the optic nerve at the back of the eyeball. The optic nerve is the trunk line to the thalamus—which sends

TABLE 14.2 Parts of the Eye	
WALL OF EYEBALL	**(three layers)**
Sensory tunic (inner layer)	*Retina.* Absorbs, transduces light energy
	Fovea. Increases visual acuity
Vascular tunic (middle layer)	*Choroid.* Blood vessels nutritionally support wall cells; pigments prevent light scattering
	Ciliary body. Muscles control lens shape; fine fibers hold lens upright
	Iris. Adjusting iris controls incoming light
	Pupil. Serves as entrance for light
	Start of optic nerve. Carries signals to brain
Fibrous tunic (outer layer)	*Sclera.* Protects eyeball
	Cornea. Covers iris; focuses light
INTERIOR OF EYEBALL	
Lens	Focuses light on photoreceptors
Aqueous humor	Transmits light, maintains pressure
Vitreous body	Transmits light, supports lens and eyeball

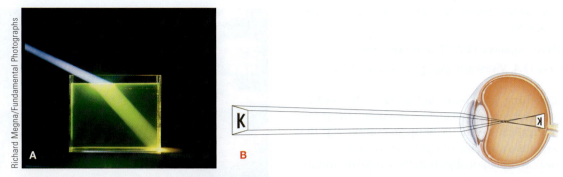

Figure 14.14 **Light entering the eye bends as it travels toward the retina. A** How light can bend. **B** How light rays reverse as they travel toward the retina. The pattern of light rays that converge on the retina is upside-down and reversed left to right. (© Cengage Learning)

signals on to the **visual cortex** in the brain. The place where the optic nerve exits the eye is a "blind spot" because there are no photoreceptors there.

The surface of the cornea is curved. This means that incoming light rays hit it at different angles and, as they pass through the cornea, their trajectories (paths) bend (Figure 14.14A). There, because of the way the rays were bent at the curved cornea, the rays converge at the back of the eyeball. They stimulate the retina in a pattern that is upside-down and reversed left to right relative to the source of the light rays. Figure 14.14B gives a simplified diagram of this process. The brain corrects the "upside-down and backwards" orientation.

Eye muscle movements fine-tune the focus

Light rays from sources at different distances from the eye strike the cornea at different angles. As a result, they will be focused at different distances behind it, and adjustments must be made so that the light will be focused precisely on the retina. Normally, the lens can be adjusted so that the focal point coincides exactly with the retina. A ciliary muscle adjusts the shape of the lens. As you can see in Figure 14.15, the muscle encircles the lens and attaches to it by ligaments. When the muscle contracts, the lens bulges, so the focal point moves closer. When the muscle relaxes, the lens flattens, so the focal point moves farther back. Adjustments like these are called **accommodation**. If they are not made, rays from distant objects will be in focus at a point just in front of the retina, and rays from very close objects will be focused behind it.

Sometimes the lens can't be adjusted enough to place the focal point on the retina. Sometimes also, the eyeball is not shaped quite right. The lens is too close to or too far away from the retina, so accommodation alone cannot produce a precise match. Eyeglasses or contact lenses can correct these problems, which we will consider more fully in Section 14.10.

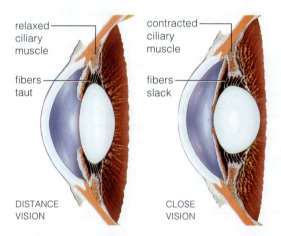

relaxed ciliary muscle

contracted ciliary muscle

fibers taut

fibers slack

DISTANCE VISION

CLOSE VISION

Figure 14.15 **Adjusting the lens focuses light on the retina.** Adjustments of the ciliary muscle focus light from near or distant sources on the retina by changing the tension of fibers that ring the lens. (© Cengage Learning)

TAKE-HOME MESSAGE

HOW DO THE EYES ALLOW PHOTORECEPTION— THE DETECTION OF LIGHT?

- In the eye's outer layer, the curved cornea covers the iris and helps focus incoming light.
- In the middle layer, the choroid prevents light scattering, the iris controls incoming light, and the ciliary body and lens aid in focusing light on photoreceptors.
- Photoreception occurs in the retina of the inner layer. Adjustments in the position or shape of the lens focus incoming visual stimuli onto the retina.

14.9   From Visual Signals to "Sight"

- Our vision sense is based on the sensory pathway from the retina to the brain.

- Links to Nerve impulses 13.2, Chemical synapses 13.3, Nerves 13.4, Visual processing in the brain 13.8

"Seeing" something is a multistep process that begins when your eyes receive raw visual information. The information then is transmitted to the brain and processed. The result is conscious awareness of light and shadows, of colors, and of near and distant objects in the world around us.

Rods and cones are the photoreceptors

Vision begins when light reaches the retina, at the back of the eyeball. Between the retina and the choroid is a layer of epithelium where visual pigments form. Millions of photoreceptors called **rod cells** and **cone cells** rest on this layer (Figure 14.16 and Table 14.3) and have visual pigments embedded in them. Rod cells are sensitive to dim light. They detect changes in light intensity across the visual field. Their signals are the start of coarse perception of motion. Cone cells detect bright light. Their signals are the start of sharp daytime vision and color perception.

cone cells Photoreceptors that detect bright light.

fovea Area near the center of the retina where visual acuity (sharpness) is the greatest.

rod cells Photoreceptors that detect dim light.

Visual pigments intercept light energy

Like sound, light energy travels in waves, and different light wavelengths correspond to different colors. As you can see in the lower part of Figure 14.16, there are stacks of membrane disks in the light-sensitive part of rods and cones. These disks are where visual pigments are found.

Visual pigments are proteins that change shape when they absorb certain wavelengths, or colors, of light. They consist of different versions of a protein called opsin together with retinal, a light-absorbing substance that is

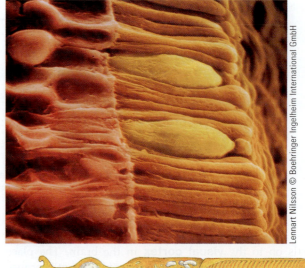

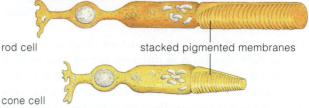

rod cell   stacked pigmented membranes

cone cell

Figure 14.16 Rods and cones contain visual pigments.
(© Cengage Learning)

derived from vitamin A. Rods contain a single type of visual pigment, called rhodopsin. It absorbs mainly blue to green light. By contrast, depending on the type of opsin in its pigment, a cone may be sensitive to red, green, or blue light. Thus we say there there are three types of cones—red, green, or blue.

Changes in visual pigments are key to our vision sense. When light stimulates a visual pigment, its opsin changes shape. The change begins a process that converts light energy to nerve impulses. In this process, a series of chemical reactions slow the release of a neurotransmitter that inhibits neurons next to the photoreceptor. When they are no longer inhibited, the neurons start sending signals about the visual stimulus on toward the brain. So-called night blindness results when a person's diet is deficient in vitamin A, so too little retinal is available to form visual pigments. The effect is most severe in rods.

Near the center of the retina is a tiny depression called the **fovea** (Figure 14.17). It is packed with cones. As a result, visual acuity, the ability to discriminate between two objects, is greatest there. For example, the fovea's dense cluster of cones enables you to distinguish between neighboring points in space—like the *e* and the period at the end of this sentence.

TABLE 14.3 Rods and Cones Compared

Cell Type	Sensitive To	Related Perception
Rod	Dim light	Coarse perception of movement
Cone	Bright light	Daytime vision and perception of color

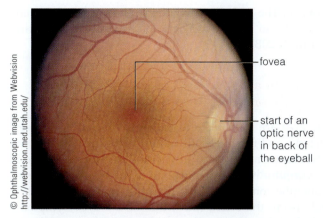

— fovea

— start of an optic nerve in back of the eyeball

Figure 14.17 **The fovea contains densely packed rods and cones.** This image shows the location of the fovea and the start of the optic nerve.

The retina begins processing visual signals

In an early embryo, its retinas arise from its developing brain. As a result, anatomically speaking, the retina is an extension of the brain. Perhaps it is not surprising, then, that cells in the retina process visual signals before they are sent on to the brain's vision centers.

Neurons in the eye are organized in layers above the rods and cones. As you can see in Figure 14.18, signals flow from rods and cones to *bipolar* interneurons, then to interneurons called *ganglion cells*. Signals also travel to *horizontal* cells and *amacrine* cells. These neurons jointly strengthen or weaken the signals before they reach ganglion cells. The axons of ganglion cells form the two optic nerves to the brain.

Signals move on to the visual cortex

The part of the outside world you actually see is called the "visual field." The right side of each retina intercepts light from the left half of the visual field and the left side intercepts light from the right half. As you can see in Figure 14.19, signals from each eye "criss-cross." The optic nerve leading out of each eye delivers signals from the left visual field to the right cerebral hemisphere, and signals from the right go to the left hemisphere.

Axons of the optic nerves end in an island of gray matter in the cerebrum (the lateral geniculate nucleus). Its layers each have a map corresponding to receptive fields of the retina. Each map's interneurons deal with one aspect of a visual stimulus—its form, movement, depth, color, texture, and so on. After initial processing all the visual signals travel rapidly, at the same time, to different parts of the visual cortex. There, final processing produces the sensation of sight.

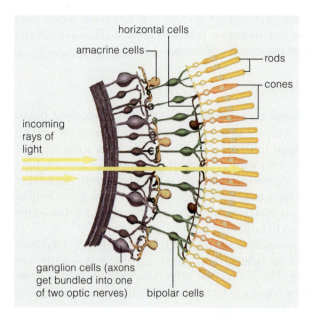

horizontal cells
amacrine cells
rods
cones

incoming rays of light

ganglion cells (axons get bundled into one of two optic nerves)
bipolar cells

Figure 14.18 **Animated!** **Photoreceptors connect with sensory neurons in the retina.** (© Cengage Learning)

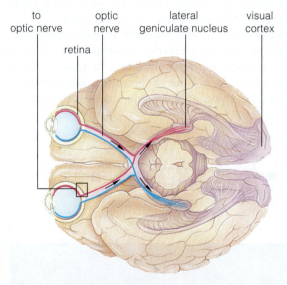

to optic nerve
optic nerve
lateral geniculate nucleus
visual cortex
retina

Figure 14.19 **Sensory signals criss-cross as they travel from the retina to the brain.** (© Cengage Learning)

TAKE-HOME MESSAGE

WHAT ARE THE EYE'S PHOTORECEPTORS?

- Rods and cones in the retina are the eye's photoreceptors.
- Rods detect dim light. Cones detect bright light and provide our sense of color.
- The eye analyzes information on the distance, shape, brightness, position, and movement of a visual stimulus.
- Visual signals move through layers of neurons in the retina before moving on to the brain.

Disorders of the Eye

Problems that disrupt normal eye functions range from injuries and diseases to inherited abnormalities and natural changes associated with aging.

Some eye disorders are inherited

Red-green color blindness is a common inherited abnormality. It shows up most often in males, for reasons you can read about in Chapter 20. The retina lacks some or all of the cone cells with pigments that normally respond to light of red or green wavelengths. Most of the time, color-blind people have trouble distinguishing red from green only in dim light. However, some cannot distinguish between the two even in bright light.

Occasionally, some or all of the cone cells that selectively respond to light of red, green, or blue wavelengths are missing. The rare people who have only one of the three kinds of cones are totally color-blind. They see the world only in shades of gray.

Some inherited vision problems are due to misshapen eye parts that affect the eye's ability to focus light. In **astigmatism**, one or both corneas curve unevenly, so they can't bend incoming light rays to the same focal point.

In **myopia**, or nearsightedness, the eyeball is wider than it is high, or the ciliary muscle responsible for adjusting the lens contracts too strongly. Then, images of distant objects are focused in front of the retina instead of on it (Figure 14.20A). **Hyperopia**, farsightedness, is the opposite

problem. The eyeball is "taller" than it is wide (or the lens is "lazy"), so close images are focused behind the retina (Figure 14.20B).

The eyes also are vulnerable to infections and cancer

The eyes are vulnerable to pathogens including viruses, bacteria, and fungi. Health authorities estimate that in the U.S., about one in every fifty visits to a doctor's office is for **conjunctivitis**, inflammation of the transparent membrane (the conjunctiva) that lines the inside of the eyelids and covers the sclera (the white of the eye). Symptoms include redness, discomfort, and a discharge. In children, conjunctivitis usually is caused by bacteria; in adults it more often is triggered by allergy (Figure 14.21). Most cases of bacterial conjunctivitis are easily treated with antibiotics.

Herpes simplex, a virus that causes cold sores and genital herpes, can infect the cornea. Because blindness can result from a **herpes infection** in the eyes, a pregnant woman who has a history of genital herpes likely will deliver by cesarean section to avoid any chance of exposing her newborn to the virus.

Malignant melanoma is the most common eye cancer. It typically develops in the choroid (the eye's middle layer) and may not trigger noticeable vision problems until it has spread to other parts of the body. About 1 in 20,000 babies is born with **retinoblastoma**, a cancer of the retina. Because

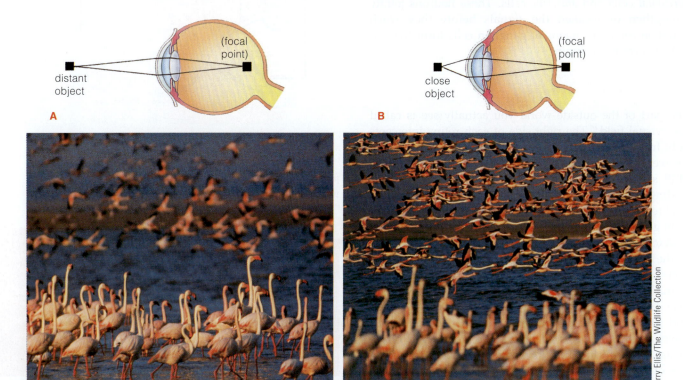

distant object   (focal point)

A

close object   (focal point)

B

Gerry Ellis/The Wildlife Collection

Figure 14.20 A Nearsighted and B farsighted vision are the most common vision problems. (© Cengage Learning)

it can spread along the optic nerve to the brain, the affected eye often is removed surgically. If both eyes are involved, radiation therapy may be used to try to save one of them.

Aging increases the risk of some types of eye disorders

Clouding of the eye's lens, or **cataracts**, is associated with aging, although an injury or diabetes can also cause them to develop. The underlying change may be an alteration in the structure of transparent proteins that make up the lens. This change in turn may scatter incoming light rays (Figure 14.22A). If the lens becomes totally opaque, no light can enter the eye.

Even a normal lens loses some of its natural flexibility as we grow older. This normal stiffening is why people over 40 years old often must start wearing eyeglasses.

In **macular degeneration** part of the retina breaks down and is replaced by scar tissue that results in a "blind spot" (Figure 14.22B). Most cases of macular degeneration are related to advancing age. Treatment is difficult unless the problem is detected early. **Glaucoma** results when too much aqueous humor builds up in the eyeball. Blood vessels that service the retina collapse under the increased pressure, and vision deteriorates as blood-starved neurons of the retina and optic nerve die. Although chronic glaucoma often is associated with advanced age, the problem really starts in a person's middle years. If detected early, the fluid pressure can be relieved by drugs or surgery before the damage becomes severe.

Medical technologies can remedy some vision problems and treat eye injuries

Today many different procedures are used to correct eye disorders. In *corneal transplant surgery*, a defective cornea is removed; then an artificial cornea (made of clear plastic) or a natural cornea from a donor (a cadaver) is stitched in place. Within a year, the patient is fitted with eyeglasses or contact lenses. Similarly, cataracts often can be surgically

Figure 14.22 Cataracts and macular degeneration obscure vision. A Cataracts produce overall fuzzy vision. **B** Macular degeneration causes a blind spot in the center of the visual field.

corrected by removing the lens and replacing it with an artificial one.

Severely nearsighted people may opt for procedures that eliminate the need for corrective lenses. So-called "Lasik" (for laser-assisted in situ keratomilieusis) and "lasek" (for laser-assisted subepithelial keratectomy) use a laser to reshape the cornea. All or part of the surface of the cornea is peeled back and then replaced into position after the defect being treated is corrected. *Conductive keratoplasty* (CK) uses radio waves to reshape the cornea and bring near vision back into focus.

Retinal detachment is the eye injury we read about most often. It may follow a blow to the head or an illness that tears the retina. As the jellylike vitreous body oozes through the torn region, the retina lifts away from the underlying choroid. In time it may leave its blood supply behind. Early symptoms include blurred vision, flashes of light that occur in the absence of outside stimulation, and loss of peripheral vision. Without medical help, the person may become totally blind in the damaged eye.

A detached retina may be treatable with *laser coagulation*, a painless technique in which a laser beam seals off leaky blood vessels and "spot welds" the retina to the underlying choroid.

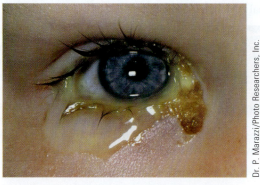

Figure 14.21 Conjunctivitis may be due to a bacterial infection or an allergy.

EXPLORE
ON YOUR OWN

As Section 14.10 described, there are various forms of color blindness.
Figure 14.23 shows simple tests, called Ishihara plates, which are standardized tests for different forms of color blindness. For instance, you may have one form of red-green color blindness if you see the numeral "7" instead of "29" in the circle in part A. You may have another form if you see a "3" instead of an "8" in the circle in part B.

If you do this exercise and have questions about your color vision, visit your doctor to determine whether additional testing is in order.

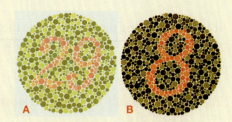

Figure 14.23 Color blindness tests.
(© Cengage Learning)

SUMMARY

Section 14.1 A stimulus is a form of energy that the body detects by means of sensory receptors. A sensation is a conscious awareness that stimulation has occurred. Perception is understanding what the sensation means.

Sensory receptors are endings of sensory neurons or specialized cells next to them. They respond to stimuli, which are specific forms of energy, such as mechanical pressure and light.

Mechanoreceptors detect mechanical energy that is associated with changes in pressure (e.g., sound waves), changes in position, or acceleration.

Thermoreceptors detect the presence of or changes in radiant energy from heat sources.

Nociceptors (pain receptors) detect tissue damage. Their signals are perceived as pain.

Chemoreceptors detect chemical substances that are dissolved in the body fluids around them.

Osmoreceptors detect changes in water volume (hence solute concentrations) in the surrounding fluid.

Photoreceptors detect light.

A sensory system has receptors for specific stimuli and nerve pathways from those receptors to processing centers in the brain. The brain assesses each stimulus based on which nerve pathway is delivering the signals, how often signals are traveling along each axon of the pathway, and the number of axons that were recruited into action. In sensory adaptation, the response to a stimulus decreases.

The special senses include taste, smell, hearing, balance, and vision. The receptors associated with these senses are in sense organs or another specific body region.

Section 14.2 Somatic sensations include touch, pressure, pain, temperature, and muscle sense. Receptors associated with these sensations occur in various parts of the body. Their signals are processed in the somatosensory cortex of the brain. The simplest receptors, including those for temperature and pain, are free nerve endings in the skin or internal tissues. Some somatic sensations arise when encapsulated receptors respond to stimuli.

Section 14.3 Taste and smell are chemical senses. Their sensory pathways travel from chemoreceptors to processing regions in the cerebral cortex and limbic system. Taste buds in the tongue and mouth contain the taste receptors. The sense of smell relies on olfactory receptors in patches of epithelium in the upper nasal passages.

Sections 14.5, 14.6 The sense of hearing requires parts of the outer, middle, and inner ear that collect, amplify, or respond to sound waves that vibrate the tympanic membrane (eardrum). The vibrations are transferred to fluid in the cochlea of the inner ear, where they in turn vibrate the tectorial membrane. The moving fluid bends sensory hair cells in the organ of Corti. The bending triggers nerve impulses that travel to the brain via the auditory nerve.

Balance organs are located in the vestibular apparatus of the inner ear. Sensory receptors in these semicircular canals (including hair cells) respond to gravity, velocity, acceleration, and other factors that affect body positions and movements.

Section 14.8 Eyes are the sensory organs associated with the sense of vision. Key eye structures include the cornea and lens, which focus light; the iris, which adjusts incoming light; and the retina, which contains photoreceptors (rods and cones). The optic nerve at the back of the eyeball transmits visual signals to the visual cortex in the brain.

Section 14.9 The rod cells and cone cells detect dim and bright light, respectively. Light detection in rods depends on changes in the shape of the visual pigment rhodopsin. The visual pigments in cones respond to colors. Visual signals are processed in the retina before being sent on to the brain. In the retina, abundant receptors in the fovea provide sharp visual acuity.

REVIEW QUESTIONS

1. When a receptor cell detects a specific kind of stimulus, what happens to the stimulus energy?

2. Name six categories of sensory receptors and the type of stimulus that each type detects.

3. How do somatic sensations differ from special senses?

4. Explain where free nerve endings are located in the body and note some functions of the various kinds.

5. What is pain? Describe one type of pain receptor.

6. What are the stimuli for taste receptors?

7. How do "smell" signals arise and reach the brain?

8. Label the parts of the ear:

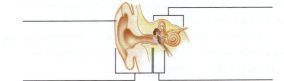

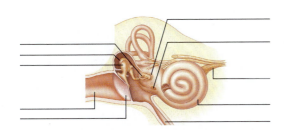

9. In the ear, sound waves cause the tympanic membrane to vibrate. What happens next in the middle ear? In the inner ear?

10. Label the parts of the eye:

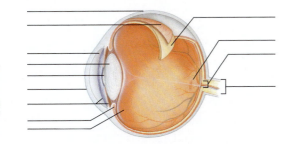

11. How does the eye focus the light rays of an image? What do *nearsighted* and *farsighted* mean?

4. A sensory system is composed of _____.
    a. nerve pathways from specific receptors to the brain
    b. sensory receptors
    c. brain regions that deal with sensory information
    d. all of the above

5. _____ detect energy associated with changes in pressure, body position, or acceleration.
    a. Chemoreceptors   c. Photoreceptors
    b. Mechanoreceptors   d. Thermoreceptors

6. Detecting substances present in the body fluids that bathe them is the function of _____.
    a. thermoreceptors   c. mechanoreceptors
    b. photoreceptors   d. chemoreceptors

7. Which of the special senses is based on the following events? Membrane vibrations cause fluid movements, which lead to bending of mechanoreceptors and firing of action potentials.
    a. taste   c. hearing
    b. smell   d. vision

8. Rods differ from cones in the following ways:
    a. They detect dim light, not bright light.
    b. They have a different visual pigment.
    c. They are not located in the retina.
    d. All of the above.
    e. a and b only

9. The outer layer of the eye includes the _____.
    a. lens and choroid   c. retina
    b. sclera and cornea   d. both a and c are correct

10. The inner layer of the eye includes the _____.
    a. lens and choroid   c. retina
    b. sclera and cornea   d. start of optic nerve

11. Your visual field is _____.
    a. a specific, small area of the retina
    b. what you actually "see"
    c. the area where color vision occurs
    d. where the optic nerve starts

12. Match each of the following terms with the appropriate description.
    _____ somatic senses
         (general senses)
    _____ special senses
    _____ variations in
         stimulus intensity
    _____ action potential
    _____ sensory receptor

    a. produced by strong stimulation
    b. endings of sensory neurons or specialized cells next to them
    c. taste, smell, hearing, balance, and vision
    d. frequency and number of action potentials
    e. touch, pressure, temperature, pain, and muscle sense

SELF-QUIZ *Answers in Appendix V*

1. Fill in the blank: A _____ is a specific form of energy that can elicit a response from a sensory receptor.

2. Fill in the blank: Awareness of a stimulus is called a _____.

3. Fill in the blank: _____ is understanding what particular sensations mean.

CRITICAL THINKING

1. Juanita started having bouts of dizziness. Her doctor asked her whether "dizziness" meant she felt lightheaded as if she were going to faint, or whether it meant she had sensations of *vertigo*—that is, a feeling that she herself or objects near her were spinning around. Why was this clarification important for the diagnosis?

2. Michael, a 3-year-old, experiences a chronic middle-ear infection, which is common among youngsters, in part due to an increase in antibiotic-resistant bacteria. This year, despite antibiotic treatment, an infection became so advanced that he had trouble hearing. Then his left eardrum ruptured and a jellylike substance dribbled out. The pediatrician told Michael's parents not to worry, that if the eardrum had not ruptured on its own she would have had to drain it. Suggest a reason why the physician concluded that this procedure would have been necessary to cure Michael's problem.

3. Jill is diagnosed with sensorineural deafness, a disorder in which sound waves are transmitted normally to the inner ear but they are not translated into neural signals that travel to the brain. Sometimes the cause is a problem with the auditory nerve, but in Jill's case it has to do with a problem in the inner ear itself. Where in the inner ear is the disruption most likely to be located?

4. Larry goes to the doctor complaining that he can't see the right side of the visual field with either eye. Where in the visual signal-processing pathway is Larry's problem occurring?

5. In a rock climber like the man pictured in Figure 14.24, which organs of equilibrium are activated?

Greg Epperson/Shutterstock.com

Figure 14.24 Dangling upside down affects several organs of equilibrium.

THE ENDOCRINE SYSTEM

LINKS TO EARLIER CONCEPTS

This chapter expands on what you have learned about the functions of the hypothalamus (13.7).

You will see more examples of how homeostatic feedback loops help regulate body functions (4.10).

You will also see how certain proteins in cell plasma membranes function in physiological processes (3.4)—in this case, by serving as receptors for hormone molecules.

KEY CONCEPTS

How Hormones Work

Hormones bind to and activate receptors on target cells. Their signals are converted into forms that work inside target cells to bring about a response. Sections 15.1–5.2

The Endocrine System

Glands and tissues of the endocrine system release most hormones. The hypothalamus and pituitary glands control much of this activity. Major health concerns can result from endocrine system disorders, including disruptions of normal bodily growth and metabolism. Sections 15.3–5.10

CONNECTIONS:
The Endocrine System and Homeostasis
Section 15.11

All images: © Cengage Learning

Diabetes mellitus is a disease that results when body cells don't take in enough insulin, a hormone that stimulates cells to take up glucose from the blood. Type 1 diabetes is a form of the disease that develops when an autoimmune response destroys insulin-making cells in the pancreas. MacKenzie Burger was diagnosed with it at age 5 and must carefully track her blood sugar. Yet MacKenzie has not let diabetes be a barrier to a full and active life. In describing her experience, she wrote:

> Growing up with diabetes taught me that I was personally responsible for maintaining my health. While classmates were learning their ABCs, I was learning to self-monitor blood glucose levels with finger pokes and to give myself insulin injections. Now, with the aid of an insulin pump and a continuous glucose monitoring system, I keep close watch on my blood glucose levels. I've been able to snorkel in the ocean, climb the Great Wall of China (below), and work toward my dream of becoming a science writer. Hardly a disability, right?

MacKenzie's story introduces this chapter's topic—how hormones help regulate many body functions that are essential to everyday health and well-being.

MacKenzie Burger

Homeostasis Preview
The endocrine system works in concert with the nervous system to adjust many body functions. Hormones generally govern long-term events such as growth and metabolism.

- Hormones are signaling molecules that help coordinate and manage the activities of the billions of body cells.

- Links to Receptor proteins in cell membranes 3.4, Neurotransmitters 13.3, Pheromones 14.3

endocrine system The glands, organs, and cells that produce hormones.

hormones Signaling molecules of the endocrine system.

target cell A cell that has receptors for a signaling molecule and so may respond to the molecule in some way.

Hormones are signaling molecules carried in the bloodstream

Previous chapters have discussed several types of signaling molecules, including neurotransmitters that carry nervous system messages. These chemical messengers are all alike in one key way: They act on target cells. A **target cell** is any cell that has receptors for the signaling molecule and that may change its activities in response. A target cell may or may not be next to the cell that sends the signal.

Hormones, our main topic here, are secreted by the body's endocrine glands, endocrine cells, and some neurons. They travel the bloodstream to target cells some distance away. Many types of cells also release "local" signaling molecules that change conditions in nearby tissues. Prostaglandins are an example (Table 15.1). Their targets include smooth muscle cells in the walls of bronchioles, which then close up or dilate and so change air flow in the lungs (Section 10.6). Prostaglandins that affect smooth muscle in the uterus cause menstrual cramps.

The word *hormone*—from the Greek *hormon*, "to set in motion"—was coined in 1900 by scientists studying food digestion in dogs. They discovered that a substance released by gland cells in a dog's GI tract could stimulate the pancreas. Later on, other researchers identified a variety of hormones and their sources (Figure 15.1).

Hormone-producing glands, organs, and cells form the **endocrine system**. For several reasons the name is misleading, however. First, it implies that there is an independent hormone-based control system for the body, when in fact almost all organ systems produce hormones. Section 15.10 will describe some major examples, including the GI tract hormones introduced in Chapter 11. In addition, as you will soon see, the functioning of endocrine glands, cells, and organs is closely allied with operations of the central nervous system.

Hormones are produced in small amounts and often interact

In general, endocrine glands usually release small amounts of hormones in short bursts. Controls usually prevent hormones from being either overproduced or underproduced. Negative feedback is the most common control mechanism.

It's not unusual for two or more hormones to affect the same process. There are three common kinds of these hormone "partnerships":

1. **Opposing interaction.** The effect of one hormone may oppose the effect of another. Insulin, for example, reduces the level of glucose in the blood, and glucagon increases it.

2. **Synergistic interaction.** The combined action of two or more "cooperating" hormones may be required to trigger a certain effect on target cells. For instance, a woman's mammary glands can't produce and secrete milk without the synergistic interaction of three other hormones: prolactin, oxytocin, and estrogen.

3. **Permissive interaction.** One hormone can exert its effect on a target cell only when a different hormone first "primes" the target cell. For example, even if one of a woman's eggs is fertilized, she can't become pregnant unless the lining of her uterus has been exposed to reproductive hormones.

TABLE 15.1 Examples of Chemical Signals in the Body

Type	Route to Target Cells
HORMONES	Carried by blood to distant targets
NEUROTRANSMITTERS	Released at synapses between neurons and target cells
PROSTAGLANDINS	Released in tissues and diffuse to target cells
PHEROMONES	Possibly reach target cells in other individuals

TAKE-HOME MESSAGE

WHAT ARE HORMONES?

- Hormones are signaling molecules secreted by endocrine glands, endocrine cells, and some neurons.

- The bloodstream carries hormones to distant target cells.

- Together, the glands and cells that secrete hormones make up the endocrine system. Their activity usually is regulated by negative feedback.

- Hormones are secreted in small amounts. Different ones may have opposing effects or may exert their effects in concert with other hormones. Still other hormone effects require a target cell to be primed by exposure to one hormone in order to respond to a second one.

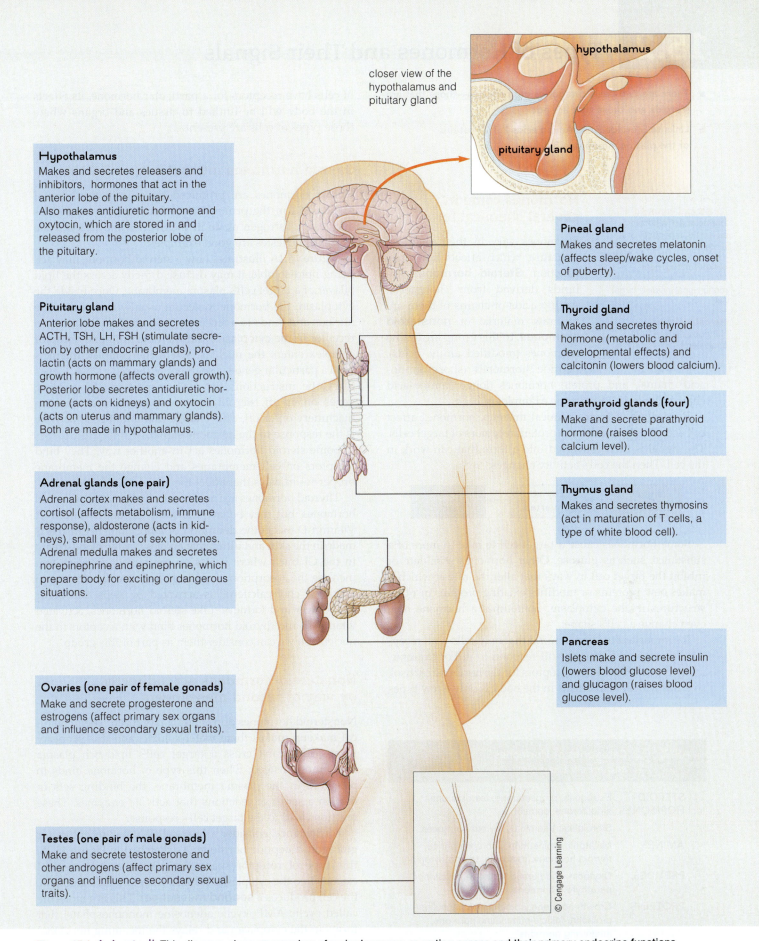

Hypothalamus
Makes and secretes releasers and inhibitors, hormones that act in the anterior lobe of the pituitary. Also makes antidiuretic hormone and oxytocin, which are stored in and released from the posterior lobe of the pituitary.

Pituitary gland
Anterior lobe makes and secretes ACTH, TSH, LH, FSH (stimulate secretion by other endocrine glands), prolactin (acts on mammary glands) and growth hormone (affects overall growth). Posterior lobe secretes antidiuretic hormone (acts on kidneys) and oxytocin (acts on uterus and mammary glands). Both are made in hypothalamus.

Adrenal glands (one pair)
Adrenal cortex makes and secretes cortisol (affects metabolism, immune response), aldosterone (acts in kidneys), small amount of sex hormones. Adrenal medulla makes and secretes norepinephrine and epinephrine, which prepare body for exciting or dangerous situations.

Ovaries (one pair of female gonads)
Make and secrete progesterone and estrogens (affect primary sex organs and influence secondary sexual traits).

Testes (one pair of male gonads)
Make and secrete testosterone and other androgens (affect primary sex organs and influence secondary sexual traits).

closer view of the hypothalamus and pituitary gland

hypothalamus

pituitary gland

Pineal gland
Makes and secretes melatonin (affects sleep/wake cycles, onset of puberty).

Thyroid gland
Makes and secretes thyroid hormone (metabolic and developmental effects) and calcitonin (lowers blood calcium).

Parathyroid glands (four)
Make and secrete parathyroid hormone (raises blood calcium level).

Thymus gland
Makes and secretes thymosins (act in maturation of T cells, a type of white blood cell).

Pancreas
Islets make and secrete insulin (lowers blood glucose level) and glucagon (raises blood glucose level).

© Cengage Learning

Figure 15.1 Animated! This diagram gives an overview of major hormone-secreting organs and their primary endocrine functions.

Types of Hormones and Their Signals

- There are two basic categories of hormones—those that are steroids and those that are not.

- Links to Steroids 2.10, Amino acids 2.11, Proteins of the plasma membrane 3.4

nonsteroid hormone
Hormone derived from an amine, a peptide, or a protein.

second messenger
Molecule that relays a hormone signal inside a target cell.

steroid hormone
Hormone derived from cholesterol.

Hormones come in several chemical forms

Hormones vary in their chemical structure, which affects how they function. **Steroid hormones** are lipids derived from cholesterol. Amino acids or chains of them are the raw material of **nonsteroid hormones**. In this group are amine hormones (modified amino acids), peptide hormones (short amino-acid chains), and protein hormones (longer amino-acid chains). Table 15.2 lists some examples of each.

Regardless of their chemical makeup, hormones affect cell activities by binding to protein receptors of target cells. The signal is then converted into a form that can work in the cell. Then the cell's activity changes:

Some hormones cause a target cell to take in more of a substance, such as glucose. Other hormones stimulate or inhibit the target cell in ways that alter the rate at which it makes new proteins or modifies existing proteins or other structures in the cytoplasm. Sometimes a hormone may even change a cell's shape.

It's important to keep in mind that only cells with receptors for a given hormone will respond to it. For example, many types of cells have receptors for the hormone cortisol, so it has widespread effects in the body. If only a few types of cells have receptors for a particular hormone, its effects in the body will be limited to tissues and organs where those types of cells are present.

Steroid hormones interact with cell DNA

Steroid hormones are produced by cells in the adrenal glands and in the primary reproductive organs—ovaries and testes. Estrogen made in the ovaries and testosterone made in the testes are good examples.

Figure 15.2A illustrates how a steroid hormone may act. Being lipid-soluble, it may diffuse directly across the lipid bilayer of a target cell's plasma membrane. Once inside the cytoplasm, the hormone molecule usually moves into the nucleus and binds to a receptor. In some cases it binds to a receptor in the cytoplasm, and then the hormone–receptor complex enters the nucleus. There the complex interacts with a particular gene—a segment of the cell's DNA. Genes carry the instructions for making proteins. By turning genes on or off, steroid hormones turn protein-making machinery on or off. This change in a target cell's activity is the response to the hormone signal.

Some steroid hormones act in another way. They bind receptors on cell membranes and change the membrane properties in ways that affect the target cell's function.

Thyroid hormones are not chemically the same as steroid hormones, but they behave the same. So does vitamin D. Vitamin D meets the definition of a hormone because it is made in the skin and ultimately arrives via the bloodstream in the GI tract, where it acts on target cells in ways that increase the absorption of calcium. En route "raw" vitamin D (called cholecalciferol) is activated by steps that occur in the liver and kidneys. Like steroid hormones, activated vitamin D and thyroid hormones bind with receptors in the nucleus, so we can consider them as part of this group.

Nonsteroid hormones act indirectly, by way of second messengers

Nonsteroid hormones don't enter a target cell. Their chemical makeup makes them water-soluble, and this property means they can't cross a target cell's lipid-rich plasma membrane. Instead, when this type of hormone binds to receptors in the plasma membrane, the binding sets in motion a series of reactions that activate enzymes. These reactions lead to the target cell's response.

For instance, consider a liver cell that has receptors for glucagon, a peptide hormone. As sketched in Figure 15.2B, this type of receptor spans the plasma membrane and extends into the cytoplasm. When a receptor binds glucagon, the cell produces a **second messenger**. This is a molecule called cyclic AMP (cyclic adenosine monophosphate) that forms in the cytoplasm and relays the incoming hormonal signal onward. (The hormone itself is the "first messenger.")

TABLE 15.2 Categories of Hormones and a Few Examples	
STEROID HORMONES	Estrogens, progesterone, testosterone, aldosterone, cortisol
	Steroidlike: Vitamin D, thyroid hormones
AMINES	Melatonin, epinephrine, norepinephrine, thyroid hormone (thyroxine, triiodothyronine)
PEPTIDES	Oxytocin, antidiuretic hormone, calcitonin, parathyroid hormone
PROTEINS	Growth hormone (somatotropin), insulin, prolactin, follicle-stimulating hormone, luteinizing hormone

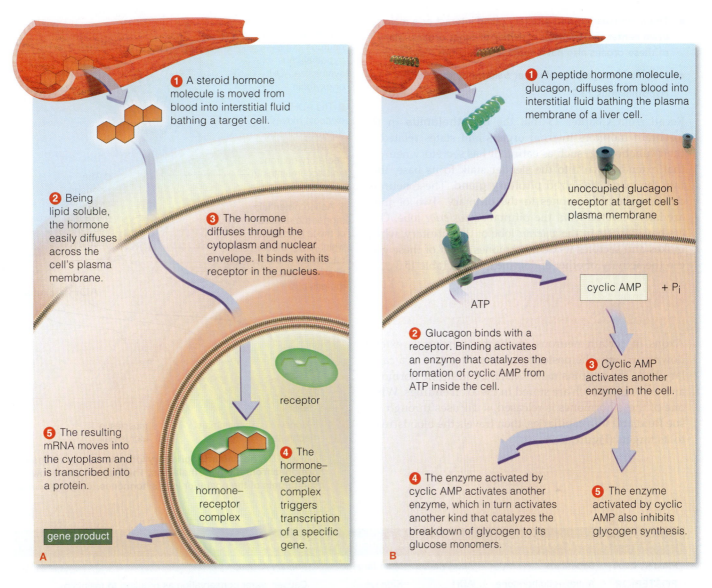

1 A steroid hormone molecule is moved from blood into interstitial fluid bathing a target cell.

2 Being lipid soluble, the hormone easily diffuses across the cell's plasma membrane.

3 The hormone diffuses through the cytoplasm and nuclear envelope. It binds with its receptor in the nucleus.

receptor

5 The resulting mRNA moves into the cytoplasm and is transcribed into a protein.

hormone–receptor complex

4 The hormone–receptor complex triggers transcription of a specific gene.

gene product

A

1 A peptide hormone molecule, glucagon, diffuses from blood into interstitial fluid bathing the plasma membrane of a liver cell.

unoccupied glucagon receptor at target cell's plasma membrane

cyclic AMP   + Pᵢ

ATP

2 Glucagon binds with a receptor. Binding activates an enzyme that catalyzes the formation of cyclic AMP from ATP inside the cell.

3 Cyclic AMP activates another enzyme in the cell.

4 The enzyme activated by cyclic AMP activates another enzyme, which in turn activates another kind that catalyzes the breakdown of glycogen to its glucose monomers.

5 The enzyme activated by cyclic AMP also inhibits glycogen synthesis.

B

Figure 15.2 Different types of hormones cause change in a target cell by different mechanisms. Part A shows an example of a mechanism by which a steroid hormone triggers changes in a target cell's activities. Part **B** is an example of how a peptide hormone triggers changes in the activity of a target cell. In this example, the hormone is glucagon. Cyclic AMP, a type of second messenger, relays the hormone's signal inside the cell. (© Cengage Learning)

An activated enzyme launches a cascade of reactions by converting ATP to cyclic AMP. Molecules of cyclic AMP are signals for the cell to activate molecules of another enzyme. These act on still other enzymes, and so forth, until a final reaction converts stored glycogen in the cell to glucose. Soon a huge number of molecules are taking part in the cell's final response to the hormone.

A slightly different example is a muscle cell that has receptors for insulin, a protein hormone. When insulin binds to the receptor, one result is that transport proteins insert themselves into the plasma membrane so that the cell can take up glucose faster. The signal also activates enzymes that catalyze reactions allowing the cell to store glucose not needed right away for its metabolism.

TAKE-HOME MESSAGE

HOW DO HORMONES INTERACT WITH TARGET CELLS?

- Hormones interact with receptors at a target cell's plasma membrane or in its nucleus. The hormone stimulates or inhibits the synthesis of proteins or the activity of enzymes in the cell.

- Most steroid hormones interact with a target cell's DNA after they enter the nucleus or bind a receptor in the cell's cytoplasm.

- Nonsteroid hormones bind to receptors in a target cell's plasma membrane. This binding activates an enzyme system. Often a second messenger relays the signal to the cell's interior, where the full response unfolds.

The Hypothalamus and Pituitary Gland

- The hypothalamus and pituitary gland interact as a major brain center that controls activities of other organs. Many of these organs also have endocrine functions.

- Links to Management of water balance by the kidneys 12.4, Hypothalamus 13.7

Recall from Chapter 13 that the **hypothalamus** in the forebrain monitors internal organs and states related to their functioning, such as eating. It has secretory neurons that extend down into the slender stalk to its base, then into the lobed, pea-sized **pituitary gland**. These neurons deliver several hormones to the pituitary. Two of them are later secreted from the pituitary's *posterior* lobe. Others have targets in the *anterior* lobe of the pituitary, which makes and secretes its own hormones. Most of these govern the activity of other endocrine glands (Table 15.3).

The posterior pituitary lobe stores and releases hormones from the hypothalamus

Axons of certain neurons in the hypothalamus extend downward into the posterior lobe, ending next to a capillary bed. The neurons make ADH (antidiuretic hormone) and oxytocin, which are stored in the axon endings. When one of these hormones is released, it diffuses through tissue fluid and into capillaries, then travels the bloodstream to its targets (Figure 15.3).

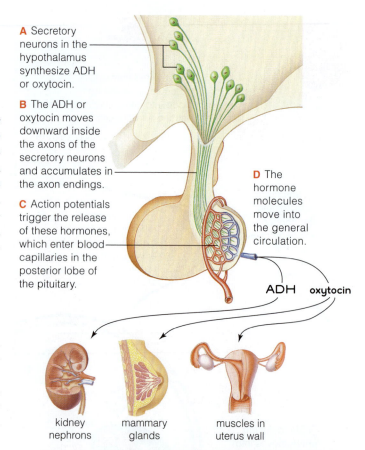

A Secretory neurons in the hypothalamus synthesize ADH or oxytocin.

B The ADH or oxytocin moves downward inside the axons of the secretory neurons and accumulates in the axon endings.

C Action potentials trigger the release of these hormones, which enter blood capillaries in the posterior lobe of the pituitary.

D The hormone molecules move into the general circulation.

ADH   oxytocin

kidney nephrons   mammary glands   muscles in uterus wall

Figure 15.3 The posterior pituitary lobe stores and releases hormones from the hypothalamus. The diagram also shows main targets of the posterior lobe's hormones. (© Cengage Learning)

TABLE 15.3 Primary Actions of Hormones Released from the Pituitary Gland

Pituitary Lobe	Secretions	Abbreviation	Main Targets	Primary Actions
POSTERIOR Nervous tissue (extension of hypothalamus)	Antidiuretic hormone (vasopressin)	ADH	Kidneys	Causes water conservation as required to maintain extracellular fluid volume and solute concentrations
	Oxytocin	OT	Mammary glands	Causes milk to move into secretory ducts
			Uterus	Causes uterine contractions during childbirth
ANTERIOR Glandular tissue, mostly	Adrenocorticotropic hormone (corticotropin)	ACTH	Adrenal glands	Stimulates release of cortisol, an adrenal steroid hormone
	Thyroid-stimulating hormone (thyrotropin)	TSH	Thyroid gland	Stimulates release of thyroid hormones
	Follicle-stimulating hormone	FSH	Ovaries, testes	In females, stimulates estrogen secretion, egg maturation; in males, helps stimulate sperm formation
	Luteinizing hormone	LH	Ovaries, testes	In females, stimulates progesterone secretion, ovulation, corpus luteum formation; in males, stimulates testosterone secretion, sperm release
	Prolactin	PRL	Mammary glands	Stimulates and sustains milk production
	Growth hormone (somatotropin)	GH	Most cells	Promotes growth in young; causes protein synthesis, cell division; roles in glucose, protein metabolism in adults

ADH acts on cells of kidney nephrons and collecting ducts. As Chapter 12 described, it promotes the reabsorption of water when the body must conserve water. The hypothalamus also releases ADH into the bloodstream when blood pressure falls below a set point. ADH causes the arterioles in some tissues to narrow, so blood pressure rises. This is why ADH is sometimes called vasopressin.

Oxytocin affects reproduction. In a pregnant woman, for example, it triggers muscle contractions in the uterus during labor and causes milk to be released when a mother nurses her infant. In sexually active people, both male and female, oxytocin apparently is a chemical trigger for feelings of satisfaction after sexual contact. Studies suggest that oxytocin is a "cuddle hormone" that helps stimulate affectionate behavior.

The anterior pituitary lobe makes hormones

Unlike the posterior pituitary lobe, the anterior pituitary lobe produces and secretes six hormones:

Corticotropin	ACTH
Thyrotropin	TSH
Follicle-stimulating hormone	FSH
Luteinizing hormone	LH
Prolactin	PRL
Growth hormone (somatotropin)	GH (or STH)

Anterior pituitary hormones have widespread effects. ACTH and TSH regulate the secretion of hormones from the adrenal glands and thyroid gland, respectively. FSH and LH influence reproduction, as described in Chapter 16. Prolactin is best known for stimulating and sustaining the production of breast milk, after other hormones have primed the tissues. There also is evidence that it promotes the synthesis of the male sex hormone testosterone.

Growth hormone (GH) affects most body tissues. It stimulates the processes by which cells divide and make new proteins, and so has a major influence on growth. GH is also important as a "metabolic hormone." It stimulates cells to take up amino acids and promotes the breakdown and release of fat stored in adipose tissues when cells require more fatty acids. GH also adjusts the rate at which cells take up glucose. In this way it helps to maintain proper blood sugar levels.

The hypothalamus regulates the anterior lobe by secreting hormones that enter blood capillaries in the pituitary stalk (Figure 15.4). The bloodstream carries those hormones to another capillary bed in the anterior lobe. There the hormones leave the blood and act on their target cells. Most of these hormones are *releasers* that spur target cells to secrete their own hormones. For example, GnRH (gonadotropin-releasing hormone) triggers the secretion of FSH and LH. These hormones are called *gonadotropins* because they affect the functioning of cells in the gonads, or reproductive organs. TRH (for thyrotropin-releasing hormone) stimulates the release of TSH. Other hypothalamic hormones

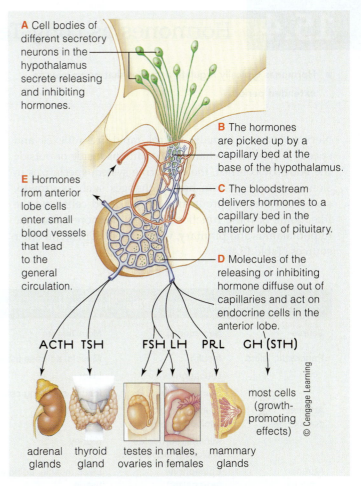

A Cell bodies of different secretory neurons in the hypothalamus secrete releasing and inhibiting hormones.

B The hormones are picked up by a capillary bed at the base of the hypothalamus.

C The bloodstream delivers hormones to a capillary bed in the anterior lobe of pituitary.

D Molecules of the releasing or inhibiting hormone diffuse out of capillaries and act on endocrine cells in the anterior lobe.

E Hormones from anterior lobe cells enter small blood vessels that lead to the general circulation.

ACTH   TSH   FSH LH   PRL   GH (STH)

adrenal glands   thyroid gland   testes in males, ovaries in females   mammary glands   most cells (growth-promoting effects)

© Cengage Learning

Figure 15.4 Animated! **The anterior pituitary lobe both makes and releases hormones.** Hormones from the hypothalamus control this activity.

are *inhibitors*. They block secretions from cells in the anterior pituitary. One of them, called somatostatin, inhibits the secretion of growth hormone and thyrotropin.

hypothalamus Forebrain region that controls processes related to homeostasis and has endocrine functions.

pituitary gland An endocrine gland that interacts with the hypothalamus to control many physiological functions, including the activity of some other glands.

15.4 Hormones as Long-Term Controllers

■ Hormones typically regulate activities that occur over an extended period.

Nervous system signals control rapid-fire reflexes and speedy responses to changing conditions inside or outside the body. By contrast, the endocrine system specializes in slower, often long-term bodily changes such as growth, sexual maturation, production of red blood cells, and the like. Some of these functions involve hormones from the hypothalamus and pituitary, while others depend on other sources (Table 15.4).

We have now completed our overview of hormones and general information about how they function. The rest of the chapter looks at how some major hormones operate in the body and how disorders arise when those key substances do not function properly.

TAKE-HOME MESSAGE

WHAT KINDS OF BODILY ACTIVITIES DO HORMONES TYPICALLY REGULATE?

- Hormones generally regulate slower, often long-term changes in the growth or functioning of body parts.

TABLE 15.4 Hormone Sources Other Than the Hypothalamus and Pituitary

Source	Secretion(s)	Main Targets	Primary Actions
Pancreatic islets	Insulin	Muscle, adipose tissue	Lowers blood-sugar level
	Glucagon	Liver	Raises blood-sugar level
	Somatostatin	Insulin-secreting cells	Influences carbohydrate metabolism
Adrenal cortex	Glucocorticoids (including cortisol)	Most cells	Promote protein breakdown and conversion to glucose
	Mineralocorticoids (including aldosterone)	Kidney	Promote sodium reabsorption; control salt–water balance
Adrenal medulla	Epinephrine (adrenaline)	Liver, muscle, adipose tissue	Raises blood level of sugar, fatty acids; increases heart rate, force of contraction
	Norepinephrine	Smooth muscle of blood vessels	Promotes constriction or dilation of blood vessel diameter
Thyroid	Triiodothyronine, thyroxine	Most cells	Regulate metabolism; have roles in growth, development
	Calcitonin	Bone	Lowers calcium levels in blood
Parathyroids	Parathyroid hormone	Bone, kidney	Elevates levels of calcium and phosphate ions in blood
Thymus	Thymosins, etc.	Lymphocytes	Have roles in immune responses
Gonads:			
Testes (in males)	Androgens (including testosterone)	General	Required in sperm formation, development of genitals, maintenance of sexual traits; influence growth, development
Ovaries (in females)	Estrogens	General	Required in egg maturation and release; prepare uterine lining for pregnancy; required in development of genitals, maintenance of sexual traits; influence growth, development
	Progesterone	Uterus, breasts	Prepares, maintains uterine lining for pregnancy; stimulates breast development
Pineal	Melatonin	Hypothalamus	Influences daily biorhythms
Endocrine cells of stomach, gut	Gastrin, secretin, etc.	Stomach, pancreas, gallbladder	Stimulate activity of stomach, pancreas, liver, gallbladder
Liver	IGFs (Insulin-like growth factors)	Most cells	Stimulate cell growth and development
Kidneys	Erythropoietin	Bone marrow	Stimulates red blood cell production
	Angiotensin*	Adrenal cortex, arterioles	Helps control blood pressure, aldosterone secretion
	Vitamin D3*	Bone, gut	Enhances calcium resorption and uptake
Heart	Atrial natriuretic peptide	Kidney, blood vessels	Increases sodium excretion; lowers blood pressure

*These hormones are not produced in the kidneys but are formed when enzymes produced in kidneys activate specific substances in the blood.

GH Growth Functions and Disorders

- Growth hormone (GH) is so important to normal bodily growth that major abnormalities develop when it does not function properly.

- Links to Bone development 5.1, Skeletal muscles 6.1

Growth hormone from the anterior pituitary affects target cells throughout the body. It acts indirectly, by triggering the synthesis of a growth factor, mainly in the liver. One major GH effect is stimulating the growth of cartilage and bone and increasing muscle mass. You may recall from Chapter 5 that GH prevents the epiphyseal plates at the ends of growing long bones from hardening during childhood and adolescence. Because this hormone has such major effects on bodily growth, if the pituitary secretes too much or too little of it, the impact can be profound.

For instance, **gigantism** results when the anterior lobe of the pituitary overproduces it during childhood. Affected adults are proportionally like an average-sized person but much larger (Figure 15.5A). If too much GH is secreted during adulthood, bones, cartilage, and other connective tissues in the hands, feet, and jaws thicken abnormally. So do epithelia of the skin, nose, eyelids, lips, and tongue. The result is **acromegaly** (Figure 15.5B). Both gigantism and acromegaly usually develop as the result of a benign (noncancerous) pituitary tumor.

Pituitary dwarfism occurs when the pituitary makes too little GH or when receptors cannot respond normally to it. Affected people are quite short but have normal proportions. Pituitary dwarfism can be inherited (Figure 15.5C) or it can result from a pituitary tumor or injury.

Human growth hormone is now made through genetic engineering (Chapter 21). Children who have a naturally low GH level may receive injections of recombinant human growth hormone (rhGH), although the treatment is expensive (up to $20,000 a year) and controversial. Some physicians and ethicists object to short stature being treated as a defect to be cured.

Injections of rhGH are also used to treat adults who have a low GH level as the result of an injury or a tumor of the pituitary or hypothalamus. The injection can help maintain healthy bone and muscle mass while reducing body fat. Entrepreneurs and others have touted rhGH injections as a means to slow normal aging or boost athletic performance. Thus far, clinical trials don't bear out this claim, and the drug is not approved for those uses. Negative side effects include increased risk of high blood pressure and diabetes.

TAKE-HOME MESSAGE

HOW DOES GROWTH HORMONE AFFECT BODILY GROWTH?

- Growth hormone (GH) stimulates the growth of bone and skeletal muscle, among other tissues.
- Excessive GH causes faster-than-normal bone growth that leads to gigantism in children and acromegaly in adults. A deficiency during childhood can cause pituitary dwarfism.

Figure 15.5 Disorders in bodily growth may result from too much or too little growth hormone. A Bao Xishun, one of the world's tallest men, stands 7 feet, 9 inches (2.36 m) tall. He is shown here with his wife. **B** A woman before and after she became affected by acromegaly. Notice how her chin became elongated. **C** Dr. Hiralal Maheshwari, *right*, with two men from a village in Pakistan where an inherited form of pituitary dwarfism is common. Men of the village average a little over 4 feet (130 cm) tall.

REUTERS/China Daily

Age 16

Courtesy of Dr. William H. Daughaday, Washington University School of Medicine, from A.I. Mendelhoff and D.E. Smith, eds., American Journal of Medicine, 20:133 (1956).

Age 52

Courtesy of G. Baumann, M.D. Northwestern University

■ Hormones from the thyroid gland are required for normal growth and metabolism. The thyroid and parathyroid glands work together to regulate calcium levels in the blood.

■ Links to Bone growth and remodeling 5.1, Autoimmune disorders 9.10, Major dietary minerals 11.13

parathyroid glands Four small endocrine glands behind the thyroid gland. Parathyroid hormone (PTH) regulates blood calcium levels.

thyroid gland Endocrine gland that produces thyroid hormone (TH), which is required for normal metabolism, growth, and development.

Thyroid hormones affect metabolism and growth

The **thyroid gland** is located at the base of the neck in front of the trachea, or windpipe (Figure 15.6). The main hormones it produces, thyroxine (T_4) and triiodothyronine (T_3), are known jointly as TH (thyroid hormone). TH affects cells throughout the body. It is largely responsible for setting a person's basal metabolic rate (Section 11.14). It also enhances the production of GH, and in this way has a major influence on growth. Adequate TH is essential in order for the central nervous system of a fetus to develop properly. Optimal functioning of an adult's CNS depends on it as well.

The thyroid also makes the hormone calcitonin, which helps lower the level of calcium (and of phosphate) in blood in response to homeostatic feedback.

TH cannot be formed without iodide, a form of iodine. Iodine-deficient diets cause one or both lobes of the thyroid gland to enlarge (Figure 15.7). The enlargement, a simple goiter, occurs after low blood levels of TH set in motion a negative feedback loop that causes the anterior pituitary to secrete TSH (the thyroid-stimulating hormone thyrotropin). The thyroid attempts to make TH but cannot do so, which leads to continued secretion of TSH, and

so on, in a sustained abnormal feedback loop. Simple goiter is no longer common in places where people use iodized salt.

Hypothyroidism is the clinical name for low blood levels of TH. Metabolism slows in affected adults, so they tend to gain weight, feel sluggish physically and mentally, and find it difficult to tolerate cold temperatures.

Graves disease and some other conditions are due to *hyperthyroidism*, in which metabolic activity "revs up" due to excess of TH in the blood. Symptoms include elevated heart rate and blood pressure and unusually heavy sweating. Some cases are autoimmune disorders, in which antibodies wrongly stimulate thyroid cells. In other cases the cause can be traced to inflammation or a tumor in the thyroid gland. Some people are genetically predisposed to the disorder.

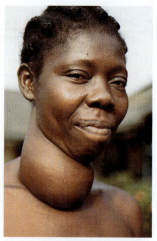

Figure 15.7 A diet low in the micronutrient iodine may cause a simple goiter. (Scott Camazine/Photo Researchers, Inc.)

PTH from the parathyroids is the main calcium regulator

Most of us have four **parathyroid glands** located on the back of the thyroid gland (Figure 15.8A). These little glands secrete parathyroid hormone (PTH), the main regulator of the calcium level in blood. Calcium is important for muscle contraction as well as for the activation of enzymes, the formation of bone, blood clotting, and

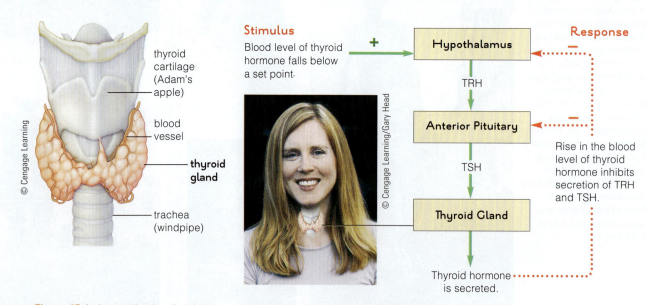

thyroid cartilage (Adam's apple)

blood vessel

thyroid gland

trachea (windpipe)

© Cengage Learning

Stimulus
Blood level of thyroid hormone falls below a set point.

Response

Hypothalamus +

TRH

Anterior Pituitary −

TSH

Thyroid Gland −

Thyroid hormone is secreted.

Rise in the blood level of thyroid hormone inhibits secretion of TRH and TSH.

© Cengage Learning/Gary Head

Figure 15.6 A negative feedback loop controls the secretion of thyroid hormone.

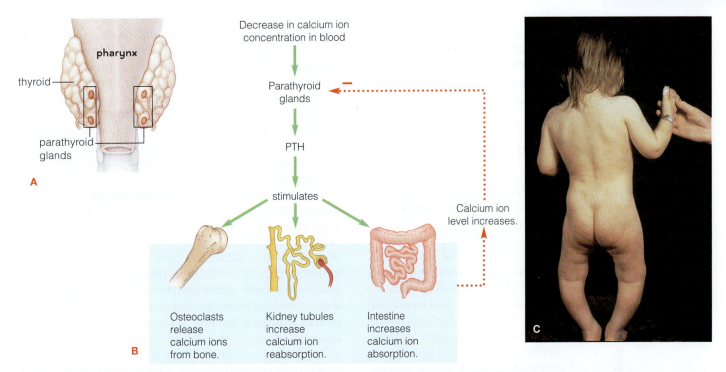

Figure 15.8 PTH regulates calcium homeostasis. A The location of the parathyroid glands. The diagram in **B** shows the feedback loop that controls the release of PTH. **C** A child with legs bowed by rickets. (© Cengage Learning; Biophoto Associates/Photo Researchers, Inc.)

Diagram labels:

- pharynx
- thyroid
- parathyroid glands
- A
- Decrease in calcium ion concentration in blood
- Parathyroid glands
- PTH
- stimulates
- B
- Osteoclasts release calcium ions from bone.
- Kidney tubules increase calcium ion reabsorption.
- Intestine increases calcium ion absorption.
- Calcium ion level increases.
- C

other tasks. The parathyroids secrete more PTH when the blood level of calcium falls below a set point, and they reduce their secretions when the calcium level rises. The hormone calcitonin from the thyroid gland contributes to processes that remove calcium from the blood.

You may remember that Section 5.1 discussed bone remodeling, the process in which bone is deposited or broken down, depending on the level of calcium in the blood. PTH is the hormone in charge of remodeling, and it acts on the skeleton and kidneys. When the blood level of calcium falls below a set point, PTH prompts the bone cells called osteoclasts to secrete enzymes that digest bone tissue (Figure 15.8B). This process releases calcium ions (and phosphate) that can be used elsewhere in the body. In the kidneys, PTH also stimulates the reabsorption of calcium from the filtrate flowing through nephrons. At the

same time, PTH helps to activate vitamin D. As described earlier, activated vitamin D is a hormone that improves the absorption of calcium in the GI tract.

In children who have vitamin D deficiency, too little calcium and phosphorus are absorbed, so the rapidly growing bones don't develop properly. Children who have the resulting bone disorder, **rickets** (Figure 15.8C), develop skeletal abnormalities such as bowed legs.

Calcium is so essential in the body that disorders related to parathyroid functioning can be quite serious. For example, excess PTH (*hyperparathyroidism*) causes so much calcium to be withdrawn from a person's bones that the bone tissue is dangerously weakened. The excess calcium in the bloodstream may cause kidney stones, and muscles don't function normally. The central nervous system's operations may be so seriously harmed that the affected person dies.

THINK OUTSIDE THE BOOK

Health experts estimate that about 16 million people in the United States have hypothyroidism. Many are unaware that this thyroid disorder is affecting their health because the main symptoms, such as fatigue and weight gain, can have many causes. Research this topic online. Who is most likely to develop hypothyroidism? What are treatment options?

TAKE-HOME MESSAGE

WHAT ARE THE BODILY ROLES OF THYROID HORMONE AND PARATHYROID HORMONE?

- Thyroid hormone (TH) influences the basal metabolic rate, growth, and the development and functioning of the nervous system.
- Parathyroid hormone (PTH) is the main regulator of calcium levels in the blood.

- Different parts of the adrenal glands make hormones that help regulate blood levels of glucose, influence blood pressure, and regulate blood circulation.

- Links to Inflammation 9.4, Nutrient processing 11.8, Urine formation 12.4, Stress responses 13.6

The adrenal cortex produces glucocorticoids and mineralocorticoids

We have two adrenal glands, one on top of each kidney. The outer part of each gland is the **adrenal cortex** (Figure 15.9). There, cells secrete two major types of steroid hormones, the glucocorticoids and mineralocorticoids.

Glucocorticoids raise the blood level of glucose. For instance, the body's main glucocorticoid, *cortisol*, is secreted when the body is stressed and glucose is in such demand that its blood level drops to a low set point. That level is an alarm signal and starts a stress response, which a negative feedback mechanism later cuts off. Among other effects, cortisol promotes the breakdown of muscle proteins and stimulates the liver to take up amino acids, from which liver cells synthesize glucose in a process called **gluconeogenesis**. Cortisol also reduces how much glucose tissues such as skeletal muscle take up from the blood. This effect is sometimes called "glucose sparing." Glucose sparing is extremely important in homeostasis, for it helps ensure that the blood will carry enough glucose to supply the brain, which usually cannot use other molecules for fuel. Cortisol also promotes the breakdown of fats and the use of the resulting fatty acids for energy.

adrenal cortex The outer portion of an adrenal gland.

adrenal medulla The inner portion of the adrenal gland.

glucocorticoids Adrenal cortex hormones that are secreted at times of physiological stress. The main one is cortisol.

gluconeogenesis The synthesis of glucose in the liver.

mineralocorticoids Adrenal cortex hormones (such as aldosterone) that adjust blood levels of mineral salts.

Figure 15.9 diagrams the negative feedback loop for cortisol. When the blood level of cortisol rises above a set point, the hypothalamus begins to produce less of the releasing hormone CRH. The anterior pituitary responds by secreting less ACTH, and the adrenal cortex secretes less cortisol. In a healthy person, daily cortisol secretion is highest when the blood glucose level is lowest, usually in the early morning. Chronic severe hypoglycemia, or low blood sugar, can develop when the adrenal cortex makes too little cortisol. Then, mechanisms that spare glucose and make new supplies in the liver don't work properly.

Glucocorticoids also reduce inflammation. The adrenal cortex pumps out more of these chemicals at times of unusual physical stress such as a painful injury, severe illness, or a strong allergic reaction. The extra cortisol and other signaling molecules helps speed recovery. That is why doctors prescribe cortisol-like drugs such as cortisone for patients with asthma or serious inflammatory disorders. Cortisone is the active ingredient in many over-the-counter products for treating skin irritations.

Unfortunately, long-term use of heavy doses of glucocorticoids has serious side effects, including suppressing the immune system. As described shortly, long-term stress has the same effect.

For the most part, **mineralocorticoids** adjust the concentrations of mineral salts, such as potassium and sodium, in the extracellular fluid. The most abundant mineralocorticoid is aldosterone. You may recall from Section 12.4 that aldosterone acts on the distal tubules of kidney nephrons, stimulating them to reabsorb sodium ions and excrete potassium ions. The reabsorption of sodium in turn promotes reabsorption of water from the tubules as urine is forming. A variety of circumstances can cause the release of aldosterone. Common triggers include falling blood pressure or falling blood levels of sodium—which reduces blood volume because water moves out of the bloodstream by osmosis.

In a fetus and early in puberty, the adrenal cortex also makes large amounts of sex hormones. The main ones are androgens (male sex hormones), but female sex hormones (estrogens and progesterone) also are produced. In adults, the reproductive organs generate most sex hormones.

Hormones from the adrenal medulla help regulate blood circulation

The **adrenal medulla** is the inner part of the adrenal gland shown in Figure 15.9. It contains neurons that release two substances, epinephrine and norepinephrine. Both act as neurotransmitters when they are secreted by neurons elsewhere in the body. When the adrenal medulla secretes them, however, their hormonelike effects help regulate blood circulation and carbohydrate use when the body is stressed or excited. For example, they increase the heart rate, dilate arterioles in some areas and constrict them in others, and dilate bronchioles. Thus the heart beats faster and harder, more blood is shunted to heart and muscle cells from other regions, and more oxygen flows to energy-demanding cells throughout the body. These are aspects of the fight–flight response noted in Chapter 13.

The operation of the adrenal medulla provides another example of negative feedback control. For example, when the hypothalamus sends the necessary signal (by way of sympathetic nerves) to the adrenal medulla, the neuron axons will start to release norepinephrine into the synapse between the axon endings and the target cells.

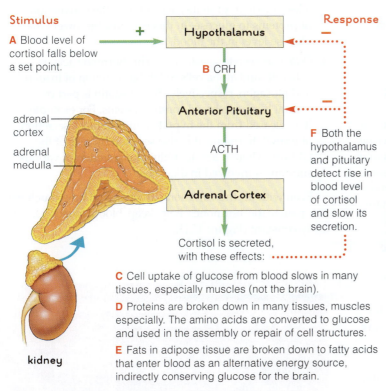

Stimulus

A Blood level of cortisol falls below a set point.

+ → **Hypothalamus** ← − **Response**

B CRH

Anterior Pituitary ← **—**

ACTH

Adrenal Cortex

F Both the hypothalamus and pituitary detect rise in blood level of cortisol and slow its secretion.

adrenal cortex

adrenal medulla

Cortisol is secreted, with these effects:

C Cell uptake of glucose from blood slows in many tissues, especially muscles (not the brain).

D Proteins are broken down in many tissues, muscles especially. The amino acids are converted to glucose and used in the assembly or repair of cell structures.

E Fats in adipose tissue are broken down to fatty acids that enter blood as an alternative energy source, indirectly conserving glucose for the brain.

kidney

Figure 15.9 The adrenal glands produce cortisol, the stress hormone. Each gland rests atop a kidney. The diagram shows a negative feedback loop that governs the secretion of cortisol. (© Cengage Learning)

Figure 15.10 Reducing stress may benefit your health. Physical exercise and social activities are good stress reducers. (Nancy Ney/Photodisc/Getty Images)

Soon, norepinephrine molecules collect in the synapse, setting the stage for a localized negative feedback mechanism. As the accumulating norepinephrine binds to receptors on the axon endings, the release of norepinephrine soon shuts down.

Long-term stress can damage health

As you've just read, when the body is stressed, nervous system commands trigger the fight–flight response and the release of cortisol, epinephrine, and norepinephrine. In daily life, most people also encounter a wide variety of psychosocial stressors—an exam, financial difficulties, a new job or romance, and the like. As you can see from this short list, some stressors are positive, others are negative. Not everyone reacts the same way to life's challenges, but there is ample evidence that being routinely "stressed out" by negative stressors may contribute to hypertension and related cardiovascular disease. And because cortisol suppresses the immune system, people who experience a lot of "bad" stress may be more susceptible to disease.

Chronic negative stress also is linked to insomnia, anxiety, and depression.

Research also shows that social connections seem to moderate the effects of stress, as does regular physical exercise (Figure 15.10). Friends, family, support groups, and counselors can not only make you feel better, they may make you healthier as well.

TAKE-HOME MESSAGE

WHAT HORMONES DO THE ADRENAL GLANDS PRODUCE?

- The adrenal cortex produces glucocorticoids such as cortisol and mineralocorticoids such as aldosterone.
- The adrenal medulla makes epinephrine and norepinephrine.
- Cortisol raises blood glucose levels and suppresses inflammation. Aldosterone helps regulate blood pressure by adjusting reabsorption of potassium and sodium in the kidneys.
- Epinephrine and norepinephrine adjust blood circulation and the use of blood sugar in the fight–flight response to stress.

■ The pancreas hormones insulin and glucagon work antagonistically—the action of one opposes the action of the other. Controls over the release of these hormones regulate the glucose level in blood.

■ Link to Accessory organs of digestion 11.5

pancreatic islets Clusters of endocrine cells in the pancreas.

The pancreas has both exocrine *and* endocrine functions. Its exocrine cells release digestive enzymes into the small intestine. It also has some 2 million scattered clusters of endocrine cells. Each cluster is a **pancreatic islet** and contains three types of hormone-secreting cells:

1. *Alpha cells* secrete *glucagon.* Between meals, cells use the glucose delivered to them by the bloodstream. When the blood glucose level decreases below a set point, secreted glucagon acts on cells in the liver and muscles. It causes glycogen (a storage polysaccharide) and amino acids to be converted to glucose. In this way glucagon raises the glucose level in the blood.

2. *Beta cells* secrete the hormone *insulin.* After meals, when a lot of glucose is circulating in the blood, insulin stimulates muscle and adipose cells to take up glucose. It also promotes synthesis of fats, glycogen, and to

a lesser extent, proteins, and inhibits the conversion of proteins to glucose. In this way insulin lowers the glucose level in the blood.

3. *Delta cells* secrete *somatostatin.* This hormone acts on beta cells and alpha cells to inhibit secretion of insulin and glucagon, respectively. Somatostatin is part of several hormone-based control systems. For example, it is released from the hypothalamus to block secretion of growth hormone; it is also secreted by cells of the GI tract, where it acts to inhibit the secretion of various substances involved in digestion.

Even with all the variations in when and how much we eat, pancreatic hormones help keep blood glucose levels fairly constant (Figure 15.11).

TAKE-HOME MESSAGE

HOW DO ENDOCRINE CELLS IN PANCREATIC ISLETS REGULATE BLOOD SUGAR?

• Alpha cells secrete glucagon when the blood level of glucose (sugar) falls below a set point. Beta cells secrete insulin when blood levels of glucose rise above the set point.

• Somatostatin from delta cells regulates the functioning of alpha and beta cells.

Figure 15.11 Animated! Cells that secrete insulin and glucagon respond to a change in the level of glucose in blood. These two hormones work antagonistically to maintain normal blood sugar levels.

1 *After* a meal, the blood level of glucose increases. In the pancreas, the increase **2** stops alpha cells from secreting glucagon and **3** stimulates beta cells to secrete insulin. In response to insulin, **4** adipose and muscle cells take up and store glucose, and liver cells make more glycogen. As a result, insulin *lowers* blood sugar **5**.

6 *Between* meals, blood sugar falls. The decrease **7** stimulates alpha cells to secrete glucagon and **8** slows the insulin secretion by beta cells. **9** In the liver, glucagon causes cells to convert glycogen back to glucose, which enters the blood. As a result, glucagon *raises* blood sugar **10**. (© Cengage Learning)

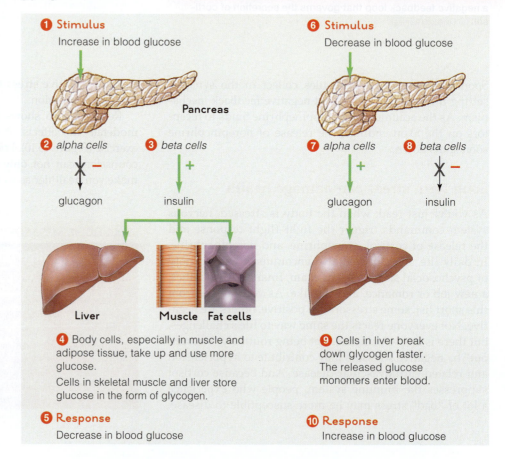

1 Stimulus
Increase in blood glucose

Pancreas

2 *alpha cells*   **3** *beta cells*

glucagon   insulin

Liver   Muscle Fat cells

4 Body cells, especially in muscle and adipose tissue, take up and use more glucose.
Cells in skeletal muscle and liver store glucose in the form of glycogen.

5 Response
Decrease in blood glucose

6 Stimulus
Decrease in blood glucose

7 *alpha cells*   **8** *beta cells*

glucagon   insulin

9 Cells in liver break down glycogen faster. The released glucose monomers enter blood.

10 Response
Increase in blood glucose

15.9 | Blood Sugar Disorders

As described in this chapter's introduction, too little insulin can lead to **diabetes mellitus**. Because target cells can't take up glucose from blood, glucose builds up in the blood (*mellitus* means "honey" in Greek). The kidneys move excess sugar into the urine, water is also lost, and the body's water–solute balance is upset. Affected people become dehydrated and extremely thirsty. They also lose weight as their glucose-starved cells break down protein and fats for energy. Fat breakdown releases ketones, so these acids build up in the blood and urine. This can lead to dangerously low blood pressure and a condition called **metabolic acidosis**—a blood pH so low (acidic) that it may harm functioning of the brain.

Like MacKenzie Burger, about one in ten diabetics has **type 1 diabetes**, in which an autoimmune response destroys pancreas beta cells. It may be caused by a viral infection in combination with genetic susceptibility. Symptoms of type 1 diabetes usually appear early in life, and affected people survive with insulin injections or insulin provided by a pump (Figure 15.12).

Type 2 diabetes is a global health crisis

In **type 2 diabetes**, insulin levels are near or above normal, but for any of several reasons target cells can't respond properly to the hormone. The beta cells break down and steadily produce less insulin. According to the World Health Organization, in developed countries type 2 diabetes has reached crisis proportions, along with its major risk factor, obesity.

Blood containing too much sugar damages capillaries. Over time, the blood supply to the kidneys, eyes, and lower limbs may be so poor that tissues die and terrible complications may develop (Table 15.5).

Diabetes also correlates strongly with cardiovascular disease. Even diabetics in their 20s and 30s are at high risk of suffering a stroke or heart attack.

"Prediabetes" is a warning sign

As many as 20 million Americans have "prediabetes"—slightly elevated blood sugar that increases the risk of developing type 2 diabetes. Metabolic syndrome, the constellation of features described in Section 11.15, is an early indicator that someone may be at risk for diabetes. These features include a fasting glucose measurement of 110 mg/dL or higher.

Type 2 diabetes can be controlled by a combination of proper diet, regular exercise, and sometimes drugs that improve insulin secretion or activity. In obese people who develop type 2 diabetes, the disease often disappears if the person loses significant weight.

Low blood sugar threatens the brain

In **hypoglycemia**, so much sugar is removed from the blood that cells in the brain and elsewhere may suddenly have too little fuel to function properly. Anything that raises the blood level of insulin, such as a miscalculated insulin injection or an insulin-secreting tumor, can cause hypoglycemia. The result can be life-threatening *insulin shock*, in which the brain essentially "stalls" as its fuel dwindles. A person experiencing insulin shock may feel dizzy and confused and have trouble talking. Anything that quickly raises blood sugar, including a shot of glucagon, solves the problem.

TABLE 15.5 Some Complications of Diabetes	
Eyes	Changes in lens shape and vision; damage to blood vessels in retina; blindness
Skin	Increased susceptibility to bacterial and fungal infections; patches of discoloration; thickening of skin on the back of hands
Digestive system	Gum disease; delayed stomach emptying that causes heartburn, nausea, vomiting
Kidneys	Increased risk of kidney disease and failure
Heart and blood vessels	Increased risk of heart attack, stroke, high blood pressure, and atherosclerosis
Hands and feet	Impaired sensations of pain; formation of calluses, foot ulcers; possible amputation of a foot or leg because of necrotic tissue that formed owing to poor circulation

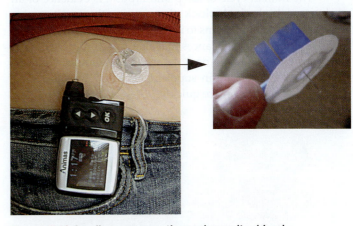

Figure 15.12 Insulin pumps continuously monitor blood glucose levels and supply insulin as needed. The device is programmed to deliver insulin through a tube inserted into the skin. It helps smooth out fluctuations in blood sugar and so reduces the risk of complications due to excessively low or high blood sugar. MacKenzie Burger's insulin pump is visible in the photograph of her in the chapter introduction. (Left: © Elizabeth Musar Right: © Manny Hernandez/Diabetes Hands Foundation, www.tudiabetes.com)

- Endocrine cells in sex organs, parts of the brain, the thymus, the heart, and the GI tract make hormones.

- Links to Heart 7.2, Controls over digestion 11.8, Lymphatic system 9.2

biological clock An internal mechanism by which the body may monitor day length.

gonads Primary sex organs—testes in males and ovaries in females.

pineal gland Endocrine gland that produces the hormone melatonin.

The gonads produce sex hormones

The human primary sex organs are called **gonads**. Most people know them as the ovaries in females and testes in males (Figure 15.13). In addition to producing sex cells—eggs in ovaries, sperm in testes—the gonads also make sex hormones. Ovaries make *estrogens* and *progesterone*. The testes make mostly *testosterone,* but they also make a little estrogen and progesterone. Small amounts of these "female" hormones are required for proper development of sperm. Similarly, a female's ovaries make small amounts of testosterone. It contributes to libido, the desire for sex.

The pineal gland makes melatonin

Many ancient vertebrates had a light-sensitive "third eye" on top of the head. In humans a version of this organ still exists, as a lump of tissue in the brain called the **pineal gland**. It releases the hormone *melatonin* into cerebrospinal fluid and the bloodstream. Melatonin influences sleep/wake cycles. It is secreted in the dark, so the amount in the bloodstream varies from day to night. It also changes with the seasons, because winter days are shorter than summer days.

The human cycle of sleep and arousal is evidence of an internal **biological clock** that apparently monitors day length. Melatonin seems to influence the clock, which can be disturbed by circumstances that alter a person's accustomed exposure to light and dark. Jet lag is an example. Some air travelers use melatonin supplements to try to adjust their sleep/wake cycles more quickly.

Depression, intense sleepiness, and other symptoms of seasonal affective disorder, or **SAD**, hit some people in winter. SAD may be due to a biological clock that is out of sync with changes in day length during winter, when days are shorter and nights longer. The symptoms get worse if a person takes melatonin. They improve when the person is exposed to intense light, which shuts down the pineal gland.

Melatonin may affect the gonads. A decline in melatonin production starts at puberty and may help trigger it. Some pineal gland disorders accelerate or delay puberty.

Skeletal muscle, the thymus, the heart, and the GI tract also produce hormones

You may recall from Section 6.8 that a newly discovered hormone, dubbed irisin (after the mythological Greek goddess Iris who acted as a messenger), is produced in skeletal muscle during and after exercise. Irisin's targets include white adipose cells, which respond by converting to a form (brown fat) that is more readily "burned" as bodily fuel. The thymus gland (see Figure 15.1) releases hormones called thymosins that help infection-fighting T cells mature. The two heart atria secrete *atrial natriuretic peptide,* or ANP. When your blood pressure rises, ANP acts to inhibit the reabsorption of sodium ions—and hence water—in the kidneys. More water is excreted, the blood volume decreases, and blood pressure falls.

Chapter 11 noted that the GI tract produces several hormones that influence appetite or have roles in digestion. For example, gastrin stimulates the release of stomach acid when proteins are being digested. Secretin stimulates the pancreas to secrete bicarbonate.

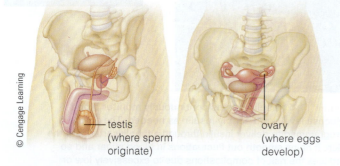

testis
(where sperm originate)

ovary
(where eggs develop)

Figure 15.13 Male and female gonads produce sex hormones as well as sex cells (sperm in males and eggs in females).

TAKE-HOME MESSAGE

WHAT ARE SOME OTHER EXAMPLES OF HORMONE SOURCES IN THE BODY?

- A female's ovaries or a male's testes are gonads that produce sex hormones as well as gametes (eggs or sperm).
- The pineal gland is in the brain and produces melatonin, which influences sleep-wake cycles and the onset of puberty.
- Skeletal muscle produces irisin, which converts white fat to brown fat.
- The thymus is in the chest and secretes thymosins that are necessary for the maturation of T cells.
- The heart atria produce ANP, which helps regulate blood pressure.
- The GI tract produces hormones that have roles in digestion.

CONNECTIONS

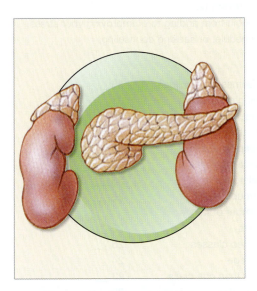

The Endocrine System

The endocrine system produces hormones, signaling molecules that travel in the bloodstream to nearly all body cells. Each kind of hormone influences the activity of its target cells. Along with signals of the nervous system, these changes adjust body functions in ways that maintain homeostasis in the body as whole.

In general, responses to hormones take longer and last longer than responses to nerve impulses. Hormones govern long-term events such as bodily growth and metabolism.

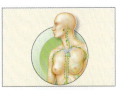

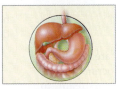

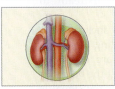

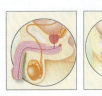

Skeletal system
Growth hormone stimulates the growth of bones. Parathyroid hormone (PTH) is the main regulator of blood calcium levels. Calcitonin stimulates uptake of calcium from blood as needed to form bone tissue.

Muscular system
Growth hormone stimulates development of skeletal muscle mass. PTH adjusts blood levels of calcium and potassium, electrolytes that are essential for muscle contraction.

Cardiovascular system and blood
Epinephrine adjusts heart rate and helps maintain blood pressure. Erythropoietin from the kidneys stimulates production of red blood cells. Aldosterone indirectly (via the kidneys) helps restore falling blood volume and pressure. PTH adjusts blood levels of calcium needed for cardiac muscle contraction.

Immunity and the lymphatic system
Thymus hormones stimulate T cells to mature. Cortisol from the adrenal cortex increases blood levels of glucose, amino acids, and other molecules used in tissue repair.

Digestive system
Insulin and GH support the delivery of nutrients to all cells by stimulating cells to take up glucose from the bloodstream.

Urinary system
Aldosterone and ANP support the urinary system's management of salt–water balance by promoting or reducing the reabsorption of sodium.

Nervous system
Epinephrine supports the sympathetic nervous system in the fight–flight response and helps the CNS regulate blood pressure. Hormones that regulate blood sugar ensure adequate fuel for brain cells.

Reproductive system
The hypothalamus regulates the release of sex hormones that govern the development and functioning of ovaries and testes (the gonads). Oxytocin triggers uterine muscle contractions during labor and (with prolactin) for milk release for a nursing infant. Luteinizing hormone (LH) and follicle-stimulating hormone (FSH) also have key roles in reproduction.

This Student Stress Scale lists a variety of life events that cause stress for young adults. The score for each event represents its relative impact on stress-related physiological responses. In general, people who score 300 points or more have the highest stress-related health risk. A score of 150–300 points indicates a moderate (50–50) stress-related health risk. A score below 150 indicates the lowest stress-related health risk, about a 1 in 3 chance of a significant, negative change in health status.

Although this test is only a general measure of stress, it can help you decide if you can benefit from adding to or improving your stress management activities, such as getting exercise, including some "down time" in your daily schedule, or seeking counseling.

Event	Points		Event	Points
Death of a close family member	100 ___		First quarter/semester in college	35 ___
Death of a close friend	73 ___		Change in living situation	31 ___
Parents' divorce	65 ___		Serious argument with instructor	30 ___
Jail term	63 ___		Lower grades than expected	29 ___
Major personal injury or illness	63 ___		Change in sleeping habits	29 ___
Marriage	58 ___		Change in social activities	29 ___
Being fired from a job	50 ___		Change in eating habits	28 ___
Failing an important course	47 ___		Chronic car trouble	26 ___
Change in health of family member	45 ___		Change in number of family	
Pregnancy (or causing one)	45 ___		get-togethers	26 ___
Sex problems	44 ___		Too many missed classes	25 ___
Serious argument with close friend	40 ___		Change of college	24 ___
Change in financial status	39 ___		Dropping more than one class	23 ___
Change of major	39 ___		Minor traffic violations	20 ___
Trouble with parents	39 ___			
New romantic interest	38 ___			
Increased workload at school	37 ___		Total _____	
Outstanding personal achievement	36 ___			

Adapted from the Holmes and Rahe Life Event Scale.

SUMMARY

Section 15.1 Hormones are produced by cells and glands of the endocrine system. They move through the bloodstream to distant target cells.

Other signaling molecules include neurotransmitters and local signaling molecules such as prostaglandins. All are chemicals released in small amounts by one cell and adjust the behavior of other target cells. Any cell with receptors for the signal is the target.

Hormones may interact in opposition, synergistically (in cooperation), or permissively (a target cell must first be primed by one hormone in order to respond to a second one).

Section 15.2 Steroid and nonsteroid hormones act on target cells by different mechanisms.

Receptors for steroid (and thyroid) hormones are inside target cells. A hormone-receptor complex binds to DNA. Binding activates genes and protein-making processes.

Amine, peptide, and protein hormones interact with receptors on the plasma membrane of target cells. Often a second messenger, such as cyclic AMP, carries their signals inside the cell.

Most nonsteroid hormones alter the activity of target cell proteins. The resulting target cell responses help maintain homeostasis in extracellular fluid or contribute to normal development or reproductive functioning.

Sections 15.3, 15.4 The hypothalamus and pituitary gland interact to integrate many body activities.

ADH and oxytocin from the hypothalamus are stored in and released from the posterior lobe of the pituitary. ADH influences fluid volume. Oxytocin affects reproductive functions such as lactation and labor.

Additional hypothalamic hormones are releasers or inhibitors of hormones secreted by the anterior lobe of the pituitary gland.

Of the six hormones produced in the anterior lobe, two (prolactin and growth hormone) have widespread effects on body cells. Four (ACTH, TSH, FSH, and LH) act on specific endocrine glands.

Hormones are released by a wide variety of organs, tissues, and cells. They typically regulate events that occur over an extended period, such as bodily growth.

Sections 15.5, 15.6 Growth hormone (GH) influences growth throughout the body, but effects are most obvious in bones and skeletal muscles.

Thyroid hormone affects overall metabolism, growth, and development. The thyroid also makes calcitonin, which helps lower blood levels of calcium and phosphate. Parathyroid hormone is the main regulator of blood calcium levels.

Section 15.7 The adrenal cortex makes two kinds of steroid hormones, the glucocorticoids and mineralocorticoids. Cortisol and other glucocorticoids raise the blood level of glucose and reduce inflammation. Mineralocorticoids adjust levels of minerals such as potassium and sodium in body fluids.

The adrenal medulla releases epinephrine and nor-epinephrine. Their hormonelike effects include the regulation of blood pressure and the metabolism of carbohydrates. (Some neurons also release them as neurotransmitters.)

Sections 15.8, 15.9 Blood levels of glucose are regulated by insulin and glucagon, which are secreted in the pancreatic islets by beta and alpha cells, respectively. Insulin stimulates muscle and adipose cells to take up glucose, while glucagon stimulates glucose-releasing reactions in muscle and the liver. Negative feedback governs both processes. Somatostatin released by islet delta cells can inhibit the release of insulin, glucagon, and some other hormones.

In blood sugar disorders, a lack of insulin or cells' inability to respond to it unbalances blood glucose levels.

Section 15.10 The gonads produce sex hormones. A female's ovaries mainly make estrogens and progesterone and a male's testes mainly make testosterone. The pineal gland in the brain produces melatonin in response to light/dark cycles. Melatonin influences sleep/wake cycles as part of an internal biological clock.

The thymus makes hormones that help T cells mature. The heart secretes ANP, which helps regulate blood pressure. The GI tract secretes several hormones that function in digestion.

REVIEW QUESTIONS

1. Distinguish among hormones, neurotransmitters, local signaling molecules, and pheromones.

2. A hormone molecule binds to a receptor on a cell membrane. It doesn't enter the cell; rather, the binding activates a second messenger inside the cell that triggers an amplified response to the hormonal signal. Is the signaling molecule a steroid or a nonsteroid hormone?

3. Which hormones produced in the posterior and anterior lobes of the pituitary gland have the targets indicated? *At right, fill in the blanks using the abbreviations noted in Section 15.3.*

4. Name the main endocrine glands and state where each is located in the body.

5. Give two examples of feedback control of hormone activity.

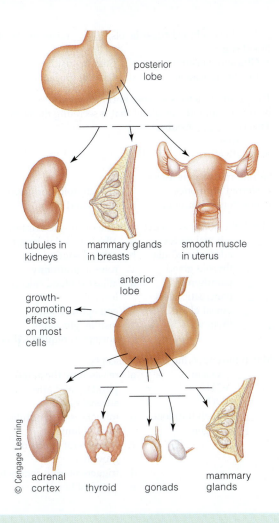

posterior lobe

tubules in kidneys | mammary glands in breasts | smooth muscle in uterus

anterior lobe

growth-promoting effects on most cells

© Cengage Learning

adrenal cortex | thyroid | gonads | mammary glands

6. _____ lowers blood sugar levels; _____ raises the level of blood sugar.
   - a. Glucagon; insulin
   - b. Insulin; glucagon
   - c. Gastrin; insulin
   - d. Gastrin; glucagon

7. The pituitary detects a rising hormone concentration in blood and inhibits the gland that is secreting the hormone. This is a _____ feedback loop.
   - a. positive
   - b. negative

8. Second messengers assist _____.
   - a. steroid hormones
   - b. nonsteroid hormones
   - c. only thyroid hormones
   - d. both a and b

9. Match the hormone source with the closest description.
   - _____ adrenal cortex
   - _____ adrenal medulla
   - _____ thyroid gland
   - _____ parathyroids
   - _____ pancreatic islets
   - _____ pineal gland
   - _____ thymus
     - a. affected by day length
     - b. cortisol source
     - c. roles in immunity
     - d. adjust(s) blood calcium level
     - e. epinephrine source
     - f. insulin, glucagon
     - g. hormones require iodine

10. Match the endocrine control concepts.
    - _____ oxytocin
    - _____ ACTH
    - _____ ADH
    - _____ growth hormone
    - _____ estrogen
      - a. released by the anterior pituitary and affects the adrenal gland
      - b. influences extracellular fluid volume
      - c. has general effects on growth
      - d. triggers uterine contractions
      - e. a steroid hormone

CRITICAL THINKING

1. Addison's disease develops when the adrenal cortex does not secrete enough mineralocorticoids and glucocorticoids. President John F. Kennedy was diagnosed with the disease when he was a young man. Before he started treatment with hormone replacement therapy, he was hypoglycemic and lost weight. Which missing hormone was responsible for his weight loss? How might Addison's disease have affected his blood pressure?

2. A physician sees a patient whose symptoms include sluggishness, depression, and intolerance to cold. After eliminating other possible causes, the doctor diagnoses a hormone problem. What disorder fits the symptoms? Why does the doctor suspect that the underlying cause is a malfunction of the anterior pituitary gland?

3. Marianne has type 1 diabetes. One day, after accidentally injecting herself with too much insulin, she starts to shake and feels confused. Following her doctor's suggestion, she drinks a glass of orange juice—a ready source of glucose—and soon her symptoms subside. What caused her symptoms? How would a glucose-rich snack help?

4. Secretion of the hormone ADH may decrease or stop if the pituitary's posterior lobe is damaged, as by a blow to the head. This is one cause of *diabetes insipidus*. People with this form of diabetes excrete so much dilute urine that they may become seriously dehydrated. Where are the target cells of ADH?

REPRODUCTIVE SYSTEMS

LINKS TO EARLIER CONCEPTS

This chapter builds on knowledge of hormones, including the steroid sex hormones estrogen and testosterone (15.1, 15.2).

You will see how negative feedback loops (4.10) regulate the production of sperm in males and the menstrual cycle in females.

You will also learn more about flagella, which propel sperm (3.9), and about chromosomes, the structures that carry genes (3.6).

We will also survey the viruses, bacteria, and other pathogens that cause sexually transmitted diseases, and gain a fuller understanding of major reproductive cancers.

KEY CONCEPTS

The Female Reproductive System

Ovaries are the primary reproductive organs of females. Hormones control their functions, such as the development of oocytes (eggs). Sections 16.1–16.2

The Male Reproductive System

A male's reproductive system consists of testes and accessory ducts and glands. Hormones control its functions, including making sperm. Sections 16.3–16.4

Sexual Intercourse and Fertility

Sexual intercourse between a male and female is the usual first step toward pregnancy. Various methods exist for limiting or enhancing fertility. Sections 16.5–16.8

Sexually Transmitted Diseases and Cancers of the Reproductive System

Sexual contact can transmit bacteria, viruses, and other disease-causing pathogens. Sections 16.9–16.12

Roughly 10 pounds. That's the combined weight of octuplets—six girls and two boys—born prematurely to a Texas mother. She had received a fertility drug, which caused many of her eggs to be ovulated at the same time. Missing from the photo below is Odera, the smallest, who weighed less than a pound (520 grams) and died of heart and lung failure after 6 days. The other newborns were in the hospital for 3 months before going home.

The incidence of triplets and other higher-order multiple births has increased sharply in recent decades, a statistic that worries some doctors. Carrying more than one embryo increases the risk of miscarriage, prematurity, and delivery complications. The babies also are more likely to have development delays.

The reproductive system is our focus in this chapter. It is the only body system that does not contribute to homeostasis. Instead, its biological role is to continue the human species.

© 1999 Dana Fineman

Top two: © Cengage Learning; Second from bottom: © David M. Phillips/Photo Researchers, Inc.
Bottom: George Musil/Visuals Unlimited

16.1 The Female Reproductive System

■ The biological function of the female reproductive system is to nurture developing offspring from the time of conception until birth.

cervix The lower portion of the uterus.

endometrium Lining of the uterus.

estrogens Female sex hormones that influence the development of female secondary sexual traits and function in the menstrual cycle.

gametes Eggs and sperm.

germ cells Cells that give rise to gametes.

menarche A female's first menstruation.

menopause The end of menstrual cycling and a female's fertility.

menstrual cycle Monthly cycle in a sexually mature female during which an oocyte matures and is released from an ovary.

menstruation Shedding of the blood-rich uterine lining (endometrium) at the start of each new menstrual cycle.

oocytes Immature eggs.

ovaries The primary reproductive organs of females.

oviduct Tube that carries oocytes (eggs) from an ovary to the uterus.

progesterone Female sex hormone that helps stimulate growth of the endometrium (uterine lining) in preparation for a possible pregnancy.

secondary sexual trait Trait that is associated with maleness or femaleness, but that is not directly involved with reproduction.

uterus Organ in which a baby can grow and develop prior to birth.

vagina Channel that receives the penis and sperm and serves as part of the birth canal.

Remember from Section 15.10 that **ovaries** are a female's primary reproductive organs—her gonads. The ovaries contain **germ cells** that produce eggs. The word *germ* comes from a Latin word that means "to sprout." A male also has germ cells, in his testes. Eggs and sperm are sometimes called **gametes** (GAM-eets), from a Greek word that means "to marry." Reproductive organs in both sexes also release hormones that guide reproduction and the development of **secondary sexual traits.**

Ovaries are a female's primary reproductive organs

Figure 16.1 shows the parts of the female reproductive system, and summarizes their functions. The ovaries release sex hormones including estrogens, and during a woman's reproductive years they also produce eggs. **Estrogens** influence the development of female secondary sexual traits, such as the "filling out" of breasts, hips, and buttocks by fat deposits. Estrogens also help govern the menstrual cycle, described shortly.

Immature eggs are called **oocytes**. When an oocyte is released from an ovary, it moves into a nearby **oviduct** (also called a *fallopian tube*). Fertilization usually occurs while an egg is in an oviduct. Regardless, an egg travels down the oviduct into the **uterus**. In this organ, a baby can grow and develop. The wall of the uterus consists of a thick layer of smooth muscle (the myometrium) and a lining, the **endometrium**. The endometrium includes epithelium, connective tissue, glands, and blood

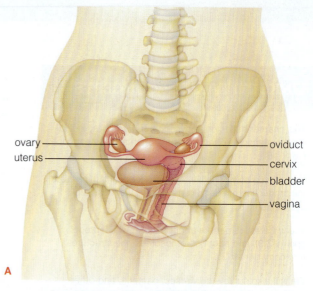

A

Figure 16.1 **The female reproductive system includes ovaries, oviducts, the uterus, the cervix, and the vagina. Part A shows where these structures are located. Part B diagrams these and related structures and summarizes their functions.** (© Cengage Learning)

vessels. The lower part of the uterus is the **cervix**. The muscular **vagina** leads from the cervix to the outside. It receives the penis and sperm and serves as part of the birth canal.

A female's outer genitals collectively form the *vulva*. Outermost are a pair of fat-padded skin folds, the *labia majora*. They enclose smaller folds, the *labia minora*, that are laced with blood vessels. The labia minora partly enclose the *clitoris*, a small organ sensitive to sexual stimulation.

A female's urethra opens about midway between her clitoris and her vaginal opening. Whereas in males the urethra carries both urine and sperm, in females it is separate and is not involved in reproduction.

During the menstrual cycle, an oocyte is released from an ovary

Like all female primates, a woman has a **menstrual cycle**. It takes about 28 days to complete one cycle, although this can vary from month to month and from woman to woman. During the cycle, an oocyte matures and is released from an ovary. Meanwhile, hormones are preparing the endometrium to receive and nourish an embryo in case the oocyte is fertilized. If the oocyte is *not* fertilized, a blood-rich fluid starts flowing out through the vaginal canal. This flow is **menstruation**, and it marks the first day of a new cycle. The disintegrating endometrium is being sloughed off, only to be rebuilt once again during the next cycle.

The menstrual cycle advances through three phases (Table 16.1). It starts with a *menstrual phase*. This is the time

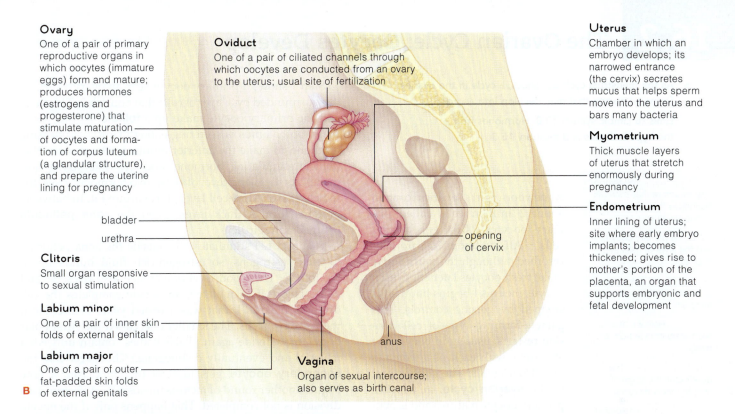

Ovary
One of a pair of primary reproductive organs in which oocytes (immature eggs) form and mature; produces hormones (estrogens and progesterone) that stimulate maturation of oocytes and formation of corpus luteum (a glandular structure), and prepare the uterine lining for pregnancy

Oviduct
One of a pair of ciliated channels through which oocytes are conducted from an ovary to the uterus; usual site of fertilization

Uterus
Chamber in which an embryo develops; its narrowed entrance (the cervix) secretes mucus that helps sperm move into the uterus and bars many bacteria

Myometrium
Thick muscle layers of uterus that stretch enormously during pregnancy

Endometrium
Inner lining of uterus; site where early embryo implants; becomes thickened; gives rise to mother's portion of the placenta, an organ that supports embryonic and fetal development

bladder
urethra

opening
of cervix

Clitoris
Small organ responsive to sexual stimulation

Labium minor
One of a pair of inner skin folds of external genitals

Labium major
One of a pair of outer fat-padded skin folds of external genitals

B

anus

Vagina
Organ of sexual intercourse; also serves as birth canal

TABLE 16.1 Phases of the Menstrual Cycle

Phase	Events	Days of the Cycle*
Menstrual phase	Menstruation; endometrium breaks down	1–5
	Follicle matures in ovary; endometrium rebuilds	6–13
Proliferative phase	Endometrium begins to thicken, ovulation occurs	14
Progestational phase	Lining of endometrium develops to receive a possible embryo	15–28

*Assumes a 28-day cycle.

of menstruation, when the endometrium disintegrates. Next comes the *proliferative phase*, when the endometrium begins to thicken again. The end of this phase coincides with ovulation—the release of an oocyte from an ovary. During the cycle's final phase, called the *progestational* ("before pregnancy") *phase*, an endocrine structure called the corpus luteum ("yellow body") forms. It secretes a flood of estrogens and another female sex hormone, **progesterone**. Together these hormones prime the endometrium for pregnancy. Feedback loops to the hypothalamus and pituitary gland from the ovaries govern the menstrual cycle.

A female's first menstruation, or **menarche**, usually occurs between the ages of 10 and 16. Menstrual cycles continue until the end of **menopause**, which usually occurs in a woman's early 50s. By then, her ovaries are making less estrogen and progesterone, and also are less sensitive to reproductive hormones from the pituitary. Falling estrogen levels may trigger a range of temporary symptoms, including moodiness, insomnia, and "hot flashes" (bouts of sweating and uncomfortable warmth). Other changes include reduced natural lubrication and thinning of the vaginal wall. The fertile phase of a woman's life ends when her menstrual cycles stop.

In the disorder called **endometriosis** endometrial tissue grows outside the uterus. Scar tissue may form on one or both ovaries or oviducts, leading to infertility. Endometriosis may develop when menstrual flow backs up through the oviducts and spills into the pelvic cavity. Or perhaps some cells became situated in the wrong place when the woman was a developing embryo, then were stimulated to grow during puberty, when her sex hormones became active. Regardless, the symptoms include pain during menstruation, sex, or urination. Treatment ranges from doing nothing in mild cases to surgery to remove the abnormal tissue or sometimes even the whole uterus.

TAKE-HOME MESSAGE

HOW DO A FEMALE'S OVARIES FUNCTION IN REPRODUCTION?

- The ovaries are primary reproductive organs that produce oocytes (immature eggs) and female sex hormones.
- Female sex hormones—estrogens and progesterone—are released as part of a recurring menstrual cycle.
- Menopause marks the natural end of a woman's fertility.

The Ovarian Cycle: Oocytes Develop

- As the menstrual cycle advances, a cycle in the ovaries forms an oocyte that may develop into an egg.
- Links to Limbic system 13.8, Hormones from the hypothalamus and pituitary 15.3

corpus luteum The temporary structure that secretes hormones that prepares the uterus for an embryo.

follicle A primary oocyte and the layer of cells that nourish it.

ovarian cycle The cycle in which a primary oocyte matures.

ovulation Release of a secondary oocyte from an ovary.

secondary oocyte The developmental stage of an oocyte that is ovulated.

zona pellucida The protein layer around an ovarian follicle.

Hormones guide ovulation

A newborn girl's ovaries contain about 2 million cells called primary oocytes ("first egg-forming cells"). All but about 300 are later resorbed, although the ovaries may make fresh oocytes later on. In each oocyte, meiosis I begins but then is stopped by genetic controls. This gamete-forming type of cell division restarts, usually in one oocyte at a time, with each of a woman's menstrual cycles. The shift is part of the **ovarian cycle**, in which a primary oocyte matures and is ovulated (Figure 16.2).

Step 1 shows a primary oocyte near an ovary's surface. It is surrounded by a layer of cells that nourish it. This layer and the primary oocyte make up a **follicle**. At this point, the hypothalamus is secreting enough GnRH, a releasing hormone, to make the anterior pituitary release more FSH (follicle stimulating hormone) and LH (luteinizing hormone). As the blood level of those two hormones rises, the follicle grows. More cell layers form around it. In between, proteins form a thick layer called the **zona pellucida** ("transparent girdle").

FSH and LH stimulate cells outside the zona pellucida to make estrogens, so estrogen-rich fluid builds up in the follicle. The blood level of estrogen also rises. Several hours before it is ovulated, an oocyte completes the cell division, meiosis I, that was arrested years before. Now, there are two cells. The smaller one, called the "first polar body," may divide again. (Polar bodies contain unneeded material and eventually disintegrate.) The larger cell, the **secondary oocyte**, gets most of the cytoplasm. It now begins another round of meiosis (meiosis II). As before, this division is not completed. That happens only if the oocyte is fertilized.

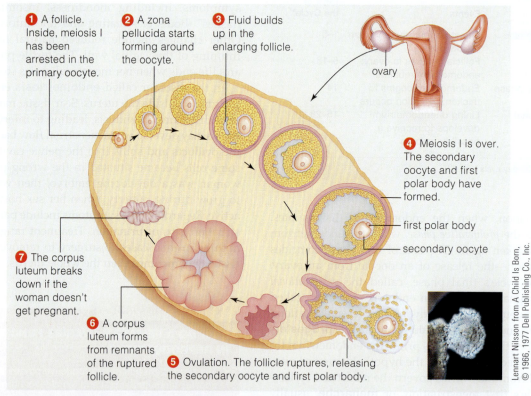

1 A follicle. Inside, meiosis I has been arrested in the primary oocyte.

2 A zona pellucida starts forming around the oocyte.

3 Fluid builds up in the enlarging follicle.

ovary

4 Meiosis I is over. The secondary oocyte and first polar body have formed.

first polar body

secondary oocyte

7 The corpus luteum breaks down if the woman doesn't get pregnant.

6 A corpus luteum forms from remnants of the ruptured follicle.

5 Ovulation. The follicle ruptures, releasing the secondary oocyte and first polar body.

Lennart Nilsson from A Child Is Born, © 1966, 1977 Dell Publishing Co., Inc.

Figure 16.2 Animated! **Oocytes develop by way of cyclic changes in the ovary.** A follicle stays in the same place in an ovary all through the ovarian cycle. It does not "move around" as in this diagram, which shows the sequence of events. In the cycle's first phase, a follicle grows and matures. The micrograph shows a secondary oocyte being released from an ovary. It will enter an oviduct, the channel to the uterus. (© Cengage Learning)

About halfway through the ovarian cycle, a woman's pituitary gland detects the rising estrogen level. It releases LH, which causes changes that make the follicle swell. The surge also causes enzymes to break down the bulging follicle wall. When the follicle ruptures—the event we call **ovulation**—fluid escapes, along with the secondary oocyte and polar body (Figure 16.2, step 5).

Once it is in the abdominal cavity, the secondary oocyte normally enters an oviduct. Long, ciliated projections from the oviduct (called *fimbriae*) extend over part of the ovary. Movements of the projections and cilia sweep the oocyte into the channel. If fertilization takes place, the oocyte will finish meiosis II and become a mature egg.

The ovarian and menstrual cycles dovetail

You may remember from Section 16.1 that estrogens released early in the menstrual cycle stimulate growth of the endometrium and its blood vessels and glands. These changes pave the way for a possible pregnancy. Just before the midcycle LH surge, cells of the follicle wall start releasing estrogens and progesterone. When ovulation occurs, the estrogens act on tissue around the cervical canal, which opens into the vagina. The cervix starts to secrete large amounts of a thin, clear mucus, which is ideal for sperm to swim through.

As diagrammed in Figure 16.3A and B, the midcycle surge of LH triggers formation of a **corpus luteum** ("yellow body"). This structure develops from cells left behind in the follicle, and it secretes some estrogen and progesterone (Figure 16.3C). The progesterone prepares the uterus for an embryo. For example, it causes mucus in the cervix to become thick and sticky, which may prevent bacteria from entering the uterus. Progesterone also maintains the endometrium during a pregnancy.

A corpus luteum lasts for about 12 days. In that time, the hypothalamus signals for a decrease in FSH, which prevents other follicles from developing. If no embryo implants in the endometrium, the corpus luteum begins to disintegrate. After it breaks down, progesterone and estrogen levels drop, so the endometrium also breaks down and menstruation begins (Figure 16.3D).

TAKE-HOME MESSAGE

WHAT HORMONAL CHANGES TRIGGER THE GROWTH AND RELEASE OF A SECONDARY OOCYTE FROM AN OVARY?

- Shifts in FSH and LH cause a follicle (primary oocyte and support cells) to grow.
- A midcycle surge of LH triggers ovulation, in which a secondary oocyte is released from the ovary.
- The cyclic release of estrogen and progesterone helps pave the way for fertilization of an egg and pregnancy.

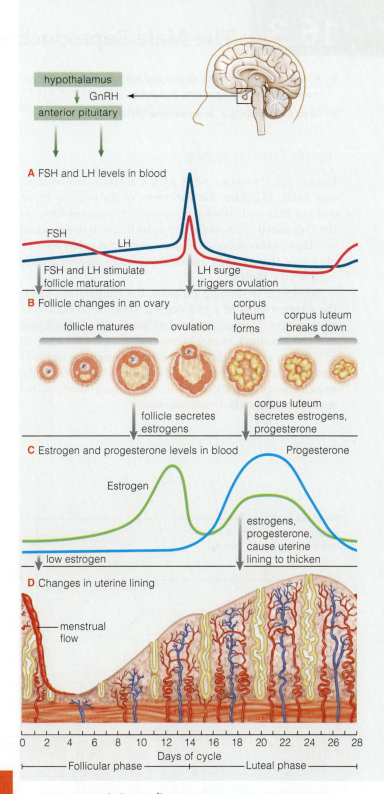

Figure 16.3 Animated! Hormones govern the menstrual and ovarian cycles. **A** GnRH from the hypothalamus stimulates the anterior pituitary to secrete FSH and LH. **B** FSH and LH stimulate a follicle to grow, an oocyte to mature, and the ovaries to secrete progesterone and estrogens that stimulate the endometrium to rebuild. **C** A midcycle LH surge triggers ovulation and the formation of a corpus luteum. **D** Progesterone and some estrogens released by the corpus luteum maintain the endometrium, but if no pregnancy occurs, they stop being released and the corpus luteum breaks down. (© Cengage Learning)

The Male Reproductive System

- A male's testes produce sperm and hormones that govern male reproductive functions and traits.
- Links to Flagella 3.9, Testosterone 15.10

Sperm form in testes

Figure 16.4 shows an adult male's reproductive system, and Table 16.2 lists the functions of its organs. In an embryo that will develop as a male, two testes form on the abdominal cavity wall. Before birth, the testes descend into the scrotum, a pouch of skin suspended below the pelvic girdle (Figure 16.5). Inside this pouch, smooth muscle encloses the **testes**.

For sperm to develop properly, the temperature inside the scrotum must be a few degrees cooler than body core temperature. To this end, a control mechanism helps assure that the scrotum's internal temperature is always close to 95°F. When a male feels cold (or afraid), muscle contractions draw his testes closer to his body. When he feels warm, the muscles relax and allow the testes to hang lower, so the sperm-making cells do not overheat.

TABLE 16.2 Male Reproductive System

REPRODUCTIVE ORGANS	
Testis (2)	Sperm, sex hormone production
Epididymis (2)	Site of sperm maturation and subsequent storage
Vas deferens (2)	Rapid transport of sperm
Ejaculatory duct (2)	Conduct sperm to penis
Penis	Organ of sexual intercourse
ACCESSORY GLANDS	
Seminal vesicle (2)	Secrete most fluid in semen
Prostate gland	Secretes some fluid in semen
Bulbourethral gland (2)	Secrete lubricating mucus

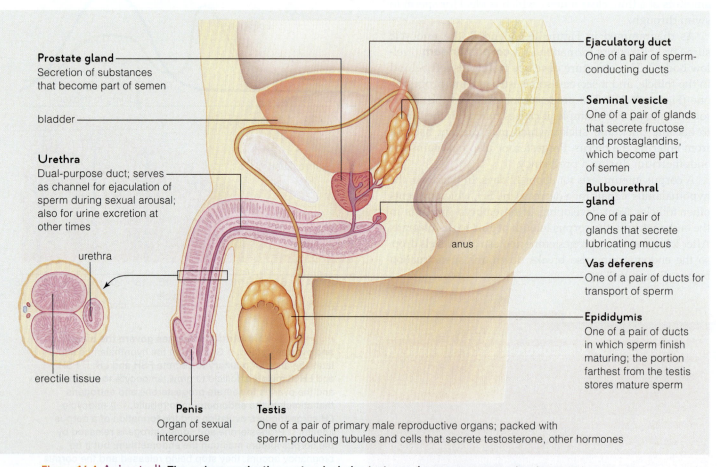

Prostate gland
Secretion of substances that become part of semen

bladder

Urethra
Dual-purpose duct; serves as channel for ejaculation of sperm during sexual arousal; also for urine excretion at other times

urethra

erectile tissue

Penis
Organ of sexual intercourse

Testis
One of a pair of primary male reproductive organs; packed with sperm-producing tubules and cells that secrete testosterone, other hormones

Ejaculatory duct
One of a pair of sperm-conducting ducts

Seminal vesicle
One of a pair of glands that secrete fructose and prostaglandins, which become part of semen

Bulbourethral gland
One of a pair of glands that secrete lubricating mucus

anus

Vas deferens
One of a pair of ducts for transport of sperm

Epididymis
One of a pair of ducts in which sperm finish maturing; the portion farthest from the testis stores mature sperm

Figure 16.4 Animated! The male reproductive system includes testes and many accessory structures. (© Cengage Learning)

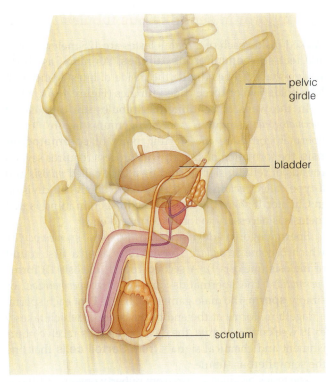

pelvic girdle

bladder

scrotum

Figure 16.5 The male reproductive system is located in the lower pelvic region. (© Cengage Learning)

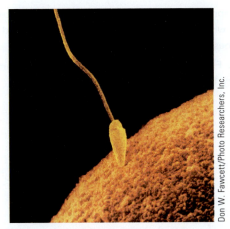

Sperm cell arriving at an egg

Sperm mature and are stored in the coiled epididymis

When sperm leave a testis they enter a long, coiled duct called an **epididymis** (ep-ih-DID-ih-muss; plural: epididymides). At this point, the sperm aren't mature. Gland cells in the walls of the ducts secrete substances that trigger final developmental changes. Until sperm leave the body, they are stored in the last stretch of each epididymis.

When a male is sexually aroused, muscle contractions propel mature sperm from each epididymis into and through a pair of thick-walled tubes. Each tube is called a **vas deferens**. From there, contractions move sperm through the two ejaculatory ducts and on through the urethra to the outside. The urethra passes through the **penis**, the male sex organ, and also carries urine.

Substances from seminal vesicles and the prostate gland help form semen

Secretions from several glands mix with sperm as they travel through the urethra. The result is **semen**, a thick fluid that is eventually expelled from the penis during sexual activity. As semen begins to form, a pair of **seminal vesicles** secrete fructose. The sperm use this sugar for energy. Seminal vesicles also secrete certain kinds of prostaglandins. You may recall from Chapter 15

that prostaglandins are signaling molecules that can trigger muscle contractions. During sex, the prostaglandins cause smooth muscles of a female's reproductive tract to contract, and so aid the movement of sperm through it toward the egg.

Substances secreted by the **prostate gland** may help buffer the acidic environment that sperm encounter in the female reproductive tract. The vaginal pH is about 3.5 to 4.0, but sperm motility improves at pH 6. **Bulbourethral glands** secrete mucus-rich fluid into the urethra when a male is sexually aroused. This fluid neutralizes acids in any traces of urine in the urethra. The more alkaline surroundings creates a more favorable chemical environment for the 150 to 350 million sperm that pass through the urethra in a typical ejaculation.

bulbourethral glands Glands of the male reproductive system that produce mucus.

epididymis Duct where sperm mature and are stored.

penis The male sex organ.

prostate gland Gland of the male reproductive system that secretes some substances in semen.

semen The fluid containing sperm that is expelled from the penis during sexual activity.

seminal vesicles Glands that secrete most of the fluid in semen.

testes A male's gonads; his primary reproductive organs.

vas deferens Tube that carries sperm from an epididymis to the ejaculatory duct.

TAKE-HOME MESSAGE

HOW DO A MALE'S TESTES FUNCTION IN REPRODUCTION?

- The testes are primary reproductive organs that produce sperm and male sex hormones.
- The male reproductive system also includes accessory glands and ducts.
- When sperm are nearly mature, they leave each testis and enter the long, coiled epididymis, where they remain until ejaculated.
- Secretions from the seminal vesicles and the prostate gland mix with sperm to form semen.

- In his reproductive years, a male continually produces sperm, which develop in a series of steps controlled by hormones.

- Links to Flagella 3.9, Hormones from the hypothalamus and pituitary 15.3

Sperm form in seminiferous tubules

Packed inside each of a male's testes are 125 meters—over 400 feet—of **seminiferous tubules**. As many as thirty wedge-shaped lobes divide the inside of a testis and each lobe holds two or three coiled tubules (Figure 16.6A, B).

In the walls of seminiferous tubules are cells called *spermatogonia* (singular: spermatogonium; Figure 16.6C). Spermatogonia are the starting point of **spermatogenesis**, the formation of sperm. This process requires several rounds of cell division, including a type called *mitosis* and a type called *meiosis*. You'll read more about cell division in Chapter 18; here the main thing to keep in mind is that meiosis is necessary to form sperm and eggs.

Spermatogonia develop into *primary spermatocytes*, which become *secondary spermatocytes* after a first round of meiosis (meiosis I). A second round (meiosis II) forms *spermatids*. The spermatids develop into *spermatozoa*, or simply **sperm**, the male gametes. The "tail" of each sperm, a flagellum, forms at the end of the process, which takes 9 to 10 weeks. Meanwhile, the developing cells receive nourishment and chemical signals from **Sertoli cells** that line the seminiferous tubule.

The testes produce sperm from puberty onward. Millions are in different stages of development on any given day. A mature sperm has a tail, a midpiece, and a head (Figure 16.7). Inside the head, a nucleus contains DNA organized into chromosomes. A cap, the **acrosome**, covers most of the head. Enzymes it releases help the sperm penetrate protective material around an egg at fertilization. In the midpiece, mitochondria supply energy for the tail's movements.

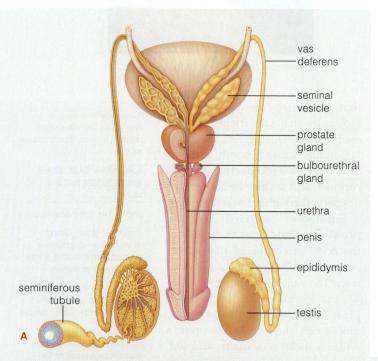

seminiferous tubule

vas deferens

seminal vesicle

prostate gland

bulbourethral gland

urethra

penis

epididymis

testis

A

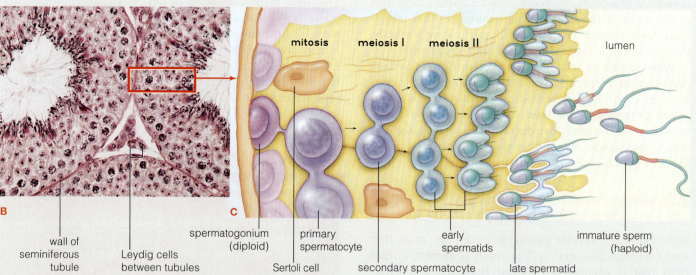

mitosis meiosis I meiosis II lumen

B

C

wall of seminiferous tubule

Leydig cells between tubules

spermatogonium (diploid)

Sertoli cell

primary spermatocyte

secondary spermatocyte

early spermatids

late spermatid

immature sperm (haploid)

Figure 16.6 Animated! Sperm form inside seminiferous tubules in the testes. **A** The male reproductive tract from behind. **B** Cells in three neighboring seminiferous tubules. Leydig cells in spaces between tubules make testosterone. **C** How sperm form, starting with a diploid germ cell. (A: © Cengage Learning; B: © Ed Reschke/Peter Arnold; C: © Cengage Learning)

Hormones control the formation of sperm

Male reproductive function depends on several hormones. **Leydig cells** (also called interstitial cells) in tissue between the seminiferous tubules in testes (Figure 16.6B), release **testosterone**. This is the hormone that governs the growth, form, and functions of the male reproductive tract. Testosterone stimulates sexual behavior, and at puberty it promotes the development of male secondary sexual traits, including facial hair and deepening of an adolescent male's voice.

When the testosterone level in a male's blood falls below a set point, the hypothalamus secretes GnRH (Figure 16.8). This releasing hormone prompts the pituitary's anterior lobe to release LH (luteinizing hormone) and FSH (follicle-stimulating hormone). These hormones are named for their functions in females, but are chemically the same in males. Both have targets in the testes. LH stimulates Leydig cells to release testosterone, which in turn stimulates the sperm-forming steps shown in Figure 16.6C. FSH acts on Sertoli cells. It is crucial to launching sperm formation at puberty.

A high level of testosterone in a male's blood inhibits the release of GnRH. Also, when a male's sperm count is high, Sertoli cells release inhibin, a hormone that acts on the hypothalamus and pituitary to inhibit the release of GnRH and FSH. Now feedback loops to the hypothalamus begin to operate, so the secretion of testosterone and the formation of sperm decline.

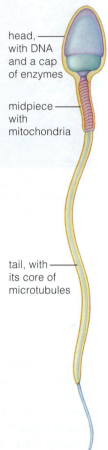

head, with DNA and a cap of enzymes

midpiece with mitochondria

tail, with its core of microtubules

Figure 16.7 A mature sperm has a head, a midpiece, and a tail.
(© Cengage Learning)

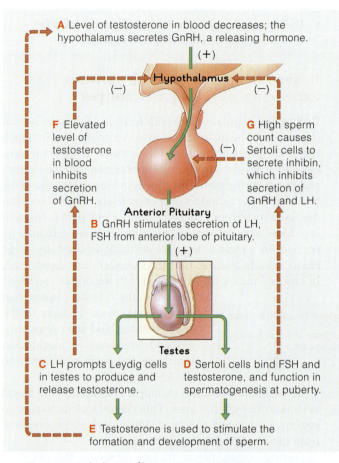

A Level of testosterone in blood decreases; the hypothalamus secretes GnRH, a releasing hormone.

(+)

Hypothalamus

(−)   (−)

F Elevated level of testosterone in blood inhibits secretion of GnRH.

(−)

G High sperm count causes Sertoli cells to secrete inhibin, which inhibits secretion of GnRH and LH.

Anterior Pituitary

B GnRH stimulates secretion of LH, FSH from anterior lobe of pituitary.

(+)

Testes

C LH prompts Leydig cells in testes to produce and release testosterone.

D Sertoli cells bind FSH and testosterone, and function in spermatogenesis at puberty.

E Testosterone is used to stimulate the formation and development of sperm.

Figure 16.8 Animated! Negative feedback loops regulate the release of male reproductive hormones.
(© Cengage Learning)

acrosome The cap of the head of a sperm cell; it contains enzymes.

Leydig cells Cells associated with seminiferous tubules and that secrete testosterone; also called interstitial cells.

seminiferous tubules Coiled tubules in the testes that contain the cells from which sperm develop.

Sertoli cells Cells in seminiferous tubules that produce substances that nourish developing sperm.

sperm Male gametes.

spermatogenesis Process in which sperm form.

testosterone Sex hormone that governs the growth, form, and functions of the male reproductive tract.

■ The penis and vagina are mechanically compatible for sexual intercourse, which may lead to pregnancy.

In sexual intercourse, both partners experience physiological changes

Coitus and copulation are both technical terms for sexual intercourse. The male sex act involves an *erection*, in which the limp penis stiffens and lengthens. It also involves *ejaculation*, the forceful expulsion of semen into the urethra and out from the penis. As shown in Figure 16.4, the penis has lengthwise cylinders of spongy tissue. The outer cylinder has a mushroom-shaped tip (the glans penis). Inside it is a dense array of sensory receptors that are activated by friction. In a male who is not sexually aroused, the large blood vessels leading into the cylinders are constricted. In aroused males, these blood vessels vasodilate, so blood flows into the cylinders faster than it flows out. Blood collects in the spongy tissue, and the organ stiffens and lengthens—a mechanism that helps the penis penetrate into the female's vagina.

coitus Sexual intercourse.

orgasm The culmination of the sex act.

In a female, arousal includes vasodilation of blood vessels in her genital area. This causes vulvar tissues to engorge with blood and swell. Mucus-rich secretions flow from the cervix, lubricating the vagina.

During coitus, pelvic thrusts stimulate the penis as well as the female's clitoris and vaginal wall. The stimulation triggers rhythmic, involuntary contractions in smooth muscle in the male reproductive tract, especially the vas deferens and the prostate. The contractions rapidly force sperm out of each epididymis. They also force the contents of seminal vesicles and the prostate gland into the urethra. The resulting mixture, semen, is ejaculated into the vagina.

During ejaculation, a sphincter closes off the neck of the male's bladder and prevents urine from being excreted. Ejaculation is a reflex response. This means that once it begins, it cannot be stopped.

Emotional intensity, heavy breathing, and heart pounding, as well as generalized contractions of skeletal muscles, accompany the rhythmic throbbing of the pelvic muscles. For both partners, **orgasm**—the culmination of the sex act—typically is accompanied by strong sensations of release, warmth, and relaxation.

Some people mistakenly believe that unless a woman experiences orgasm, she cannot become pregnant. This is not true, however. A female can become pregnant from intercourse regardless of whether she experiences orgasm, and even if she is not sexually aroused. All that is required is that a sperm meet up with a secondary oocyte that is traveling down one of her oviducts.

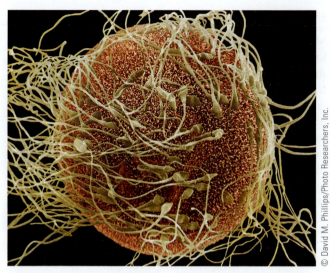

© David M. Phillips/Photo Researchers, Inc.

Figure 16.9 This image shows a secondary oocyte surrounded by sperm. If fertilization occurs, it will set the stage for a new individual to develop, continuing the human life cycle.

Intercourse can produce a fertilized egg

If sperm enter the vagina a few days before or after ovulation or anytime between, an ovulated egg may be fertilized. Within thirty minutes after ejaculation, muscle contractions in the uterus move the sperm deeper into the female reproductive tract. Only a few hundred sperm will actually reach the upper portion of the oviduct, which is where fertilization usually takes place. The remarkable micrograph in Figure 16.9 shows living sperm around a secondary oocyte.

TAKE-HOME MESSAGE

WHAT SORTS OF PHYSIOLOGICAL CHANGES OCCUR DURING SEXUAL INTERCOURSE?

- Sexual intercourse, or coitus, typically involves a series of physiological changes in both partners.
- During arousal, blood vessels dilate so that more blood flows to the penis (males) and vulva (females). Orgasm involves muscular contractions (including those leading to ejaculation of semen into the vagina) and sensations of release, warmth, and relaxation.
- Intercourse may lead to a pregnancy even if the female is not sexually aroused or she does not experience orgasm.

16.6 Fertilization

■ **Fertilization combines the genetic material in the father's sperm with that in the mother's egg.**

Fertilization is the fusion of an egg cell's nucleus and a sperm's nucleus. It begins when a sperm enters a secondary oocyte. After several steps, fertilization produces a **zygote** (ZYE-goat, "yoked together"), the first cell of the new individual. Figure 16.10 shows these steps.

As sperm swim through the cervix and uterus and into the oviducts, *capacitation* occurs. In this process, chemical changes weaken the membrane over the sperm's acrosome. Only a sperm that is capacitated ("made able") can fertilize an oocyte. Of the millions of sperm in the vagina after an ejaculation, just several hundred reach the upper part of an oviduct, where fertilization usually occurs. Contractions of smooth muscle in the uterus help move sperm toward the oviducts.

When a capacitated sperm contacts an oocyte, enzymes are released from the now-weakened membrane over the acrosome. These enzymes clear a path through the zona pellucida. Many sperm can reach and bind to the oocyte, Usually, however, only one sperm fuses with the oocyte. Rapid chemical changes in the oocyte's cell membrane prevent more sperm from entering.

Fusion with a sperm prompts the completion of the cell division process that began when the oocyte was being formed in an ovary (Section 16.2). The result is a mature egg, or **ovum** (plural: ova), plus another polar body. (Recall from Section 16.2 that one or two polar bodies are produced in the cell division step that gives rise to the secondary oocyte. Often three tiny polar bodies eventually are packaged with the ovum.) The nuclei of the sperm and ovum swell up, then fuse.

Each sperm and oocyte has twenty-three chromosomes, half the number in other body cells. Fertilization combines them into a full set of forty-six chromosomes. Thus a zygote has all the DNA required to guide development of the embryo.

fertilization Fusion of the nuclei of a egg and a sperm.

ovum A mature egg.

zygote The first cell of a new individual.

TAKE-HOME MESSAGE

WHAT IS THE ROLE OF FERTILIZATION IN REPRODUCTION?

- At fertilization, a sperm's nucleus fuses with the nucleus of an egg cell the result is a zygote, a single cell with a full set of chromosomes—half from the mother and half from the father.

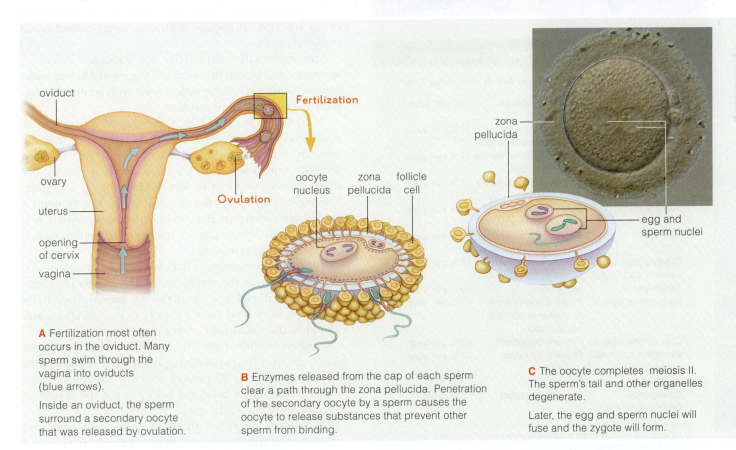

A Fertilization most often occurs in the oviduct. Many sperm swim through the vagina into oviducts (blue arrows).

Inside an oviduct, the sperm surround a secondary oocyte that was released by ovulation.

B Enzymes released from the cap of each sperm clear a path through the zona pellucida. Penetration of the secondary oocyte by a sperm causes the oocyte to release substances that prevent other sperm from binding.

C The oocyte completes meiosis II. The sperm's tail and other organelles degenerate.

Later, the egg and sperm nuclei will fuse and the zygote will form.

Figure 16.10 Animated! Fertilization unites a sperm and oocyte. (A, B: © Cengage Learning; C: Courtesy of Elizabeth Sanders, Women's Specialty Jackson, MS)

- People who choose to control whether their sexual activity produces a child have a variety of options.

The most effective method of birth control is complete *abstinence*—no sexual intercourse whatsoever. A modified form of abstinence is the *rhythm method,* also called the "fertility awareness" or *sympto-thermal method.* The idea is to refrain from intercourse during the woman's fertile period, starting a few days before ovulation and ending a few days after. Her fertile period is identified and tracked by keeping records of the length of her menstrual cycles

iStockphoto.com/Marcus Clackson

TABLE 16.3 Common Methods of Contraception

Method	Mechanism of Action	Pregnancy Rate*
Abstinence	Avoid intercourse entirely	0% per year
Rhythm method	Avoid intercourse when female is fertile	25% per year
Withdrawal	End intercourse before male ejaculates	27% per year
Vasectomy	Cut or close off male's vasa deferentia	<1% per year
Tubal ligation	Cut or close off female's oviducts	<1% per year
Condom	Enclose penis, block sperm entry to vagina	15% per year
Diaphragm, cervical cap	Cover cervix, block sperm entry to uterus	16% per year
Spermicides	Kill sperm	29% per year
Intrauterine device	Prevent sperm entry to uterus or prevent implantation	<1% per year
Oral contraceptives	Prevent ovulation	<1% per year
Hormone patches, implants, or injections	Prevent ovulation	<1% per year
Emergency contraception pill	Prevent ovulation	15–25% per use**

* Percent of users who get pregnant despite consistent, correct use
** Not meant for regular use

and sometimes by examining her cervical secretions and taking her temperature each morning when she wakes up. (Core body temperature rises by one-half to one degree just after ovulation.) The method is not very reliable (Table 16.3). Ovulation can be irregular, and it can be easy to miscalculate. Also, sperm already in the vaginal tract may survive until ovulation.

Withdrawal, removing the penis from the vagina before ejaculation, also is not very effective because fluid released from the penis before ejaculation may contain sperm. *Douching,* or rinsing out the vagina with a chemical right after intercourse, is next to useless. It takes less than 90 seconds for sperm to move past the cervix into the uterus.

Surgery and barrier methods are the most effective options

Controlling fertility by surgery is less chancy but is usually an irreversible step. In *vasectomy,* a physician makes a tiny incision in a man's scrotum, then severs and ties off each vas deferens (Figure 16.11A). Afterward, sperm can't leave the testes and so can't be present in the man's semen. A vasectomy does not change a man's sex hormones or sex drive. An alternative is the Vasclip, a device about the size of a rice grain that simply closes off the vas deferens.

In *tubal ligation,* a woman's oviducts are cauterized or cut and tied off (Figure 16.11B), so sperm cannot reach ovulated oocytes.

Spermicides kill sperm. They are packaged inside an applicator and placed in a woman's vagina just before intercourse. Neither is reliable unless used with another device, such as a diaphragm or condom.

A *diaphragm* is a flexible, dome-shaped device that is positioned over the cervix before intercourse. It must be fitted by a doctor, used with foam or jelly, and inserted correctly with each use. A *cervical cap* is a similar but smaller device and can be left in place for up to 3 days. The *contraceptive sponge* is a disposable disk that contains a spermicide and covers the cervix. After being wetted, it is inserted up to 24 hours before intercourse. No prescription or special fitting is required.

The *intrauterine device,* or IUD, is a plastic or metal device that is placed into the uterus, where it hampers implantation of a fertilized egg. Available by prescription, IUDs have been associated with a variety of complications and should be discussed fully with a physician.

Condoms are thin, tight-fitting sheaths of latex or animal skin worn over the penis during intercourse. Good brands may be as much as 95 percent effective when used with a spermicide. Latex condoms help prevent the spread of sexually transmitted diseases.

A *birth control pill,* with its synthetic estrogens and progesterone-like hormones, blocks the maturation and ovulation of oocytes. Oral contraceptives are one of the

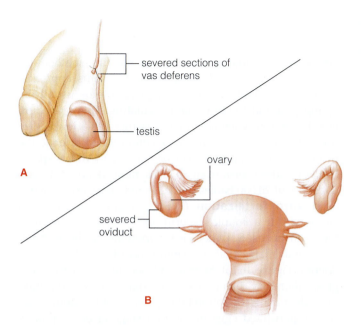

Figure 16.11 Both men and women may opt for surgical methods of birth control. A Vasectomy and **B** tubal ligation.
(© Cengage Learning)

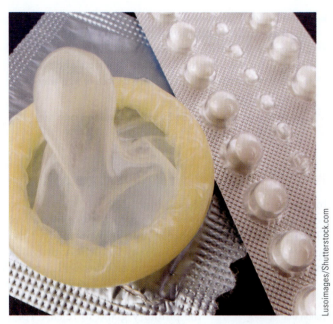

Lusoimages/Shutterstock.com

Figure 16.12 Birth control pills and condoms are among the most common approaches to preventing pregnancy.

most common methods of contraception. Some users experience (usually temporary) side effects, including nausea and weight gain. Continued use may lead to blood clots in at-risk women. Complications are more likely in women who smoke, and most physicians won't prescribe an oral contraceptive for a smoker.

A *birth control patch* is a small, flat, adhesive patch applied to the skin. It delivers the same hormones as a birth control pill and blocks ovulation the same way. It also has the same risks as oral contraceptives.

Progestin injections or *implants* prevent ovulation or implantation of an embryo. They may cause heavier menstrual periods, and implants can be difficult to remove. Even so, implants are convenient and have become increasing popular, especially among younger women.

Some women use *emergency contraception* after a condom tears, or after unprotected consensual sex or rape. These *"morning-after pills"* suppress ovulation and in most places are available without a prescription to women 18 and older. They work best taken right away but may be effective up to 5 days after intercourse.

Abortion is highly controversial

An induced or surgical **abortion** removes or dislodges an embryo or fetus from the womb. In the United States about half of unplanned pregnancies end in induced

abortion. The difficult legal and ethical conflict over legalized abortion rages on.

During the first trimester (12 weeks), abortions performed in a clinical setting usually are fast, painless, and free of complications. Even so, polls show that for both medical and moral reasons, most people in the U.S. prefer sexually responsible behavior over abortion. Aborting a late-term fetus is quite controversial unless the mother's life is threatened.

This science book can't offer any "right" answers to questions about the morality of abortion or any other reproductive decision. It can only offer an explanation of how a new individual develops to help you objectively assess the biological basis of human life. We discuss development in Chapter 17.

abortion The removal or separation of an embryo or fetus from the womb.

16.8 | Options for Coping with Infertility

- In the United States, about one in every six couples is infertile—unable to conceive a child after a year of trying. Causes run the gamut from hormonal imbalances that prevent ovulation, oviducts blocked by effects of disease, a low sperm count, or sperm that are defective in a way that impairs fertilization.

- Link to Hormones from the hypothalamus and pituitary gland 15.3

Fertility drugs stimulate ovulation

In about one-third of cases, infertility can be traced to poor quality oocytes or to irregular or absent ovulation. These situations are most common in women over the age of 37. A couple's first resort may be fertility drugs, in the hope that one or more ovarian follicles will produce a healthy oocyte. One commonly used drug, clomiphene, stimulates the pituitary gland to release FSH. As noted in Section 15.3, this hormone triggers ovulation. A drug called *human menopausal gonadotropin* (hMG) is basically a highly purified form of FSH. Injected directly into the bloodstream, it stimulates ovulation in 70 to 90 percent of women who receive it.

> **in vitro fertilization**
> Fertilization that occurs when oocytes and sperm are placed in a prepared laboratory dish.

Although fertility drugs have been used with great success since the 1970s, they can cause undesirable side effects, including the fertilization of several eggs at once. The result is a high-risk pregnancy that can result in babies with neurological and other problems.

Assisted reproductive technologies include artificial insemination and IVF

Artificial insemination was one of the first methods of *assisted reproductive technology*, or ART. In this approach, semen is placed into a woman's vagina or uterus, usually by syringe, around the time she is ovulating. This procedure may be chosen when a woman's partner has a low sperm count, because his sperm can be concentrated prior to the procedure. In *artificial insemination by donor* (AID), a sperm bank provides sperm from an anonymous donor. AID produces about 20,000 babies in the United States every year.

In vitro fertilization (IVF) is literally fertilization "in glass." If a couple's sperm and oocytes are normal, they can be used. Otherwise, variations of the technology are available that use sperm, oocytes, or both, from donors (Figure 16.13). Sperm and oocytes are placed in a glass laboratory dish in a solution that simulates the fluid in oviducts. If fertilization takes place, about 12 hours later *zygotes* (fertilized eggs in the first stage of development) are transferred to a chemical solution that will support further development. Two to four days later, one or more embryos are transferred to the woman's uterus. An embryo implants in about 20 percent of cases. In vitro fertilization often produces more embryos than can be used in a given procedure. The fate of unused embryos (which are stored frozen) has prompted ethical debates, such as whether such embryos should be used as a source of embryonic stem cells (Chapter 4).

A procedure called ICSI is a variation on IVF. ICSI stands for *intracytoplasmic sperm injection*. A single sperm is injected into an egg with a tiny glass needle. Although

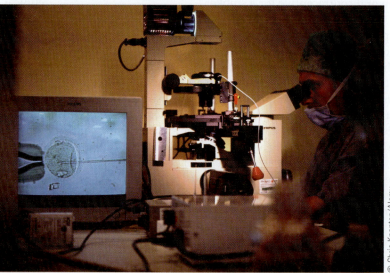

Figure 16.13 ART may allow a couple to overcome infertility. *Above*: Doctor inserting a human sperm into an egg during in vitro fertilization. He is viewing the cell through a microscope. The procedure is magnified on a monitor. The egg, held in place by the tip of a pipette, is being pierced by a micromanipulator (the thin "line" on the right).

Photodisc/Getty Images

© Chris Knapton/Alamy

NEW WAYS TO MAKE BABIES	
Artificial Insemination and Embryo Transfer	**In Vitro Fertilization**
1. Father is infertile. Mother is inseminated by donor and carries child.	1. Mother is fertile but unable to conceive. Egg from mother and sperm from father are combined in laboratory. Embryo is placed in mother's uterus.
2. Mother is infertile but able to carry child. Egg donor is inseminated with father's sperm. Then embryo is transferred and mother carries child.	2. Mother is infertile but able to carry child. Egg from donor is combined with sperm from father and implanted in mother.
3. Mother is infertile and unable to carry child. Egg donor is inseminated with father's sperm and carries child.	3. Mother is fertile, but father is infertile. Egg from mother is combined with sperm from donor.
4. Both parents are infertile, but mother is able to carry child. Egg donor is inseminated by sperm donor. Then embryo is transferred and mother carries child.	4. Both parents are infertile, but mother can carry child. Egg and sperm from donors are combined in laboratory; then embryo is transferred to mother.
	5. Mother is infertile and unable to carry child. Egg of donor is combined with sperm from father. Embryo is transferred to donor who carries child.
	6. Both parents are fertile, but mother is unable to carry child. Egg from mother and sperm from father are combined. Embryo is transferred to surrogate.
	7. Father is infertile. Mother is fertile but unable to carry child. Egg from mother is combined with sperm from donor. Embryo is transferred to surrogate mother.

LEGEND:

Sperm from father

Egg from mother

Baby born of mother

Sperm from donor

Egg from donor

Baby born of donor (surrogate)

Figure 16.14 Various options exist for assisted reproductive technologies.
(© Cengage Learning)

© Andy Walker, Midland Fertility Services/Photo Researchers, Inc.

IVF and ICSI are both in common use, evidence is mounting that babies conceived through any form of in vitro fertilization have a much higher risk of low birth weight and related developmental problems later on.

In *artificial insemination with embryo transfer* (Figure 16.14), a fertile female volunteer is inseminated with sperm from a man whose female partner is infertile. If a pregnancy results, the developing embryo is transferred to the infertile woman's uterus or to a "surrogate mother." This approach is technically difficult and has major legal complications. It isn't a common solution to infertility.

In a technique called GIFT (*gamete intrafallopian transfer*) sperm and oocytes are collected and placed into an oviduct (fallopian tube). About 20 percent of the time, a normal pregnancy follows. An alternative is ZIFT (*zygote intrafallopian transfer*). First, oocytes and sperm are placed in a laboratory dish. If fertilization occurs, the zygote is placed in a woman's oviducts. GIFT and ZIFT have about the same success rate as in vitro fertilization.

TAKE-HOME MESSAGE

HOW MAY ASSISTED REPRODUCTIVE TECHNOLOGIES HELP OVERCOME INFERTILITY?

- Fertility drugs include hormones that stimulate ovulation.
- In vitro fertilization, intrafallopian transfers, and artificial insemination are techniques for producing an embryo that may be transferred to a prospective mother's body.

Chlamydia infections and PID are most common in young sexually active people

One of the most common **sexually transmitted diseases (STDs)** is caused by the bacterium *Chlamydia trachomatis* (Figure 16.15A). This infection is often called **chlamydia** for short. Each year an estimated 3 million Americans are infected, about two-thirds of them under age 25. Around the world, *C. trachomatis* infects roughly 90 million people annually. At least 30 percent of newborns who are treated for eye infections and pneumonia were infected with *C. trachomatis* during birth.

The bacterium infects cells of the genital and urinary tract. Infected men may have a discharge from the penis and a burning sensation when they urinate. Women may have a vaginal discharge as well as burning and itching. Often, however, *C. trachomatis* is a "stealth" STD with no outward signs of infection. About 80 percent of infected women and 40 percent of infected men don't have obvious symptoms—yet they can still pass the bacterium to others.

Once a bout of chlamydia is under way, the bacteria will migrate to the person's lymph nodes, which become enlarged and tender. Impaired lymph drainage can cause swelling in the surrounding tissues.

Chlamydia can be treated with antibiotics. However, because so many people are unaware they're infected, this STD does a lot of damage. Between 20 and 40 percent of women with genital chlamydial infections develop **pelvic inflammatory disease (PID)**. PID strikes about 1 million women each year, most often sexually active women in their teens and 20s.

Although PID can arise when microorganisms that normally inhabit the vagina ascend into the pelvic region (typically as a result of too much douching), it is also a serious complication of both chlamydia and gonorrhea. Usually, a woman's uterus, oviducts, and ovaries are affected. Pain may be so severe that infected women often think they are having an attack of acute appendicitis. If the oviducts become scarred, additional complications, such as chronic pelvic pain and even sterility, can result. PID is the leading cause of infertility among young women. An affected woman may also develop chronic menstrual problems.

As soon as PID is diagnosed, a woman usually will be prescribed antibiotics. Advanced cases can require removal of the uterus (hysterectomy). A woman's partner should also be treated, even if there are no symptoms.

Gonorrhea may have no symptoms at first

Like chlamydia, **gonorrhea** can be cured if it is diagnosed soon after the infection starts. Gonorrhea is caused by *Neisseria gonorrhoeae* (Figure 16.15B). This bacterium, also called gonococcus, can infect epithelial cells of the genital tract, the rectum, eye membranes, and the throat. Each year in the United States there are about 650,000 new cases reported; there may be up to 10 million unreported cases.

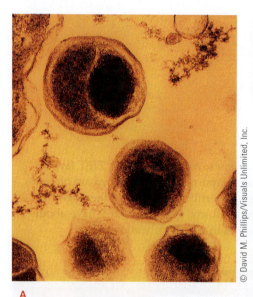
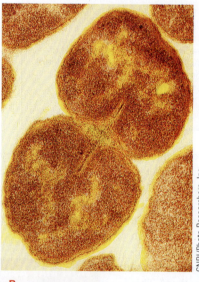
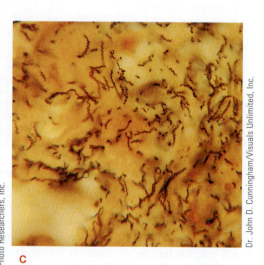

A   B   C

© David M. Phillips/Visuals Unlimited, Inc.   CNRI/Photo Researchers, Inc.   Dr. John D. Cunningham/Visuals Unlimited, Inc.

Figure 16.15 **Chlamydia, gonorrhea, and syphilis all are caused by bacteria.** **A** Color-enhanced micrograph of *Chlamydia trachomatis* bacteria. **B** *Neisseria gonorrhoeae*, or gonococcus, a bacterium that typically is seen as paired cells, as shown here. **C** The bacteria that cause syphilis.

Part of the problem is that the initial stages of the disease can be so uneventful that, as with chlamydia, a carrier may be unaware of being infected.

Early symptoms in males usually are easy to see. Pus begins to ooze from the penis and urinating becomes painful and more frequent. A man can become sterile if untreated gonorrhea leads to inflammation of his testicles or scarring of the vas deferens.

In females, the early stages of gonorrhea can be much more difficult to notice. For example, a woman may not experience burning while urinating, and she may not have an abnormal vaginal discharge. As a result, a woman's gonorrhea infection may well go untreated while the gonococcus is spreading into her oviducts. Eventually, she may experience violent cramps, fever, and vomiting. She may even become sterile if PID develops and her oviducts become blocked with scar tissue.

Antibiotics can kill the gonococcus and thus prevent complications of gonorrhea. Penicillin was once the most commonly used drug treatment. Unfortunately, antibiotic-resistant strains of gonococcus have developed. As a result, many doctors now order testing to determine the strain responsible for a particular patient's illness and then treat the infection with an appropriate antibiotic—if one is available. In 2011 medical researchers in Japan announced the discovery of a strain called H041 that is resistant to all known antibiotic treatment.

Many people believe that once cured of gonorrhea, they can't be reinfected. That is not true, partly because there are many different strains of *N. gonorrhoeae*.

Syphilis eventually affects many organs

Syphilis is caused by the bacterium *Treponema pallidum* (Figure 16.15C). The bacterium is transmitted by sexual contact. Once it reproduces, an ulcer called a chancre ("shanker," Figure 16.16A) develops. Usually the chancre is flat rather than bumpy, is not painful, and teems with treponemes. It becomes visible 1 to 8 weeks after infection and is a symptom of the *primary stage* of syphilis. Syphilis can be diagnosed in a cell sample taken from a chancre. By then, however, bacteria have already moved into the person's bloodstream.

The *secondary stage* of syphilis begins a couple of months after the chancre appears. Lesions can develop in mucous membranes, the eyes, bones, and the central nervous system. A blotchy rash breaks out over much of the body (Figure 16.16B). After the rash heals, the infection enters a latent stage that can last for years. During that time, the disease does not produce major outward symptoms and can be detected only by laboratory tests.

The *tertiary stage* of syphilis usually begins from 5 to 20 years after infection. Lesions may develop in the skin

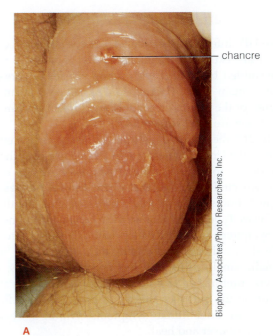

chancre

Biophoto Associates/Photo Researchers, Inc.

A

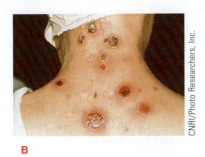

CNRI/Photo Researchers, Inc.

B

Figure 16.16 Skin ulcers are signs of syphilis. A An ulcer called a chancre ("shanker"), a sign of the first stage of syphilis. It appears any time from about 9 days to 3 months after infection, on the genitals or near the anus or the mouth. **B** This photograph shows chancres typical of secondary syphilis.

and internal organs, including the liver, bones, and aorta. Scars form; the walls of the aorta can weaken. Treponemes also damage the brain and spinal cord in ways that lead to various forms of insanity and paralysis. Infected women who become pregnant typically have miscarriages, stillbirths, or sickly infected infants.

Penicillin may cure syphilis during the early stages, although antibiotic-resistant strains have now developed.

Genital herpes is a lifelong infection

Like HIV infections (Section 9.11), herpesvirus infections are extremely contagious. Herpes simplex is transmitted by contact with active viruses or sores that contain them (Figure 16.17A). Mucous membranes of the mouth or genitals and broken or damaged skin are most susceptible.

In 2005 the National Institutes of Health estimated that in the United States, one in five people over the age of 12—roughly 45 million people—have one of the two viral strains of that cause **genital herpes**. Type 1 strains infect mainly the lips, tongue, mouth, and eyes. Type 2 strains cause most genital infections.

Symptoms usually develop within 2 weeks after infection, although sometimes they are mild or absent. Usually, small, painful blisters erupt on the penis, vulva, cervix, urethra, or around the anus. The sores can also occur on the buttocks, thighs, or back. The first flare-up may cause brief flulike symptoms. Within 3 weeks the sores crust over and heal.

Every so often the virus may be reactivated. Then it produces new, painful sores at or near the original site of infection. Recurrences can be triggered by stress, sexual intercourse, menstruation, a rise in body temperature, or other infections.

There is no cure for herpes. Between flare-ups, the virus simply is latent in nervous tissue. However, several antiviral drugs inhibit its ability to reproduce. They also reduce the shedding of virus particles from sores, and sores are often less painful and heal faster.

Human papillomavirus can lead to cancer

Worldwide, an estimated 20 million people become infected with the human papillomavirus (HPV) each year (Table 16.4). One form of HPV causes the growths called **genital warts**, which can develop months or years after a person is exposed to the virus. Usually they occur in clusters on the penis, the cervix, or around the anus (Figure 16.17B).

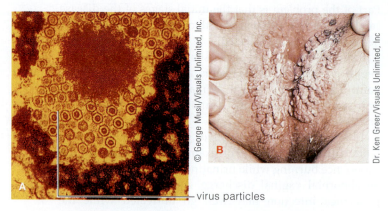
— virus particles

Figure 16.17 Viruses cause herpes and genital warts. A Herpes virus particles in an infected cell. **B** Genital warts caused by HPV (human papillomavirus).

Certain forms of HPV cause most cases of invasive cervical cancer, as well as some anal and oral cancers. Any woman having a history of genital warts should tell her physician, who may recommend an annual *Pap smear*—a test for abnormal growth of cervix cells. Several years ago an anti-HPV vaccine became available, and the Centers for Disease Control recommends that both girls and boys (who can transmit the virus) be vaccinated before they become sexually active.

Hepatitis can be sexually transmitted

Two types of hepatitis can be transmitted through sex. Like HIV, the **hepatitis B** virus (HBV) is transmitted in blood or body fluids such as saliva, vaginal secretions, and semen. However, HBV is far more contagious than HIV. The number of sexually transmitted cases is growing; in the United States about 80,000 new cases are reported each year. The virus attacks the liver. A key symptom is jaundice, yellowing of the skin and whites of the eyes as the liver loses its ability to process bilirubin pigments produced when liver cells break down hemoglobin from red blood cells. In some cases the infection becomes chronic and can lead to liver cirrhosis or cancer. The only treatment is rest. However, people at known risk for getting the disease (such as health care workers and anyone who requires repeated blood transfusions) can be vaccinated against HBV.

The **hepatitis C** virus (HCV) causes liver cirrhosis and sometimes cancer. It is carried in the blood and can reside in the body for years before symptoms develop. A blood-borne disease, HCV can be transmitted sexually if contaminated blood enters a sex partner's body through cut or torn skin.

TABLE 16.4 New STD Cases Annually*		
STD	U.S. Cases	Global Cases
HPV infection	5,500,000	20,000,000
Trichomoniasis	5,000,000	174,000,000
Chlamydia	3,000,000	92,000,000
Genital herpes	1,000,000	20,000,000
Gonorrhea	650,000	62,000,000
Syphilis	70,000	12,000,000
AIDS	40,000	4,900,000

* Global data on HPV and genital herpes were last compiled in 1997.

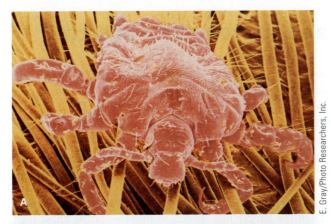

Figure 16.18 Animal parasites also cause STDs. A A crab louse, magnified 120 times. Crab lice may be visible as moving brownish dots. **B** The protozoan *Trichomonas vaginalis*, which causes trichomoniasis.

E. Gray/Photo Researchers, Inc.

Dr. Dennis Kunkel/Visuals Unlimited, Inc.

Parasites cause some STDs

Several animal parasites can be transmitted by close body contact. One is **pubic lice**, also called crab lice or simply "crabs" (Figure 16.18A). These tiny relatives of spiders usually turn up in the pubic hair, although they can make their way to any hairy spot on the body. They cling to hairs and attach their small, whitish eggs ("nits") to the base of the hair shaft. Itching and irritation can be intense when the parasites bite into the skin and suck blood. Antiparasitic drugs get rid of pubic lice.

Many microorganisms may live inside the vagina, although its rather acidic pH usually keeps pathogens in check. When certain vaginal infections do occur, they can be transmitted to a sex partner during intercourse. Any factor that alters the vagina's usual chemistry (such as taking an antibiotic) can trigger overgrowth of *Candida albicans*, a type of yeast (a fungus) that often lives in the vagina. A vaginal yeast infection, or **candidiasis**, causes a "cottage cheesy" discharge and itching and irritation of the vulva. A male may notice itching, redness, and flaky skin on his penis. Yeast infections are easily treated by over-the-counter and prescription medications, but both partners may need to be treated to prevent reinfection.

Trichomonas vaginalis, a protozoan parasite (Figure 16.18B), can cause the severe vaginal inflammation called **trichomoniasis**. The symptoms include a foul-smelling vaginal discharge and burning and itching of the vulva. An infected male may experience painful urination and have a discharge from the penis, both due to an inflamed urethra. Usually both partners are treated with antibiotics.

16.11 Eight Steps to Safer Sex

The only people who are not at risk of STDs are those who are celibate (never have sex) or who are in a long-term, mutually monogamous relationship in which both partners are disease-free. The following guidelines can help you minimize your risk of acquiring or spreading an STD.

1. Use a latex condom during either genital or oral sex to greatly reduce your risk of being exposed to HIV, gonorrhea, herpes, and other diseases. With the condom, use a spermicide that contains nonoxynol-9, which may help kill virus particles. Condoms are available for men and women.
2. Limit yourself to one partner who also has sex only with you.
3. Get to know a prospective partner before you have sex. A friendly but frank discussion of your sexual histories, including any previous exposure to an STD, is very helpful.
4. If you decide to become sexually intimate, be alert to the presence of sores, a discharge, or any other sign of possible trouble in your partner's genital area.

5. Avoid abusing alcohol and drugs. Studies show that alcohol and drug abuse both are correlated with unsafe sex practices.
6. Learn about and be alert for symptoms of STDs. If you have reason to think you have been exposed, abstain from sex until a medical checkup rules out any problems. Self-treatment won't help. See a doctor or visit a clinic.
7. Take all prescribed medication and don't share it with a partner. Unless both of you take a full course of medication, your chances of reinfection will be great. Your partner may need to be treated even if he or she does not have symptoms.
8. If you do become exposed to an STD, avoid sex until medical tests confirm that you are not infected.

© Jupiterimages

In the United States, breast cancer is a major killer of women. In both females and males, reproductive system cancers also are major health concerns.

Breast cancer is a major cause of death

In the United States, about one woman in eight, and a small number of men, develop **breast cancer**. Of all cancers in women, breast cancer currently ranks second only to lung cancer as a cause of death.

Obesity, early puberty, late childbearing, late menopause, excessive estrogen levels, and a fatty diet are risk factors for women. The risk is much greater for women with a family history of the disease. They may carry a faulty version of a gene such as BRCA1 or BRCA2. (Cancer is more likely when such genes are mutated.) Only 20 percent of breast lumps are cancer, but a woman should see a doctor about any breast lump, thickening, dimpling, breast pain, or discharge.

Chances for cure are excellent if breast cancer is detected early. Hence a woman should examine her breasts every month (about a week after her menstrual period, during her reproductive years). Figure 16.19 shows the steps of a self-exam. Low-dose mammography (breast X-ray) combined with ultrasound is the most effective method for detecting small breast cancers. The American Cancer Society recommends an annual mammogram for women over 40 and for younger women at high risk.

Early breast cancer often is treated by *lumpectomy*, which removes the tumor but leaves nearly all of the breast tissue. In *modified radical mastectomy*, the affected breast tissue, overlying skin, and nearby lymph nodes are removed. When the cancer has spread to muscles of the chest wall, they also must be taken out (*radical mastectomy*). In all cases, lymph nodes are examined because they reveal whether the cancer has begun to spread.

Various chemotherapy drugs are also used in the fight against breast cancer. A few can sometimes shrink tumors.

Uterine and ovarian cancer affect women

Cancers of the uterus most often affect the endometrium (uterine lining) and the cervix. Various types are treated by surgery, radiation, or both. The incidence of uterine cancers is falling, in part because precancerous phases of cervical cancer can be easily detected by the *Pap smear* that is part

❶ Lie down and put a folded towel under your left shoulder, then put your left hand behind your head. With the right hand (fingers flat), begin the examination of your left breast by following the outer circle of arrows shown. Gently press the fingers in small, circular motions to check for any lump, hard knot, or thickening. Next, follow the inner circle of arrows. Continue doing this for at least three more circles, one of which should include the nipple. Then repeat the procedure for the right breast. For a complete examination, repeat the procedure while standing in a shower. Hands glide more easily over wet skin.

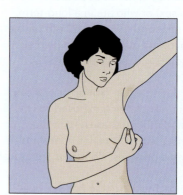

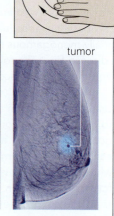

tumor

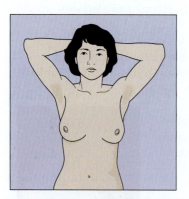
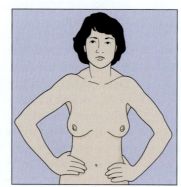

❷ Stand before a mirror, lift your arms over your head, and look for any unusual changes in the contour of your breasts, such as a swelling, dimpling, or retraction (inward sinking) of the nipple. Also check for any unusual discharge from the nipple.

If you discover a lump or any other change during a breast self-examination, it's important to see a physician at once. Most changes are not cancerous, but let the doctor make the diagnosis.

Figure 16.19 Women should perform monthly breast self-examination. The diagram below shows how to perform a breast self-examination. The mammogram shown at right has revealed a breast cancer tumor. (© Cengage Learning)

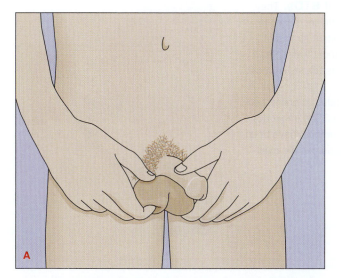

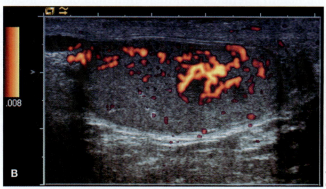

Figure 16.20 Men should perform testicular self-examination monthly. The instructions on this page are the method recommended by the American Cancer Society. Do the exam when the scrotum is relaxed, as it is after a warm bath or shower. **A** Simply roll each testicle between the thumb and forefinger, feeling for any lumps or thickening. As with breast lumps, most such changes are not cancer, but a doctor must make the diagnosis. **B** In this colored ultrasound of a testicle, the sites of a spreading cancer appear orange. (© Cengage Learning)

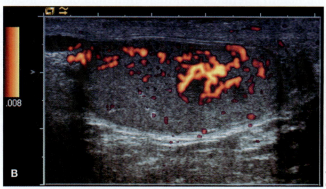 is labelled with credit text: Du Cane Medical Imaging Ltd./ Photo Researchers, Inc.

of a routine gynecological examination. The risk factors for cervical cancer include having many sex partners, early age of first intercourse, cigarette smoking, and genital warts (Section 16.10). **Endometrial cancer** is more common during and after menopause.

Ovarian cancer is often lethal because its chief symptom, an enlarged abdomen, doesn't show up until the cancer is advanced and has spread. The first sign may be abnormal vaginal bleeding or abdominal discomfort. Risk factors include family history of the disease, having breast cancer, and not bearing children. Surgery to remove the ovaries and other affected tissue is the usual first step in treatment. Patients often also receive chemotherapy, which is moderately successful, especially in early stages of the disease.

Testicular and prostate cancer affect men

Several thousand cases of **testicular cancer** are diagnosed annually in the United States. In early stages this cancer is painless. However, it can spread to lymph nodes in a man's abdomen, chest, neck, and, eventually, his lungs. Once a month from high school onward, men should examine each testicle separately after a warm bath or shower (when

the scrotum is relaxed). The testis should be rolled gently between the thumb and forefinger to check for any unusual lump, enlargement, or hardening (Figure 16.20). Because the epididymis may be confused with a lump, the important thing is to compare the two testes. A lump may not be painful, but only a physician can rule out the possibility of disease. Surgery is the usual treatment, and the success rate is high when the cancer is caught before it can spread.

Prostate cancer is second only to lung cancer in causing cancer deaths in men. There are no definite risk factors other than having a family history of the disease. Symptoms include various urinary problems, although these can also signal simply a noncancerous enlarged prostate. For men over 40, an annual digital rectal examination, which enables a physician to feel the prostate, is the first step in detecting unusual lumps. A physician who suspects cancer may order the PSA blood test, which can detect suspiciously large amounts of that tumor marker. Depending on the results of these two tests, the next step may be a biopsy—removing a small tissue sample for microscopic analysis. Over 90 percent of prostate tumors detected early are cured.

Many prostate cancers grow slowly and cause few problems. In such cases a physician may recommend simply monitoring the tumor for worrisome changes.

Public health agencies maintain statistics on the incidence of STDs. They use the numbers to measure the success of public education efforts, to identify increases in reported cases of various STDs, and to monitor the appearance of drug-resistant strains of disease-causing organisms. Infection by human papillomavirus (HPV) is the most widespread and fastest growing STD in the United States. Table 16.4 in Section 16.10 lists the seven most prevalent STDs globally.

To explore how these health concerns are affecting your community or state, go online and find out if your local or state public health department maintains statistics on STDs (most do). Then see which are the most prevalent STDs in your area and whether the numbers have been rising or declining. If someone thinks they may have been exposed, what resources are available for confidential testing?

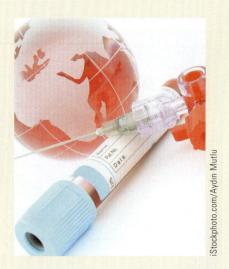

iStockphoto.com/Aydin Mutlu

SUMMARY

Section 16.1 The paired ovaries, which produce eggs, are a female's primary reproductive organs. Accessory glands and ducts, such as the oviducts, are also part of the female reproductive system. Oviducts open into the uterus, which is lined by the endometrium.

Unless a fertilized egg begins to grow in the uterus, the endometrium proliferates, then is shed in the three-phase menstrual cycle, which averages about 28 days.

Section 16.2 The menstrual cycle overlaps with an ovarian cycle. At the end of each menstrual period, a follicle (containing an oocyte) matures in an ovary. Under the influence of hormones, the endometrium starts to rebuild.

A midcycle peak of LH triggers ovulation, the release of a secondary oocyte from the ovary.

A corpus luteum forms from the rest of the follicle. It secretes progesterone that prepares the endometrium to receive a fertilized egg and helps maintain the endometrium during pregnancy. When no egg is fertilized, the corpus luteum degenerates, and the endometrial lining is shed through menstruation.

Estrogen, progesterone, FSH, and LH control the maturation and release of eggs, as well as changes in the endometrium. They are part of feedback loops involving the hypothalamus, anterior pituitary, and ovaries.

Section 16.3 Testes are a male's primary reproductive organs. The male reproductive system also includes accessory ducts and glands.

Sperm develop mostly in the seminiferous tubules and mature in the epididymis. The seminal vesicles, bulbourethral glands, and prostate gland produce fluids that mix with sperm, forming semen.

A vas deferens leading from each testis transports sperm outward when a male ejaculates.

Section 16.4 The hormones testosterone, LH (luteinizing hormone), and FSH (follicle-stimulating hormone) control the formation of sperm. They are part of feedback loops among the hypothalamus, anterior pituitary, and testes. Sertoli cells, which line the seminiferous tubules, nourish sperm. Leydig cells in tissue between the tubules secrete testosterone.

A mature sperm cell has a head, midpiece, and tail. Covering much of the head is the acrosome, which contains enzymes that help a sperm penetrate an egg.

In both males and females, gonadotropin-releasing hormone (GnRH) from the hypothalamus stimulates the anterior pituitary to release LH and FSH.

Sections 16.5, 16.6 Sexual intercourse (coitus) is the usual way an egg (a secondary oocyte) and sperm meet for fertilization. It typically involves a sequence of physiological changes in both partners and may culminate in orgasm. Fertilization produces a zygote, a cell with chromosomes from both the mother and father.

Sections 16.7, 16.8 Physical, chemical, surgical, and behavioral strategies are available for controlling unwanted pregnancies and helping infertile couples. Efforts to control fertility raise important ethical questions.

Sections 16.9–16.11 Sexually transmitted diseases (STDs) are passed by sexual activity. Bacteria cause chlamydia, gonorrhea, and syphilis. In addition to AIDS, viral STDs include genital herpes, genital warts (HPV), and viral hepatitis. Untreated STDs can seriously harm health. Only people who abstain from sexual contact or who are in an infection-free monogamous relationship can be sure of not being exposed to an STD.

Section 16.12 Major reproductive cancers in females include cancers of the breast, ovaries, cervix, and uterus. In males major reproductive cancers are cancers of the testes and prostate.

REVIEW QUESTIONS

1. Distinguish between:
   a. seminiferous tubule and vas deferens
   b. sperm and semen
   c. Leydig cells and Sertoli cells
   d. primary oocyte and secondary oocyte
   e. follicle and corpus luteum
   f. the three phases of the menstrual cycle

2. Which hormones influence the development of sperm?

3. Which hormones influence the menstrual and ovarian cycles?

4. List four events that are triggered by the surge of LH at the midpoint of the menstrual cycle.

5. What changes occur in the endometrium during the ovarian cycle?

6. Label the parts of the female reproductive system and list their functions.

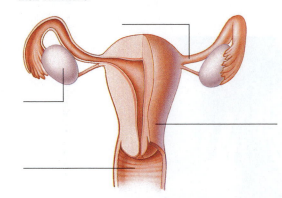

© Cengage Learning

7. Label the parts of the male reproductive system and state their functions.

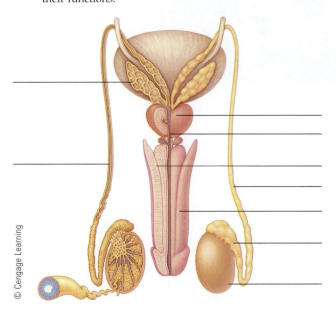

© Cengage Learning

Lennart Nilsson from A Child Is Born, © 1966, 1977 Dell Publishing Co., Inc.

Figure 16.21 The entrance to an oviduct, the tubelike channel to the uterus.

8. Figure 16.21 shows the billowing opening to the oviduct into which an ovulated oocyte is swept. Which oocyte stage is ovulated? What happens to it if it encounters a sperm cell there? What happens if it does not meet up with sperm?

SELF-QUIZ *Answers in Appendix V*

1. Besides producing gametes (sperm and eggs), the primary male and female reproductive organs also produce sex hormones. The _____ and the pituitary gland control secretion of both.

2. _____ production is continuous from puberty onward in males; _____ production is cyclic and intermittent in females.
   a. Egg; sperm   c. Testosterone; sperm
   b. Sperm; egg   d. Estrogen; egg

3. The secretion of _____ controls the formation of sperm.
   a. testosterone   c. FSH
   b. LH   d. all of the above are correct

4. During the menstrual cycle, a midcycle surge of _____ triggers ovulation.
   a. estrogen   c. LH
   b. progesterone   d. FSH

5. Ovulation releases _____.
   a. the corpus luteum
   b. a primordial follicle
   c. a primary oocyte
   d. a secondary oocyte and first polar body

6. Which is the correct order for one turn of the mentrual cycle?
   a. corpus luteum forms, ovulation, follicle forms
   b. follicle grows, ovulation, corpus luteum forms

7. In order for sexual intercourse to produce a pregnancy, both partners must experience _____.
   a. orgasm   c. affection
   b. ejaculation   d. none of the above

8. Fertilization _____.
   a. unites the nuclei of a sperm and egg cell
   b. forms a zygote that has chromosomes of both parents
   c. usually occurs in an oviduct
   d. all of the above
   e. a and b but not c

CRITICAL THINKING

1. Fertility specialists sometimes advise a couple who wish to conceive a child to use an alkaline (basic) douche immediately before intercourse. Speculate about what the reasoning behind this advice might be.

2. In the "fertility awareness" method of birth control, a woman gauges her fertile period each month by monitoring changes in the consistency of her vaginal mucus. What kind of specific information does such a method provide? How does it relate to the likelihood of getting pregnant?

3. Some women experience premenstrual syndrome (PMS), which can include a distressing combination of mood swings, fluid retention (edema), anxiety, backache and joint pain, food cravings, and other symptoms. PMS usually develops after ovulation and lasts until just before or just after menstruation begins. A woman's doctor can recommend strategies for managing PMS, which often include diet changes, regular exercise, and use of diuretics or other drugs. Many women find that taking vitamin B_6 and vitamin E helps reduce pain and other symptoms. Although the precise cause of PMS is unknown, it seems clearly related to the cyclic production of ovarian hormones. After reviewing Figure 16.3, suggest which hormonal changes may trigger PMS.

4. Some infertile couples are willing to go to considerable lengths to have a baby. From your reading of Section 16.8, which of the variations of reproductive technologies produces a child that is least related (genetically) to the infertile couple? Would you view having a child by that method as preferable to adopting a baby? Why or why not?

5. The absence of menstrual periods, or amenorrhea, is normal in pregnant and postmenopausal women and in girls who have not yet reached puberty. However, in females of reproductive age, amenorrhea can result from tumors of the pituitary or adrenals. Based on discussions in this chapter and Chapter 15, speculate about why such tumors might disrupt monthly menstruation.

Your Future

A diagnosis of ovarian cancer, even in early stages, often means a treatment regimen of surgery, chemotherapy, and radiation. Nearly always, so much ovarian tissue is lost that menopause results and the patient can no longer become pregnant. Recently, an experimental procedure has shown promise for restoring this lost fertility. Before cancer treatment begins, a portion of the patient's still-healthy ovarian tissue is removed and frozen. If the cancer treatment is successful, the frozen tissue is thawed and transplanted back into the patient. As of this writing, nine babies have been conceived by ovarian cancer survivors who received this type of transplant. Advocates have expressed the hope that the method may soon become more widely available.

© Cengage Learning

AJPhoto/Photo Researchers, Inc.

LINKS TO EARLIER CONCEPTS

This chapter builds on the principles of reproduction introduced in Chapter 16, including how hormones influence the menstrual cycle (16.1, 16.2).

We will discuss how organ systems (4.8) start to develop in an embryo and learn about special features of the cardiovascular system in a developing fetus (7.1).

We will also see how hormones from the hypothalamus and pituitary set the stage for birth (15.3).

KEY CONCEPTS

Development Begins

A new individual begins to develop in steps that produce a multicellular embryo in which tissues and organs form. Each step is guided by genes and builds on body structures that are formed in the preceding stage. Sections 17.1–17.4

Prenatal Development

As an embryo develops, so do vital support structures such as the placenta. In the fetal phase, organs and other structures grow and mature. Sections 17.5–17.8

Birth to Adulthood

Birth launches life outside the womb. Body structures and their functioning change from infancy onward. Sections 17.9–17.12

Later Life

Over time, age-related changes begin to affect virtually all body organs and tissues. Section 17.13

A lifetime of change—that's human development in a nutshell. Your own biological development began when you were a zygote and will continue until the day you die. Like everyone else, you own a body that is the result of the gene-guided formation of tissues and organ systems, not to mention years of growth. The proportions of your body have altered dramatically from infancy to adulthood. Puberty brought other major changes as sex hormones kicked in. If you're under the age of thirty, you are at your biological prime. If you've passed thirty, you may be noticing new shifts—possibly the beginning of wrinkles, for starters. If you've ever watched a baby transform into a toddler, or a child become an adolescent, or a parent make the transition to "senior" status, you've had your real-life introduction to human development.

With this chapter, we consider the many aspects of how a human develops. We start with principles that govern how all the specialized cells and tissues of an adult come into being—a biological journey we all have made.

17.1 Overview of Early Human Development

- Like all animals, humans begin life as a single cell from which tissues and organs soon begin to develop.
- Links to Organ systems 4.8, How eggs and sperm form 16.2, 16.4, Fertilization 16.6

After fertilization, the zygote soon becomes a ball of cells

Section 16.6 described how fertilization produces a zygote, the first cell of a new individual. Within a day or two after fertilization, cell divisions convert the zygote to a ball of cells (Figure 17.1). This process is called **cleavage**. It occurs while the zygote is traveling down the oviduct toward the uterus. By the time it reaches the uterus, the zygote is a cluster of sixteen cells called a *morula* (MOE-roo-lah, from a Latin word for mulberry).

Each new cell that forms during cleavage is called a **blastomere**. Each one ends up with a different portion of the egg's cytoplasm. Which bit of cytoplasm a blastomere receives helps determine the developmental fate of cells that arise from it later on. For example, its fate may be to become the forerunner of some kind of nervous tissue or perhaps of epithelium.

Three primary tissues form

After cleavage comes **gastrulation** (gas-tru-LAY-shun), a process that rearranges the morula's cells. It lays out the basic organization for the body as cells are arranged into three primary tissues, the **germ layers** (Table 17.1). The outer one is called *ectoderm*, the middle one *mesoderm*, and the inner one *endoderm*. Body tissues and organs will develop from groups of cells in each layer.

Next, cells become specialized

In the germ layers, different genetic instructions begin to operate in different groups of cells. This is the start of **cell differentiation**. In simple terms, this process makes cells "specialists" for a particular function. For example, you may remember from Chapter 9 how various subsets of T cells, each with a different function, differentiate in the thymus.

An adult has about 200 differentiated cell types. In each type, the genetic instructions needed for its specialized function operate. For instance, as your eyes developed, certain cells turned on genes for a transparent protein called crystallin. These differentiated cells then formed the lens of each eye. They are the only body cells that make crystallin.

Although each type of differentiated cell has its own particular genetic marching orders, a differentiated cell still has all the genes that parents pass to an embryo. Efforts to "reverse engineer" adult cells back into stem cells, as described in the introduction to Chapter 4, rely on this fact. It also is why scientists have been able to clone adult animals—that is, to create a genetic copy—from some types of differentiated cells. Chapter 21 discusses some examples of cloning.

Morphogenesis forms organs

Morphogenesis ("the beginning of form") is the process by which body tissues and organs form. Several factors influence these changes. They include cell division in certain areas and the growth and movement of cells and tissues from one place to another.

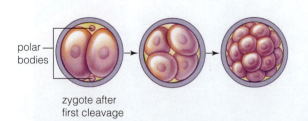

polar bodies

zygote after first cleavage

Figure 17.1 Cleavage in a zygote creates a ball of cells. In the diagram of the first cleavage, on the left, the small spheres are the polar bodies described in Section 16.2. They contain unneeded material and will disintegrate. Cleavage produces a morula by the fourth day after fertilization. (© Cengage Learning)

TABLE 17.1 The Three Germ Layers and Tissues and Organs That Form from Them	
Germ Layer	**Body Parts in an Adult**
ECTODERM	Nervous system and sense organs Pituitary gland Outer layer of skin (epidermis) and its associated structures, such as hair
MESODERM	Cartilage, bone, muscle, and various connective tissues Cardiovascular system and blood Lymphatic system Urinary system Reproductive system Outer layers of the digestive tube and of structures that develop from it, including parts of the respiratory system
ENDODERM	Lining of the digestive tube and of structures that develop from it, such as the lining of the respiratory airways

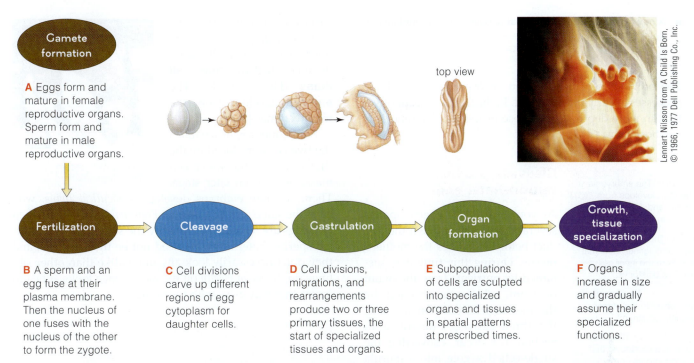

A Eggs form and mature in female reproductive organs. Sperm form and mature in male reproductive organs.

top view

Gamete formation

Fertilization → Cleavage → Gastrulation → Organ formation → Growth, tissue specialization

B A sperm and an egg fuse at their plasma membrane. Then the nucleus of one fuses with the nucleus of the other to form the zygote.

C Cell divisions carve up different regions of egg cytoplasm for daughter cells.

D Cell divisions, migrations, and rearrangements produce two or three primary tissues, the start of specialized tissues and organs.

E Subpopulations of cells are sculpted into specialized organs and tissues in spatial patterns at prescribed times.

F Organs increase in size and gradually assume their specialized functions.

Figure 17.2 An early embryo begins to develop soon after fertilization. This flow chart shows the stages from fertilization to about 6 weeks. For clarity, membranes surrounding the embryo are not shown, and several stages are shown in cross section. (© Cengage Learning)

For example, most of the bones of your face descended from cells that migrated from the back of your head when you were an early embryo. In a similar way, neurons in the center of the developing brain creep along parts of glial cells or axons of other neurons until they reach their final destination. Morphogenesis also requires sheets of tissue to fold and certain cells to die on cue. You will read about some other examples of these events in Section 17.4.

Figure 17.2 summarizes the stages of development we have been discussing. It is important to remember that by the end of each stage, the embryo is more complex than it was before. Normal development requires that each stage be completed before the next one begins.

blastomere A cell that forms during cleavage of a zygote.

cell differentiation Process by which newly formed cells become specialized for a certain function.

cleavage Rounds of cell division that transform a zygote into a ball of cells.

gastrulation The process of early development that produces the three germ layers.

germ layers The three primary tissues that form as an early embryo develops. The outer tissue is ectoderm, the middle one is mesoderm, and the inner one is endoderm.

morphogenesis The process by which tissues and organs form.

TAKE-HOME MESSAGE

WHAT ARE THE FIRST STEPS IN EMBRYONIC DEVELOPMENT?

- Cleavage of the zygote into a ball of cells occurs shortly after fertilization.
- Gastrulation forms the three germ layers.
- Cell differentiation specializes cells for their final roles in the body. Tissues and organs form during morphogenesis.
- Each stage of early embryonic development builds on structures that were formed during the preceding stage.

From Zygote to Implantation

■ **A newly formed embryo cannot survive unless it implants in the mother's uterus.**

■ **Link to Fertilization 16.6**

The previous section gave you an overview of the basic processes of early development. In the next few sections we look in more depth at how these processes culminate in a newborn baby.

blastocyst The embryonic stage that develops from a morula and eventually implants in the uterine wall.

embryo The stage of a newly forming individual that arises from the inner cell mass of a blastocyst.

implantation Attachment of a blastocyst to the uterine lining (endometrium).

inner cell mass The mass of cells in a blastocyst that develops into an embryo.

Cleavage produces a multicellular embryo

As you've read, the zygote spends several days moving down the oviduct before it reaches the mother's uterus. During this time it is sustained by nutrients from the ovum or from substances secreted by the mother's tissues. On the way, the three cleavages described in Section 17.1 occur, converting the single-celled zygote into a morula (Figure 17.3, stages A and B).

When the morula finally reaches the uterus, a cavity filled with fluid begins to open up inside it. This change transforms the morula into a **blastocyst** (*blast-* means "bud"). The blastocyst has two tissues: a surface epithelium called the *trophoblast* (*tropho-* means "to nourish") and a small clump of cells called the **inner cell mass** (Figure 17.3E). The **embryo** develops from the inner cell mass.

Sometimes a split separates the two cells produced by the first cleavage, the inner cell mass, or an even later stage.

Identical twin sisters
(© Dennis Degnan/Corbis)

Then, separate embryos develop as *identical twins*, who have the same genetic makeup. *Fraternal twins* result when two eggs are fertilized at roughly the same time by different sperm. Fraternal twins need not be the same sex, and they don't necessarily look any more alike than other siblings do. *Focus on Health* on the facing page looks at health issues that may arise with twinning.

Implantation secures the embryo in the uterus

About a week after fertilization, **implantation** begins as the blastocyst breaks out of the zona pellucida. Cells of the blastocyst's epithelium then invade the endometrium (the lining of the uterus) and cross into the underlying connective tissue (Figure 17.4). This gives the blastocyst a foothold in the

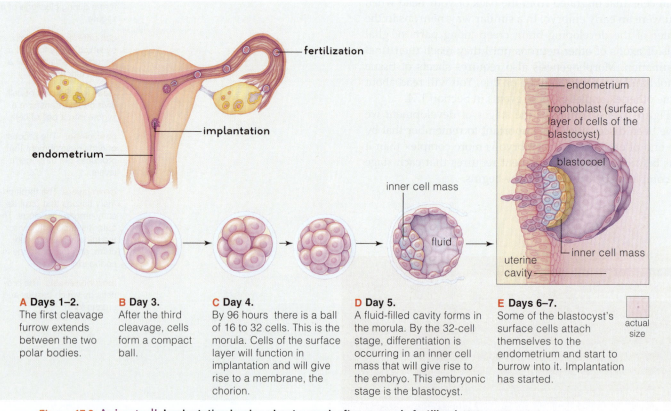

A Days 1–2. The first cleavage furrow extends between the two polar bodies.

B Day 3. After the third cleavage, cells form a compact ball.

C Day 4. By 96 hours there is a ball of 16 to 32 cells. This is the morula. Cells of the surface layer will function in implantation and will give rise to a membrane, the chorion.

D Day 5. A fluid-filled cavity forms in the morula. By the 32-cell stage, differentiation is occurring in an inner cell mass that will give rise to the embryo. This embryonic stage is the blastocyst.

E Days 6–7. Some of the blastocyst's surface cells attach themselves to the endometrium and start to burrow into it. Implantation has started.

actual size

Figure 17.3 Animated! Implantation begins about a week after an egg is fertilized. (© Cengage Learning)

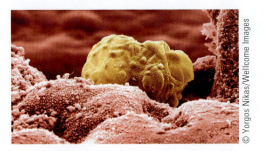

Figure 17.4 A blastocyst implants in the endometrium of the uterus. In this image the blastocyst is colored gold.

© Yorgos Nikas/Wellcome Images

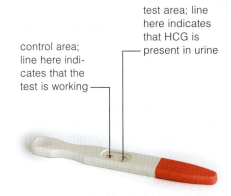

control area; line here indicates that the test is working

test area; line here indicates that HCG is present in urine

Figure 17.5 A home pregnancy test measures HCG.
(iStockphoto.com/RonTech2000)

uterus. As time passes it will sink into the connective tissue of the uterus, and the endometrium will close over it.

Occasionally a blastocyst implants in the wrong place—in the oviduct or in the outer surface of the ovary or in the abdominal wall. The result is an *ectopic (tubal) pregnancy.* It can't go to full term and must be terminated by surgery.

Implantation is complete two weeks after the original oocyte was ovulated. Menstruation, which would begin at this time if the woman were not pregnant, does not occur because the implanted blastocyst secretes HCG (human chorionic gonadotropin). HCG stimulates the corpus luteum to continue secreting estrogen and progesterone, which prevent the uterus lining from being shed. By the third week of pregnancy, HCG can be detected in the mother's blood or urine. Home pregnancy tests use chemicals that change color when the woman's urine contains HCG (Figure 17.5).

TAKE-HOME MESSAGE

AFTER CLEAVAGE, WHAT NEXT DEVELOPMENTAL STEPS LEAD TO IMPLANTATION OF THE EMBRYO?

- A blastocyst develops from the morula.
- About seven days after fertilization, the blastocyst implants as cells of its outer layer of epithelium invade the endometrium of the uterus.
- Implantation usually is complete by two weeks after the original oocyte was ovulated.

17.3 A Baby Times Two

FOCUS ON HEALTH

As Section 17.2 mentioned, fraternal twins result when two eggs are fertilized at once, and identical twins result when a very early stage embryo splits into two. We now know more about some patterns of twinning, such as twins "running in families" and cases of conjoined twins.

Conjoined twins form when an embryo partially splits after day twelve of development. The twins remain joined, usually at the chest or abdomen, although other configurations also are possible (Figure 17.6A). Doctors usually try to surgically separate conjoined twins early in life so that each twin can develop as normally as possible.

A high level of the hormone FSH, which stimulates the maturation of woman's eggs (Section 16.2), increases the likelihood of fraternal twins. Genetic quirks explain why twinning runs in some families and is more common in some ethnic groups. A woman who herself is a fraternal twin has double the average chance of giving birth to fraternal twins, and once she does her odds triple for having a second set. The Yoruba people of Africa have the world's highest fraternal twinning rate. Yoruba women have unusually high levels of FSH. Unfortunately, many Yoruba mothers lack access to good medical care and half of twins die soon after birth (Figure 17.6B).

Top left: Courtesy of © Amy Waddell/UCLA's Mattel Children's Hospital Bottom left: JORGE SILVA/Reuters/Landov; Right: © Marilyn Houlberg

A   B

Figure 17.6 Twinning may bring health issues. A These Guatemalan sisters were joined at the head until doctors at UCLA's Mattel Hospital separated them. **B** This grieving Yoruba mother of twins carries a doll as a ritual point of contact for one of her infants, who died.

How the Early Embryo Develops

■ The embryonic period lasts for eight weeks. During that time, the basic body plan of the embryo takes shape.

A developing baby is considered an embryo for most of the first trimester, or three months, of the nine months of gestation. When the three germ layers—the ectoderm, mesoderm, and endoderm—are in place, morphogenesis begins and the embryo's organ systems start to develop.

First, the basic body plan is established

By the time a woman has missed her first menstrual period, the embryo has implanted and the inner cell mass has been transformed into a pancake-shaped **embryonic disk**. Around day fifteen, gastrulation has rearranged cells so that a faint "primitive streak" appears at the midline of the disk (Figure 17.7A). Now, ectoderm along the midline thickens to establish the beginnings of a **neural tube** (Figure 17.7B). This tube is the forerunner of the embryo's brain and spinal cord. Some of its cells also give rise to a flexible rod of cells called a *notochord*. The vertebral column will form around this rod.

embryonic disk Pancake-shaped stage of an early embryo.

neural tube The forerunner of an embryo's brain and spinal cord.

These events establish the body's long axis and its bilateral symmetry. In other words, the embryonic disk is reshaped in ways that provide the body with the basic form we see in all vertebrates.

On the surface of the embryonic disk near the neural tube, the third primary tissue layer—mesoderm—also has been forming. Toward the end of the third week, some mesoderm gives rise to *somites* (SOE-mites). These are paired blocks of mesoderm, and they will be the source of most bones and skeletal muscles of the neck and trunk. The dermis overlying these regions comes from somites as well. Structures called pharyngeal arches start to form (Figure 17.7C). They will contribute to development of the face, neck, mouth, and associated parts. In other mesodermal tissues, spaces open up. Eventually, these spaces will merge

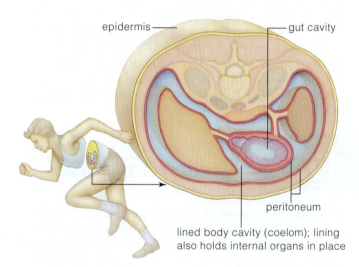

epidermis — gut cavity

peritoneum

lined body cavity (coelom); lining also holds internal organs in place

Figure 17.7 Several important steps mark the embryonic period of development. These steps include the appearance of a primitive streak foreshadowing the brain and spinal cord and the formation of somites and pharyngeal arches. These are dorsal views (of the embryo's back) except for days 24–25, which is a side view. (© Cengage Learning)

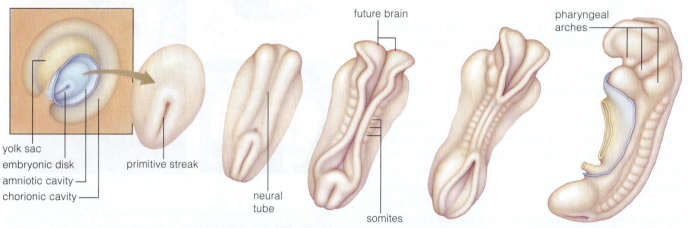

future brain

pharyngeal arches

yolk sac
embryonic disk
amniotic cavity
chorionic cavity

primitive streak

neural tube

somites

A Day 15. A primitive streak appears along the axis of the embryonic disk. This thickened band of cells marks the onset of gastrulation.

B Days 19–23. Cell migrations, tissue folding, and other morphogenetic events lead to the formation of a hollow neural tube and to somites (bumps of mesoderm). The neural tube gives rise to the brain and spinal cord. Somites give rise to most of the axial skeleton, skeletal muscles, and much of the dermis.

C Days 24–25. By now, some cells have given rise to pharyngeal arches, which contribute to the face, neck, mouth, nasal cavities, larynx, and pharynx.

to form the cavity (called the *coelom*, SEE-lahm) between the body wall and the digestive tract.

Next, organs develop and take on the proper shape and proportions

After gastrulation, organs and organ systems begin to form. An example is *neurulation*, the first stage in the development of the nervous system. Figure 17.8 shows how cells of ectoderm at the embryo's midline elongate and form a neural plate. This is the first sign that a region of ectoderm is starting to develop into nervous tissue. Next, cells near the middle become wedge-shaped. The changes in cell shape cause the neural plate to fold over and meet at the embryo's midline to form the neural tube.

The folding of sheets of cells is extremely important in morphogenesis. The folding takes place as microtubules lengthen and rings of microfilaments in cells tighten like purse strings.

Section 17.1 mentioned that morphogenesis also requires cells to move from one place to another. Migrating cells find their way in part by following so-called adhesive cues. For instance, as the nervous system is developing, migrating Schwann cells stick to adhesion proteins on the surface of axons but not on blood vessels. Adhesive cues also tell the cells when to stop. Cells migrate to places where the signals are strongest, then stay there once they arrive.

Successful embryonic development requires that body parts form according to normal patterns, in a certain sequence. Genetically programmed cell death, the process called *apoptosis* (a-poh-TOE-sys) introduced in Chapter 9, helps sculpt body parts. Inside cells that are destined to die, enzymes begin digesting cell parts. For instance, morphogenesis at the ends of limb buds first produced paddle-shaped hands at the ends of your arms (Figure 17.9A). Then epithelial cells between the lobes in the paddles died on cue, leaving separate fingers (Figure 17.9B). Figure 17.9C shows what can happen when apoptosis doesn't occur normally while a hand is forming.

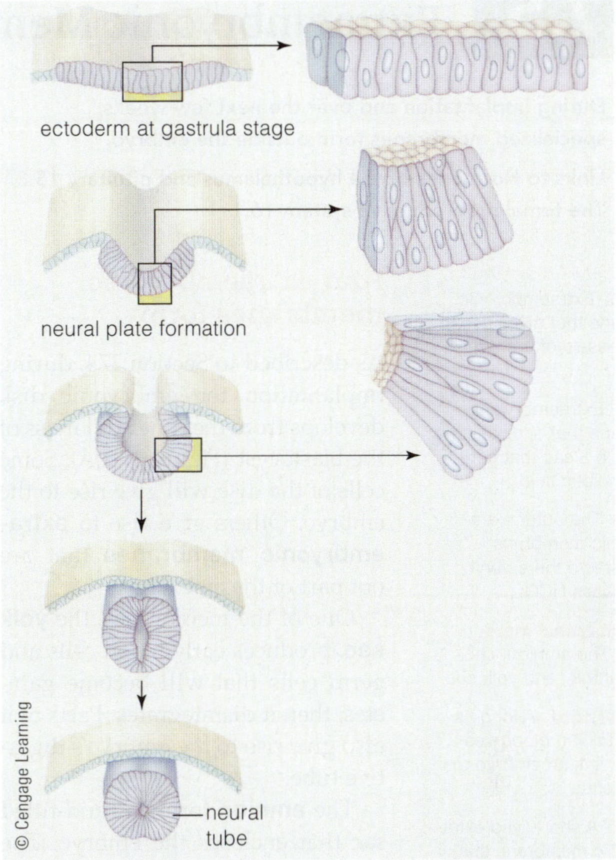

ectoderm at gastrula stage

neural plate formation

© Cengage Learning

neural tube

Figure 17.8 Animated! Formation of the neural tube is an example of morphogenesis. A neural tube is the forerunner of the brain and spinal cord. It forms as certain ectodermal cells change shape. In some cells, microtubules lengthen, and the elongating cells form a neural plate. In other cells, microfilament rings at one end constrict and the cells become wedge-shaped. Their part of the ectodermal sheet folds over the neural plate to form the tube.

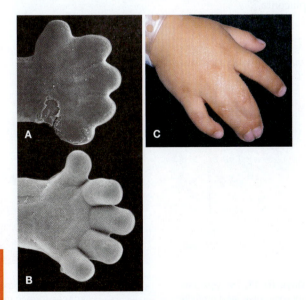

A and B: Courtesy of Kathleen K. Sulick, Bowles Center for Alcohol Studies, The University of North Carolina at Chapel Hill; C: © John Dasiai, MD/Custom Medical Stock Photo

Figure 17.9 Programmed cell death separates the digits when fingers form. A At first, webs of tissue connect the digits. **B** Then, cells in the webs die by apoptosis and the digits are separated. **C** These fingers remained attached when embryonic cells did not die on cue.

■ During implantation and over the next few weeks, specialized membranes form outside the embryo.

■ Links to Hormones of the hypothalamus and pituitary 15.3, The female reproductive system 16.1

allantois Extraembryonic membrane that gives rise to blood vessels of the umbilical cord.

amnion Extraembryonic membrane that encloses the embryo in a sac that contains amniotic fluid.

chorion The outer extraembryonic membrane that protects the embryo and releases HCG.

extraembryonic membranes The amnion, chorion, allantois, and yolk sac.

umbilical cord A long tissue containing blood vessels that link an embryo to the placenta.

yolk sac A short-lived extraembryonic membrane that produces early blood cells, germ cells, and parts of the embryo's digestive tube.

Four extraembryonic membranes form

As described in Section 17.4, during implantation the embryonic disk develops from the inner cell mass of the blastocyst (Figure 17.10A). Some cells of the disk will give rise to the embryo. Others give rise to **extraembryonic membranes** that are not part of the embryo.

One of the membranes, the **yolk sac**, produces early blood cells and germ cells that will become gametes; then it disintegrates. Parts of it also give rise to the embryo's digestive tube.

The **amnion** forms a fluid-filled sac that encloses the embryo. The amniotic fluid insulates the embryo, absorbs shocks, and prevents the embryo from drying out. Just outside it is the **allantois**, which gives rise to blood vessels that will invade the **umbilical cord**. These vessels are the embryo's contribution to circulatory "plumbing" that will link the embryo with its lifeline, the placenta.

As a blastocyst is implanting, a new cavity opens up around the amnion and yolk sac (Figure 17.10B). The lining of this cavity becomes the **chorion**, a membrane that is folded into fingerlike projections called *chorionic villi*. While these changes take place, the erosion of the endometrium that began with implantation continues. As capillaries in the endometrium break down, spaces in the disintegrating tissue fill with the mother's blood. The chorionic villi extend into these spaces. Inside each villus are small blood vessels, as shown in Figure 17.11.

Eventually the chorion wraps around the embryo and the other three membranes (Figure 17.10C). It continues the secretion of HCG that began when the embryo implanted. HCG will prevent the lining of the uterus (the endometrium) from breaking down until the placenta can produce enough estrogen and progesterone to maintain the lining.

TAKE-HOME MESSAGE

WHAT ARE THE EXTRAEMBRYONIC MEMBRANES?

• The extraembryonic membranes are the yolk sac, the amnion, the allantois, and the chorion.

• The yolk sac produces early blood cells and some other parts, including portions of the GI tract. The amnion forms a sac that encloses the embryo and produces amniotic fluid.

• The allantois gives rise to blood vessels of the umbilical cord, while the chorion forms fingerlike villi that contain blood capillaries. The villi grow into the endometrium and become part of the placenta.

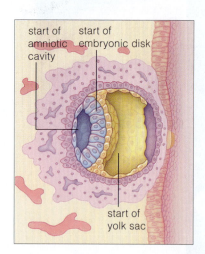

A Days 10–11. The yolk sac, embryonic disk, and amniotic cavity have started to form from parts of the blastocyst.

actual size

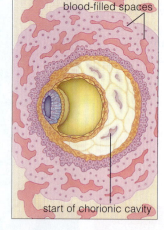

B Day 12. Blood-filled spaces form in maternal tissue. The chorionic cavity starts to form.

actual size

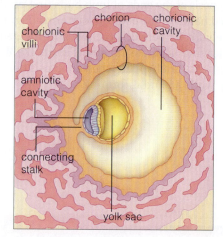

C Day 14. A connecting stalk has formed between the embryonic disk and chorion. Chorionic villi, which will be features of a placenta, start to form.

actual size

Figure 17.10 Extraembryonic membranes begin to form during the first two weeks of life. (© Cengage Learning)

17.6 The Placenta: A Pipeline for Oxygen, Nutrients, and Other Substances

- **Three weeks after fertilization, nearly a fourth of the inner surface of the uterus has become a spongy tissue, the developing placenta.**

The **placenta** is the link through which nutrients and oxygen pass from the mother to the embryo and waste products from the embryo pass back to the mother's bloodstream. The placenta is a way of sustaining a developing baby while allowing its blood vessels to develop apart from the mother's.

A fully developed placenta is considered an organ, but it's useful to think of it as a close association of the chorion and the upper cells of the endometrium where the embryo implanted (Figure 17.11). The mother's side of the placenta is endometrial tissue that contains arterioles and venules.

As the embryo develops, very little of its blood ever mixes with that of its mother. Oxygen and nutrients simply diffuse out of the mother's blood vessels, across the blood-filled spaces in the endometrium, then into the embryo's blood vessels. Carbon dioxide and other wastes diffuse in the opposite direction, leaving the embryo.

Besides nutrients and oxygen, many other substances taken in by the mother—including alcohol, caffeine, drugs, pesticide residues, and toxins in cigarette smoke—can cross the placenta, as can HIV.

> **placenta** A temporary organ in which blood vessels of the mother and an embryo (later the fetus) are in contact; it provides nutrients and oxygen to an embryo and carries away wastes.

TAKE-HOME MESSAGE

WHAT IS THE PLACENTA?

- The placenta is an organ in which maternal and embryonic blood vessels are in close contact.
- The placenta provides nutrients and oxygen to the embryo from the mother's bloodstream. The embryo's bloodstream also discharges wastes that the mother's bloodstream will transport away.

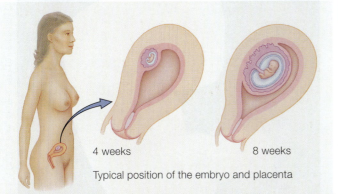

4 weeks   8 weeks

Typical position of the embryo and placenta

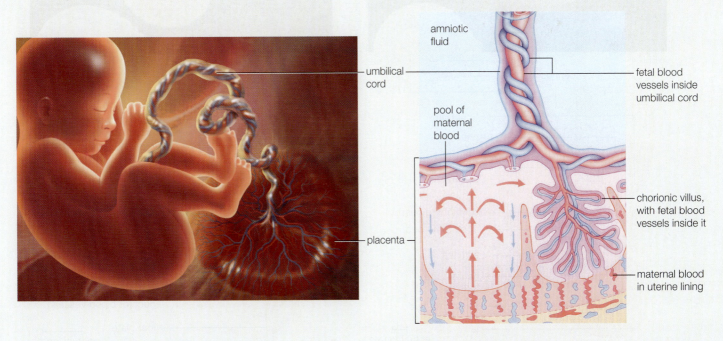

Artist's depiction of the view inside the uterus, showing a fetus connected by an umbilical cord to the pancake-shaped placenta.

amniotic fluid

umbilical cord

pool of maternal blood

placenta

fetal blood vessels inside umbilical cord

chorionic villus, with fetal blood vessels inside it

maternal blood in uterine lining

The placenta consists of maternal and fetal tissue. Fetal blood flowing in vessels of chorionic villi exchanges substances by diffusion with maternal blood around the villi. The bloodstreams do not mix.

Figure 17.11 Animated! Blood vessels of the mother and fetus are in close contact in a full-term placenta. (© Cengage Learning)

The Second Four Weeks: Human Features Appear

- By the end of four weeks, the embryo has grown to 500 times its original size. Over the next several weeks it will develop recognizable human features.

In an embryo's first few weeks of life, it grows rapidly and its cells begin to specialize. Morphogenesis begins to sculpt limbs, fingers, and toes. The circulatory system becomes more intricate, and the umbilical cord forms. Growth of the all-important head now surpasses that of any other body region (Figure 17.12A–B). The embryonic period ends as the eighth week draws to a close. The embryo is no longer merely "a vertebrate." As you can see in Figure 17.12C it now clearly looks like a human being.

As the second half of the first trimester begins, gonads begin to develop. In an embryo that has inherited X and Y sex chromosomes, a sex-determining portion of the Y chromosome now triggers development of testes (Figure 17.13). Sex hormones made by the testes then influence the devel-

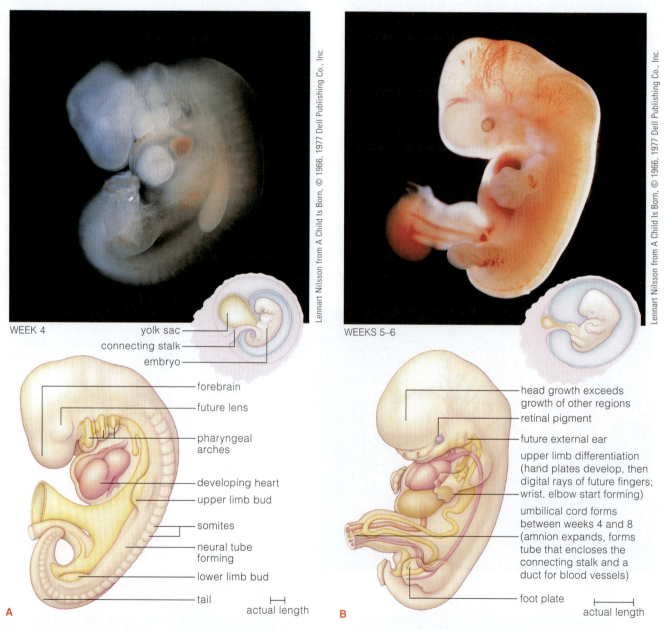

WEEK 4

yolk sac
connecting stalk
embryo

forebrain
future lens
pharyngeal arches
developing heart
upper limb bud
somites
neural tube forming
lower limb bud
tail

actual length

A

WEEKS 5–6

head growth exceeds growth of other regions
retinal pigment
future external ear
upper limb differentiation (hand plates develop, then digital rays of future fingers; wrist, elbow start forming)
umbilical cord forms between weeks 4 and 8 (amnion expands, forms tube that encloses the connecting stalk and a duct for blood vessels)
foot plate

actual length

B

Lennart Nilsson from A Child Is Born, © 1966, 1977 Dell Publishing Co., Inc.

Figure 17.12 **Amazing photographs show the development of a human embryo. A** Human embryo at four weeks. It has a tail and pharyngeal arches, which all vertebrate embryos have. **B** The embryo at five to six weeks after fertilization. **C** An embryo at the boundary between the embryonic and fetal periods. It is floating in fluid within the amniotic sac. The chorion, which normally covers the amniotic sac, has been opened and pulled aside. (© Cengage Learning)

opment of the entire reproductive system. An embryo with XX sex chromosomes will be female, and female reproductive structures begin to form in her body. Notice that no hormones are required to stimulate development of female gonads—all that is necessary is the absence of testosterone.

After eight weeks the embryo is just over 1 inch long, its organ systems are formed, and it is designated a **fetus**.

As the first trimester ends, a heart monitor can detect its heartbeat. Its genitals are well formed, and an ultrasound image often will reveal the baby's sex.

fetus A term used to designate a human embryo after it has completed eight weeks of development.

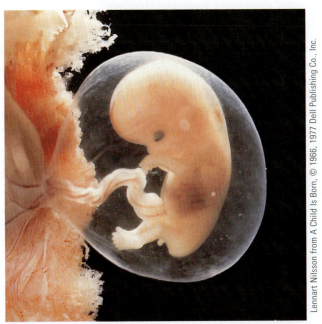

WEEK 8

Lennart Nilsson from A Child Is Born, © 1966, 1977 Dell Publishing Co., Inc.

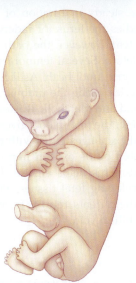

final week of embryonic period; embryo looks distinctly human compared to other vertebrate embryos

upper and lower limbs well formed; fingers and then toes have separated

early tissues of all internal, external structures now developed

tail has become stubby

actual length

C

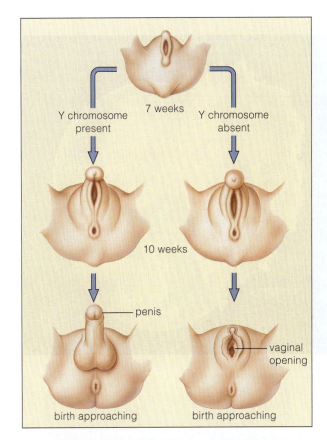

Figure 17.13 **Genitals of all embryos start out the same.** The male sex hormone testosterone must be present in order for male genitals to develop. (© Cengage Learning)

TAKE-HOME MESSAGE

HOW DOES AN EMBRYO CHANGE DURING THE SECOND FOUR WEEKS OF DEVELOPMENT?

- During its second four weeks, an embryo begins to take on the appearance of a human as features of its head, face, limbs, and genitals develop.

17.8 Development of the Fetus

- In the second and third trimesters, organs and organ systems gradually mature in preparation for birth.

- Links to Cardiovascular system 7.1, Red blood cells 8.1, Gas exchange in the lungs 10.2

In the second trimester, movements begin

When the fetus is 3 months old, it is about 4.5 inches long. Soft, fuzzy hair called the lanugo covers its body. Its reddish skin is wrinkled and protected from chafing by a thick, cheesy coating called the *vernix caseosa*.

The second trimester of development extends from the start of the fourth month to the end of the sixth. Figure 17.14 shows what the fetus looks like at 16 weeks. Its tiny facial muscles now produce frowns, squints, and sucking movements—evidence of a sucking reflex. Before the second trimester ends, the mother can easily feel her fetus's arms and legs move. During the sixth month, its eyelids and eyelashes form.

Organ systems mature during the third trimester

The third trimester extends from the seventh month until birth. At seven months the fetus is about 11 inches long, and soon its eyes will open. Although the fetus is growing larger and rapidly becoming "babylike," it will not be able to survive on its own until the middle of the third trimester. At seven months few fetuses can maintain a normal body temperature or breathe normally. However, with intensive medical care, fetuses as young as 23 to 25 weeks have survived early delivery. A baby born before seven months' gestation is at high risk of *respiratory distress syndrome* (described in Chapter 10) because its lungs lack surfactant and so can't expand adequately. The longer the baby can stay in its mother's uterus, the better. By the ninth month, its survival chances are about 95 percent.

The blood and circulatory system of a fetus have special features

The steady maturation of its organs and organ systems readies the fetus for independent life. For the circulatory system, however, the path toward independence requires a detour. Several temporary bypass vessels form and will function until birth. As Figure 17.15A shows, two umbilical arteries inside the umbilical cord transport deoxygenated blood and metabolic wastes from the fetus to the placenta. There, the fetal blood gives up wastes, takes on nutrients, and exchanges gases with the mother's blood. Fetal hemoglobin binds oxygen more easily than adult hemoglobin does. This helps ensure that enough oxygen will reach developing fetal tissues. The oxygenated blood, enriched with nutrients, returns from the placenta to the fetus in the umbilical vein.

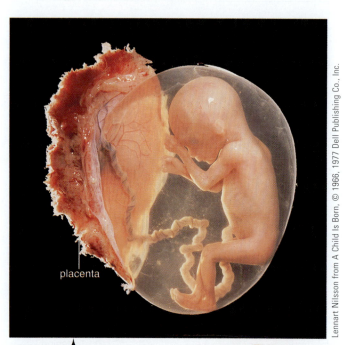

placenta

Lennart Nilsson from A Child Is Born, © 1966, 1977 Dell Publishing Co., Inc.

WEEK 16
Length: 16 centimeters
         (6.4 inches)
Weight: 200 grams
         (7 ounces)

WEEK 29
Length: 27.5 centimeters
         (11 inches)
Weight: 1,300 grams
         (46 ounces)

WEEK 38 (full term)
Length: 50 centimeters
         (20 inches)
Weight: 3,400 grams
         (7.5 pounds)

During fetal period, length measurement extends from crown to heel (for embryos, it is the longest measurable dimension, as from crown to rump).

Figure 17.14 At 16 weeks a fetus is well formed and can move. Movements begin as soon as nerves establish functional connections with developing muscles. Legs kick, arms wave, fingers grasp, the mouth puckers. These reflex actions will be vital skills in the world outside the uterus. The drawing shows a baby at full term—ready to be born. (© Cengage Learning)

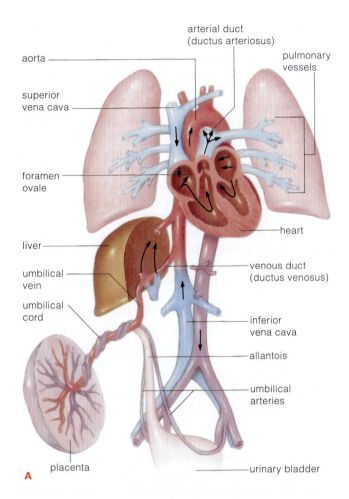

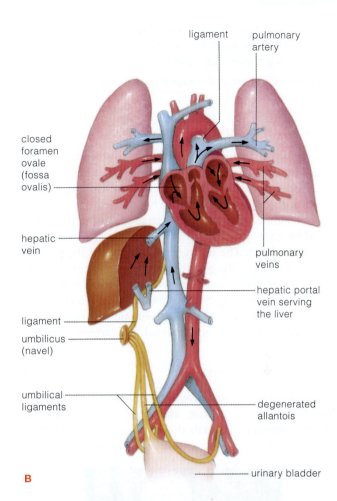

arterial duct
(ductus arteriosus)

pulmonary
vessels

aorta

superior
vena cava

foramen
ovale

liver

umbilical
vein

umbilical
cord

heart

venous duct
(ductus venosus)

inferior
vena cava

allantois

umbilical
arteries

placenta

urinary bladder

A

ligament

pulmonary
artery

closed
foramen
ovale
(fossa
ovalis)

hepatic
vein

ligament

umbilicus
(navel)

umbilical
ligaments

pulmonary
veins

hepatic portal
vein serving
the liver

degenerated
allantois

urinary bladder

B

Figure 17.15 Blood circulates in a special pattern in a fetus
(*arrows*). **A** Umbilical arteries carry deoxygenated blood and
wastes from fetal tissues to the placenta. Blood in the umbilical vein picks up oxygen and nutrients from the mother's
bloodstream and returns to the fetus. Blood mainly bypasses
the lungs, moving through the foramen ovale and the arterial
duct. It bypasses the liver by moving through the venous duct.

B At birth the foramen ovale closes, and the pulmonary and
systemic circuits of blood flow become completely separate.
The arterial duct, venous duct, umbilical vein, and portions
of the umbilical arteries become ligaments, and the allantois
degenerates. (© Cengage Learning)

Other temporary vessels divert blood past the lungs and
liver. These organs don't develop as rapidly as some others,
because (by way of the placenta) the mother's body can perform their functions. The lungs of a fetus are collapsed and
won't begin gas exchange until the newborn takes its first
breaths. Until then, its lungs receive only enough blood to
sustain their development.

Some of the blood entering the heart's right atrium flows
into the right ventricle and on to the lungs. Most of it, however, travels through a gap in the interior heart wall (called
the *foramen ovale*, or "oval opening") or into an arterial duct
(*ductus arteriosus*) that bypasses the lungs.

Similarly, most blood bypasses the fetal liver because
the mother's liver performs most liver functions until birth.
Nutrient-rich blood from the placenta travels through a
venous duct (the *ductus venosus*) past the liver and on to

the heart, which pumps it to body tissues. At birth, blood
pressure in the heart's left atrium increases. This causes
a flap of tissue to close off the foramen ovale, which then
gradually seals and separates the pulmonary and systemic
circuits of blood flow (Figure 17.15B). The temporary vessels that have formed in a fetus gradually close during the
first few weeks after birth.

TAKE-HOME MESSAGE

**WHAT ARE THE BASIC DEVELOPMENTAL EVENTS OF
THE SECOND AND THIRD TRIMESTERS?**

- The second and third trimesters of gestation are when the
organs and organ systems of a fetus enlarge and mature.
- Because the fetus exchanges gases and receives nutrients via
its mother's bloodstream, its circulatory system develops temporary vessels that bypass the lungs and liver until birth.

Birth and Beyond

- Birth, or parturition, takes place about 39 weeks after fertilization—about 280 days from the start of the woman's last menstrual period.

- Link to Hormones of the hypothalamus and pituitary 15.3

Hormones trigger birth

Usually within two weeks of a pregnant woman's "due date," the birth process, "labor," begins when smooth muscle in her uterus starts to contract. These contractions are the indirect result of a cascade of hormones from the fetus's hypothalamus, pituitary, and adrenal glands, which is triggered by an as-yet-unknown signal that says, in effect, it's time to be born. The hormonal flood causes the placenta to produce more estrogen. Rising estrogen in turn calls for a rush of oxytocin and of prostaglandins (also produced by the placenta), which jointly stimulate the uterine contractions. For about the next 2 to 18 hours, the contractions will become stronger, more painful, and more frequent.

Labor has three stages

Labor is divided into three stages that we can think of loosely as "before, during, and after." In the first stage, uterine contractions push the fetus against its mother's cervix. Initial contractions occur about every 15 to 30

Petro Feketa/Shutterstock.com

minutes and are relatively mild. As the cervix gradually dilates to a diameter of about 10 centimeters (4 inches, or "5 fingers"), contractions become more frequent and intense. Usually, the amniotic sac ruptures during this stage, which can last 12 hours or more.

The second stage of labor, actual birth of the fetus, typically occurs less than an hour after the cervix is fully dilated. This stage is usually brief—under 2 hours. Strong contractions of the uterus and abdominal muscles occur every 2 or 3 minutes, and the mother feels an urge to push. Her efforts and the intense contractions move the soon-to-be newborn through the cervix and out through the vaginal canal, usually head first (Figure 17.16). Complications can develop if the baby begins to emerge in a "bottom-first" or *breech* position. In that case the doctor may use hands or forceps to aid the delivery.

After the baby is born, the third stage of labor gets under way. More uterine contractions force fluid, blood, and the placenta (now called the afterbirth) from the mother's body. The umbilical cord—the lifeline to the mother—is now severed. A lifelong reminder of this separation is the scar we call the navel, the site where the umbilical cord was attached.

Without the placenta to remove wastes, carbon dioxide builds up in the baby's blood. Together with other factors, including handling by medical personnel, this stimulates control centers in the brain, which respond by triggering inhalation—the newborn's crucial first breath.

As the infant's lungs begin to function, the bypass vessels of the fetal circulation begin to close, soon to shut completely. The fetal heart opening, the foramen ovale, normally closes slowly during the first year of life.

Most full-term pregnancies end in the birth of a healthy infant. Yet babies born prematurely—especially before about eight months of intrauterine life—can suffer complications because their organs have not developed to the point where they can function independently. Then, attempts to sustain the baby's life under conditions that will permit the necessary additional development may require a variety of advanced medical technologies. Even then, the majority of extremely premature infants do not survive.

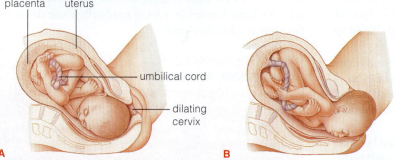

placenta   uterus

umbilical cord

dilating cervix

A

B

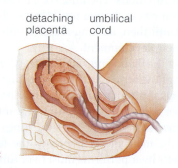

detaching placenta   umbilical cord

C

Figure 17.16 During birth, the fetus is pushed out of the uterus. The afterbirth—the placenta, fluid, and blood—is expelled shortly afterward. (© Cengage Learning)

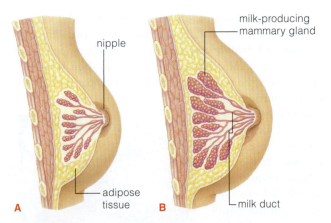

Figure 17.17 **Female breasts contain milk glands. A** The anatomy of the breast of a non-lactating woman. **B** Breast of a lactating woman. (© Cengage Learning)

nipple

milk-producing mammary gland

adipose tissue

milk duct

A

B

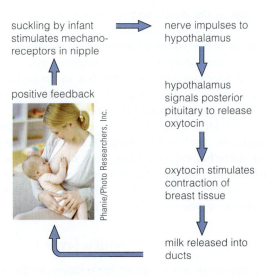

suckling by infant stimulates mechano-receptors in nipple

nerve impulses to hypothalamus

positive feedback

hypothalamus signals posterior pituitary to release oxytocin

oxytocin stimulates contraction of breast tissue

milk released into ducts

Phanie/Photo Researchers, Inc.

Figure 17.18 **Positive feedback keeps milk flowing to a suckling infant.** (© Cengage Learning)

Hormones also control milk production in a mother's mammary glands

Milk production provides an excellent example of how hormones may interact in the body. In this case, a total of four hormones are involved. During pregnancy, estrogen and progesterone stimulate the growth of mammary glands and ducts in the mother's breasts (Figure 17.17). For the first few days after birth, those glands produce colostrum, a pale fluid that is rich in proteins, antibodies, minerals, and vitamin A. Then prolactin secreted by the pituitary stimulates milk production, or **lactation**.

The "let-down" or flow of milk from a nursing mother's mammary glands is a reflex and an example of positive feedback (Figure 17.18). When a newborn nurses, mechanoreceptors in the nipple send nerve impulses to the hypothalamus, which in turn stimulates the mother's pituitary to release oxytocin, which causes the mother's breast tissues to contract. This forces milk into the ducts. This response continues as long as the baby suckles. Oxytocin also triggers contractions of uterine muscle that will help to "shrink" the uterus back to its normal size.

lactation The production of milk by mammary glands in a female's breasts.

THINK OUTSIDE THE BOOK

In recent years there has been a significant increase in births by cesarean section, in which a full-term fetus is surgically removed from the mother's uterus instead of via a vaginal birth. In 2009 in the United States, to the growing alarm of some health authorities, more than 30 percent of babies were born this way. Use Web resources such as www.childbirthconnection.org to research this topic. What are some possible reasons for the increase in cesarean births? Why are critics concerned about the rise?

TAKE-HOME MESSAGE

WHAT ARE BASIC FEATURES OF LABOR AND LACTATION?

- The mother's cervix dilates during the first stage of labor. The baby is born during the second stage. In the third stage, contractions of the uterus expel the placenta.

- Lactation, or milk production, begins a few days after birth. It is stimulated by the hormone prolactin, which is released from the mother's pituitary and acts on her breast (mammary gland) tissues.

- Suckling triggers a reflex in which oxytocin from the pituitary acts to force milk into mammary ducts. The response continues as long as the infant suckles.

17.10 Disorders: Miscarriage, Stillbirths and Birth Defects

Miscarriage is the death and spontaneous expulsion of an embryo or fetus before 20 weeks of gestation. It occurs in more than 20 percent of all conceptions, usually during the first trimester. Many factors can trigger a miscarriage (also called spontaneous abortion), but in up to half of cases the embryo (or the fetus) has one or more genetic disorders that prevent normal development.

Death of a fetus after 20 weeks is called a **stillbirth**. Some stillbirths result from complications during delivery, but most occur before labor begins. The risk is greatest for twins (or other multiples) who share a placenta and for the fetuses of women who are diabetic, smokers, have a history of drug abuse, or who are of African descent.

Poor maternal nutrition puts a fetus at risk

A pregnant woman must nourish her unborn child as well as herself. In general, the same balanced diet that is good for her should also provide her developing baby with all the carbohydrates, lipids, and proteins it needs. Vitamins and minerals are a different story, however. Physicians recommend that a pregnant woman take supplemental vitamins and minerals, not only for her own benefit but also to meet the needs of her fetus. This is particularly true for the nutrient folic acid (folate), which is required for the neural tube to develop properly. If too little folic acid is available, a birth defect called **spina bifida** ("split spine") may develop, in which the neural tube doesn't close and separate from ectoderm. The infant may be born with

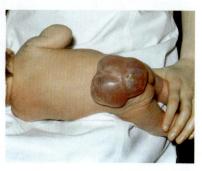

An exposed spinal cord due to spina bifida (Biophoto Associates/Photo Researchers, Inc.)

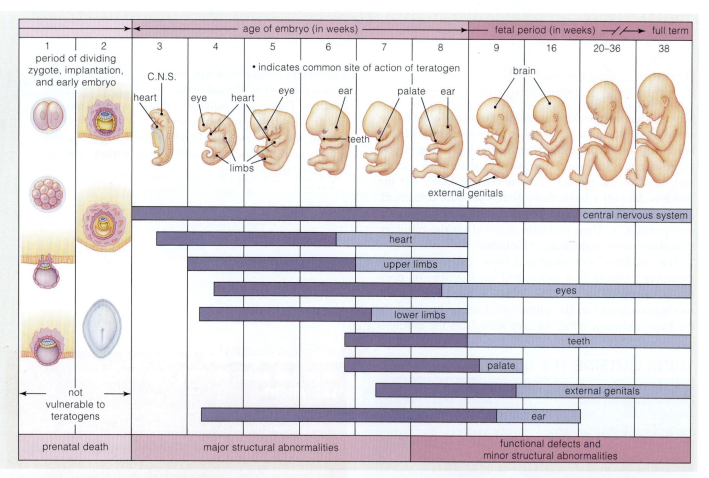

Figure 17.19 Animated! Many factors can cause birth defects. Light blue bars indicate periods when developing organs are most sensitive to damage from alcohol, viral infection, and so on. Dark blue bars indicate the time when major structural abnormalities due to other causes become apparent. Numbers signify the week of development. (© Cengage Learning)

part of its spinal cord exposed inside a cyst. Infection is a serious danger, and the resulting neurological problems can include poor bowel and bladder control. To prevent neural tube defects, folic acid now is added to wheat flour and other widely used foods.

A pregnant woman must eat enough to gain between 20 and 35 pounds, on average. If she gains much less than that, she may be putting her fetus at risk. Infants who are severely underweight have more complications after delivery. As birth approaches, the growing fetus demands more and more nutrients from the mother's body. For example, a fetus's brain grows the most in the weeks just before and after birth. Poor nutrition during that time, especially protein deficiency, can have repercussions on intelligence and other brain functions later in life.

Infections present serious risks

A pregnant woman's IgG antibodies cross the placenta. They can help protect her developing infant from all but the most severe bacterial infections. Other *teratogens*—agents that can cause birth defects—are more serious threats. Some viral diseases can be dangerous during the first six weeks of pregnancy, when the organs of a fetus are forming (Figure 17.19). For example, if a pregnant woman contracts **rubella** (German measles) during this time, there is a 50 percent chance that some organs of the embryo won't form properly. If she contracts the virus when the embryo's ears are forming, her newborn may be deaf. With time, the risk of damage diminishes, and getting vaccinated before pregnancy can eliminate it.

Drugs of all types may do harm

During its first trimester in the womb, an embryo is extremely sensitive to drugs the mother takes. In the 1960s many women using the tranquilizer thalidomide gave birth to infants with missing or deformed arms and legs. Although it wasn't known at the time, thalidomide alters the steps required for normal limbs to develop. When the connection became clear, thalidomide was withdrawn from the market (although it now has other medical uses).

Other commonly used tranquilizers, as well as some sedatives and barbiturates, may cause similar, although less severe, damage. Anti-acne drugs such as retinoic acid increase the risk of facial and cranial deformities. The antibiotic streptomycin causes hearing problems and may adversely affect the nervous system. A pregnant woman who uses the antibiotic tetracycline may have a child whose teeth are yellowed.

Like many other drugs, alcohol crosses the placenta and affects the fetus. **Fetal alcohol syndrome** (FAS) is a

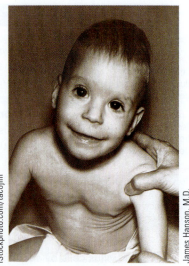

iStockphoto.com/tacojim

James Hanson, M.D.

Figure 17.20 Fetal alcohol syndrome may cause mental retardation. Outward symptoms are low and prominent ears, poorly developed cheekbones, and a long, smooth upper lip. The child may have growth problems and abnormalities of the nervous system.

constellation of defects that can result from alcohol use by a pregnant woman. Babies born with FAS typically have a smaller than normal brain and head, facial deformities, poor motor coordination, and, sometimes, heart defects (Figure 17.20). The symptoms can't be reversed, and FAS children never catch up physically or mentally. Between 60 and 70 percent of alcoholic women give birth to infants with FAS. Many doctors urge near or total abstinence from alcohol during pregnancy.

A pregnant woman who uses cocaine, especially crack, prevents her child's nervous system from developing normally. As a result, the child may be chronically irritable as well as abnormally small.

Research evidence suggests that tobacco smoke reduces the level of vitamin C in a pregnant woman's blood, and in that of her fetus as well. Cigarette smoke also harms the growth and development of a fetus in other ways. A pregnant woman who smokes daily will give birth to an underweight newborn even if her own weight, nutrition, and all other relevant variables are the same as those of pregnant nonsmokers. As noted at the beginning of this section, a pregnant smoker also has a greater risk of stillbirth, as well as of miscarriage and premature delivery. No one knows just how cigarette smoke damages a fetus. However, its demonstrated effects are additional evidence that the placenta cannot protect a developing fetus from every danger.

SCIENCE COMES TO LIFE

amniocentesis Testing fetal cells in a sample of amniotic fluid for evidence of birth defects.

chorionic villus sampling Testing of fetal cells removed from chorionic villi for birth defects.

preimplantation diagnosis Testing for birth defects in an early embryo that was conceived by in vitro fertilization.

A growing number of technologies now enable us to detect more than 100 genetic disorders before a child is born.

Chorionic villus sampling (CVS) uses tissue from the chorionic villi of the placenta. CVS can be used as early as the eighth week of pregnancy, but it is tricky. Using ultrasound, the physician guides a tube through the vagina, past the cervix, and along the uterine wall, then removes a small sample of chorionic villus cells by suction. Results are available within days.

Fran Heyl Associates

Amniocentesis is performed during the fourteenth to sixteenth weeks of pregnancy. It samples the amniotic fluid that surrounds the fetus (Figure 17.21). The thin needle of a syringe is inserted through the mother's abdominal wall, into the amnion. The physician must take care that the needle doesn't puncture the fetus and that no infection occurs. Amniotic fluid contains sloughed fetal cells, and as the syringe withdraws fluid, some of those cells are included. They are then cultured and tested for genetic abnormalities.

Methods of embryo screening also are available. In **preimplantation diagnosis**, an embryo conceived by in vitro fertilization (Section 16.8) is analyzed for genetic defects using recombinant DNA technology. The testing occurs at the eight-cell stage (*left*), which by one view is a *pre*-pregnancy stage. Like unfertilized eggs discarded during monthly menstruation, the ball is not implanted in the uterus. Its cells all have the same genes and are not yet committed to giving rise to specialized cells of a heart, lungs, or other organs. Doctors take one of the undifferentiated cells and analyze its genes for suspected disorders. If the cell has no detectable genetic defects, the ball is inserted into the uterus. Embryo screening is designed to help parents who are at high risk of having children with a genetic birth defect. Even so, for some people it raises questions of morality.

It is now possible to see a live, developing fetus with the aid of an endoscope, a fiberoptic device. In *fetoscopy*, sound waves are pulsed across the mother's uterus. Images of parts of the fetus, umbilical cord, or placenta show up on a computer screen that is connected to the endoscope (Figure 17.22). A sample of fetal blood often is drawn at the same time in order to diagnose blood cell disorders such as sickle-cell anemia and hemophilia.

All three procedures pose some risk for the fetus, including infections, punctures, or miscarriage. With CVS there also is a slight chance the forthcoming child will have missing or underdeveloped fingers or toes.

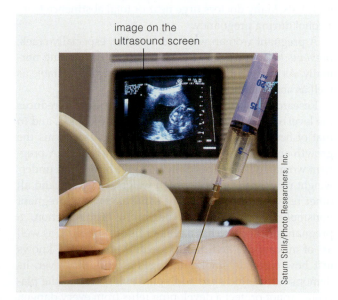

image on the ultrasound screen

Saturn Stills/Photo Researchers, Inc.

Figure 17.21 Animated! Amniocentesis is a common prenatal diagnostic tool.

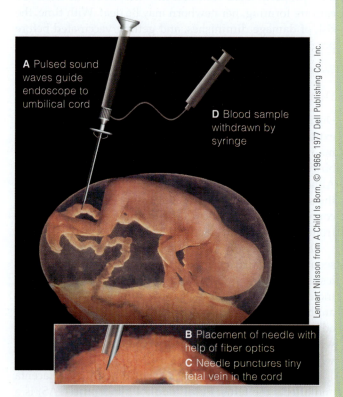

A Pulsed sound waves guide endoscope to umbilical cord

D Blood sample withdrawn by syringe

B Placement of needle with help of fiber optics
C Needle punctures tiny fetal vein in the cord

Lennart Nilsson from A Child Is Born, © 1966, 1977 Dell Publishing Co., Inc.

Figure 17.22 Fetoscopy gives a direct view of a developing fetus in the womb.

■ A gene-dictated course of growth and development leads to adulthood.

There are many transitions from birth to adulthood

Table 17.2 summarizes the prenatal (before birth) and postnatal (after birth) stages of life. A newborn is called a *neonate*. During infancy, which lasts until about 15 months of age, the child's nervous and sensory systems mature rapidly, and a series of growth spurts makes its body longer. Figure 17.23 shows how body proportions change during childhood and adolescence. **Puberty** marks the arrival of sexual maturity as a person's reproductive organs begin to function. Sex hormones trigger the appearance of secondary sex characteristics, such as pubic and underarm hair, and behavior changes. A mix of hormones triggers another growth spurt at this time. Boys usually grow most rapidly between the ages of 12 and 15, whereas girls tend to grow most rapidly between the ages of 10 and 13. After several

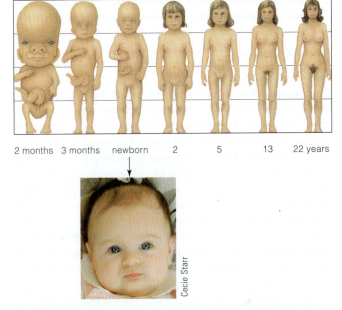

2 months   3 months   newborn   2   5   13   22 years

Cecie Starr

Figure 17.23 Body proportions change during prenatal and postnatal growth. (© Cengage Learning)

TABLE 17.2 Stages of Human Development	
PRENATAL PERIOD	
1. Zygote	Single cell resulting from fertilization
2. Morula	Ball of cells produced by cleavage
3. Blastocyst	Ball of cells with surface layer and inner cell mass
4. Embryo	All developmental stages from two weeks after fertilization until end of eighth week
5. Fetus	All developmental stages from the ninth week until birth (about 39 weeks after fertilization)
POSTNATAL PERIOD	
6. Newborn (neonate)	The first two weeks after birth
7. Infancy	From two weeks to about fifteen months after birth
8. Childhood	From infancy to about 12 or 13 years
9. Pubescence	Puberty, when secondary sexual traits develop; girls between 10 and 16 years, boys between 13 and 16 years
10. Adolescence	From puberty until about three or four years later; physical, mental, emotional maturation occur
11. Adulthood	Early adulthood (between 18 and 25 years); bone formation and growth completed; changes proceed very slowly afterward
12. Old age	Aging culminates in general body deterioration

years, the influence of sex hormones causes the cartilaginous plates near the ends of long bones to harden into bone. Humans stop growing by their early twenties.

puberty Onset of functioning of reproductive organs.

senescence Aging.

Adulthood is also a time of bodily change

Although in the United States the average life expectancy is 74 years for males and 79 years for females, we reach the peak of our physical potential in adolescence and early adulthood. A healthy diet, regular exercise, and other good lifestyle habits can help keep people vigorous for decades of adult life. Even so, after about age 40, body parts and their functioning begin to deteriorate. This process, called **senescence** or aging, is a natural part of the life cycle of all organisms that have highly specialized cells. We take a brief look at the possible causes and well-known effects of aging in the following section.

TAKE-HOME MESSAGE

WHAT ARE THE STAGES OF HUMAN DEVELOPMENT?

• Following birth, development proceeds through childhood and adolescence, which includes the arrival of sexual maturity at puberty.

• Puberty is the gateway to the adult phase of life, including changes associated with aging.

Time's Toll: Everybody Ages

■ Time takes a toll on body tissues and organs. To some extent, our genes determine how long each of us will live.

Genes may determine the maximum life span

Each species has a maximum life span. For example, we know the maximum is about twenty years for dogs and twelve weeks for butterflies. So far as we can document, no human has lived beyond 122 years. The consistency of life span within species is a sign that genes help govern aging.

One idea is that each type of cell, tissue, and organ is like a clock that ticks at its own genetically set pace. When researchers investigated this possibility, they grew normal human embryonic cells, all of which divided about fifty times, then died. In the body, most cells divide eighty or ninety times, at most. As discussed in Chapter 18, a cell copies its chromosomes before it divides. The ends of chromosomes are capped by numerous segments of DNA called telomeres. A bit of each telomere is lost each time a cell divides. The cell dies when only a nub remains.

Cancer cells, and cells in gonads that give rise to sperm and oocytes, make an enzyme that causes telomeres to lengthen. Apparently, that is why such cells can divide over and over, without dying.

© Jose Carillo/Photo Edit

Cumulative damage to DNA may also play a role in aging

A "cumulative assaults" hypothesis proposes that aging results from mounting damage to DNA combined with a decline in DNA's mechanisms of self-repair. Chapter 2 described how free radicals can damage DNA and other biological molecules. If changes in DNA aren't fixed, they may prevent cells from making needed enzymes and other proteins required for normal cell operations.

Ultimately, it's quite possible that aging involves processes in which genes, free radical damage, a decline in DNA repair mechanisms, and even other factors all come into play.

Visible changes occur in skin, muscles, and the skeleton

Aging is a gradual loss of vitality as cells, tissues, and organs function less and less efficiently. Starting at about age 40, all of us begin to see and feel the effects of such changes.

Changes in structural proteins may contribute to many of the more obvious aging-related characteristics. Remember from Chapter 4 that many connective tissues contain large amounts of the protein collagen, and some also contain the flexible protein elastin. As we age, chemical changes make collagen molecules more rigid and reduce the amount of elastin in many tissues. For example, as the elastin fibers that give skin its flexibility are slowly replaced with more rigid collagen, the skin thins, sags and wrinkles. It also becomes drier as sweat and oil glands begin to break down and are not replaced.

As hair follicles die or become less active, there is a general loss of body hair. And as pigment-producing cells die and are not replaced, the remaining body hair begins to appear gray or white.

In general, aging muscles lose mass and strength. The lost muscle tends to be replaced by fat and, with time, by collagen. Bone cells become less efficient at taking up calcium and generating new bone tissue, so the risk of osteoporosis rises. Osteoarthritis also is more common in older people. With the passing years, intervertebral disks gradually deteriorate, reducing the distance between vertebrae. This is why people tend to get shorter by about a centimeter (half an inch) every ten years from middle age onward. Staying physically active can help slow most of these changes.

Most other organ systems also decline

The heart, lungs, and kidneys also function less well with increasing age (Table 17.3). In the lungs, alveoli break down, so there is less respiratory surface available for gas exchange. In a person who does not have cardiovascular

TABLE 17.3 Some Physiological Changes in Aging

Maximum Age	Lung Heart Rate	Muscle Capacity	Kidney Strength	Efficiency
25	100%	100%	100%	100%
45	94%	82%	90%	88%
65	87%	62%	75%	78%
85	81%	50%	55%	69%

Note: Age 25 is the benchmark for maximal efficiency of physiological functions.

disease (which may cause an enlarged heart), the heart muscle shrinks slightly and so its strength and blood-pumping ability decline. The waning blood supply may be a factor in aging-related changes throughout the body. Blood transport is also affected by changes in aging blood vessels. Elastin fibers in blood vessel walls are replaced with connective tissue containing collagen or become hardened with calcium deposits, and so vessels stiffen. Cholesterol plaques often cause further narrowing of arteries and veins (Section 7.8). This is why resting blood pressure may rise as people get older. However, as with the muscular and skeletal systems, lifestyle choices such as not smoking, eating a healthy diet, and getting regular exercise can help each of us maintain a vigorous respiratory and cardiovascular system well past middle age.

In the immune system, the total number of T cells falls and B cells become less active. Older people also are more likely to develop autoimmune diseases. It is possible that faltering DNA repair mechanisms no longer fix genetic changes that alter self-markers. This could provoke immune responses against the body's own cells.

In the aging GI tract, mucous glands in the stomach and intestines gradually break down, and the pancreas secretes fewer digestive enzymes. Although it is vital for older people to eat a healthy diet, we require fewer calories as we age. By age 50, your basal metabolic rate will be only 80 to 85 percent of what it was in childhood and will keep declining about 3 percent every decade. This is why middle-aged people tend to gain weight unless they compensate by eating less, increasing their physical activity, or both.

Levels of most hormones stay steady throughout life. Sex hormones are exceptions, however. As Chapter 16 noted, falling levels of estrogens and progesterone trigger menopause in women. In older men, falling levels of testosterone reduce fertility. That said, men have fathered children into their 80s. Men and women both retain their capacity for sexual response well into old age.

Aging also alters the brain and senses

Brain neurons die throughout life, so the brain shrinks slightly over time. It loses about 10 percent of its mass after eighty years. Brain neurons also are damaged by the cumulative effects of free radicals. In addition, in most people who live to old age, tangled clumps of a protein called *tau* develop in the cell bodies of many brain neurons. These *neurofibrillary tangles* can disrupt normal cell operations. Clotlike plaques containing misfolded proteins called *beta amyloid* also develop between neurons.

In people who develop **Alzheimer's disease** (AD), the brain is riddled with masses of neurofibrillary tangles and

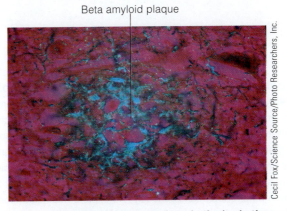

Beta amyloid plaque

Figure 17.24 Beta amyloid plaques form in the brain tissue of Alzheimer's patients.

beta amyloid plaques (Figure 17.24). AD symptoms include progressive loss of normal mental functions, including short-term memory.

Some cases of AD are inherited. The increased risk is significant for people who inherit a version of a gene that codes for a protein called apolipoprotein E. Around 16 percent of the U.S. population has one or two copies of this gene. Those with two copies have a 90 percent chance of developing AD. Other AD-related genes also are known. Most are associated with early-onset Alzheimer's, which develops before the age of 65.

Treatments for AD are limited, although drugs can temporarily help alleviate some symptoms or slow the progression of the disease. Even so, researchers are making major strides in understanding the causes of AD. There is hope that this growing knowledge will lead to strategies for preventing the disease.

After about age 60 even otherwise healthy people begin to have "senior moments," or occasional difficulty with short-term memory. Older neurons do not conduct action potentials as efficiently, and neurotransmitters such as acetylcholine may be released more slowly. Such changes are why older people tend to move more slowly, have slower reflexes, and have more problems with muscle coordination.

Our sensory organs also become less efficient at detecting or responding to stimuli. The taste buds become less sensitive over time, and as Chapter 14 noted, people also tend to become farsighted as they grow older because the eye lens loses elasticity and is altered in other ways that prevent it from properly focusing incoming light. Most people over 60 also have some hearing loss due to "worn out" sensory nerve cells in the ear.

TAKE-HOME MESSAGE

HOW DOES AGING AFFECT BODY SYSTEMS?

• Virtually every body system undergoes age-related declines in structure and functioning.

• Aging may result from several factors, including an internal biological clock that ticks out the life spans of cells, and the accumulation of DNA damage.

EXPLORE
ON YOUR OWN

Housecats can carry the parasite that causes *toxoplasmosis*. In an otherwise healthy person the disease may only produce flulike symptoms, but it is dangerous for a pregnant woman and her fetus. A mother-to-be may suffer a miscarriage, and if the parasite infects a fetus it causes birth defects. An infected cat may not appear to be ill, but its feces will contain infectious cysts. This is why some physicians advise pregnant women to avoid contact with cats and not to clean sandboxes, take care of housecat "accidents," or empty a litterbox.

All that said, toxoplasmosis is not especially common—so is it something a cat-loving expectant mother should take seriously? To explore this health concern, research the kinds of birth defects caused by toxoplasmosis and find out what stance (if any) public health authorities in your community take on this issue. Is the disease more common in some regions or settings than in others? Can cat owners have their pets tested for the disease?

Lauren and Homer. Photography by Gary Head

SUMMARY

Section 17.1 The fertilization of an oocyte by a sperm launches several key stages in early development.

During cleavage, the fertilized egg undergoes cell divisions that form the early multicellular embryo. The destiny of various cell lines is established in part by the portion of cytoplasm inherited at this time.

Gastrulation lays out the organizational framework of the whole body. Endoderm, ectoderm, and mesoderm form; all the tissues of the adult body will develop from these three germ layers.

Differentiation is the process during which cells come to have specific structures and functions.

In morphogenesis, tissues form and become organized into organs. Tissues and organs continue to mature as the fetus develops and even after birth.

Sections 17.2, 17.3 During the first week or so after fertilization, cell divisions and other changes transform the zygote into a multicellular blastocyst, which attaches to the mother's uterus during implantation. The blastocyst includes the inner cell mass, a small clump of cells from which the embryo develops.

Section 17.4 Gastrulation and morphogenesis shape the body's basic plan. A key step is the formation of the neural tube, the forerunner of the brain and spinal cord, from ectoderm. The skeleton and most muscles develop from blocks of cells called somites that arise from mesoderm. In morphogenesis, sheets of cells fold and cells migrate to new locations in the developing embryo.

Sections 17.5, 17.6 During implantation, the inner cell mass is transformed into an embryonic disk. Some of its cells give rise to four extraembryonic membranes: the yolk sac, the allantois, the amnion, and the chorion. Table 17.4 summarizes their functions. The embryo and its mother exchange nutrients, gases, and wastes by way of the placenta, a spongy organ that is a combination of endometrium and extraembryonic membranes.

Section 17.7 The first eight weeks of development are the embryonic period; thereafter the developing individual is considered a fetus. By the ninth week of development, the fetus clearly looks human.

Section 17.8 During the last three months of gestation (the third trimester), the fetus grows rapidly and many organs mature. However, because the fetus exchanges gases and receives nourishment via its mother's bloodstream, its own circulatory system routes blood flowing to the lungs and liver through temporary blood vessels.

Sections 17.9, 17.10 Birth takes place approximately 39 weeks after fertilization. Labor advances through three stages; a baby is born at the end of stage two, and the afterbirth (placenta) is expelled in stage three. At birth, contractions of the uterus dilate expel the fetus and afterbirth. After delivery, nursing causes the secretion of hormones that stimulate lactation—the production and release of milk.

Miscarriage is the spontaneous expulsion of a developing embryo before 20 weeks of gestation. The term "stillbirth" refers to the death and delivery of a fetus after 20 weeks. Poor maternal nutrition, certain infections, and use of drugs including nicotine, alcohol, and therapeutic drugs may have serious harmful effects on her developing fetus.

Sections 17.12, 17.13 Human development can be divided into a prenatal period before birth, followed by the neonate (newborn) stage, childhood, adolescence, and adulthood. The process of aging is called senescence. As the body ages, changes occur in the structure and functional efficiency of many organ systems. These changes are due to multiple factors.

TABLE 17.4 Extraembryonic Membranes

Membrane	Function
YOLK SAC	Source of digestive tube; helps form blood cells and forerunners of gametes
ALLANTOIS	Source of umbilical blood vessels and vessels of the placenta
AMNION	Sac of fluid that protects the embryo and keeps it moist
CHORION	Forms part of the placenta; protects the embryo and the other extraembryonic membranes

REVIEW QUESTIONS

1. Define and describe the main features of the following developmental stages: fertilization, cleavage, gastrulation.

2. Label the following stages of early development:

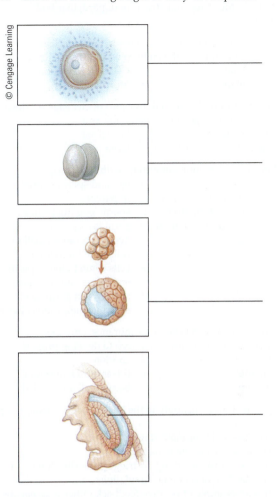

© Cengage Learning

3. Define cell differentiation and morphogenesis, two processes that are critical for development.

4. Summarize the development of an embryo and a fetus. When are body parts such as the heart, nervous system, and skeleton largely formed?

SELF-QUIZ *Answers in Appendix V*

1. Development cannot proceed properly unless each of the following processes is successfully completed before the next begins, starting with _____.
   a. gamete formation d. gastrulation
   b. fertilization e. organ formation
   c. cleavage f. growth, tissue specialization

2. During cleavage, the _____ is converted to a ball of cells, which in turn is transformed into the _____.
   a. zygote; blastocyst c. ovum; embryonic
   b. trophoblast; embryonic disk
   disk d. blastocyst; embryonic disk

3. In the week following implantation, cells of the _____ will give rise to the embryo.
   a. blastocyst c. embryonic disk
   b. trophoblast d. zygote

4. Fill in the blank: The developmental process called _____ produces the shape and structure of particular body regions.

5. _____ is the gene-guided process by which cells in different locations in the embryo become specialized.
   a. Implantation   c. Cell differentiation
   b. Neurulation   d. Morphogenesis

6. In a human zygote, the cell divisions of cleavage produce an embryonic stage known generally as a _____.
   a. zona pellucida   c. blastocyst
   b. gastrula   d. larva

7. Match each developmental stage with its description.
   _____ cleavage
   _____ gamete formation
   _____ organ formation
   _____ cell differentiation
   _____ gastrulation
   _____ fertilization

   a. egg and sperm mature in parents
   b. sperm, egg nuclei fuse
   c. germ layers form
   d. zygote becomes a ball of cells called a morula
   e. cells come to have specific structures and functions
   f. starts when germ layers split into subgroups of cells

8. Of the four extraembryonic membranes, only the _____ is not needed in order for an embryo to develop properly.
   a. yolk sac   d. chorion
   b. allantois   e. this is a trick question, because all are needed
   c. amnion

9. All of the following apply to the birth process (labor) except _____.
   a. It consists of four main stages.
   b. Hormones from the fetus trigger it.
   c. In a newborn, breathing air triggers the shutdown of the "bypass" vessels of fetal circulation.
   d. It is an example of positive feedback rather than negative feedback.

10. Of the following, _____ cannot cross the placenta.
    a. alcohol
    b. the mother's antibodies
    c. antibiotics
    d. toxic substances in tobacco smoke
    e. all can cross the placenta

CRITICAL THINKING

1. How accurate is the statement "A pregnant woman must do everything for two"? Give some specifics to support your answer.

2. A renowned developmental biologist, Lewis Wolpert, once observed that birth, death, and marriage are not the most important events in human life—rather, Wolpert said, the most important moment in life is gastrulation. Given the discussion in Section 17.1, what do you think he meant?

3. One of your best friends tells you that she and her husband think she might be pregnant. She feels she can wait until she's several months along before finding an obstetrician. You think she could use some medical advice sooner, and you suggest she discuss her plans with a physician as

Your Future

Cecil Fox/Science Source/Photo Researchers, Inc.

Researchers studying Alzheimer's disease may be making significant progress in understanding how the disease spreads in the brain. Recent studies suggest that faulty proteins that form beta amyloid plaques may infiltrate still-healthy brain tissue by following pathways of neuron circuits in the brain. The research tracked the spread of AD lesions in transgenic mice that develop the disease. The hope is that as more is learned about how plaques spread, it may be possible to develop drug therapy that can halt the process before a patient's AD becomes severe.

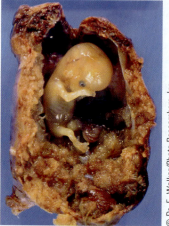

© Dr. E. Walker/Photo Researchers, Inc.

Figure 17.25 This embryo, the result of an ectopic pregnancy, could not survive implantation outside its mother's uterus.

soon as possible. What kinds of health issues might you be concerned about?

4. The complications of ectopic pregnancy (Section 17.2) are life-threatening for the mother, and in fact each year in the United States a few pregnant women die when their situation is not diagnosed in time. Tragically, the only option is to surgically remove the embryo, which was doomed from the beginning. Based on what you know about where an embryo normally develops, explain why the ectopic embryo in Figure 17.25 could not have long survived.

© Gary Head

CELL REPRODUCTION

18

LINKS TO EARLIER CONCEPTS

This chapter builds on the discussion in Chapter 3 of the cell nucleus and chromosomes and explains how microtubules assist in cell division. (3.6, 3.9)

You will also learn more about how sperm and eggs form during the processes of spermatogenesis and oogenesis (16.2 and 16.4).

You will gain a fuller understanding of how the union of sperm and egg at fertilization (16.6) provides a zygote with the full set of parental chromosomes required for normal development.

KEY CONCEPTS

Basic Principles of Cell Division

Cells reproduce by duplicating their chromosomes and then dividing the chromosomes and cell cytoplasm among the daughter cells. Sections 18.1–18.2

Mitosis: Body Growth and Repair

The body grows and tissues are repaired when cells divide by the mechanism called mitosis. This mechanism divides the nucleus so that each newly formed cell has the same number of chromosomes as the parent cell. Sections 18.3, 18.4, 18.9

Meiosis: Cells for Sexual Reproduction

Sperm and oocytes form by the cell division mechanism of meiosis. Meiosis reduces the number of chromosomes so that each gamete has half the number of chromosomes of the parent cell. Sections 18.6–18.9

Top: L. Willatt, East Anglian Regional Genetics Service/Photo Researchers, Inc. Middle and bottom: © Cengage Learning

Body cells multiply by dividing. By the time you were born, cell division had given you a body of about a trillion cells.

One type of cell division is the basis for the growth that transforms a zygote—the first cell of a new individual—into a newborn, then a child, then an adult. Many types of cells, such as epithelial cells of the skin, continue to divide throughout adult life, replacing cells that are damaged or are simply worn out.

Another type of cell division occurs only in the germ cells of the testes or ovaries. It produces gametes—a male's sperm or a female's eggs—that allow our species to produce each new generation.

Learning about cell division begins to help us understand how each of us is put together in the image of our parents. Our study of this topic will lead to the answers to three basic questions. First, what kind of information guides inheritance? Second, how is the information copied in a parent cell before being distributed into daughter cells? And third, what kinds of mechanisms parcel out the information to daughter cells?

iStockphoto.com/Rosemarie Gearhart

Reproduction: Continuing the Life Cycle

autosomes Chromosomes other than the sex chromosomes.

chromosome A DNA molecule together with the proteins attached to it.

chromosome number The sum of the chromosomes in a species' body cells.

diploid Having two of each type of chromosome occurring in a species; body cells are diploid.

haploid Having one of each type of chromosome occurring in a species; gametes are haploid.

homologous chromosomes Chromosomes with the same length, shape, and set of genes.

life cycle The recurring series of events in which individuals grow, develop, maintain themselves, and produce a new generation.

■ The continuity of life depends on the ability of cells to faithfully reproduce themselves.

■ Links to Life characteristics 1.1, Cell nucleus 3.6, Formation of eggs and sperm 16.2, 16.4

In biology, **reproduction** is when a "parent" cell produces a new generation of cells, or when parents produce a new individual. Reproduction is part of a **life cycle**, a recurring series of events in which individuals grow, develop, maintain themselves, and reproduce a new generation. The instructions for the human life cycle are encoded in our DNA, which we inherit from our parents.

Remember from Section 3.6 that DNA is organized into structures called **chromosomes** in the nucleus of a cell. When a cell is not dividing, its threadlike chromosomes are dispersed in the nucleus. As a cell prepares to divide, however, each chromosome is copied and each copy is coiled and packed tightly. Each copy is condensed as it coils back on itself again and again (Figure 18.1). Notice that the DNA loops around some

proteins (called histones), forming beadlike structures. The "beads" then coil up into a long fiber.

As a cell nucleus starts to divide, the DNA coils tighten up even more to form the condensed chromosome shown at the upper right in Figure 18.1. This "supercoiling" may help keep chromosomes from getting tangled up when they are moved and sorted into parcels for daughter cells. When the coiling is complete, a chromosome has its typical size and shape.

Body cells have two sets of chromosomes

Every species of organism has a characteristic number of chromosomes. In humans, DNA is carried on 23 different types of chromosomes (Figure 18.2). In all cells but gametes, however—that is, in all somatic ("of the body") cells—each type of chromosome comes as a paired set, in which one member of the pair comes from each parent. The sum of the chromosomes in a species' cells is called the **chromosome number**. In humans the chromosome number is 46.

A cell that has two of each type of chromosome is called a **diploid** cell. The shorthand $2n$ indicates that a cell is diploid. The n stands for the number of chromosomes in one full set. All the body's somatic cells are diploid. When a soon-to-divide diploid cell copies its chromosomes, it has four sets of them. The division process called *mitosis*, which we consider shortly, puts half of this doubled genetic material in each new cell. As a result, each cell ends up with the diploid number of chromosomes.

1 The DNA inside the nucleus of a cell is typically divided up into a number of chromosomes. Inset: a duplicated human chromosome.

2 At its most condensed, a duplicated chromosome is packed tightly into an X shape.

3 A chromosome unravels as a hollow cylinder formed by coiled coils.

4 The coiled coils consist of a long molecule of DNA (blue) and the proteins that are associated with it (purple).

5 At regular intervals, the DNA molecule is wrapped twice around a core of proteins. In this "beads-on-a-string" structure, the "string" is the DNA, and each "bead" is called a nucleosome.

6 The DNA molecule itself has two strands that are twisted into a double helix. Chapter 21 discusses DNA's structure.

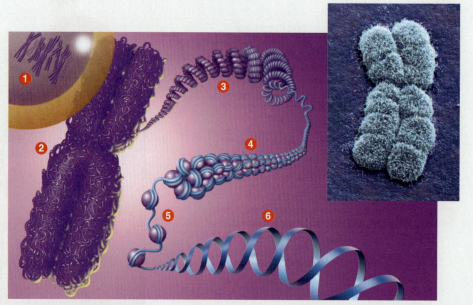

Figure 18.1 Animated! DNA in a chromosome is looped and coiled. (© Cengage Learning; Andrew Syred/Photo Researchers, Inc.)

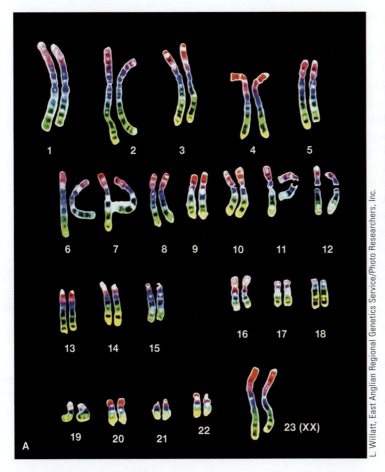

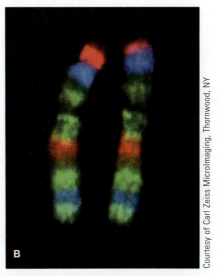

Figure 18.2 **Human somatic cells have 46 chromosomes. A** In this image, the presence of pairs of chromosomes (two of each type) tells you that they came from a diploid cell. One member of each pair contains genetic instructions inherited from the father. The other member contains instructions from the mother. **B** Close-up of a pair of homologous chromosomes from an animal cell.

L. Willatt, East Anglian Regional Genetics Service/Photo Researchers, Inc.

Courtesy of Carl Zeiss Microlmaging, Thornwood, NY

Gametes have one set of chromosomes

Unlike a "regular" body cell, a gamete has just one set of 23 chromosomes, instead of 46 paired chromosomes. This is because the cell division process that creates gametes, *meiosis*, halves the diploid number of chromosomes ($2n$) to a **haploid** number (n). And not just any half. Each haploid gamete ends up with one partner from each pair of homologous parent chromosomes.

In Figure 18.2, each type of human chromosome has been lined up with its partner from the other parent. Pairs 1 through 22 are **autosomes**: the two members of each pair are about the same length and both carry hereditary instructions for the same traits. Pair 23 consists of the **sex chromosomes**, which determine a person's biological sex. The two types of sex chromosomes are denoted X and Y. A female has two X chromosomes, while a male is XY.

The paired corresponding chromosomes, one from each parent, are called **homologous chromosomes** (Figure 18.2B), or simply *homologues* (from a Greek word meaning "to agree"). The X and Y sex chromosomes are considered to be homologues, even though they are of different size and shape and for the most part they carry genetic information for different traits. We will return to

the topic of sex chromosomes in Chapter 20. The rest of this chapter deals with how cells divide, beginning with an overview of the cell cycle, mitosis, and meiosis.

reproduction The process in which a parent cell or organism produces a new individual cell or whole organism.

sex chromosomes Chromosomes that determine a person's biological sex.

TAKE-HOME MESSAGE

WHAT ARE THE TWO TYPES OF CHROMOSOMES?

- A cell's DNA is organized into autosomes and sex chromosomes.
- In humans there are 22 kinds of autosomes and two possible types of sex chromosomes, X and Y.
- Body (somatic) cells have a diploid number of chromosomes —that is, they have paired sets of 23 chromosomes, making 46 chromosomes in all. The chromosomes in each paired set are called homologous chromosomes. Gametes have half the diploid chromosome number—23—which is called the haploid number of chromosomes.
- In a diploid cell the paired set of sex chromosomes may be either two Xs or an X and a Y. Females have an XX pair. Males have an XY pair.

- Mitosis occurs during the "lifetime" of a somatic cell, called the cell cycle. Meiosis occurs in germ cells only.

- Links to Cell nucleus 3.6, Cytoskeleton 3.9

cell cycle The series of events from the time a cell forms until it divides.

centromere The constricted area where sister chromatids attach to each other.

meiosis Process that divides the nucleus in a dividing germ cell in a way that halves the chromosome number in daughter cells.

mitosis Process that divides the nucleus of a dividing somatic cell in a way that maintains the chromosome number in daughter cells.

sister chromatids A duplicated chromosome and its copy.

The **cell cycle** starts every time a new cell is produced, and it ends when the cell completes its own division. Usually, the longest phase of the cell cycle is interphase, as sketched in Figure 18.3. This phase has three parts in which a cell grows larger, more or less doubles the number of components in its cytoplasm, then copies its DNA. This copying process duplicates the chromosomes. Then comes **mitosis** (my-TOE-sis), the division of the cell nucleus into two nuclei, each with a full diploid set of chromosomes (Figure 18.4). The parts of the cell cycle are

G1 The part of interphase when the cell grows

S The part of interphase when a cell's DNA is copied and its chromosomes are duplicated

G2 The part of interphase after chromosomes are duplicated, when other events prepare the cell to divide

M Mitosis, when chromosomes (copied DNA) are sorted into two sets and the cytoplasm divides

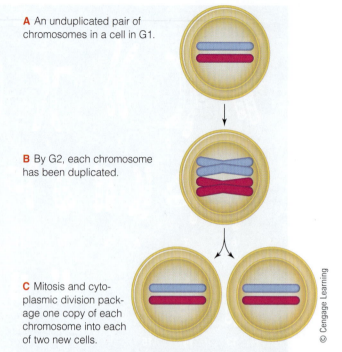

A An unduplicated pair of chromosomes in a cell in G1.

B By G2, each chromosome has been duplicated.

C Mitosis and cytoplasmic division package one copy of each chromosome into each of two new cells.

© Cengage Learning

Figure 18.4 **Mitosis maintains the chromosome number.** After mitosis is complete, the parent cell's cytoplasm divides. Section 18.4 explains this final step in cell division.

1 A cell spends most of its life in interphase, which includes three stages: G1, S, and G2.

2 G1 is the interval of growth before DNA replication. The cell's chromosomes are unduplicated.

3 S is the time of synthesis, during which the cell copies its DNA.

4 G2 is the interval after DNA replication and before mitosis. The cell prepares to divide during this stage.

5 The nucleus divides during mitosis, the four stages of which are detailed in Section 18.3. After mitosis, the cytoplasm may divide. The cycle begins anew, in interphase, for each descendant cell.

6 Built-in checkpoints ☑ stop the cycle from proceeding until certain conditions are met.

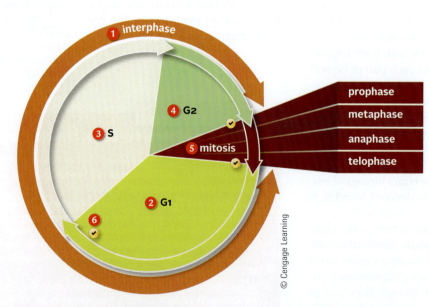

prophase
metaphase
anaphase
telophase

© Cengage Learning

Figure 18.3 **The cell cycle has two main stages, interphase (*orange arrow*) and mitosis (*brown arrow*).** The duration of each interval differs among cells.

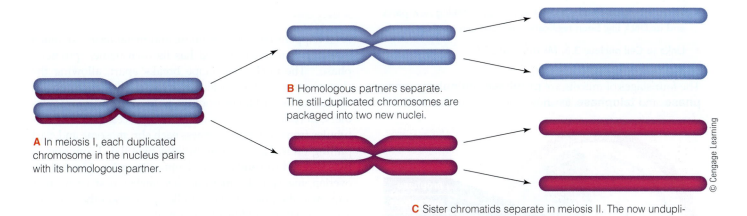

A In meiosis I, each duplicated chromosome in the nucleus pairs with its homologous partner.

B Homologous partners separate. The still-duplicated chromosomes are packaged into two new nuclei.

C Sister chromatids separate in meiosis II. The now undupli-cated chromosomes are packaged into four new nuclei.

© Cengage Learning

Figure 18.5 Meiosis halves the chromosome number.

The length of the cell cycle varies depending on the type of cell. For instance, the cycle lasts 25 hours in epithelial cells in your stomach lining and 18 hours in bone marrow cells. New red blood cells form and replace your worn-out ones at an average rate of 2 to 3 million each second.

Mitosis occurs in cells that are dividing as the body grows or repairs itself. It assures that each new "daughter" cell is diploid, with a full complement of 46 parental chromosomes. By contrast, **meiosis** (my-OH-sis) occurs in germ cells in the ovaries and testes, and it halves the number of parental chromosomes (Figure 18.5). Two rounds of division, called meiosis I and meiosis II, produce gametes that are haploid cells, with half the diploid number of chromosomes. Table 18.1 summarizes the differences between meiosis and mitosis, which we'll look at in more detail in the following sections.

A chromosome undergoes changes in preparation for cell division

As already mentioned, a chromosome consists of DNA and some proteins that are attached to it. (Together, the DNA and protein are called *chromatin*.) Before a cell begins to divide by either mitosis or meiosis, its chromosomes are duplicated:

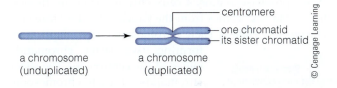

centromere
one chromatid
its sister chromatid

a chromosome (unduplicated)   a chromosome (duplicated)

© Cengage Learning

For much of the cell division process, each chromosome and its copy stay together. During this time they are called **sister chromatids**. A sister chromatid has at least one "pinched in" region called a **centromere**. It will provide attachment sites for microtubules that move chromosomes when a cell nucleus is dividing.

With this background we now begin a closer look at cell division, starting with mitosis.

TABLE 18.1 Overview of Mitosis and Meiosis		
	Mitosis	**Meiosis**
FUNCTION	Growth, including repair and maintenance	Gamete production (sperm/eggs)
OCCURS IN	Somatic (body) cells	Germ cells in gonads (testes and ovaries)
MECHANISM	Chromosomes are duplicated once, then the cytoplasm is divided	Chromosomes are duplicated twice, then the cytoplasm is divided
OUTCOME	Maintains the diploid chromosome number ($2n \rightarrow 2n$)	Halves the diploid chromosome number ($2n \rightarrow n$)
EFFECT	Two diploid daughter cells	Four haploid daughter cells

TAKE-HOME MESSAGE

WHAT ARE THE CELL CYCLE, MITOSIS AND MEIOSIS?

- The cell cycle is the sequence of events in a somatic cell during which the cell grows, doubles the components of its cytoplasm, then copies its DNA, forming duplicates of its chromosomes.

- A somatic cell divides its nucleus and cytoplasm at the end of the cycle. The division of a somatic cell's nucleus is called mito-sis. Mitosis gives each daughter cell the same diploid number of chromosomes as were in the parent cell.

- Meiosis occurs in germ cells that give rise to gametes. Meiosis provides each gamete with a haploid number of chromosomes, one half the number that were in the parent cell.

- When interphase ends, a cell has stopped making new parts and its DNA has been replicated. Mitosis can begin.

- Links to Cell nucleus 3.6, Microtubules 3.9

The four stages of mitosis are **prophase**, **metaphase**, **anaphase**, and **telophase**, as shown here and in Figure 18.6.

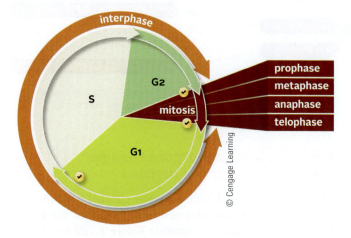

© Cengage Learning

Mitosis begins with prophase

When prophase begins, a cell's chromosomes are thread-like. ("Mitosis" comes from the Greek *mitos,* for "thread.") During interphase each chromosome was duplicated, forming two sister chromatids joined at the centromere. The sister chromatids of each chromosome twist and fold into a more compact form. By the end of prophase, all chromosomes will be condensed into thick rod shapes.

Meanwhile, in the cytoplasm, most microtubules of the cell's cytoskeleton are breaking apart into their subunits (Section 3.9) and new microtubules are forming near the nucleus. The nuclear envelope physically prevents them from making contact with the chromosomes inside the nucleus, but not for long: The nuclear envelope starts to break up as prophase ends.

In both mitosis and meiosis, chromosomes move into new positions with help of a **spindle**. This structure consists of microtubules that extend from centrioles. Each barrel-shaped centriole was duplicated in interphase, so there are two pairs of them in prophase. In prophase, microtubules start moving each pair of centrioles to the opposite ends, or "poles," of the spindle.

anaphase Stage of mitosis when sister chromatids of each chromosome separate and move to opposite spindle poles.

metaphase Stage of mitosis when duplicated chromosomes are lined up midway between the poles of the spindle.

prophase The first stage of mitosis, when a cell's chromosomes condense.

spindle An array of microtubules that moves chromosomes during mitosis or meiosis.

telophase The last stage of mitosis, when a nuclear envelope forms around each of the two nuclei of forthcoming daughter cells.

Next comes metaphase

A lot happens between prophase and metaphase—so much that this transitional period has its own name, "prometaphase." The nuclear envelope breaks apart, allowing the chromosomes in the nucleus to interact with microtubules extending toward them from the poles of the forming spindle. Cell biologists have watched this process occur. Microtubules from both poles harness each chromosome and start pulling on it. The two-way pulling tugs the chromosome's two sister chromatids toward opposite poles. Meanwhile, overlapping spindle microtubules ratchet past each other and push the poles of the spindle apart. Soon the chromosomes reach the middle of the spindle.

When all the duplicated chromosomes are lined up midway between the poles of a spindle, we say the cell is in metaphase (*meta-* means "midway between"). This alignment sets the stage for anaphase, the next stage of mitosis.

Anaphase, then telophase follow

During anaphase, the sister chromatids of each chromosome separate from each other and move to opposite spindle poles. Two mechanisms produce this movement. First, the microtubules attached to the centromeres pull the chromosomes toward the poles. Second, the spindle elongates as overlapping microtubules continue to ratchet past each other and push the two spindle poles even farther apart. Once each chromatid is separated from its sister, it is an independent chromosome.

Telophase begins as soon the two clusters of chromosomes each arrive at a spindle pole. The chromosomes are no longer connected to microtubules, and they return to threadlike form. Bit by bit, a new nuclear envelope forms around each cluster, separating it from the cytoplasm. Now there are two new nuclei. Each new nucleus has the same chromosome number as the parent nucleus. Once two nuclei form, telophase is over—and so is mitosis.

TAKE-HOME MESSAGE

WHAT ARE THE STAGES OF MITOSIS?

- Mitosis occurs in consecutive stages, called prophase, metaphase, anaphase, and telophase.

- Before mitosis, each chromosome in a cell's nucleus is duplicated, so that it consists of two sister chromatids.

- A spindle of microtubules moves the sister chromatids of each chromosome apart, to opposite spindle poles. A new nuclear envelope forms around the two chromosome clusters. Both daughter nuclei have the same number of chromosomes as the parent cell's nucleus.

centrosome

1 Early prophase
Mitosis begins. In the nucleus, the DNA begins to condense. The centrosome gets duplicated.

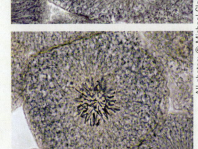

2 Prophase
The duplicated chromosomes become visible as they condense. One of the two centrosomes moves to the opposite side of the nucleus. The nuclear envelope breaks up.

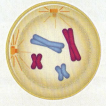

3 Transition to metaphase
The nuclear envelope is gone, and the chromosomes are at their most condensed. Spindle microtubules assemble and bind to chromosomes at the centromere. Sister chromatids are attached to opposite spindle poles.

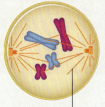

microtubule of spindle

4 Metaphase
All of the chromosomes are aligned midway between the spindle poles.

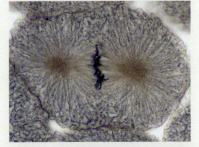

5 Anaphase
Spindle microtubules separate the sister chromatids and move them toward opposite spindle poles. Each sister chromatid has now become an individual, unduplicated chromosome.

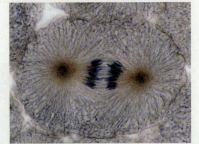

6 Telophase
The chromosomes reach the spindle poles and decondense. A nuclear envelope forms around each cluster, and mitosis ends.

All photos: © Michael Clayton, University of Wisconsin

Figure 18.6 Animated! Mitosis ensures that daughter cells will have the same chromosome number as the parent cell. For clarity, the diagram shows only two pairs of chromosomes from a diploid (2n) animal cell. (© Cengage Learning)

■ After mitosis produces two new cell nuclei, each with a set of the parent cell's chromosomes, the parent cell's cytoplasm also must be divided between the two daughter cells.

cleavage furrow The area of a dividing cell's plasma membrane where microfilaments pull the cell surface inward until the cell is pinched in two.

cytokinesis Division of the cytoplasm of a dividing cell.

Division of the cytoplasm, or **cytokinesis**, usually begins toward the end of anaphase. By this time, the two sister chromatids of each chromosome have been separated and are independent chromosomes. In an animal cell, about midway between the cell's two poles, a patch of plasma membrane sinks inward, forming a **cleavage furrow**. Microfilaments made of the contractile protein actin steadily pull the plasma membrane inward all around the cell, until the cell is pinched in two (Figure 18.7). Now there are two new cells, each with a nucleus, cytoplasm, and a plasma membrane.

This concludes our tour of mitosis. Now we turn to meiosis, the nucleus-dividing mechanism that forms gametes. It is difficult not to be in awe of the astonishing precision with which both mitosis and meiosis take place.

TAKE-HOME MESSAGE

WHAT IS CYTOKINESIS?

- Cytokinesis completes cell division following mitosis. It divides the parent cell's cytoplasm in a way that allots cytoplasm, one of the two newly formed nuclei, and plasma membrane to each daughter cell.

- In a diploid cell the paired set of sex chromosomes may be either two Xs or an X and a Y. Females have an XX pair. Males have an XY pair.

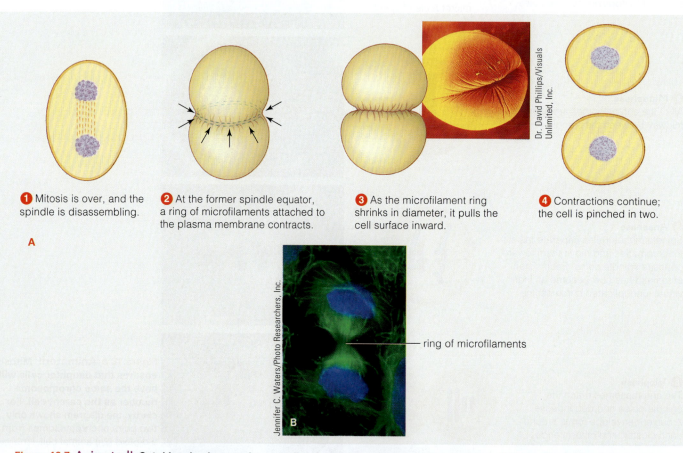

Dr. David Phillips/Visuals Unlimited, Inc.

① Mitosis is over, and the spindle is disassembling.

② At the former spindle equator, a ring of microfilaments attached to the plasma membrane contracts.

③ As the microfilament ring shrinks in diameter, it pulls the cell surface inward.

④ Contractions continue; the cell is pinched in two.

A

Jennifer C. Waters/Photo Researchers, Inc.

ring of microfilaments

B

Figure 18.7 Animated! Cytokinesis gives each new cell a share of the parent cell's cytoplasm. A Steps of cytokinesis in an animal cell. **B** The beltlike contracting ring of microfilaments inside a dividing animal cell that is undergoing cytokinesis. (© Cengage Learning)

Concerns and Controversies over Irradiation

What do a routine dental X-ray and an irradiated side of beef have in common? Both are examples of ways we use ionizing radiation. Like some other technologies, this one can be a double-edged sword and can even fuel serious controversy.

Irradiation effects on the body

Ionizing radiation includes various potentially harmful types of electromagnetic energy—for instance, radio waves, visible light, microwaves, cosmic rays from outer space, and radioactive radon gas in rocks and soil. Forms that can harm living cells, including radon and X-rays, have enough energy to remove electrons from atoms and change them to positively charged ions (Section 2.4).

When ionizing radiation enters an organism, it may break apart chromosomes, alter genes, or both. If the chromosomes in an affected cell have been broken into fragments, the spindle apparatus will not be able to harness and move the fragments when the cell divides. The cell or its descendants may then die. When ionizing radiation damage occurs in germ cells, the resulting gametes can carry damaged DNA. Therefore, an infant who inherits the DNA may have a genetic defect. If only somatic cells are affected, only the person exposed to the radiation will suffer damage.

When a person receives a sudden, large dose of ionizing radiation, it typically destroys cells of the immune system, epithelial cells of the skin and intestinal lining, and red blood cells, among other cell types. The results are serious infections, intestinal hemorrhages, anemia, and wounds that do not heal.

Small doses of ionizing radiation over a long period of time apparently cause less damage than the same total dosage given all at once. This may be due in part to the body's ability to repair damaged DNA. Even so, ionizing radiation is associated with miscarriages, eye cataracts, and various cancers (Chapter 22).

On the other hand, medical X-rays and diagnostic technologies such as magnetic resonance imaging (MRI) and PET scanning (Section 2.1) are valuable uses of ionizing radiation in health care. So is radiation therapy used in treating some cancers. Despite these positives, recent studies have raised concerns about the relatively high level of ionizing radiation that patients are exposed to during MRI procedures, among others.

Irradiated food

Just as living body cells can be damaged or killed by radiation, so can harmful bacteria, fungi, and other microorganisms. As a result, foods ranging from grains and potatoes to fruits, spices, beef, pork, and other meats may be irradiated. Irradiated food sold in the United States carries an identifying logo (Figure 18.8).

Irradiated food is not radioactive, and some people are quite comfortable eating it because there is no scientific evidence that it presents a health hazard. In addition, irradiation limits spoilage, and proponents argue that it may reduce the incidence of food-borne illnesses. On the other hand, opponents worry that irradiation might promote the development of radiation-resistant microbes. Some also are concerned that irradiation may chemically change food in ways that could harm consumers. For the time being, there is no scientific evidence to support that fear.

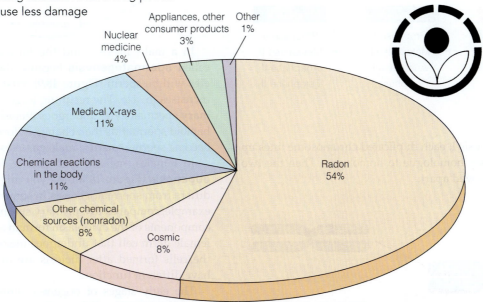

Figure 18.8 Ionizing radiation can break apart chromosomes. The sources of radiation exposure for people in the United States. (*far right*) Logo required to be placed on irradiated food products in the United States. (© John W. Gofman and Arthur R. Tamplin. From Poisoned Power: The Case Against Nuclear Power Plants Before and After Three Mile Island, Rodale Press, 1979)

18.6 Meiosis: The Beginnings of Eggs and Sperm

- Meiosis divides the nuclei of germ cells in a way that halves the number of chromosomes in daughter cells. It is the first step toward the gametes required for sexual reproduction.

- Links to Oocyte formation 16.2, Sperm formation 16.4

In meiosis the parent cell nucleus divides twice

As described in Sections 18.1 and 18.2, meiosis is like mitosis in some ways, but the outcome is different. During interphase, a germ cell copies its DNA, forming duplicated chromosomes. Each duplicated chromosome consists of two sister chromatids attached to one another:

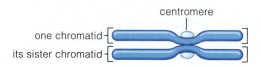

one chromosome in the duplicated state

As in mitosis, a spindle moves chromosomes into the proper position for the formation of daughter nuclei. In meiosis, however, there are *two consecutive* divisions of the chromosomes. The result will be four haploid nuclei. There is no interphase between the two nuclear divisions, which we call meiosis I and meiosis II:

Interphase	Meiosis I	Meiosis II
DNA is replicated prior to meiosis I	Prophase I Metaphase I Anaphase I Telophase I	Prophase II Metaphase II Anaphase II Telophase II

During meiosis I, each duplicated chromosome lines up with its partner, homologue to homologue. Then the two partners are moved apart:

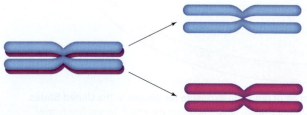

Unless otherwise noted, all art on this page is © Cengage Learning.

The cytoplasm typically divides after each homologue is separated from its partner. The two daughter cells are haploid, with only one of each type of chromosome.

Later, during meiosis II, the two sister chromatids of each chromosome are separated from each other:

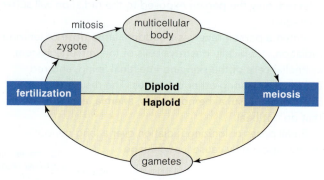

Each sister chromatid is now a separate chromosome. After the four nuclei form, the cytoplasm divides again. The result is four haploid cells, either sperm or oocytes.

Meiosis leads to the formation of gametes

The human life cycle starts with meiosis. Next come the formation of gametes, fertilization, then growth of the new individual by way of mitosis:

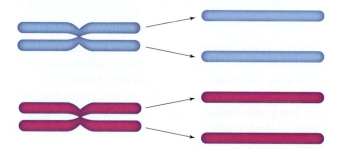

In a male, meiosis and the formation of gametes are called **spermatogenesis** because the forthcoming gametes will be sperm (Figure 18.9). First, a diploid germ cell increases in size. The resulting large, immature cell (a primary spermatocyte) undergoes meiosis. The resulting four haploid spermatids then change in form, develop tails, and become sperm—mature male gametes.

In females, meiosis and gamete formation are called **oogenesis** (Figure 18.10). As you might expect, oogenesis differs from spermatogenesis in some important ways. For example, compared to a primary spermatocyte, many more components of the cytoplasm form in a primary oocyte, the female germ cell that undergoes meiosis. Also, in females the cells formed after meiosis are of different sizes and have different functions.

The early stages of oogenesis unfold in a developing female embryo. Recall from Chapter 16, however, that until a girl reaches puberty, her primary oocytes are arrested in prophase I. Then, each month, meiosis resumes in (usually)

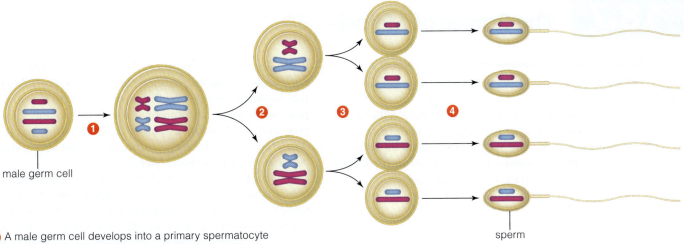

1 A male germ cell develops into a primary spermatocyte as it replicates its DNA. Both types of cell are diploid.

2 Meiosis I in the primary spermatocyte results in two secondary spermatocytes, which are haploid.

3 Four haploid spermatids form when the secondary spermatocytes undergo meiosis II.

4 Spermatids mature as sperm (haploid male gametes).

Figure 18.9 **Spermatogenesis is the process that forms sperm.** (© Cengage Learning)

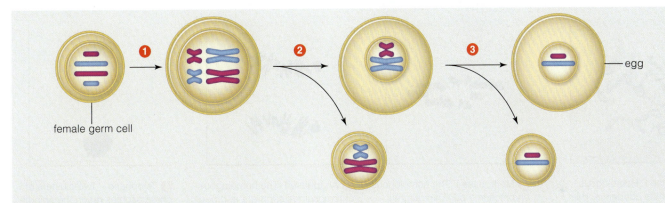

1 A female germ cell (an oogonium) develops into a primary oocyte as it replicates its DNA. Both types of cell are diploid.

2 Meiosis I in the primary oocyte results in a secondary oocyte and a first polar body. The polar body is much smaller than the oocyte. Both cells are haploid. Polar bodies typically degenerate.

3 Meiosis II in the secondary oocyte results in a polar body and an ovum, or egg. Both cells are haploid.

Figure 18.10 **Animated! Oogenesis is the process that forms eggs. For clarity, only one polar body is shown.** (© Cengage Learning)

one oocyte that is ovulated. This cell, the secondary oocyte, receives nearly all the cytoplasm; the other, much smaller cell is a polar body. Both cells enter meiosis II, but the process is arrested again at metaphase II. If the secondary oocyte is fertilized, meiosis II continues. It results in one large cell and one to three small polar bodies. The polar bodies are "dumping grounds" for three sets of chromosomes so that the egg will end up with only the necessary haploid number. The large cell develops into the mature egg (ovum). Its cytoplasm contains components that will help guide the development of an embryo.

Oogenesis and spermatogenesis make gametes that are available for fertilization. As you may remember from Section 16.6, fertilization restores the diploid number of chromosomes in a zygote.

oogenesis The formation of a secondary oocyte.

spermatogenesis The formation of sperm.

TAKE-HOME MESSAGE

WHAT IS THE FUNCTION OF MEIOSIS?

• Meiosis is the first step in the formation of gametes—sperm and eggs—for sexual reproduction.

• Meiosis reduces the parental chromosome number by half, to the haploid number.

• In males, meiosis and gamete formation are called spermatogenesis. In females, these two processes are called oogenesis.

The Stages of Meiosis

■ During meiosis, chromosomes of a diploid nucleus of a germ cell are distributed into four haploid nuclei of gametes.

Before meiosis begins, DNA is replicated and a cell's chromosomes are duplicated, so each chromosome consists of two sister chromatids. The starting nucleus is diploid—it contains two sets of chromosomes, one from each parent. Figure 18.11 takes us step by step through meiosis.

Meiosis I

The first stage of meiosis I is prophase I. During this phase, the chromosomes condense, align, and exchange segments. (This "swap" is discussed in Section 18.8). The centrosome is duplicated along with its two centrioles.

Pairs of centrioles are positioned at the opposite sides of the nucleus as the nuclear envelope breaks up (step 1). Microtubules of the spindle link the chromosomes to the spindle poles so that each chromosome is attached to one spindle pole, and its homologous partner is attached to the other. At metaphase I, all the chromosomes are aligned midway between the spindle poles (step 2).

In anaphase I, the spindle microtubules separate the homologous chromosomes and pull them toward opposite spindle poles (step 3). During telophase I, the chromosomes reach the spindle poles (step 4). New nuclear envelopes form around the two clusters of chromosomes. Each of the two haploid nuclei that form contains one set of duplicated chromosomes. The cytoplasm divides at this point (by cytokinesis), forming two haploid cells, but the cells' DNA is not copied before meiosis II begins.

1 Prophase I. Homologous chromosomes condense, pair up, and swap segments. Spindle microtubules attach to them as the nuclear envelope breaks up.

2 Metaphase I. The homologous chromosome pairs are aligned midway between spindle poles.

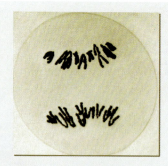

3 Anaphase I. The homologous chromosomes separate and begin heading toward the spindle poles.

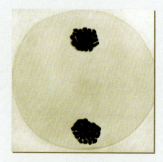

4 Telophase I. Two clusters of chromosomes reach the spindle poles. A new nuclear envelope forms around each cluster, so two haploid nuclei form.

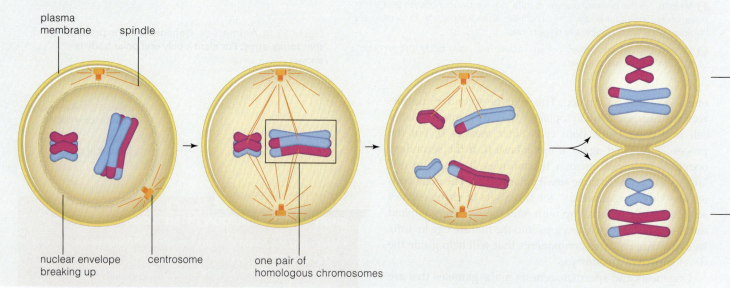

plasma membrane   spindle

nuclear envelope breaking up   centrosome   one pair of homologous chromosomes

Figure 18.11 Animated! The two consecutive divisions in meiosis reduce the parental number of chromosomes by half (to the haploid number) for forthcoming gametes. Only two pairs of homologous chromosomes (*blue* and *pink*) are shown.

Meiosis II

In prophase II and metaphase II, the same events occur in each new nucleus as in prophase I and metaphase I (steps 5 and 6). In anaphase II, the spindle microtubules pull the sister chromatids apart (step 7). Each chromosome now consists of one DNA molecule. During telophase II, the chromosomes (which are no longer duplicated) cluster around each spindle pole. There are four clusters in all. Each one is a set of unduplicated chromosomes. New nuclear envelopes form around the four clusters, forming four haploid nuclei—two in each cell (step 8). Cytokinesis then divides the cytoplasm of each cell, producing four haploid cells.

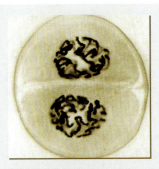

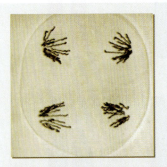

5 Prophase II. The chromosomes condense. Spindle microtubules attach to each sister chromatid as the nuclear envelope breaks up.

6 Metaphase II. The (still duplicated) chromosomes are aligned midway between poles of the spindle.

7 Anaphase II. All sister chromatids separate. The now unduplicated chromosomes head to the spindle poles.

8 Telophase II. A cluster of chromosomes reaches each spindle pole. A new nuclear envelope encloses each cluster, so four haploid nuclei form.

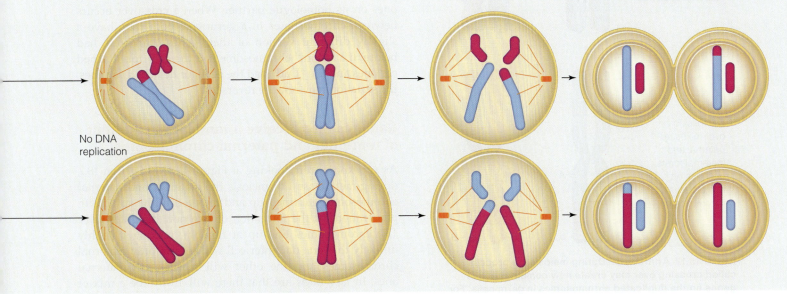

No DNA replication

■ Events that happen during meiosis explain why no person is genetically identical to either parent.

Pieces of chromosomes may be exchanged

No one ever looks, or has a body that operates, exactly like his or her parents. Most of the trait variations we take for

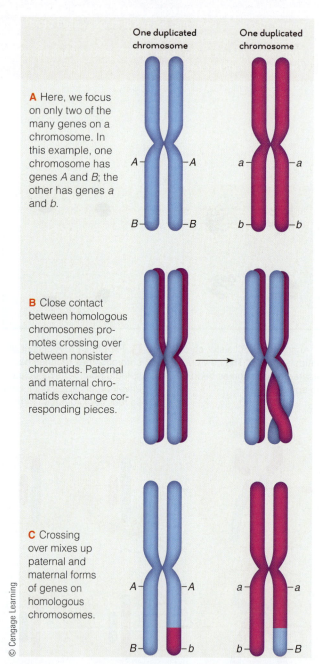

A Here, we focus on only two of the many genes on a chromosome. In this example, one chromosome has genes *A* and *B*; the other has genes *a* and *b*.

B Close contact between homologous chromosomes promotes crossing over between nonsister chromatids. Paternal and maternal chromatids exchange corresponding pieces.

C Crossing over mixes up paternal and maternal forms of genes on homologous chromosomes.

© Cengage Learning

Figure 18.12 Animated! During meiosis I, a process called crossing over may create new combinations of genes on the duplicated chromosomes in germ cells. For clarity, this diagram of a cell shows only one pair of homologous chromosomes and one crossover. Here, the paternal chromosome is *blue* and its maternal homologue is *pink*.

granted result from changes to chromosomes that occurred during meiosis, when germ cells were forming sperm in a father's testes or eggs in a mother's ovaries.

Some genetic variations come about during prophase I of meiosis. This is a time when parts of the duplicated chromosomes—the sister chromatids—in gonad germ cells are rearranged. Remember that germ cells contain sets of homologous chromosomes, one set that came originally from a person's mother and a corresponding set from the father. These homologues therefore can be called "maternal" and "paternal" chromosomes and they carry genes for the same traits. During meiosis I, the two pairs of sister chromatids (one maternal, the other paternal) line up very closely.

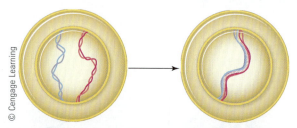

© Cengage Learning

This close alignment favors **crossing over**. In a crossover, nonsister chromatids break at the same places along their length and exchange corresponding segments. The X-shaped areas shown in Figure 18.12 are crossovers. Each segment in the exchange includes one or more genes.

The exchange of chromosome pieces is called *genetic recombination*. It leads to variation in the details of inherited traits because a gene may have several chemical forms. For example, as described in Chapter 19, the gene for earlobe shape has two forms, one for attached earlobes and the other for detached earlobes. Often, particular forms of genes on one chromosome differ from corresponding ones on its homologue partner. When a crossover occurs between chromosomes in a germ cell, both then have a slightly different version of their genes than they had before. If a "Mom" chromosome had the gene for attached earlobes and the "Dad" had the gene for detached earlobes, the situation may now be reversed.

Gametes also receive a random assortment of maternal and paternal chromosomes

As you know from looking at Figure 18.11 in Section 18.7, during metaphase I of meiosis the maternal and paternal chromosomes get lined up and tethered to the spindle in preparation for the formation of two new daughter cells. The chromosomes line up at random, making it highly unlikely that one daughter cell will receive only maternal chromosomes and the other will receive only paternal ones. In fact, odds are that there will always be a mix of maternal and paternal chromosomes in each cell that forms during meiosis. Figure 18.13 shows the possibilities when there are only three pairs of homologues. In this case, eight

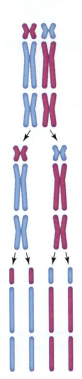

1 The four possible alignments of three pairs of chromosomes in a nucleus at metaphase I.

2 Resulting combinations of maternal and paternal chromosomes in the two nuclei that form at telophase I.

3 Resulting combinations of maternal and paternal chromosomes in the four nuclei that form at telophase II. Eight different combinations are possible.

Figure 18.13 Maternal and paternal chromosomes are assorted randomly into gametes. The diagram shows possible outcomes of the random alignment of three pairs of homologous chromosomes at metaphase I of meiosis. For simplicity, no crossing over occurs in this example. (© Cengage Learning)

Jeff Greenberg/Visuals Unlimited, Inc.

crossing over The exchange of corresponding segments between homologous chromosomes during meiosis.

disjunction Separation of homologous chromosomes during meiosis, so that gametes receive only a haploid set of chromosomes.

combinations (2^3) of maternal and paternal chromosomes are possible for the forthcoming gametes.

Of course, a human germ cell has a full 23 pairs of homologous chromosomes, not just three. So a grand total of 2^{23}, or 8,388,608, combinations of maternal and paternal chromosomes are possible every time meiosis in a germ cell produces a gamete! This is why striking mixes of traits can show up even in the same family.

During meiosis II, each homologue normally is separated from its partner so that gametes receive only the required haploid set of chromosomes. This separation is called **disjunction**. As Chapter 20 describes, birth defects can result when this process does not occur as usual.

TAKE-HOME MESSAGE

HOW DOES MEIOSIS CREATE VARIATION IN THE CHROMOSOMES GAMETES RECEIVE?

- Meiosis moves maternal and paternal chromosomes into gametes at random.
- This random pattern creates new combinations of chromosomes in sperm and eggs, so children have varied combinations of their parents' traits.
- Also, crossing over of segments between pairs of homologous chromosomes creates new combinations of genes on the chromosomes.

18.9 Meiosis and Mitosis Compared

■ Miitosis occurs in somatic cells, while meiosis takes place in germ cells. The diagram presented here summarizes the similarities and key differences of these mechanisms.

The end results of mitosis and meiosis differ in a crucial way (Figure 18.14). Mitosis produces genetically identical copies of a parent cell. Meiosis is an important source of genetic variation in the traits offspring will have.

Figure 18.14 Animated! This diagram compares the similarities and differences between mitosis and meiosis. As in other diagrams in this chapter, maternal chromosomes are *pink*, and paternal chromosomes are *blue*. (© Cengage Learning)

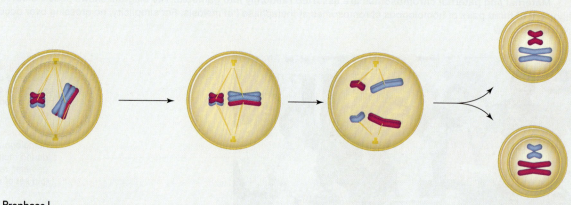

Prophase I
- Chromosomes condense.
- Homologous chromosomes pair.
- Crossovers occur (not shown).
- Spindle forms and attaches chromosomes to spindle poles.
- Nuclear envelope breaks up.

Metaphase I
- Chromosomes align midway between spindle poles.

Anaphase I
- Homologous chromosomes separate and move toward opposite spindle poles.

Telophase I
- Chromosome clusters arrive at spindle poles.
- New nuclear envelopes form.
- Chromosomes decondense.

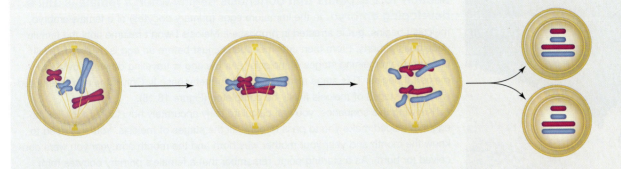

Prophase
- Chromosomes condense.
- Spindle forms and attaches chromosomes to spindle poles.
- Nuclear envelope breaks up.

Metaphase
- Chromosomes align midway between spindle poles.

Anaphase
- Sister chromatids separate and move toward opposite spindle poles.

Telophase
- Chromosome clusters arrive at spindle poles.
- New nuclear envelopes form.
- Chromosomes decondense.

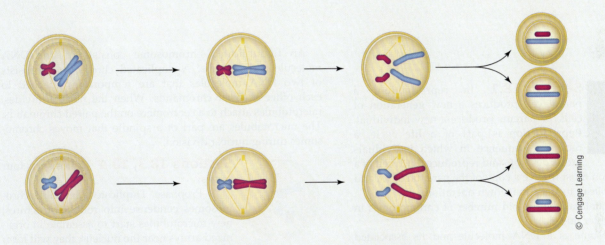

Prophase II
- Chromosomes condense.
- Spindle forms and attaches chromosomes to spindle poles.
- Nuclear envelope breaks up.

Metaphase II
- Chromosomes align midway between spindle poles.

Anaphase II
- Sister chromatids separate and move toward opposite spindle poles.

Telophase II
- Chromosome clusters arrive at spindle poles.
- New nuclear envelopes form.
- Chromosomes decondense.

EXPLORE
ON YOUR OWN

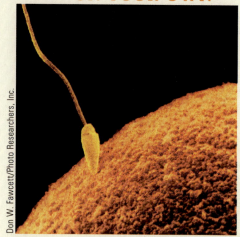

Figure 18.15 **This photograph shows an egg with a sperm that will penetrate and fertilize it—marking the end of meiosis in the egg and the beginning of bodily growth by mitosis.**

Don W. Fawcett/Photo Researchers, Inc.

Section 16.2 explains that oogenesis begins when a female is still a developing embryo. In the immature eggs (primary oocytes) of a female embryo, meiosis I begins, but is arrested in prophase I. Meiosis I won't resume until the female undergoes puberty. From then until menopause, just before an egg is ovulated, it will undergo the remaining stages of meiosis I. As the egg is traveling down an oviduct, meiosis II begins. This stage also is arrested, in metaphase II. Only if the egg is fertilized will all the stages of meiosis finally be completed (Figure 18.15).

Knowing this sequence, you can calculate fairly accurately how long it took for the egg that helped make you to pass through all the stages of meiosis. You only need to know the month and year your mother was born and the month and year you were conceived (or born). As a starting point, remember that a female's primary oocytes form during the third month of embryonic development, about 6 months before birth. If you have siblings, do the same calculation for them.

SUMMARY

Section 18.1 In reproduction, a parent cell produces a new generation of cells, or parents produce a new individual. Reproduction is part of a life cycle, a recurring sequence in which individuals grow, develop, and reproduce. Each cell of a new generation must receive a copy of the parental DNA and enough cytoplasm to start up its own operation.

Mitosis maintains the diploid number of chromosomes in the two daughter nuclei.

A chromosome is a DNA molecule and its associated proteins. The sum of the chromosomes in a given cell type is the chromosome number. Human somatic cells have a diploid chromosome number of 46, or two copies of 23 types of chromosomes. Except for the sex chromosomes, pairs of homologous chromosomes are the same length and carry similar genes.

Section 18.2 The cell cycle begins when a new cell is produced and ends when that cell divides. The longest part of the cycle is interphase, which includes a growth stage (G1), when the cell roughly doubles the number of organelles and other components in its cytoplasm. In the S stage the cell's chromosomes are duplicated. There is also a final, short growth stage (G2). The duration of the cell cycle varies in different types of cells.

Human cells divide the cell nucleus by either mitosis or meiosis. Mitosis is the division mechanism in somatic cells—body cells that are not specialized to make gametes. It functions in growth and tissue repair. Meiosis divides the nucleus in germ cells, cells in the gonads that give rise to gametes.

An unduplicated chromosome consists of one DNA molecule and proteins. A duplicated chromosome consists of two DNA molecules that are temporarily attached to each other as sister chromatids. When the nucleus divides, microtubules attach to a centromere on the paired chromatids. The microtubules are part of a spindle that moves chromosomes during nuclear division.

Sections 18.3, 18.4 Mitosis has four stages:

1. Prophase. Duplicated, threadlike chromosomes condense into rodlike structures; new microtubules start to assemble in organized arrays near the nucleus; they will form a spindle. The nuclear envelope disappears.

2. Metaphase. Spindle microtubules orient the sister chromatids of each chromosome toward opposite spindle poles. The chromosomes line up at the spindle equator.

3. Anaphase. Sister chromatids of each chromosome separate. Both are now independent chromosomes, and they move to opposite poles.

4. Telophase. Chromosomes become threadlike again and a new nuclear envelope forms around the two clusters of chromosomes. Mitosis is completed. Cytokinesis divides the cytoplasm; the result is two diploid cells.

Division of the cytoplasm, called cytokinesis, occurs toward the end of mitosis or shortly afterward.

Sections 18.6, 18.7 Meiosis reduces the total number of chromosomes in daughter cells. Two consecutive divisions of a germ cell cut the parental diploid chromosome number in half. Meiosis I distributes the pairs of homologous chromosomes. Meiosis II separates sister chromatids. The result, after cytokinesis, is four haploid cells.

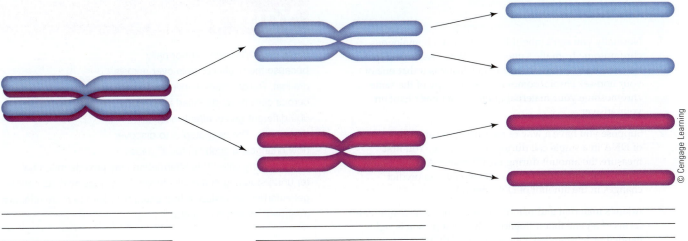

_____
_____
_____

_____
_____
_____

_____
_____
_____

 Section 18.8 Meiosis contributes to genetic variation. Crossing over during prophase I creates new combinations of genes in chromosomes. Random alignment of pairs of homologues at metaphase I results in new combinations of maternal and paternal traits.

REVIEW QUESTIONS

1. Define somatic cell and germ cell.

2. What is a chromosome? What is the difference between a diploid cell and a haploid cell?

3. What are homologous chromosomes?

4. Name the four main stages of mitosis, and describe what happens in each stage.

5. In the diagram above, summarize the main events in meiosis.

6. In a paragraph, summarize the similarities and differences between mitosis and meiosis.

SELF-QUIZ _Answers in Appendix V_

1. Fill in the blanks: DNA, packaged in chromosomes, is distributed to daughter cells by _____ or _____.

2. Fill in the blanks: Each kind of organism contains a characteristic number of _____ in each cell; each of those structures is composed of a _____ molecule and proteins.

3. A pair of chromosomes that are similar in length and the traits they govern are called _____.
    a. diploid chromosomes
    b. mitotic chromosomes
    c. homologous chromosomes
    d. germ chromosomes

4. Fill in the blank: Somatic cells have a _____ number of chromosomes.

5. Interphase is the stage when _____.
    a. a cell ceases to function
    b. a germ cell forms its spindle
    c. a cell grows and duplicates its DNA
    d. mitosis takes place

6. After mitosis, each daughter cell contains genetic instructions that are _____ and _____ chromosome number of the parent cell.
    a. identical to the parent cell's; the same
    b. identical to the parent cell's; one-half the
    c. rearranged; the same
    d. rearranged; one-half the

7. All of the following are stages of mitosis except _____.
    a. prophase
    b. interphase
    c. metaphase
    d. anaphase

8. A duplicated chromosome has _____ chromatids.
    a. one   b. two   c. three   d. four

9. Crossing over in meiosis _____.
    a. occurs between sperm DNA and egg DNA at fertilization
    b. leads to genetic recombination
    c. occurs only rarely

10. Because of the _____ alignment of homologous chromosomes during meiosis, gametes can end up with _____ mixes of maternal and paternal chromosomes.
    a. unvarying; different   c. random; duplicate
    b. unvarying; duplicate   d. random; different

11. Match the stage of mitosis with the following key events.
    _____ metaphase   a. sister chromatids separate
    _____ prophase   and move to opposite poles
    _____ telophase   b. chromosomes condense and a
    _____ anaphase   microtubular spindle forms
                           c. chromosomes decondense,
                              daughter nuclei re-form
                           d. chromosomes line up at a
                              spindle equator

CRITICAL THINKING

1. Normally you can't inherit both copies of a homologous chromosome from the same parent. Why? Assuming that no crossing over has occurred, how likely is it that one of your non-sex chromosomes is an exact copy of the same chromosome your maternal grandmother had? Explain your answer.

2. Suppose you have a way of measuring the amount of DNA in a single cell during the cell cycle. You first measure the amount during the G1 phase. At what points during the remainder of the cycle would you predict changes in the amount of DNA per cell?

3. Adam's maternal and paternal chromosomes have alternate forms of a gene that influences whether a person is right-handed or left-handed. One form says "right" and its partner says "left." Visualize one of his spermatogonia, in which chromosomes are being duplicated prior to meiosis. Visualize what happens to the chromosomes during anaphase I and II. (It might help to use toothpicks as models of the sister chromatids of each chromosome.) What fraction of Adam's sperm will carry the gene for right-handedness? For left-handedness?

4. Fresh out of college, Maria has her first job teaching school. When she goes for a pre-employment chest X-ray required by the school district, the technician places a lead-lined apron over her abdomen but not over any other part of Maria's body. The apron prevents electromagnetic radiation from penetrating into the protected body area. What cells is the lead shield designed to protect, and why?

INTRODUCTION TO GENETICS

LINKS TO EARLIER CONCEPTS

Your reading in this chapter will flesh out the concept of inheritance introduced at the beginning of this textbook (1.1).

You will also draw on what you have learned about diploid and haploid sets of chromosomes (18.1) and how gametes form during meiosis (18.7).

KEY CONCEPTS

Genes and Inheritance

Genes are segments of DNA that code for inherited traits. Genes have different forms called alleles. A gamete carries one copy of each gene. Sections 19.1–19.2

Probability Rules

Chance determines which sperm will fertilize which egg. This means that rules of probability apply to the inheritance of traits coded by a single gene. Section 19.3

Sorting Genes into Gametes

The paired copies of a gene on one chromosome are sorted into gametes independently of genes on other chromosomes. Section 19.4

Gene Effects

Genetic traits are not always predictable. Examples are traits determined by more than one gene, and instances in which one gene affects several traits. Sections 19.5–19.6

Human skin color can range from very pale to to very dark brown. As Chapter 4 described, skin color comes from the pigment melanin made in skin cells called melanocytes. Some human traits are governed by single genes, but not this one. Geneticists have identified more than one hundred genes that may influence the amount and chemical makeup of the melanin in a person's skin cells—and therefore the skin's natural shade.

This complex genetic picture explains why Kian and Remee, the little girls shown below, can have such different skin and hair colors even though they are fraternal twins. The girls' parents both are of mixed African and European descent, and each carries skin color genes common in both groups. As you will read in this chapter, chance determined the exact combinations of skin color genes each sister inherited.

With this chapter we start to explore the topics of genes and basic principles of inheritance. In the following pages you will gain a much fuller understanding of the genetic events that ultimately produced all your biological traits.

© Gary Roberts/worldwidefeatures.com

Basic Concepts of Heredity

- Genes provide the instructions for all human traits, including physical features and how body parts function. Each person inherits a particular mix of maternal and paternal genes.

- Link to Reproduction and the chromosome number 18.1

allele Each chemical version of a gene.

gene A unit of DNA that provides chemical instructions for building a protein.

genotype The alleles a person inherits.

heterozygous Having two different alleles for a trait.

homozygous Having the same two alleles for a trait.

locus The location of a gene on a chromosome.

phenotype Observable functional or physical genetic traits.

Having read Chapter 18, you already know a bit about chromosomes. The following list and Figure 19.1 explain some basic concepts about the genes chromosomes carry.

1. **Genes** are units of DNA that provide chemical instructions for building proteins. They are passed from parents to offspring and provide the information necessary to produce specific traits. Humans have about 21,500 genes. Each gene has a specific location, or **locus**, on a given chromosome.

2. Diploid cells have two copies of each gene, one on each member of a pair of homologous chromosomes.

3. All copies of a gene deal with the same trait, but their information about it may vary a little due to chemical differences between them. Each version of the gene is called an **allele**. Contrasting alleles produce much of the variation we see in traits (Figure 19.2).

4. If the two copies of a gene are identical alleles, a person is **homozygous** for the trait (*homo-*: same; *zygo*: joined together). If the two allele copies are different, the person is **heterozygous** (*hetero-*: different) for the trait.

5. An allele is *dominant* when its effect on a trait masks that of any recessive allele paired with it. A dominant allele is represented by an uppercase letter, a recessive one by a lowercase (for instance, *A* and *a* or *C* and *c*).

A A *pair of homologous chromosomes*, each in the unduplicated state (most often, one from a male parent and its partner from a female parent)

B A *gene locus* (plural, loci), the location for a specific gene on a specific type of chromosome

C A *pair of alleles* (each being one chemical form of a gene) at corresponding loci on a pair of homologous chromosomes

D Three *pairs of genes* (at three loci on this pair of homologous chromosomes); same thing as three pairs of alleles

© Cengage Learning

Figure 19.1 Animated! Knowing a few basic genetic terms will help you understand principles of inheritance.

6. Someone who is *homozygous dominant* has a pair of dominant alleles (*AA*) for the trait. Someone who is *homozygous recessive* for the trait has a pair of recessive alleles (*aa*). A *heterozygous* person has a pair of nonidentical alleles (*Aa*).

7. The alleles a person inherits are his or her **genotype**. Observable functional or physical traits, such as attached earlobes, are the **phenotype** (Table 19.1).

TABLE 19.1	Genotype and Phenotype Compared	
Genotype	**Described as**	**Phenotype**
EE	homozygous dominant	detached ear lobe
Ee	heterozygous (one of each allele; dominant form of trait observed)	detached ear lobe
ee	homozygous recessive	attached earlobe

Table by Lisa Starr

TAKE-HOME MESSAGE

HOW MANY GENES FOR A TRAIT DOES A PERSON INHERIT?

- Each of us inherits two copies of each gene, one on each member of a pair of homologous chromosomes.
- If the two inherited alleles of a gene are identical, the person is homozygous for the trait. If the two inherited alleles are different, the person is heterozygous for the trait.

A: © Lisa O'Connor/Zuma Press
B: © Fabian Cevallos/Sygma/Corbis

Figure 19.2 Many genetic traits have dominant and recessive forms. **A** Actor Tom Cruise has attached earlobes. **B** Actress Joan Chen's are detached. The detached version is dominant.

19.2 One Chromosome, One Copy of a Gene

- We inherit pairs of genes (alleles) on pairs of chromosomes, but a gamete receives only one gene from each pair.

- Links to Stages of meiosis 18.7, 18.8

For geneticists working with nonhuman organisms, a "monohybrid" experiment (*mono-* means one) is one way to learn more about the genotypes of their test subjects. In this kind of experiment, the two parents have different alleles for the gene being studied. Although a scientist can't ethically do genetic experiments on human beings, monohybrid matings do occur naturally and they show patterns of inheritance at work.

Consider, for example, the "dimple" in actor Viggo Mortensen's chin, a trait called a chin fissure (Figure 19.3). The configuration of a person's chin is governed by a gene that has two allele forms. One allele, which calls for a chin fissure (actually an indentation in the skin), is dominant when it is present. We can use *C* to represent this allele. The recessive allele, which codes for a smooth chin, is *c*. The diagram below shows the genotypes gametes can have when one parent is homozygous for the *C* form of the gene and has a chin fissure, and the other is homozygous for the *c* form and has a smooth chin. The *CC* parent's gonads make only *C* gametes while the *cc* parent makes only *c* gametes:

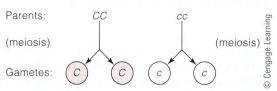

Each gamete gets only one allele for the trait because each gamete has only one copy of each chromosome. This process was described in Section 18.7, and Figure 19.4 will remind you how it works. To summarize what the diagram shows, when meiosis separates homologous chromosomes—as it must do to reduce the diploid number of chromosomes to the haploid number—the pairs of alleles on those chromosomes also are separated and each one ends up in a different gamete. This separation of gamete pairs is called **segregation**.

© Lisa O'Connor/ZUMA/Corbis

Figure 19.3 The trait called a chin fissure arises from one allele of a gene. A Actor Viggo Mortensen received a gene that influences this trait from each of his parents. At least one of those genes was dominant. **B** What Mr. Mortensen's chin might have looked like if he had inherited identical alleles for "no chin fissure" instead.

segregation The separation of pairs of gametes during meiosis.

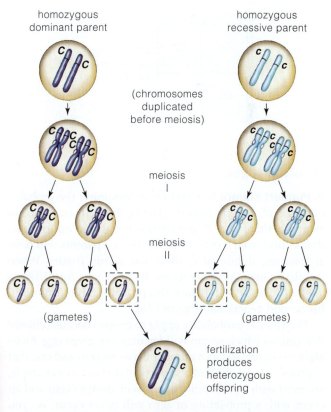

Figure 19.4 Each pair of gene alleles is separated and the two alleles end up in different gametes. Due to this segregation, two parents that are each homozygous for a different version of a trait will have only offspring who are heterozygous for that trait. (© Cengage Learning)

TAKE-HOME MESSAGE

WHY DOES A GAMETE GET ONLY ONE COPY OF EACH GENE?

- The two copies of each chromosome in a diploid organism separate from each other (segregate) during meiosis in germ cells, so each copy ends up in a different gamete.

- Because a gamete has only one copy of each chromosome, it contains only one copy of the genes on chromosomes.

Genetic Tools: Testcrosses and Probability

- When potential parents are concerned about passing a harmful trait to a child, genetic counselors must try to predict the likely outcome of the mating.

probability A measure of the chance that a given outcome will occur.

Punnett square A grid for determining the probable outcome of genetic crosses.

Geneticists and genetic counselors are very concerned about **probability**—a measure of the chance that some particular outcome will occur. Probability is a factor in the inheritance of single-gene traits. A bit like a lottery, that chance depends on the number of possible outcomes.

To begin to get a feel for how probability works, let's again use the chin fissure trait as our example. Each child will inherit a pair of differing alleles for the trait, one from each homozygous parent. The children will thus each be heterozygous for the chin genotype, or *Cc*. Because C is dominant, each child will have a chin fissure.

Suppose now that one of the *Cc* children grows up and has a family with another *Cc* person. Because half of each parent's gametes (sperm or eggs) are C and half are c (due to segregation at meiosis), four outcomes are possible every time a sperm fertilizes an egg:

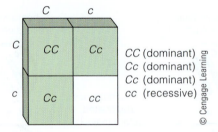

CC (dominant)	
Cc (dominant)	
Cc (dominant)	
cc (recessive)	

© Cengage Learning

A Punnett square can be used to predict the result of a genetic cross

A **Punnett square** is a grid for determining the probable outcome of genetic crosses (Figure 19.5). In our current example, there is a 75 percent chance (three out of four) that a child from a cross between two *Cc* parents will have at least one dominant C allele and a chin fissure. When the first generation parents are homozygous for different alleles (*CC* and *cc*), in theory this probable ratio turns up in the second generation (Figure 19.6).

The rules of probability apply to crosses because chance determines which sperm will fertilize any given egg. Probability is expressed as a number between zero and one that expresses the likelihood of a particular event. For example, an event with a probability of one will always occur and an event with a probability of zero will never occur. As you might guess, an event that has a probability of one-half (or 50 percent) is likely to occur in about half of all situations (Figure 19.7).

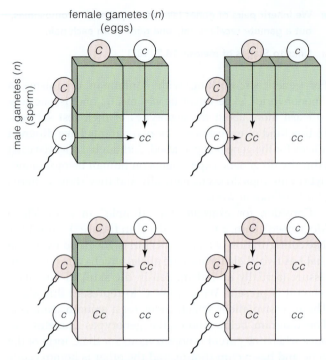

female gametes (*n*)
(eggs)

male gametes (*n*) (sperm)

Figure 19.5 Making a Punnett square is one way to determine the likely outcome of a genetic cross. Here the cross is between two heterozygous individuals. Circles are gametes. Letters represent gene alleles. Genotypes of the resulting offspring are inside the squares. (© Cengage Learning)

Having a chin fissure doesn't affect a person's health. However, quite a few genetic disorders result from single-gene defects and so follow the laws of probability. As you think about probability, two ideas are important:

1. The most probable outcome doesn't have to occur. For instance, it's common to see families in which the parents have produced a string of boys or girls when probability says that parents with two or more children will have equal numbers of girls and boys. Predicted ratios usually only turn up with a large number of events. You can test this for yourself by flipping a coin. Probability predicts that heads and tails should each come up about half the time, but you may have to flip the coin a hundred times to end up with a ratio close to 1:1.

2. In a given genetic situation, *probability doesn't change*. The chance that a certain genotype will occur—say, a baby with cystic fibrosis—is the same for every child no matter how many children a couple has. Based on the parents' genotypes, if the probability that a child will inherit a given genotype is one in four, then each child of those parents has a one-in-four (25 percent) chance of inheriting the genotype. If the parents have three children without the trait, the fourth child still has only a one-in-four chance of inheriting it.

A testcross also can reveal genotypes

Until the advent of high-tech genetic analysis, a testcross was used to learn the genotype of a (nonhuman) organism. In this method, an individual with an unknown genotype is crossed with an individual that is homozygous recessive for the trait being studied (say, aa). Then the phenotypes of the offspring are observed.

If all offspring have a dominant form of the trait, the "mystery" parent's genotype must include at least one dominant allele (the parent must be *Aa* or *AA*). If some offspring have the dominant phenotype and some have the recessive one, then the parent with the unknown genotype must be a heterozygote, or *Aa*.

Similar situations can shed light on the genotype of a human parent. Suppose that a woman has smooth cheeks and her husband has dimpled cheeks. "No dimples" is a recessive trait, so the woman is *dd*. If a child is born with no dimples, then the father must be a heterozygote for this trait, with a genotype of *Dd*. That is the only way he could himself have dimples and also father a *dd* child. If, on the other hand, the child has dimples, the father can be either *DD* or *Dd*. If he is *Dd*, the probability that he will have a dimpled child is 1/2, a 50–50 chance every time. If he were *DD*, every child of his would have dimples.

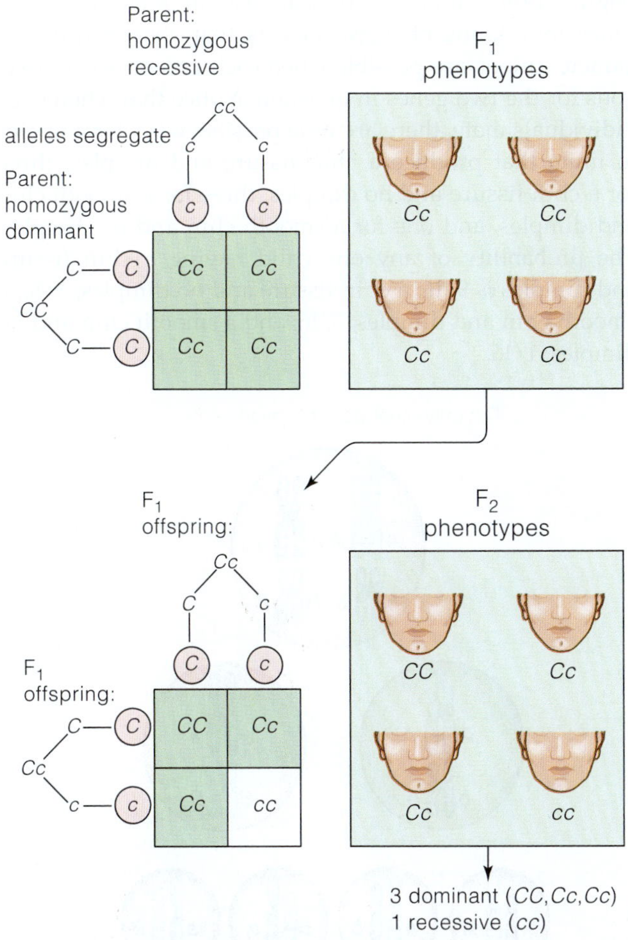

Figure 19.6 With a testcross performed to learn parent genotypes for a single trait, a different set of genetic results is possible in the second generation. Notice that the dominant-to-recessive ratio is 3:1 for the second generation of offspring.
(© Cengage Learning)

How to Calculate Probability

Step 1. Actual genotypes of parental gametes

In the cross *Cc* × *Cc*, gametes have a 50–50 chance of receiving either allele (*C* or *c*) from each parent. Said another way, the probability that a particular sperm or egg will be *C* is 1/2, and the probability that it will be *c* is also 1/2:

probability of *C*:	1/2
probability of *c*:	1/2

Step 2. Probable genotypes of offspring

Offspring receive one allele from each parent. Three different combinations of alleles are possible in this cross. To figure the probability that a child will receive a particular allele combination, simply multiply the probabilities of the individual alleles:

probability of *CC*:	1/2 × 1/2 = 1/4
probability of *Cc*:	1/2 × 1/2 = 1/4 ⎫
probability of *cC*:	1/2 × 1/2 = 1/4 ⎭ 1/2
probability of *cc*:	1/2 × 1/2 = 1/4

Step 3. Probable phenotypes

Chin fissure: (*CC,Cc,cC*)	1/4 + 1/4 + 1/4 = 3/4
Smooth chin: (*cc*)	1/4

Figure 19.7 Simple multiplication lets you figure the probability that a child will inherit alleles for a particular phenotype.
(© Cengage Learning)

How Genes for Different Traits Are Sorted into Gametes

- When we consider more than one trait, we see that the gene for each trait is inherited independently of the gene for other traits.

- Links to Meiosis 18.7, Crossing over 18.8

independent assortment
The mechanism in meiosis by which the alleles of a given gene segregate into gametes independently of the alleles of other genes.

Section 19.2 explained how segregation separates the pairs of gene alleles for a single trait so that only one allele of each pair ends up in forming sperm or eggs. It also is important to understand how people inherit genes for different traits. Among other uses, such knowledge can help explain the genetic basis for some disorders. For example, is someone who inherits the faulty gene that causes hemophilia likely to also have a gene for color blindness? As you will read later on, the answer in that case is yes, because sometimes certain genes on the same chromosome are closely linked. Crossing over doesn't often separate them, so in general they are inherited together. Otherwise, however, we inherit each of our traits independently of all others.

The reason most traits are inherited independently is a mechanism called **independent assortment** that occurs during meiosis. As you can see in Figure 19.8, there are two ways the two members of each pair of homologous parent chromosomes can line up before the pairs become separated. Exactly how each lines up is a random process. This means that a given chromosome and its genes may end up in any of eight gametes—and the same rule applies to all the parent chromosomes. As a result, meiosis can yield gametes having all the possible combinations of parental genes. Adding to this the fact that which sperm fertilizes which egg is random makes it easy to understand why we see so much variety in the mix of traits we humans have.

To get an idea of how independent assortment works, let's look at a case in which two traits, the chin fissure and cheek dimples, are inherited. As with earlier examples, there are two contrasting alleles of each gene, one dominant and one recessive. The dominant alleles are *C* for a chin fissure and *D* for cheek dimples and *c* and *d* for recessive forms of the genes. Let's also say that both parents are heterozygous for both traits. That is, they are both *CcDd*. As a result, they can each produce equal proportions of four types of gametes:

$$1/4\ CD \quad 1/4\ Cd \quad 1/4\ cD \quad 1/4\ cd$$

A Punnett square can show the probabilities that a child will inherit a particular combination of single-gene traits (Figure 19.9). Simple multiplication (four kinds of sperm times four kinds of eggs) tells us that sixteen different gamete unions are possible when each parent is heterozygous for the two genes in question. Notice that when such individuals mate, there are nine possible ways for gametes to unite that produce a chin fissure and dimples, three for a chin fissure and no dimples, three for a smooth chin and dimples, and one for a smooth chin and no dimples. The probability of any one child having a chin fissure and dimples is 9/16; a chin fissure and no dimples, 3/16; a smooth chin and dimples, 3/16; and a smooth chin and no dimples, 1/16.

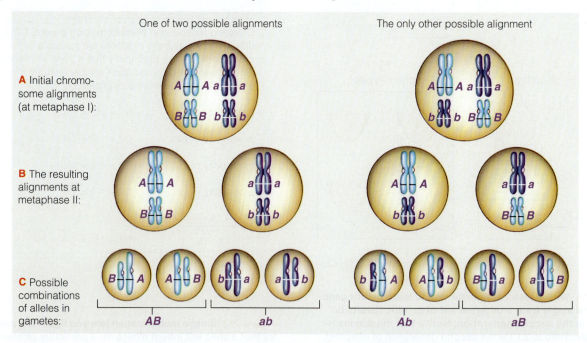

A Initial chromosome alignments (at metaphase I):

B The resulting alignments at metaphase II:

C Possible combinations of alleles in gametes:

One of two possible alignments — AB, ab

The only other possible alignment — Ab, aB

Figure 19.8 Animated! In independent assortment, chromosomes and the genes they carry are moved at random into forming gametes. (© Cengage Learning)

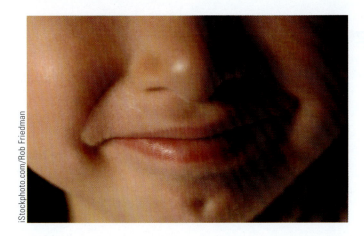

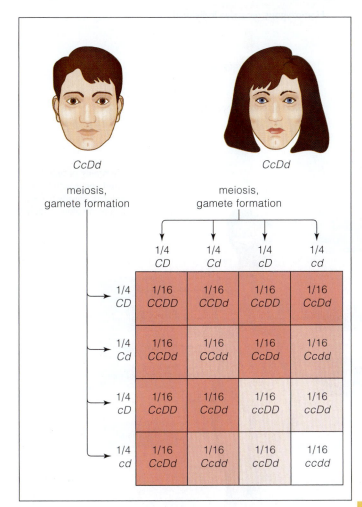

CcDd   CcDd

meiosis,   meiosis,
gamete formation   gamete formation

	1/4 CD	1/4 Cd	1/4 cD	1/4 cd
1/4 CD	1/16 CCDD	1/16 CCDd	1/16 CcDD	1/16 CcDd
1/4 Cd	1/16 CCDd	1/16 CCdd	1/16 CcDd	1/16 Ccdd
1/4 cD	1/16 CcDD	1/16 CcDd	1/16 ccDD	1/16 ccDd
1/4 cd	1/16 CcDd	1/16 Ccdd	1/16 ccDd	1/16 ccdd

Adding up the combinations possible:

9/16 or 9 chin fissure, dimples

3/16 or 3 chin fissure, no dimples

3/16 or 3 smooth chin, dimples

1/16 or 1 smooth chin, no dimples

Figure 19.9 **Tracking two traits shows the results of independent assortment.** Both parents are heterozygous for both genes. Rules of probability predict that certain combinations of phenotypes among offspring of this type of cross occur in a 9:3:3:1 ratio, on average. (© Cengage Learning)

Probability in a Mating Where Both
Parents Are Heterozygous at Two Loci

If both parents are heterozygous for two traits as shown in Figure 19.9, a testcross produces the following 9:3:3:1 phenotype ratio:

9/16 or 9 chin fissure, dimples

3/16 or 3 chin fissure, no dimples

3/16 or 3 smooth chin, dimples

1/16 or 1 smooth chin, no dimples

Individually, these phenotypes have the following probabilities:

probability of chin fissure (12 of 16)   = 3/4
probability of dimples (12 of 16)   = 3/4
probability of smooth chin (4 of 16)   = 1/4
probability of no dimples (4 of 16)   = 1/4

To figure the probability that a child will show a particular *combination* of phenotypes, multiply the probabilities of the individual phenotypes in each possible combination:

Trait combination		Probability
Chin fissure, dimples	3/4 × 3/4	9/16
Chin fissure, no dimples	3/4 × 1/4	3/16
Smooth chin, dimples	1/4 × 3/4	3/16
Smooth chin, no dimples	1/4 × 1/4	1/16

Figure 19.10 **Probability rules apply to independent assortment.** (© Cengage Learning)

Figure 19.10 shows how to calculate the probability that a child will inherit genes for a particular set of two traits on different chromosomes.

TAKE-HOME MESSAGE

WHAT IS INDEPENDENT ASSORTMENT?

- Usually, the two members of each pair of homologous chromosomes are assorted into gametes without regard to how other chromosome pairs are assorted.

- As a result of this independent assortment, the individual gene alleles on chromosomes also are sorted into gametes without regard to which other genes the gamete may get.

- Because of independent assortment, the probability is high that gametes—sperm and egg—will carry every possible combination of parental genes.

Single Genes, Varying Effects

- Some traits have clearly dominant and recessive forms. For most traits, however, the story is not so simple.
- Links to Red blood cells 8.1, ABO blood groups 8.4

Section 19.1 noted that genes are chemical instructions for building proteins. A gene is "expressed" when its instructions are carried out and the cell makes the protein. In some cases, the expression of a gene leads to a single phenotype, or observable trait. Usually, however, the genetic foundation of traits is more complicated.

One gene may affect several traits

Sometimes expression of a gene affects two or more traits. This wide-ranging effect of a single gene is called **pleiotropy** (ply-AH-trow-pee, after the Greek *pleio-*, meaning "more," and *-tropic*, meaning "to change"). Much of what

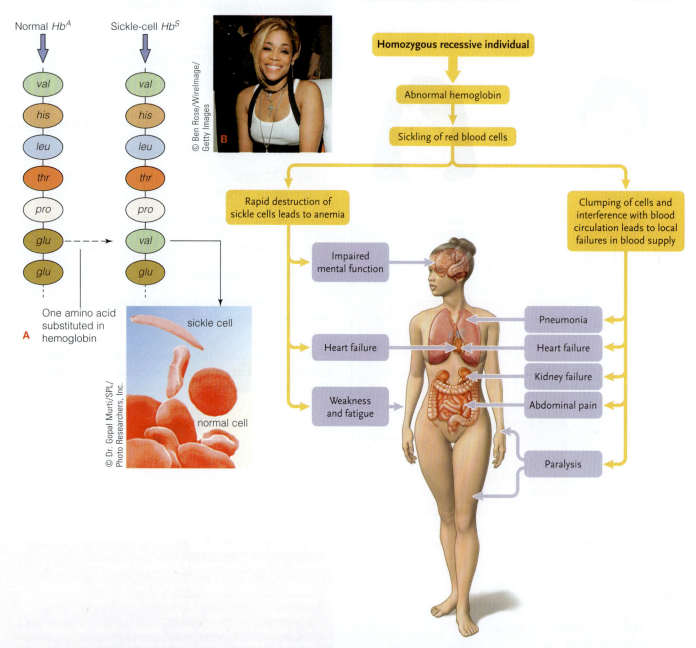

Figure 19.11 A single genetic change leads to the many physical effects of sickle-cell anemia. A shows how an incorrect amino acid has been substituted in the chain of amino acids making up the hemoglobin protein. The inset shows how the shape of a sickled red blood cell differs from that of a normal red blood cell. **B** Tionne "T-Boz" Watkins, celebrity spokesperson for sickle-cell anemia organizations. *Right*: The range of symptoms for a person who has inherited the mutated gene for hemoglobin's beta chain from both parents. There may be other effects as well. (© Cengage Learning)

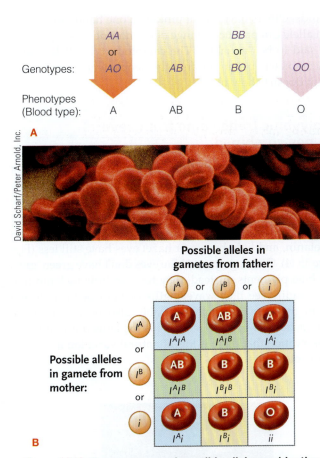

Genotypes: AA or AO | AB | BB or BO | OO

Phenotypes (Blood type): A | AB | B | O

David Scharf/Peter Arnold, Inc.

A

Possible alleles in gametes from father:

I^A or I^B or i

Possible alleles in gamete from mother:

I^A or I^B or i

	I^A	I^B	i
I^A	A $I^A I^A$	AB $I^A I^B$	A $I^A i$
I^B	AB $I^A I^B$	B $I^B I^B$	B $I^B i$
i	A $I^A i$	B $I^B i$	O ii

B

Figure 19.12 There are several possible allele combinations for ABO blood types. **A** An overview of the genotypes that produce the phenotypes of the human ABO blood group—an example of codominance. **B** Figuring out ABO phenotypes using a Punnett square. (© Cengage Learning)

researchers have learned about how such genes function has come from studies of genetic diseases. An example is **sickle-cell anemia**, which was introduced in the discussion of malaria in Chapter 8. This disabling and painful disease is most common in people of African and Mediterranean descent. It arises when a person is homozygous for a recessive allele. The normal allele, Hb^A, has instructions for building normal hemoglobin, the oxygen-transporting protein in red blood cells. When a person inherits two copies of the recessive mutant allele, Hb^S, he or she develops sickle-cell anemia.

Red blood cells, which normally are biconcave disks, take on a sickle shape when the oxygen content of blood falls below a certain level (Section 8.8). The sickled cells clump in blood capillaries and can rupture. The flow of blood can be so disrupted that the person's oxygen-deprived tissues are severely damaged (Figure 19.11B). Homozygotes for the mutated hemoglobin gene (Hb^S/Hb^S) often die relatively young. Heterozygotes (Hb^A/Hb^S), on the other hand, have *sickle-cell trait*. They generally have few

symptoms because the one Hb^A allele provides enough normal hemoglobin to prevent red blood cells from sickling.

During a crisis, sickle-cell anemia patients may receive blood transfusions, oxygen, antibiotics, and painkilling drugs. There is evidence that the food additive butyrate can reactivate "dormant" genes responsible for fetal hemoglobin, an efficient oxygen carrier that normally is produced only before birth. For this reason, some states require hospitals to screen newborn infants for sickle-cell anemia so that appropriate action can begin right away.

In codominance, more than one allele of a gene is expressed

As you now know, people who are heterozygous for a trait have two contrasting alleles for that trait. Usually, one is dominant and one is recessive. In some cases, however, *both* alleles are expressed. We see an example of this **codominance** in people who are heterozygotes for alleles that confer A and B blood types (Figure 19.12). Remember from Section 8.4 that the alleles you carry for the *ABO* gene determines your blood type. The gene provides instructions for making an enzyme that helps build a polysaccharide (a sugar) on the surface of red blood cells. Each *ABO* allele provides slightly different instructions for building the sugar. The sugar in turn gives each person's red blood cells their particular chemical identity—which we call blood type.

Two *ABO* alleles, *IA* and *IB*, are codominant when paired with each other. Someone who inherits them has type AB blood. A third allele, *O*, is recessive. When paired with either *IA* or *IB*, the *O*'s effect is masked. A person who has it plus an *IA* allele has type A blood, while someone who has it paired with the *IB* allele has type B blood. Someone who inherits two *O*s has type O blood. A gene that has three or more alleles is called a **multiple allele system**. There are many such human genes.

> **codominance** The expression of both contrasting alleles of a gene in heterozygotes.
>
> **multiple allele system** A gene that has three or more different alleles.
>
> **pleiotropy** The expression of a gene that affects more than one trait.

TAKE-HOME MESSAGE

WHY DO MOST GENETIC TRAITS NOT HAVE CLEAR DOMINANT AND RECESSIVE FORMS?

- Pleiotropy is common in humans. In pleiotropy, one gene affects more than one trait. The effects may not be simultaneous, but may have repercussions over time as one altered trait changes another trait and so on, as in sickle-cell anemia.
- Some genes have codominant alleles—that is, both are expressed.
- Some genes have more than two alleles. These multiple allele systems include the alleles for the ABO blood group.

■ Many phenotypes, such as eye color, can't be predicted with certainty. Biologists have uncovered several underlying causes for these variations.

continuous variation
Variation in a genetic trait that shows up as a range of phenotypes.

multifactorial trait   A phenotype shaped by multiple genes and one or more nongenetic factors.

penetrance   The degree to which someone who inherits an allele has the phenotype associated with it.

polygenic trait
Phenotypes produced by the combined expression of several genes.

A gene or multiple allele system may have an all-or-nothing effect on a trait. Either you have dimples or type A blood or you don't. But in other cases, the expression of a gene varies due to gene interactions or nongenetic factors in the environment.

The term **penetrance** refers to the likely degree to which someone who inherits an allele will have the phenotype associated with it. For example, the recessive allele that causes cystic fibrosis is completely penetrant. A full 100 percent of people who are homozygous for it develop CF. The dominant allele for having extra fingers or toes (called *polydactyly*) is incompletely penetrant. Some people who inherit it have the usual ten digits, while others have more (Figure 19.13). *Camptodactyly* is caused by the abnormal attachment of muscles to bones of the little finger. Some people who inherit the allele for it have a stiff, bent little finger on both

hands. Others have a bent pinkie on one hand only. When an allele can produce a range of phenotypes, its expression is said to be "variable." The camptodactyly allele also is incompletely penetrant. In some people who inherit it, the trait doesn't show up.

Polygenic traits come from several genes combined

Polygenic traits result from the combined expression of several genes. For example, like skin color, eye color is the cumulative result of many genes involved in the stepwise production and distribution of melanin. Black eyes have abundant melanin in the iris. Dark-brown eyes have less melanin, and light-brown or hazel eyes have still less (Figure 19.14). Green, gray, and blue eyes don't have green, gray, or blue pigments. Instead, they have so little melanin that we readily see blue wavelengths of light being reflected from the iris. Hair color probably also results from the interactions of several genes. This explains our real-world observation that there is lots of natural variation in human hair color.

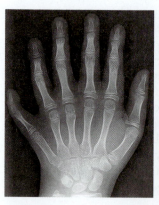
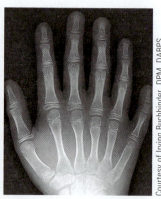

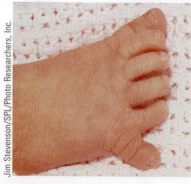

Figure 19.13 People with polydactyly have extra digits on their hands or feet. Usually the extra digits are duplicates. These X-rays reveal two "middle" fingers on each hand. The photograph shows the same pattern with a baby's toes.

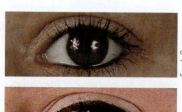

Figure 19.14 Eye color is just one of many human polygenic traits. Alleles of more than one gene interact to produce and deposit melanin, the pigment that helps color the eye's iris (and skin, too).

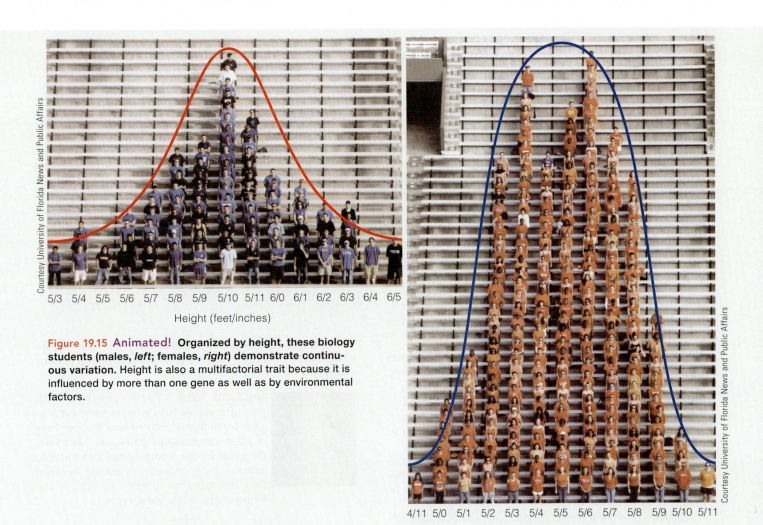

Figure 19.15 **Animated!** **Organized by height, these biology students (males, *left*; females, *right*) demonstrate continuous variation. Height is also a multifactorial trait because it is influenced by more than one gene as well as by environmental factors.**

For many polygenic traits a population as a whole may show **continuous variation**. That is, its members show a range of continuous, rather than incremental, differences in some trait. Continuous variation is especially evident in traits that are easily measurable, such as height (Figure 19.15). Height is also a **multifactorial trait**. This term applies to complex phenotypes that are shaped by more than one gene and also by some aspect of the individual's environment.

The environment can affect phenotypes

There are many examples of human traits in which the environment can help determine the exact phenotype we observe. As just described, a person's adult height is programmed to a great extent by genes, but diet also may play a role. If a young child's diet is deficient in protein, or a disease or injury prevents the normal release of growth hormone, then the person will be shorter than his or her genetic potential would allow. Tanning darkens the natural color of a person's skin, a change that may become permanent in extreme cases.

Many of us inherit genetic predispositions for certain conditions, such as some cancers, obesity, hypertension, even alcoholism and depression. Today a great deal of scientific research is aimed at discovering the degree to which genes and environmental factors contribute to developing one of these health problems. Because it can be difficult to sort out these issues, many of the "good lifestyle choices" you hear so much about these days—from not smoking to eating a healthy diet and reducing stress—are recommendations for ways each of us can limit the chances a harmful gene or genes will be expressed.

TAKE-HOME MESSAGE

WHY ARE SOME PHENOTYPES UNPREDICTABLE?

- Oberved traits can be difficult to predict when expression of the underlying gene or genes can vary.
- Examples include alleles that are incompletely penetrant and polygenic traits that result from the combined expression of two or more genes.
- With complex, multifactorial traits, the phenotype is shaped by more than one gene as well as by environmental factors.

© Cengage Learning/Gary Head

Identical twins have identical genes and look alike. In addition, they have many behaviors in common. Are these parallels coincidence, clever hoaxes, or evidence that aspects of behavior are inherited characteristics, like hair and eye color? Although the question is intriguing, clear answers have proven difficult to come by.

There is strong evidence that some basic behaviors, such as smiling to indicate pleasure and the crying of an infant when it is hungry, are genetically programmed. Scientists have also begun to look for links between genes and alcoholism, some types of mental illness, violent behavior, and even sexual orientation. Such studies raise controversial social issues. So far their greatest impact has been to point out how little we know about the biological basis of human behavior. To learn more about these efforts and find links to some other fascinating and reputable human genetics websites, visit the Behavior Genetics page on the Personality Research website at personalityresearch.org.

SUMMARY

Section 19.1 Genes are specific units of inheritance that are passed to offspring. Each gene has a specific location on a chromosome. Chromosomes come in pairs, so each person has two copies of each gene. The copies are called alleles, and they may or may not be identical.

Someone who is homozygous for a trait (such as *AA*) has two identical alleles for the trait. Having two different alleles (*Aa*) is a heterozygous condition. Alleles (and traits) may be dominant (*A*) or recessive (*a*).

The term *genotype* refers to the particular alleles an individual has. *Phenotype* is the term used to refer to an individual's actual observable traits.

Section 19.2 As diploid organisms, humans have two copies (alleles) of each gene, one on each of the two chromosomes of a homologous pair. The two copies of each gene segregate from each other during meiosis, so each gamete formed ends up with one gene or the other.

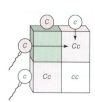

Section 19.3 A Punnett square is a grid for figuring the probable outcome of a genetic cross. Probability expresses the likelihood of a particular event. Matings between two heterozygous individuals (*Cc* × *Cc*) produce the following combinations of alleles in offspring:

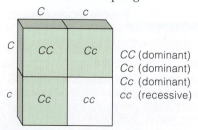

	C	*c*
C	*CC*	*Cc*
c	*Cc*	*cc*

CC (dominant)
Cc (dominant)
Cc (dominant)
cc (recessive)

This results in a probability of 3/4 that any one child will have the dominant phenotype and 1/4 that the child will have the recessive phenotype.

A testcross is a tool in which the phenotypes of offspring are interpreted to identify a parent's genotype. For ethical reasons testcrosses are not used with humans.

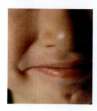

Section 19.4 The members of each pair of homologous chromosomes sort into gametes independently of how the members of other chromosome pairs sort. Therefore, the genes on the chromosomes also sort into gametes independently. A mating between two heterozygous parents results in the following probable phenotypes:

CcDd × *CcDd*

9 dominant for both traits
3 dominant for *C*, recessive for *d*
3 dominant for *D*, recessive for *c*
1 recessive for both traits

Thus in this situation, sixteen genotypes and four phenotypes are possible.

Section 19.5 In cases of pleiotropy, a single gene can influence many seemingly unrelated traits (as in sickle-cell anemia). In codominance, two contrasting alleles of a gene are both expressed. *ABO* blood types provide an example: people who are type AB have codominant alleles for type A and type B blood. Someone who has two copies of the *O* allele has type O blood. A gene that has three or more alleles is called a multiple allele system.

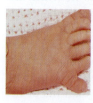

Section 19.6 Various factors can influence the expression of genes. Penetrance refers to the probability that someone who inherits an allele will have the phenotype associated with it. Polygenic traits, such as eye color and height, are due to the expression of several genes. For polygenic traits we may see continuous variation in populations. In multifactorial inheritance the phenotype associated with a polygenic trait can be influenced by nongenetic environmental factors.

REVIEW QUESTIONS

1. Define the difference between (*a*) gene and allele, (*b*) dominant allele and recessive allele, (*c*) homozygote and heterozygote, and (*d*) genotype and phenotype.

2. State the theory of segregation. Does segregation occur during mitosis or during meiosis?

3. What is a testcross, and why can it be useful in genetic analysis?

4. What is independent assortment? Does independent assortment occur during mitosis or during meiosis?

SELF-QUIZ *Answers in Appendix V*

1. Alleles are _____.
   a. alternate forms of a gene
   b. different molecular forms of a chromosome
   c. always homozygous
   d. always heterozygous

2. A heterozygote has _____.
   a. only one of the various forms of a gene
   b. a pair of identical alleles
   c. a pair of contrasting alleles
   d. a haploid condition, in genetic terms

3. The observable traits of an organism are its _____.
   a. phenotype   c. genotype
   b. pedigree   d. multiple allele system

4. Offspring of a monohybrid cross *AA* × *aa* are _____.
   a. all *AA*   d. 1/2 *AA* and 1/2 *aa*
   b. all *aa*   e. none of the above
   c. all *Aa*

5. Second-generation offspring from a cross between two homozygotes are the _____.
   a. F₁ generation   c. hybrid generation
   b. F₂ generation   d. none of the above

6. Assuming complete dominance, offspring of the cross *Aa* × *Aa* will show a phenotypic ratio of _____.
   a. 1:2:1   c. 9:1
   b. 1:1:1   d. 3:1

7. Which statement best fits the principle of segregation?
   a. Units of heredity are transmitted to offspring.
   b. Two genes of a pair separate from each other during meiosis.
   c. Members of a population become segregated.
   d. A segregating pair of genes is sorted out into gametes independently of how gene pairs located on other chromosomes are sorted out.

8. Crosses involving parents that are heterozygous for two traits (*AaBb* × *AaBb*) lead to F₂ offspring with phenotypic ratios close to _____.
   a. 1:2:1   c. 3:1
   b. 1:1:1:1   d. 9:3:3:1

9. Reasons that many phenotypes can't be predicted with certainty include:
   a. There is continuous variation in a gene's effects.
   b. A trait is shaped by more than one gene and by some aspect of a person's environment.
   c. The gene responsible is recessive.
   d. The trait is polygenic.

10. Match each genetic term appropriately.
    _____ testcross involving two traits   a. *Aa* × *Aa*
    _____ testcross involving one trait   b. *Aa*
    _____ homozygous condition   c. *AABB* × *aabb*
    _____ heterozygous condition   d. *aa*

CRITICAL THINKING

1. One gene has alleles *A* and *a*. Another has alleles *B* and *b*. For each genotype listed, what type(s) of gametes can be produced? (Assume independent assortment occurs.)
   a. *AABB*   c. *Aabb*
   b. *AaBB*   d. *AaBb*

2. Still referring to Problem 1, what will be the possible genotypes of offspring from the following matings? With what frequency will each genotype show up?
   a. *AABB* × *aaBB*   c. *AaBb* × *aabb*
   b. *AaBB* × *AABb*   d. *AaBb* × *AaBb*

3. Go back to Problem 1, and assume you now study a third gene having alleles *C* and *c*. For each genotype listed, what type(s) of gametes can be produced?
   a. *AABBCC*   c. *AaBBCc*
   b. *AaBBcc*   d. *AaBbCc*

4. The young woman shown at right has albinism—very pale skin, white hair, and pale blue eyes. This phenotype is due to the absence of melanin, which imparts color to the skin, hair, and eyes. It typically is caused by a recessive allele. In the following situations, what are the probable genotypes of the father, the mother, and their children?
   a. Both parents have normal phenotypes; some of their children are albino and others are not.
   b. Both parents and all their children are albino.
   c. The mother is not albino, the father is albino, and one of their four children is albino.

© Rick Guidotti, Positive Exposure

5. When you decide to breed your Labrador retriever Molly and sell the puppies, you discover that two of Molly's four siblings have developed a hip disorder that is traceable to the action of a single recessive allele. Molly herself shows no sign of the disorder. If you breed Molly to a male Labrador that does not carry the recessive allele, can you assure a purchaser that the puppies will also be free of the condition? Explain your answer.

6. The ABO blood system has been used to settle cases of disputed paternity. Suppose, as a geneticist, you must testify during a case in which the mother has type A blood, the child has type O blood, and the alleged father has type B blood. How would you respond to the following statements?
   a. *Man's attorney*: "The mother has type A blood, so the child's type O blood must have come from the father. Because my client has type B blood, he could not be the father."
   b. *Mother's attorney*: "Further tests prove this man is heterozygous, so he must be the father."

7. Soon after a couple marries, tests show that both the man and the woman are heterozygotes for the recessive allele that causes sickling of red blood cells; they are both Hb^A/Hb^S. What is the probability that any of their children will have sickle-cell trait? Sickle-cell anemia?

8. A man is homozygous dominant for ten different genes that assort independently. How many genotypically different types of sperm could he produce? A woman is homozygous recessive for eight of these genes and is heterozygous for the other two. How many genotypically different types of eggs could she produce? What can you conclude about the relationship between the number of different gametes possible and the number of heterozygous and homozygous gene pairs that are present?

9. As is the case with the mutated hemoglobin gene that causes sickle-cell anemia, certain dominant alleles are crucial to normal functioning (or development). Some are so vital that when the mutant recessive alleles are homozygous, the combination is lethal and death results before birth or early in life. However, such recessive alleles can be passed on by heterozygotes (Ll). In many cases, these are not phenotypically different from homozygous normals (LL). If two heterozygote parents mate ($Ll \times Ll$), what is the probability that any of the children will be heterozygous?

10. Bill and Marie each have flat feet, long eyelashes, and "achoo syndrome" (chronic sneezing). All are dominant traits. The genes for these traits each have two alleles, which we can designate as follows:

	Dominant	Recessive
Foot arch	A	a
Sneezing	S	s
Eyelash length	E	e

Bill is heterozygous for each trait. Marie is homozygous for all of them. What is Bill's genotype? What is Marie's genotype? If they have four children, what is the probability that each child will have the same phenotype as the parents? What is the probability that a child will have short lashes, high arches, and no achoo syndrome?

11. You decide to breed a pair of guinea pigs, one black and one white. In guinea pigs, black fur is caused by a dominant allele (B) and white is due to homozygosity for a recessive allele (b) at the same locus. Your guinea pigs have seven offspring, four black and three white. What are the genotypes of the parents? Why is there a 1:1 ratio in this cross?

Your Future

In diabetes, severe asthma, many cancers, and many other common diseases, affected people typically show multiple gene-based (multifactorial) traits associated with the disease. For instance, medical researchers have identified several dozen separate traits commonly seen in patients with hard-to-manage asthma. Such complex scenarios have made it extremely difficult to sort out the genetic basis of many diseases. Researchers in the field of computational biology now are developing computer programs that can analyze the genetic underpinnings of several correlated disease traits at one time. This approach may provide a much clearer picture of the "gene networks" that collectively cause disease.

KEY CONCEPTS

Chromosomes and Genes

Humans have two types of chromosomes, autosomes and sex chromosomes. Each of the chromosomes carries genes. Each gene is located at a particular place on a specific chromosome. Sections 20.1–20.3

Patterns of Inheritance

Studies of genetic disorders can reveal how traits in general are inherited. Some disease traits arise from dominant or recessive alleles on an autosome or sex chromosome. Sections 20.4–20.6

Changes in Chromosomes

Many genetic disorders arise from rare changes in the number or structure of chromosomes. Such changes can cause harmful or lethal effects. Sections 20.8–20.9

Top: Omikron/Photo Researchers, Inc. Middle: © Cengage Learning Bottom: Lauren Shear/Photo Researchers, Inc.

Cystic fibrosis, or CF, is caused by a faulty gene on chromosome 7. The gene provides instructions for building a plasma membrane transport protein called CFTR. The protein helps chloride and water move into and out of exocrine cells, which secrete mucus or sweat. Mutations of the gene can make the CFTR protein abnormal. CF results when a child inherits a flawed CFTR gene from both parents. CF is most common in people of European ancestry, including the six young people pictured here.

In CF, dry, thickened mucus clogs the airways and makes it hard to breathe. Lung infections develop when bacteria grow in the mucus. Each day patients must undergo physiotherapy that includes thumping on the chest and back to loosen the mucus so it can be expelled. In most cases the lungs eventually fail. Every day in the United States at least one person dies of this disease. Most are under the age of thirty.

This chapter's topic is the link between chromosomes, the genes they carry, and inheritance. As you will discover, researchers are learning more and more about how our genes operate, including the causes of many inherited diseases.

Cody, 23

Jeff, 21

Lindsay, 22

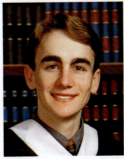

Ben, 23

Savannah, 19

Brandon, 18

A Review of Genes and Chromosomes

- Chapters 18 and 19 provided general information about chromosomes and what happens to them during meiosis. We can relate this information to some common patterns of heredity.

- Links to Chromosome structure 18.1, Meiosis 18.6–18.8, Concepts of heredity 19.1

Understanding inheritance starts with gene–chromosome connections

Chapters 18 and 19 discussed how chromosomes carry genes, and outlined basic "rules" for how genes are passed from one generation to another. To recap some of what you have learned up to this point:

1. Each gene has a particular location (its *locus*) on a specific chromosome.

2. A diploid cell (2*n*) has pairs of homologous chromosomes, one from the mother and one from the father. Except for the sex chromosomes (*X* and *Y*), the chromosomes of each pair are alike in length, shape, and the genes they include.

3. During meiosis in germ cells, homologous chromosomes line up together, then later separate from each other. While they are lined up, they may exchange corresponding segments. This exchange of segments and their genes is called crossing over.

4. There may be two or more versions (alleles) of given gene, but a diploid cell can have only two of them, one on each member of a pair of homologous chromosomes.

5. In general, each gene on a chromosome is sorted into gametes independently of the chromosome's other genes. Geneticists call this process independent assortment.

The concept of independent assortment helps explain why even close relatives have such a varied mix of genetic traits. Geneticists could not help noticing, however, that some traits often seemed to go together. It turns out that there are exceptions to the "rule" of independent assortment, and they result in some traits often being inherited together.

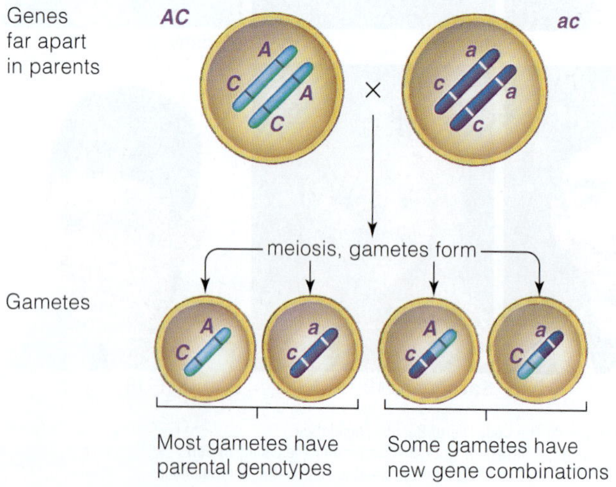

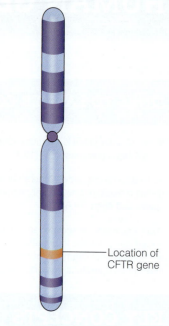

Human chromosome 7

Figure 20.2 **The gene for cystic fibrosis has been mapped to human chromosome 7.** (© Cengage Learning)

Some traits often are inherited together because their genes are physically linked

Although most genes on a chromosome do sort into gametes independently, others are physically connected. When the distance between two genes is short, we say there is close *linkage* between them. Closely linked genes nearly always end up in the same gamete. On the other hand, when two genes on a chromosome are far apart, it is more likely that crossing over will break up the linkage. Those genes are less likely to stay together as gametes form (Figure 20.1). The patterns in which genes are distributed into gametes are so regular they can be used to map the positions of the genes on a chromosome. The simplified map of human chromosome 7 in Figure 20.2 shows the location of the cystic fibrosis gene (CFTR).

Genes far apart in parents   *AC*   ×   *ac*

— meiosis, gametes form —

Gametes

Most gametes have parental genotypes   Some gametes have new gene combinations

Figure 20.1 **Closely linked genes tend to stay together when meiosis sorts genes into gametes.** (© Cengage Learning)

TAKE-HOME MESSAGE

WHAT IS GENE LINKAGE?

- Linkage refers to the physical distance between genes on a chromosome. It is an important factor in determining whether genes move into the same or different gametes during meiosis.

Picturing Chromosomes with Karyotypes

A diagram called a **karyotype** can help answer questions about a person's chromosomes. Chromosomes are the most condensed and easiest to identify at the phase of mitosis called metaphase (Section 18.3).

A technician who wants to make a karyotype doesn't assume that it will be possible to find a body cell that is dividing. Instead, cells are cultured in the laboratory along with chemicals that stimulate the cells to grow and to divide by mitosis. Blood cells are often used for this purpose.

Once the cell culture is established, a chemical called colchicine is added to stop mitosis at metaphase. After the colchicine treatment, the "soup" of cultured cells is placed into glass tubes that are whirled in a centrifuge. The spinning force moves the cells to the bottom of the test tubes.

> **karyotype** A preparation of an individual's metaphase chromosomes arranged by length, shape, and the location of the centromere.

Next the cells are transferred to a saline (salt–water) solution. They swell (by osmosis) and separate, as do the metaphase chromosomes. At this point the cells are placed on a microscope slide, "fixed" (stabilized by air-drying or some other method), and stained so that they are easy to see.

The chromosomes are photographed through the microscope, and the image is enlarged. Then the photograph is cut apart, one chromosome at a time. The cutouts are arranged in order, essentially from largest to smallest (Figure 20.3). The sex chromosomes are placed last. All pairs of homologous chromosomes are aligned horizontally, by their centromeres. Figure 20.3F shows a karyotype diagram prepared this way.

A Add cells from a small blood sample to a medium that has a chemical stimulator for mitosis. Add colchicine to arrest mitosis at metaphase. Transfer culture to a centrifuge. (This motor-driven rotary device spins test tubes at high speed. Tube contents respond to the centrifugal force according to their mass, density, and shape.)

Charles D. Winters/Photo Researchers, Inc.

B Centrifugation forces cells to bottom of tube. Draw off culture medium. Add a dilute saline solution to tube. Add a fixative.

C Prepare and stain cells for microscopy.

D Put cells on a microscope slide. Observe.

E Photograph one cell through microscope. Enlarge image of its chromosomes. Cut the image apart. Arrange chromosomes as a set.

Figure 20.3 A karyotype gives a portrait of the chromosomes in a cell. **A–E** The diagrams show how to prepare a karyotype. **F** A human karyotype. Human somatic cells have twenty-two pairs of autosomes and one pair of sex chromosomes (XX or XY). These are metaphase chromosomes from a female; and each has been duplicated. In the orange box at the far right are the two sex chromosomes (XY) of a male.
(© Cengage Learning)

1 2 3 4 5 6 7 8 9 10 11 12

13 14 15 16 17 18 19 20 21 22 XX (or XY)

Omikron/Photo Researchers, Inc.

F

The Sex Chromosomes

■ Sex chromosomes carry genes associated with sexual traits.

■ Links to Formation of eggs and sperm 16.2, 16.4

Gender is a question of X or Y

It has been said that "men are from Mars, women are from Venus." While that topic goes far beyond biology class, there *are* biological differences between the sexes. You already know that the **X chromosome** is the female sex chromosome and the **Y chromosome** is the male sex chromosome. A female's diploid cells each have two X chromosomes, so females are said to be XX. A male's diploid cells each have one X chromosome and one Y, so males are said to be XY. Each X chromosome carries an estimated 1,400 genes, but a Y chromosome is much smaller and carries an estimated 200 genes, at most.

Despite their differences, the X and Y chromosomes can be joined together briefly in a small region along their length. This allows the X and Y to function as homologues during meiosis.

A mother's egg always carries an X chromosome, so the father's sperm determines a baby's gender. If an X-bearing sperm fertilizes an egg, the embryo will be XX and develop into a female. On the other hand, if the sperm has a Y chromosome, the embryo will be XY and develop into a male (Figure 20.4A).

The genes on a Y chromosome include the master gene for male sex determination, which has been dubbed *SRY*. When the gene is expressed, testes form in the embryo (Figure 20.4B). When that gene is "missing" because no Y chromosome is present, ovaries form, and the developing embryo is female.

Although the X chromosome has some genes associated with sexual traits, such as the distribution of body fat, most of its genes deal with nonsexual traits such as blood clotting. Males have one X chromosome, so these genes can be expressed in males as well as in females. The genes on X and Y chromosomes are sometimes called **X-linked genes** and **Y-linked genes**, respectively.

In females, one X is inactivated

Since females have two X chromosomes and males have only one, do females have twice as many X-linked genes and therefore a double dose of their gene products? Not really, because a compensating mechanism called

Figure 20.4 **The father's sperm determines whether a baby will be male or female. A** Males transmit their Y chromosome to sons but not daughters. Males get their X chromosome from their mother. **B** How the duct system in an early embryo develops into a male or a female reproductive system. (© Cengage Learning)

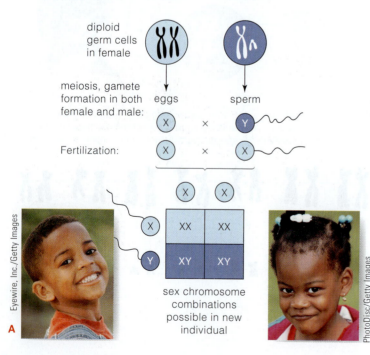

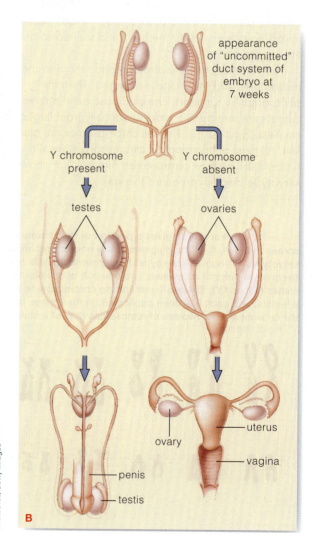

X inactivation occurs in females. Apparently, most or all of the genes on one of a female's X chromosomes are turned off soon after the first cleavages of the zygote. In a given cell, either of the two X chromosomes can be inactivated. The inactivated X is condensed into a **Barr body** (Figure 20.5A). From then on, the same X chromosome will be inactivated in all the descendants of the cell. After X inactivation takes place, the embryo continues to develop. Typically, a female's body has patches of tissue where the genes of the maternal X chromosome are expressed, and other patches where the genes of the paternal X chromosome are expressed.

Some females have a mutation that makes the mosaic tissues visible (Figure 20.5C). For example, *incontinentia pigmenti* is an X-linked disorder that affects the skin, teeth, nails, and hair. In females who are heterozygous for the trait (one X chromosome has the mutation and the other X is normal), mosaic tissue shows up as lighter and darker patches of skin.

Some genes are expressed differently in males and females

You may have noticed that many more men than women have pattern baldness. This common form of hair loss in men is an example of a *sex-influenced* trait. Such traits appear much more often in one sex than in the other, or else the phenotype differs depending on whether the person is male or female. Genes for sex-influenced traits are on autosomes, not on sex chromosomes. The gene allele that causes pattern baldness acts like a dominant gene in males but behaves like a recessive gene in females. That is, a male needs only to inherit one copy in order to become bald. A woman will develop pattern baldness only if she has two copies of the allele, and usually much later in life than a male who has the allele. The difference is due to differences in the effects of male sex hormones in females and males.

Secondary sexual characteristics such as the growth of a man's beard and the development of a woman's breasts are governed by *sex-limited* genes. Both males and females inherit the same genes (on the X chromosome), but only the genes appropriate to a person's biological sex are turned on when a youngster reaches puberty. Sex hormones once again are the "switch."

Rivera Collection/SuperStock, Inc.

Barr body An X chromosome in which the genes are inactivated.

X chromosome The female sex chromosome.

X inactivation Inactivation of genes on one of the two X chromosomes in a female's somatic cells.

X-linked gene Any of the genes on an X chromosome.

Y chromosome The male sex chromosome.

Y-linked gene Any of the genes on a Y chromosome.

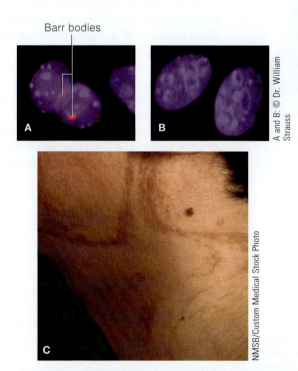

Barr bodies

A and B: © Dr. William Strauss

NMSB/Custom Medical Stock Photo

Figure 20.5 X inactivation halts the functioning of one of a female's two X chromosomes. A Light micrographs show Barr bodies (condensed X chromosomes) in the nuclei of several cells. **B** The X chromosome in cells of males is not condensed this way. **C** A "mosaic" tissue effect that shows up in females who have incontinentia pigmenti. In darker patches of this girl's skin, the mutated X chromosome is active.

TAKE-HOME MESSAGE

HOW DO THE FEATURES AND FUNCTIONS OF THE X AND Y SEX CHROMOSOMES COMPARE?

• The X chromosome, the female sex chromosome, is much larger and carries more genes than the Y chromosome—the male sex chromosome.

• A person's sex is determined by the father's sperm, which can carry either an X chromosome or a Y chromosome. XY embryos develop as males, and XX embryos as females.

• In females, one of the two X chromosomes is inactivated soon after embryonic development begins.

• Sex chromosomes do not carry genes for sex-influenced traits, such as pattern baldness.

• Genes that govern secondary sexual characteristics are on the X chromosome. Male and female sex hormones determine which ones are expressed in a given individual.

• Sex-influenced traits are governed by genes on autosomes but are expressed differently in males and females.

20.4 Human Genetic Analysis

- In some cases prospective parents can assess their risk of conceiving a child with an inherited disorder. A first step can be to construct a genetic family history.

- Links to Chromosomes 18.1, Dominant and recessive conditions 19.1

A pedigree shows genetic connections

In nonhuman organisms, geneticists use experimental crosses to do genetic analysis. Since we can't experiment with humans, however, a basic tool is a genetic family history called a **pedigree**. This is a chart that tracks several generations of a family, showing who exhibited the trait being investigated. The example shown in Figure 20.6 includes definitions of some of the symbols used.

When analyzing a pedigree, geneticists use their knowledge of probability and of basic inheritance patterns, which may yield clues to a trait's genetic basis. For instance, they might figure out that the allele that causes a disorder is dominant or recessive or that it is located on an autosome or a sex chromosome. Pedigrees often are used to identify those at risk of transmitting or developing the trait in question—including any children that a couple may have.

Gathering numerous family pedigrees increases the numerical base for analysis. Figure 20.7 shows a series of pedigrees for **Huntington's disease**, in which the nervous system progressively degenerates. Genetic researcher Nancy Wexler constructed the pedigrees for a huge extended family in Venezuela that includes more than 10,000 people.

Someone who is heterozygous for a recessive disease trait can be designated as a *carrier*. A carrier shows the dominant phenotype (no disease symptoms) but still can produce sperm or eggs with the recessive allele and potentially pass it on to a child. If both parents are carriers for a disorder, a child has a 25 percent chance of being homozygous for the harmful recessive allele.

In thinking about genetics, it's good to keep in mind the difference between an abnormality and a disorder. A **genetic abnormality** is simply a deviation from the average, such as having six toes on each foot instead of five. An abnormality doesn't necessarily cause health problems. A **genetic disorder** causes mild to severe medical problems. We use the word **syndrome** to mean a set of symptoms that usually occur together and characterize a disorder.

Each of us carries an average of three to eight harmful recessive alleles. Why don't alleles that cause severe disorders simply disappear from human populations? One reason is that new gene alleles can come about by way of mutation, as described in Section 20.8. Also, in heterozygotes, a recessive allele is paired with a normal dominant one that prevents the recessive phenotype from showing up. Even so, the recessive allele still can be transmitted to offspring.

It's not uncommon for some genetic disorders to be described as diseases (for example, Huntington's disease and

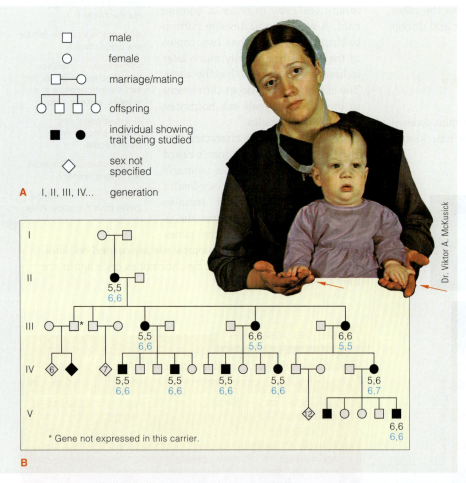

Dr. Viktor A. McKusick

Figure 20.6 Animated! A pedigree is the family history of a genetic trait. **A** Here you see some symbols used in constructing pedigree diagrams. **B** A pedigree for polydactyly, in which affected people have extra fingers, extra toes, or both. Polydactyly and several other genetic conditions are relatively common in the closed society of the Old Order Amish. As described in Section 19.6, expression of the gene governing polydactyly can vary. Here, *black* numerals designate the number of fingers on each hand. *Blue* ones designate the number of toes on each foot. (© Cengage Learning)

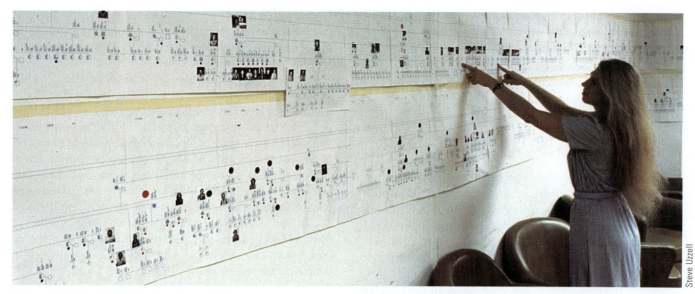

Figure 20.7 **Genetic researcher Nancy Wexler made a pedigree for Huntington's disease (HD).** Wexler has a special interest in HD because it runs in her family.

cystic fibrosis), but in other situations the terms are not interchangeable. For instance, a person's genes may increase her or his susceptibility or weaken the response to infection by a virus, bacterium, or some other pathogen. Strictly speaking, however, the resulting illness isn't a genetic disease. A common example is a genetic predisposition to develop allergies or asthma that runs in some families.

Genetic analysis may predict disorders

Some prospective parents suspect that they are likely to produce a severely afflicted child. Their first child, a close relative, or they themselves may have a genetic disorder, and they wonder how likely it is that future children also will be affected. Psychologists, geneticists, and other specialists may be called in to provide answers.

A common first step is determining the genotype of each parent. Family pedigrees can aid the diagnosis. For disorders that follow one of the basic inheritance patterns described in Chapter 19, it's possible to predict the chances a given child will be affected. But not all follow basic patterns. And those that do can be influenced by other factors, some identifiable, others not. Even when the extent of risk has been determined with confidence, prospective parents must understand that the risk is the same for *each* pregnancy. If a pregnancy has one chance in four of producing a child with a genetic disorder, the same odds apply to every subsequent pregnancy.

genetic abnormality A deviation from the usual phenotype that does not harm health.

genetic disorder Genetic condition that causes medical problems.

pedigree Family history of a genetic trait.

syndrome A set of symptoms that usually occur together and that are associated with a disorder.

THINK OUTSIDE THE BOOK

Various types of genetic tests are available for prospective parents. One, called carrier identification testing, does what its name suggests. Using the Web or library, find out who is most likely to receive this type of test. Which diseases are the most common concerns?

TAKE-HOME MESSAGE

WHAT CAN WE LEARN FROM PEDIGREE ANALYSIS?

- Pedigree analysis may reveal basic inheritance patterns that provide information about the probability that certain genes may be transmitted to children.
- A genetic abnormality is an uncommon version of an inherited trait.
- A genetic disorder is an inherited condition that produces mild to severe health problems.

- Research on genetic disorders and abnormalities has revealed patterns in the way dominant and recessive genes on autosomes are inherited.

- Links to Blood cholesterol 7.8, Basal nuclei and Parkinson's disease 13.11, Basic heredity concepts and probability 19.1, 19.3

Inherited recessive traits on autosomes cause a variety of disorders

For some traits, inheritance patterns reveal two clues that point to a recessive allele on an autosome. *First*, if both parents are heterozygous, any child of theirs will have a 50 percent chance of being heterozygous and a 25 percent chance of being homozygous recessive (Figure 20.8). *Second*, if both parents are homozygous recessive, any child of theirs will be too.

Cystic fibrosis, which you read about in the chapter introduction, is an **autosomal recessive** condition. So is phenylketonuria (PKU), which results from abnormal buildup of the amino acid phenylalanine. Affected people are homozygous for a recessive allele that fails to provide instructions for an enzyme that is needed to convert phenylalanine to another amino acid, tyrosine. Excess phenylalanine builds up and may be used by cells to make phenylpyruvic acid. At high levels, phenylpyruvic acid can cause mental retardation. Fortunately, a diet low in phenylalanine will prevent PKU symptoms. Many diet products are sweetened with aspartame, which contains phenylalanine. They must carry a warning label so people with PKU can avoid using them.

In some autosomal recessive disorders, the defective gene product is an enzyme needed to metabolize lipids. Infants born with **Tay-Sachs disease** lack hexosaminidase A, which is an enzyme required for the metabolism of sphingolipids, a type of lipid that is especially abundant in the plasma membrane of cells in nerves and the brain. Affected babies seem normal at birth, but over time they lose motor functions and also become deaf, blind, and mentally retarded (Figure 20.8). Most die in early childhood.

Tay-Sachs disease is most common among children of Eastern European Jewish descent. Biochemical tests before conception can determine whether either member of a couple carries the recessive allele.

Some disorders are due to dominant genes

Other kinds of clues indicate that an **autosomal dominant** allele is responsible for a trait. First, the trait usually appears in each generation because the dominant allele generally is expressed even in heterozygotes. Second, if one parent is heterozygous and the other is homozygous for the normal, recessive allele, there is a 50 percent chance that any child of theirs will be heterozygous (Figure 20.9).

A few dominant alleles that cause severe genetic disorders persist in populations. Some result from spontaneous mutations. In other cases, expression of a dominant allele may not prevent reproduction, or affected people have children before the disorder's symptoms become severe.

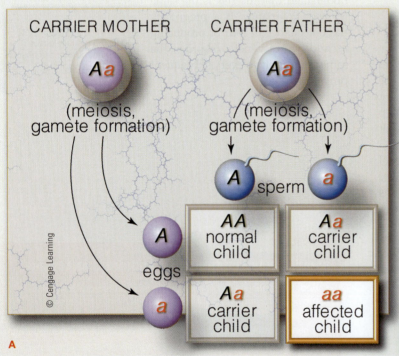

Figure 20.8 Animated! Both parents may be carriers in autosomal recessive inheritance. A The pattern in which both parents are heterozygous carriers of the recessive allele (*red*). **B** Conner Hopf, who was diagnosed with Tay-Sachs disease as an infant. He died at 22 months.

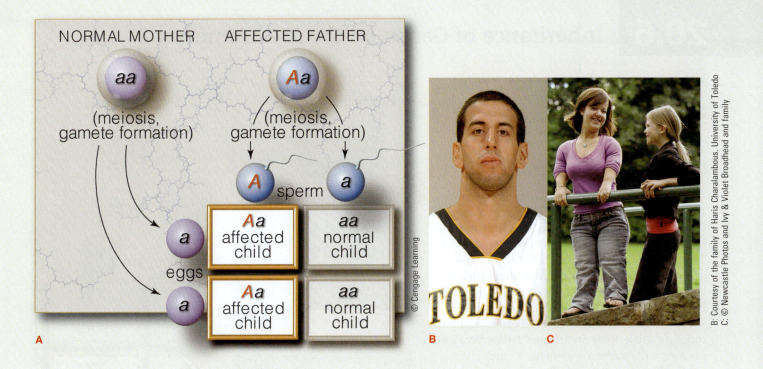

Figure 20.9 Animated! **Autosomal dominant inheritance underlies a variety of genetic disorders. A** Here the phenotype for the dominant allele (*red*) is expressed in the carriers. **B** College basketball star Haris Charalambous, who died when his aorta burst during warmup exercises at the University of Toledo in 2006. Charalambous was very tall and lanky, with long arms and legs, traits that are associated with Marfan syndrome. **C** Achondroplasia affects Ivy Broadhead (*left*) as well as her brother, father, and grandfather.

Huntington's disease (Section 20.4) is a prime example of an autosomal dominant disorder that doesn't cause symptoms until well into adulthood. In about half the cases, symptoms emerge after age thirty, when the person may already have had children (so the allele may be passed on). Homozygotes for the Huntington's allele die as embryos, so affected adults are always heterozygous. Today testing can reveal the disease-causing allele, which is on chromosome 4. There is no cure for HD. Some at-risk people opt not to be tested for the disease, and many don't have children to avoid passing on the disorder.

Marfan syndrome (Figure 20.9B) is another autosomal dominant condition. The responsible allele codes for a defective form of the protein fibrillin, which is found in connective tissue. Abnormal fibrillin causes a range of problems, including disrupting both the structure and the function of smooth muscle cells in the wall of the aorta, the large artery that carries blood away from the heart. Over time the wall weakens, and it can rupture suddenly during strenuous exercise. Marfan syndrome affects 1 in 10,000 people throughout the world. Until recent medical advances, it killed most affected people before age fifty.

Another example, **achondroplasia**, also affects about 1 in 10,000 people. Homozygous dominant infants usually are stillborn. Heterozygotes can reproduce, but while they are young and their limb bones are forming, the cartilage elements of those bones cannot form properly. For that reason, at maturity affected people have abnormally short arms and legs (Figure 20.9C). Adults with achondroplasia don't grow taller than about 4 feet, 4 inches. In many cases, the dominant allele has no other effects.

About 1 person in 500 is heterozygous for a dominant autosomal allele that causes a condition called **familial hypercholesterolemia**. This allele leads to dangerously elevated blood cholesterol because it fails to encode the normal number of cell receptors for LDLs (low-density lipoproteins). You may remember from Chapter 7 that LDLs bind cholesterol in the blood, the first step in removing it from the body. A person who is homozygous for the allele may develop severe cholesterol-related heart disease as a child. Affected people usually die in their twenties and thirties.

autosomal dominant
Caused by a dominant allele on an autosome.

autosomal recessive
Caused by a recessive allele on an autosome.

TAKE-HOME MESSAGE

WHICH GENES CAUSE GENETIC DISORDERS?

- Recessive and dominant genes on autosomes both may cause genetic disorders.
- If both parents are heterozygous carriers of a recessive allele on an autosome, there is a 25 percent chance that a child of theirs will be homozygous for the trait and exhibit the recessive phenotype.
- A dominant autosomal trait may appear in each generation, because the dominant allele is expressed even in heterozygotes.

20.6 Inheritance of Genes on the X Chromosome

- Genes on the X chromosome also are inherited according to predictable patterns.
- Links to Muscle contraction 6.3, Blood clotting 8.7, Eye disorders 14.10

Some disorders are recessive X-linked traits

When a recessive allele on an X chromosome causes a trait, two clues often point to this source. First, many more males than females are affected. This is because a recessive allele can be masked in females, who may inherit a dominant allele on their other X chromosome. It cannot be masked in males, who have only one X chromosome (Figure 20.10). Second, only a daughter can inherit the recessive allele from an affected father, because his sons will receive a copy of his Y chromosome, not the X.

Two forms of the bleeding disorder **hemophilia** are X-linked recessive traits. The most common one, hemophilia A, is caused by a mutation in the gene for factor VIII, a blood-clotting protein (Section 8.7). A male with a recessive allele on his X chromosome is always affected, and he risks death from anything that causes bleeding, even a bruise. The blood of a heterozygous female clots normally, because the normal gene on her normal X chromosome makes enough factor

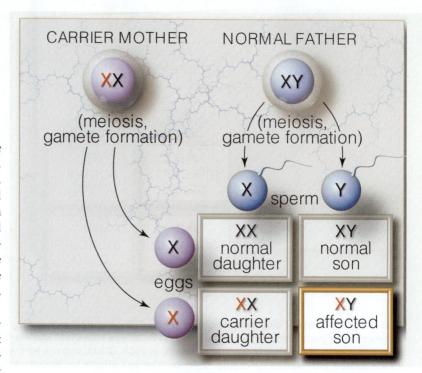

Figure 20.10 Animated! Here is one pattern for X-linked inheritance. It shows the outcomes possible when the mother carries a recessive allele on one of her X chromosomes (*red*).
(© Cengage Learning)

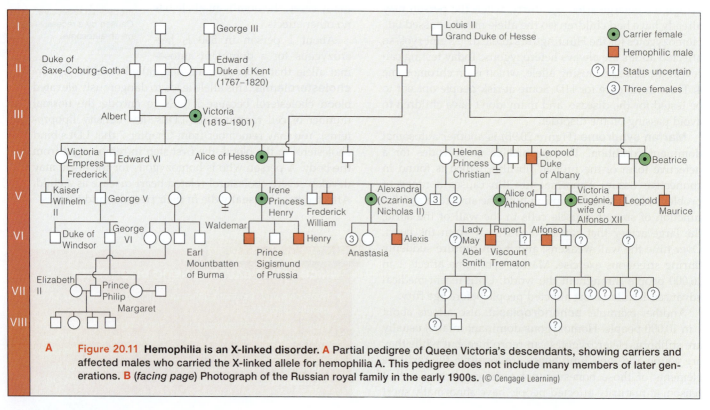

A **Figure 20.11 Hemophilia is an X-linked disorder. A** Partial pedigree of Queen Victoria's descendants, showing carriers and affected males who carried the X-linked allele for hemophilia A. This pedigree does not include many members of later generations. **B** (*facing page*) Photograph of the Russian royal family in the early 1900s. (© Cengage Learning)

VIII. (Hemophilia B, which involves clotting factor IX, has similar symptoms.)

Hemophilia A affects about 1 in 7,000 males. Among nineteenth-century European royal families, however, the frequency was unusually high because close relatives often married. Queen Victoria of England and two of her daughters were carriers. In a pedigree developed some years ago, more than 15 of her 69 descendants at that time were affected males or female carriers (Figure 20.11).

You may recall from Chapter 6 that diseases lumped under "muscular dystrophy" involve progressive wasting of muscle tissue. **Duchenne muscular dystrophy (DMD)** is X-linked. It affects 1 in 3,500 males. As muscles degenerate, affected boys become weak and unable to walk. They usually die by age twenty from cardiac or respiratory failure. The gene that is mutated in DMD normally encodes the protein dystrophin, which gives structural support to muscle fibers. In DMD, muscle fibers lack dystrophin, so they can't withstand the physical stress of contraction and break down. In time the whole muscle is destroyed.

Red-green color blindness is an X-linked recessive trait. About 8 percent of males in the United States have this condition. It arises from mutation of an allele that codes for the protein opsin, which binds visual pigments in cone cells of the retina. Females also can have red-green color blindness, but this occurs rarely because a girl must inherit the recessive allele from both parents. The genes for opsin and blood clotting factor VIII are closely linked, so often hemophilia and red-green color blindness are inherited together.

Some X-linked abnormalities are quite rare

Amelogenesis imperfecta is one of a few known examples of a trait that may be caused by a dominant mutant allele on the X chromosome. With this disorder, the hard, thick enamel coating that normally protects teeth doesn't develop properly (Figure 20.12).

On rare occasions, someone whose sex chromosomes are XY develops as a female. The result is **testicular feminizing syndrome**, or "androgen insensitivity." In people who have this syndrome, a gene mutation on the X chromosome produces defective receptors for male sex hormones (androgens), including testosterone. Normally, cells in the testes and other male reproductive organs bind one or more of the hormones and then develop further. With defective receptors, however, they can't bind the hormones. As a result, the embryo develops externally as a female but has no uterus or ovaries.

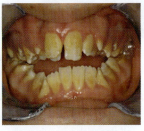

Figure 20.12 Discolored, abnormal tooth enamel is typical of a person with amelogenesis imperfecta. ("Amelogenesis imperfecta" by Peter JM Crawford, Michael Aldred, and Agnes Bloch-Zupan. *Orphanet Journal of Rare Diseases* © 2007 Crawford et al; licensee BioMed Central Ltd. 2:17.)

Many factors complicate genetic analysis

This chapter's examples of autosomal and X-linked traits give a general idea of the clues that geneticists look for. Genetic analysis is usually a difficult task, however. These days few people have large families, so it may be necessary to pool several pedigrees. Typically the geneticist will make detailed analyses of clinical data and keep abreast of current research, in part because more than one gene may be responsible for a given phenotype. For example, we know of dozens of conditions that can arise from a mutated gene on an autosome or from a mutated gene on the X chromosome. Also, some genes on autosomes are dominant in males but recessive in females—so they may initially appear to be due to X-linked recessive inheritance, even though they are not.

Appendix VI at the back of this textbook provides more information on genetic disorders and a database of human genetic traits that you can access online.

Czarina Alexandra (a carrier, descendant of Queen Victoria)

Czar Nicholas II (free of allele for hemophilia A)

Alexis (hemophilic son)

© Bettmann/Corbis

The Russian royal family members. All are believed to have been executed near the end of the Russian Revolution. They were recently exhumed from their hidden graves, but DNA analysis indicated that the remains of Alexis and one daughter, Anastasia, were not among them. The search has continued for graves where the two missing children might have been buried.

20.7 Personalized Medicine

Everybody's different. And thanks to our genes, so is every body. Because each of us has our own personal mix of alleles—the varying chemical forms of genes—we also may respond differently to therapeutic drugs. A field of study called pharmacogenetics aims to pinpoint genetic variations that influence how individuals respond to medications. The idea is to allow physicians to custom-prescribe the drugs that will be safest and most effective for each patient.

All medicines work for some people, but none is a perfect fit for all. Blood pressure drugs are an example. There are more than one hundred different ones, partly because there are many individual differences in how well each one controls high blood pressure. The medication and dose that work well for one patient may be only modestly effective for another, or may cause dangerous or unpleasant side effects for someone else. Figuring out the best course often is a matter of trial and error.

How can we take the guesswork out of prescribing drugs? A first step is identifying the genes that control common reactions to various drugs. That research is happening now (Figure 20.13). The technology for rapid, low-cost genetic screening is becoming widely available. With these tools, a physician could order a genetic profile for a patient and use it to select the best medicine to deal with that person's illness.

AP Photo/Al Behrman

Pharmacogenetics promises to improve patient care by allowing doctors to tailor treatments to their patients' genes. As medical treatment moves in this direction, it will be important to have safeguards in place to protect the privacy of patients' genetic records.

Figure 20.13 Dr. Stephen Liggett is a researcher in the field of pharmacogenetics. He is studying genes that control how patients respond to drug treatments for asthma and congestive heart failure.

20.8 Changes in a Chromosome or Its Genes

- The structure of a chromosome isn't "written in stone." It can change in a variety of ways.

- Link to Eye disorders 14.10, Varying effects of genes 19.5

You may recall that DNA in chromosomes consists of various types of nucleotides linked by chemical bonds (Section 2.13). A *gene mutation* is a change in one or more of the nucleotides that make up a gene. Such mutations can arise in several ways that we discuss in detail in Chapter 21. In this section we are concerned with changes in the structure of whole chromosomes. During meiosis, pieces of chromosomes can be deleted, duplicated, or moved around in other ways. The result often is harmful.

Various changes in a chromosome's structure may cause a genetic disorder

A chromosome region may be deleted spontaneously, or by a virus, irradiation, chemical assault, or some other environmental factor:

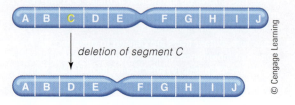

deletion of segment C

© Cengage Learning

Any part of a chromosome can be lost. Wherever such a **deletion** happens, it permanently removes one or more of the chromosome's genes. The loss of a gene can lead to serious problems. For example, one deletion from human chromosome 5 leads to abnormal mental development and an abnormally shaped larynx. When an affected infant cries, the sounds produced resemble meowing—hence the name of the disorder, **cri-du-chat** (French, meaning

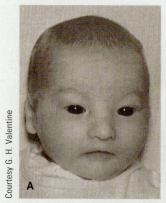

Courtesy G. H. Valentine

Courtesy of the Aniridia Foundation International, www.aniridia.net

Figure 20.14 Animated! Some genetic disorders are due to deletion of part of a chromosome. **A** This child developed cri-du-chat syndome due to a deletion on chromosome 5. Having ears low on the head relative to the eyes is a typical sign of the disorder. **B** A deletion on chromosome 11 causes aniridia, the lack of an iris in the eye, as shown in this photograph.

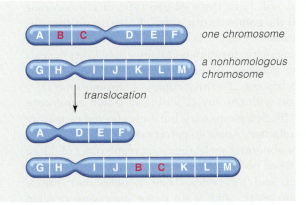

Figure 20.15 Animated! A translocation is one kind of chromosome change. (© Cengage Learning)

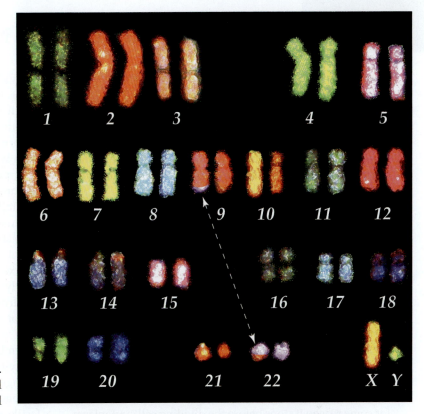

Figure 20.16 A karyotype may reveal damage to a chromosome. Like the karyotype in Figure 18.2, this one displays the 46 chromosomes in a diploid human cell. The arrow indicates the bit of chromosome 22 fused to the Philadelphia chromosome (chromosome 9). (From "Multicolor Spectral Karyotyping of Human Chromosomes" by E. Schrock, T. Ried, et al., Science, 26 July 1996, Volume 273, pp. 495. Used by permission of E. Schrock and T. Ried and The American Association for the Advancement of Science)

"cat cry"). Figure 20.14A shows an affected child. Figure 20.14B shows an eye in which the colored iris did not develop properly. This disorder, called aniridia, is the result of a deletion on chromosome 11. People with aniridia usually have other eye abnormalities as well, because the deleted gene governs several aspects of eye development.

Given that humans have diploid cells, it might seem logical that genes on the affected chromosome's homologue would make up for the loss. In fact, this often happens if a segment deleted from one chromosome is present—and normal—on the homologous chromosome. However, if the remaining, homologous segment is abnormal or carries a harmful recessive allele, nothing will mask its effects.

Another kind of chromosome change is called a **translocation**. Here, part of one chromosome exchanges places with a corresponding part of another chromosome that is *not* its homologous partner (Figure 20.15). This sort of change to a chromosome's structure is virtually sure to be harmful. For instance, in some people a region of chromosome 8 has been translocated to chromosome 14—and the result can be several rare types of cancer. The disease develops because genes in that region are no longer properly regulated, a topic that we consider again in Chapter 22.

Patients who have a chronic type of leukemia (a blood cell cancer) have an abnormally long chromosome 9—called the Philadelphia chromosome after the city where it was discovered. The extra length is actually a piece of chromosome 22 (Figure 20.16). By chance, both chromosomes break in a stem cell in bone marrow. Then, each broken piece reattaches to the *wrong* chromosome—and a gene located at the end of chromosome 9 becomes fused with a gene in chromosome 22. Instructions from this altered gene lead to the synthesis of an abnormal protein. In some way that researchers do not yet understand, that protein promotes the runaway multiplication of white blood cells.

Even normal chromosomes contain the changes called **duplications**, which are sequences of nucleotides that are repeated. Often the same sequence is repeated thousands of times. You might guess that so much duplicate DNA would be harmful, but no genetic disorder has yet been linked to duplication. In the next section we will look at some genetic effects that can arise when a gamete or new embryo receives an abnormal number of chromosomes—either too few or too many.

deletion The loss of part of a chromosome.

duplication The repetition of a sequence of nucleotides in a chromosome.

translocation The movement of part of a chromosome to a nonhomologous chromosome.

20.9 Changes in Chromosome Number

■ Several kinds of events can increase or decrease the number of chromosomes in gametes.

Sometimes gametes—and, later on, embryos—end up with the wrong chromosome number. The effects range from minor physical ones to deadly disruption of body function. More often, an affected fetus is miscarried, or spontaneously aborted before birth.

nondisjunction The failure of pairs of chromosomes to separate during mitosis or meiosis.

About half of fertilized eggs have a lethal condition called *aneuploidy* (AN-yoo-ploy-dee). In this situation, the embryo doesn't have an exact multiple of the normal haploid set of 23 chromosomes. A *polyploid* embryo has three, four, or more sets of the normal haploid set of 23 chromosomes. All but 1 percent of human polyploids die before birth, and the rare newborns die soon afterward.

Chromosome numbers can change during mitosis or meiosis or even at fertilization. For instance, a cell cycle might advance through DNA duplication and mitosis; then for some reason it stops before the dividing cell's cytoplasm divides. The cell then is polyploid—it has four of each type of chromosome.

Nondisjunction is a common cause of abnormal numbers of autosomes

In **nondisjunction**, one or more pairs of chromosomes fail to separate during mitosis or meiosis. Here again, some or all of the resulting cells end up with too many or too few chromosomes (Figure 20.17).

If fertilization involves a gamete that has an extra chromosome ($n + 1$), the result will be *trisomy*: the new individual will have three of one type of chromosome ($2n + 1$). If the gamete is missing a chromosome, then the result is *monosomy* ($2n - 1$). Most changes in the number of autosomes arise through nondisjunction when meiosis is forming gametes. About 1 in every 1,000 children is born with trisomy 21—three copies of chromosome 21.

A person with trisomy 21 will have **Down syndrome** (Figure 20.18). Symptoms vary, but most affected people are mentally retarded. About 40 percent develop heart defects. Because of abnormal skeletal development, older children have shortened body parts, loose joints, and poorly aligned hip, finger, and toe bones. Their muscles and muscle reflexes are weak, and their motor functions develop slowly. With special training, though, people with Down syndrome often engage in normal activities.

For women, the incidence of nondisjunction increases with age. The probability that a woman will conceive an embryo with Down syndrome rises steeply after age thirty-five. Yet 80 percent of trisomic 21 infants are born to younger mothers. This statistic reflects the fact that between the ages of eighteen and thirty-five women are the most fertile, so more babies are born to mothers in this age range.

Nondisjunction also can change the number of sex chromosomes

Most sex chromosome abnormalities come about as a result of nondisjunction as gametes are forming. Let's look at a few of the resulting phenotypes, which are listed in Figure 20.19.

About 1 in every 5,000 newborns has **Turner syndrome**, in which a nondisjunction has reduced the chromosome number to 45. Most people with Turner syndrome are missing an X chromosome (in most cases, the one that would

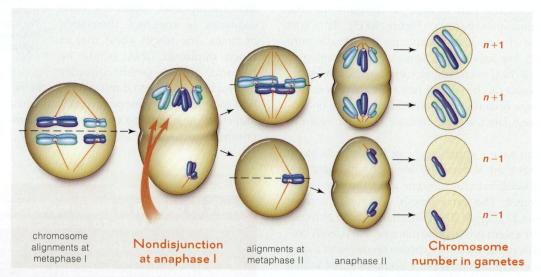

chromosome alignments at metaphase I — **Nondisjunction at anaphase I** — alignments at metaphase II — anaphase II — **Chromosome number in gametes**

$n + 1$
$n + 1$
$n - 1$
$n - 1$

Figure 20.17 Animated! In nondisjunction, chromosomes don't separate properly during meiosis. In this example, chromosomes fail to separate during anaphase I of meiosis and so there is a change in the chromosome number in gametes that form. (© Cengage Learning)

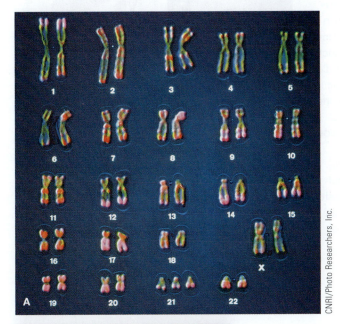

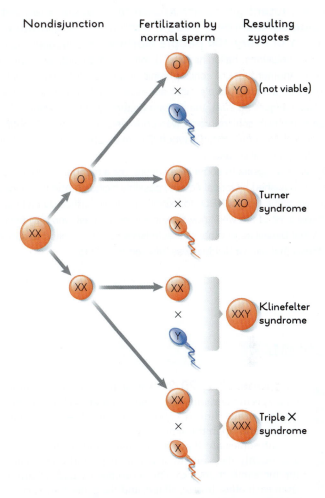

Figure 20.19 Genetic disorders can result from nondisjunction of X chromosomes followed by fertilization by normal sperm. (© Cengage Learning)

Figure 20.18 In Down syndrome there are three copies of chromosome 21. A In this karyotype of a girl with Down syndrome, note the extra copy of chromosome 21. B A boy with Down syndrome.

have come from the father), and the condition is symbolized as XO. Turner syndrome occurs less often than other sex chromosome abnormalities, probably because most XO embryos are miscarried early in the pregnancy. Affected people are female and have a webbed neck and other phenotypic abnormalities. Their ovaries don't function, they are sterile, and secondary sexual traits don't develop at puberty. People who have Turner syndrome often age prematurely and have shortened life expectancies.

Roughly 1 in 1,000 females has three X chromosomes. Two of these X chromosomes are condensed to Barr bodies, and most XXX females develop normally.

In **Klinefelter syndrome**, nondisjunction produces the genotype XXY. This sex chromosome abnormality occurs in about 1 in 500 males. XXY males have low fertility and many have some mental retardation. They have abnormally small testes, sparse body hair, and may develop enlarged

breasts. Testosterone injections can reverse some aspects of the phenotype.

About 1 in every 1,000 males has one X and two Y chromosomes, a condition due to nondisjunction of duplicated Y chromosomes during meiosis. XYY males tend to be taller than average, but otherwise they have a normal male phenotype.

Several mutant genes are known to be associated with neurobiological disorders (NBDs) such as schizophrenia, which affects 1 of every 100 people worldwide. Schizophrenia is characterized by delusions, hallucinations, disorganized speech, and abnormal social behavior. Another facet of schizophrenia and other NBDs (such as bipolar disorder) is that affected people often are exceptionally creative. One example is John Nash (Figure 20.20A), the brilliant mathematician and Nobel Prize winner whose battle with schizophrenia was portrayed in the film *A Beautiful Mind*. Another was the writer Virginia Woolf (Figure 20.20B), who committed suicide after a long mental breakdown.

Evidence suggests to some researchers that a number of other highly creative, distinguished historical figures, possibly including Abraham Lincoln, suffered from some type of NBD. To explore this topic further, do an Internet search and make a list of ten well-known people from the past who may have had an NBD. What behaviors or other characteristics have been cited to support the hypothesis that each individual was affected by an NBD?

Figure 20.20 **Some highly creative people have suffered from a genetic neural disorder.** Two famous cases are **A** mathematician John Nash and **B** writer Virginia Woolf.

SUMMARY

Sections 20.1, 20.2 A gene has a specific location on a specific chromosome. The genes on a chromosome are physically linked. Those that are closest together usually end up in the same gamete.

Autosomes are the same in males and females. They are roughly the same in size and shape and carry genes for the same traits. Sex chromosomes (X and Y) differ from each other in size, shape, and the genes they carry.

Section 20.3 A child's gender is determined by the father's sperm, which can have either an X or a Y chromosome. Males have an XY genotype; females are XX. Genes on the X and Y chromosomes are X-linked and Y-linked, respectively. In a female, X inactivation shuts down the expression of genes carried on one of her X chromosomes. Sex-influenced traits, such as pattern baldness, appear more frequently in one sex. They may reflect the varying influences of sex hormones.

Section 20.4 A pedigree chart can help establish inheritance patterns and track genetic abnormalities through several generations of a family. Table 20.1 lists some common genetic disorders and abnormalities.

Section 20.5 Genetic disorders provide information about patterns of gene inheritance. In disorders that involve autosomal recessive inheritance, a person who is homozygous for a recessive allele has the recessive phenotype. Heterozygotes generally have no symptoms.

In autosomal dominant inheritance, a dominant allele usually is expressed to some degree.

Section 20.6 Many genetic disorders are X-linked—the mutated gene occurs on the X chromosome. Males, who inherit only one X chromosome, typically are affected.

Sections 20.8, 20.9 A chromosome's structure can be changed by deletions, duplications, or translocations. Such changes usually lead to harmful changes in traits.

Chromosome number can be altered by nondisjunction, in which one or more pairs of chromosomes do not separate during meiosis or mitosis. In trisomy, an individual inherits an extra copy of one type of chromosome. If one copy of a given chromosome is missing, the condition is called monosomy.

REVIEW QUESTIONS

1. How do X and Y chromosomes differ?

2. What is a "carrier" of a genetic trait?

3. What evidence indicates that a trait is coded by a dominant allele on an autosome?

4. Explain the difference between an X-linked trait and a sex-influenced trait.

5. Explain what nondisjunction is, and give two examples of phenotypes that can result from it.

TABLE 20.1 Examples of Human Genetic Disorders and Genetic Abnormalities

Disorder or Abnormality	Main Symptoms	Disorder or Abnormality	Main Symptoms
AUTOSOMAL RECESSIVE INHERITANCE		**X-LINKED RECESSIVE INHERITANCE**	
Albinism	Absence of pigmentation	Androgen insensitivity some syndrome	XY individual but having female traits; sterility
Hereditary methemoglobinemia	Blue skin coloration	Red–green color blindness	Inability to distinguish among some or all shades of red and green
Cystic fibrosis	Abnormal glandular secretions leading to tissue, organ damage	Fragile X syndrome	Mental impairment
Ellis–van Creveld syndrome	Dwarfism, heart defects, polydactyly	Hemophilia	Impaired blood clotting ability
Fanconi anemia	Physical abnormalities, bone marrow failure	Muscular dystrophies	Progressive loss of muscle function
Galactosemia	Brain, liver, eye damage		
Phenylketonuria (PKU)	Mental impairment	**CHANGES IN CHROMOSOME STRUCTURE**	
Sickle-cell anemia	Adverse pleiotropic effects on organs throughout body	Aniridia	Eyes lacking an iris, other abnormalities
AUTOSOMAL DOMINANT INHERITANCE		Chronic myelogenous leukemia (CML)	Overproduction of white blood cells in bone marrow; organ malfunctions
Achondroplasia	One form of dwarfism		
Camptodactyly	Rigid, bent fingers	Cri-du-chat syndrome	Mental impairment; abnormally shaped larynx
Familial hypercholesterolemia	High cholesterol levels in blood; eventually clogged arteries		
Huntington's disease	Nervous system degenerates progressively, irreversibly	**CHANGES IN CHROMOSOME NUMBER**	
Marfan syndrome	Abnormal or no connective tissue	Down syndrome	Mental impairment; heart defects
Polydactyly	Extra fingers, toes, or both	Turner syndrome (XO)	Sterility; abnormal ovaries, abnormal sexual traits
Progeria	Drastic premature aging	Klinefelter syndrome	Sterility; mild mental impairment
Neurofibromatosis	Tumors of nervous system, skin	XXX syndrome	Minimal abnormalities
		XYY condition	Mild mental impairment or no effect

SELF-QUIZ   *Answers in Appendix V*

1. _____ segregate during _____.
   a. Homologues; mitosis
   b. Genes on one chromosome; meiosis
   c. Homologues; meiosis
   d. Genes on one chromosome; mitosis

2. The alleles of a gene on homologous chromosomes end up in separate _____.
   a. body cells
   b. gametes
   c. nonhomologous chromosomes
   d. offspring
   e. both b and d are possible

3. Genes on the same chromosome tend to stay together during _____ and end up in the same _____.
   a. mitosis; body cell   d. meiosis; gamete
   b. mitosis; gamete   e. both a and d
   c. meiosis; body cell

4. The probability of a crossover occurring between two genes on the same chromosome _____.
   a. is unrelated to the distance between them
   b. increases the closer they are
   c. increases the farther apart they are
   d. zero

5. A chromosome's structure can be altered by _____.
   a. deletions   c. translocations
   b. duplications   d. all of the above

6. Nondisjunction can be caused by _____.
   a. crossing over in meiosis
   b. segregation in meiosis
   c. failure of chromosomes to separate during meiosis
   d. multiple independent assortments

7. A gamete affected by nondisjunction could have _____.
   a. a change from the normal chromosome number
   b. one extra or one missing chromosome
   c. the potential for a genetic disorder
   d. all of the above

8. Genetic disorders can be caused by _____.
   a. gene mutations
   b. changes in chromosome structure
   c. changes in chromosome number
   d. all of the above

9. A person who is a carrier for a genetic trait _____.
   a. is heterozygous for a dominant trait
   b. is heterozygous for a recessive trait
   c. is homozygous for a recessive trait
   d. could be either a or b but not c

10. Match the following chromosome terms appropriately.
   _____ crossing over   a. a chemical change in DNA
   _____ deletion   that may affect genotype
   _____ nondisjunction   and phenotype
   _____ translocation   b. movement of a chromosome
   _____ gene mutation   segment to a nonhomologous
                                 chromosome
                             c. disrupts gene linkages during
                                 meiosis
                             d. causes gametes to have abnor-
                                 mal chromosome numbers
                             e. loss of a chromosome segment

CRITICAL THINKING

1. If a couple has six boys, what is the probability that a seventh child will be a girl?

2. Human sex chromosomes are XX for females and XY for males.
   a. With respect to an X-linked gene, how many different types of gametes can a male produce?
   b. If a female is homozygous for an X-linked allele, how many different types of gametes can she produce with respect to this allele?
   c. If a female is heterozygous for an X-linked allele, how many different types of gametes can she produce with respect to this allele?

3. People with Down syndrome have an extra copy of chromosome 21, for a total of 47 chromosomes. However, in a few cases of Down syndrome, 46 chromosomes are present. This total includes two normal-looking chromosomes 21, one normal chromosome 14, and

From "Multicolor Spectral Karyotyping of Human Chromosomes" by E. Schrock, T. Ried, et al., *Science*, 26 July 1996, Volume 273, pp. 495. Used by permission of E. Schrock and T. Ried and The American Association for the Advancement of Science

Your Future

Medical treatment tailored to a patient's genetic makeup is still in its infancy. One obstacle is the length of time it currently takes to analyze genes. Much faster testing methods are being developed, however. It may not be long before personalized medicine is much more widely available.

a longer-than-normal chromosome 14. Interpret this observation. How can these individuals have 46 chromosomes?

4. If a trait appears only in males, is this good evidence that the trait is due to a Y-linked allele? Explain why you answered as you did.

5. A woman unaffected by hemophilia A whose father had hemophilia A marries a man who also has hemophilia A. If their first child is a boy, what is the probability he will have the disorder?

6. About 4 percent of people of Northern European descent have a cystic fibrosis allele, but only about 1 in 2,500 of these people actually has the disorder. What is the most likely reason for this finding?

7. The following pedigree shows the pattern of inheritance of red-green color blindness in a family. Females are shown as circles and males as squares; the squares or circles of individuals affected by the trait are filled in black. What is the chance that a son of the third-generation female indicated by the arrow will be color-blind if the father is not color-blind? If he is color-blind?

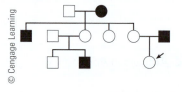

© Cengage Learning

DNA, GENES, AND BIOTECHNOLOGY

LINKS TO EARLIER CONCEPTS

This chapter expands on Chapter 2's discussion of amino acids and the primary structure of proteins (2.11).

The chapter's discussion of how DNA is copied will deepen your understanding of events in the cell cycle, before chromosomes are duplicated and assorted into new cells (18.2, 18.3, 18.6).

Here you will learn more about RNA's role in making proteins. You will also gain a fuller understanding of the key role a cell's ribosomes (3.6) play in protein synthesis.

KEY CONCEPTS

Genetic Instructions in DNA

DNA consists of two strands of nucleotides twisted into a double helix. A gene is a sequence of nucleotides in DNA. Sections 21.1–21.2

Making Proteins

Genes are the genetic code for proteins. Cells build proteins in two steps. First, an mRNA molecule is transcribed from DNA. Then mRNA is translated into a string of amino acids, the primary structure of proteins. Sections 21.3–21.6

Engineering and Exploring Genes

Biotechnology is a tool for changing genes and studying their effects. Practical applications range from uses in agriculture to gene therapy, DNA fingerprinting, and studying the human genome. Sections 21.7–21.12

Top and middle: © Cengage Learning Bottom: TEK Image/ Photo Researchers, Inc.

DNA is the genetic material—the molecule that stores information about inherited traits. More than one hundred years of scientific study have taught us how DNA is built and how it can provide instructions for building and operating the body. These days our growing understanding of DNA has begun to allow us to manipulate the inheritance of traits in organisms ranging from bacteria to food crops, livestock, and humans.

For example, in the United States a whopping 38 percent of soybeans and 25 percent of corn crops have been genetically modified (GM) to withstand weed killers or make their own pesticides. GM corn and soybeans are used in breakfast cereals, soy sauce, vegetable oils, soft drinks, and lots of other food products. They also are used in feed for livestock.

To some people, the ability to genetically modify organisms holds the promise of crucial advances such as curing genetic disorders and developing more nutritious food crops. To others, genetic modification is a slippery slope toward harmful or even unethical outcomes. The student in the upper photograph at right is studying the structure of DNA. The protesters pictured in the bottom photograph are ripping up a GM crop.

This chapter discusses DNA and how our cells use it to guide life processes. We will also look at examples of how that knowledge is being applied in biotechnology.

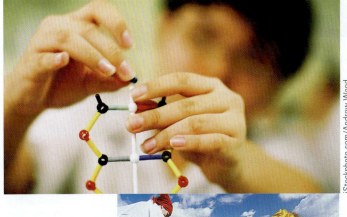

iStockphoto.com/Andrew Wood

©Adrian Arbib/Still Pictures

- DNA is built of nucleotides—the subunits of the biological molecules called nucleic acids. DNA nucleotides are arranged to form a double helix.

- Links to Nucleic acids 2.13, Genes 19.1

DNA is built of four kinds of nucleotides

As you know from Chapter 18, a chromosome consists of a DNA molecule and proteins. A DNA molecule, in turn, is built from four kinds of nucleotides (Figure 21.1), the building blocks of nucleic acids introduced in Section 2.13.

A DNA nucleotide is built of a five-carbon sugar (deoxyribose), a phosphate group, and one of these four nitrogen-containing bases:

adenine	guanine	thymine	cytosine
A	G	T	C

Many researchers were in the race to discover DNA's structure, but James Watson and Francis Crick were the first to realize that DNA consists of two strands of nucleotides

twisted into a double helix. Nucleotides in a strand are linked together, like boxcars in a train, by strong covalent bonds. Weaker hydrogen bonds link the bases of one strand with bases of the other. The two strands run in opposite directions, as shown in Figure 21.2.

Chemical "rules" determine which nucleotide bases in DNA can pair up

The bases in the four DNA nucleotides have different shapes, and different sites where hydrogen bonds can form. These factors determine which bases can pair up. Adenine pairs with thymine, and guanine pairs with cytosine. Therefore, two kinds of **base pairs** occur in DNA: A—T and G—C. In a double-stranded DNA molecule, the amount of adenine equals the amount of thymine, and the amount of guanine equals the amount of cytosine.

While base pairs must form as we've just described—A with T and G with C—the nucleotides can line up in any order. For example, these are just three possibilities of the pattern one might find in DNA:

© Cengage Learning

Figure 21.1 Animated! There are four kinds of nucleotides in DNA. Here the five-carbon sugars are *orange*. Each one has a phosphate group attached to its ring structure (*on the left*). Small numerals on the structural formulas identify the carbon atoms where various parts of the molecule are attached. The photograph shows James Watson (*left*) and Francis Crick, who figured out the structure of DNA. (© Cengage Learning)

A. Barrington Brown/Photo Researchers, Inc.

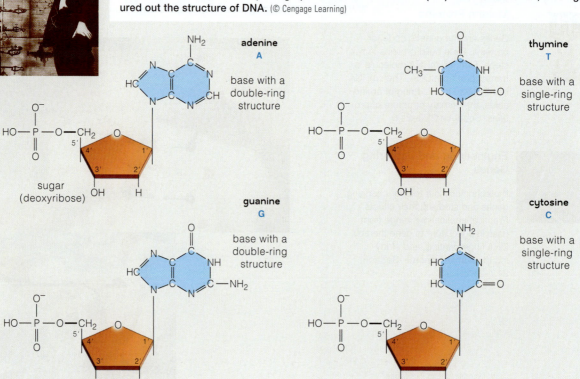

Figure 21.2 Animated!

This diagram shows how nucleotide bases are arranged in the DNA double helix. Three different models are combined here. Notice that the two sugar–phosphate backbones run in opposite directions. It may help to think of the sugar units of one strand as being upside down. By comparing the numerals used to identify each carbon atom of the deoxyribose molecule (1′, 2′, 3′, and so on), you can see that the strands run in opposing directions. (© Cengage Learning)

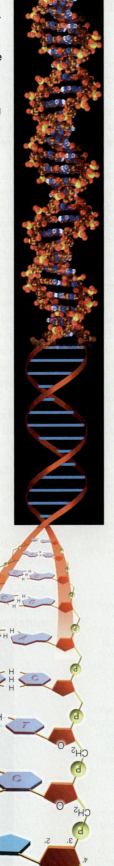

The pattern of base pairing (A with T, and G with C) is consistent with the known composition of DNA (A = T, and G = C).

Human DNA

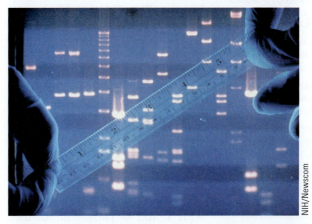

Figure 21.3 **DNA fragments can be visualized through a process called gel electrophoresis.** *Above*: DNA extracted from human cells. The extraction process has given the DNA a "cotton candy" appearance.

A gene is a sequence of nucleotides

As you already know, genes are the units of heredity. Chemically, a **gene** is a sequence of nucleotides in a DNA molecule. The **nucleotide sequence** of each gene codes for a specific polypeptide chain. Polypeptide chains, remember, are the basic structural units of proteins. We'll see how they form later on. Figure 21.3 shows one way researchers can visualize small fragments of DNA.

We now turn to a key feature of each DNA molecule—how it can be copied, or replicated, so that the information it contains is faithfully passed on to new generations.

base pair Two bases (A—T or G—C) held together by hydrogen bonds.

gene A sequence of nucleotides in a DNA molecule.

nucleotide sequence The order of nucleotides in a gene; the sequence codes for a specific polypeptide chain.

TAKE-HOME MESSAGE

WHAT IS A DNA MOLECULE?

- A DNA molecule consists of two strands of nucleotides held together at their bases by hydrogen bonds. The two strands run in opposite directions and twist into a double helix.
- DNA's nucleotides are built of the sugar deoxyribose, a phosphate group, and one of the nitrogen-containing bases adenine (A), guanine (G), thymine (T), and cytosine (C).
- In a DNA molecule, two kinds of base pairs occur: A—T and G—C.

21.2 Passing on Genetic Instructions

- Each DNA molecule must be faithfully copied so parents may pass on their traits to offspring.
- Links to aging 17.13, Cell cycle 18.2, Genes on chromosomes 20.1

base-pair substitution Type of gene mutation in which the wrong nucleotide is paired with a base during DNA replication.

DNA polymerases Enzymes that function in DNA replication and repair.

DNA repair The repair of DNA that has been miscopied during replication.

DNA replication The copying of DNA in a cell before it divides.

gene mutation Change in a gene's nucleotide sequence.

semiconservative replication Mechanism of DNA replication, in which one strand of an existing DNA molecule is the template for the formation of a new second strand.

How is a DNA molecule copied?

Chapter 19 noted that DNA must be copied before a cell divides. This process is called **DNA replication**. During replication, enzymes called **DNA polymerases** and other proteins unwind the DNA molecule, keep the two unwound strands separated, and assemble and seal a new strand on each one. The unwinding exposes stretches of the molecule's nucleotide bases. Cells contain unattached nucleotides that can pair with the exposed bases. **A** pairs with **T** and **G** pairs with **C**; they are linked by hydrogen bonds. Each parent strand stays intact while a new strand is assembled on it, nucleotide by nucleotide.

As replication occurs, the newly formed double-stranded molecule twists back into a double helix. Because one strand is from the starting molecule, that strand is said to be conserved. Only the second strand has been freshly assembled, so each DNA molecule is really half new and half "old." The DNA replication mechanism is thus called **semiconservative replication** (Figure 21.4).

Every time a cell replicates its DNA, at least 3 billion nucleotides must be assembled properly. DNA is copied rapidly—between ten and twenty nucleotides per second are added at a replication site. It's no wonder, then, that mistakes happen. By some estimates a human cell must fix breaks in a strand of its DNA up to 2 million times every hour! Fixing broken or otherwise damaged DNA is called **DNA repair**. It is another task carried out by DNA polymerases and other enzymes.

DNA can be damaged by certain chemicals, ionizing radiation, and ultraviolet light (as in sunlight or the rays of tanning lamps). In one type of damage, UV light causes two neighboring thymine bases to link together. The new structure distorts the affected DNA molecule in a way that prevents effective DNA repair. **Xeroderma pigmentosum** (Figure 21.5) is one type of genetic disorder that results when radiation damage can't be fixed. Patients are at high risk for lethal skin cancer.

A mutation is a change in the sequence of a gene's nucleotides

Every so often, DNA is altered in a way that changes a gene. Sometimes one base gets substituted for another in the nucleotide sequence. At other times, an extra base is inserted into the sequence or a base is deleted from it. Such a small-scale change in the nucleotide sequence of a gene is a **gene mutation**.

A Parent DNA molecule; two complementary strands of base-paired nucleotides.

B Replication begins; the two strands unwind and separate from each other at specific sites along the length of the DNA molecule.

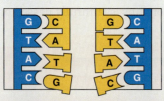
C Each "old" strand serves as a structural pattern (a template) for the addition of bases according to the base-pairing rule.

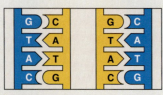
D Bases positioned on each old strand are joined together into a "new" strand. Each half-old, half-new DNA molecule is just like the parent molecule.

Figure 21.4 Animated!
One strand stays the same when DNA is replicated. Here the original two-stranded DNA molecule is shown in blue. A new strand (*yellow*) is assembled on each parent strand. (© Cengage Learning)

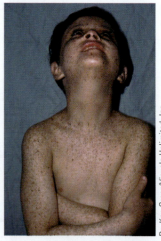
Dr. Ken Greer/Visuals Unlimited, Inc.

Figure 21.5 Skin lesions like these are typical of xeroderma pigmentosum.

Section 2.13 introduced ribonucleic acid, or RNA. When a cell makes a protein, the DNA instructions for the order of its amino acids are first rewritten, or *transcribed*, into a form of RNA called mRNA (Figure 21.6A). Section 21.3 explains this rewriting process. For now, the point to remember is that when a gene is mutated, the change alters mRNA instructions for building the protein the normal gene encodes.

One common kind of gene mutation is **base-pair substitution**, in which the wrong nucleotide is paired with an exposed base while DNA is being replicated (Figure 21.6B). Proofreading enzymes may fix the error. But if they don't, a mutation will be established in the DNA in the next round of replication. As a result of this mutation, one amino acid might replace another during protein synthesis. This is what has happened in people who have sickle-cell anemia (Section 19.5). In another common type of mutation, a *deletion*, a base is lost (Figure 21.6C). This loss is smaller than the deletion of part of a chromosome described in Section 20.8, but both may have major impacts on body functioning.

In an *expansion mutation*, a nucleotide sequence is repeated over and over, sometimes many hundreds of times. Expansion mutations cause several genetic disorders, including Huntington's disease and **fragile X syndrome** (Figure 21.7). In this disorder, a sequence of three nucleotides is repeated in an X chromosome gene. The result is an abnormal protein needed for proper brain development. Boys tend to be more seriously affected, because unlike girls they have only one X, so there is no "backup" X with a normal gene to code for a normal version of the affected protein.

While mutations can occur in the DNA at any time in in any cell, they are inherited only when they take place in germ cells that give rise to gametes. Whether a mutation turns out to be harmful, neutral, or beneficial depends on a variety of factors, including how the resulting protein affects body functions.

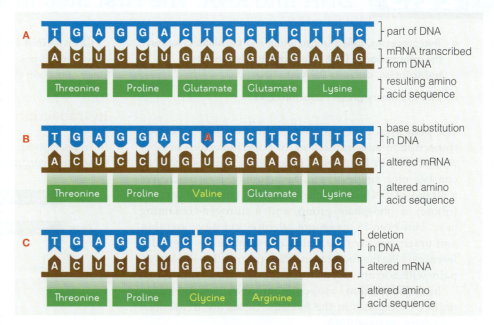

Figure 21.6 **Animated!** **There are several types of gene mutations. A** Part of a gene (*blue*), the mRNA (*brown*), and the specified amino acid sequence. **B** A base-pair substitution. **C** A deletion in DNA. (© Cengage Learning)

C. J. Harrison

Figure 21.7 **Gene mutations cause numerous disorders.** This X chromosome shows the constriction that occurs with fragile X syndrome (*arrow*).

THINK OUTSIDE THE BOOK

Some genetic defects result in "inborn errors of metabolism." These disorders arise when an enzyme vital to some aspect of metabolism is faulty. A fairly common example is alkaptonuria. In people with this condition, the urine turns black when it is exposed to the air. Use Web or library resources to learn more about inborn errors of metabolism. Why does alkaptonuria make urine turn black? What are some other inborn errors of metabolism?

TAKE-HOME MESSAGE

HOW IS DNA REPLICATED?

- In DNA replication, enzymes and other proteins unwind the double helix, assemble a new, complementary strand on it, then seal the strands together.

- DNA replication is said to be semiconservative because each newly formed molecule contains one of the parent strands.

- Enzymes active in DNA replication also may repair damage in DNA.

- A gene mutation is a change in one or more bases in the nucleotide sequence of DNA.

- A mutation may be harmful, neutral, or beneficial depending on how it affects body structures and functions.

21.3 DNA into RNA: The First Step in Making Proteins

- Two processes, called transcription and translation, convert the information in DNA into proteins. The first step, transcription, converts DNA into RNA.

- Link to Primary protein structure 2.11

The path from genes to proteins involves two processes, transcription and translation. In both, molecules of ribonucleic acid, or RNA, have major roles. Most often, RNA consists of a single strand. Structurally, it is much like a strand of DNA. Its nucleotides each consist of a sugar (ribose), a phosphate group, and a nitrogen-containing base. However, its bases are adenine, cytosine, guanine, and uracil, not thymine. Table 21.1 summarizes these differences. Like the thymine in DNA, the uracil in RNA base-pairs with adenine.

In **transcription**, molecules of RNA are assembled on DNA templates in the nucleus. Genes are transcribed into three types of RNA:

rRNA (ribosomal RNA)	a nucleic acid chain that combines with proteins to form a ribosome, a structure on which a polypeptide chain is assembled
mRNA (messenger RNA)	a sequence of nucleotides that carries instructions for building proteins to ribosomes in the cell cytoplasm
tRNA (transfer RNA)	a nucleic acid chain that can (transfer-bring a specific amino acid to the mRNA /ribosome complex during the translation step of protein synthesis

Only mRNA eventually is translated into a protein. The other two types of RNA operate in *translation*, the second stage of protein synthesis, which is described in Section 21.6.

In transcription, DNA is decoded into RNA

In transcription, a strand of RNA is assembled on a DNA template according to the base-pairing rules:

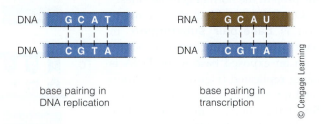

base pairing in DNA replication

base pairing in transcription

© Cengage Learning

Transcription takes place in the cell nucleus, but it isn't the same as DNA replication. Only the gene that is being transcribed serves as the template, not the whole DNA strand, and the enzymes involved are **RNA polymerases**. Also, transcription makes a single-stranded molecule, not one with two strands.

Transcription starts at a *promoter*, a sequence of bases that signals the start of a gene. As transcription starts, a nucleotide "cap" is added to the beginning of the mRNA for protection. This capped end (designated 5′) is also where the mRNA will bind to a ribosome when the time comes for translation.

As RNA polymerase moves along the DNA, it joins nucleotides together (Figure 21.8). When it reaches a

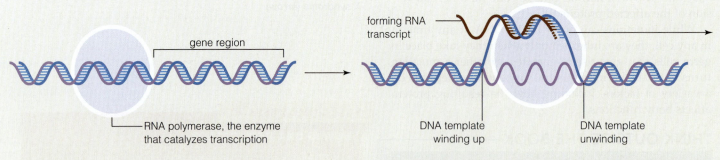

gene region

RNA polymerase, the enzyme that catalyzes transcription

forming RNA transcript

DNA template winding up

DNA template unwinding

A RNA polymerase binds to a promoter in the DNA. The binding positions the polymerase near a gene in the DNA.

In most cases, the nucleotide sequence of the gene occurs on only one of the two strands of DNA. Only the complementary strand will be translated into RNA.

B The polymerase begins to move along the DNA and unwind it. As it does, it links RNA nucleotides into a strand of RNA in the order specified by the base sequence of the DNA.

The DNA double helix winds up again after the polymerase passes. The structure of the "opened" DNA molecule at the transcription site is called a transcription bubble, due to its appearance.

Figure 21.8 Animated! In transcription, an mRNA molecule is assembled on a DNA template. **A** A gene in part of a DNA double helix. The base sequence of one of the two nucleotide strands is used as the template. **B–D** Transcribing that gene results in a molecule of mRNA. (© Cengage Learning)

TABLE 21.1 DNA and RNA

	DNA	RNA
SUGAR	Deoxyribose	Ribose
BASES	Adenine, cytosine, guanine, thymine	Adenine, cytosine, guanine, uracil

termination sequence of bases, the RNA strand—called a transcript—is released, but it is not yet finished. It must be modified before its protein-building instructions can be used. For example, when many human genes are transcribed, this "pre-mRNA" contains sections called **introns**—in some cases including as many as 100,000 nucleotides! Introns may be a sort of genetic gibberish. Researchers have not discovered any that code for proteins.

All new mRNA transcripts also contain regions called **exons**. Unlike introns, exons are the nucleotide sequences that carry DNA's protein-building instructions. Before an mRNA leaves the nucleus, its introns are snipped out and its exons are spliced together. Now the mRNA is ready to enter the cell cytoplasm and be translated into a protein. Some genetic disorders, such as the blood disorder thalassemia (Chapter 8) arise when the splicing step goes awry.

Gene transcription can be turned on or off

Most cells of your body carry the same genes. Many of those genes carry instructions for making proteins that are essential to any cell's structure and functioning. Yet each type of cell also uses a small subset of genes in specialized ways. For example, every cell carries the genes for hemoglobin, but only the precursors of red blood cells activate those genes. Each cell determines which genes are active and which gene products appear, when, and in what amounts. Some genes might be switched on and off throughout a person's life. Others might be turned on only in certain cells and only at certain times.

Genes are regulated by molecules that interact with DNA, RNA, or other substances. For example, **regulatory proteins** speed up or halt transcription. Some also may bind with noncoding DNA sequences and in this way trigger or shut down the transcription of a neighboring gene. For example, this is how steroid hormones, such as cortisol, estrogen, and testosterone, act.

exons Nucleotide sequences in mRNA that carry the protein-building instructions of DNA.

introns Sections of mRNA that do not carry DNA's protein-building instructions.

mRNA Messenger RNA; its sequence of nucleotides is a DNA-based template for building proteins.

regulatory proteins Proteins that speed up or stop gene transcription.

RNA polymerases Enzymes that catalyze transcription of DNA to RNA.

rRNA Ribosomal RNA; it combines with proteins to form ribosomes.

transcription The assembly of RNA with a base sequence that corresponds to the sequence of the DNA region where it was assembled.

tRNA Transfer RNA; it picks up amino acids and pairs them with mRNAs.

TAKE-HOME MESSAGE

HOW IS DNA CONVERTED INTO RNA?

- The process of transcription converts DNA into RNA. A sequence of bases in one strand of a DNA molecule is the template for assembling a strand of RNA. Transcription is the first step in protein synthesis.
- Transcription occurs in the nucleus. Before leaving the nucleus, new RNA transcripts are modified into their final form.
- Gene transcription can be turned on or off to produce specialized cell structures or functions.

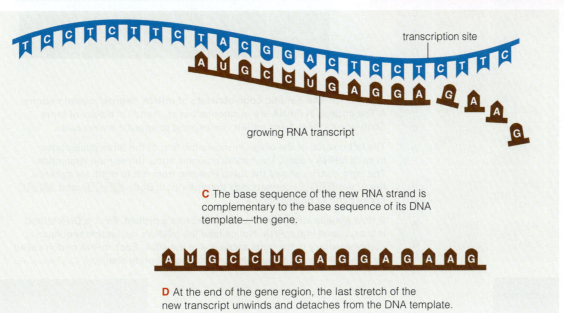

C The base sequence of the new RNA strand is complementary to the base sequence of its DNA template—the gene.

growing RNA transcript

transcription site

D At the end of the gene region, the last stretch of the new transcript unwinds and detaches from the DNA template.

21.4 The Genetic Code

- The sequence of nucleotides in an mRNA molecule is like a string of three-letter protein-building "words."
- Link to Primary structure of proteins 2.11

codon A set of three nucleotide bases in mRNA; they are the three-letter "words" of the genetic code.

genetic code The array of codons that provide instructions for making proteins.

start codon A codon that marks the start of a new amino acid chain during protein synthesis.

stop codon A codon that marks the end of a new polypeptide chain.

Codons are mRNA "words" for building proteins

Each "word" in the mRNA instructions for building a protein is a set of three nucleotide bases that are "read" by a cell's protein-making machinery. These *base triplets* are called **codons**. There are sixty-four kinds of codons (Figure 21.9A). Together they are the **genetic code**—a cell's direct instructions for making proteins. The order of different codons in an mRNA molecule determines the order of amino acids that are assembled into a protein (Figure 21.9B).

Gene by gene, mRNAs carry protein building instructions from the DNA in a cell's nucleus to the cytoplasm where amino acids are. A **start codon** marks the first amino acid of a new polypeptide chain. Most of the twenty kinds of amino acids can be ordered up by more than one start codon. (For example, glutamate corresponds to the code words GAA *or* GAG.) The codon AUG sets the "reading frame" for translation. That is, ribosomes start their "three-bases-at-a-time" selections at an AUG that is the start signal in an mRNA strand. Three different **stop codons** (UAA, UAG, and UGA) can signal ribosomes to stop adding amino acids to the growing chain.

TAKE-HOME MESSAGE

WHAT IS THE GENETIC CODE?

- The genetic code is a set of sixty-four different groups of three mRNA bases called codons.
- A cell's protein-making machinery "reads" codons, which specify different amino acids.
- Accordingly, mRNA provides the instructions for assembling amino acids into the polypeptide chains of proteins.

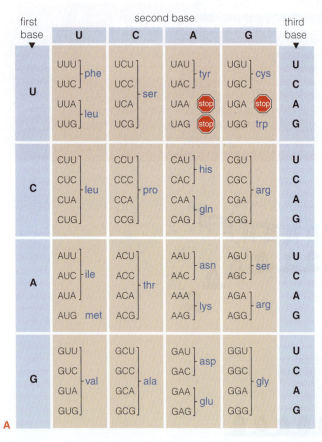

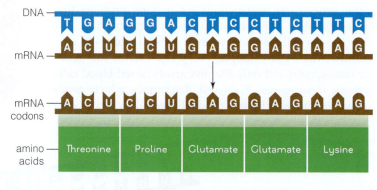

B

Figure 21.9 The genetic code consists of mRNA "words" called codons.
A The codons in mRNA are nucleotide bases, "read" in blocks of three. Sixty-one of these base triplets correspond to specific amino acids.

The left column of the diagram shows the first of the three nucleotides in each mRNA codon. The middle columns show the second nucleotide. The right column shows the third. Reading from left to right, for instance, the triplet U G G corresponds to tryptophan. Both U U U and U U C correspond to phenylalanine.

B How genetic information is converted into a protein. First, a DNA strand is transcribed into mRNA. Notice how the mRNA's nucleotide sequence is complementary to the gene sequence in the DNA. Each mRNA codon called for one amino acid in a growing protein (polypeptide chain). (© Cengage Learning)

- Transfer RNA and ribosomal RNA carry out the protein-building instructions of mRNA.
- Links to Primary protein structure 2.11 and The parts of a eukaryotic cell, 3.2

tRNAs match amino acids with mRNA "code words"

A cell's cytoplasm contains amino acids and tRNA molecules. Each tRNA has a "hook" site where it can attach to a specific amino acid. As shown in Figure 21.10, a tRNA also has an **anticodon**, a nucleotide triplet that can base-pair with codons. When a series of tRNAs bind to a series of codons, the matching up of codons and anticodons automatically lines up the amino acids attached to tRNAs in the order specified by mRNA.

A cell has more than sixty kinds of codons but fewer kinds of tRNAs. The needed match-ups occur anyway. Remember that by the base-pairing rules, adenine pairs with uracil, and cytosine with guanine. However, for codon–anticodon interactions, the rules loosen up for the third base. For example, only two tRNAs are needed to hook onto the codons CCU, CCC, CCA, and CCG, which all have instructions for making the amino acid proline.

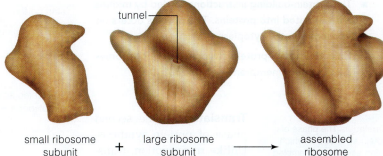

small ribosome subunit + large ribosome subunit → assembled ribosome

Figure 21.11 **Ribosomes consist of small and large subunits formed by rRNA.** Chains of amino acids—the primary structure of proteins, called polypeptide chains—are assembled on part of the small subunit. Newly forming chains move through a tunnel in the large subunit. (© Cengage Learning)

rRNAs are ribosome building blocks

Our discussion of cell organelles in Section 3.2 introduced you to ribosomes. In a cell's cytoplasm, tRNAs come into contact with mRNAs at binding sites on the surfaces of ribosomes. Each ribosome has two subunits, as you can see in Figure 21.11. The subunits are built in the nucleus from rRNA and proteins; then they are shipped to the cytoplasm. There they will combine into ribosomes. These temporary structures are the workbenches where proteins are assembled during translation, as described in the next section.

> **anticodon** A nucleotide triplet that can base-pair with codons. The matching of codons and anticodons lines up amino acids in the proper order in a new polypeptide chain.

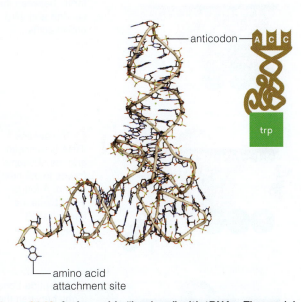

anticodon — A C C

trp

amino acid attachment site

Figure 21.10 **Amino acids "hook up" with tRNAs.** The model above represents a tRNA that carries the amino acid tryptophan. Each tRNA's anticodon is complementary to an mRNA codon. Each also carries the amino acid that is specified by the codon. (© Cengage Learning)

TAKE-HOME MESSAGE

WHAT DO rRNAS AND tRNAS DO?

- rRNAs and proteins form subunits that combine into ribosomes.
- Different kinds of tRNAs pick up amino acids that are used to build polypeptide chains.
- Different tRNAs bind specific codons in mRNA. In this way amino acids line up in the order specified by mRNA.
- Ribosomes are temporary structures where proteins are assembled.

The Three Stages of Translation

- The protein-building instructions carried by mRNAs are translated into proteins. This process occurs on ribosomes in the cytoplasm.

- Links to Primary protein structure 2.11, Ribosomes 3.2, The endomembrane system, 3.7

translation The phase of protein synthesis in which amino acids are assembled into a chain, the primary structure of a protein.

Translation is the second phase of protein synthesis. Unlike transcription, translation occurs in the cell cytoplasm. During its three stages, called initiation, elongation, and termination, mRNA transcribed from DNA is "translated" into a linear sequence of amino acids that forms a polypeptide chain. Figure 21.12 shows the steps involved.

To begin *initiation*, a tRNA that can start transcription is loaded onto a ribosome subunit. This initiator tRNA binds with the small ribosome subunit. AUG, the start codon for the mRNA transcript, matches up with this tRNA's anticodon. The AUG also binds with the small subunit, which in turn binds with a large ribosome subunit (Step 1). Now the next stage can begin.

In the *elongation* stage of translation, a polypeptide chain forms as the mRNA strand passes between the ribosome subunits, like a thread moving through the eye of a needle. Some proteins in the ribosome are enzymes. They join amino acids together in the sequence dictated by the codon sequence in the mRNA molecule. In Steps 2 through 6 you can see that a peptide bond forms between the most recently attached amino acid and the next amino acid being delivered to the ribosome. (Section 2.11 explains how a peptide bond forms.)

During the last stage of translation, *termination*, a stop codon in the mRNA moves onto the ribosome platform. No tRNA has a corresponding anticodon, so translation stops. Enzymes then detach the mRNA *and* the new chain from the ribosome.

Some cells, such as unfertilized oocytes, may have to be ready to make many copies of different proteins very quickly. As "preparation," they stockpile transcribed mRNA in their cytoplasm. In cells that are already rapidly using or secreting proteins (such as endocrine cells that are making hormones), *polysomes* are often present. A polysome is a cluster of ribosomes, all translating the same mRNA transcript at the same time. The transcript threads through all of them, one after another (Figure 21.13A and B).

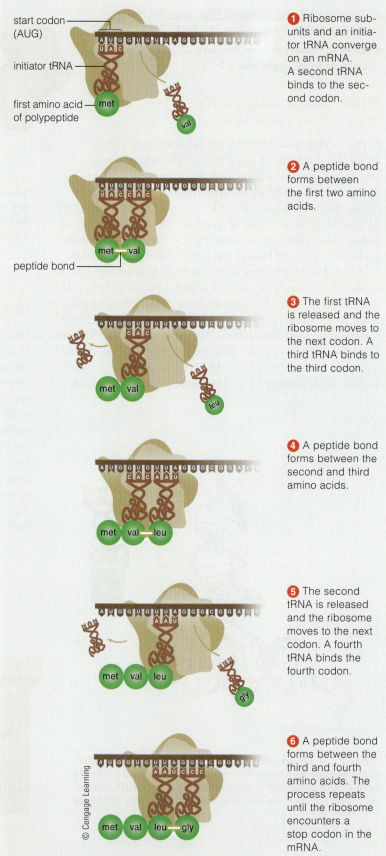

① Ribosome sub-units and an initiator tRNA converge on an mRNA. A second tRNA binds to the second codon.

② A peptide bond forms between the first two amino acids.

③ The first tRNA is released and the ribosome moves to the next codon. A third tRNA binds to the third codon.

④ A peptide bond forms between the second and third amino acids.

⑤ The second tRNA is released and the ribosome moves to the next codon. A fourth tRNA binds the fourth codon.

⑥ A peptide bond forms between the third and fourth amino acids. The process repeats until the ribosome encounters a stop codon in the mRNA.

Figure 21.12 Animated! Translation occurs in a series of steps.

© Cengage Learning

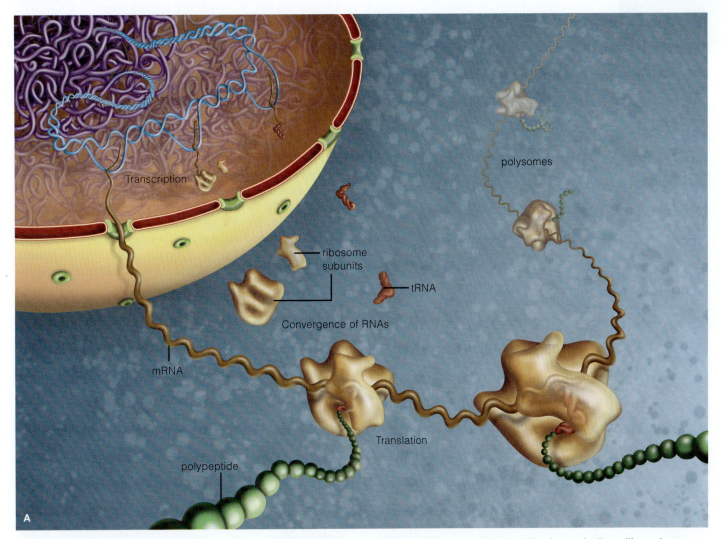

Transcription

ribosome subunits

tRNA

Convergence of RNAs

mRNA

polysomes

Translation

polypeptide

A

Figure 21.13 Translation takes place in a cell's cytoplasm. A Translation begins while transcription is still going on in the cell's nucleus. **B** The micrograph below shows clusters of ribosomes (polysomes) all translating DNA in a cell. (© Cengage Learning)

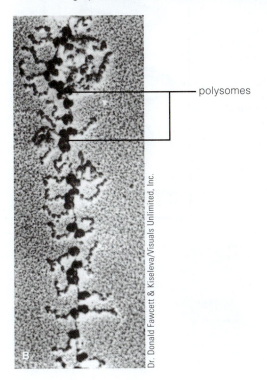

polysomes

B

Many newly forming polypeptide chains carry out their functions in the cytoplasm. Others have a "shipping label," a special sequence of amino acids. The label allows them to enter the rough ER of the endomembrane system (Section 3.7). There they are modified into their final form before being shipped to their ultimate destinations inside or outside the cell.

TAKE-HOME MESSAGE

HOW IS mRNA TRANSLATED INTO PROTEIN?

- Translation begins when a small ribosome unit and an initiator tRNA both arrive at an mRNA's start codon and a large ribosome subunit binds to them.

- tRNAs deliver amino acids to the ribosome in the order dictated by the sequence of mRNA codons. A chain of amino acids (a polypeptide chain) grows as peptide bonds form between the amino acids.

- Translation ends when a stop codon triggers events that cause the chain and the mRNA to detach from the ribosome.

- Researchers can use advanced technology to change genetic traits.
- Link to Cell division 18.2

genetic engineering Insertion of purposely modified genes into an organism.

genome All the DNA in one (haploid) set of a species' chromosomes.

PCR The polymerase chain reaction.

recombinant DNA A DNA molecule that contains DNA from more than one species.

Today researchers routinely cut and splice DNA from different species, then insert the modified molecules into cells such as bacteria that can replicate genetic material and divide. The cells copy the foreign DNA along with their own. Copying produces large quantities of **recombinant DNA** molecules. This recombinant DNA technology is used for **genetic engineering**, in which genes are altered and inserted back into the original organism or into a different one.

Enzymes and plasmids from bacteria are basic tools

Recombinant DNA technology depends on the genetic workings of bacteria. Many bacteria have small circular molecules of "extra" DNA called *plasmids*, which contain a few genes. The bacterium's replication enzymes can copy plasmid DNA. Bacteria also have *restriction enzymes*—enzymes that can detect and cut apart specific short sequences of bases in DNA. Today, plasmids and restriction enzymes are basic parts of a tool kit for doing genetic recombination in the laboratory.

Many restriction enzymes make staggered cuts that leave single-stranded "tails" on the end of DNA fragments. Depending on the molecule being cut, the fragments may be long enough to be useful for studying the organization of a genome. A **genome** is all the DNA in a haploid set of a species' chromosomes.

DNA fragments with staggered cuts have so-called "sticky" ends. This means that a restriction fragment's single-stranded tail can base-pair with a complementary tail of any other DNA fragment or molecule cut by the same restriction enzyme. If you mix together some DNA fragments cut by the same restriction enzyme, the sticky ends of fragments that have complementary base sequences will base-pair and form a recombinant DNA molecule. Another enzyme seals the nicks.

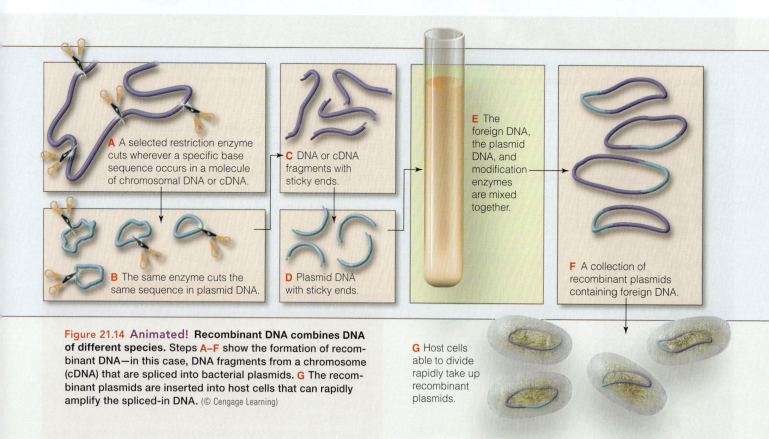

A A selected restriction enzyme cuts wherever a specific base sequence occurs in a molecule of chromosomal DNA or cDNA.

B The same enzyme cuts the same sequence in plasmid DNA.

C DNA or cDNA fragments with sticky ends.

D Plasmid DNA with sticky ends.

E The foreign DNA, the plasmid DNA, and modification enzymes are mixed together.

F A collection of recombinant plasmids containing foreign DNA.

G Host cells able to divide rapidly take up recombinant plasmids.

Figure 21.14 Animated! Recombinant DNA combines DNA of different species. Steps **A–F** show the formation of recombinant DNA—in this case, DNA fragments from a chromosome (cDNA) that are spliced into bacterial plasmids. **G** The recombinant plasmids are inserted into host cells that can rapidly amplify the spliced-in DNA. (© Cengage Learning)

Volker Steger/SPL/Photo Researchers, Inc.

Figure 21.15 Animated! PCR makes quick work of copying DNA. The photograph shows rows of PCR systems that are copying human DNA. *Right*: The steps of the polymerase chain reaction (PCR). (© Cengage Learning)

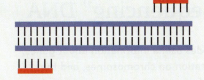

A DNA (blue) is mixed with primers (red), nucleotides, and heat-tolerant DNA polymerase.

B When the mixture is heated, DNA strands separate. When it is cooled, some primers bond to the template DNA.

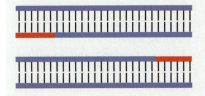

C DNA polymerase uses the primers to begin synthesis, and complementary strands of DNA form. The first round of PCR is now complete.

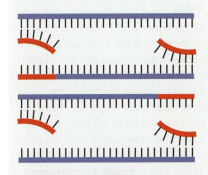

D The mixture is heated again, and all of the DNA separates into single strands. When the mixture is cooled, some of the primers bond to the DNA.

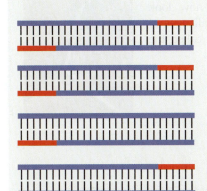

E DNA polymerase uses the primers to begin DNA synthesis, and complementary strands of DNA form. The second round of PCR is complete.

Each round can double the number of DNA molecules. After 30 rounds, the mixture contains huge numbers of DNA fragments, all copies of the starting DNA.

Using the necessary enzymes, it's possible to splice foreign DNA into bacterial plasmids. The result is called a DNA clone, because when a bacterium copies a plasmid, it makes many identical, "cloned" copies of it.

A DNA clone carries foreign DNA into a host cell that can divide rapidly (such as a bacterium or a yeast cell). This can be the start of a cloning "factory"—a population of rapidly dividing cloned cells, all with identical copies of the foreign DNA (Figure 21.14). As they divide they make much more of, or amplify, the foreign DNA.

PCR is a super-fast way to copy DNA

The *polymerase chain reaction*, or **PCR**, is an even faster way to copy DNA. These reactions occur in test tubes, and primers get them started.

A *primer* is a manmade, short nucleotide sequence that base-pairs with any complementary sequences in DNA. The workhorses of DNA replication—the DNA polymerases—chemically recognize primers as start tags. Following a computer program, machines make a primer one step at a time.

PCR also uses a DNA polymerase that is not destroyed at the high temperatures that are required to unwind a DNA double helix. (Such high temperatures will denature and destroy the activity of most DNA polymerases.)

In the lab, primers, the polymerase, DNA from an organism, and nucleotides are all mixed together. Next, the mixture is exposed to precise temperature cycles. During each temperature cycle, the two strands of all the DNA molecules in the mixture unwind from each other.

Primers line up on exposed nucleotides at the targeted site according to base-pairing rules (Figure 21.15). Each round of reactions doubles the number of DNA molecules amplified from the target site. For example, if there are 10 such molecules in the test tube, there soon will be 20, then 40, 80, 160, 320, and so on. Very soon there will be billions of copies of a target piece of DNA.

PCR can amplify samples that contain tiny amounts of DNA, and it is used in laboratories all over the world. As you'll read shortly, it can copy DNA from even a single hair follicle or a drop of blood left at a crime scene.

<table>
<tr><td>

21.8   "Sequencing" DNA

</td><td>

21.9   Mapping the Human Genome

</td></tr>
</table>

■ DNA sequencing provides useful information about genes, including their size, location on chromosomes, and the order of their nucleotides.

To study how a particular gene functions, what kind of mutations occur in it, and how it interacts with other genes,

DNA sequencing
Determining the order of nucleotides in DNA, typically using a supercomputer.

genomics The study of whole genomes.

it's useful to have information such as where the gene is on its chromosome, how many nucleotides are in it, and the order of the nucleotides. **DNA sequencing** gives this sort of information. Powerful supercomputers now can sequence DNA with astonishing speed.

The sequencing method uses standard and modified versions of the four nucleotides—A, T, G, and C. Each modified version has been attached to another molecule that fluoresces (lights up) a preselected color during the sequencing process. All the nucleotides are mixed with millions of copies of the DNA for which the sequence is to be determined, along with a primer and DNA polymerase. Next a series of chemical steps produce a "soup" containing millions of copies of DNA fragments, each one tagged with a fluorescing molecule. More processing produces fragments in which each nucleotide in the "mystery" base sequence fluoresces. After several more steps the computer program interprets the information from all the "marked" nucleotides in the sample and assembles the original DNA's sequence (Figure 21.16).

DNA sequencing is a major tool in the field of **genomics**, the study of genomes. We will now look at the promise and challenges of the multinational effort called the Human Genome Project.

TEK Image/Photo Researchers, Inc.

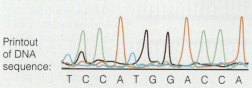

Printout of DNA sequence:

T C C A T G G A C C A

Figure 21.16 Animated! **The printout from a DNA sequencing machine matches colors with DNA nucleotides.** (© Cengage Learning)

TAKE-HOME MESSAGE

WHAT IS DNA SEQUENCING?

- DNA sequencing determines the order of nucleotides in a DNA fragment.
- Researchers who study genomes use the information from DNA sequencing, among other tools.

■ The Human Genome Project is providing valuable insights into the genetic basis of many disorders and diseases.

Thanks to the DNA sequencing of the **Human Genome Project**, we now know that the human genome consists of about 3.2 billion nucleotide bases—As, Ts, Gs, and Cs (Figure 21.17). The bases are subdivided into roughly 21,500 genes, which provide instructions to build and operate the body Researchers have not discovered what all the genes encode, only where they are in the genome.

One interesting discovery is that only about 1.5 percent of human DNA is devoted to the protein-coding parts of our genes, the exons. One challenge biologists now face is figuring out what noncoding gene regions do.

It also turns out that our DNA is sprinkled with SNPs ("snips"). Each SNP (for single *n*ucleotide *p*olymorphism) is a change in one nucleotide in a sequence. It appears there are around 1.4 million SNPs in the human genome. Many result in different gene alleles—the different versions of a gene that encode slightly different traits.

Mapping shows where genes are located

Unraveling the human genome has extremely important implications for human biology and medicine, in part because genome sequencing can identify where specific genes are located on chromosomes. For instance, we know that genes on chromosome 21 are responsible for early-onset Alzheimer's disease, some forms of epilepsy, and one type of **amyotrophic lateral sclerosis**, or ALS, a disease that destroys motor neurons (Figure 21.18). More than sixty disorders have been mapped to chromosome 14. A consortium of public and private laboratories has mounted a $100 million effort to correlate disorders with individual genetic

Volker Steger/SPL/Photo Researchers, Inc.

Figure 21.17 Supercomputers are used in human genome research. These gene-sequencing computers are at Celera Genomics in Maryland. Many new genetic tests make use of Celera's database.

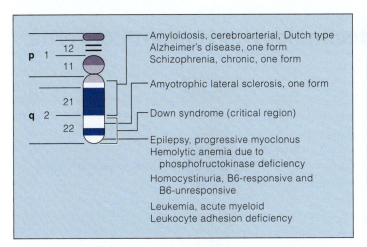

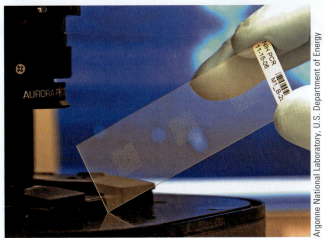

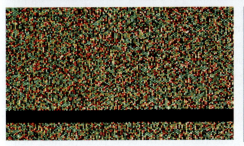

Figure 21.18 **A map of human chromosome 21 shows genes correlated with several diseases.** The upper arm of the chromosome is marked p and the lower arm q. In this case, each arm has a single region, labeled 1 and 2, respectively. In other chromosomes, each arm may have two or more regions. Stains used in chromosome analysis produce a series of bands, indicated by the small numbers to the left of each arm. A combination of letters and numbers indicates the chromosome region where a given gene is found; for instance, the gene for one form of amyotrophic lateral sclerosis (ALS) is 21q2.22. A total of 225 genes have been mapped to this chromosome. (© Cengage Learning)

Figure 21.19 **DNA chips, like those stamped on this glass plate, can be used to analyze thousands of genes at once.** Below you see a small portion of a "SNP-chip," a type of microarray that reveals SNPs. Analysis of SNPs is useful in screening individuals for mutations associated with increased risk of certain diseases.

differences. Appendix VI at the end of this book shows maps for all 23 human chromosomes and provides information you can use to research human genetic diseases on your own.

DNA chips help identify mutations and diagnose diseases

One key tool is the **DNA chip**, a microscopic array (microarray) of thousands of DNA sequences that are stamped onto a small glass plate (Figure 21.19). The chips can quickly pinpoint which genes are being expressed in a tissue, including tissues such as cancerous tumors. DNA chips, including so-called "SNP-chips," are now being used to locate mutations, diagnose genetic diseases, and test how drugs or other therapies affect the functioning of genes.

As new genes are identified, biologists are exploring the roles of proteins they encode. We already know how having a particular allele can set the course of diseases such as sickle-cell anemia and forms of breast cancer. SNP-chips now are used to test patients for genetic mutations associated with increased breast cancer risk. Soon it will be possible to apply the deeper understanding of human genes to many more health concerns. For instance, a simple blood test might be able to provide a complete genetic profile of a person's inborn predisposition to heart disease, asthma, diabetes, and certain cancers. Armed with your profile, your doctor might be able to diagnose and treat problems earlier and more effectively. Drugs could be customized for various genetic situations, as described in Section 20.7. This sort of personalized medicine is rapidly becoming a reality.

Along with the promise of genome sequencing come serious cautions. Many people worry that genetic profiling could lead to discrimination against people seeking employment or insurance, based solely on their genetic makeup. New issues could arise in the genetic screening of embryos, already a major ethical concern to some. Clearly, a challenging genetic future awaits us.

DNA chip A glass plate containing a microscopic array of DNA sequences.

Human Genome Project A research project that determined the approximate number of human genes and the order of nucleotides in them.

- Practical applications of biotechnology are coming along almost as fast as the leaps in our understanding of human genetics. They include genetic fixes for diseases and using DNA as a personal identifier.

- Link to Inheritance of genes on autosomes 20.5

Researchers are exploring gene therapy

There are about 15,500 known human genetic disorders. Some, such as cystic fibrosis and hemophilia, develop when one single gene mutates, producing a malfunctioning protein. Many cancers also arise following changes in single genes. **Gene therapy** aims to replace mutated genes with normal ones that will encode functional proteins, or to insert genes that restore normal controls over gene activity. Scientists around the world are exploring these possibilities.

Genes can be inserted two ways

The size of a gene (how many base pairs it has) helps determine how it might be inserted into a host cell. Smaller genes can be carried into animal cells by a *vector* such as a virus. Larger genes must enter a host cell some other way.

DNA profile A set of repeated DNA fragments unique to each person.

DNA fingerprint A DNA profile used in criminal investigations.

gene therapy Inserting one or more normal genes into body cells to correct a genetic defect or enhance the activity of specific genes.

In *transformation*, cells grown in the laboratory are exposed to DNA that contains a gene of interest, and some of the foreign DNA may become integrated into the host cell's genome. Exposing the host cells to a weak electric current seems to help. Even so, sometimes only one cell in 10 million takes up a new gene.

In *transfection*, a gene is inserted into a virus. The first step is to remove from the virus its genetic instructions that would allow it to replicate and cause disease. In their place goes the gene to be transferred. Next, the virus is allowed to infect target cells (Figure 21.20). Once the virus is inside the host cell, the desired foreign DNA usually becomes integrated into the host cell's DNA.

Transfection can be used only with genes that are expressed in the tissues into which the new DNA will be inserted. In addition, many introduced genes turn off within a few days or weeks. A few patients have died from complications of the procedure, which raises serious safety questions.

Gene therapy results have been mixed

The first federally approved gene therapy test on humans began in the early 1990s. It aimed to replace a defective gene that causes one type of **severe combined immune deficiency**, called SCID-X1. In this disorder, stem cells

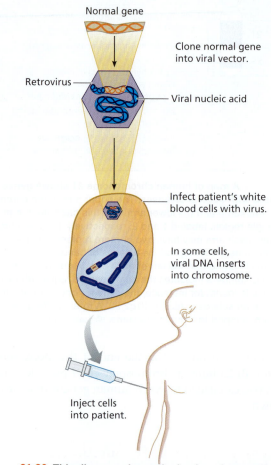

Figure 21.20 This diagram shows the basics of one gene therapy method. Here, white blood cells are removed from a patient with a genetic disorder. Next, a retrovirus is used as a vector to insert a normal gene into the DNA. After the normal gene begins to produce its protein, the cells are placed back into the patient. If enough of the normal protein is present in the patient, the disorder's symptoms diminish. Because the genetically altered blood cells live but a few months, however, the procedure must be repeated regularly. (© Cengage Learning)

in the affected person's bone marrow fail to make the immune system's infection-fighting lymphocytes (Section 9.1). Affected children (sometimes called "bubble babies") must live in germ-free isolation tents. Some years ago, a virus was used to insert copies of a normal allele into stem cells taken from the bone marrow of eleven children with the disorder. Then the genetically modified stem cells were infused back into the marrow, which began to produce lymphocytes. Seven of the treated children, including Rhys Evans, the boy shown in Figure 21.21, eventually developed working immune systems. Later on, however, two of the children developed leukemia as a side effect of of their treatment.

Cystic fibrosis has been another target. The idea is to introduce normal copies of the defective CFTR gene into cells of the respiratory system, using a viral vector in a nasal spray. In the trials thus far, only about 5 percent of

© Jeans for Gene Appeal

Figure 21.21 **Gene therapy helped Rhys Evans, born with SCID-X1.** His immune system did not develop properly, so his body couldn't fight infections. A gene transfer restored his immune system.

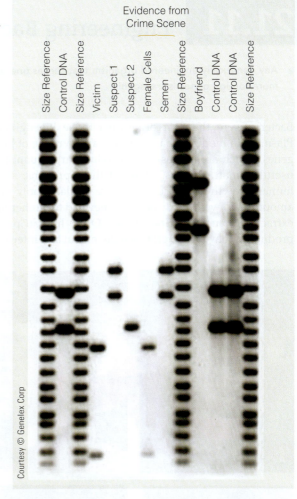

Figure 21.22 **Animated!** These DNA fingerprints compare DNA gathered during the investigation of a sexual assault. The assay compared DNA from the victim and from the semen of two suspects and the victim's boyfriend. Can you see how only one exactly matches the pattern from the crime scene? Three control samples were included to confirm that the test procedure was working correctly.

Courtesy © Genelex Corp

affected cells have taken up the normal gene. The retrovirus vectors used are not effective enough at delivering the new genes to cells that need them.

More recently, patients with a congenital genetic disorder that causes blindness underwent gene therapy that restored limited vision. To date, however, gene therapy has been most successful not for "curing" genetic disorders but for treating cancer. Trials have targeted malignant melanoma, leukemia, a fast-growing form of lung cancer, and cancers of the brain, ovaries, and other organs. In some approaches, tumor cells are first removed from a patient and grown in the laboratory. Genes for an interleukin (which helps activate T cells of the immune system) are then introduced into the cells, and the cells are returned to the body. In theory, interleukins produced by the tumor cells may act as "suicide tags" that stimulate T cells to recognize cancerous cells and attack them.

At present, gene therapy is still experimental, costly, and available to only a few carefully selected patients. It will probably be years before it is widely used for treating and curing disease.

Genetic analysis also is used to read DNA fingerprints

Other than identical twins, no two people have exactly the same sequence of bases in their DNA. Thus each of us has a **DNA profile**—a unique set of certain DNA fragments. When used in the context of a criminal investigation, DNA profiling produces a **DNA fingerprint**. DNA fingerprints are so accurate that they can easily distinguish between tissues from full siblings.

More than 99 percent of human DNA is exactly the same in all people, regardless of race and gender. Thus DNA fingerprinting focuses only on the part that tends to differ from one person to the next. Throughout the human genome are short regions of repeated DNA that are very different from person to person. Each person has a unique combination of repeats.

Forensic scientists use DNA fingerprinting to identify criminals and crime victims from the DNA in blood, semen, or bits of tissue left at a crime scene. Figure 21.22 shows DNA fingerprints that were analyzed in an effort to identify the perpetrator of a rape. DNA fingerprinting was a key tool in identifying remains of people killed at the World Trade Center on September 11, 2001.

THINK OUTSIDE THE BOOK

DNA analysis has suggested that Thomas Jefferson, third president of the United States, fathered one or more children by his slave Sally Hemings. A 2007 article in the *University of Virginia Magazine* looks at both sides of this controversial issue. Check out "Anatomy of a Mystery" at archives.uvamagazine .org. Are you convinced by the arguments on either side of the issue?

TAKE-HOME MESSAGE

WHAT ARE SOME WAYS SCIENTISTS APPLY BIOTECHNOLOGY TO ISSUES OF CONCERN TO HUMANS?

• Biotechnology is used in experimental gene therapy and DNA profiling.

21.11 Engineering Bacteria, Animals, and Plants

- Any genetically engineered organism that carries one or more foreign genes is transgenic.

Bacteria were the first organisms to be bioengineered. Plasmids in modified bacteria can carry a range of human genes, which are expressed to produce large quantities of useful human proteins. Many of these proteins, such as human growth hormone, once were available only in tiny amounts and were costly because they had to be chemically extracted from endocrine tissues. Other human proteins produced today by bacteria include insulin and interferons.

Many types of animal cells can be "micro-injected" with foreign DNA. For instance, when researchers introduced the gene for human growth hormone into mice, the result was the "super mouse" shown in Figure 21.23A. Recombinant DNA technology also has been used to transfer human and cow growth hormones into pigs, which then grow much faster. Transgenic goats (Figure 21.23B) produce CFTR protein (used to treat cystic fibrosis), as well as drugs that prevent dangerous blood clots. These and other medically useful engineered proteins show up in the goats' milk.

People with hemophilia A now can obtain the needed blood-clotting factor VIII from a drug that is produced by hamster ovary cells in which genes for human factor VIII have been inserted. Factor VIII produced in this way eliminates the need to obtain it from human blood, and thus the risk of transmitting blood-borne diseases.

Plants have been intriguing genetic engineers for a long time. Researchers now routinely grow crop plants and many other plant species from cells cultured in the laboratory. They use a variety of methods to pinpoint genes that confer useful traits, such as resistance to salt, a pathogen, or an herbicide (Figure 21.23C). Later, whole plants with the trait can be grown from the cultured cells.

Figure 21.23 Animated! These are just a few of the organisms with new traits bestowed by genetic engineering. **A** Mouse littermates. The larger mouse grew from a fertilized egg into which the gene for human growth hormone had been inserted. **B** Myra, a goat transgenic for human antithrombin III, an anticlotting factor. **C** A bacterial gene conferred insect resistance to the corn in the photograph at left. Farmers who grow this GM corn don't need to apply as much pesticide to their fields. Corn in the photograph at right was not genetically modified and is more vulnerable to pests. **D** The chart lists a few genetically modified crop plants approved by the USDA.
(A: R. Brinster and R. E. Hammer, School of Veterinary Medicine, University of Pennsylvania B: Transgenic goat produced using nuclear transfer at Biotherapeutics. Photo used with permission C: Bt and Non-Bt corn, part of field trial conducted on the main campus of Tennessee State University at the Institute of Agricultural and Environmental Research. The work was supported by a competitive grant from the CSREES, USDA titled *Southern Agricultural Biotechnology Consortium for Underserved Communities* (2000–2005). Dr. Fisseha Tegegne and Dr. Ahmad Aziz served as Principal and Co-principal Investigators respectively to conduct the portion of the study in the State of Tennessee.)

USDA-approved Crop Plant	Modified Trait
Tomato, potato, corn, rice, sugar beet, canola bean, cotton, flax	Resistance to weed-killing herbicides used in agriculture croplands
Potato, squash, papaya	More resistance to harmful viruses, bacteria, and fungi
Tomato	Delayed ripening; easier to ship, with less bruising
Corn, chicory **D**	Plants cannot interbreed with wild stocks

TAKE-HOME MESSAGE

WHAT TYPES OF NONHUMAN ORGANISMS HAVE BEEN BIOENGINEERED?

- Scientists have created transgenic bacteria, plants, and numerous nonhuman animals.
- Some of these genetic engineering efforts aim to develop transgenic animals or animal cells capable of producing medically useful substances.
- Genetically engineered plants have a variety of traits, such as resistance to disease and herbicides.

Many people believe that genetic technologies are a double-edged sword: they can be used for good or ill. Some say that no matter what the species of organism, DNA should never be altered—even though natural mutations change DNA all the time. Cloning, or making a genetic copy of a cell or organism, has become another issue. Geneticists have become increasingly expert in applying this technology, from the cloned bacteria used in recombinant DNA technology to embryos cloned to obtain stem cells and cloned adult animals.

Cloning of bacteria, plants, and nonhuman animals raises concerns

There are pros and cons associated with many current uses of biotechnology. For instance, most recombinant bacteria or viruses are altered in ways that will prevent them from reproducing outside a laboratory. In theory, though, transgenic bacteria or viruses could mutate and possibly become new pathogens in the process. On the other hand, genetically engineered "oil-eating" bacteria have been used to help clean up oil spills, an example of *bioremediation*.

The chapter introduction mentioned some objections to bioengineered plants. Critics also point out that such plants could escape from test plots and become "superweeds" that are resistant to herbicides and other controls. It is also possible that crop plants with engineered insect resistance could trigger the evolution of new, even worse pests. Although experts deem many, even most, of these possibilities to be unlikely, there are documented cases of engineered plant genes turning up in wild plants.

Controversy swirls over cloning

Scientists have achieved considerable success in cloning embryos and whole animals of several species. The 1997 cloning of a fully grown sheep named Dolly has been followed by a string of similar experiments that have produced clones of cattle, rabbits, pigs, mules, goats, and other animals, including cats (Figure 21.24). Uses for adult animal clones range from saving endangered species from extinction to cloning livestock with desirable traits to "replacing" a beloved pet that has died.

To make a clone of an adult animal, the nucleus of an unfertilized egg cell is removed and replaced with the nucleus from some type of adult cell. An embryo then may develop from the cell. In theory, the embryo can become a source of stem cells (Chapter 4), or it may grow into an adult. Either way, the clone contains only the DNA of the original adult cell.

To date many clones have developed serious health problems and have aged much more rapidly than usual. This happens when the substituted adult nucleus does not revert to an embryonic state (when a cell is not yet differentiated for its final function in the body). Getting a nucleus to "reprogram" in this way is tricky and often is not successful.

Even so, efforts are under way in many laboratories to perfect methods for making cloned embryos as well as healthy adult animal clones. Such embryos might be used for **therapeutic cloning**—that is, as a source of embryonic stem cells that can be used to grow replacement human tissues and organs. Some companies have announced plans for **reproductive cloning**—creating a cloned embryo that can be implanted in a woman's uterus and allowed to develop into a baby. So-called germline engineering would take the technology a step further by altering the cloned embryo's genetic makeup in some desired way before implanting it.

We humans have been doing genetic experiments for centuries by manipulating matings to create modern crop plants and new breeds of animals. Going forward, the real issue will be how to bring about beneficial changes without doing harm.

reproductive cloning
Creating a cloned embryo to produce a pregnancy.

therapeutic cloning
Creating a cloned embryo that will be used as a source of embryonic stem cells.

Photos by Victor Fisher, courtesy Genetic Savings & Clone

Figure 21.24 Cloned cats and other animals are a modern-day reality. The owner of Tahini, a Bengal cat (*top*), founded Genetic Savings and Clone, a biotechnology company. The two kittens (*bottom*) are exact clones of Tahini. As they grow older their eye color will become exactly like Tahini's.

EXPLORE
ON YOUR OWN

In some parts of the world about 140 million children under the age of six suffer blindness and other serious health problems due to a vitamin A deficiency.
Eating one cup a day of genetically modified Golden Rice (*right*) can correct the deficiency. This GM rice has been engineered to contain beta carotene, a pigment formed in some plants that is a precursor for the formation of vitamin A. Rice is a cheap staple food in many regions where poor people suffer the most from vitamin A deficiency. For them Golden Rice might make the difference between relative good health and a life of misery. Even so, opponents of GM food crops have campaigned strongly against allowing human consumption of Golden Rice.

To explore this topic further, do some Web research on GM foods. Do you buy products that are likely to contain material from genetically modified plants? If you live in an agricultural area, do farmers there plant GM crops?

Golden Rice

SUMMARY

Section 21.1 A gene is a sequence of nucleotide bases in DNA. These bases are adenine, thymine, guanine, and cytosine (A, T, G, and C). The nucleotide sequence of most genes codes for the sequence of amino acids in a protein (polypeptide chain).

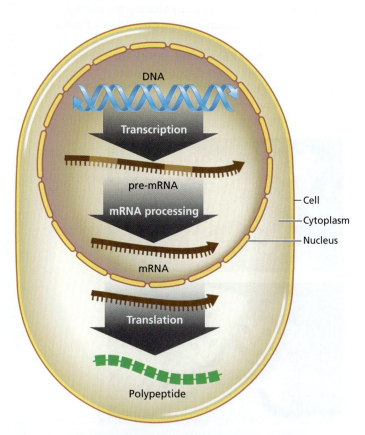

Figure 21.25 This diagram is a summary of transcription and translation, the two steps leading to protein synthesis in cells. DNA is transcribed into RNA in the nucleus. RNA is translated into proteins in the cytoplasm. (© Cengage Learning)

Section 21.2 A DNA molecule consists of two strands of nucleotides twisted together in a double helix. DNA is copied (duplicated) by semiconservative replication. One strand in each new DNA molecule is new and one is from the parent molecule. DNA polymerases and other enzymes unwind the existing DNA molecule, keep the strands apart, and assemble a new strand on each one.

A gene mutation is a change in the DNA nucleotide sequence. Mutations may be harmful, neutral, or helpful. Mutations in germ cells (cells that produce gametes) can be passed to the next generation.

Section 21.3 Converting genetic information into a protein requires steps called transcription and translation, which are summarized in Figure 21.25. In transcription, DNA instructions guide the formation of RNA from nucleotides in the cell. The double-stranded DNA is unwound at a gene region, then RNA polymerases use the exposed bases as a template to build a corresponding strand of RNA. Base-pairing rules govern which bases pair up. In RNA, guanine pairs with cytosine, and uracil (not thymine) pairs with adenine.

DNA encodes three kinds of RNA molecules. Messenger RNA (mRNA) carries instructions for building proteins. Ribosomal RNA (rRNA) forms the subunits of ribosomes, the structures on which amino acids are assembled into polypeptide chains. Different kinds of transfer RNA (tRNA) pick up amino acids and deliver them to ribosomes in the order specified by mRNAs.

A new mRNA transcript consists of introns (nucleotide sequences that do not code for proteins) and exons. Exons are the mRNA sequences that carry protein-building instructions. Regulatory proteins stimulate or suppress gene transcription and so control gene activity.

Section 21.4 In translation, RNAs link amino acids in the sequence required to produce a specific polypeptide chain.

Translation follows the genetic code, a set of sixty-four base triplets—nucleotide bases that ribosome proteins "read" three at a time.

A base triplet in an mRNA molecule is a codon. A given combination of codons specifies the amino acid sequence of a polypeptide chain.

Sections 21.5, 21.6 Translation has three stages. In initiation, a small ribosome subunit and an initiator tRNA bind with an mRNA transcript and move along it until they reach an AUG start codon. The small subunit binds with a large ribosome subunit.

In elongation, tRNAs deliver amino acids to the ribosome. Their anticodons base-pair with mRNA codons. The amino acids are joined (by peptide bonds) to form a new polypeptide chain.

In chain termination, an mRNA stop codon moves onto the ribosome; then the polypeptide chain and mRNA detach from the ribosome.

Section 21.7 Recombinant DNA technology is the foundation for genetic engineering. Restriction enzymes are used to cut DNA molecules into fragments, which are inserted into a cloning vector (such as a plasmid) and then multiplied in rapidly dividing cells.

A DNA clone is a foreign DNA sequence that has been introduced and amplified in dividing cells. DNA sequences also can be amplified in test tubes by the polymerase chain reaction.

The genome of a species is all the DNA in a haploid set of chromosomes. The human genome, including our species' 21,500 genes, consists of about 3.2 billion nucleotides. Some tools of recombinant DNA technology can be used to identify genes in the genome.

Sections 21.8, 21.9 Automated DNA sequencing can quickly determine the sequence of nucleotides in segments of DNA—and, accordingly, in genes. The human genome has been sequenced, and researchers are identifying genes responsible for a variety of traits and genetic disorders.

Sections 21.10, 21.11 Recombinant DNA technology and genetic engineering have enormous potential for research and applications in medicine, agriculture, and industry. Both also may pose ecological and social risks.

REVIEW QUESTIONS

1. Why is DNA replication called "semiconservative"?

2. Name one kind of mutation that produces an altered protein. What determines whether the altered protein will have beneficial, neutral, or harmful effects?

3. How are the polypeptide chains of proteins that are specified by DNA assembled?

4. How does RNA differ from DNA?

5. Name the three classes of RNA and describe their functions.

6. Distinguish between a codon and an anticodon.

7. Describe the three steps of translation.

8. What is a restriction enzyme?

9. What is a "gene sequence"?

SELF-QUIZ   Answers in Appendix V

1. Fill in the blank: Nucleotide bases, read _____ at a time, serve as the "code words" of genes.

2. DNA contains genes that are transcribed into _____.
   a. proteins   d. tRNAs
   b. mRNAs   e. b, c, and d
   c. rRNAs

3. mRNA is produced by _____.
   a. replication   c. transcription
   b. duplication   d. translation

4. _____ carries coded instructions for an amino acid sequence to the ribosome.
   a. DNA   c. mRNA
   b. rRNA   d. tRNA

5. tRNA _____.
   a. delivers amino acids to ribosomes
   b. picks up genetic messages from rRNA
   c. synthesizes mRNA
   d. all of the above

6. An anticodon pairs with the bases of _____.
   a. mRNA codons   c. tRNA anticodons
   b. DNA codons   d. amino acids

7. The loading of mRNA onto the small ribosomal subunit occurs during _____.
   a. initiation of transcription c. translation
   b. transcript processing   d. chain elongation

8. Use the genetic code (Figure 21.9) to translate the mRNA sequence AUGcgcaccucaggaugagau. (Human reading frames start with AUG.) Which amino acid sequence is being specified?
   a. meth-arg-thr-ser-gly-stop-asp . . .
   b. meth-arg-thr-ser-gly . . .
   c. meth-arg-tyr-ser-gly-stop-asp . . .
   d. none of the above

9. Match the terms related to protein building.
   _____ alters genetic   a. initiation, elongation,
           instructions   termination
   _____ codon   b. conversion of genetic
   _____ transcription   messages into
   _____ translation   polypeptide chains
   _____ stages of   c. base triplet for an amino
           transcription,   acid
           translation   d. RNA synthesis
                                e. mutation

10. Rejoined cut DNA fragments from different organisms are best known as _____.
    a. cloned genes
    b. mapped genes
    c. recombinant DNA
    d. conjugated DNA

CRITICAL THINKING

1. Which mutation would be more harmful: a mutation in DNA or one in mRNA? Explain your answer.

2. Jimmie's Produce put out a bin of tomatoes having beautiful red color and looking "just right" in terms of ripeness. A sign above the bin identified them as genetically modified produce. Most shoppers selected unmodified tomatoes in the neighboring bin, even though those tomatoes were pale pink and hard as rocks. Which ones would you pick? Why?

3. Previous chapters have discussed various types of cloning, including the cloning of stem cells and of embryos. Dolly, the ewe shown with her first lamb in Figure 21.26, grew from a cloned cell. Is cloning the same as genetic engineering? Explain your answer.

Figure 21.26 Transgenic animals have become important research subjects. The cloned sheep Dolly is shown here with her first lamb. The lamb was conceived the old-fashioned way.

Your Future

The Human Genome Project examined the DNA of only a few people. In fact, about 80 percent of the gene sequences it identified came from just eight individuals. Now an effort called the 1,000 Genomes Project aims to sequence the DNA of 1,000 anonymous participants from around the world. Already, several thousand new DNA sequences have been discovered. They represent genetic variations that weren't in the DNA sequenced by the Human Genome Project. As the 1,000 Genomes Project continues, it will provide a much more complete picture of the genetic variety in humans—information that is almost sure to translate into better understanding of the roles genes play in health and disease.

KEY CONCEPTS

Cancer: Uncontrolled Cell Division

Cancer cells are abnormal in both structure and function. Cancer develops when gene changes remove the normal controls over cell division. Sections 22.1–22.4

Diagnosis and Screening

Cancer is diagnosed by biopsy and other tools. Early detection increases the chances of successful treatment. Section 22.5

Treatment and Prevention

Cancer treatments include surgery, chemotherapy, radiation, and immunotherapy. Lifestyle decisions that promote health can limit a person's risk of developing cancer. Section 22.6

Cancer strikes one in three people in the United States and kills one in four. According to the American Cancer Society, there are about 1,500 cancer deaths every day, over half a million annually.

Overall, more males than females develop cancer, but the pattern varies depending on the cancer type. Henrietta Lacks, shown in the old photograph at right, had cervical cancer. In 1951, researchers George and Margaret Gey at Johns Hopkins University were trying to grow a line of cancer cells that could be kept alive indefinitely in the lab and used for scientific study. Their efforts failed until they obtained a sample of Henrietta Lacks's cancerous cells, dubbed HeLa cells for short. Runaway cell division is the hallmark of cancer, but the HeLa cells divided especially aggressively. They were still dividing in the Geys' laboratory a few months later when Mrs. Lacks's cancer killed her at age thirty-one.

Today, more than half a century later, descendants of the cancerous cells of Henrietta Lacks's are still dividing and being used in research all over the world. The longevity of the cell line is a testament to cancer's capacity to sidestep normal genetic controls that regulate cell division.

Intensive research is rapidly increasing our understanding of many kinds of cancer, including those of the breast, ovary, colon, and skin. Most cancers are treatable, and many are curable if the disease is discovered early.

Henrietta's son David with a photograph of his parents

© Bill Denison Photography

The Characteristics of Cancer

- As genes switch on and off, they determine when and how fast the cell will grow and divide, when it will stop dividing, and even when it will die. Cancer can result when controls over cell division are lost.

- Links to Cell structure 3.2, Cell differentiation 17.1, Mutations in DNA 21.2

Some tumors are cancer, others are not

If cells in a tissue overgrow—an abnormal enlargement called **hyperplasia**—the result is a defined mass of tissue called a **tumor**. Technically, a tumor is a *neoplasm*, which means "new growth."

A tumor may not be "cancer." As Figure 22.1A shows, the cells of a *benign* tumor are often enclosed by a capsule of connective tissue, and inside the capsule they are organized in an orderly way. They also tend to grow slowly and to be well differentiated (structurally specialized for

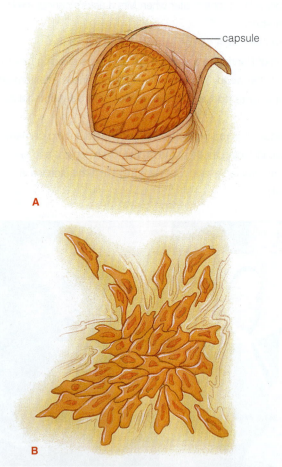

— capsule

A

B

Figure 22.1 Cancer cells are abnormal in their growth and appearance. A Sketch of a benign neoplasm. Cells appear nearly normal, and connective tissue encapsulates the tumor. **B** A cancerous neoplasm. Due to the abnormal growth of cancer cells, the tumor is a disorganized heap of cells. Some of the cells may break off and invade surrounding tissues, a process called metastasis. (© Cengage Learning)

TABLE 22.1 Main Features of Benign and Malignant Tumors		
	Malignant Tumor	**Benign Tumor**
RATE OF GROWTH	Rapid	Slow
NATURE OF GROWTH	Invades surrounding tissue	Expands in the same tissue
SPREAD	Metastasizes via the bloodstream and the lymphatic system	Does not spread
CELL DIFFERENTIATION	Usually poor	Nearly normal

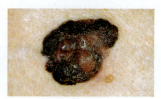

A Benign mole   B Melanoma

Figure 22.2 Normal moles are examples of benign growths. A Harmless moles, like this one, are all one color, symmetrical, and have a smooth edge. **B** Malignant melanomas are asymmetrical (they look blobby), have a ragged edge, and often have differently colored areas. A "mole" with these characteristics is suspicious and should be evaluated right away by a doctor. (A: National Cancer Institute/Photo Researchers, Inc. B: James Stevenson/Photo Researchers, Inc.)

a particular function), much like normal cells of the same tissue (Section 17.1). Benign tumors usually stay put in the body, push aside but don't invade surrounding tissue, and generally can be easily removed by surgery. Benign tumors *can* threaten health, as when they occur in the brain. Nearly everyone has at least several of the benign tumors we call *moles*. Most of us also have or have had some other type of benign neoplasm, such as a cyst. A *malignant* growth, by contrast, is potentially harmful. Table 22.1 compares the main features of malignant and benign tumors, and Figure 22.2 shows the outward differences between a harmless mole and a malignant melanoma, the most dangerous skin cancer.

Dysplasia ("bad form") is an *abnormal* change in the sizes, shapes, and organization of cells in a tissue. Such change is often an early step toward **cancer**. Under the microscope, the edges of a cancerous tumor usually look ragged (Figure 22.1B), and its cells form a disorganized clump. Most cancer cells also have characteristics that enable them to behave differently from normal cells.

A cancer cell's structure is abnormal

Cancer is the result of a series of mutations in a cell's genes, as you will read in Section 22.2. One effect of these

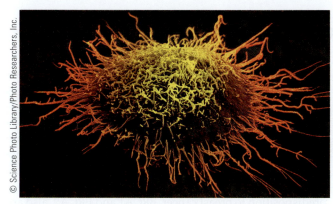

Figure 22.3 **This cervical cancer cell has the threadlike "false feet" that are a common feature of cancerous cells.** Color has been added to this photograph.

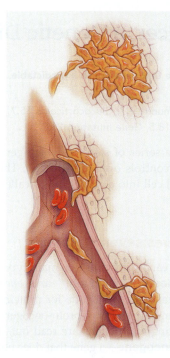

A Cancer cells break away from their home tissue.

B The metastasizing cells become attached to the wall of a blood vessel or lymph vessel. They secrete enzymes that break down part of the wall. Then they enter the vessel.

C Cancer cells creep or tumble along inside blood vessels, then leave the bloodstream the same way they got in. They start new tumors in new tissues.

Figure 22.4 **Animated!** **Cancer spreads step by step.**
(© Cengage Learning)

changes is that a cancer cell's structure is abnormal. Often, the nucleus is much larger than usual and there is much less cytoplasm. Cancer cells also often do not have the structural specializations of healthy cells in mature body tissues. As a general rule, the less specialized cancer cells are, the more likely they are to break away from the primary tumor and spread the disease.

When a normal cell is transformed into a cancerous one, more changes take place. The cytoskeleton shrinks, becomes disorganized, or both. Proteins that are part of the plasma membrane are lost or altered, and new, different ones appear. These changes are passed on to the cell's descendants: When a transformed cell divides, its daughter cells are cancerous cells too.

Cancer cells also do not divide normally

Contrary to popular belief, cancer cells don't necessarily divide more rapidly than normal cells do, but they do increase in number faster. This is because the death of normal cells usually closely balances the production of new ones by mitosis. In a cancer, however, at any given moment more cells are dividing than are dying. As this runaway cell division continues, the cancer cells do not respond to crowding, as normal cells do. A normal cell stops dividing once it comes into contact with another cell, so the arrangement of cells in a tissue remains orderly. By contrast, a cancer cell keeps on dividing. Therefore, cancer cells pile up in a disorganized heap. This is why cancer tumors are often lumpy.

Cancer cells also do not stay well connected physically to the cells next to them in a tissue, and they may form extensions (pseudopodia, "false feet") that enable them to move about (Figure 22.3). These extensions allow cancer cells to break away from the parent tumor and invade other tissues, including the lymphatic and circulatory systems (Figure 22.4). The spread of cancer is called **metastasis**. It is what makes a cancer malignant.

Some kinds of cancer cells produce the hormone HCG, human chorionic gonadotropin. (Recall from Chapter 17 that HCG maintains the uterus lining when a pregnancy

begins.) The presence of HCG in the blood can serve as a red flag that a cancer exists somewhere in a person's body.

Some cancer cells produce a chemical that stimulates cell division, and the cells themselves have receptors for that chemical. Cancer cells also secrete a growth factor called angiogenin that encourages new blood vessels to grow around the tumor. The blood vessels can "feed" the tumor with the large supply of nutrients and oxygen it needs to continue growing. Patients with certain types of cancer may receive drugs that essentially starve tumors to death by blocking the effects of angiogenin, although the drugs can have serious side effects.

cancer Disease state in which cells divide in an uncontrolled manner and develop other abnormal biological features.

dysplasia Abnormal change in the sizes, shapes, and organization of cells in a tissue; often a step toward cancer.

hyperplasia Overgrowth of cells in a tissue.

metastasis The process in which cancer cells spread from one part of the body to another.

tumor A defined mass of tissue formed as cells of the tissue overgrow.

Cancer, a Genetic Disease

- Cancer is a genetic disease that develops in a predictable sequence of steps.

- Links to Cell-mediated immunity 9.7, Glucocorticoids 15.7, Cell cycle 18.2, Radiation 18.5, Gene mutation 21.2

Cancer develops through a series of steps in which gene mutation removes normal controls over cell division. The transformation of a normal cell into a cancer cell is called **carcinogenesis**.

Cancer usually involves several genes

As a rule, the beginning of cancer involves two main types of genes. **Proto-oncogenes** (*proto,* "before," and *onco,* "mass") are genes in normal cells. They code for proteins that act in cell division. If a mutation alters a proto-oncogene or the way its protein-making instructions are read out, it may be converted into an **oncogene**—a gene that does not respond to the control signals that regulate cell division.

By itself, an oncogene does not cause malignant cancer. That usually requires mutations in several other genes (Figure 22.5). At least one of the altered genes is likely to be a **tumor suppressor gene**. These are genes that can halt abnormal cell growth and division, preventing cancers from developing. They also may prevent oncogenes from being expressed.

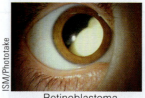

Retinoblastoma

We now understand how some tumor suppressor genes operate. For example, the childhood eye cancer **retinoblastoma** (*left*) is likely to develop when a child has only one functional copy of a tumor suppressor gene on chromosome 13. The genes associated with a predisposition to breast cancer, BRCA1 and BRCA2, also are tumor suppressor genes. People who inherit mutant forms of these genes are at high risk of developing breast cancer.

Researchers know a lot about a tumor suppressor gene called p53. This gene codes for a regulatory protein that stops cell division when cells are stressed or damaged. When p53 mutates, the controls turn off. Then an affected cell may begin runaway division. Even worse, a mutated p53 gene's faulty protein may turn on an oncogene. Half or more of cancers involve a mutated or missing p53 gene.

Various factors can cause mutations leading to cancer

Inherited susceptibility to cancer Heredity plays a major role in about 5 percent of cancers, including cases of familial breast cancer, lung cancer, and colorectal cancer. If

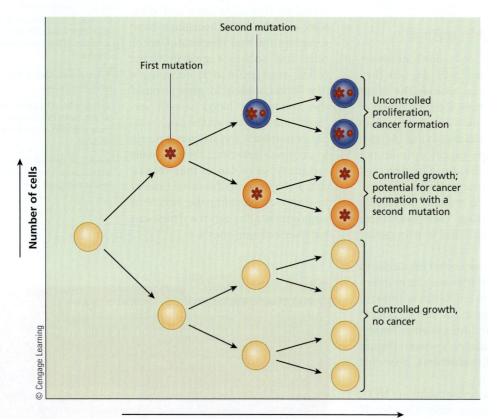

Number of cells

Second mutation

First mutation

Uncontrolled proliferation, cancer formation

Controlled growth; potential for cancer formation with a second mutation

Controlled growth, no cancer

Cell division over time

Figure 22.5 Carcinogenesis occurs in steps as cells acquire mutations. Only two mutation events may be enough to launch the development of some cancers. Three or more separate mutations may underlie other malignancies, such as colon cancer (see Figure 22.6).

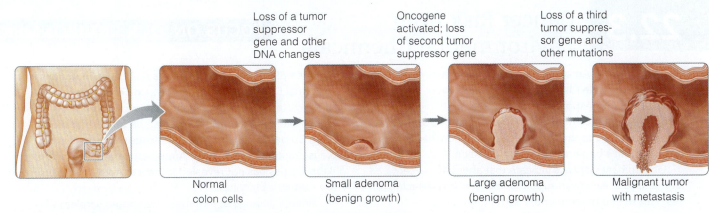

Loss of a tumor suppressor gene and other DNA changes

Oncogene activated; loss of second tumor suppressor gene

Loss of a third tumor suppressor gene and other mutations

Normal colon cells

Small adenoma (benign growth)

Large adenoma (benign growth)

Malignant tumor with metastasis

Figure 22.6 A common type of colorectal cancer may develop by these steps. (© Cengage Learning)

a mutation in a germ cell or a gamete (sperm or egg) alters a proto-oncogene or tumor suppressor gene, the defect can be passed on from parent to child. An affected person may be more likely to develop cancer if later mutations occur in other proto-oncogenes, in tumor suppressor genes, or in genes that control aspects of cell metabolism and responses to hormones. Figure 22.6 diagrams the steps toward a common type of colorectal cancer.

Viruses Viruses cause some cancers. For example, a viral infection may switch on a proto-oncogene when the viral DNA is inserted at a certain location in the host cell's DNA. Other viruses carry oncogenes as part of their genetic material and insert them into the host's DNA.

Chemical carcinogens Thousands of **carcinogens**—cancer-causing substances—can lead to a mutation in DNA. The list of known carcinogens includes many chemicals that are by-products of industrial activities, such as asbestos, vinyl chloride, and benzene. The list also includes hydrocarbons in cigarette smoke and on the surfaces of charred meats, and substances in dyes and pesticides. Some plants and fungi also produce carcinogens. Aflatoxin, which is produced by a fungus that attacks stored grain, peanuts, and other seeds, causes liver cancer. For this reason, some authorities advise against eating unprocessed peanut butter.

Radiation Section 18.5 noted that radiation can cause cancer-related mutations in DNA. Common sources include ultraviolet radiation from sunlight and tanning lamps, medical and dental X-rays, and some radioactive materials used to diagnose diseases. Other sources are radon gas in soil and water, background radiation from cosmic rays, and the gamma rays emitted from nuclear reactors and radioactive wastes. Sun exposure is probably the greatest radiation risk factor for most people (Figure 22.7).

Breakdowns in immunity When a normal cell turns cancerous, altered proteins at its surface function like foreign antigens—the "nonself" tags that mark a cell for destruction by cytotoxic T cells and NK cells. A healthy immune system can destroy some types of cancer cells,

but this protection wanes as a person ages. This is why the risk of cancer rises with age.

A person's cancer risk may rise whenever the immune system is suppressed for a long time. In addition to factors such as infection by HIV, anxiety and severe depression can suppress immunity. So can some therapeutic drugs, such as the glucocorticoids discussed in Section 15.7.

Finally, for various reasons, the cells of a growing cancer may not trigger an immune response. When this happens, the immune system is "blind" to the cancer threat.

carcinogens Substances that cause cancer by mutating DNA.

carcinogenesis Transformation of a normal cell into a cancer cell.

oncogene Gene that doesn't respond to control signals for cell division.

proto-oncogene Normal gene that codes for a protein that stimulates cell division.

tumor suppressor gene Gene that can halt cell growth and division.

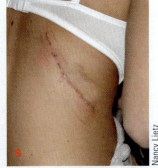

Figure 22.7 Tanning can lead to malignant melanoma. A Former Miss Maryland Brittany Lietz Cicala developed malignant melanoma at age 20. Before her cancer was diagnosed she had regularly used a tanning bed since the age of 16. Since her diagnosis she has had more than twenty-five tumors or precancerous lesions removed and must be checked for new lesions every three months. **B** A scar on Brittany's back where a cancerous tumor was removed.

TAKE-HOME MESSAGE

WHAT TRIGGERS CANCER?

- Cancer typically develops due to mutations in oncogenes or proto-oncogenes.
- Carcinogenesis usually also requires the absence or mutation of at least one tumor suppressor gene.
- Inherited susceptibility, viruses, chemical carcinogens, radiation, and weakened immunity all may trigger carcinogenesis.

According to the American Cancer Society, factors in our environment lead to about half of all cancers. This statistic includes exposure to UV light and radiation, and it also includes agricultural and industrial chemicals. How are people exposed to these chemicals? And how dangerous are they? Let's begin with the first question.

Government statistics indicate that about 40 percent of the food in American supermarkets contains detectable residues of one or more of the active ingredients in commonly used pesticides. The residues are especially likely to be found in tomatoes, grapes, apples, lettuce, oranges, potatoes, beef, and dairy products. Imported crops, such as fruits, vegetables, and coffee beans, can also carry significant pesticide residues—sometimes including pesticides, such as DDT, that are banned in the United States. The pesticide category includes roughly 600 chemicals used as fungicides, insecticides, and herbicides, which are used alone or in combination.

Avoiding exposure to pesticides is difficult. Although residues of some pesticides can be removed from the surfaces of fruits and vegetables by washing, it can be difficult to avoid coming into contact with pesticides used in community spraying programs to control mosquitoes and other pests, or used to eradicate animal and plant pests on golf courses and along roadsides. We have more control over chemicals we use in gardens and on lawns (Figure 22.8).

Agricultural chemicals are not the only potential threats to human health. Industrial chemicals also have been linked to cancer. In one way or another, the industrial chemicals in Table 22.2 all can cause carcinogenic mutations in DNA.

Biochemist Bruce Ames developed a test that could be used to assess the ability of chemicals to cause mutations. This Ames test uses *Salmonella* bacteria as the "guinea pigs," because chemicals that cause mutations in bacterial DNA may also have the same effect on human DNA. After extensive experimentation, Ames arrived at some interesting conclusions. First, he found that more than 80 percent of known cancer-causing chemicals do cause mutations. However, Ames testing at many different laboratories has not revealed a "cancer epidemic" caused by synthetic chemicals.

Ames's findings do not mean we should carelessly expose ourselves to environmental chemicals. The National Academy of Sciences has warned that the active ingredients in 90 percent of all fungicides, 60 percent of all herbicides, and 30 percent of all insecticides used in the United States have the potential to cause cancer in humans. At the same time, responsible scientists recognize that it is virtually impossible to determine that a certain level of a specific chemical caused a particular cancer or some other harmful effect. Given these facts, it seems wise to be cautious and limit our exposure to the potential carcinogens in an increasingly chemical world.

TABLE 22.2 Some Industrial Chemicals Linked to Cancer

Chemical/Substance	Type of Cancer
Benzene	Leukemias
Vinyl chloride	Liver, various connective tissues
Various solvents	Bladder, nasal epithelium
Ether	Lung
Asbestos	Lung, epithelial linings of body cavities
Arsenic	Lung, skin
Radioisotopes	Leukemias
Nickel	Lung, nasal epithelium
Chromium	Lung
Hydrocarbons in soot, tar smoke	Skin, lung

Figure 22.8 Home garden chemicals are just one way people can come into contact with potentially carcinogenic substances. Most commercially produced oranges are sprayed with fungicides that are potential human carcinogens (*right*). Agricultural workers have the greatest risk of significant exposure.

■ **In general, a cancer is named according to the type of tissue in which it first forms.**

Although there are dozens of specific types of cancer, they can be sorted into more general categories based on the tissue where the primary (first) cancer develops.

For example, cancers of connective tissues such as muscle and bone are **sarcomas**. Types of **carcinomas** arise from cells in epithelium, including cells of the skin and the epithelial linings of internal organs. **Lymphomas** are cancers of lymphoid tissues in organs such as lymph nodes, and cancers arising in blood-forming regions—mainly stem cells in bone marrow—are **leukemias**. **Gliomas** develop in glial cells of the brain.

Scientists use Latin prefixes to indicate the particular tissue or organ where cancer develops (Figure 22.9). For example, the prefix *adeno-* refers to gland cells. A woman whose breast cancer develops in milk ducts will therefore be diagnosed with an adenocarcinoma of the breast.

Previous chapters have discussed examples of cancers that strike different body systems. Figure 22.10 summarizes American Cancer Society statistics on where the most common types of cancer form.

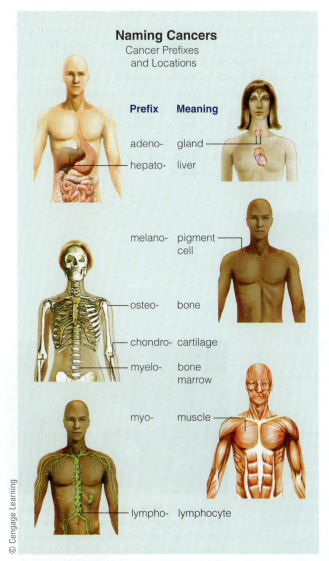

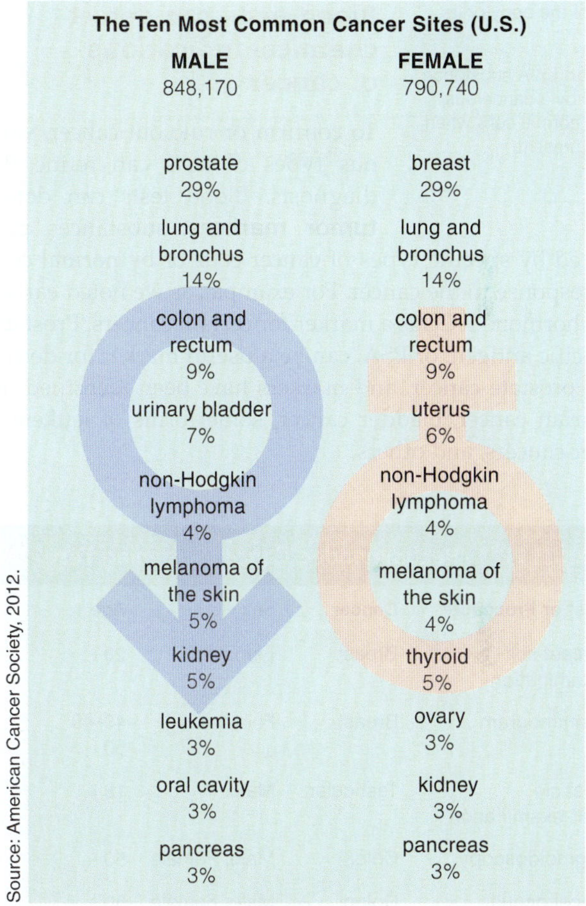

Naming Cancers
Cancer Prefixes and Locations

Prefix	Meaning
adeno-	gland
hepato-	liver
melano-	pigment cell
osteo-	bone
chondro-	cartilage
myelo-	bone marrow
myo-	muscle
lympho-	lymphocyte

© Cengage Learning

Figure 22.9 Cancer is named for the site in the body where it first develops. This list shows Latin prefixes used to indicate the location of various forms of carcinoma.

The Ten Most Common Cancer Sites (U.S.)

MALE 848,170	FEMALE 790,740
prostate 29%	breast 29%
lung and bronchus 14%	lung and bronchus 14%
colon and rectum 9%	colon and rectum 9%
urinary bladder 7%	uterus 6%
non-Hodgkin lymphoma 4%	non-Hodgkin lymphoma 4%
melanoma of the skin 5%	melanoma of the skin 4%
kidney 5%	thyroid 5%
leukemia 3%	ovary 3%
oral cavity 3%	kidney 3%
pancreas 3%	pancreas 3%

Source: American Cancer Society, 2012.

Figure 22.10 In the United States, more than 1.6 million people are diagnosed with cancer each year. This chart shows the American Cancer Society's estimates of the incidence of common cancers in the United States in 2012.

TAKE-HOME MESSAGE

HOW IS CANCER NAMED?

• Cancer is categorized and named according to the tissue where it first develops.

22.5 Cancer Screening and Diagnosis

- Early and accurate diagnosis of cancer is important to maximize the chances that a cancer can be cured.
- Links to Gene mutation 21.2, DNA profiling, 21.10

biopsy Microscopic examination of tissue for evidence of cancer cells; biopsy is the definitive test for cancer.

medical imaging Methods such as MRI, X-rays, and ultrasound that are used to obtain an internal view of the body or its parts.

tumor marker A substance produced by a cancerous cell or by normal cells when cancer is present.

Routine screening is important for people with a family history of cancer or whose risk is elevated for some other reason, including simply getting older. Table 22.3 lists some recommended cancer screening tests.

Blood tests can detect chemical indications of cancer

To confirm or rule out cancer, various types of tests can refine the diagnosis. Blood tests can detect **tumor markers**, substances produced by specific types of cancer cells or by normal cells in response to the cancer. For example, as we noted earlier, the hormone HCG is a marker for certain cancers. Prostate-specific antigen, or PSA, can be a useful marker for detecting prostate cancer, and markers have been identified for ovarian cancer, bladder cancer, some forms of leukemia, liver cancers, and others.

Medical imaging can reveal the site and size of tumors

Medical imaging includes methods such as MRI (magnetic resonance imaging), X-rays, ultrasound, and CT (computed tomography). Unlike a standard X-ray, an MRI scan can reveal tumors that are obscured by bone, such as in the brain (Figure 22.11).

You may remember from Section 2.1 that radioactive tracers (substances with a radioisotope attached to them) are another important tool for diagnosing cancer. A doctor administers the tracer, then uses a tracking device such as a PET scanner to see where the tracer ends up in the body. For example, thyroid cancer be diagnosed using a tracer that includes a radioactive isotope of iodine (Figure 22.12).

Radioactively labeled monoclonal antibodies, which home in on tumor antigens, are useful for pinpointing the location and sizes of tumors of the colon, brain, bone, and some other tissues. A *DNA probe* (radioactively labeled DNA) can be used to locate mutated genes, such as the p53 tumor suppressor gene, or genes associated with some inherited cancers. This type of screening can allow people with increased genetic susceptibility to make medical and lifestyle choices that may reduce their cancer risk. The procedure is expensive, however, and relatively few people have insurance that covers it.

TABLE 22.3 Recommended Cancer Screening Tests

Test or Procedure	Cancer	Sex	Age	Frequency
Breast self-examination	Breast	Female	20+	Monthly
Mammogram	Breast	Female	40–49 50+	Every 1–2 years Yearly
Testicle self-examination	Testicular	Male	18+	Monthly
Sigmoidoscopy	Colon	Male, Female	50+	Every 3–5 years
Fecal occult blood test	Colon	Male, Female	50+	Yearly
Digital rectal	Prostate, colorectal	Male, Female	40+	Colorectal: Yearly examination Prostate: Yearly up to age 75
Pap test	Uterus, cervix	Female	18+ and all sexually active women	Every other year until age 35; yearly thereafter
Pelvic examination	Uterus, ovaries, cervix	Female	18–39 40+	Every 1–3 years w/ Pap Yearly
General checkup		Male, Female	20–39 40+	Every 3 years Yearly

Figure 22.11 MRI scanning is a noninvasive tool for diagnosing cancer. The patient is placed in a chamber that is surrounded by a magnet. The machine produces a magnetic field in which nuclei of common atoms in the body align and absorb energy. A computer analyzes the information and uses it to generate an image of soft tissues. (Paul Shambroom/Photo Researchers, Inc.)

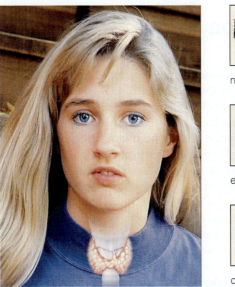

normal thyroid

enlarged

cancerous

Figure 22.12 Radioactive tracers also can reveal cancer tumors. Shown above are scans of the thyroid gland from three patients who have ingested radioactive iodine, which is taken up by the thyroid. (© Cengage Learning/Gary Head)

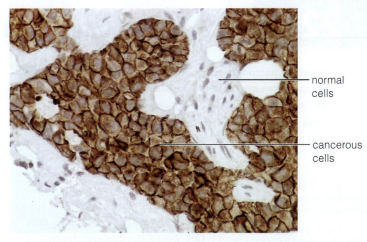

normal cells

cancerous cells

Figure 22.13 This light microscope image shows cancerous cells in breast tissue. The cancer cells are stained brown. Normal cells are the ones with lighter staining. (From the archives of www.breastpath.com, courtesy Askew, Jr., M.D., P.A. Reprinted with permission, 2004 Breastpath.com)

Biopsy is the surest way to diagnose cancer

When a test or exam suggests that a patient has cancer, the usual next step to confirm the diagnosis is **biopsy**. A small piece of suspect tissue is removed from the body through a hollow needle or exploratory surgery. A pathologist then microscopically examines cells of the tissue sample to see if cancer cells are present (Figure 22.13).

Table 22.4 lists the American Cancer Society's seven common cancer warning signs. Notice that the first letters of the signs spell CAUTION. Watching for these signs can help people spot cancer in its early stages, when treatment is most effective. Anyone who has one or more of these signs should be evaluated by a doctor as soon as possible.

Today sophisticated tools of personal genetic profiling may be used to identify an inherited predisposition for some forms of cancer. Prime examples are the mutated forms of BRCA1 and BRCA2 associated with familial breast cancer. The genes are located on different chromosomes (Figure 22.14), and a female who inherits one or both of them has a dramatically increased risk of developing breast cancer.

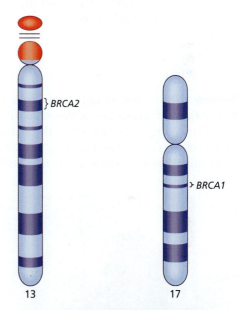

} BRCA2

> BRCA1

13   17

Figure 22.14 Personal genetic profiling helps identify women at risk of inherited breast cancer. Faulty BRCA1 and BRCA2 genes underlie the majority of cases of familial breast cancer. (© Cengage Learning)

TABLE 22.4 The Seven Warning Signs of Cancer*
Change in bowel or bladder habits and function
A sore that does not heal
Unusual bleeding or bloody discharge
Thickening or lump
Indigestion or difficulty swallowing
Obvious change in a wart or mole
Nagging cough or hoarseness

*Notice that the first letters of the signs spell the word CAUTION.
Source: American Cancer Society.

TAKE-HOME MESSAGE

HOW IS CANCER DIAGNOSED?

- Preliminary procedures for diagnosing cancer include blood tests for substances produced by cancer cells and various types of medical imaging.
- Only a biopsy can definitely diagnose cancer.
- Everyone should be aware of the seven common cancer warning signs.

- When a person is diagnosed with cancer, a variety of weapons are available to combat it. And anyone can adopt an "anticancer lifestyle."
- Links to Monoclonal antibodies and immunotherapy 9.8, Cell cycle 18.2

adjuvant therapy Cancer therapy that combines surgery with chemotherapy.

chemotherapy Use of anticancer drugs to treat cancer.

radiation therapy Irradiation of relatively small, localized cancer tumors. The radiation comes from radioisotopes.

Patients understandably dread a diagnosis of cancer, but today many forms of cancer can be treated successfully. Even if a complete cure is not possible, modern treatment approaches may prolong a patient's life and improve the quality of life for years. The major weapons against cancer are chemotherapy drugs, radiation therapy, and surgery.

Surgery may even be a complete cure when a tumor is fully accessible and has not spread.

Chemotherapy and radiation kill cancer cells

Chemotherapy uses drugs to kill cancer cells (Figure 22.15). Most anticancer drugs are designed to kill dividing cells by disrupting some aspect of the normal cell cycle. Unfortunately, chemotherapy drugs are also toxic to rapidly dividing healthy cells such as hair cells, stem cells in bone marrow, immune system lymphocytes, and epithelial cells of the intestinal lining. This is why chemotherapy patients may suffer side effects such as hair loss, nausea, vomiting, and reduced immune responses. A treatment option called **adjuvant therapy** (*adjuvant* means "helping") combines surgery and a less toxic dose of chemotherapy. A cancer patient might receive enough chemotherapy to shrink a tumor, for instance, then have surgery to remove what's left.

Drugs used in chemotherapy typically have been matched to the organ in which a cancer occurs—this drug for breast cancer, that one for lung cancer, and so on. A promising new strategy instead matches chemotherapy with the genetic characteristics of a patient's cancer. This approach recognizes that there are hundreds of genetically different subgroups of cancer, and that some subgroups have the same gene mutations—and chemical features—regardless of where the cancer develops. For example, the drug Gleevec works well against some types of leukemia and also against some sarcomas.

Like surgery, **radiation therapy** may be used when the cancer is small and has not spread (Figure 22.16). The radiation comes from radioisotopes such as radium 226 and cobalt 60. Like traditional chemotherapy, it is something of a "shotgun" approach to cancer treatment because it kills both cancer cells and healthy cells in the irradiated area.

Because chemotherapy and radiation both damage or kill healthy body cells, cancer researchers have looked for

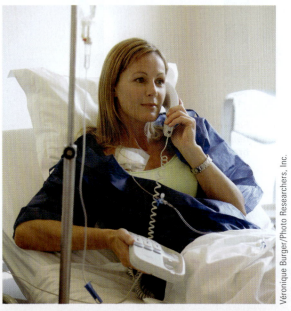

Figure 22.15 Chemotherapy uses cell-killing drugs. In some cases, the drug is delivered through a tube that connects with a port inserted into a vein, as shown here. Patients usually receive chemotherapy over a period of weeks. Imaging tests are used to determine whether the patient's condition has improved.

Véronique Burger/Photo Researchers, Inc.

Larry Mulvehill/Photo Researchers, Inc.

Figure 22.16 Radiation therapy kills cells with a targeted dose of lethal radiation.

more precise cancer treatments. Section 9.8 described how monoclonal antibodies can be used to deliver lethal doses of anticancer drugs to tumor cells while sparing healthy cells. The antibodies target cell surface markers (antigens) on various types of tumors. The idea is to link monoclonal antibodies that bind to tumor cell markers with lethal doses of cytotoxic (cell-killing) drugs. Experiments have shown promising results in some patients with one form of leukemia, one type of breast cancer, and certain gliomas (cancers that arise in glia in the brain).

Interferons also can activate cytotoxic T cells and NK (natural killer) cells, which then may recognize and kill some types of cancer cells. So far, interferon therapy has been useful only against some rare forms of cancer.

Good lifestyle choices can limit cancer risk

None of us can control factors in our heredity or biology that might lead one day to cancer, but we can all make lifestyle decisions that promote health. The National Cancer Institute estimates that 40 percent of cancers are related to lifestyle factors, such as smoking, suntanning, and obesity that is due to improper diet and sedentary habits. The American Cancer Society recommends the following strategies for limiting cancer risk:

1. Avoid tobacco in any form, including secondary smoke from others (Figure 22.17).

2. Maintain a desirable weight. Being more than 40 percent overweight increases the risk of several cancers.

3. Eat a low-fat diet that includes plenty of vegetables and fruits. These foods contain antioxidants, such as vitamin E, that may help prevent some kinds of cancer.

4. Drink alcohol in moderation. Heavy alcohol use, especially in combination with smoking, increases risk for cancers of the mouth, larynx, esophagus, and liver.

5. Learn whether your job or residence exposes you to such industrial agents as nickel, chromate, vinyl chloride, benzene, asbestos, and agricultural pesticides, which are associated with various cancers.

6. Protect your skin from excessive sunlight.

THINK OUTSIDE THE BOOK

Four-dimensional computed tomography—4D CT, a type of CT scanning—is an advanced medical imaging technology used to precisely locate cancerous tumors prior to radiation treatments. Visit the website of the University of Pittsburgh Cancer Institute at www.upmccancercenters.com. How might using 4DCT improve the treatment of a cancer patient slated to receive radiation therapy?

Photos.com/Jupiter Images

Billy E. Barnes/Jeroboam

Figure 22.17 Lifestyle choices have a major impact on cancer risk. Eating a healthy diet and choosing not to smoke are just two ways everyone can improve their chances of avoiding cancer.

TAKE-HOME MESSAGE

HOW IS CANCER TREATED?

- Surgery, chemotherapy, radiation, and other treatment strategies are used to fight cancer.

- The best defense against cancer is making lifestyle choices that promote good health.

EXPLORE
ON YOUR OWN

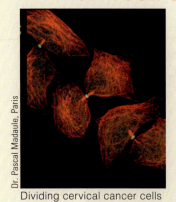

Dr. Pascal Madaule, Paris

Dividing cervical cancer cells

Most families have been touched by cancer in one way or another. The American Cancer Society website is a portal to a huge amount of reliable information on the risks for, causes of, and treatments for virtually any cancer. Choose a cancer to investigate and see how much you can learn about it in just 15 minutes. Does your research give you any new insights into your own risk for the cancer? What is your reaction to the stories of cancer survivors that are posted on the website?

SUMMARY

Section 22.1   Overgrowing cells lead to a tissue mass called a tumor. In dysplasia, a common precursor to cancer, cells develop abnormalities in size, shape, and organization. Cancer results when the genetic controls over cell division are lost completely. Cancer cells (Figure 22.18) differ from normal cells in both structure and function. They usually have an overly large nucleus and altered surface proteins, and lack features of a normal, specialized body cell. Cancer cells also grow uncontrolled and can invade surrounding tissues, a process called metastasis.

Sections 22.2, 22.3   Cancer develops during carcinogenesis, a process that involves a series of genetic changes. Initially, mutation may alter a proto-oncogene into a cancer-causing oncogene. Infection by a virus can also insert an oncogene into a cell's DNA or disrupt normal controls over a proto-oncogene. One or more tumor suppressor genes must be missing or become mutated before a normal cell can be transformed into a cancerous one (Table 22.5).

TABLE 22.5	Cancer Causes and Contributing Factors
Cause/Factor	**Impact**
Oncogene	May alter control of cell division
Faulty tumor suppressor gene	Fails to halt runaway cell division
Viral infection	Switches proto-oncogene to oncogene or inserts an oncogene into the host cell DNA
Carcinogen	Damages DNA
Radiation	Damages DNA
Faulty immunity	Fails to tag cancer cells for destruction

A predisposition to a certain type of cancer can be inherited. Other causes of carcinogenesis are viral infection, chemical carcinogens, radiation, faulty immune system functioning, and possibly a breakdown in DNA repair.

Section 22.4   In general, a cancer is named according to the type of tissue in which it arises. Common ones include sarcomas (connective tissues such as muscle and bone), carcinomas (epithelium), adenocarcinomas (glands or their ducts), lymphomas (lymphoid tissues), and leukemias (blood-forming regions).

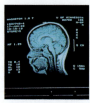

Section 22.5   Common methods for cancer diagnosis include blood testing for the presence of substances produced either by specific types of cancer cells or by normal cells in response to the cancer. Medical imaging (such as magnetic resonance imaging) also can aid diagnosis. Biopsy provides a definitive diagnosis.

Section 22.6   Cancer treatments include surgery, chemotherapy, and tumor irradiation. Under development are target-specific monoclonal antibodies and immune therapy using interferons.

Lifestyle choices such as the decision not to use tobacco, to maintain a low-fat diet, and to avoid overexposure to direct sunlight and chemical carcinogens can help limit personal cancer risk.

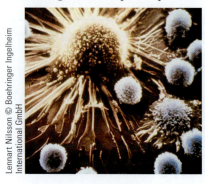

Lennart Nilsson © Boehringer Ingelheim International GmbH

Figure 22.18 This cancer cell is surrounded by white blood cells that may or may not be able to destroy it. The cancer cell has extended dozens of "false feet" (pseudopodia) that help it move about in tissues.

REVIEW QUESTIONS

1. How are cancer cells structurally different from normal cells of the same tissue? What is the relevance of altered surface proteins to uncontrolled growth?

2. What are the differences between a benign tumor and a cancerous one?

3. Write a short paragraph that summarizes the roles of proto-oncogenes, oncogenes, and tumor suppressor genes in carcinogenesis.

4. List the four main categories of cancer tumors.

5. What are the seven warning signs of cancer? (Remember the American Cancer Society's clue word, CAUTION.)

6. Using the diagram below as a guide, indicate the major steps in cancer metastasis.

Step 1.

_____
_____
_____

Step 2.

_____
_____
_____
_____
_____
_____

Step 3.

_____
_____
_____

1. A tumor is _____.
   a. malignant by definition
   b. always enclosed by connective tissue
   c. a mass of tissue that may be benign or malignant
   d. always slow-growing

2. Cancer cells _____.
   a. lack normal controls over cell division
   b. secrete the growth factor angiogenin
   c. display altered surface proteins
   d. are not inhibited by contact with other cells
   e. all of the above

3. Fill in the blank: The onset of cancer seems to require the activity of an oncogene plus the absence or mutation of at least one _____.

4. Chemical carcinogens _____.
   a. include viral oncogenes
   b. can damage DNA and cause a mutation
   c. must be ingested in food
   d. are not found in foods

5. So far as we know, carcinogenesis is *not* triggered by _____.
   a. breakdowns in DNA repair   d. protein deficiency
   b. a breakdown in immunity   e. inherited gene
   c. radiation   defects

6. Tumor suppressor genes _____.
   a. occur normally in cells
   b. promote metastasis
   c. are brought into cells by viruses
   d. only rarely affect the development of cancer

7. _____ is the definitive method for detecting cancer.
   a. Blood testing   c. Biopsy
   b. Physician examination   d. Medical imaging

8. The most common therapeutic approaches to treating cancer include all of the following except _____.
   a. chemotherapy
   b. irradiation of tumors
   c. surgery to remove cancerous tissue
   d. administering doses of vitamins

9. The goal of immune therapy is to _____.
   a. cause defective T cells in the thymus to disintegrate
   b. activate cytotoxic T cells
   c. dramatically increase the numbers of circulating macrophages
   d. promote the secretion of monoclonal antibodies

10. Currently in the United States, _____ cancer is the leading cancer diagnosed among adult females; _____ cancer is the leading cancer most often diagnosed among adult males.
    a. lung; colon   c. colon; lung
    b. breast; prostate   d. breast; lung

CRITICAL THINKING

1. Look back at the discussion of aging and DNA repair in Chapter 17 and Chapter 21, then propose an explanation of the observation that higher rates of cancer are associated with increasing age.

2. A textbook on cancer contains the following statement: "Fundamentally, cancer is a failure of the immune system." Why does this comment make sense?

3. Ultimately, cancer kills because it spreads and disturbs homeostasis. Consider, for example, a kidney cancer that metastasizes to the lungs and liver. What are some specific aspects of homeostasis that the spreading disease could affect?

4. Over the last few months, your best friend, Mark, has noticed a small, black-brown, raised growth developing on his arm. When you suggest that he have it examined by his doctor, he says he's going to wait and see if it gets any larger. You know that's not a very smart answer. Give three arguments that you can use to try to convince Mark to seek medical advice as soon as possible.

5. Some desperate cancer patients consume pills or other preparations containing shark cartilage, which the manufacturers tout as an anti-angiogenesis compound. The basis for these claims is the fact that blood vessels do not grow into cartilage. Responsible researchers point out that, regardless of the properties of cartilage, there is no way that *eating* it could provide any anticancer benefit. Why is this counterargument correct?

Your Future

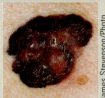

James Stevenson/Photo Researchers, Inc.

In 2009 the World Health Organization (WHO) reported that people under age thirty who regularly use tanning beds increase their risk of developing skin cancer by 75 percent. Although critics have cited some flaws in the WHO study, lawmakers in the United States and some other nations are proposing stricter restrictions on tanning bed use. At this writing, the states of California and Vermont have banned the use of tanning beds by anyone under eighteen years of age. The indoor tanning industry says more regulations aren't needed. Regardless of how the current debate turns out, in the future it's likely that the safety of tanning bed use—especially by those under age eighteen—will continue to receive increased scrutiny.

PRINCIPLES OF EVOLUTION

23

LINKS TO EARLIER CONCEPTS

This chapter returns to the evolution of life on Earth (1.2). The discussion in Chapter 1 of the use of the term *theory* in science (1.6) also applies to a key topic of this chapter, the theory of evolution by natural selection.

This chapter also draws on your understanding of genes (19.1, 20.1), of genetic mutations (20.8, 20.9), and of how environmental factors can alter the expression of genes (19.6).

The section on life's origins builds on your knowledge of enzymes and organic compounds (2.8), amino acids (2.11), and cell membranes (3.1).

KEY CONCEPTS

Basics of Evolution

The theory of evolution by natural selection draws on observations and ideas about how populations of organisms interact with their environment. Evolution occurs through inherited changes passed through the generations. Sections 23.1–23.3

Evidence for Evolution

Evidence for evolution comes from biogeography, fossils, and comparisons of body form, development, and biochemistry. Sections 23.4–23.7

Human Evolution

Trends in the evolution of humans include upright walking, refined vision and hand movements, and development of a complex brain and behaviors. Sections 23.8–23.9

Life's Origins

Molecules of life and the first living cells are thought to have emerged on Earth around 3.8 billion years ago. Section 23.10

The Barringer crater in Arizona, shown below, is immense—a full one mile wide. Geologists estimate that this huge hole in the ground was created by a 300,000-ton asteroid that collided with Earth 50,000 years ago. Geological evidence also tells us that an even larger asteroid struck Earth about 65 million years ago. It may well have altered the course of evolution on our planet. Many scientists are convinced that the impact led to the extinction of the dinosaurs and many other life forms.

It has been only about 100,000 years since modern humans evolved. In fact, dozens of humanlike species arose in Africa during the 5 million years before our own species—*Homo sapiens*—even showed up. Have asteroid impacts changed the course of human evolution? It's possible, but we will never know for sure. Thinking about time and the changes it can bring is our task in this chapter. We will start with the basic principles of evolution and the kinds of evidence evolutionary biologists use in their work. Later we'll survey key trends in human evolution, and we'll close the chapter with a look at what is currently known about the origins of life on our planet.

© Brad Snowder

A Little Evolutionary History

■ In Latin, the word *evolutio* means "unrolling." That image is an apt way to begin thinking of biological evolution.

evolution Genetic change in a line of descent over time.

In biology, **evolution** is genetic change in a line of descent over time. The evolution of life on Earth has resulted in many hundreds of thousands of species. Of those living today, some are closely related, others more distantly. As described later in this chapter, genetic evidence reveals that humans share a common ancestor with life forms as different from us as bacteria and corn plants. Shared ancestry is a basic idea in evolution.

In the early 1800s, the source of Earth's remarkable diversity of life forms was hotly disputed. Many people believed that all species had come into existence at the same time in the distant past, at the same center of creation, and had not changed since. By the mid-1800s, however, centuries of exploration and advances in the sciences of geology and comparative anatomy had raised questions. Why were some species found only in particular isolated regions? Why, on the other hand, were similar (but not identical) species found in widely separated parts of the world? Why did species as different as humans, whales, and bats have some strikingly similar body features? What was the significance of discoveries of similar fossil organisms in similar layers of Earth's sedimentary rocks—regardless of where in the world the layers were?

In 1831 Charles Darwin was a 22-year-old with a degree in theology—though he wasn't interested in being a clergyman. He enjoyed hunting, fishing, collecting shells, or simply watching wildlife. A botanist friend arranged for him to work as a naturalist aboard the HMS *Beagle* during a five-year voyage around the world (Figure 23.1). The *Beagle* sailed first to South America, and during the long Atlantic crossing Darwin studied geology and collected marine life. During stops along the coast and at various islands, he observed other species of organisms in environments ranging from sandy shores to high mountains. After returning to England in 1836, Darwin began talking with other naturalists about a topic that was on many scholars' minds—the growing body of evidence that life forms evolve, changing over time.

A clue as to how this might happen came from Thomas Malthus, a British clergyman and economist, who proposed that a population will tend to outgrow its resources, and so in time its members must compete for what is available. Darwin's observations also suggested that any population can produce more individuals than the environment can support, yet populations tend to be stable over time. For instance, a starfish can release 2.5 million eggs a year, but the oceans are not filled with starfish. What determines who lives and who dies as predators, starvation, and environmental events take their toll? Chance could be a factor, but Darwin and others came up with a second major clue— that even members of the same species vary in their traits.

Darwin's melding of his observations of the natural world with the ideas of others led him to propose that evolution could occur by way of a process called *natural selection*. Wide acceptance of this theory would not come until nearly 70 years later, when the then-new field of genetics provided insights into how traits could vary. Now let's consider some current views of the mechanisms of evolution.

route of *Beagle*

EQUATOR

Galápagos Islands

A

B

C

Figure 23.1 Animated! Darwin's voyage on the *Beagle* spurred his thinking about evolution. A The *Beagle*'s route from England to South America, where the ship called at **B** the Galápagos Islands. **C** Charles Darwin. (A: © Cengage Learning B: Dieter & Mary Plage/Survival Anglia/Oxford Scientific C: Charles Darwin (w/c on paper), English School, (19th century)/Down House, Downe, Kent, UK/© English Heritage Photo Library/The Bridgeman Art Library)

TAKE-HOME MESSAGE

WHAT IS BIOLOGICAL EVOLUTION?

- In biology, evolution is genetic change in a line of descent through successive generations.
- Combining observations of the natural world with ideas about interactions of populations with their environment, Darwin forged his theory of evolution by natural selection.

A Key Evolutionary Idea: Individuals Vary

- Evolution occurs in populations of organisms. It begins when the genetic makeup of a population changes.

- Links to Concepts of heredity 19.1, Genes 20.1

The history of life on Earth spans nearly 4 billion years. It is a story of how species originated, survived or went extinct, and stayed put or spread into new environments. The overall "plot" of the story is evolution, genetic change in lines of descent over time. **Microevolution** is the name for cumulative genetic changes that may give rise to new species. **Macroevolution** is the name for the large-scale patterns, trends, and rates of change among *groups* of species. Later on you'll get a fuller picture of these two patterns. Here, we begin with a fundamental principle of evolution: The traits of individuals in a population vary.

Individuals don't evolve—populations do

An individual fish, flower, or person doesn't evolve. Evolution occurs only when there is change in the genetic makeup of whole *populations of organisms*. In biology, a **population** is a group of individuals of the same species occupying a given area. As you know from your own experience with other people, there is plenty of genetic variation within and among populations of the same species (Figure 23.2).

Members of a population have similar traits—that is, phenotypes. They have the same general form and appearance (*morphological* traits), their body structures function in the same way (*physiological* traits), and they respond the same way to certain basic stimuli (*behavioral* traits). However, the details of traits vary quite a bit. For instance, individual humans vary in the color of their body hair, as well as in its texture, amount, and distribution—an example that only hints at the great genetic variation in human populations. Populations of most other species show the same kinds of variation.

Genetic differences produce variation

In theory, the members of a population have inherited the same number and kinds of genes. These genes make up the population's **gene pool**. Remember, though, that each kind of gene in the pool may have slightly different forms, called alleles. Variations in traits in a population—skin or hair color, say—result when individuals inherit different combinations of alleles. Whether your hair is black, brown, red, or blond depends on which alleles of certain genes you inherited from your mother and father.

If you go through a bag of chocolate candies that have different-colored sugar coatings, you'll see that some colors turn up more or less often than others do. The same is true for the gene alleles in a population. Some are much more common than others. The manufacturer can adjust the number of "reds" or "blues" in the overall mix of candy pieces, but where genes are concerned, changes come about by mutation, natural selection, and other processes. Those changes can lead to microevolution, which we consider next.

gene pool The total number and kinds of genes present in a given population.

macroevolution Large-scale patterns, trends, and rates of evolutionary change among groups of organisms.

microevolution Genetic changes that may give rise to new species.

population A group of individuals of the same species living in a given area.

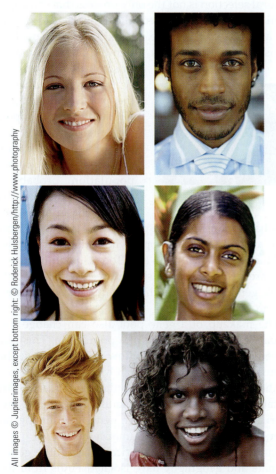

All images © Jupiterimages, except bottom right: © Roderick Hulsbergen/http://www.photography

Figure 23.2 **The traits of individuals in a population vary.** These photos show a small sample of the outward variation in our own species, *Homo sapiens*.

TAKE-HOME MESSAGE

WHAT CAUSES EVOLUTION?

- Evolution results from changes in the genetic makeup of populations of organisms.
- Members of a population share many traits overall, but in most populations there also is a great deal of underlying genetic variation.

23.3 Microevolution: How New Species Arise

- From the perspective of evolution, genetic mutations are important because they are the source of alleles—the alternative forms of genes.
- Links to Gene mutations 20.8, Changes in chromosomes 20.9

Mutation produces new forms of genes

A mutation is harmful when it alters a trait such that an affected individual can't survive or reproduce as well as other individuals. For example, for us humans, small cuts are common. But before modern medical treatments were available, people with hemophilia, whose blood does not clot properly, could die young from such minor injuries. As a result, the various hemophilias were rare, because affected people rarely lived long enough to pass on the faulty genes. By contrast, a *beneficial trait* improves some aspect of an individual's functioning in the environment and so improves chances of surviving and reproducing. A *neutral trait*, such as attached earlobes in humans, doesn't help nor hurt survival.

Natural selection can reshape the genetic makeup of a population

Changes in genes are the raw material of microevolution, the process that leads to new species. Actually, several processes are included in this category, but the one that probably accounts for most changes in the mix of alleles in a gene pool is natural selection. As Section 23.1 noted, Darwin formulated his theory of evolution by **natural selection** by correlating his understanding of inheritance with certain features of populations. In 1859 he published his ideas in a classic book, *On the Origin of Species*. We can express the main points of Darwin's insight as follows:

1. **The individuals of a population vary in their body form, functioning, and behavior.**

2. **Many variations can be passed from generation to generation.** This simply means that different versions of genes—alleles—can pass from parents to offspring.

3. **In every set of circumstances, some versions of a trait are more advantageous than others.** That is, some traits impart a better chance of surviving and reproducing. The expression "survival of the fittest" is verbal shorthand for this advantage.

4. **Natural selection is the difference in survival and reproduction** that we observe in individuals who have different versions of a trait. The "fittest" traits—and the gene alleles that govern them—are more likely to be "selected" for survival.

5. **A population is evolving when some forms of a trait are becoming more or less common** relative to the other forms. The shifts are evidence that the corresponding versions of genes are becoming more or less common.

6. **Over time, shifts in the makeup of gene pools** have been responsible for the amazing diversity of life forms on Earth.

As natural selection occurs over time, organisms come to have characteristics that suit them to the conditions in a particular environment. We call this trend **adaptation**.

Recall from Section 1.6 that the accumulated evidence of evolution and natural selection has elevated both ideas to the status of fundamental principles of the living world. Later we will consider a few examples of this evidence.

Chance can also change a gene pool

Natural selection is not the only process that can adjust the relative numbers of different alleles in a gene pool. Chance can also play a major role. This kind of gene pool tweaking is called **genetic drift**. Often the change is most rapid in small populations. In one type of genetic drift, called the *founder effect*, a few individuals leave a population and establish a new one. By chance, the relative numbers of various alleles in the new population probably will differ from those in the old group. For example, geneticists have strong evidence that ethnic Finns are descended from a small band of people who settled in what is now Finland about 4,000 years ago. Today, blond hair and blue eyes are distinctive Finnish features. In addition, at least thirty genetic disorders that are rare elsewhere are common in Finland.

Finnish boy

© Layne Kennedy/Corbis

The makeup of a gene pool also can change as individuals migrate into or out of a population. This physical movement of alleles, or **gene flow**, helps keep neighboring populations genetically similar. Over time it tends to counter the differences between populations that arise through mutation, genetic drift, and natural selection. Now that international travel is common, gene flow among human populations has increased dramatically.

In Finland, there has historically been little gene flow. Climate and geography have isolated Finns from the rest of Europe for centuries. Even now, Finns have much less genetic variation than other Europeans.

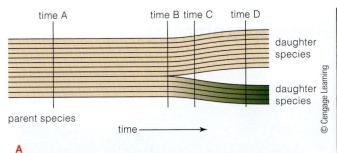

Figure 23.3 **Divergence is the first step toward the formation of new species. A** Horizontal lines in this diagram represent different populations. Because evolution is gradual, we cannot say at any one point in time that there are now two species rather than one. At time A there is only one species. At D there are two. At B and C the split has begun but isn't complete. **B** Artist's view of *Homo habilis*.

The ability to interbreed defines a species

For humans and other sexually reproducing organisms, a **species** is a genetic unit consisting of one or more populations of organisms that usually closely resemble each other physically and physiologically. Members of the same species can interbreed and produce fertile offspring under natural conditions. It doesn't matter how diverse their traits are, so long as they can interbreed successfully and so share a common gene pool. From this perspective, a female lawyer in India may never meet up with an Icelandic fisherman, but there's no biological reason why the two couldn't mate and produce children. But neither of them could mate successfully with a chimpanzee, even though chimps and humans are closely related species and have more than 90 percent of their genes in common. Humans and chimps are "reproductively isolated." This means that their genetic differences ensure that they can't mate and produce fertile offspring.

Reproductive isolation develops when gene flow between two populations stops. This often occurs when two populations are separated geographically. When populations are in different environments, mutation, natural selection, and genetic drift begin to operate independently in each one. These processes can change the gene pools of each in different ways. Eventually the differences can result in changes in body structure, function, or behavior that reduce the chances of successful interbreeding. For example, the two populations may breed in different seasons, or there may be bodily changes that physically interfere with mating. Other changes may prevent zygotes or hybrid offspring from developing properly.

The buildup of genetic differences between isolated populations is called *divergence* (Figure 23.3A). When the genetic differences are so great that members of the two populations can't interbreed, **speciation** has occurred: the populations have become separate species. Figure 23.3B is an artist's rendition of an extinct human species, *Homo habilis*. Clearly, there are many differences between

H. habilis and our own species, *Homo sapiens*.

Biologists disagree about the pace and timing of microevolution. Some hold to a model of gradual speciation, in which new species emerge through many small changes in form over long spans of time. Other scholars favor a model in which most evolutionary change occurs in bursts. That is, each species undergoes a spurt of changes in form when it first branches from the parental lineage, then changes little from then on.

The driving force behind rapid changes in species may be dramatic changes in climate or some other aspect of the physical environment. This type of change alters the physical conditions to which populations of organisms are adapted. On the other hand, bursts of evolutionary change could help explain why the fossil record has scanty evidence of a continuum of microevolution—the "missing links" between closely related species. Both models probably have a place in explaining the history of life.

adaptation Accumulation of adaptive traits.

gene flow Physical movement of gene alleles into or out of a population's gene pool.

genetic drift Chance change in a gene pool.

natural selection Process of evolution in which individuals of a species are more or less likely to survive and reproduce, depending on the details of their inherited traits.

speciation Accumulation of so many genetic differences that members of two populations can't interbreed successfully.

species A group of populations in which the individuals share enough traits that they can successfully interbreed.

23.4 Looking at Fossils and Biogeography

■ **The fossil record and biogeography are tools for studying the evolutionary journey of life on Earth.**

A **fossil** is the remains or traces (such as tracks) of an organism of a past geologic age embedded in the Earth's crust (Figure 23.4). Similarities among fossils and living organisms—or differences between them—provide strong evidence of evolution by natural selection as populations adapted to their surroundings.

Fossils are found in sedimentary rock

When an organism dies, its soft parts usually decompose first. As a result, the most common fossils are bones, teeth, shells, seeds, and other hard parts. *Fossilization* begins when an organism is buried in sediments or volcanic ash. With time, water seeps into the organic remains, infusing them with dissolved inorganic compounds. As more and more sediments accumulate above the burial site, the remains are subjected to increasing pressure. Over long spans of time, the chemical changes and growing pressure transform them to stony hardness.

Organisms are more likely to be preserved when they are buried quickly in the absence of oxygen. Entombment by volcanic ash or anaerobic mud (which lacks oxygen) satisfies this condition very well. Preservation also is favored when a burial site is not disturbed. Usually, though, fossils are broken, crushed, deformed, or swept away by erosion, rock slides, and other geologic events.

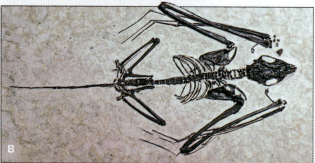

Figure 23.4 High quality fossils such as these are rare. A Fossilized footprint of one of the flesh-eating dinosaurs called theropods. **B** The skeleton of a bat. (A: Igor Karasi/Shutterstock.com B: Department of Geosciences, Princeton University)

Figure 23.5 The Grand Canyon of the American Southwest reveals sedimentary rock layers that formed over hundreds of millions of years.

Fossil-containing layers of sedimentary rock formed long ago, when silt, volcanic ash, and other materials were gradually deposited, one above the other (Figure 23.5). This layering of sediments is called *stratification*. Although most sedimentary layers form horizontally, earthquakes or other geologic disturbances can tilt or break them.

The fossil record is spotty

We currently have fossils of about 250,000 species. However, judging from present diversity, there must have been millions of ancient, now-extinct species. For several reasons we will not be able to recover fossils for most of them. Most of the important, large-scale movements in the Earth's crust have wiped out evidence from crucial periods in the history of life. In addition, most members of ancient communities simply have not been preserved. For example, plenty of hard-shelled mollusks and bony fishes are represented in the fossil record. Jellyfishes are not, even though they may have been common. Population density and body size skew the record more. A population of ancient plants may have produced millions of pollen grains in each growing season, while the earliest human ancestors lived in small groups and produced few young. Therefore, the chance of finding a fossilized skeleton of an early human is small compared to the chance of finding spores of plant species that lived at the same time.

The fossil record is also biased toward certain environments and locations. Most species for which we have fossils lived on land or in shallow seas that, through geologic uplifting, became part of continents. We have only a few fossils from deep ocean sediments. Also, most fossils have been discovered in the Northern Hemisphere, probably because most geologists have lived and worked there.

How do we know a fossil's age? Sedimentary rocks that contain fossils are dated by way of their position relative

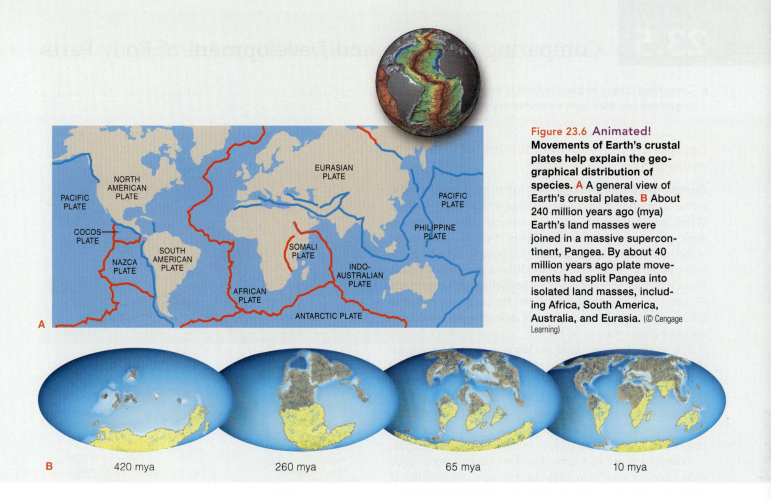

Figure 23.6 Animated!
Movements of Earth's crustal plates help explain the geographical distribution of species. A A general view of Earth's crustal plates. **B** About 240 million years ago (mya) Earth's land masses were joined in a massive supercontinent, Pangea. By about 40 million years ago plate movements had split Pangea into isolated land masses, including Africa, South America, Australia, and Eurasia. (© Cengage Learning)

Plate labels: NORTH AMERICAN PLATE, PACIFIC PLATE, COCOS PLATE, NAZCA PLATE, SOUTH AMERICAN PLATE, AFRICAN PLATE, EURASIAN PLATE, SOMALI PLATE, INDO-AUSTRALIAN PLATE, PHILIPPINE PLATE, PACIFIC PLATE, ANTARCTIC PLATE

A

B   420 mya   260 mya   65 mya   10 mya

to nearby volcanic rocks. The age of the volcanic rocks is determined by **radiometric dating**. This method tracks the radioactive decay of an isotope of some element that had been trapped inside the rock when the rock formed. Like the ticking of a perfect clock, the decay rate is constant. Radiometric dating is about 90 percent accurate.

Biogeography provides other clues

Biogeography—the study of the world distribution of plants and animals—also can shed light on past evolutionary changes. Biogeography asks why certain species (and higher groupings) occur where they do. For example, why do Australia, Tasmania, and New Guinea have species of monotremes (egg-laying mammals such as the duckbilled platypus), while such animals are absent from other parts of the world where the living conditions are similar? And why do the tropics have the greatest diversity of life forms? The simplest explanation for such biogeographical patterns is that species occur where they do either because they evolved there from ancestral species or because they dispersed there from elsewhere.

Charles Darwin probably would have been fascinated to learn of modern *plate tectonics*, the movement of plates of Earth's crust (Figure 23.6A). From studying evidence of such movements, we know that early in our planet's history all present-day continents, including Africa and

South America, were parts of a massive "supercontinent" called Pangea (Figure 23.6B). By determining the locations of plates at different times in Earth's history, researchers can shed light on possible dispersal routes for some groups of organisms and when (in geological history) the movements took place.

biogeography The study of the world distribution of plants and animals.

fossil Remains or traces of an organism of a past geologic age embedded and preserved in Earth's crust.

radiometric dating Method that uses the rate of decay of an isotope to determine the age of rock samples.

TAKE-HOME MESSAGE

HOW DOES THE STUDY OF FOSSILS AND BIOGEOGRAPHY CONTRIBUTE TO KNOWLEDGE OF EVOLUTION?

• Fossils and biogeography both provide evidence of evolution.

• The completeness of the fossil record varies, depending on the kinds of organisms represented, where they lived, and how stable their burial sites have been.

• Biogeographical patterns can provide clues to where a species arose. Along with evidence from plate tectonics, the patterns also may shed light on the routes by which some groups of organisms spread to new areas.

23.5 Comparing the Form and Development of Body Parts

- Comparing stages of development in major groups of organisms can shed light on evolutionary history.

Comparing body forms may reveal evolutionary connections

Comparative morphology uses information contained in patterns of body form to reconstruct evolutionary history. When populations of a species diverge, they diverge in their appearance, or in the functions of certain body parts, or both. Yet the related species also remain alike in many ways, because their evolution modifies a shared body plan. In such species we very often see **homologous structures**. These are the same body parts that have been modified in different ways in different lines of descent from a common ancestor (*homo-* means "same").

For example, most land vertebrates have homologous structures and probably share a common ancestor that had four five-toed limbs. The limbs diverged in form and became wings in pterosaurs, birds, and bats (Figure 23.7). All these wings are homologous—they have the same parts. The five-toed limb also evolved into the flippers of porpoises and the anatomy of your own forearms and fingers.

Body parts in organisms that *don't* have a recent common ancestor may also come to resemble one another in form and function. These **analogous structures** (from *analogos*, meaning "similar") arise when different lineages evolve in the same or similar environments. Different body parts, which were put to similar uses, were modified through natural selection and ended up resembling one another. For example, a dolphin, a fast-swimming marine mammal, has a sleek, torpedo-shaped torso—and so does a tuna, a fast-swimming fish.

Development patterns also provide clues

Vertebrates include fishes, amphibians, reptiles, birds, and mammals. Yet despite how different these groups are, comparing the ways in which their embryos develop provides strong evidence of their evolutionary links.

Early in development, the embryos of all the different vertebrate lineages go through strikingly similar stages (Figure 23.8). During vertebrate evolution, mutations that disrupted an early stage of development would have had devastating effects on the organized interactions required for later stages. Evidently, embryos of different groups remained similar because mutations that altered early steps in development were selected against.

How did the *adults* of different vertebrate groups come to be so different? At least some differences resulted from mutations that altered the onset, rate, or time of completion of certain developmental steps. For instance,

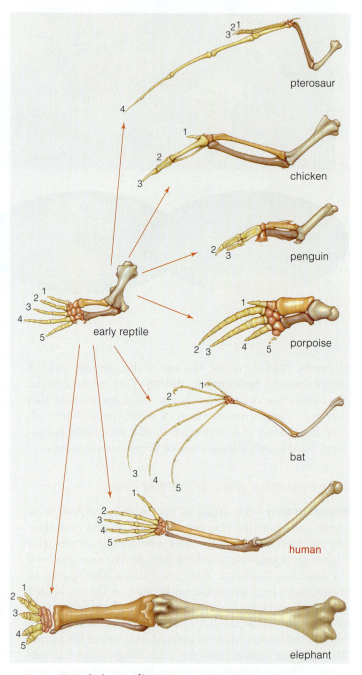

Figure 23.7 Animated! The form of forelimbs of humans and other vertebrates diverged as different groups evolved. This diagram starts with a generalized form of ancestral early reptiles. Diverse forms evolved even as similarities in the number and position of bones were preserved. The drawings are not to the same scale. (© Cengage Learning)

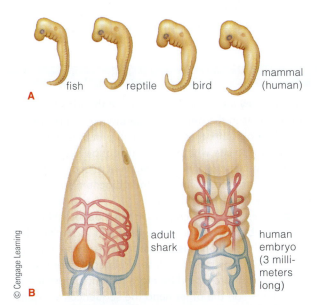

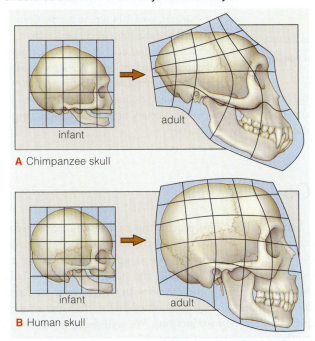

Figure 23.8 Comparing embryos also provides evidence of evolutionary relationships. A reveals that the early embryos of vertebrates are similar, even though the adults are very different. This is evidence of change in a common program of development. **B** Fishlike structures still form in the early embryos of reptiles, birds, and mammals. For example, a two-chambered heart (*orange*), certain veins (*blue*), and portions of arteries called aortic arches (*red*) develop in the embryos of sharks and other fishes. Adult fishes have them, too. These structures also form in an early human embryo.

Figure 23.9 Animated! The skulls of a chimpanzee and a human start out with similar proportions but diverge as growth continues. A shows a chimp's skull, **B** a human skull. Imagine that the skulls represented here are paintings on a blue rubber sheet divided into a grid. Stretching the sheet deforms the grid's squares. For the adult skulls, differences in size and shape in corresponding grid sections reflect differences in growth patterns. (© Cengage Learning)

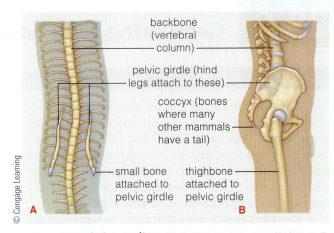

Figure 23.10 Animated! Vestigial body parts are "left over" from ancestral species. A Some python bones correspond to the pelvic girdle of other vertebrates, including humans. A snake has small vestigial hind limbs that are remnants from a limbed ancestor. **B** The human coccyx is a similar vestige from an ancestral species that had a bony tail, although muscles still attach to it.

Figure 23.9 depicts how change in the growth rate at a key point in development may have produced differences in the proportions of chimpanzee and human skull bones, which are alike at birth. Later on, they change dramatically for chimps but only slightly for humans. Chimps and humans have almost identical genes. Yet on the separate evolutionary road leading to humans, some regulatory genes probably mutated in ways that proved adaptive. From then on, instead of promoting the rapid growth required for dramatic changes in skull bones, the mutated genes have blocked it.

As Figure 23.10 indicates, the bodies of humans, pythons, and other organisms can have what seem like useless *vestigial* structures. For example, consider your own ear-wiggling muscles—which four-footed mammals (such as dogs) use to orient their ears. In humans, such body parts are left over from a time when more functional versions were important for an ancestor.

analogous structures
Body parts that are similar in different lineages because they are adaptations to a similar environment.

comparative morphology
Comparing the body form of different groups to help reconstruct evolutionary history.

homologous structures
Body parts that are similar in different lineages because the lineages have a common ancestor.

TAKE-HOME MESSAGE

HOW DOES THE STUDY OF PATTERNS OF BODY FORM HELP OUR UNDERSTANDING OF EVOLUTION?

- Comparing the patterns of body form in different groups of organisms provides clues about how body parts may have become modified in the course of evolution.

23.6 | Comparing Genetics

The kinds and numbers of outward traits species do (or don't) share are clues to how closely they are related. The same holds true for genes and proteins. Remember, the DNA of each species contains instructions for making RNAs and then proteins. This means that comparisons of DNA, RNA, and proteins from different species are additional ways of evaluating evolutionary relationships.

If you didn't already know that apes, monkeys, humans, and chimpanzees are all primates (Section 1.2), you might guess that they are related. You can test this idea by looking for differences in the amino acid sequence of a protein, such as hemoglobin, that occurs in all primates. You also could decide to see whether the nucleotide sequences in their DNA match closely or not much at all. Logically, the species that are most similar in their biochemistry are the most closely related.

Over time, mutations crop up in most genes. When two species both have the same gene and its nucleotide sequence is the same or nearly so in both, they must be closely related. Otherwise there would be more genetic difference. If on the other hand the sequences are quite different, many neutral mutations must have occured in them. A very long time must have passed since the species shared a common ancestor.

For instance, many organisms produce cytochrome c, a protein of electron transport chains. Studies show that the gene coding for the protein has changed very little over vast spans of time. Human cytochrome c has a primary structure of 104 amino acids. Chimps have the identical amino acid sequence. It differs by one amino acid in rhesus monkeys, by thirty-one in turtles, and fifty-six in yeasts, a type of fungus (Figure 23.11). Given this biochemical information, would you assume humans are more closely related in evolutionary time to a chimpanzee, a rhesus monkey, or a turtle?

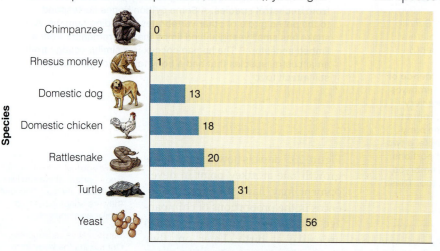

Figure 23.11 Comparing the structure of an amino acid in different organisms can shed light on evolutionary relationships. The human version of the protein cytochrome c is built from 104 amino acids. This graph shows the amino acid count for the same protein in some other organisms. (© Cengage Learning)

23.7 | How Species Come and Go

- The history of life on Earth is marked by extinction of some species and by the evolution of new species.

In extinction, species are lost forever

Extinction is the permanent loss of a species. Overall, species disappear at a fairly steady rate of "background extinction." A **mass extinction** is a sudden, widespread rise in extinctions above the background level (Figures 23.12 and 23.13). Major groups are wiped out simultaneously and the overall number of living species plummets. About 65 million years ago dinosaurs and many marine groups died out during a mass extinction, possibly due to environmental changes that occurred after one or more large meteorites struck Earth in a short time. The fossil record indicates that it may take as long as 100 million years for the overall number of species to recover after a mass extinction.

You may be aware that in the past few hundred years human activities have caused the extinction of thousands of species. The extinction rate is speeding up as we cut down forests, fill in wetlands, and otherwise destroy habitats of other animals and plants with which we share Earth. Global climate change is another factor that has contributed to patterns of extinctions. We will delve more deeply into these concerns in Chapter 25.

In adaptive radiation, new species arise

In **adaptive radiation**, new species of a lineage move into a wide range of habitats during bursts of microevolution. Many adaptive radiations have occurred during the first few million years after a mass extinction. Fossil evidence suggests this happened after dinosaurs went extinct. Many

Era	Period	Major extinction under way
CENOZOIC	QUATERNARY — 1.8 mya — TERTIARY	With high population growth rates and cultural practices (e.g., agriculture, deforestation), humans become major agents of extinction.
		Major extinction event
	— 65.5 —	
MESOZOIC	CRETACEOUS — 145.5 — JURASSIC — 199.6 — TRIASSIC	Slow recovery after Permian extinction, then adaptive radiations of some marine groups and plants and animals on land. Asteroid impact at K–T boundary, 85% of all species disappear from land and seas.
	— 251 —	*Major extinction event*
PALEOZOIC	PERMIAN — 299 — CARBONIFEROUS	Pangea forms; land area exceeds ocean surface area for first time. Asteroid impact? Major glaciation, colossal lava outpourings, 90%–95% of all species lost.
	— 359 —	*Major extinction event*
	DEVONIAN — 416 — SILURIAN	More than 70% of marine groups lost. Reef builders, trilobites, jawless fishes, and placoderms severely affected. Meteorite impact, sea level decline, global cooling?
	— 443 —	*Major extinction event*
	ORDOVICIAN — 488 — CAMBRIAN	Second most devastating extinction in seas; nearly 100 families of marine invertebrates lost.
	— 542 —	*Major extinction event*
	(Precambrian)	Massive glaciation; 79% of all species lost, including most marine microorganisms.

Figure 23.12 Many mass extinctions have occurred during Earth's history. Each extinction has been followed by a period of slow recovery, and the resulting mix of species is not the same as before. (© Cengage Learning)

Figure 23.13 Many marine species died out during the mass extinction at the end of the Devonian period. One of them was *Dunkleosteus*, a massive placoderm ("plate skin"). Here you see a cast of its skull, which was reconstructed from fossils. (Man: Photo by Lisa Starr; Skull: Courtesy of John McNamara, www.paleodirect.com)

new species of mammals arose and radiated into habitats where dinosaurs had once lived.

Adaptive radiations also have occurred in the human lineage. The ancestors of modern humans, including the tool-using species *Homo habilis* ("handy man," pictured in Figure 23.3B), apparently remained in Africa until about 2 million years ago. Around that time, genetic divergence led to new species, including *Homo erectus*, a human species that the fossil record places on the evolutionary road to modern humans. *H. erectus* coexisted for a time with *H. habilis*. But while members of its sister species also were upright walkers, *H. erectus* really earned the name. Its populations walked out of Africa, going left into Europe or right into

Asia. Judging from Middle Eastern fossils, our species, *Homo sapiens*, had evolved by 100,000 years ago.

What kinds of selection pressures triggered this adaptive radiation? Although we don't know for sure, this *was* a time of physical changes. One group of early humans, the Neanderthals, had large brains and were massively built (Figure 23.14). Some Neanderthal populations were the first humans to adapt to colder climates, such as in Europe. Later genetic changes resulted in anatomically modern humans. Scientists now consider Neanderthals our closest extinct relatives.

adaptive radiation The movement of new species of a lineage into a wide range of habitats.

extinction Permanent loss of a species.

mass extinction A sudden, widespread rise in extinctions above the typical background level.

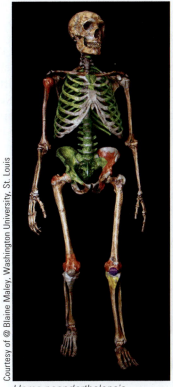

Courtesy of @ Blaine Maley, Washington University, St. Louis

Homo neanderthalensis   *Homo sapiens*

Figure 23.14 Adaptive radiation of humans included the evolution of several sister species, including Neanderthals and our own species, *Homo sapiens*. These skeletons are both from males. Scientists reconstructed the Neanderthal skeleton from fossils found in several locations, designated by different colors.

TAKE-HOME MESSAGE

HOW DO EXTINCTION AND THE RISE OF NEW SPECIES FIT INTO THE HISTORY OF LIFE ON EARTH?

- Extinction and the appearance of new species are part of the natural course of evolution.
- Species are always going extinct. In a mass extinction major groups of species go extinct in a short period of geological time.
- In adaptive radiation, new species rapidly (on a geological time scale) fill a range of habitats.

■ Like other life forms, we humans have a well-defined place in the evolutionary scheme of things.

The scientific name for the human species is *Homo sapiens*. Scientific names for organisms are always shown in italic type. In a binomial system devised centuries ago, the name has two parts. The first part is the genus name. A **genus** (plural: genera) encompasses all the species that are similar to one another and distinct from others in certain traits. The second part of the name indicates the particular species within the genus.

Species are organized into a hierarchy of groupings. Table 23.1 lists them, using humans as the example. Each group above the genus level includes a larger array of organisms that share more general features.

Each lineage of life forms has its defining traits. We humans share certain characteristics with other primates (such as being land-dwellers). At the same time, we differ in major ways from other primate lineages, such as the New World monkeys. We are genetically closer to the great apes and closest to the chimpanzees and bonobos (Figure 23.15).

Five trends mark human evolution

Primates evolved from ancestral mammals more than 60 million years ago. Fossils suggest that the first primates resembled small rodents. They may have foraged in the forest for insects, seeds, buds, and eggs. Between 54 and 38 million years ago, some primates were living in the trees—a habitat where natural selection would strongly favor some traits over others.

Precision grip and power grip The first mammals spread their toes apart to help support the body as they walked or ran on four legs. Primates still spread their toes or fingers. Many also make cupping motions, as when a monkey lifts food to its mouth. Other hand movements also developed in our ancient tree-dwelling relatives. Changes in hand bones allowed fingers to be wrapped around objects (that is, *prehensile* hand movements were possible), and the thumb and tip of each finger could touch (opposable movements).

TABLE 23.1 Classification of Humans

DOMAIN	Eukarya
KINGDOM	Animalia
PHYLUM	Chordata
CLASS	Mammalia
ORDER	Primates
FAMILY	Homininae
GENUS	*Homo*
SPECIES	*sapiens* (only living species of this genus)

Bruce Coleman Ltd./Photoshot

Table by Lisa Starr

Lemurs, lorises, galagos — Tarsiers — New World monkeys — Old World monkeys — Gibbons — Orangutans — Gorillas — Chimpanzees, bonobos — Humans

Hominins

Hominoids

Anthropoids

Figure 23.15 The primates include several nonhuman species. As the diagram shows, bonobos and chimpanzees are our closest primate relatives. Like us, bonobos walk upright.
(© Cengage Learning)

In time, hands also began to be freed from load-bearing functions—they were not needed to support the body. Much later, refinements in hand movements led to the precision grip and the power grip:

These hand positions would enable early humans to make and use tools. They helped form the foundation for the development of early technologies and culture.

Improved daytime vision Early primates had an eye on each side of the head. Later ones had forward-directed eyes, which is better for detecting shapes and movements in three dimensions. Later modifications allowed the eyes to respond to variations in color and light intensity (dim to bright)—another advantage for life in the trees.

Jaw shape and teeth of an early primate

Changes in dentition Changes in dentition—the teeth and jaws—of early primates accompanied a shift from eating insects to fruits and leaves, and on to an *omnivorous* (mixed) diet. Rectangular jaws and long canine teeth came to be further defining features of monkeys and apes. On the road leading to humans, a bow-shaped jaw and teeth that were smaller and all about the same length evolved.

Changes in the brain and behavior Living on tree branches also favored shifts in reproductive and social behavior. Imagine the advantages of single births over litters, for example, or of clinging longer to the mother. In many primate lineages, parents started to invest more effort in fewer offspring. They formed strong bonds with their young, maternal care became more intense, and the learning period grew longer.

Brain regions, such as the cerebral cortex, that are involved in information processing began to expand, and the brain case enlarged. New behavior promoted more brain development—which in turn stimulated more new behavior. In other words, brain modifications and behavioral complexity became closely linked. We see the links clearly in the parallel evolution of the human brain and culture. *Culture* is the sum total of behavior patterns of a social group, passed between generations by learning and symbolic behavior—especially language. The capacity

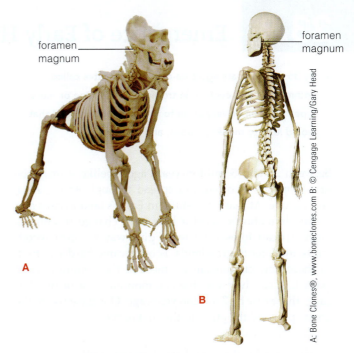

Figure 23.16 Compare the skeletal organization of two primates. Pictured are **A** a gorilla and **B** a human.

for language arose among ancestral humans through changes in the skull bones and expansion of parts of the brain.

Upright walking Of all primates, only humans can stride freely on two legs for long periods of time. This two-legged gait, called **bipedalism**, emerged as elements of the ancestral primate skeleton were reorganized. As shown in Figure 23.16, humans have a shorter, S-shaped backbone as compared with apes. In addition, in apes (and monkeys) the foramen magnum, the opening at the base of the skull where the spinal cord can connect with the brain, is at the back of the skull. In humans, it is close to the center of the base of the skull. These and other features, such as the position and shape of the knee and ankle joints and pelvic girdle, make bipedalism possible. By current thinking, the evolution of bipedalism was the key change in the origin of human and humanlike species, both past and present.

> **bipedalism** A two-legged gait.
>
> **genus** All the species that are similar to one another and distinct from others with respect to certain traits.

TAKE-HOME MESSAGE

WHAT TRENDS MARKED THE EVOLUTION OF HUMANS?

- Trends associated with the evolution of humans include refined hand movements, improved vision, dentition for an omnivorous diet, the interconnected development of a more complex brain and cultural behavior, and upright walking.

Emergence of Early Humans

■ **By 36 million years ago tree-dwelling primates called anthropoids had evolved in tropical forests. One or more types were on or very close to the evolutionary road that would lead to monkeys, apes, and humans.**

Between 23 and 5 million years ago, apelike forms—the first *hominins*—evolved and spread through Africa, Asia, and Europe. At that time, shifts in Earth's land masses and ocean circulation caused a long-term change in climate. Lush African forests began to give way to open woodlands and later to grasslands. Food became harder to find. In these new circumstances, most of the hominins went extinct. A survivor was the common ancestor of two lineages that arose by 7 million years ago. One gave rise to the great apes, and the other to the first *hominins*.

Early hominins lived in central Africa

Sahelanthropus tchadensis may have been a hominin. It lived in central Africa 6 to 8 million years ago, when the ancestors of humans were becoming distinct from the apes (Figure 23.17A). By 3.9 million years ago a clearly bipedal hominin, *Australopithecus afarensis*, lived in Africa. The individual whose skeleton is shown in Figure 23.17B, dubbed "Lucy," had a slight build, unlike some other African hominins. About 2 million years earlier, at what is now Laetoli, Tanzania, *A. afarensis* walked across fresh volcanic ash during a light rain, which turned the ash to quick-drying cement. We know little about how various hominins were related or if they used tools, but the footprints at Laetoli (Figure 23.17C), as well as fossil hip and limb bones, confirm that they walked upright.

Is *Homo sapiens* "out of Africa"?

By a little over 2 million years ago, species of *humans*—members of the genus *Homo*—were living in woodlands of eastern Africa. One was *Homo habilis* (Figure 23.17D). Compared to other hominins, these early humans had a larger brain, smaller face, and thickly enameled teeth. They ate a mixed diet of plant and animal foods, and used tools. Fossil hunters have found many stone tools dating to the time of *H. habilis*.

Divergence produced *Homo erectus* (Figure 23.17E), a species related to modern humans. Its name means "upright man." *Homo erectus* coexisted for a time with *H. habilis*. Fossils reveal that between 2 million and 500,000 years ago, *H. erectus* began leaving Africa. Some still lived in Southeast Asia as recently as 37,000 years ago. Some populations evolved into a new species, *H. neanderthalensis*, the Neanderthals mentioned in Section 23.8. Although we don't know exactly when Neanderthals went extinct, evidence from caves in southern Europe indicates Neanderthals were living there as recently as 18,000 years ago. Genetic evidence shows Neanderthals and modern humans mated at some time during their shared history.

A

Sahelanthropus tchadensis
7–6 million years

Australopithecus afarensis
3.6–2.9 million years

A left: HO/AFP/Newscom A right: © Friedrich Saurer/Alamy B: Dr. Donald Johanson, Institute of Human Origins C: © Images of Africa Photobank/Alamy D: Colin Keates/ Dorling Kindersley/Getty Images E: Pascal Goetgheluck/Photo Researchers, Inc.

B

C

Figure 23.17 We have fossil evidence of African hominins and early humans. **A** Representative fossils of African hominins. **B** Remains of "Lucy" (*Australopithecus afarensis*), who was a hominin. **C** At Laetoli in Tanzania, Mary Leakey found these footprints made 3.7 million years ago. **D** A skull of *H. habilis* (see also Figure 23.3B), an early species of human. **E** A skull of *H. erectus*, a relative of modern humans that walked upright.

D

E

H. habilis
1.9–1.6 million years

H. erectus
1.9 million to 53,000 years ago

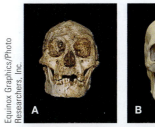

Figure 23.18 These fossil skulls are from early modern human forms. **A** An *H. floresiensis* skull, and **B** the skull of a modern human, *H. sapiens*. **C** Svante Paabo with a reconstructed Neanderthal skull. Paabo headed a team that sequenced the Neanderthal genome in 2010.

So *where* did our species, *Homo sapiens*, originate? Researchers dispute this point, although all base their hypotheses on measurements of the small genetic differences among modern human populations. We know that by 1 million years ago, *H. erectus* was living in many regions. The *multiregional model* holds that *H. erectus* evolved along different paths in different regions, in response to local selection pressures. Subpopulations of *H. sapiens* may have evolved from these groups, with gene flow preventing speciation. In 2003, fossils of early humans that date to 18,000 years ago turned up on the Indonesian island of Flores. The species was named *H. floresiensis* (Figure 23.18), but some researchers believe the fossil bones may be from modern humans who suffered a nutritional deficiency.

According to the *African emergence model*, humans arose in sub-Saharan Africa between 200,000 and 100,000 years ago, *then* moved out of Africa (Figure 23.19). Wherever they settled, they replaced *Homo erectus* populations that had preceded them. Phenotype differences that we associate with races evolved later.

About 15,000 years ago one small band of humans crossed a now-submerged land bridge that connected Siberia with North America. Some 14,000-year old fossil feces from a cave in Oregon is reportedly the oldest evidence of humans in North America.

For the past 40,000 years, cultural evolution has outpaced biological evolution of the human species—and so we leave our story. In thinking about this subject, we can keep in mind that humans spread rapidly over the planet by devising cultural means to deal with a wide range of environments. They also developed cultural features such as art and religious beliefs. People in some parts of the world still live as "stone age" hunters and gatherers even as other groups have moved to the age of high-tech. These differences are testimony to the remarkable behavioral flexibility and depth of human adaptations.

Cave painting made by early modern humans at Lascaux, France

TAKE-HOME MESSAGE

WHEN AND WHERE DID THE GENUS HOMO ARISE?

- The genus *Homo* arose about 2 million years ago, evidently in Africa.
- *Homo sapiens*, modern humans, is the only remaining species in this genus.

40,000 years ago

15,000–30,000 years ago

60,000 years ago

160,000 years ago

35,000–60,000 years ago

Figure 23.19 The map shows estimated times when populations of early *Homo sapiens* were colonizing different regions of the world, based on radiometric dating of fossils. The presumed dispersal routes (*white* lines) seem to support the African emergence model, but other hypotheses have been proposed. (© Cengage Learning)

- Experiments provide indirect evidence of how life may have emerged on Earth.

- Link to Characteristics of life 1.1

Four billion years ago, Earth was a thin-crusted, fiery inferno (Figure 23.20). Yet within 200 million years, life had originated on its surface! Geological changes have wiped out all physical traces of life's origin. Still, researchers have been able to put together a plausible explanation of how life began.

Conditions on early Earth were intense

What were the physical and chemical conditions on Earth at the time of life's origin? To answer this question, we need to know a little bit about what the young Earth was like. When patches of its crust were forming, heat and gases blanketed Earth. This first atmosphere probably consisted of gaseous hydrogen (H_2), nitrogen (N_2), carbon monoxide (CO), and carbon dioxide (CO_2). Were gaseous oxygen (O_2) and water also present? Probably not. Rocks don't release much oxygen during volcanic eruptions. Even if oxygen *were* released, those small amounts would have reacted at once with other elements, and any water would have evaporated because of the intense heat.

> **chemical evolution** The evolution of Earth's first biological molecules.

When the crust finally cooled and solidified, water condensed into clouds and the rains began. For millions of years, runoff from rains stripped mineral salts and other compounds from Earth's parched rocks. Salt-laden waters collected in depressions in the crust and formed the early seas.

The foregoing events were crucial to the beginning of life. Without an oxygen-free atmosphere, the organic compounds that started the story of life never would have formed on their own. Why? Oxygen would have attacked them and disrupted their functioning (as free radicals of oxygen do in cells). Without liquid water, cell membranes would not have formed, because cell membranes take on their bilayer organization only in water.

As you know, cells are the basic units of life. Each has a capacity for independent existence. Cells could never simply have appeared one day on the early Earth. Their emergence required the existence of biological molecules built from organic compounds. It also required metabolic pathways, which could be organized and controlled inside the confines of a cell membrane. Let's look at some current ideas about these essential events.

Biological molecules paved the way for cells to evolve

Some scientists believe that the structure shown in Figure 23.21 is a fossilized string of cells that is 3.5 billion years old. The first living cells probably emerged around 3.8 billion years ago and may have resembled certain modern bacteria that do not require or use oxygen. Before something as complex as a cell was possible, however, biological molecules must have come about through **chemical evolution**. Researchers have been able to put together several reasonable scenarios by which life on Earth could have emerged.

Rocks collected from Mars, meteorites, and Earth's moon—which all formed at the same time as Earth, from the same cosmic cloud—contain precursors of biological molecules. Possibly, sunlight, lightning, or heat escaping from Earth's crust supplied enough energy to drive chemical reactions that yielded more complex organic molecules. In various experiments that re-created conditions on the early Earth, molecules such as amino acids, glucose, ribose, deoxyribose, and other sugars were produced from formaldehyde. Adenine was produced from hydrogen cyanide. Adenine plus ribose occur in ATP, NAD, and other nucleotides vital to cells.

Painting by William K. Hartmann

Figure 23.20 The primordial Earth, about 4 billion years ago, may have looked something like this. Within another 500 million years, various types of living cells would be present on the surface.

THINK OUTSIDE THE BOOK

Based on years of research and new findings made possible by high-tech telescopes, astronomers now believe that many more Earthlike planets are likely to exist in other solar systems. It's possible that life could or does exist elsewhere in the universe. Why haven't we discovered other planets similar to our own? Research this topic to discover how astronomers answer this question.

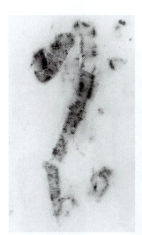

Figure 23.21 Is this a 3.5 billion-year-old fossil? Some researchers believe that this is a string of walled cells. It was unearthed in the Warrawoona rocks of Western Australia. (Stanley W. Awramik)

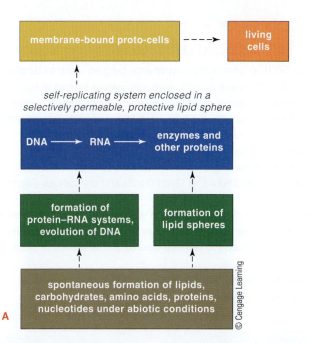

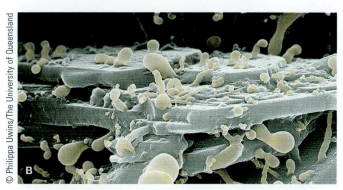

A

Figure 23.22 There are several hypotheses about how life began on Earth. A One possible sequence of events that may have led to the first self-replicating systems, then to the first living cells. B Nanobes—possible models for the first "pre-cells" on Earth.

How did complex compounds such as proteins form? In one scenario, these kinds of molecules could have assembled on clay in the muck of tidal flats. Clay is formed of thin, stacked layers of aluminosilicates with metal ions at its surface. Clay and metal ions attract amino acids. From experiments, we know that when clay is warmed by sunlight, then alternately dried out and moistened, it actually promotes reactions that produce complex organic compounds.

In another hypothesis, a variety of complex organic compounds formed near deep-sea hydrothermal vents. Today, species of primitive bacteria thrive in and around the vents. In laboratory tests, when amino acids were heated and immersed in water, they ordered themselves into small proteinlike molecules.

You may recall from Chapter 1 that metabolism and reproduction are two basic characteristics of life. During the first 600 million years of Earth history, enzymes, ATP, and other molecules that are important in metabolism could have assembled spontaneously in places where they were near one another. If so, their proximity would have promoted chemical interactions—and possibly the beginning of metabolic pathways.

In the area of reproduction, it's possible that the first "molecule of life" was not DNA but RNA. Simple self-replicating systems of RNA, enzymes, and coenzymes have been created in some laboratories. If so, how DNA entered the picture is still a mystery.

Figure 23.22A summarizes some key events that may have led to the first cells. Before cells could exist, there must have been membrane-bound sacs that sheltered molecules such as DNA (or RNA), key amino acids, and the like. Here again, this kind of structure has formed in the laboratory. Working with hot rocks from nearly 4 kilometers (2.5 miles) below Earth's surface, researchers in Australia also have found threadlike and bloblike structures that contain DNA and other organic compounds enclosed in a membrane (Figure 23.22B). These "nanobes" appear to grow and take up substances from outside. They are simpler than living cells, but some scientists think they could be like the forerunners of the first living cells. With time we may discover if that is correct.

EXPLORE
ON YOUR OWN

About 50,000 years ago humans began domesticating wild dogs. By about 14,000 years ago, people started to favor new breeds of dogs using artificial selection. Dogs having desired forms of traits were selected from each litter and later encouraged to breed. Those with undesired forms of traits were passed over.

This process has produced scores of domestic dog breeds, including sheep-herding border collies, sled-pulling huskies, and dogs as strikingly different as Great Danes and chihuahuas (Figure 23.23).

With a little bit of sleuthing on the Web or in a library, you should be able to discover numerous other examples of how humans have used artificial selection to develop desired animal breeds or plant varieties.

How has artificial selection affected aspects of your own life, such as pets, foods you eat, and ornamental garden plants?

Figure 23.23 The Great Dane (legs, *left*) and the chihuahua are both "designer" dog breeds.

SUMMARY

Section 23.1 Evolution modifies existing species. Therefore, broadly speaking, all species, past and present, share a common ancestry. The naturalist Charles Darwin proposed that evolution could occur by way of a process he named natural selection.

Section 23.2 In biology, a population is a group of individuals of the same species occupying a given area. The totality of genes in a population make up its gene pool. Each kind of gene may have different alleles, and these genetic variations produce variations in traits.

The relative numbers of different alleles—that is, the different versions of genes—can change as a result of four processes of microevolution: mutation, genetic drift, gene flow, and natural selection (Table 23.2). The large-scale patterns, trends, and rates of change among groups of species over time are called macroevolution.

Section 23.3 The theory of evolution by natural selection holds that there may be a difference in the survival and reproduction among members of a population that vary in one or more traits. That is, under prevailing conditions, one form of a trait may be favored because individuals that have it tend to survive and therefore to reproduce more often, so it becomes more common than other forms of the trait.

Organisms tend to come to have characteristics that suit them to the conditions in a particular environment. This trend is called adaptation.

A species is a genetic unit consisting of populations of organisms that closely resemble each other and that can interbreed and produce fertile offspring under natural conditions. New species come into being when the differences between isolated populations become so great that their members are not able to interbreed successfully in nature.

TABLE 23.2 Major Processes of Microevolution

MUTATION	A heritable change in DNA
GENETIC DRIFT	Random fluctuation in allele frequencies over time, due to chance occurrences alone
GENE FLOW	Movement of genes (alleles) among populations as individuals migrate
NATURAL SELECTION	Change or stabilization of allele frequencies due to differences in survival and reproduction among variant members of a population

Sections 23.4, 23.6 Evidence of evolutionary relationships comes in part from fossils and studies of biogeography. Fossils are dated using radiometric dating.

Comparative morphology often reveals similarities in development that indicate an evolutionary relationship. Similarities may reveal homologous structures, shared as a result of descent from a common ancestor. Alternatively, analogous structures arise when different lineages evolve in the same or similar environments. In comparative biochemistry, gene mutations that have accumulated in different species provide evolutionary clues.

Section 23.7 In a mass extinction, major lineages perish abruptly. Adaptive radiation is a burst of evolutionary activity; a lineage rapidly produces many new species. Both kinds of events have changed the course of biological evolution many times.

Section 23.8 We see five major evolutionary trends in the primate lineage leading to *H. sapiens*. These are (1) a transition to bipedalism, with related changes in the skeleton; (2) increased motor skills related to structural modification of the hands; (3) more reliance on daytime vision, including color vision and depth perception; (4) transition away from specialized eating habits, with corresponding modification of dentition; and (5) the enlargement and reorganization of the brain.

Section 23.9 Humans (*Homo sapiens*) are classified in the hominin family of the primate order, and are members of the only existing species of the genus *Homo*. In human evolution, the development of a larger, more complex brain correlated with increasingly sophisticated technology and with the development of complex behaviors and culture. For about the last 40,000 years, human cultural evolution has outpaced our species' biological evolution.

Section 23.10 Life originated on Earth about 3.8 billion years ago. Various experiments provide indirect evidence that life originated under conditions that presumably existed on the early Earth.

Comparisons of the composition of cosmic clouds and of rocks from other planets and Earth's moon suggest that precursors of the complex molecules associated with life were available.

When researchers simulated primordial conditions, chemical precursors assembled into sugars, amino acids, and other organic compounds.

Metabolic pathways could have evolved as a result of chemical competition for the limited supplies of organic molecules that had accumulated in the seas.

Self-replicating systems of RNA, enzymes, and coenzymes have been synthesized in the laboratory. How DNA entered the picture is not yet understood.

1. Distinguish between microevolution and macroevolution.

2. As shown in Figure 23.24, there is considerable variation in the facial features of humans. Explain this fact in terms of the concept of the gene pool.

3. Explain how natural selection differs from adaptation.

4. Explain the difference between
   a. divergence and gene flow
   b. homologous and analogous structures

5. Explain the difference between a primate and a hominin.

6. Describe the chemical and physical environment in which the first living cells may have evolved.

Top: Tomas Rodriguez/Fancy/Jupiterimages Bottom left: © Owen Franken/Corbis Bottom right: Christopher Briscoe/Photo Researchers, Inc.

Figure 23.24 Humans show a great deal of variation in their outward appearance.

SELF-QUIZ *Answers in Appendix V*

1. Fill in the blank: A _____ is a genetic unit consisting of one or more populations of organisms that usually closely resemble one another physically and physiologically.

2. Fill in the blanks: The relative numbers of different genes (alleles) in a gene pool change as a result of four processes of microevolution: _____, _____, _____, and _____.

3. Fill in the blank: A difference in survival and reproduction among members of a population that vary in one or more traits is called _____.

4. The fossil record of evolution correlates with evidence from _____.
   a. the geologic record
   b. radiometric dating
   c. comparing development patterns and morphology
   d. comparative biochemistry
   e. all of the above

5. Comparative biochemistry _____.
   a. is based mainly on the fossil record
   b. often reveals similarities in embryonic development stages that indicate evolutionary relationships
   c. is based on mutations that have accumulated in the DNA of different species
   d. compares the proteins and the DNA from different species to reveal relationships
   e. both c and d are correct

6. Comparative morphology _____.
   a. is based mainly on the fossil record
   b. shows evidence of divergences and convergences in body parts among certain major groups
   c. compares the proteins and the DNA from different species to reveal relationships
   d. both b and c are correct

7. In _____, new species of a lineage move into a wide range of habitats by way of bursts of microevolutionary events.
   a. an adaptive radiation
   b. natural selection
   c. genetic drift
   d. punctuated equilibrium

8. The pivotal modification in hominin evolution was _____.
   a. the transition to bipedalism
   b. hand modification that increased manipulative skills
   c. a shift from omnivorous to specialized eating habits
   d. less reliance on smell, more on vision
   e. expansion and reorganization of the brain

CRITICAL THINKING

1. The cheetahs in Figure 23.25 are among the roughly 20,000 of these sleek, swift cats left in the world. One reason cheetahs have become endangered is that about 10,000 years ago cheetahs experienced a severe loss in their numbers, and since then the survivors have been inbreeding. As a result, today's cheetah gene pool has very little variation and includes alleles that reduce the sperm counts of males and impair the animals' resistance to certain diseases. Based on what you've read in this chapter about genetic variation, does it seem likely that a "gene therapy"

program might be able to correct the genetic problems and help cheetahs make a comeback? Explain your answer.

2. Humans can inherit various alleles for the liver enzyme ADH (alcohol dehydrogenase), which breaks down alcohol. People of Italian and Jewish descent commonly have a form of ADH that detoxifies alcohol very rapidly. People of northern European descent have forms of ADH that are moderately effective in alcohol breakdown, while people of Asian descent typically have ADH that is less efficient at processing alcohol. Explain why researchers have been able to use this information to help trace the origin of human use of alcoholic beverages.

3. In 1992 the frozen body of a Stone Age man was discovered in the Austrian Alps. Although the "Iceman" died about 5,300 years ago, his body is amazingly well preserved and researchers have analyzed DNA extracted from bits of his tissue. Can these studies tell us something about early human evolution? Explain your reasoning.

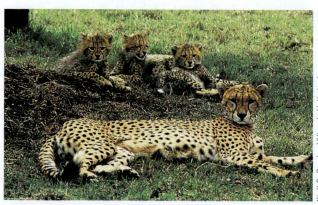

Figure 23.25 Genetically, the world's remaining cheetahs vary only a little—the result of inbreeding.

Kjell B. Sandved/Visuals Unlimited

PRINCIPLES OF ECOLOGY

LINKS TO EARLIER CONCEPTS

This chapter places your study of human biology in the broader context of the whole living world (1.3).

You will gain a global perspective on the energy and raw materials that sustain living organisms (2.1, 3.13). You will also learn how a global water cycle makes water available for life processes in organisms (2.5).

KEY CONCEPTS

Principles of Ecology

Ecology is the study of interactions of organisms with one another and the environment. Energy flows through ecosystems, passing from organism to organism by way of food webs. Sections 24.1–24.3

Chemical Cycles

Nutrients cycle in ecosystems. In biogeochemical cycles, water and nutrients such as carbon and nitrogen move from the physical environment to organisms, then back to the environment. Sections 24.4–24.6

Firefighters in dry Western states gear up for wildfire season, from July to October. In 2008 California had its worst fire season in fifty years, with more than 1 million acres and dozens of homes burned in a series of massive wildfires. Several firefighters died in the line of duty.

A series of dry winters set the stage for the infernos. For several years running, winter and spring rains had been few and far between. While occasional droughts have always been part of California's weather picture, the current "dry" is more ominous. Although the winter of 2009–2010 was fairly rainy, dry conditions returned the following year. Fire officials are bracing for a future marked by increasing fire danger, because Earth is warming, a trend that most climate experts expect will change climates and weather patterns for the long term. In the mountains of western North America, average temperatures will rise, so there will be less snow, drier forests and grasslands, and more intense wildfires.

Elsewhere, climate change is expected to have other effects—more severe hurricanes in the Atlantic and cyclones in the Pacific, melting glaciers and Arctic ice, rising sea levels that cause coastal flooding around the globe. All these changes will ripple through Earth's ecosystems and affect a variety human activities, from where people choose to live to the availability of reliable energy and water supplies and the kinds of crops farmers can grow.

In this chapter we start thinking about how we and other organisms interact with our environment, the science of ecology. This topic will lead to our survey of human impacts on Earth's ecosystems in Chapter 25.

24.1 Some Basic Principles of Ecology

- All life on Earth is part of some type of ecosystem.

- Link to Life's organization 1.3

Ecology is the study of the interactions of organisms with one another and with the physical environment. The general type of place in which a species normally lives is its **habitat**. For example, muskrats live in a stream habitat, damselfish in a coral reef habitat. The habitat of any organism has certain characteristic physical and chemical features. Every species also interacts with others that occupy the same habitat. Humans live in "disturbed" habitats, which we have deliberately altered for purposes such as agriculture and urban development.

Directly or indirectly, the populations of all species in a habitat interact with one another as a **community**. The large-scale community of organisms in a major ecological setting such as a desert or coral reef may be called a **biome** (Figure 24.1). Different land biomes have different dominant types of vegetation. For example, the plant species of a tropical rain forest are different from those of a rain forest in North America.

An **ecosystem** consists of one or more communities of organisms interacting with one another and with the physical environment through a flow of energy and a cycling of materials. Figure 24.2 shows some typical organisms of an arctic tundra ecosystem.

A species' **niche** (NITCH) consists of the various physical, chemical, and biological conditions the species needs to live and reproduce in an ecosystem. Examples of those conditions include the amount of water, oxygen, and other nutrients a species needs, the temperature ranges it can tolerate, the places it finds food, and the type of food it consumes. *Specialist* species have narrow niches. They may be able to use only one or a few types of food or live only in one type of habitat. For example, the red-cockaded woodpecker builds its nest mainly in longleaf pines that are at least seventy-five years old. Humans and houseflies are examples of *generalist* species with broad niches. Both can live in a range of habitats and eat many types of food.

Communities of organisms make up the *biotic*, or living, parts of an ecosystem. New communities may develop in habitats that were once empty of life, such as land exposed

biome The large-scale community of organisms in a major ecological setting, such as a desert.

community The interacting populations of all species in a habitat.

ecology The study of the interactions of organisms with one another and the physical environment.

ecosystem One or more communities of organisms interacting with one another and their physical environment through a flow of energy and a cycling of materials.

habitat The type of place where a species normally lives.

niche The physical, chemical, and biological conditions a species needs in order to live and reproduce in an ecosystem.

succession Sequence in which the first species to inhabit an area are then replaced by others.

© George H.H. Huey/Corbis

© John Easley, www.johneasley.com

Figure 24.1 Examples of biomes are **A** the hot desert near Tucson, Arizona, and **B** tropical aquatic realms such as this coral reef in the South Pacific.

by a retreating glacier, or in a previously disturbed inhabited area, such as an abandoned pasture. Through a process called **succession**, the first species to thrive in the habitat are then replaced by others, which are replaced by others in a predictable sequence. Eventually the composition of species stabilizes as long as other conditions remain the same. This more or less stable array of species is called a *climax community*.

In *primary* succession, changes begin when pioneer species colonize a newly available habitat, such as a recently deglaciated region (Figure 24.3). In *secondary* succession, a community develops toward the climax state after parts of a habitat have been disturbed. For example, this pattern occurs in abandoned fields, where wild grasses and other plants spring up when cultivation stops. Changing climate, natural disasters (such as forest fires), and other factors often interfere with succession so we rarely see truly stable climax communities.

TAKE-HOME MESSAGE

WHY LEARN ABOUT ECOLOGY AND ECOSYSTEMS?

- Humans and other organisms interact with one another in ecosystems.

- The general type of place where a species lives is its habitat. The populations of all species in a habitat make up a community.

- In all ecosystems there is a flow of energy and a cycling of materials.

Figure 24.2 **Animals, plants, and other organisms interact in ecosystems.** These photographs show some of the organisms you might see in the boreal forest of the Canadian Rockies **A**, including shelf fungi growing on a large conifer **B**; a pine grosbeak **C**; and a red fox **D**.

Figure 24.3 **Animated!** **Alaska's Glacier Bay provides examples of primary succession. A** A glacier is receding, leaving newly exposed soil. **B** The first plants are lichens, mosses, and small flowering plants that can grow and spread over glacial till. **C** Within twenty years, young alders begin to flourish. After eighty years, a climax community of spruces dominates. (A: © Douglas Peebles/Corbis B: © Pat O'Hara/Corbis C: © Tom Bean/Corbis)

■ Although there are many different types of ecosystems, they are all alike in many aspects of their structure and function.

Nearly every ecosystem runs on energy from the sun. Plants and other photosynthetic organisms are **producers** (or *autotrophs*, which means "self-feeders"). They capture sunlight energy and use it to build organic compounds from inorganic raw materials (Figure 24.4).

consumers The organisms in an ecosystem that eat producers (plants and other photosynthesizers).

food chain Linear sequence of who eats whom in an ecosystem.

food web Network of inter-linked food chains.

producers Photosynthetic organisms that capture energy and use it to make organic compounds.

All other organisms in an ecosystem are **consumers** (or *heterotrophs*, "other-feeders"). One way or another, consumers take in energy that has been stored in the tissues of producers. For the most part, consumers fall into four categories:

Herbivores, such as grazers and insects, eat plants; they are *primary* consumers.

Carnivores, such as lions, eat animals. Carnivores are *secondary* or *tertiary* (third-level) consumers.

Omnivores, such as humans, dogs, and grizzly bears, feed on a variety of foods, either plant or animal.

Decomposers, such as fungi, bacteria, and worms, get energy from the remains or products of organisms.

Producers obtain an ecosystem's nutrients and its initial pool of energy. As they grow, they take up water and carbon dioxide (which provide oxygen, carbon, and hydrogen), as well as minerals such as nitrogen. These materials, recall, are the building blocks for biological molecules. When decomposers get their turn at this organic matter, they can break it down to inorganic bits. If those bits are not washed away or otherwise removed from the system, producers can reuse them as nutrients.

It is important to remember that ecosystems must have an ongoing input of energy; in most cases, this energy source is the sun. Ecosystems often depend on outside sources of nutrients as well, as when erosion carries minerals into a lake. Ecosystems also *lose* energy and nutrients. Most of the energy that producers capture eventually is lost to the environment in the form of metabolic heat. Though nutrients generally are recycled, some also are lost, as when minerals are leached out of soil by water seeping down through it.

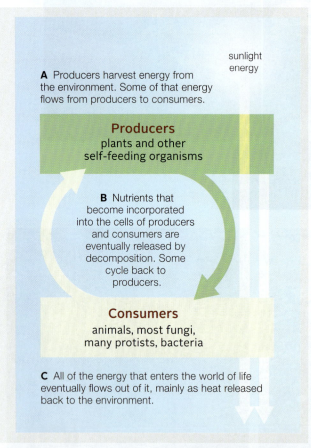

A Producers harvest energy from the environment. Some of that energy flows from producers to consumers.

sunlight energy

Producers
plants and other self-feeding organisms

B Nutrients that become incorporated into the cells of producers and consumers are eventually released by decomposition. Some cycle back to producers.

Consumers
animals, most fungi, many protists, bacteria

C All of the energy that enters the world of life eventually flows out of it, mainly as heat released back to the environment.

Figure 24.4 Animated! In ecosystems, there is a flow of energy and cycling of materials. **A** Energy from the environment flows through producers, then consumers. All energy that entered this ecosystem eventually flows out of it, mainly as heat. **B** Producers and consumers concentrate nutrients in their tissues. Some nutrients released by decomposition get cycled back to producers. (© Cengage Learning)

A simple food chain

Marsh Hawk

Upland Sandpiper

Garter Snake

Cutworm

Plants

Figure 24.5 A food chain is a simple hierarchy of feeding levels. (© Cengage Learning)

Energy moves through a series of ecosystem feeding levels

Each species in an ecosystem has its own position in a hierarchy of feeding levels (also called *trophic levels;* *troph-* means "nourishment"). A key factor in how any ecosystem functions is the transfer of energy from one of its feeding levels to another.

Primary producers, which gain energy directly from sunlight, make up the first feeding level. Corn plants in a field or waterlilies in a pond are examples. Snails and other herbivores that feed on the producers are at the next feeding level. Birds and other primary carnivores that prey on the herbivores form a third level. A hawk that eats a snake is a secondary carnivore. Decomposers, humans, and many other organisms can obtain energy from more than one source. For this reason they can't be assigned to a single feeding level.

human (Inuk)
arctic wolf
arctic fox
HIGHER TROPHIC LEVELS
A sampling of carnivores that feed on herbivores and one another
gyrfalcon
snowy owl
ermine
SECOND TROPHIC LEVEL
Major parts of the buffet of primary consumers (herbivores)
vole
arctic hare
lemming
Major parasitic consumers that feed at different trophic levels
mosquito
flea
tick
FIRST TROPHIC LEVEL
This is just part of the buffet of primary producers.
grasses, sedges
purple saxifrage
arctic willow

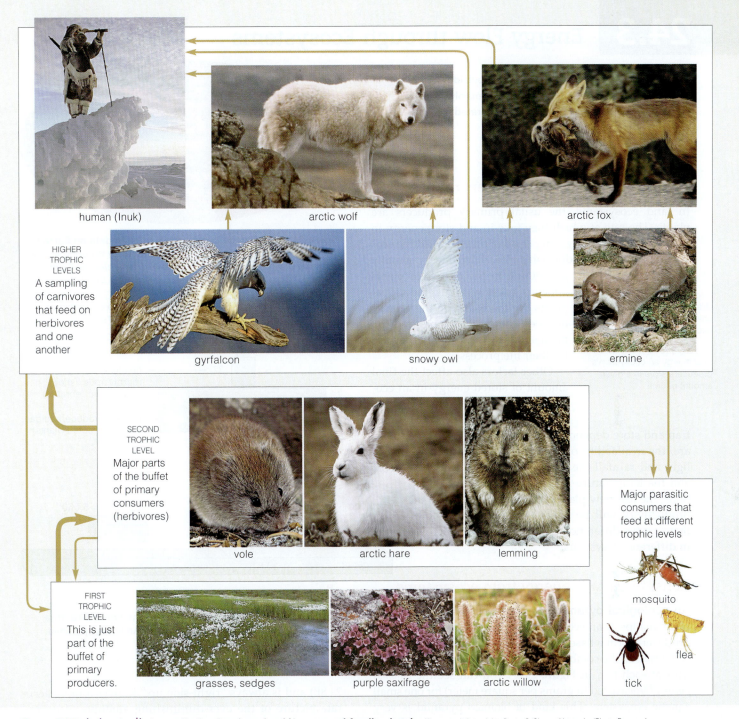

Figure 24.6 Animated! An arctic food web on land has several feeding levels. (Top row, left to right: Bryan & Cherry Alexander/Photo Researchers, Inc.; © Dave Mech; © Imagestate Media Partners Limited - Impact Photos/Alamy Second row, left to right: © Tom Wakefield/Photoshot; © Paul J. Fusco/Photo Researchers, Inc.; E. R. Degginger/Photo Researchers, Inc. Third row, left to right: © Tom J. Ulrich/Visuals Unlimited; © Dave Mech; Tom McHugh/Photo Researchers, Inc. Bottom row, left to right: © Jim Steinborn; © Jim Riley; © Matt Skalitzky Inset box: Photo by James Gathany, Centers for Disease Control and Prevention)

Food chains and webs show who eats whom

A linear sequence of who eats whom in an ecosystem is sometimes called a **food chain**. However, you won't often find such a simple, isolated chain as the one shown in Figure 24.5. Most species belong to more than one food chain, especially when they are at a low feeding level. It's more accurate to view food chains as cross-connecting with one another in **food webs**. Figure 24.6 shows a typical food web in an arctic ecosystem.

TAKE-HOME MESSAGE

WHAT IS A FOOD WEB?

- A food web is a complex system of feeding relationships in an ecosystem.
- Producers and various consumers and decomposers make up a food web.
- By way of food webs, different species in an ecosystem are interconnected.

24.3 Energy Flow through Ecosystems

- Energy flows into food webs from an outside source, usually the sun. Energy leaves mainly when organisms lose heat that is generated by their metabolism.
- Link to Metabolism 3.13

Producers capture and store energy

In land ecosystems, the usual primary producers are plants. The rate at which they take in and store energy in their tissues during a given period of time is called the ecosystem's **primary productivity**. How much energy actually gets stored in the tissues of plants depends on how many individual plants live there, and on the balance between energy the plants trap (by photosynthesis) and energy they use for their life processes.

ecological pyramid Graphic that represents the energy relationships in an ecosystem.

primary productivity The total energy an ecosystem's producers store in a given amount of time.

Other factors also affect the final amount of stored energy in an ecosystem at any given time. For example, how much energy ecosystems trap and store depends partly on how large the producers are, the availability of mineral nutrients, how much sunlight and rainfall are available during a growing season, and the temperature range. The harsher the conditions, the less new plant growth per season—so the productivity will be lower. As Figure 24.7 shows, there are big differences in the primary productivity in different ecosystems.

Figure 24.7 This image summarizes satellite data on Earth's primary productivity during 2002. Productivity is coded as *red* (highest) down through *orange*, *yellow*, *green*, *blue*, and *purple* (lowest). (Image by Reto Stöckli, based on data provided by the MODIS Science Team/ NASA Earth Observatory)

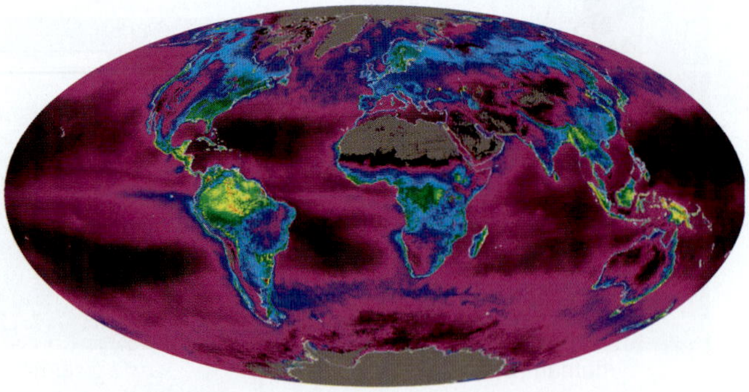

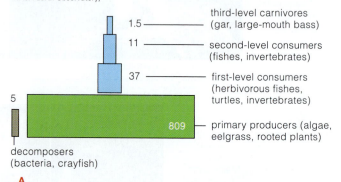

1.5
third-level carnivores (gar, large-mouth bass)
11
second-level consumers (fishes, invertebrates)
37
first-level consumers (herbivorous fishes, turtles, invertebrates)
5
809
primary producers (algae, eelgrass, rooted plants)
decomposers (bacteria, crayfish)
A
top carnivores
21
carnivores
383
herbivores
3,368
producers
20,810
decomposers = 5,060
B

Figure 24.8 Ecological pyramids show the tiered structure of ecosystems. A A biomass pyramid. B An energy pyramid. (© Cengage Learning)

Consumers take energy from ecosystems

An **ecological pyramid** is a way to represent the energy relationships of an ecosystem. In these pyramids, primary producers form a base for tiers of consumers above them.

Biomass is the combined weight of an ecosystem's organisms at each tier. In Figure 24.8A you can see the *biomass pyramid* (measured in grams per square meter) for one small aquatic ecosystem. This kind of biomass pyramid is very common in nature. There are lots of primary producers (aquatic plants) and few top carnivores such as large fishes.

Some biomass pyramids have the smallest tier on the bottom. A pond or the sea is like this. In those ecosystems primary producers have less biomass (they collectively weigh less) than consumers feeding on them. The producers consist of phytoplankton (tiny floating photosynthetic organisms), which grow and reproduce fast enough to provide a steady supply of food for a much greater biomass of zooplankton (small floating animals). Zooplankton in turn become food for larger animals.

An *energy pyramid* also shows how usable energy declines as it flows through an ecosystem. An energy pyramid has a large energy base at the bottom (Figure 24.8B) and is always "right-side up." It gives a more accurate picture of the ever-diminishing amounts of energy flowing through the ecosystem's feeding levels.

TAKE-HOME MESSAGE

HOW DOES ENERGY FLOW THROUGH ECOSYSTEMS?

- Energy enters ecosystems from outside—usually the sun. Energy leaves ecosystems mainly by the loss of metabolic heat from organisms.
- Ecological pyramids depict the tiered feeding (trophic) structure of ecosystems.
- A biomass pyramid shows the combined weight of organisms in each tier. An energy pyramid shows the loss of usable energy from the base tier to higher levels.

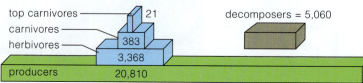

- Ecosystems depend on primary productivity. This is why the availability of water, carbon dioxide, and the mineral ions that serve as nutrients for producers have such an important impact on ecosystems.

In a **biogeochemical cycle**, ions or molecules of a nutrient are moved from the environment to organisms, then back to the environment—part of which serves as a reservoir for them. They generally move slowly through the reservoir, compared to their rapid movement between organisms and the environment. There are three types of biogeochemical cycles, based on the part of the environment that has the largest supply of the ion or molecule being considered. In *atmospheric cycles*, much of the nutrient is in the form of a gas. *Sedimentary cycles* move chemicals that do not occur as gases, such as phosphorus. Such solid nutrients move from land to the seafloor and return to dry land only through geological uplifting, which may take millions of years. The Earth's crust is the main storehouse for these substances.

The global **water cycle** is a third type of biogeochemical cycle (Figure 24.9). In this cycle, oxygen and hydrogen move in the form of water molecules.

Driven by solar energy, Earth's waters move slowly, on a vast scale, from the ocean into the atmosphere, to land, and back to the ocean—the main reservoir. Water evaporating into the lower atmosphere initially stays aloft in the form of vapor, clouds, and ice crystals. It returns to Earth as precipitation, mostly rain and snow. Ocean currents and prevailing wind patterns influence this global cycling.

biogeochemical cycle Movement of nutrients from the environment to organisms, then back to a reservoir in the environment.

water cycle The cycling of water molecules between bodies of water (the reservoirs) to the atmosphere and back to Earth, where water is available to organisms.

TAKE-HOME MESSAGE

WHAT IS A BIOGEOCHEMICAL CYCLE?

- In a biogeochemical cycle, nutrients move from the environment to organisms, then back to their reservoir in the environment.
- Nutrients usually move slowly through the environment but rapidly *between* organisms and the environment.
- In the water cycle, the world ocean is the main reservoir.

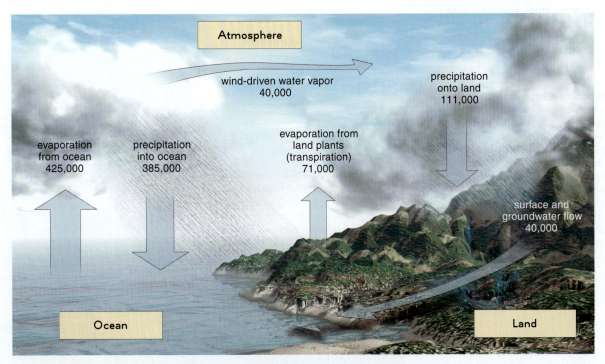

Figure 24.9 In the water cycle, water moves from oceans to the atmosphere, the land, and back to the ocean. *Yellow* boxes indicate the main reservoirs. *Arrow* labels identify the processes involved in the movement of water between reservoirs, measured in cubic kilometers per year. (© Cengage Learning)

24.5 The Carbon Cycle

- Carbon moves through the atmosphere and food webs on its way to and from the oceans, sediments, and rocks.

carbon cycle Movement of carbon molecules from organisms and the Earth's crust to the atmosphere and oceans, then back into organisms.

Figure 24.10 sketches the global **carbon cycle**. Sediments and rocks hold most of the carbon, followed by the ocean, soil, atmosphere, and land biomass. Carbon enters the atmosphere as cells engage in aerobic respiration, as fossil fuels burn, and when volcanoes erupt and release it from rocks in Earth's crust. Most atmospheric carbon occurs as carbon dioxide (CO_2). Most carbon dissolved in the ocean is in the forms of bicarbonate and carbonate.

You've likely seen bubbles of CO_2 rising to the surface of a glass of carbonated soda. Why doesn't the CO_2 in warm ocean surface waters escape to the atmosphere? Driven by winds and regional differences in water den-

sity, water makes a gigantic loop from the surface of the Pacific and Atlantic oceans to the Atlantic and Antarctic seafloors. There, its CO_2 moves into deep "storage" before bottom water loops up again (Figure 24.11). This looping movement is a factor in carbon's distribution in the biosphere and the global carbon "budget."

Photosynthesizers capture billions of metric tons of carbon atoms in organic compounds every year. However, the average length of time that a carbon atom is held in any given ecosystem varies quite a bit. For example, organic wastes and remains decompose rapidly in tropical rain forests, so not much carbon accumulates at the surface of soils. In marshes, bogs, and other places where there is little or no oxygen, decomposers cannot break down organic compounds completely, so carbon gradually builds up in peat and other forms of compressed organic matter. Also, in ancient aquatic ecosystems, carbon was incorporated in shells and other hard parts. The shelled organisms died and sank, then were buried in sediments. The same

Figure 24.10 Animated! Carbon cycles through the oceans, atmosphere, and the bodies of organisms. Part **A** shows the cycle through typical marine ecosystems. Part **B** shows how carbon cycles through land ecosystems. *Yellow* boxes indicate the main carbon reservoirs. The vast majority of carbon atoms are in sediments and rocks, followed by ever lesser amounts in ocean water, soil, the atmosphere, and biomass. Here are typical annual fluxes in the global distribution of carbon, in gigatons:

From atmosphere to plants by carbon fixation	120
From atmosphere to ocean	107
To atmosphere from ocean	105
To atmosphere from plants	60
To atmosphere from soil	60
To atmosphere from fossil-fuel burning	5
To atmosphere from net destruction of plants	2
To ocean from runoff	0.4
Burial in ocean sediments	0.1

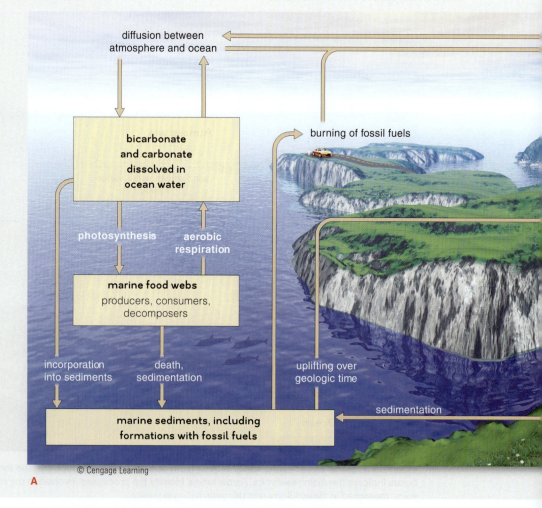

diffusion between atmosphere and ocean

burning of fossil fuels

bicarbonate and carbonate dissolved in ocean water

photosynthesis   aerobic respiration

marine food webs
producers, consumers, decomposers

incorporation into sediments

death, sedimentation

uplifting over geologic time

sedimentation

marine sediments, including formations with fossil fuels

© Cengage Learning

A

things are happening today. Carbon remains buried for many millions of years in deep sediments until part of the seafloor is uplifted above the ocean surface through geologic forces. Other buried carbon is slowly converted to long-standing reserves of gas, petroleum, and coal, which we humans have been tapping for use as fossil fuels.

Human activities, including the burning of fossil fuels, are putting more carbon into the atmosphere than can be cycled to the ocean reservoir. This factor is contributing to global warming and climate change, topics we will consider in some detail in Chapter 25.

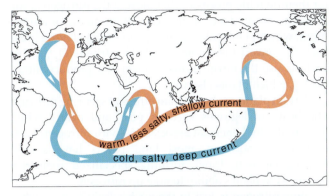

Figure 24.11 A "loop" of moving ocean water delivers carbon dioxide to carbon's deep ocean reservoir. The CO_2 sinks in the cold, salty North Atlantic and rises in the warmer Pacific. (© Cengage Learning)

TAKE-HOME MESSAGE

WHAT IS THE GLOBAL CARBON CYCLE?

- In the global carbon cycle, carbon is released by the metabolic processes of organisms, by decomposing organic material, by fossil fuel burning, and by geologic events.
- Much of Earth's carbon moves into the atmosphere (as carbon dioxide gas) and becomes dissolved in the oceans.
- Carbon may be buried in deep sea or land-based sediments for millions of years before cycling back into organisms.

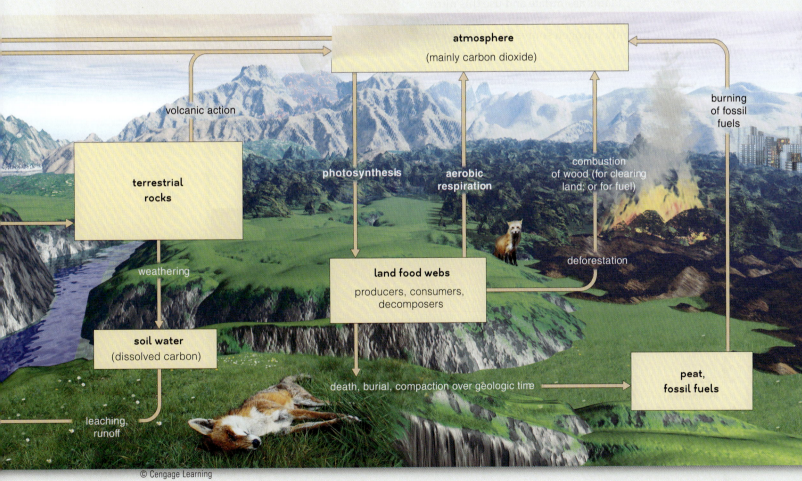

- Nitrogen, a component of our proteins and nucleic acids, moves in both an atmospheric and a sedimentary cycle called the nitrogen cycle.

- Link to Proteins and nucleic acids 2.11–2.13

Gaseous nitrogen (N_2) makes up about 80 percent of the atmosphere, the largest nitrogen reservoir. Among organisms, only a few kinds of bacteria can break the triple covalent bonds that hold its two atoms together.

Figure 24.12 shows the **nitrogen cycle**. As you can see, bacteria play key roles. They convert nitrogen to forms plants can use, and they also release nitrogen to complete the cycle. Land ecosystems lose more nitrogen through leaching of soils, although leaching provides nitrogen inputs to aquatic ecosystems such as streams, lakes, and the oceans.

nitrogen cycle Movement of nitrogen from the atmosphere, through nitrogen-fixing organisms, into plants, then back to the atmosphere.

nitrogen fixation Process in which certain bacteria and fungi convert nitrogen into forms that can be taken up by plants.

Nitrogen fixation is the process in which certain bacteria convert N_2 to ammonia (NH_3), which then dissolves to form ammonium (NH_4^+). Plants assimilate and use this nitrogen to make amino acids, proteins, and nucleic acids. Plant tissues are the only nitrogen source for humans and other animals.

Bacteria and some fungi also break down nitrogen-containing wastes and remains of organisms. The decomposers use part of the released proteins and amino acids for their own life processes. But most of the nitrogen is still in the decay products, in the form of ammonia or ammonium, which plants take up. In a process called *nitrification*, bacteria convert these compounds to nitrite (NO_2^-). Other bacteria use the nitrite in metabolism and produce nitrate (NO_3^-), which plants also use.

Certain plants are better than others at securing nitrogen. The best are legumes such as peas and beans, which have mutually beneficial associations with nitrogen-fixing bacteria. In addition, most land plants have similar associations with fungi, forming specialized roots that enhance the plant's ability to take up nitrogen.

Some nitrogen is lost to the air by *denitrification*, when bacteria convert nitrate or nitrite to N_2 and a bit of nitrous oxide (N_2O). Much of the N_2 escapes into the atmosphere, completing the nitrogen cycle.

TAKE-HOME MESSAGE

WHAT IS THE NITROGEN CYCLE?

- Nitrogen cycles from the atmosphere, through nitrogen-fixing organisms in soil and water, into plants and then to consumers, and back into the atmosphere.

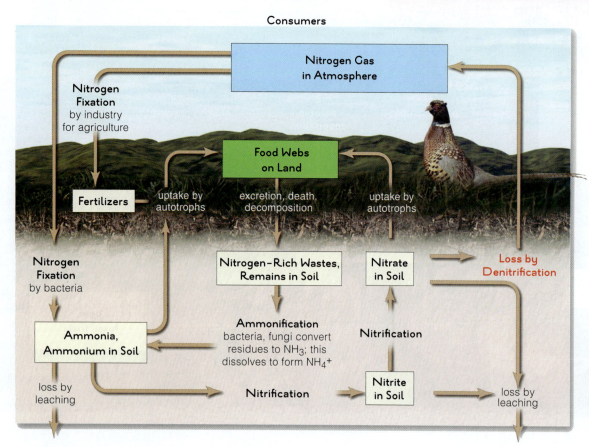

Consumers

Nitrogen Gas in Atmosphere

Nitrogen Fixation by industry for agriculture

Food Webs on Land

Fertilizers

uptake by autotrophs

excretion, death, decomposition

uptake by autotrophs

Nitrogen Fixation by bacteria

Nitrogen-Rich Wastes, Remains in Soil

Nitrate in Soil

Loss by Denitrification

Ammonia, Ammonium in Soil

Ammonification
bacteria, fungi convert residues to NH_3; this dissolves to form NH_4^+

Nitrification

loss by leaching

Nitrification

Nitrite in Soil

loss by leaching

Figure 24.12 In the nitrogen cycle bacteria play key roles. The atmosphere is the largest nitrogen reservoir.
(© Cengage Learning)

EXPLORE
ON YOUR OWN

E.F. Benfield, Virginia Tech

Benjamin Franklin once remarked, "It's not until the well runs dry that we know the worth of water."
Everybody needs water, and as Chapter 25 describes more fully, that essential fluid is in increasingly short supply. Have you ever wondered about your own "water impact"? You can get an idea using the following average statistics:

In a typical U.S. home, flushing toilets, washing hands, and bathing account for about 78 percent of the water used.

Nearly all toilets installed in the United States since 1994 use 1.6 gallons for each flush. Older toilets use about 4 gallons.

A shower uses about 5 gallons per minute (less if you have a low-flow shower head). Brushing teeth with the water running uses about 2 gallons. Shaving with the water running full blast can use up to 20 gallons.

Washing dishes with an automatic dishwasher uses about 15 gallons; handwashing dishes with the water running doubles that water use.

Using these numbers as a guide, keep track of your water use for a typical day. Are there ways you could conserve and still meet your basic needs?

SUMMARY

Section 24.1 Ecology is the study of interactions of organisms with one another and with their physical environment. The regions of Earth's crust, waters, and atmosphere where organisms live make up the biosphere. Every kind of organism has a habitat where it normally lives. The species in a given habitat associate with each other as a community.

An ecosystem encompasses producers, consumers, and decomposers and their physical environment. All interact with their environment and with one another through a flow of energy and a cycling of materials. A species' niche consists of the combined physical, chemical, and biological conditions it needs to live and reproduce in an ecosystem.

In ecological succession, the first species that take hold in a habitat are later replaced by others which themselves are replaced until conditions support a more or less stable array of species in the habitat.

Section 24.2 Ecosystems gain and lose energy and nutrients. Sunlight is the main energy source. Primary producers (photosynthesizing plants or algae) convert solar energy to forms that consumers can use. Primary producers also assimilate many of the nutrients that are transferred to other members of the system.

Consumers include herbivores, which feed on plants; carnivores, which feed on animals; and omnivores, which have combination diets. Consumers also include decomposers, such as fungi and bacteria that feed on particles of dead or decomposing material.

An ecosystem's energy supply is transferred through feeding (trophic) levels. Primary producers make up the first feeding level, herbivores make up the next, carnivores the next, and so on.

Organisms that get energy from more than one source cannot be assigned to a single feeding level.

A food chain is a straight-line sequence of who eats whom in an ecosystem. Food chains usually cross-connect in intricate food webs.

Section 24.3 Primary productivity is the rate at which producers capture and store a given amount of energy in a given time period in an ecosystem. The amount of energy flowing through consumer levels drops at each energy transfer through the loss of metabolic heat and as food energy is shunted into organic wastes. Energy relationships in an ecosystem can be represented as an ecological pyramid in which producers form a base for successive levels of consumers.

Sections 24.4, 24.6 In a biogeochemical cycle, a substance moves from the physical environment into organisms, then back to the environment.

Water enters and leaves ecosystems through the water cycle. In the carbon cycle and the nitrogen cycle, where the elements exist mainly as atmospheric gases, the elements move through food webs and then ultimately return to the atmosphere.

REVIEW QUESTIONS

1. Explain what an ecosystem is, and name the central roles that producers play in all ecosystems.

2. Explain the difference between an organism's habitat and its niche.

3. Define and give examples of the different feeding levels in ecosystems. Which feeding level is most likely to include most humans?

4. Explain the difference between a food chain and a food web in an ecosystem.

5. Describe the reservoirs and organisms involved in the carbon cycle and the nitrogen cycle.

SELF-QUIZ *Answers in Appendix V*

1. _____ can be thought of as an ecosystem.
   a. A freshwater spring   c. A city
   b. A rain forest   d. All of the above

2. Ecosystems have _____.
   a. energy gains and losses   c. one feeding level
   b. nutrient cycling but not losses   d. a and b

3. Fill in the blank: _____ is the study of how organisms interact with one another as well as with their physical and chemical environment.

4. Feeding levels can be described as _____.
   a. structured feeding relationships
   b. who eats whom in an ecosystem
   c. a hierarchy of energy transfers
   d. all of the above

5. A feeding relationship that proceeds from algae to a fish, then to a fisherman, and then to a shark is _____.
   a. a food chain   c. a and b
   b. a food web

6. Primary productivity is affected by _____.
   a. photosynthesis and energy use by plants
   b. how many plants are neither eaten nor decomposed
   c. rainfall
   d. temperatures
   e. all of the above

7. Match the following terms with the suitable description.
   _____ ecological pyramid
   _____ biogeochemical cycle
   _____ ecosystem parts
   _____ primary productivity

   a. water or nutrients moving from the environment to organisms, then back
   b. producers, consumers, decomposers
   c. producers capturing and storing energy in their tissues
   d. energy relationships in an ecosystem

Your Future

© Cengage Learning

Changes that are already occuring due to global warming include melting polar ice and rising sea levels in several coastal areas around the globe, including the eastern and southeastern United States and parts of Southeast Asia. In the future we can expect to see more government efforts to protect coastal ecosystem features, including wetlands and sandy shorelines. In some vulnerable areas plans are under way to change building codes and even move residents to higher ground.

CRITICAL THINKING

1. Imagine and describe a situation whereby you would be a link (not the top predator) in a food chain.

2. The use of off-road recreational vehicles may double in the next twenty years. Enthusiasts would like increased access to government-owned deserts like the one shown in Figure 24.1A. Some argue that deserts are the perfect places for off-roaders because "there's nothing there." Explain whether you agree, and why.

3. If you were growing a vegetable garden, what variables might affect its net primary production?

HUMAN IMPACTS ON THE BIOSPHERE

LINKS TO EARLIER CONCEPTS

In this chapter, discussions of climate change and thinning of the ozone layer will expand on your knowledge of the carbon cycle (24.5).

Learning about human impacts on natural resources and biodiversity will draw on your understanding of ecosystems and biogeochemical cycles (24.1, 24.4).

KEY CONCEPTS

Human Population Growth

Humans have sidestepped some natural limits on population growth. Soaring human population size has led to many impacts on the natural world. Sections 25.1–25.3

Global Climate Change

Warming of Earth's lower atmosphere due to increases in greenhouse gases is changing climate patterns. Sections 25.4–25.5

Impacts on Resources and Biodiversity

Human activities and increasing demand for natural resources are creating pollution and straining ecosystems. Sections 25.6–25.10

Seven billion humans make a lot of trash. Stuff we toss out ends up everywhere—in huge landfills, along roadsides, in remote places like Antarctica and the slopes of Mount Everest, in the open ocean thousands of miles from shore. An unlucky seabird died after eating the trash pictured in the photograph below, including more than 300 pieces of discarded plastic.

A planet littered with trash is just one of many impacts our species is having on the biosphere. Our growing population fuels demand for land, water, and other natural resources at a faster and faster clip. Efforts to meet these demands are a major force driving the rapid pace of extinction of wild species noted in Chapter 23. The fossil fuels we burn are one major cause of the warming of Earth's lower atmosphere and global climate change. More and more people are becoming aware that human activities are changing planet Earth in ways that do serious harm not just to other species but to our own as well.

This chapter will help give you the tools to think critically about human impacts on Earth's environments. Decisions we make today about climate change and other environmental issues are likely to shape Earth's ecosystems and the quality of human life far into the future.

Claire Fackler/NOAA

Human Population Growth

■ Advances in agriculture, industrialization, sanitation, and health care have fueled ever-faster growth of the human population. This growth is a major factor in changes that are occuring in Earth's ecosystems.

age structure The relative number of individuals of each age in a population.

population density The total number of individuals in a given area of habitat.

TFR Total fertility rate; the average number of children born to women of a population during their reproductive years.

The human population has grown rapidly

In 2011, there were 7 billion people on Earth. It took a long time, 2.5 million years, for the human population to reach the 1 billion milestone (Figure 25.1). It took less than 200 years more to reach 6 billion!

In any population, the growth rate is determined by the balance between births and deaths, plus gains and losses from immigration and emigration. Populations in different parts of the world grow at different rates, but overall, birth rates have been coming down worldwide. Death rates are falling, too, mainly because improved nutrition and health care are lowering infant mortality rates (the number of infants per 1,000 who die in their first year).

In order, the six countries expected to show the most growth are India, China, Pakistan, Nigeria, Bangladesh, and Indonesia. China and India combined dwarf all other countries in population size. They make up 38 percent of the world population. The United States is next in line. But with about 309 million people, it represents less than 5 percent of the world population.

The **total fertility rate (TFR)** is the average number of children born to women of a population during their reproductive years. In 1950, the worldwide TFR averaged 6.5. Currently it is 2.8, which is still far above replacement level of 2.1—the number of children a couple must have to replace themselves. These numbers are averages. In many developed nations TFRs are at or below replacement levels. Rates are highest in developing countries in western Asia and Africa. Figure 25.2 has some examples of the differences in the population distribution.

Even if every couple decides to bear no more than two children, the world population will keep growing for sixty years. It is projected to reach nearly 9 billion by 2050. Can we grow enough food and find enough drinkable water, energy sources, and all the wood, steel, and other materials to meet everyone's basic needs? That seems like a tall order, especially because billions of people do not have those necessities even now.

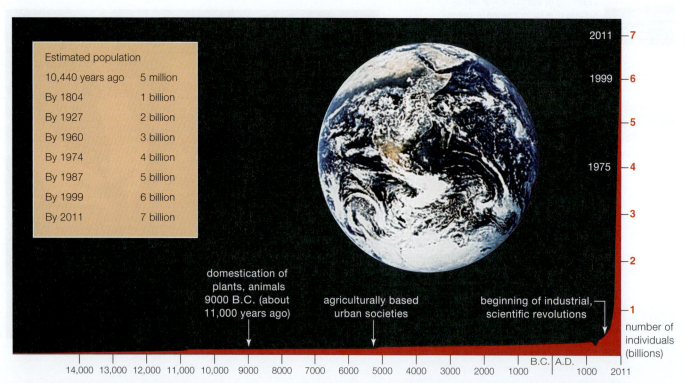

Estimated population

10,440 years ago	5 million
By 1804	1 billion
By 1927	2 billion
By 1960	3 billion
By 1974	4 billion
By 1987	5 billion
By 1999	6 billion
By 2011	7 billion

domestication of plants, animals 9000 B.C. (about 11,000 years ago)

agriculturally based urban societies

beginning of industrial, scientific revolutions

number of individuals (billions)

14,000 13,000 12,000 11,000 10,000 9000 8000 7000 6000 5000 4000 3000 2000 1000 | B.C. | A.D. | 1000 2011

Figure 25.1 Agricultural revolutions, industrialization, and improvements in health care have sustained the accelerated growth of the human population over the last two centuries. In the growth curve (*red*) the vertical axis represents world population, in billions. The dip between the years 1347 and 1351 is the time when 60 million people died from bubonic plague in Asia and Europe. (NASA)

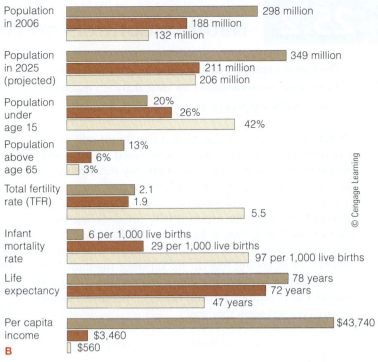

Figure 25.2 **Demographic information provides insight into the future growth of a population.** The graph at right shows key demographic indicators for three countries, mainly in 2006. The United States (*gold* bar) is highly developed, Brazil (*brown* bar) is moderately developed, and Nigeria (*ivory* bar) is less developed. The photograph shows just a few of India's more than 1.2 billion people.

Population statistics help predict growth

A population's demographics—its vital statistics, such as its size, density, and age structure—strongly influence its growth and its impact on ecosystems. Population "size" is the number of individuals in the population's gene pool. **Population density** is the number of individuals in a given area of habitat, such as the number of people who live within a given area of land. Another demographic is the general pattern in which a population's members are distributed in their habitat. We humans tend to cluster in towns and cities, where we interact socially, find jobs, and access other resources.

Age structure is the relative number of individuals of each age in a population. Often, the population is divided into prereproductive, reproductive, and postreproductive age categories. In theory, people in the first category will be able to produce offspring when they are sexually mature. Along with the people in the reproductive category, they

help make up the population's *reproductive base*. The population of the United States has a narrow base and is an example of slow growth. This pattern contrasts with the age structure in rapidly growing populations, which have a broad reproductive base (Figure 25.3).

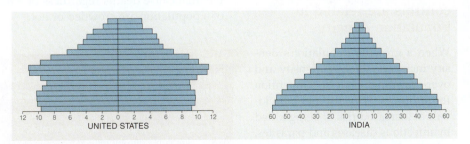

Figure 25.3 **Animated!** An age structure diagram provides a snap-shot of the number of people in a population who are of reproductive age. These diagrams show age structure of the populations of the United States and India in 1997. (© Cengage Learning)

■ **The human population can't continue to grow faster and faster. Even when conditions are ideal, there is a maximum rate at which any population can grow.**

Human populations can grow at a maximum rate of 2 to 5 percent per year. The rate is determined by how soon people begin to reproduce, how often they reproduce, and how many offspring are born each time.

Even when a population isn't reproducing at its full potential, it can grow exponentially (in doubling increments from 2 to 4, then 8, 16, and so on). For instance, it's biologically possible for human females to bear twenty children or more, but few do so. Yet since the mid-1700s, our population has been growing exponentially. This rapid growth can't continue forever, however, because "Mother Nature" doesn't work that way.

There is a limit on how many people Earth can support

Environmental factors prevent most populations of organisms from reaching their full biotic potential. For instance, when a basic resource such as food or water is in short supply, it becomes a **limiting factor** on growth. Other kinds of limiting factors include predation (as by pathogens) and competition for living space.

The concept of limiting factors is important because it defines the **carrying capacity**—how many individuals of a species that the resources in a given area can sustain on an ongoing basis. Some experts believe that Earth's resources can support from 7 to 12 billion humans, with a reasonable standard of living for many. Others believe that the current human population of 6.9 billion is already exceeding Earth's carrying capacity. Both viewpoints share the premise that overpopulation is the root of many, if not most, of today's environmental problems.

A low-density population grows slowly at first, then grows more rapidly. Growth levels off once the carrying capacity is reached. This pattern, called *logistic growth*, gives us an S-shaped curve (Figure 25.4). This curve is a simple approximation of what goes on in nature.

Some natural population controls are related to population density

When a growing population's density increases, the high density and over-crowding result in competition for resources. They also put individuals at increased risk of being killed by infectious diseases and parasites, which are more easily spread in crowded living conditions. These are *density-dependent controls* on population growth. Once such factors take their

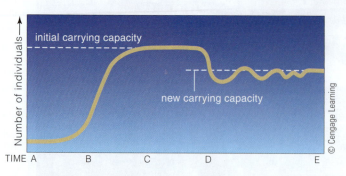

Figure 25.4 The S-shaped curve of logistic growth. The curve flattens out as the carrying capacity is reached. Changed environmental conditions can reduce the carrying capacity. This happened to the human population in Ireland in the late 19th century, when a fungus wiped out potatoes, which were the mainstay of the diet for all but the wealthy. An estimated 1 million people starved to death and half a million more emigrated to the United States and elsewhere in search of a better life. *Right*: Sculpture commemorating Irish immigration in the 19th century. (Keith Murphy/Shutterstock.com)

toll on a population and its density decreases, the pressures ease and the population may grow once more.

A classic example is the bubonic plague that killed 60 million Asians and Europeans—about one-third of the population—during the Middle Ages. *Yersinia pestis*, the bacterium responsible, normally lives in wild rodents, and fleas transmit it to new hosts. It spread like wildfire through the cities of medieval Europe because dwellings were crowded together, sanitation was poor, and rats were everywhere. In 1994, bubonic plague and a related disease, pneumonic plague, raced through rat-infested cities in India where garbage and animal carcasses had piled up for months in the streets. Only crash efforts by public health officials averted a public health crisis.

Density-independent controls can also operate. These are events such as floods, earthquakes, or other natural disasters that cause deaths regardless of whether the members of a population are crowded or not.

carrying capacity The number of individuals of a species that can be sustained indefinitely by the resources in a given area.

limiting factor Any factor that limits the growth of a population, such as the supply of food or the availability of living space.

TAKE-HOME MESSAGE

WHY IS THERE AN UPPER LIMIT TO THE NUMBER OF HUMANS THE EARTH'S RESOURCES CAN SUPPORT?

• Earth's resources are not unlimited, so over the long term given areas can only support a finite number of individuals of any species, including humans.

• Disease organisms and effects of pollution also limit population growth.

■ Population growth and the unsustainable use of natural resources are two major root causes of current and future environmental problems.

Today we hear constantly about environmental challenges facing humanity. Many of these problems are due directly or indirectly to the growing human population and its demand for natural resources.

Everyone has an ecological footprint

As societies strive to become more affluent, they replace natural landscapes such as forests with cropland, factories, and housing developments. Their citizens also want the conveniences of affluent life—cars, labor-saving appliances, computers, cell phones, and other electronic gadgets, to name a few. Having more money also allows people to travel over long distances and to buy more food—often, more meat and fish—sometimes from sources thousands of miles away. The net result of these and other changes is a growing **ecological footprint**—a shorthand term for the total resources a population consumes and the resulting wastes that are returned to the environment. Each of us has a personal ecological footprint as well.

Ecological footprints vary widely. For example, in a year the average consumer in the United States uses 100 times the resources consumed by someone in the poorest regions of Africa and Asia (Figure 25.5). The ecological footprints of people in rapidly developing nations such as China and India are growing at the same rapid pace.

Resources are renewable or nonrenewable

Overall, our planet's resources fall into two categories. **Renewable resources**, such as fresh water and forests, can be tapped indefinitely if they are replenished. On the

other hand, fossil fuels and minerals such as copper are for all intents and purposes **nonrenewable resources**. Earth's crust contains finite, limited amounts of them. More may form over many millions of years, but not on a human time scale. Therefore nonrenewable resources can be depleted, leaving us to our technological ingenuity to find usable, cost-effective replacements.

U.S. Department of Agriculture

Pollution can result from human activities

Often, using a resource produces wastes that the user must dispose of. All too often in technologically advanced societies, wastes or chemical by-products of industry and agriculture create pollution. A **pollutant** is a substance that in some way harms the health, activities, or survival of a population.

Natural events such as a volcanic eruption can release pollutants, but today most pollution comes from human activities. There are two basic sources for pollution. A *point source* is a single place or outlet, such as a leaky toxic waste dump. A *nonpoint source* is not tied to a particular location and so is harder to pin down. Pesticide runoff from farms and homes into a river or bay is an example.

In the remainder of this chapter we'll consider the effects of human activities on natural resources, starting with the impact of industry and fossil fuel use on the air we breathe.

ecological footprint The total resources a population consumes and the wastes it produces.

nonrenewable resources Resources, such as fossil fuels, that can't be restored within a human time frame.

pollutant A substance that harms the health, activities, or survival of a population.

renewable resources Resources that can be replenished.

Figure 25.5 Who has the largest ecological footprint? In Africa the rural poor rely heavily on homegrown crops and typically lack access to conveniences people in wealthier regions take for granted. (Left: © Adrian Arbib/Corbis Right: © Don Mason/Corbis)

TAKE-HOME MESSAGE

WHAT IS AN ECOLOGICAL FOOTPRINT?

- An ecological footprint is a measure of a population's impact on its environment—the total resources it consumes and the wastes it produces.
- Renewable resources can be replenished and so in theory may be available indefinitely. There are only finite amounts of nonrenewable resources.

■ A variety of human activities are polluting the air, with major consequences for ecosystems and human health.

Air pollutants include carbon dioxide, oxides of nitrogen and sulfur, and chlorofluorocarbons (CFCs). Others are photochemical oxidants formed as the sun's rays interact with certain chemicals. The United States releases more than 700,000 metric tons of air pollutants every day.

In some weather conditions, a layer of cool, dense air gets trapped under a warmer air layer (Figure 25.6). As a result of this *thermal inversion*, the atmospheric condition called **smog** develops. Some of the worst air pollution disasters have been due to thermal inversions.

Where winters are cold and wet, industrial smog forms as a gray haze over industrialized cities that burn coal and other fossil fuels for manufacturing, heating, and generating electric power. The burning releases airborne dust, smoke, ashes, soot, asbestos, oil, bits of lead and other heavy metals, and sulfur oxides. Most industrial smog forms in cities of developing countries, including China and India, as well as in eastern Europe.

In warm climates, photochemical smog forms as a brown, smelly haze over large cities. The key culprit is nitric oxide. After it is released from vehicles, nitric oxide reacts with oxygen in the air to form nitrogen dioxide. When exposed to sunlight, nitrogen dioxide can react with hydrocarbons (such as partly burned gasoline) to form photochemical oxidants. Some of those in smog resemble tear gas; even traces can sting the eyes, irritate lungs, and damage crops.

acid rain Rain or snow that contains large amounts of acid, usually sulfuric and nitric acids from industrial emissions.

smog Haze that develops when dust, smoke, acids, and other air pollutants accumulate in the lower atmosphere.

Oxides of sulfur and nitrogen are among the worst air pollutants. These substances come mainly from power plants and factories fueled by coal, oil, and gas, as well as from motor vehicles. Dissolved in atmospheric water, they form weak sulfuric and nitric acids that winds may disperse over great distances. If they fall to Earth in rain and snow, they form **acid rain**. Acid rain can be much more acidic than normal rainwater, sometimes as acidic as lemon juice (pH 2.3). The acids eat away at marble, metals, even nylon (Figure 25.7A). They also seriously damage the chemistry of ecosystems (Figure 25.7B).

Canadian researchers have reported that inhaled soot particles from a steel mill caused DNA mutations that showed up in the sperm of male mice and were passed on to offspring. More study is needed to learn whether such pollution-spurred mutations can cause disease in mice—or in humans who also breathe the polluted air.

Figure 25.7 Animated! **Acid rain causes widespread damage.** **A** A coal-burning power plant (*left*) and a stone sculpture eroded by acid rain (*right*). **B** Part of a forest in Great Smoky Mountains National Park, where nitrogen oxides and other forms of air pollution have killed trees. (A left: Michael Grecco Photography/Getty Images A right: Fletcher & Baylis/Photo Researchers, Inc. B: © Frederica Georgia/Photo Researchers, Inc.)

Figure 25.6 Thermal inversions set the stage for smog. **A** shows how air normally would circulate in smog-forming regions. **B** shows how air pollutants become trapped under a thermal inversion layer.
(© Cengage Learning)

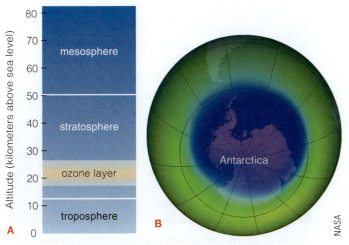

Figure 25.8 **Ozone concentrated in the stratosphere helps shield life on Earth from UV radiation. A** The location of the ozone layer. **B** Seasonal ozone thinning above Antarctica in 2007. Darkest blue indicates the area with the lowest ozone level.
(© Cengage Learning)

NASA

TABLE 25.1 Some Predicted Effects of Ozone Depletion
More skin cancers, eye cataracts, and worse sunburns
Increased acid rain and photochemical smog
Reduced ocean phytoplankton, harming ocean food webs and human seafood supplies
Increased global warming due to CFCs in the troposphere

Air pollution has damaged the ozone layer

Ozone is a molecule of three oxygen atoms (O_3). It occurs in two regions of Earth's atmosphere. In the troposphere, the region closest to Earth's surface, ozone is part of smog and can damage the respiratory system (as well as other organisms). On the other hand, ozone in the next atmospheric layer—the stratosphere (17–48 kilometers, or 11–30 miles, above Earth)—intercepts harmful ultraviolet radiation that can cause skin cancer and eye cataracts. **Ozone thinning** has damaged this protective screen. September through mid-October, an ozone "hole" appears over the Antarctic, extending over an area about the size of the continental United States. Since 1987 the ozone layer over the Antarctic has been thinning by about half every year; a new hole, over the Arctic, appeared in 2001. Several years later the Antarctic hole was the biggest ever, covering an area greater than North America (Figure 25.8).

Chlorofluorocarbons (CFCs) are the main ozone depleters. These gases (compounds of chlorine, fluorine, and carbon) are used as coolants in refrigerators and air conditioners and in solvents and plastic foams. They slowly escape into the air and do not break down easily. Through a series of chemical steps, each of their chlorine molecules can destroy over 10,000 molecules of ozone. A widely used fungicide called methyl bromide is even worse. It will account for 15 percent of ozone thinning in future years unless production stops.

CFC production in developed countries has been phased out. Some developing countries have announced plans to phase it out within the next few years. Methyl bromide production may also end soon. Even if these efforts succeed, it will be decades before ozone thinning is reversed. Current scientific models project that ozone depletion over the poles will be severe until at least 2019. In the meantime we can expect some ongoing negative impacts (Table 25.1). One of these repercussions, the loss of ocean phytoplankton (microscopic floating plantlike organisms), will worsen global warming, the topic we turn to next.

chlorofluorocarbons (CFCs) Compounds of chlorine, fluorine, and carbon that are used as coolants and that destroy ozone molecules when they escape into the atmosphere.

ozone thinning Reduction in the usual amount of ozone (O_3) in the stratosphere.

TAKE-HOME MESSAGE

WHAT ARE IMPACTS OF AIR POLLUTION?

- Air pollution can have local, regional, and global impacts.
- These impacts include smog, acid rain, and thinning of Earth's protective ozone layer as CFCs and other compounds destroy ozone molecules.

■ Human activities have increased the concentrations of greenhouse gases in the lower atmosphere so that the average temperature is warming. This warming is driving major changes in Earth's climate.

A variety of gases in Earth's atmosphere play a key role in shaping the average temperature near its surface. Temperature, in turn, has huge effects on global and regional climates.

Atmospheric molecules of carbon dioxide, water, ozone, methane, nitrous oxide, and chlorofluorocarbons are the major players in interactions that affect global temperature. Collectively, the gases act like the panes of glass in a greenhouse—hence their name, "greenhouse gases." Wavelengths of visible light pass through these gases to Earth's surface, which absorbs them and then emits them as heat. Greenhouse gases slow the escape of this heat back into space. Instead, much of it radiates back toward Earth's surface (Figure 25.9).

With time, heat builds up in the lower atmosphere and the air temperature near the surface rises. This warming action is known as the **greenhouse effect**. Without it, Earth's surface would be too cold to support life.

In the 1950s, researchers in a laboratory on Hawaii's highest volcano began a long-term program of measuring the concentrations of greenhouse gases in the atmosphere. They found that carbon dioxide concentrations follow the annual cycle of plant growth (that is, primary production) in the Northern Hemisphere. This includes the growth of phytoplankton at the ocean surface. CO_2 levels drop in summer, when photosynthesis rates are highest (and plants use lots of CO_2). They rise in winter, when photosynthesis by plants slows.

© Yann Arthus-Bertrand/Corbis

Figure 25.10 **Human activities produce large amounts of greenhouse gases.** Recorded changes in global temperature between 1875 and 2008 (*facing page*). At this writing, 2005 and 2010 were tied as the hottest years on record in the Northern Hemisphere. *Above*, Mexico City on a smoggy morning. With more than 10 million residents, it's the world's largest city.

The troughs and peaks in Figure 25.10 (*facing page*) are annual lows and highs of global carbon dioxide concentrations. For the first time, scientists have been able to see the big picture regarding effects of the carbon balances for an entire hemisphere. Notice the midline of the troughs and peaks in the cycle. It shows that the concentration of carbon dioxide is steadily increasing, as are the concentrations of other greenhouse gases.

Greenhouse gas levels are far higher than they were in most of the past. Carbon dioxide may be at its highest level since 420,000 years ago, perhaps even longer.

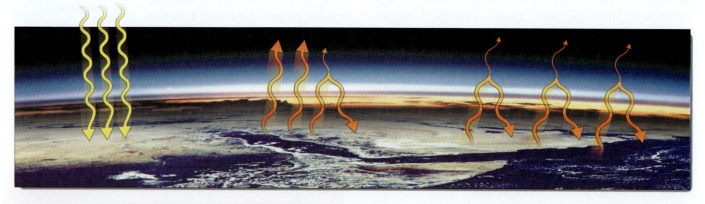

A Rays of sunlight penetrate the lower atmosphere and warm Earth's surface.

B The surface radiates heat (infrared wavelengths) to the lower atmosphere. Some heat escapes into space. But greenhouse gases and water vapor absorb some infrared energy and radiate a portion of it back toward Earth.

C Increased concentrations of greenhouse gases trap more heat near Earth's surface. Sea surface temperature rises, more water evaporates into the atmosphere, and Earth's surface temperature rises.

Figure 25.9 **The greenhouse effect warms Earth's surface.** (© Cengage Learning)

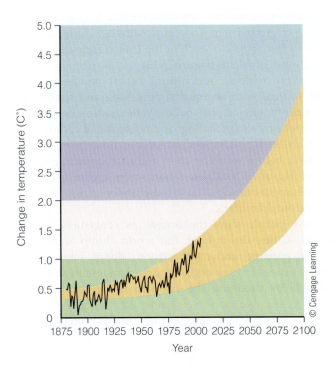

Figure 25.11 Glaciers are in retreat in high mountains all over the world. In Glacier Bay, Alaska, rocks and gravel now are exposed where ice covered the land ten years earlier (*left*). (Roger K. Burnard)

Today the great majority of climate scientists agree that the increase in greenhouse gases is a factor in **global warming**, a long-term rise in temperature near Earth's surface. Since direct measurements started in 1861, the lower atmosphere's temperature has risen by more than 1°F, mostly since 1946. Nine of the ten hottest years occurred between 1990 and the present. In 2007, the combined findings of nineteen different climate research programs and the Intergovernmental Panel on Climate Change (IPCC) were clear: It is almost certain that Earth's surface will warm by 3.6–8.1°F by the year 2100. Irreversible **global climate change** is under way, and it is happening much faster than expected.

What will climate change mean for us?

Continued temperature increases will have drastic effects on the climate of every region on Earth. Here we can mention only a few examples of the projected impacts.

To begin with, climate change may benefit agriculture in regions where winters become milder. In the Arctic, where sea ice once made the ocean impassable for much of the year, warming will ensure open channels and opportunities for fisheries. Several nations already are planning to expand commercial fishing operations there.

Elsewhere the impacts may be much less desirable. As evaporation increases, so will overall precipitation. Intense rains and flooding are expected to become more frequent in some regions, while other regions likely will experience more frequent and intense droughts. Current research suggests that severe, extended droughts may wipe out agriculture in parts of Africa where many people are desperately poor and farm small plots for food.

Glaciers all over the world are retreating (Figure 25.11), and polar ice is rapidly melting. One result of these changes is rising sea level in many coastal areas. The IPCC projects an average rise of 0.6 to 1.9 feet by 2100. If this estimate is accurate, in your lifetime the sea may submerge up to a third of the world's coastal wetlands and coral reefs, begin to flood many large coastal cities and agricultural lands, and erode large chunks of coastlines such as the Atlantic coast of the United States.

Disappearing glaciers also will wipe out a major source of fresh water for domestic and agricultural use in India, parts of South America and Africa, as well as western North America.

Although this discussion may seem depressing, it is important to remember that many smart people are working on strategies for slowing global warming and climate change. *Explore on Your Own* at the end of this chapter suggests ways you can assess your personal contribution to both these phenomena.

global climate change A shift in the Earth's climates.

global warming A long-term rise in temperature near the Earth's surface.

greenhouse effect The build-up of heat in the lower atmosphere due largely to the accumulation of carbon dioxide, methane, water vapor, and other gases.

TAKE-HOME MESSAGE

HOW DOES GLOBAL WARMING AFFECT THE EARTH?

- Global warming is the long-term rise in Earth's surface temperature due to the rapid accumulation of greenhouse gases in the atmosphere.
- Due to warming, a major shift in Earth's climate may be under way.

Problems with Water and Wastes

■ Three of every four humans do not have enough clean water to meet basic needs. Most of Earth's water is salty (in oceans). Of every million liters of water on our planet, only 6 liters are readily usable for human activities.

Water issues affect 75 percent of humans

As the human population grows exponentially, so do the demands and impacts on Earth's limited supply of fresh water.

About a third of the world's food grows on land that is irrigated with water piped in from groundwater, lakes, or rivers. Irrigation water often contains large amounts of mineral salts. Where soil drainage is poor, evaporation may cause salt buildup, or *salinization*. Globally, salinization is estimated to have reduced yields on 25 percent of all irrigated cropland. Large-scale irrigation (Figure 25.12) is depleting groundwater stored in the Ogallala aquifer, which extends from South Dakota to Texas and has been providing about 30 percent of the groundwater used for irrigation in the United States. In some areas the farmers are working to reduce their water use by switching to more efficient irrigation systems.

Communities located in deserts have been notorious water wasters. In a 1999 estimate, Las Vegas, Nevada—desert home of golf courses, swimming pools, and lush lawns—was said to use more water per resident than any other city in the world. Since then, dwindling supplies have forced the city to launch a program of water conservation. In coastal areas, overuse of groundwater can cause saltwater intrusion into human water supplies. Much of the United States is experiencing water problems (Figure 25.13).

In many regions, agricultural runoff pollutes public water sources with sediments, pesticides, and fertilizers. Power plants and factories pollute water with excess heat as well as chemicals (including carcinogens) and radioactive materials. Such pollutants may accumulate in lakes, rivers, and bays before reaching their ultimate destination, the oceans (Figure 25.14). Contaminants from human activities have begun to turn up even in supposedly "pure" water in underground aquifers.

Many people view the oceans as convenient refuse dumps. Cities throughout the world dump untreated sewage, garbage, and other noxious debris into coastal waters. Cities along rivers and harbors maintain shipping channels by dredging the polluted muck and barging it out to sea. We don't yet know the full impact such practices may have on fisheries that provide human food.

Managing solid wastes is another challenge

In 2008, the United States generated 505 million tons of garbage. In natural ecosystems, solid wastes are recycled, but we humans bury them in landfills or incinerate them. Incinerators can add heavy metals and other pollutants to the air and leave a highly toxic ash that must be disposed

Figure 25.12 Large-scale irrigation is common in many U.S. agricultural areas. A center-pivot sprinkler system is about 70 to 80 percent efficient. (© Craig Aurness/Corbis)

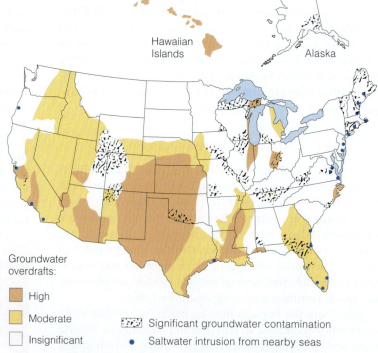

Groundwater overdrafts:

■ High
■ Moderate
□ Insignificant

▨ Significant groundwater contamination
● Saltwater intrusion from nearby seas

Figure 25.13 Many areas of the United States have groundwater troubles. (© Cengage Learning)

Laurence Lowry/Photo Researchers, Inc.

Figure 25.14 Industrial wastes polluted Lake Erie. A steel mill in Lackawanna, New York, discharged industrial wastes into Lake Erie until 1983, when the mill was closed. Major efforts to clean up Lake Erie have been under way.

Bill Bachmann/Photo Researchers, Inc.

© Alan Schein Photography/Corbis

of safely. Land that is both available and acceptable for landfills is scarce and becoming scarcer. All landfills eventually leak, posing a threat to groundwater supplies. That is one reason why communities increasingly take the "not in my back yard" (NIMBY) approach to landfills. For some highly populated areas, the "solution" has been to ship much of their garbage to rural areas (often in other states or countries) or dump it in the ocean.

Some components of trash may stay around for decades, especially if they contain plastic. Plastic bags may still be intact after forty or fifty years, and parts of a disposable diaper will last more than a hundred years.

On the plus side, about one-third of the total U.S. trash is recycled (Figure 25.15). That still leaves many millions of tons of plastic containers and glass bottles for society to deal with. To reduce the impact of plastic trash, consumers can buy fewer disposable items such as razors and avoid buying plastic when other, less environmentally harmful products exist. If plastic is used, it can be recycled. These days most communities make recycling easy with curbside pickups or recycling centers. Many grocery stores and other businesses provide free bins where plastic bags also can be recycled.

Figure 25.15 Recycling is an important part of efforts to manage solid wastes. A Recyclable material includes many plastics, glass containers, metal cans, and often other items. Many communities have special facilities for recycling items such as batteries, computers, and paint cans, which contain hazardous substances such as mercury and lead. **B** A garbage barge laden with trash. Most unfortunately, this barge is probably headed for the open ocean where the trash will be dumped.

TAKE-HOME MESSAGE

HOW DO HUMAN ACTIVITIES AFFECT WATER SUPPLIES AND SOLID WASTES?

- Human water supplies are threatened by overuse and pollution. Our activities also generate trillions of tons of solid wastes each year.
- Landfills are rapidly filling up, and land for new ones is becoming scarcer.
- Recycling helps reduce solid waste and conserve resources.

- The demand for food, housing, and forest products has led to widespread deforestation, agricultural practices that harm the environment, and the loss of habitat for other species.

- Link to Genetic engineering of plants 21.10

Like other ecological problems, those related to land use are linked to the rapid growth of the human population. People must have places to live and grow food, as well as materials to build homes and use as fuel.

Feeding and housing billions of humans requires land and other scarce resources

Only 25 percent of Earth is dry land, and only a fraction of that is available for human use. Today, almost a quarter of Earth's land is being used for agriculture. Scientists have made valiant efforts to improve crop production on existing land. As part of the "green revolution," research has been geared to improving the varieties of crop plants for higher yields and exporting modern agricultural practices and equipment to developing countries.

deforestation The removal of trees from large tracts of land.

desertification The conversion of agricultural land or grassland to a desertlike state, often due to overgrazing of livestock.

Unfortunately, the green revolution is based on huge inputs of fertilizers and pesticides and ample irrigation to sustain high-yield crops (Figure 25.16). It is based also on fossil fuel energy to drive farm machines. Crop yields *are* four times as high as from traditional methods. But modern practices use up to 100 times more energy and minerals such as nitrogen and phosphorus. Also there are signs that limiting factors are coming into play to slow down further increases in crop yields.

Overgrazing of livestock on marginal lands is a prime cause of **desertification**—the conversion of grasslands or cropland to a desertlike state that does not readily support useful plants (Figure 25.17). Worldwide, this has happened to about 10 million square kilometers over the past fifty years, and the trend is continuing.

Development of land for housing, towns, and cities also has an ecological price. One of the most important is the loss of habitat used by other species. In areas where water is scarce, such as much of the western United States, housing developments also increase the pressure on limited water supplies and political wrangling over access to them. As climate change alters conditions in these areas, these problems will be increasingly difficult to solve. We will return to the topic of habitat losses in Section 25.9.

Deforestation has global repercussions

The world's great forests influence the biosphere in many ways. Like giant sponges, forested watersheds absorb, hold,

A

© Charles O'Rear/Corbis

Figure 25.16 Agriculture requires large amounts of land and water. Rice must grow in shallow water of continuously flooded fields. It is a major staple food crop for tens of millions of people in Asia.

Figure 25.17 Desertification has become a major problem in parts of Africa. A A satellite photo taken in 2005. Red lines mark the huge area where desertification has occurred. **B** People who live in this area can barely eke out a living on the land.
(A: © Cengage Learning/NASA B: © ROMANO CAGNONI/Peter Arnold, Inc.)

and gradually release water. Forests also help control soil erosion, flooding, and sediment buildup in rivers, lakes, and reservoirs.

Deforestation is the name for removal of all trees from large tracts of land for logging, agricultural, or grazing operations. The loss of vegetation exposes the soil, and this promotes leaching of nutrients and erosion, especially on steep slopes. Cleared plots soon become infertile and are abandoned. The photograph in Figure 25.18A shows forest destruction in the Amazon basin of South America.

Deforestation is linked with several ecological problems. One of the most troubling effects relates to the global carbon cycle. Tropical forests absorb much of the sunlight reaching equatorial regions of Earth's surface. When the forests are cleared, the land becomes "shinier" and reflects more incoming energy back into space. The many millions of photosynthesizing trees in these vast forests help sustain the global cycling of carbon and oxygen. When trees are harvested or burned, carbon stored in their biomass is released to the atmosphere in the form of carbon dioxide—and this may be boosting the greenhouse effect.

About half the world's tropical forests have been cut down for cropland, fuel wood, grazing land, and timber. Deforestation is greatest in Brazil, Indonesia, Colombia, Mexico, Nigeria, and Cote d'Ivoire (Figure 25.18B). If clearing continues at present rates, within a few years only Brazil and the Democratic Republic of the Congo will have large tropical forests. By 2035, most of *their* forests will be gone.

Conservation biologists are attempting to reverse the trend. For example, in Brazil, a coalition of 500 groups is working to preserve the country's remaining tropical forests. In Kenya, women have planted millions of trees. Their success has inspired similar programs in more than a dozen countries in Africa. In eastern North America, forested land has increased in recent years due to regrowth in logged areas and the creation of commercial tree plantations. Elsewhere, such as the western United States, British Columbia (western Canada), and Siberia, logging is rapidly clearing vast tracts of old growth temperate forests (Figure 25.18C).

Figure 25.18 Huge tracts of tropical and temperate lands are being deforested. A Forest destruction in South America. **B** *Orange* shading on this map indicates countries where about 2,000 to 14,800 square kilometers of tropical forests are removed annually. *Gold* indicates sites of "moderate" tropical deforestation (100 to 1,900 square kilometers). **C** A huge clear-cut in a temperate conifer forest in British Columbia. (A: R. Bieregaard/Photo Researchers, Inc.; B: © Cengage Learning; C: Ted Kerasote/Photo Researchers, Inc.)

■ **Paralleling the growth of the human population is a steep rise in energy consumption.**

In theory, our planet has ample energy supplies to meet human needs for transportation and other uses. For example, world reserves of coal can meet our energy needs for at least several centuries. There are pros and cons associated with all energy sources. But as society wrestles with the rising costs and environmental risks associated with nonrenewable forms of energy such as coal and petroleum, pressure is building for a large-scale shift to renewable energy sources such as solar and wind energy.

Figure 25.19 shows the percentages of different energy sources used in the United States in 2006. Overall, developed countries use far more energy per person. This is a major reason why consumers in developed countries have a much larger ecological footprint.

There are growing issues with fossil fuels

Oil, coal, and natural gas are **fossil fuels**, the fossilized remains of ancient forests. Although there is plenty of coal

in the ground, known oil and gas reserves may be used up in this century. As the reserves run out in accessible areas—with political and market forces increasing the strain—pressure has mounted to drill in wilderness areas in Alaska, offshore, and other fragile environments. Due in part to the associated environmental costs, many people oppose the idea.

Increased coal burning is not popular either. It has been a major source of air pollution, because most reserves contain low-quality, high-sulfur coal. In addition, extraction and transportation of these fuels have harmful impacts. Oil harms many species when it leaks from pipelines or from ships. Strip mining for coal degrades the immediate area and often lowers the water quality of nearby streams. And as we have seen, burning fossil fuels also contributes to acid rain and adds to the greenhouse effect.

Can "green" energy sources meet the need?

Fossil fuels are the main source of power for all forms of transportation. The search for environmentally acceptable alternatives has gone in several directions.

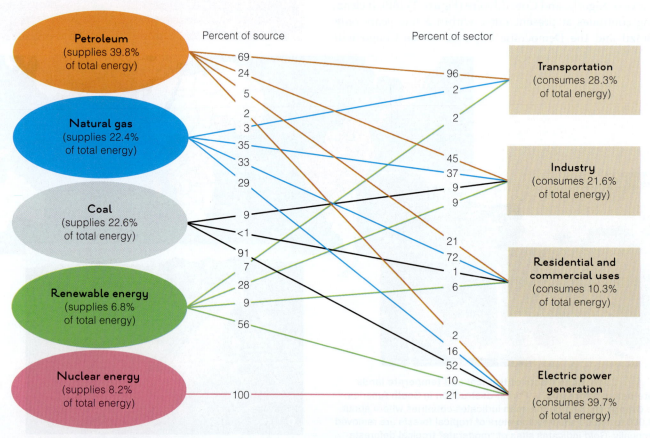

Figure 25.19 Currently, most of our energy comes from nonrenewable sources. This diagram links the sources of energy used in the United States in 2006 to different sectors of users. For example, petroleum supplies 96 percent of the transportation sector's energy needs, and this amounts to 69 percent of all the petroleum consumed in the United States. (© Cengage Learning)

Figure 25.20 **Electricity-producing photovoltaic cells are placed in panels that collect sunlight, a renewable source of energy.**

Figure 25.21 **This field of turbines is harvesting wind energy.**

Today solar power, produced by photovoltaic cells in panels such as those shown in Figure 25.20, is becoming a viable alternative in places where there is plenty of sunlight much of the year. Technological advances are bringing down the cost, but a homeowner who wants to install enough solar panels to supply all of a home's electricity often must spend tens of thousands of dollars.

As you know, automakers now sell hybrid vehicles that run on a combination of gasoline and electric power generated by batteries. There also has been much research aimed at developing all-electric vehicles and hydrogen fuel cells to power cars. So far, though, both are quite expensive and have other problems as well. Few electric-car "charging stations" exist, and the process for making hydrogen fuel cells generates high levels of greenhouse gases.

Another renewable option involves liquid **biofuels**— products such as ethanol and biodiesel made from plants and organic wastes. Unfortunately, growing most biofuel species (such as corn) and manufacturing and delivering the liquid fuel to consumers actually costs *more* in energy than it produces.

Hydropower from dams is a renewable energy source, but it too has drawbacks. For example, dams in rivers of the Pacific Northwest and California generate a great deal of electricity, but they also prevent endangered salmon from returning to streams above the dam to breed. As the salmon populations have suffered, so have endangered whales that feed on salmon in the ocean.

There are other alternatives, although none has yet proven itself for widespread use. In windy places, the mechanical energy of wind generates electricity as wind turns giant turbines (Figure 25.21). Analysts have estimated that a "wind power corridor" stretching from Texas north through South and North Dakota would generate enough electricity to meet the needs of 80 percent of the continental United States. Some people object to the turbines as eyesores and point out that the turbine blades kill birds.

What about nuclear power?

Today, nuclear power generates electricity at a relatively low cost. Increasing its use would decrease dependence on oil from the politically unstable Middle East. Also, nuclear power does not contribute to global warming, acid rain, or smog. So why hasn't there been a "nuclear power revolution"? Safety concerns are a major reason. For instance, there is always the potential for a meltdown if the reactor core becomes overheated. This happened at Chernobyl in Ukraine in 1986, and in 2011 at the Fukushima Dai-ichi power plant in Japan, where three reactors melted down in the aftermath of a tsunami. In both cases the meltdowns released potentially deadly radiation and millions of people were exposed to dangerous levels of radioactive fallout

biofuels Fuel products made from plants and organic wastes.

fossil fuels Oil, coal, and natural gas, all the fossilized remains of ancient forests.

Radioactive wastes also are highly dangerous. Some must be isolated for 10,000 years or longer, and there is little agreement on the best way to store them. Debate swirled around the decision to store nuclear wastes generated in the United States in a remote area of Nevada.

In the final analysis, all commercially produced energy has some kind of negative environmental impact. Experts agree on one thing: The best way to minimize that impact is for all of us to find ways to use less energy, now and in the future.

TAKE-HOME MESSAGE

WHAT DOES OUR ENERGY FUTURE LOOK LIKE?

- Demand is increasing for safe, cost-effective, renewable energy alternatives, but all of the technologies now available or being developed have drawbacks.
- Energy conservation is the best way to limit negative environmental impacts.

- Today human activities are a major cause of the rapid loss of other species.

- Link to Mass extinction 23.7

A major mass extinction is under way, and we humans are largely responsible for it. By one estimate our actions are leading to the premature extinction of six species an hour! At present the major culprits are destruction of wildlife habitats and the overexploitation of wild species for food or profit.

Habitat loss pushes species to the brink

endangered species A species native to an area that is vulnerable to extinction.

sustainability An approach for managing human population growth, resource use, and the preservation of wild habitats in ways that will help ensure the long-term survival of the human species.

The underlying causes of today's rapid pace of species extinctions are human population growth and economic policies that promote unsustainable exploitation. As our population grows, we clear, occupy, and damage more land to supply food, fuel, timber, and other resources (Figure 25.22A). In some regions, the combination of rapid population growth and poverty pushes the poor to cut forests, grow crops on marginal land, overgraze grasslands, and poach endangered animals (Figure 25.22B).

Globally, tropical deforestation is the greatest killer of species. The loss of plant species is extremely important because most animals depend directly or indirectly on plants for food, and often for shelter as well. We humans also have traditionally depended on plants as sources of medicines. For example, we get two anticancer drugs from a tropical plant called the rosy periwinkle. Unfortunately, most of its native forest habitat on the island of Madagascar has been cleared away to make room for human enterprises. Climate change also has begun affecting species such as the polar bear, which recently was listed as endangered by the U.S. Environmental Protection Agency (Figure 25.23). No giant asteroids have hit Earth for 65 million years. Yet a major extinction event is under way. Throughout the world, human activities are swiftly driving many species to extinction. An **endangered species** is an endemic species extremely vulnerable to extinction. *Endemic* means it originated in only one geographic region and lives nowhere else.

In the United States we are destroying the habitats of wild species at a dizzying pace. For instance, we have logged more than 90 percent of old-growth forests and drained half the wetlands, which filter human water supplies and provide homes for waterfowl and juvenile fishes. Hundreds of native species in these areas have gone extinct and dozens more are endangered.

Marine resources are being overharvested

People have been looking to the seas as a major source of food for the expanding human population, but there, too, little attention has been given to sustainable use of marine resources. One recent study estimated that for a third of seafood species, the annual catch has plummeted by 90 percent. The threatened species include Atlantic cod, black sea bass, and several shark species (Figure 25.24).

Figure 25.22 Habitat loss and overexploitation of wildlife is a major threat to the survival of many species. A Cities displace wild species and require huge amounts of resources to sustain the people who live in them. **B** Confiscated products made from endangered species. It's estimated that 90 percent of the illegal wildlife trade, which threatens hundreds of species of animals and plants, goes undetected.

James Marshall/Image Works/Time Life Pictures/Getty Images

Steve Hillebrand, U.S. Fish & Wildlife Service

Figure 25.23 Melting Arctic ice has put polar bears at risk. These polar bears are checking out a U.S. Navy submarine in the Arctic Ocean. The bears hunt from ice shelves and floes, but the ice appears to be melting rapidly due to climate change. (U.S. Navy photo by Chief Yeoman Alphanso Braggs)

The principle of sustainability is the answer

We humans are the dominant species on our planet, but many people have begun to wonder how long that will be the case. As you have read in this chapter, by our sheer numbers and many of our activities we have disturbed many, if not virtually all, ecosystems on the planet. That course cannot continue for much longer. Fortunately, governments and individuals are beginning to embrace the principle of **sustainability**. This principle is simple: By controlling our population growth, using resources wisely, embracing renewable energy sources, and protecting the wild places where other species live, we will be taking steps that help ensure our own survival as well.

Figure 25.24 Worldwide, sharks are among the animals increasingly at risk of overexploitation by human enterprises. (FAUP/Shutterstock.com)

25.10 Biological Magnification

SCIENCE COMES TO LIFE

Wildlife can be harmed and even driven to extinction by the process called **biological magnification**. This term refers to an increase in concentration of a substance in organisms as it is passed upward through food chains.

An example is the peregrine falcon, which nearly went extinct as a result of biological magnification of DDT. This synthetic pesticide was first used in the tropics during World War II to kill mosquitos that carried malaria. Later in Europe it helped control body lice that were transmitting a bacterium that causes typhus. After the war, it seemed like a good idea to use DDT against agricultural, garden, or forest pests.

DDT can build up in the tissues of animals that come into contact with it in the air, water, or food. After the war, it began to move through the global environment, infiltrate food webs, and affect organisms in ways that no one had predicted. In cities where DDT was sprayed to control Dutch elm disease, songbirds started dying. In streams flowing through forests where DDT was sprayed to control spruce budworms, salmon started dying. In croplands sprayed to control one kind of pest, new kinds of pests moved in. DDT was killing off natural predators that had been keeping pest populations in check.

Much later, side effects of biological magnification started showing up far from the areas where DDT was applied. Most devastated were species at the top of food chains, including bald eagles, brown pelicans, ospreys, and peregrine falcons (*below*). These predators fell prey to a product of DDT breakdown that interferes with physiological processes. The birds produced eggs with thin, brittle shells—and many of the chick embryos didn't survive to hatching time. Some species, including the peregrine falcon, were facing extinction.

DDT now has been banned in the United States for decades, except for limited applications where public health is endangered. Populations of peregrine falcons and other birds have begun to recover. Even today, however, some birds lay thin-shelled eggs. They pick up DDT at their winter ranges in Latin America, where DDT is still widely used.

biological magnification
An increase in the concentration of a substance in organisms as it is passed upward through food chains.

Peregrine falcon

© David T. Grewcock; Frank Lane Picture Agency/Corbis

© Cengage Learning

SUMMARY

Section 25.1 Recent, rapid growth of the human population has been due mainly to advances in agriculture, industrialization, sanitation, and health care. Differences in growth among countries correlate partly with levels of economic development and partly with demographics such as population density and age structure. Populations of countries with a large reproductive base generally grow the fastest.

Section 25.2 Carrying capacity is the number of individuals of a species that can be sustained long-term by the available resources in an area.

Populations initially show logistic growth—rapid growth that levels off when carrying capacity is reached. Carrying capacity, competition, and other factors limit population growth. Density-dependent controls on growth include competition for resources, disease, and predation.

Section 25.3 An ecological footprint is the sum total of resources used by a population or person, together with the resulting wastes. If renewable resources are replenished, they may be available indefinitely. Nonrenewable resources (such as fossil fuels) can't be readily replenished.

A pollutant is a substance that harms the health, activities, or survival of a population.

Sections 25.4, 25.9 Exponential growth of the human population has brought increased pollution and demands for energy, water, food, and waste disposal sites.

Air pollutants are present in industrial and photochemical smog and acid rain. Long-lasting chlorofluorocarbons (CFCs), once widely used as coolants and propellants in aerosol sprays, have played a major role in the seasonal thinning of the Earth's protective ozone later. Widespread deforestation is associated with leaching of soil nutrients, erosion, and possible disruption of the global carbon cycle.

Fossil fuels are nonrenewable and are major sources of air pollution. To meet the increasing demand for energy, conservation and the development of alternative, renewable energy sources will be crucial.

Human population growth and related activities also are contributing to the rapid loss of other species.

REVIEW QUESTIONS

1. Describe at least two ways that human activities are altering air quality.

2. Explain the difference between global warming and global climate change.

3. Describe the greenhouse effect. Make a list of twenty agricultural and other products that you depend on. Are any implicated in global warming?

4. What is deforestation, and how does large-scale deforestion damage an ecosystem?

SELF-QUIZ *Answers in Appendix V*

1. The number of individuals that can be sustained indefinitely by the resources in a given region is the _____.
   a. biotic potential
   b. carrying capacity
   c. environmental resistance
   d. density-dependent control

2. _____ shields organisms against the sun's UV wavelengths.
   a. A thermal inversion   c. The ozone layer
   b. Acid rain   d. The greenhouse effect

3. Acid rain is one outcome of _____.
   a. coal burning   c. some industrial processes
   b. gas and oil burning   d. all of the above

4. Greenhouse gases _____.
   a. slow the escape of heat from Earth into space
   b. are produced by natural and human activities
   c. are at higher levels than they were 100 years ago
   d. all of the above

5. All the following are renewable energy sources except _____.
   a. hydropower   c. natural gas
   b. wind   d. solar energy

6. Earth's biodiversity is threatened mainly by _____.
   a. the spread of disease organisms due to climate change
   b. habitat destruction by human activities
   c. disasters such as wildfires and hurricanes
   d. water pollution
   e. all of the above

CRITICAL THINKING

1. The map below shows the availability of direct solar energy in North America. Some areas are good candidates for solar heating systems and use of solar cells to produce electricity (see legend). What is the potential for using more solar energy where you live?

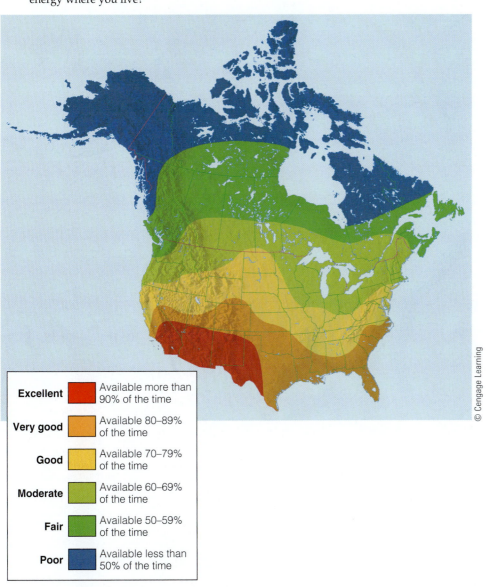

Excellent		Available more than 90% of the time
Very good		Available 80–89% of the time
Good		Available 70–79% of the time
Moderate		Available 60–69% of the time
Fair		Available 50–59% of the time
Poor		Available less than 50% of the time

© Cengage Learning

2. Some researchers and policymakers are proposing that we should protect entire ecosystems, such as forests, rather than just endangered species. Opponents say that this approach might endanger property values. Would you favor setting aside large tracts of land for restoration? Or would you consider such efforts too intrusive on individual rights of property ownership?

Your Future

Given the many urgent ecological issues raised in this chapter, it's no wonder that many people and businesses are making efforts to "go green." If each of us finds ways to live more sustainably, the future will be brighter for all of us.

3. Some people have proposed that all nuclear power plants in the United States be phased out over the next twenty years. Explain why you agree or disagree.

Concepts in Cell Metabolism

The nature and uses of energy

Any time an object is not moving, it has a store of **potential energy**—a capacity to do work, simply owing to its position in space and the arrangement of its parts. If a stationary runner springs into action, some of the runner's potential energy is transformed into **kinetic energy**, the energy of motion.

Energy on the move does work when it imparts motion to other things—for example, when you throw a ball. In skeletal muscle cells in your arm, the energy currency ATP (adenosine triphosphate, Sections 2.13, 3.8, 3.13–14) gave up some of its potential energy to molecules of contractile units and set them in motion. The combined motions in many muscle cells resulted in the movement of whole muscles. The transfer of energy from ATP also released another form of kinetic energy called **heat**, or *thermal energy*.

The potential energy of molecules is called **chemical energy** and is measured as kilocalories. A **kilocalorie** is the amount of energy it takes to heat 1,000 grams of water from 14.5°C to 15.5°C at standard pressure.

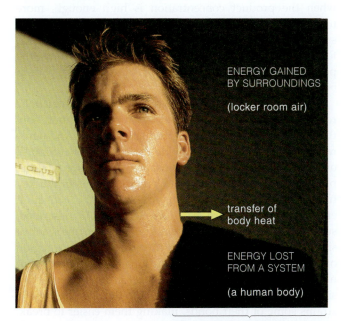

ENERGY GAINED BY SURROUNDINGS

(locker room air)

transfer of body heat

ENERGY LOST FROM A SYSTEM

(a human body)

NET ENERGY CHANGE = 0

Figure A.1 The body uses energy for life processes and loses a portion of it as heat. (© Cengage Learning; photo: Evan Cerasoli)

As noted in Chapter 3, cells use energy for chemical work, to stockpile, build, rearrange, and break apart substances. They channel it into *mechanical work*—to move cell structures and the whole body or parts of it. They also channel it into *electrochemical work*—to move charged substances into or out of the cytoplasm or an organelle compartment.

Laws of thermodynamics

We cannot create energy from scratch; we must first get it from someplace else. Why? According to the **first law of thermodynamics**, the total amount of energy in the universe remains constant. More energy cannot be created; existing energy cannot vanish or be destroyed. It can only be converted from one form to some other form. For instance, when you eat, your cells extract energy from food and convert it to other forms, such as kinetic energy for moving about.

With each metabolic conversion, some of the energy escapes to your surroundings, as heat. Even when you "do nothing," your body gives off about as much heat as a 100-watt lightbulb because of conversions in your cells. The energy being released is transferred to atoms and molecules that make up the air, and in this way it heats up the surroundings, as shown in Figure A.1. In general, the body cannot recapture energy lost as heat, but the energy still exists in the environment outside the body. Overall, there is a one-way flow of energy in the universe.

The human body obtains its energy mainly from the covalent bonds in organic compounds, such as glucose and glycogen. When the compounds enter metabolic reactions, specific bonds break or are rearranged. For example, your cells release usable energy from glucose by breaking all of its covalent bonds. After many steps, six molecules of carbon dioxide and six of water remain. Compared with glucose, these leftovers have more stable arrangements of atoms, but chemical energy in their bonds is much less. Why? Some energy was lost at each breakdown step leading to their formation. This is why glucose is a much better source of usable energy than, for example, water is.

As the molecular events just described take place, some heat is lost to the surroundings and cannot be recaptured. Said another way, no energy conversion can ever be 100

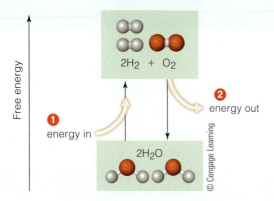

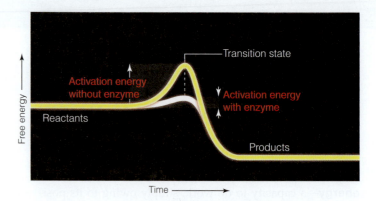

Figure A.2 Energy inputs and outputs in chemical reactions.
1 Endergonic reactions convert molecules with lower energy to molecules with higher energy, so they require a net energy input in order to proceed.
2 Exergonic reactions convert molecules with higher energy to molecules with lower energy, so they end with a net energy output.

Figure A.3 An enzyme enhances the rate of a reaction by lowering its activation energy. (© Cengage Learning)

percent efficient. Therefore, the total amount of energy in the universe is spontaneously flowing from forms rich in energy (such as glucose) to forms having less and less of it. This is the main point of the **second law of thermodynamics**.

Examples of energy changes

When cells convert one form of energy to another, there is a change in the amount of potential energy that is available to them. Cells of photosynthetic organisms, notably green plants, convert energy in sunlight into chemical energy, which is stored in the bonds of organic compounds. The outcome is a net increase in energy in the product molecule (such as glucose), as diagrammed in Figure A.2. A reaction in which there is a net increase in energy in the product compound is an **endergonic reaction** (meaning energy in). By contrast, reactions in cells that break down glucose (or another energy-rich compound) release energy. They are called **exergonic reactions** (meaning energy out).

The role of enzymes in metabolic reactions

The catalytic molecules called **enzymes** are crucial actors in metabolism. To better understand why, it helps to begin with the idea that in cells, molecules, or ions of substances are always moving at random. As a result of this random motion, they are constantly colliding. Metabolic reactions may take place when participating molecules collide—but only if the energy associated with the collisions is great enough. This minimum amount of energy required for a chemical reaction is called **activation energy**. Activation

energy is a barrier that must be surmounted one way or another before a reaction can proceed.

Nearly all metabolic reactions are reversible. That is, they can run "forward," from starting substances to products, or in "reverse," from a product back to starting substances. Which way such a reaction runs depends partly on the ratio of reactant to product. When there is a high concentration of reactant molecules, the reaction is likely to run strongly in the forward direction. On the other hand, when the product concentration is high enough, more molecules or ions of the product are available to revert spontaneously to reactants. Any reversible reaction tends to run spontaneously toward **chemical equilibrium**—the point at which it will be running at about the same pace in both directions.

As just described, before reactants enter a metabolic reaction they must be activated by an energy input; only then will the steps leading to products proceed. And while random collisions might provide the energy for reactions, our survival depends on thousands of reactions taking place with amazing speed and precision. This is the key function of enzymes, for enzymes lower the activation energy barrier (Figure A.3). As Section 3.13 described, substrates and enzymes interact at the enzyme's active site. According to the **induced fit model**, a surface region of each substrate has chemical groups that are almost but not quite complementary to chemical groups in an active site. However, as substrates settle into the site, the contact strains some of their bonds, making them easier to break. There also are interactions among charged or polar groups that prime substrates for conversion to an activated state. With these changes, substrates fit precisely in the enzyme's

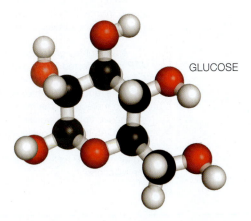

Figure A.4 A glucose molecule has six carbon atoms, which are colored black here. (© Cengage Learning)

active site. They now are in an activated state, in which they will react spontaneously.

Glycolysis: the first stage of the energy-releasing pathway

Energy that is converted into the chemical bond energy of adenosine triphosphate—ATP—fuels cell activities. Cells make ATP by breaking down carbohydrates (mainly glucose), fats, and proteins. During the breakdown reactions, electrons are stripped from intermediates, then energy associated with the liberated electrons drives the formation of ATP.

Recall that cells rely mainly on **aerobic respiration**, an oxygen-dependent pathway of ATP formation. The main energy-releasing pathways of aerobic respiration all start with the same reactions in the cytoplasm. During this initial stage of reactions, called **glycolysis**, enzymes break apart and rearrange a glucose molecule into two molecules of pyruvate, which has a backbone of three carbon atoms. Following up on the discussion in Section 3.14, here you can track in a bit more detail on what happens to a glucose molecule in the first stage of aerobic respiration.

Glucose is one of the simple sugars. Each molecule of glucose contains six carbon, twelve hydrogen, and six oxygen atoms, all joined by covalent bonds (Figure A.4). The carbons make up the backbone. With glycolysis, glucose or some other carbohydrate in the cytoplasm is partially broken down, the result being two molecules of the three-carbon compound pyruvate:

$$\text{glucose} \rightarrow \text{glucose-6-phosphate} \rightarrow 2 \text{ pyruvate}$$

The first steps of glycolysis require energy. As diagrammed in Figure A.5 on page A-4, they advance only when two ATP molecules each transfer a phosphate group to glucose and so donate energy to it. Such transfers, recall, are phosphorylations. In this case, they raise the energy content of glucose to a level that is high enough to allow the *energy-releasing* steps of glycolysis to begin.

The first energy-releasing step breaks the activated glucose into two molecules. Each of these molecules is called PGAL (phosphoglyceraldehyde). Next, each PGAL is converted to an unstable intermediate that allows ATP to form by giving up a phosphate group to ADP. The next intermediate in the sequence does the same thing. Thus, a total of four ATP form by **substrate-level phosphorylation**. This metabolic event is the direct transfer of a phosphate group from a substrate of a reaction to some other molecule—in this case, ADP. Remember, though, two ATP were invested to jump-start the reactions. So the *net* energy yield is only two ATP.

Meanwhile, the coenzyme NAD$^+$ picks up electrons and hydrogen atoms liberated from each PGAL, thus becoming NADH. When the NADH gives up its cargo at a different reaction site, it reverts to NAD$^+$. Said another way, like other coenzymes NAD$^+$ is reusable.

In sum, glycolysis converts energy stored in glucose to a transportable form of energy, in ATP. NAD$^+$ picks up electrons and hydrogen that are removed from each glucose molecule. The electrons and hydrogen have roles in the next stage of reactions. So do the end products of glycolysis—the two molecules of pyruvate.

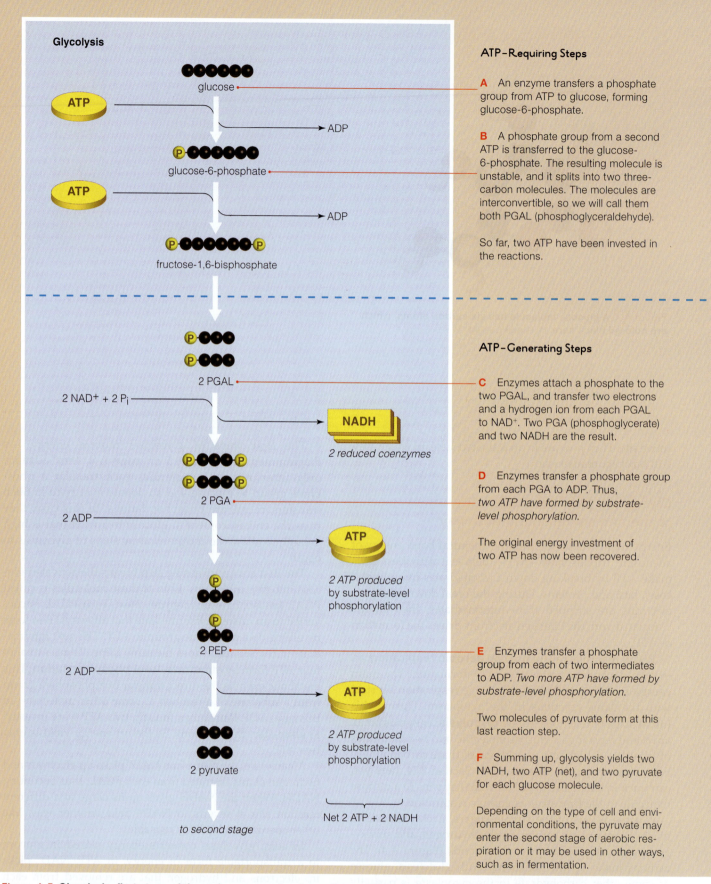

Glycolysis

glucose

ATP

→ ADP

glucose-6-phosphate

ATP

→ ADP

fructose-1,6-bisphosphate

2 PGAL

2 NAD⁺ + 2 Pᵢ

NADH

2 reduced coenzymes

2 PGA

2 ADP

ATP

2 ATP produced by substrate-level phosphorylation

2 PEP

2 ADP

ATP

2 ATP produced by substrate-level phosphorylation

2 pyruvate

to second stage

Net 2 ATP + 2 NADH

ATP–Requiring Steps

A An enzyme transfers a phosphate group from ATP to glucose, forming glucose-6-phosphate.

B A phosphate group from a second ATP is transferred to the glucose-6-phosphate. The resulting molecule is unstable, and it splits into two three-carbon molecules. The molecules are interconvertible, so we will call them both PGAL (phosphoglyceraldehyde).

So far, two ATP have been invested in the reactions.

ATP–Generating Steps

C Enzymes attach a phosphate to the two PGAL, and transfer two electrons and a hydrogen ion from each PGAL to NAD⁺. Two PGA (phosphoglycerate) and two NADH are the result.

D Enzymes transfer a phosphate group from each PGA to ADP. Thus, *two ATP have formed by substrate-level phosphorylation.*

The original energy investment of two ATP has now been recovered.

E Enzymes transfer a phosphate group from each of two intermediates to ADP. *Two more ATP have formed by substrate-level phosphorylation.*

Two molecules of pyruvate form at this last reaction step.

F Summing up, glycolysis yields two NADH, two ATP (net), and two pyruvate for each glucose molecule.

Depending on the type of cell and environmental conditions, the pyruvate may enter the second stage of aerobic respiration or it may be used in other ways, such as in fermentation.

Figure A.5 Glycolysis, first stage of the main energy-releasing pathways. The reaction steps proceed inside the cytoplasm of every living prokaryotic and eukaryotic cell. In this example, glucose is the starting material. By the time the reactions end, two pyruvate, two NADH, and four ATP have been produced. Cells invest two ATP to start glycolysis, however, so the net energy yield of glycolysis is two ATP.

Depending on the type of cell and on environmental conditions, the pyruvate may enter the second set of reactions of the aerobic pathway, which includes the Krebs cycle. Or it may be used in other reactions, such as a fermentation pathway. (© Cengage Learning)

Second stage of the aerobic pathway

When pyruvate molecules formed by glycolysis leave the cytoplasm and enter a mitochondrion, the scene is set for both the second and the third stages of the aerobic pathway. Figure A.6 diagrams these steps in detail.

A An enzyme splits a pyruvate molecule into a two-carbon acetyl group and CO_2. Coenzyme A binds the acetyl group (forming acetyl–CoA). NAD^+ combines with released hydrogen ions and electrons, forming NADH.

B The Krebs cycle starts as one carbon atom is transferred from acetyl–CoA to oxaloacetate. Citrate forms, and coenzyme A is regenerated.

C A carbon atom is removed from an intermediate and leaves the cell as CO_2. NAD^+ combines with released hydrogen ions and electrons, forming NADH.

D A carbon atom is removed from another intermediate and leaves the cell as CO_2, and another NADH forms.

In summary, pyruvate's three carbon atoms have now exited the cell, in CO_2.

H The final steps of the Krebs cycle regenerate oxaloacetate.

G NAD^+ combines with hydrogen ions and electrons, forming NADH.

F The coenzyme FAD combines with hydrogen ions and electrons, forming $FADH_2$.

E One ATP forms by substrate-level phosphorylation.

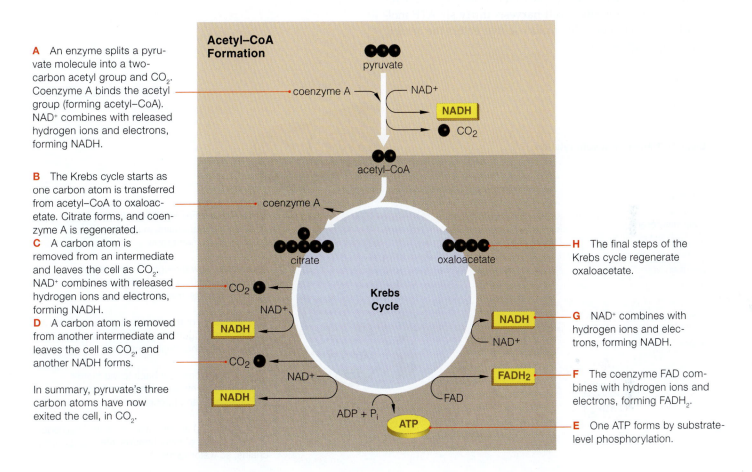

Figure A.6 **Second stage of aerobic respiration: the Krebs cycle and reaction steps that precede it.** For each three-carbon pyruvate molecule entering the cycle, three CO_2, one ATP, four NADH, and one $FADH_2$ molecules form. The steps shown proceed *twice*, because each glucose molecule was broken down earlier to *two* pyruvate molecules. (© Cengage Learning)

Third stage of aerobic cellular respiration

Most ATP is produced in the third stage of the aerobic pathway. Electron transport systems and neighboring proteins called ATP synthases serve as the production machinery. They are embedded in the inner membrane that divides the mitochondrion into two compartments (Figure A.7). They interact with electrons and H⁺ ions, which coenzymes deliver from reaction sites of the first two stages of the aerobic pathway. Typically, a cell harvests thirty-six ATP molecules for each molecule of glucose that enters the aerobic pathway (Figure A.8).

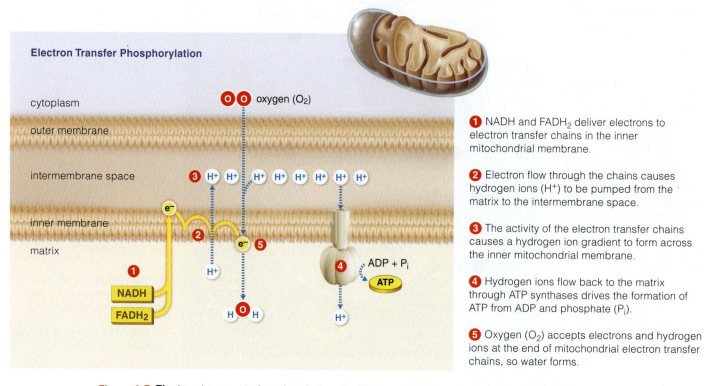

Electron Transfer Phosphorylation

1 NADH and FADH$_2$ deliver electrons to electron transfer chains in the inner mitochondrial membrane.

2 Electron flow through the chains causes hydrogen ions (H⁺) to be pumped from the matrix to the intermembrane space.

3 The activity of the electron transfer chains causes a hydrogen ion gradient to form across the inner mitochondrial membrane.

4 Hydrogen ions flow back to the matrix through ATP synthases drives the formation of ATP from ADP and phosphate (P$_i$).

5 Oxygen (O$_2$) accepts electrons and hydrogen ions at the end of mitochondrial electron transfer chains, so water forms.

Figure A.7 Electron transport phosphorylation, the third and final stage of aerobic respiration. (© Cengage Learning)

Summary of glycolysis and aerobic cellular respiration

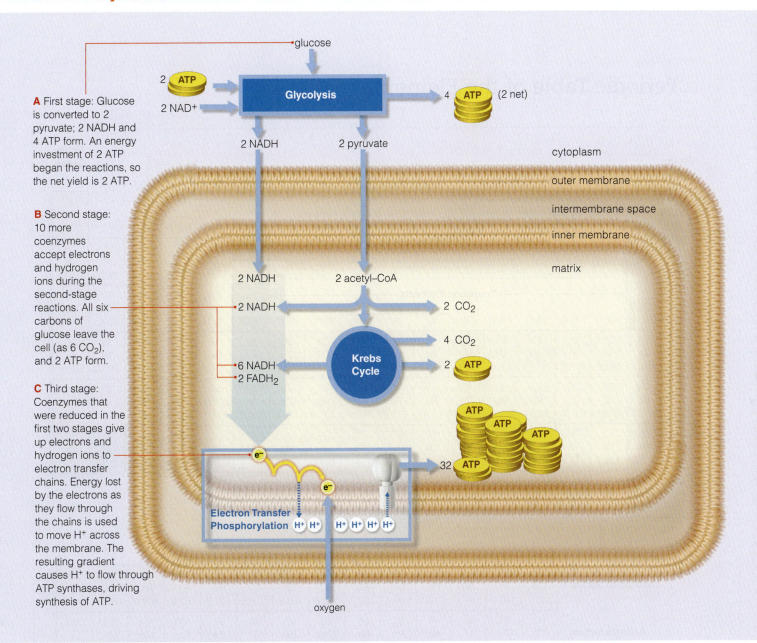

A First stage: Glucose is converted to 2 pyruvate; 2 NADH and 4 ATP form. An energy investment of 2 ATP began the reactions, so the net yield is 2 ATP.

B Second stage: 10 more coenzymes accept electrons and hydrogen ions during the second-stage reactions. All six carbons of glucose leave the cell (as 6 CO_2), and 2 ATP form.

C Third stage: Coenzymes that were reduced in the first two stages give up electrons and hydrogen ions to electron transfer chains. Energy lost by the electrons as they flow through the chains is used to move H^+ across the membrane. The resulting gradient causes H^+ to flow through ATP synthases, driving synthesis of ATP.

glucose

2 ATP

Glycolysis

2 NAD$^+$

4 ATP (2 net)

2 NADH   2 pyruvate

cytoplasm

outer membrane

intermembrane space

inner membrane

matrix

2 NADH   2 acetyl–CoA

2 NADH   2 CO_2

4 CO_2

Krebs Cycle

6 NADH

2 FADH$_2$

2 ATP

32 ATP

e$^-$ e$^-$

Electron Transfer Phosphorylation   H$^+$ H$^+$ H$^+$ H$^+$ H$^+$ H$^+$

oxygen

Figure A.8 Summary of the harvest from the energy-releasing pathway of aerobic respiration. Commonly, thirty-six ATP form for each glucose molecule that enters the pathway. But the net yield varies according to shifting concentrations of reactants, intermediates, and end products of the reactions. It also varies among different types of cells.

Cells differ in how they use the NADH from glycolysis, which cannot enter mitochondria. At the outer mitochondrial membrane, these NADH give up electrons and hydrogen to transport proteins, which shuttle the electrons and hydrogen across the membrane. NAD$^+$ or FAD already inside the mitochondrion accept them, thus forming NADH or FADH$_2$.

Any NADH inside the mitochondrion delivers electrons to the highest possible entry point into a transport system. When it does, enough H^+ is pumped across the inner membrane to make three ATP. By contrast, any FADH$_2$ delivers them to a lower entry point. Fewer hydrogen ions can be pumped, so only *two* ATP can form.

In liver, heart, and kidney cells, for example, electrons and hydrogen from glycolysis enter the highest entry point of transport systems, so the energy harvest is thirty-eight ATP. More commonly, as in skeletal muscle and brain cells, they are transferred to FAD— so the harvest is thirty-six ATP. (© Cengage Learning)

APPENDIX II

Periodic Table of the Elements

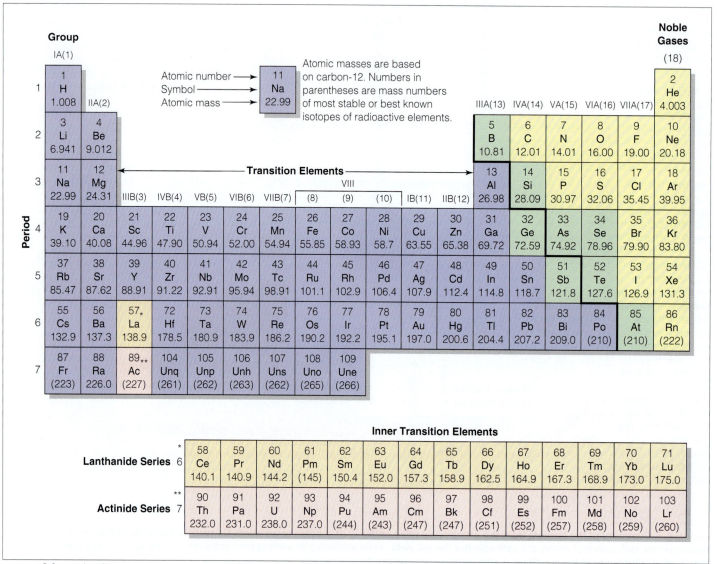

Group

| | IA(1) | | | | | | | | | | | | | | | | | | Noble Gases (18) |

Atomic number → 11
Symbol → Na
Atomic mass → 22.99

Atomic masses are based on carbon-12. Numbers in parentheses are mass numbers of most stable or best known isotopes of radioactive elements.

Transition Elements

Period	IA(1)	IIA(2)	IIIB(3)	IVB(4)	VB(5)	VIB(6)	VIIB(7)	(8) VIII (9)	(10)	IB(11)	IIB(12)	IIIA(13)	IVA(14)	VA(15)	VIA(16)	VIIA(17)	(18)
1	1 H 1.008																2 He 4.003
2	3 Li 6.941	4 Be 9.012										5 B 10.81	6 C 12.01	7 N 14.01	8 O 16.00	9 F 19.00	10 Ne 20.18
3	11 Na 22.99	12 Mg 24.31										13 Al 26.98	14 Si 28.09	15 P 30.97	16 S 32.06	17 Cl 35.45	18 Ar 39.95
4	19 K 39.10	20 Ca 40.08	21 Sc 44.96	22 Ti 47.90	23 V 50.94	24 Cr 52.00	25 Mn 54.94	26 Fe 55.85 / 27 Co 58.93	28 Ni 58.7	29 Cu 63.55	30 Zn 65.38	31 Ga 69.72	32 Ge 72.59	33 As 74.92	34 Se 78.96	35 Br 79.90	36 Kr 83.80
5	37 Rb 85.47	38 Sr 87.62	39 Y 88.91	40 Zr 91.22	41 Nb 92.91	42 Mo 95.94	43 Tc 98.91	44 Ru 101.1 / 45 Rh 102.9	46 Pd 106.4	47 Ag 107.9	48 Cd 112.4	49 In 114.8	50 Sn 118.7	51 Sb 121.8	52 Te 127.6	53 I 126.9	54 Xe 131.3
6	55 Cs 132.9	56 Ba 137.3	57* La 138.9	72 Hf 178.5	73 Ta 180.9	74 W 183.9	75 Re 186.2	76 Os 190.2 / 77 Ir 192.2	78 Pt 195.1	79 Au 197.0	80 Hg 200.6	81 Tl 204.4	82 Pb 207.2	83 Bi 209.0	84 Po (210)	85 At (210)	86 Rn (222)
7	87 Fr (223)	88 Ra 226.0	89** Ac (227)	104 Unq (261)	105 Unp (262)	106 Unh (263)	107 Uns (262)	108 Uno (265) / 109 Une (266)									

Inner Transition Elements

	*	58 Ce 140.1	59 Pr 140.9	60 Nd 144.2	61 Pm (145)	62 Sm 150.4	63 Eu 152.0	64 Gd 157.3	65 Tb 158.9	66 Dy 162.5	67 Ho 164.9	68 Er 167.3	69 Tm 168.9	70 Yb 173.0	71 Lu 175.0
Lanthanide Series 6															
Actinide Series 7	**	90 Th 232.0	91 Pa 231.0	92 U 238.0	93 Np 237.0	94 Pu (244)	95 Am (243)	96 Cm (247)	97 Bk (247)	98 Cf (251)	99 Es (252)	100 Fm (257)	101 Md (258)	102 No (259)	103 Lr (260)

Units of Measure

Metric-English Conversions

Length

English		Metric
inch	=	2.54 centimeters
foot	=	0.30 meter
yard	=	0.91 meter
mile (5,280 feet)	=	1.61 kilometer

To convert	multiply by	to obtain
inches	2.54	centimeters
foot	30.00	centimeters
centimeters	0.39	inches
millimeters	0.039	inches

Weight

English		Metric
grain	=	64.80 milligrams
ounce	=	28.35 grams
pound	=	453.60 grams
ton (short) (2,000 pounds)	=	0.91 metric ton

To convert	multiply by	to obtain
ounces	28.3	grams
pounds	453.6	grams
pounds	0.45	kilograms
grams	0.035	ounces
kilograms	2.2	pounds

Volume

English		Metric
cubic inch	=	16.39 cubic centimeters
cubic foot	=	0.03 cubic meter
cubic yard	=	0.765 cubic meters
ounce	=	0.03 liter
pint	=	0.47 liter
quart	=	0.95 liter
gallon	=	3.79 liters

To convert	multiply by	to obtain
fluid ounces	30.00	milliliters
quart	0.95	liters
milliliters	0.03	fluid ounces
liters	1.06	quarts

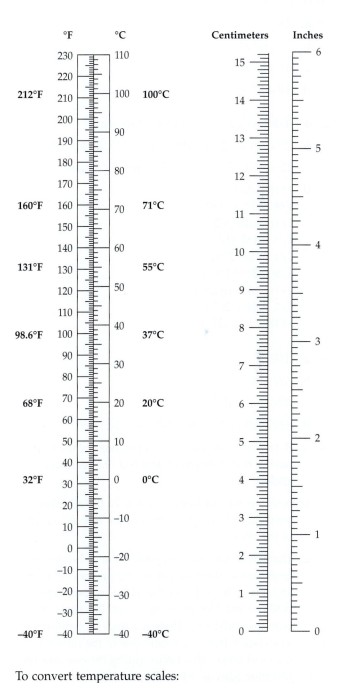

To convert temperature scales:

Fahrenheit to Celsius: $°C = 5/9(°F - 32)$

Celsius to Fahrenheit: $°F = 9/5(°C + 32)$

APPENDIX IV

Answers to Critical Thinking Genetics Problems

CHAPTER 19

1. a: *AB*

   b: *AB* and *aB*

   c: *Ab* and *ab*

   d: *AB, aB, Ab,* and *ab*

2. a: *AaBB* will occur in all the offspring.

   b: 25% *AABB*; 25% *AaBB*; 25% *AABb*; 25% *AaBb*

   c: 25% *AaBb*; 25% *Aabb*; 25% *aaBb*; 25% *aabb*

   d: 1/16 *AABB* (6.25%)
   1/8 *AaBB* (12.5%)
   1/16 *aaBB* (6.25%)
   1/8 *AABb* (12.5%)
   1/4 *AaBb* (25%)
   1/8 *aaBb* (12.5%)
   1/16 *AAbb* (6.25%)
   1/8 *Aabb* (12.5%)
   1/16 *aabb* (6.25%)

3. a: *ABC*

   b: *ABc, aBc*

   c: *ABC, ABc, aBC, aBc*

   d: *ABC, ABc, AbC, Abc, aBC, aBc, abC, abc*

4. a: Both parents must be heterozygous for the trait, having one dominant allele and one recessive allele.

   b: Both parents are homozygotes for the albinism trait—both have two copies of the recessive form of the allele.

   c: The parent with typical pigmentation is heterozygous for the albinism allele. The probability that any one child will have the albinism phenotype is 50 percent. However, with a small sample of only four offspring, there is a high probability of a deviation from a 1:1 ratio due to the random mixes of alleles that occur during meiosis and fertilization (discussed in Chapter 18).

5. Because Molly does not exhibit the recessive hip disorder, she must be either homozygous dominant (*HH*) for this trait, or heterozygous (*Hh*). If the father is homozygous dominant (*HH*), then he and Molly cannot produce offspring that are homozygous recessive (*hh*), and so none of their offspring will have the undesirable phenotype. However, if Molly is a heterozygote for the trait, notice that the probability is 1/2 (50 percent) that a puppy will be heterozygous (*Hh*) and so carry the trait.

6. a: The mother must be heterozygous ($I^A i$). The man having type B blood could have fathered the child if he were also heterozygous ($I^B i$).

   b: If the man is heterozygous, then he *could be* the father. However, because any other type B heterozygous male could also be the father, one cannot say that this particular man absolutely must be. Actually, any male who could contribute an O allele (*i*) could have fathered the child. This would include males with type O blood (*ii*) or type A blood who are heterozygous.

7. The probability is 1/2 (50 percent) that a child of this couple will be a heterozygote and have sickle cell trait. The probability is 1/4 (25 percent) that a child will be homozygous for the sickling allele and so will have sickle cell anemia.

8. For these ten traits, all the man's sperm will carry identical genes. He cannot produce genotypically different sperm. The woman can produce eggs with four genotypes. This example underscores the fact that the more heterozygous gene pairs that are present, the more genetically different gametes are possible.

9. The mating between two carriers of a lethal trait is *Ll × Ll*.

Progeny genotypes: 1/4 *LL* + 1/2 *Ll* + 1/4 *ll*. Phenotypes:

1/4 homozygous survivors (*LL*)

1/2 heterozygous survivors (*Ll*)

1/4 lethal (*ll*) nonsurvivors

10. Bill's genotype: *Aa Ss Ee*
    Marie's genotype: *AA SS EE*

No matter how many children Bill and Marie have, the probability is 100 percent that each child will have the parents' phenotype. Because Marie can produce only dominant alleles, there is no way that a child could inherit a *pair* of recessive alleles for any of these three traits—and that is what would be required in order for the child to show the recessive phenotype. Thus the probability is zero that a child will have short lashes, high arches, and no achoo syndrome.

11. The white-furred parent's genotype is *bb*; the black-furred parent must be *Bb*; because if it were *BB* all offspring would be heterozygotes (*Bb*) and would have black fur. A monohybrid cross between a heterozygote and a homozygote typically yields a 1:1 phenotype ratio. Four black and three white guinea pigs is close to a 1:1 ratio.

CHAPTER 20

1. The probability is the same for each child: 1/2, or 50 percent.

2. a: A male can produce two types of gametes with respect to an X-linked gene. One type will possess only a Y chromosome and so lack this gene; the other type will have an X chromosome and will have the X-linked gene.

   b: A female homozygous for an X-linked gene will produce just one type of gamete containing an X chromosome with the gene.

   c: A female heterozygous for an X-linked gene will produce two types of gametes. One will contain an X chromosome with the dominant allele, and the other type will contain an X chromosome with the recessive allele.

3. Most of chromosome 21 has been translocated to chromosome 14. While this individual has forty-six chromosomes, there are in fact three copies of chromosome 21. The third copy of chromosome 21 is attached to chromosome 14.

4. No. Many traits are sex-influenced and controlled by genes on autosomes.

5. Fifty percent. His mother is heterozygous for the allele, so there is a 50 percent chance that any male offspring will inherit the allele. Since males do not inherit an X chromosome from their fathers, the genotype of the father is irrelevant to this question.

6. The allele for cystic fibrosis is recessive. Most of the carriers are heterozygous for the allele and do not have the cystic fibrosis phenotype.

7. Males inherit their X chromosome from their mother, so there is no chance that a son of this female will be color-blind because she doesn't carry this X-linked trait.

APPENDIX V

Answers to Self-Quizzes

CHAPTER 1: 1. DNA, 2. cell, 3. Homeostasis, 4. vertebrates, mammals, 5. nine, 6. c, 7. d, 8. c, 9. b, 10. b

CHAPTER 2: 1. carbon, 2. a, 3. e, 4. c, 5. d, 6. b, 7. b, 8. c, 9. c, e, b, d, a, 10. a (plus R group interactions)

CHAPTER 3: 1. d, 2. cytoskeleton, 3. c, 4. a, 5. d, 6. b, g, f, d, c, a, e, 7. d, 8. d, 9. c, e, d, a, b, 10. b, 11. b, a, c, 12. The electron transport systems and carrier proteins required for ATP formation are embedded in the membrane between the inner and outer compartment of the mitochondrion.

CHAPTER 4: 1. d, 2. c, 3. b, 4. b, 5. a, 6. c, 7. d, 8. c, 9. Receptors, integrator, effectors, 10. d, e, c, b, a

CHAPTER 5: 1. Skeletal; muscular, 2. d, 3. d, 4. c, 5. b, 6. a, 7. c, 8. skull, rib cage, and vertebral column; pectoral girdles, pelvic girdle, and bones of extremities, 9. connect bones at joints, 10. g, d, f, e, c, b, h, a

CHAPTER 6: 1. Skeletal; muscular, 2. skeletal, cardiac, smooth, 3. b, 4. b, 5. d, 6. c, 7. d, 8. d, 9. b, 10. f, d, e, a, b, g, c

CHAPTER 7: 1. a, 2. systole; diastole, 3. b, 4. c, 5. b, 6. a, 7. a, 8. a, 9. d, 10. d, b, a, c, 11. c, a, b

CHAPTER 8: 1. a, c, d, 2. d, 3. c, 4. b, 5. e, 6. b, 7. c, 8. d, a, c, e, b

CHAPTER 9: 1. d, 2. f, 3. d, 4. d, 5. b, 6. d, 7. d, 8. a, 9. Tears are surface barriers that may wash away pathogens, 10. c, b, a, e, d, f

CHAPTER 10: 1. d, 2. b, 3. d, 4. b, 5. c, 6. e, 7. c, 8. d, 9. d, 10. d

CHAPTER 11: 1. digesting, absorbing, eliminating, 2. caloric, energy, 3. carbohydrates, 4. essential amino acids, essential fatty acids, 5. c, 6. d, 7. c, 8. a, 9. b, 10. e, d, a, b, c

CHAPTER 12: 1. d, 2. f, 3. a, 4. b, 5. c, 6. d, 7. a, 8. d, 9. d, 10. b, d, e, c, a

CHAPTER 13: 1. stimuli, 2. an action potential or nerve impulse, 3. neurotransmitter, 4. c, 5. b, 6. c, 7. b, 8. e, 9. e, d, b, c, a

CHAPTER 14: 1. stimulus, 2. sensation, 3. Perception, 4. d, 5. b, 6. d, 7. c, 8. e, 9. b, 10. c, 11. b, 12. e, c, d, a, b

CHAPTER 15: 1. f, 2. d, 3. b, 4. d, 5. e, 6. b, 7.b, 8. b, 9. b, e, g, d, f, a, c, 10. d, a, b, c, e

CHAPTER 16: 1. hypothalamus, 2. b, 3. d, 4. c, 5. d, 6. b, 7. d, 8. d

CHAPTER 17: 1. a, 2. a, 3. c, 4. morphogenesis, 5. c, 6. c, 7. d, a, f, e, c, b, 8. e, 9. a, 10. e

CHAPTER 18: 1. mitosis; meiosis, 2. chromosomes; DNA, 3. c, 4. diploid, 5. c, 6. a, 7. b, 8. b, 9. b, 10. d, 11. d, b, c, a

CHAPTER 19: 1. a, 2. c, 3. a, 4. c, 5. b, 6. d, 7. b, 8. d, 9. a, b, d, 10. c, a, d, b

CHAPTER 20: 1. c, 2. e, 3. e, 4. c, 5. d, 6. c, 7. d, 8. d, 9. d, 10. c, e, d, b, a

CHAPTER 21: 1. three, 2. b, 3. c, 4. c, 5. a, 6. a, 7. c, 8. a, 9. e, c, d, a, b, 10. c

CHAPTER 22: 1. c, 2. e, 3. tumor suppressor gene, 4. b, 5. d, 6. a, 7. c, 8. d, 9. b, 10. b

CHAPTER 23: 1. species, 2. mutation, genetic drift, gene flow, and natural selection, 3. natural selection, 4. e, 5. e, 6. b, 7. a, 8. a

CHAPTER 24: 1. d, 2. d, 3. Ecology, 4. d, 5. a, 6. e, 7. d, a, b, c

CHAPTER 25: 1. b, 2. c, 3. d, 4. d, 5. c, 6. b

A Plain English Map of the Human Chromosomes and Some Associated Traits

Figure A.9 Haploid set of human chromosomes. The banding patterns characteristic of each type of chromosome appear after staining with a reagent called Giemsa. **The locations of some of the 20,065 known genes (as of November 2008) are indicated.** Also shown are locations that, when mutated, cause some of the genetic diseases discussed in the text.

Figure by Lisa Starr (based on information from Susan Offner)

The OMIM: A resource for exploring human genetic disorders

Researchers at Johns Hopkins University have developed a Web-based catalog of human genetic disorders called Online Mendelian Inheritance in Man, or OMIM (Figure A.10).

Physicians, scientists, and the public can use the OMIM to learn more about more than 10,000 inherited diseases and disorders, including those discussed in this textbook.

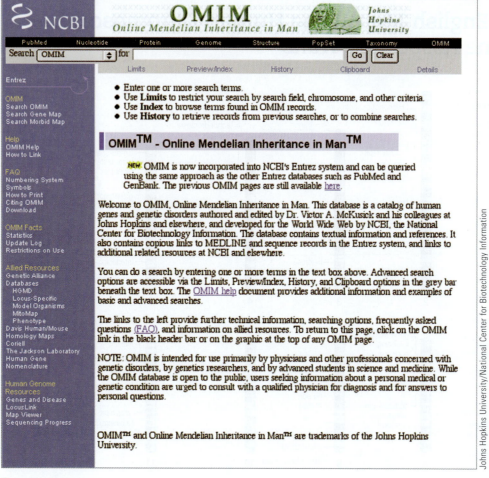

Figure A.10

abdominal cavity Body cavity that holds the stomach, liver, pancreas, most of the intestine, and several other organs.

ABO blood typing Method of characterizing an individual's blood according to whether one or both of two protein markers, A and B, are present at the surface of red blood cells. The O signifies that neither marker is present.

abortion Spontaneous or induced expulsion of the embryo or fetus from the uterus.

absorption The movement of nutrients, fluid, and ions across the gastrointestinal tract lining and into the internal environment.

accommodation In the eye, adjustments of the lens position that move the focal point forward or back so that incoming light rays are properly focused on the retina.

acetylcholine (ACh) A neurotransmitter that can excite or inhibit various target cells in the brain, spinal cord, glands, and muscles.

acid A substance that releases hydrogen ions in water.

acid–base balance State in which extracellular fluid is neither too acidic nor too basic, an outcome of controls over its concentrations of dissolved ions.

acid rain Wet acid deposition; falling of rain (or snow) rich in sulfur and nitrogen oxides.

acrosome An enzyme-containing cap that covers most of the head of a sperm and helps the sperm penetrate an egg at fertilization.

actin (AK-tin) A globular contractile protein. In muscle cells, actin interacts with another protein, myosin, to bring about contraction.

action potential An abrupt, brief reversal in the steady voltage difference (resting membrane potential) across the plasma membrane of a neuron.

activation energy The minimum amount of energy required for a chemical reaction.

active immunity Immunity that develops after a person receives a vaccine, which stimulates the immune system to produce antibodies against a particular pathogen.

active site A crevice on the surface of an enzyme molecule where a specific reaction is catalyzed.

active transport The pumping of one or more specific solutes through a transport protein that spans the lipid bilayer of a cell membrane. Most often, the solute is transported against its concentration gradient. The protein is activated by an energy boost, as from ATP.

adaptation [L. *adaptare*, to fit] In evolutionary biology, the process of becoming suited (or more suited) to a given set of environmental conditions. Of sensory neurons, a decrease in the frequency of action potentials (or their cessation) even when a stimulus is maintained at constant strength.

adaptive immunity Immune responses that the body develops in response to antigens of specific pathogens, toxins, or abnormal body cells.

adaptive radiation A burst of speciation events, with lineages branching away from one another as they partition the existing environment or invade new ones.

adenine (AH-de-neen) A purine; a nitrogen-containing base in certain nucleotides; a building block of DNA.

ADH Antidiuretic hormone. Produced by the hypothalamus and released by the posterior pituitary, it stimulates reabsorption in the kidneys and so reduces urine volume.

adhering junctions Cell junctions that cement cells together.

adipose tissue A type of connective tissue having an abundance of fat-storing cells and blood vessels for transporting fats.

adjuvant therapy Cancer therapy that combines surgery with chemotherapy.

adrenal cortex (ah-DREE-nul) Outer portion of the adrenal gland; its hormones have roles in metabolism, inflammation, maintaining extracellular fluid volume, and other functions.

adrenal medulla Inner region of the adrenal gland; its hormones help control blood circulation and carbohydrate metabolism.

aerobic exercise Exercise that works muscles at a rate that does not exceed the body's ability to keep them supplied with oxygen (in blood).

aerobic respiration (air-OH-bik) [Gk. *aer*, air, and *bios*, life] An oxygen-dependent pathway of ATP formation.

afferent arteriole In the urinary system, an arteriole that delivers blood to each nephron.

age structure Of a population, the number of individuals in each of several or many age categories.

agglutination (ah-glue-tin-AY-shun) The clumping together of foreign cells that have invaded the body (as pathogens or in tissue grafts or transplants). Clumping is caused by cross-linking between antibody molecules that have already bound an antigen at the surface of the foreign cells.

aging *See* senescence.

aldosterone (al-DOSS-tuh-roan) Hormone secreted by the adrenal cortex that helps regulate sodium reabsorption by the kidneys.

allantois (ah-LAN-twahz) [Gk. *allas*, sausage] One of four extraembryonic membranes that form during embryonic development. In humans, it functions in early blood formation and development of the urinary bladder.

allele (uh-LEEL) For a given location on a chromosome, one of two or more slightly different chemical forms of a gene that code for different versions of the same trait.

allergen Any normally harmless substance that provokes inflammation, excessive mucus secretion, and other immune responses.

allergy An immune response made against a normally harmless substance.

alveolus (al-VEE-uh-lus), plural: alveoli [L. *alveus*, small cavity] Any of the many cup-shaped, thin-walled outpouchings of respiratory bronchioles. A site where oxygen diffuses from air in the lungs to the blood, and carbon dioxide diffuses from blood to the lungs.

amine hormone A hormone derived from the amino acid tyrosine.

amino acid (uh-MEE-no) A small organic molecule having a hydrogen atom, an amino group, an acid group, and an R group covalently bonded to a central carbon atom. The subunit of polypeptide chains, which represent the primary structure of proteins.

amnesia A loss of fact memory.

amniocentesis Test of fetal cells in a sample of amniotic fluid for evidence of birth defects.

amnion (AM-nee-on) One of four extraembryonic membranes. It becomes a fluid-filled sac in which the embryo (and fetus) can grow, move freely, and be protected from sudden temperature shifts and impacts.

anabolism A metabolic activity that assembles small molecules into more complex molecules that store energy.

anal canal The canal from the rectum to the anus through which feces pass.

analogous structures Body parts, once different in separate lineages, that were put to comparable uses in similar environments and that came to resemble one another in form and function. They are evidence of morphological convergence.

anaphase (AN-uh-faze) The stage at which microtubules of a spindle apparatus separate sister chromatids of each chromosome and move them to opposite spindle poles. During anaphase I of meiosis, the two members of each pair of homologous chromosomes separate. During anaphase II, sister chromatids of each chromosome separate.

anaphylactic shock A whole-body allergic response in which a person's blood pressure plummets, among other symptoms.

aneuploidy (AN-yoo-ploy-dee) A change in the chromosome number following inheritance of one extra or one fewer chromosome than usual.

aneurysm A pouchlike weak spot in an artery.

anorexia nervosa Eating disorder in which a person deliberately starves and may become dangerously thin.

antibiotic [Gk. *anti*, against] A substance that kills or inhibits the growth of microorganisms.

antibody Any of a variety of Y-shaped receptor molecules with binding sites for specific antigens. Only B cells produce antibodies, then position them at their surface or secrete them.

antibody-mediated immune response The B cell defensive response to pathogens in the body wherein antibodies are produced.

anticodon In a tRNA molecule, a sequence of three nucleotide bases that can pair with an mRNA codon.

antigen (AN-tih-jen) [Gk. *anti*, against, and *genos*, race, kind] Substance that is recognized as foreign to the body and that triggers an immune response. Most antigens are protein molecules at the surface of infectious agents or tumor cells.

antigen–MHC complex Unit consisting of fragments of an antigen molecule bound to MHC proteins. MHC complexes displayed at the surface of an antigen-presenting cell such as a macrophage promote an immune response by lymphocytes.

antigen-presenting cell A macrophage or other cell that displays antigen–MHC complexes at its surface and so promotes an immune response by lymphocytes.

antioxidant A chemical that can give up an electron to a free radical before the free radical damages DNA or some other cell constituent.

anus Terminal opening of the gastrointestinal tract.

aorta (ay-OR-tah) [Gk. *airein*, to lift, heave] Main artery of systemic circulation; carries oxygenated blood away from the heart to all body regions except the lungs.

aortic body Any of several receptors in artery walls near the heart that respond to changes in levels of carbon dioxide and oxygen in arterial blood.

aortic valve Valve that opens from the left ventricle into the aorta.

apoptosis (a-poh-TOE-sis) Genetically programmed cell death. Molecular signals lead to self-destruction in body cells that have finished their prescribed functions or have become altered, as by infection or transformation into a cancerous cell.

appendicular skeleton (ap-en-DIK-yoo-lahr) Bones of the limbs, hips, and shoulders.

appendix A slender projection from the cup-shaped pouch (cecum) at the start of the colon. It may function in defense.

appetite The desire to eat, apart from the physical need for food.

arteriole (ar-TEER-ee-ole) Any of the blood vessels between arteries and capillaries. They are control points where the volume of blood delivered to different body regions can be adjusted.

artery Any of the large-diameter blood vessels that conduct deoxygenated blood to the lungs and oxygenated blood to all body tissues. The thick, muscular artery wall allows arteries to smooth out pulsations in blood pressure caused by heart contractions.

atom The smallest unit of matter that is unique to a particular element.

atomic number The number of protons in the nucleus of each atom of an element; it differs for each element.

ATP Adenosine triphosphate (ah-DEN-uh-seen try-FOSS-fate) A nucleotide composed of adenine, ribose, and three phosphate groups. As the main energy carrier in cells, it directly or indirectly delivers energy to or picks up energy from nearly all metabolic pathways.

ATP/ADP cycle In cells, a mechanism of ATP renewal. When ATP donates a phosphate group to other molecules (and so energizes them), it reverts to ADP, then forms again by phosphorylation of ADP.

atrioventricular (AV) node In the septum dividing the heart atria, a site that contains bundles of conducting fibers. Stimuli arriving at the AV node from the cardiac pacemaker (sino-atrial node) pass along the bundles and continue on via Purkinje fibers to contractile muscle cells in the ventricles.

atrioventricular valve One-way flow valve between the atrium and ventricle in each half of the heart.

atrium (AY-tree-um) Upper chamber in each half of the heart; the right atrium receives deoxygenated blood (from tissues) entering the pulmonary circuit of blood flow, and the left atrium receives oxygenated blood from pulmonary veins.

autoimmunity Misdirected immune response in which lymphocytes mount an attack against normal body cells.

autonomic nerves (ah-toe-NOM-ik) Those nerves leading from the central nervous system to the smooth muscle, cardiac muscle, and glands of internal organs and structures—that is, to the visceral portion of the body.

autosomal dominant Condition caused by a dominant allele on an autosome (not a sex chromosome). The allele is always expressed to some extent, even in heterozygotes.

autosomal recessive Condition caused by a recessive allele on an autosome (not a sex chromosome). Only recessive homozygotes show the resulting phenotype.

autosome Any chromosome that is not a sex (gender-determining) chromosome.

autotroph (AH-toe-trofe) [Gk. *autos*, self, and *trophos*, feeder] An organism able to build its own large organic molecules by using carbon dioxide and energy from the physical environment. Compare *heterotroph*.

axial skeleton (AX-ee-uhl) The skull, backbone, ribs, and breastbone (sternum).

axon Of a neuron, a long, cylindrical extension from the cell body, with finely branched endings. Action potentials move rapidly, without alteration, along an axon; their arrival at axon endings may trigger the release of neurotransmitter molecules that influence an adjacent cell.

baroreceptor reflex The short-term control over arterial pressure. It keeps blood pressure within normal limits in the face of sudden changes in blood pressure.

Barr body In the cells of females, a condensed X chromosome that was inactivated during early embryonic development.

basal body A centriole that, after having given rise to the microtubules of a flagellum or cilium, remains attached to its base in the cytoplasm.

basal metabolic rate (BMR) Amount of energy required to sustain body functions when a person is resting, awake, and has not eaten for 12–18 hours.

base A substance that accepts H^+ when it dissolves in water.

base pair A pair of hydrogen-bonded nucleotide bases in two strands of nucleic acids. In a DNA double helix, adenine pairs with thymine, and guanine with cytosine. When an mRNA strand forms on a DNA strand during transcription, uracil (U) pairs with the DNA's adenine.

base-pair substitution A mutation occurring in a replicating DNA molecule (a chromosome) when one base is wrongly substituted for another in a base pair.

basement membrane Noncellular layer of mostly proteins and polysaccharides that is sandwiched between an epithelium and underlying connective tissue.

basophil Fast-acting white blood cell that secretes histamine and other substances during inflammation.

B cell B lymphocyte; the only white blood cell that produces antibodies, then positions them at the cell surface or secretes them as weapons in immune responses.

B cell receptor Antigen receptor on a B cell.

bicarbonate–carbon dioxide buffer system A system used to restore the body's normal pH level by neutralizing excess H^+ and allowing for the exhalation of carbon dioxide formed during the reaction. It does not eliminate the excess H^+ and therefore has only a temporary effect.

bile A yellowish fluid made in the liver, stored in the gallbladder, and released into the upper small intestine where it aids in the digestion and absorption of fats.

binge eating Eating an abnormally large quantity of food in a few hours at least twice a week for six months or more.

biofuel An alternative, renewable fuel made from plants and organic wastes.

biogeochemical cycle The movement of an element such as carbon or nitrogen from the environment to organisms, then back to the environment.

biogeography [Gk. *bios*, life, and *geographein*, to describe the surface of Earth] The study of major land regions, each having distinguishing types of plants and animals.

biological clock Internal time-measuring mechanism that has a role in adjusting an organism's daily activities, seasonal activities, or both in response to environmental cues.

biological magnification The increasing concentration of a nondegradable or slowly degradable substance in body tissues as it is passed along food chains.

biological molecule A molecule that contains carbon and that is formed in a living organism.

biomass The combined weight of all the organisms at a particular feeding (trophic) level in an ecosystem.

biome A broad, vegetational subdivision of a biogeographic realm shaped by climate, topography, and composition of regional soils.

biopsy Diagnostic procedure in which a small piece of tissue is removed from the body through a hollow needle or exploratory surgery, and then examined for signs of a particular disease (often cancer).

biosphere [Gk. *bios,* life, and *sphaira,* globe] All regions of Earth's waters, crust, and atmosphere in which organisms live.

bipedalism A habitual standing and walking on two feet, as by humans.

bladder *See* urinary bladder.

blastocyst (BLASS-tuh-sist) [Gk. *blastos,* sprout, and *kystis,* pouch] In embryonic development, a blastula stage consisting of a hollow ball of surface cells and an inner cell mass.

blastomere One of the small, nucleated cells that form during cleavage of a zygote.

blood A fluid connective tissue composed of water, solutes, and formed elements (blood cells and platelets); it carries substances to and from cells and helps maintain an internal environment that is favorable for cell activities.

blood pressure Fluid pressure, generated by heart contractions, that keeps blood circulating.

blood–brain barrier Modified structure of brain capillaries that helps control which blood-borne substances reach neurons in the brain.

BMI Body mass index, a measure of the ratio of weight to height.

BMR *See* basal metabolic rate.

bolus Softened, lubricated ball of food, created by chewing and mixing of food with saliva.

bone Connective tissue that functions in movement and locomotion, protection of other organs, mineral storage, and (in some bones) blood cell production.

bone marrow A connective tissue where blood cells are formed.

bone remodeling Process of ongoing calcium deposits and withdrawals from bone that adjusts bone strength and maintains levels of calcium and phosphorus in blood.

bone tissue Mineral-hardened connective tissue; the main tissue in bone.

Bowman's capsule Cup-shaped portion of a nephron that receives water and solutes being filtered from blood.

brain Organ that receives, integrates, stores, and retrieves information, and coordinates appropriate responses by stimulating and inhibiting the activities of different body parts.

brain case The eight bones that together surround and protect the brain.

brain stem The midbrain, pons, and medulla oblongata, the core of which contains the reticular formation that helps govern activity of the nervous system as a whole.

bronchiole A component of the finely branched bronchial tree inside each lung.

bronchus (BRONG-cuss, BRONG-kee), plural: bronchi [Gk. *bronchos,* windpipe] Tubelike branchings of the trachea that lead into the lungs.

brush border The collective array of microvilli on epithelial cells lining the intestinal mucosa.

buffer A substance that can stabilize the pH of a solution by donating or accepting hydrogen ions.

bulbourethral glands Two glands of the male reproductive system that secrete mucus-rich fluid into the urethra when the male is sexually aroused.

bulimia nervosa Eating disorder in which a person alternately binges and purges.

bulk A volume of fiber and other undigested material that absorption processes in the colon cannot decrease.

bursa Fluid-filled sac between a tendon and a bone.

cancer A malignant tumor, the cells of which show profound abnormalities in the plasma membrane and cytoplasm, abnormal growth and division, and weakened capacity for adhesion within the parent tissue (leading to metastasis). Unless eradicated, cancer is lethal.

capillary [L. *capillus,* hair] A thin-walled blood vessel that functions in the exchange of gases and other substances between blood and interstitial fluid.

capillary bed Dense capillary networks containing true capillaries where exchanges occur between blood and tissues, and also thoroughfare channels that link arterioles and venules.

carbaminohemoglobin A hemoglobin molecule that has carbon dioxide bound to it; $HbCO_2$.

carbohydrate [L. *carbo,* charcoal, and *hydro,* water] A biological molecule built of carbon, hydrogen, and oxygen atoms, usually in a 1:2:1 ratio. All cells use carbohydrates as structural materials, energy stores, and transportable forms of energy. The three classes of carbohydrates include monosaccharides, oligosaccharides, and polysaccharides.

carbon cycle A biogeochemical cycle in which carbon moves from its reservoir in the atmosphere, through oceans and organisms, then back to the atmosphere.

carbonic anhydrase Enzyme in red blood cells that catalyzes the conversion of unbound carbon dioxide to carbonic acid and its dissociation products, thereby helping maintain the gradient that keeps carbon dioxide diffusing from interstitial fluid into the blood.

carcinogen (kar-SIN-uh-jen) An environmental agent or substance, such as ultraviolet radiation, that can trigger cancer.

carcinogenesis The transformation of a normal cell into a cancerous one.

cardiac conduction system (KAR-dee-ak) Set of noncontractile cells in heart muscle that spontaneously produce and conduct the electrical events that stimulate heart muscle contractions.

cardiac cycle The sequence of muscle contraction and relaxation constituting one heartbeat.

cardiac muscle Type of muscle found only in the heart wall; cardiac muscle cells contract as a single unit.

cardiac output The amount of blood each ventricle of the heart pumps in one minute.

cardiac pacemaker Sinoatrial (SA) node; the source of the normal rate of heartbeat. The self-excitatory cardiac muscle cells that spontaneously generate rhythmic waves of excitation over the heart chambers.

cardiovascular system Organ system that is composed of the heart and blood vessels and that functions in the rapid transport of blood to and from tissues.

carotid artery Artery in the neck that contains baroreceptors, which monitor arterial pressure.

carotid body Any of several sensory receptors that monitor carbon dioxide and oxygen levels in blood; located at the point where carotid arteries branch to the brain.

carrier An organism in which a pathogen is living without causing disease symptoms.

carrying capacity The maximum number of individuals in a population (or species) that can be sustained indefinitely by a given environment.

cartilage A type of connective tissue with solid yet pliable intercellular material that resists compression.

cartilaginous joint Type of joint in which cartilage fills the space between adjoining bones; only slight movement is possible.

catabolism Metabolic activity that breaks down large molecules into simpler ones, releasing the components for use by cells.

cell [L. *cella*, small room] The smallest living unit; an organized unit that can survive and reproduce on its own, given DNA instructions and suitable environmental conditions, including appropriate sources of energy and raw materials.

cell body The part of a neuron that contains its nucleus and organelles.

cell count The number of cells of a given type in a microliter of blood.

cell cycle Events during which a cell increases in mass, roughly doubles its number of cytoplasmic components, duplicates its DNA, then undergoes nuclear and cytoplasmic division. It extends from the time a new cell is produced until it completes its own division.

cell differentiation The gene-guided process by which cells in different locations in the embryo become specialized.

cell-mediated immune response The T cell defensive response to pathogens in the body, wherein cytotoxic T cells attack the invaders directly.

cell theory A fundamental theory in biology, which states that (1) all organisms are composed of one or more cells, (2) the cell is the smallest unit that still retains a capacity for independent life, and (3) all cells arise from pre-existing cells.

cell-to-cell junction A point of contact that physically links two cells or that provides functional links between their cytoplasm.

cellular respiration The process by which cells break apart carbohydrates, lipids, or proteins to form ATP.

central nervous system The brain and spinal cord.

centriole (SEN-tree-ohl) A cylinder of triplet microtubules that gives rise to the microtubules of cilia and flagella.

centromere (SEN-troh-meer) [Gk. *kentron*, center, and *meros*, a part] A small, constricted region of a chromosome having attachment sites for microtubules that help move the chromosome during nuclear division.

cerebellum (ser-ah-BELL-um) [L. diminutive of *cerebrum*, brain] Hindbrain region with reflex centers for maintaining posture and refining limb movements.

cerebral cortex Thin surface layer of the cerebral hemispheres. Some regions of the cortex receive sensory input, others integrate information and coordinate appropriate motor responses.

cerebral hemispheres The left and right sides of the cerebrum, which are separated by a deep fissure.

cerebrospinal fluid CSF; clear extracellular fluid that surrounds and cushions the brain and spinal cord.

cerebrum (suh-REE-bruhm) Part of the forebrain; the most complex integrating center.

cervix The lower part of the uterus.

chemical bond A union between the electron structures of two or more atoms.

chemical energy The potential energy of molecules.

chemical equilibrium The point at which a reversible reaction will be running at about the same pace in both directions.

chemical evolution Process by which biological molecules evolved.

chemical senses Senses, such as taste and smell, that detect substances dissolved in fluid that is in contact with chemoreceptors.

chemical synapse (SIN-aps) [Gk. *synapsis*, union] A small gap, the synaptic cleft, that separates two neurons (or a neuron and a muscle cell or gland cell) and that is bridged by neurotransmitter molecules released from the presynaptic neuron.

chemoreceptor (KEE-moe-ree-sep-tur) Sensory receptor that detects chemical energy (ions or molecules) dissolved in the surrounding fluid.

chemotherapy The use of therapeutic drugs to kill cancer cells.

chlorofluorocarbon (CFC) (klore-oh-FLOOR-oh-car-bun) One of a variety of odorless, invisible compounds of chlorine, fluorine, and carbon, widely used in commercial products, that are contributing to the destruction of the ozone layer above Earth's surface.

chorion (CORE-ee-on) One of four extraembryonic membranes; it encloses the embryo and the three other membranes. Absorptive structures (villi) that develop at its surface are crucial for the transfer of substances between the embryo and mother.

chorionic villus sampling (CVS) Test of fetal cells removed from chorionic villi for evidence of birth defects.

chromatid *See* sister chromatids.

chromatin A cell's DNA and all of the proteins associated with it.

chromosome (CROW-moe-soam) [Gk. *chroma*, color, and *soma*, body] A double-stranded DNA molecule that carries genetic information.

chromosome number The number of each type of chromosome in all cells except dividing germ cells or gametes.

chyme (KIME) The thick mixture of swallowed food boluses and acidic gastric fluid in the stomach that enters the small intestine during digestion.

cilia (SILL-ee-uh), singular: cilium [L. *cilium*, eyelid] Of eukaryotic cells, short, hairlike projections that contain a regular array of microtubules. Cilia serve as motile structures, help create currents of fluids, or are part of sensory structures.

clavicle Long, slender collarbone that connects the pectoral girdle with the sternum (breastbone).

cleavage Stage of development when mitotic cell divisions convert a zygote to the ball of cells called the blastula.

cleavage furrow Of a cell undergoing cytoplasmic division, a shallow, ringlike depression that forms at the cell surface as contractile microfilaments pull the plasma membrane inward. It defines where the cytoplasm will be cut in two.

climax community Following primary and secondary succession, the array of species that remains more or less steady under prevailing conditions.

cochlea Coiled, fluid-filled chamber of the inner ear. Sound waves striking the eardrum become converted to pressure waves in the cochlear fluid, and the pressure waves ultimately cause a membrane to vibrate and bend sensory hair cells. Signals from bent hair cells travel to the brain, where they may be interpreted as sound.

codominance Condition in which a pair of nonidentical alleles are both expressed, even though they specify two different phenotypes.

codon One of a series of base triplets in an mRNA molecule, most of which code for a sequence of amino acids of a specific polypeptide chain. (Of 64 codons, 61 specify different amino acids and three of these also serve as start signals for translation; one other serves only as a stop signal for translation.)

coenzyme A type of nucleotide that transfers hydrogen atoms and electrons from one reaction site to another. NAD$^+$ is an example.

coitus Sexual intercourse.

colon (CO-lun) The large intestine.

community The populations of all species occupying a habitat; also applied to groups of organisms with similar lifestyles in a habitat.

compact bone Type of dense bone tissue that makes up the shafts of long bones and outer regions of all bones. Narrow channels in compact bone contain blood vessels and nerves.

comparative morphology [Gk. *morph*, form] Anatomical comparisons of major evolutionary lineages.

complement system A set of about 30 proteins circulating in blood plasma with roles in nonspecific defenses and in immune responses. Some trigger lysis of pathogens, others promote inflammation, and others stimulate phagocytes to engulf pathogens.

compound A substance of two or more elements, whose relative proportions never vary. Organic compounds have a backbone of carbon atoms arranged as a chain or ring structure. The simpler, inorganic compounds do not have comparable backbones.

concentration gradient A difference in the number of molecules (or ions) of a substance between two adjacent regions, as in a volume of fluid.

conclusion In scientific reasoning, a statement that evaluates a hypothesis based on test results.

condensation reaction Chemical reaction in which two molecules become covalently bonded into a larger molecule, and water often forms as a by-product.

cone cell In the retina, a type of photoreceptor that responds to intense light and contributes to sharp daytime vision and color perception.

connective tissue A tissue type that consists of cells in a matrix that contains a ground substance and protein fibers. This category includes fibrous connective tissues, cartilage, bone tissue, blood, and adipose (fat) tissue.

consumer [L. *consumere*, to take completely] Of ecosystems, a heterotrophic organism that obtains energy and raw materials by feeding on the tissues of other organisms. Herbivores, carnivores, omnivores, and parasites are examples.

continuous variation A more or less continuous range of small differences in a given trait among all the individuals of a population.

control group In a scientific experiment, a group that differs from the experimental group only with respect to the variable being studied.

controlled experiment An experiment that tests only one prediction of a hypothesis at a time.

core temperature The body's internal temperature, as opposed to temperatures of the tissues near its surface. Normal human core temperature is about 37°C (98.6°F).

cornea Transparent tissue in the outer layer of the eye, which causes incoming light rays to bend.

coronary artery Either of two arteries leading to capillaries that service cardiac muscle.

coronary circulation Arteries and veins that service the heart.

corpus callosum (CORE-pus ka-LOW-sum) A band of 200 million axons that functionally link the two cerebral hemispheres.

corpus luteum (CORE-pus LOO-tee-um) A glandular structure that develops from cells of a ruptured ovarian follicle. It secretes progesterone and some estrogen, both of which maintain the lining of the uterus (endometrium).

cortex [L. *cortex*, bark] In general, a rindlike outer layer; the kidney (renal) cortex and cerebral cortex are examples.

covalent bond (koe-VAY-lunt) [L. *con*, together, and *valere*, to be strong] A sharing of electrons between atoms. When electrons are shared equally, the bond is nonpolar. When electrons are shared unequally, the bond is polar—slightly positive at one end and slightly negative at the other.

cranial cavity Body cavity that houses the brain.

creatine phosphate Organic compound that transfers phosphate to ADP in a rapid, short-term pathway that generates ATP.

critical thinking Objective evaluation of information; evidence-based thinking.

cross-bridge The interaction between actin and myosin filaments that is the basis of muscle cell contraction.

crossing over During prophase I of meiosis, an interaction between a pair of homologous chromosomes. Their nonsister chromatids break at the same place along their length and exchange corresponding segments at the break points. Crossing over breaks up old combinations of alleles and puts new ones together in chromosomes.

culture The sum total of behavior patterns of a social group, passed between generations by learning and by symbolic behavior, especially language.

cutaneous membrane A dry, sturdy epithelial membrane; the skin.

cyclic AMP Cyclic adenosine monophosphate. A nucleotide that has roles in intercellular communication, as when it serves as a second messenger (a cytoplasmic mediator of a cell's response to signaling molecules).

cytokine Any of the chemicals released by white blood cells that help muster or strengthen defense responses.

cytokinesis (sigh-toe-kih-NEE-sis) [Gk. *kinesis*, motion] Cytoplasmic division; the splitting of a parent cell into two daughter cells.

cytoplasm (SIGH-toe-plaz-um) [Gk. *plassein*, to mold] All cellular parts, particles, and semifluid substances enclosed by the plasma membrane except for the nucleus.

cytosine (SIGH-toe-seen) A pyrimidine; one of the nitrogen-containing bases in nucleotides.

cytoskeleton A cell's internal "skeleton." Its microtubules and other components structurally support the cell and organize and move its internal components.

cytosol The jellylike fluid portion of the cytoplasm.

cytotoxic T cell Type of T lymphocyte that directly kills infected body cells and tumor cells.

decomposer [L. *de-*, down, away, and *companere*, to put together] A heterotroph that obtains energy by chemically breaking down the remains, products, or wastes of other organisms. Decomposers help cycle nutrients to producers in ecosystems. Certain fungi and bacteria are examples.

deforestation The removal of all trees from a large tract of land, such as the Amazon Basin or the Pacific Northwest.

deletion At the cellular level, loss of a segment from a chromosome. In a DNA molecule, loss of one to several base pairs.

demographics A population's vital statistics.

denaturation (deh-nay-chur-AY-shun) Of a protein, the loss of three-dimensional shape following disruption of hydrogen bonds and other weak bonds.

dendrite (DEN-drite) [Gk. *dendron*, tree] A short, slender extension from the cell body of a neuron that receives messages.

dendritic cell A type of white blood cell that alerts the adaptive immune system when an antigen is present in tissue fluid of the skin or body linings.

dense connective tissue A type of fibrous connective tissue with more collagen fibers than loose connective tissue; it is strong but not very flexible.

density-dependent controls Factors, such as predation, parasitism, disease, and competition for resources, that limit population growth by reducing the birth rate, increasing the rates of death and dispersal, or all of these.

density-independent controls Factors such as storms or floods that increase a population's death rate more or less independently of its density.

dermis The layer of skin underlying the epidermis, consisting mostly of dense connective tissue.

desertification (dez-urt-ih-fih-KAY-shun) The conversion of grasslands, rain-fed cropland, or irrigated cropland to desert-like conditions, with a drop in agricultural productivity of 10 percent or more.

development Of complex multicellular species, a series of stages that begins with the formation of gametes, followed by fertilization and subsequent embryonic and adult phases.

diaphragm (DIE-uh-fram) [Gk. *diaphragma*, to partition] Muscular partition between the thoracic and abdominal cavities, the contraction and relaxation of which contributes to breathing.

diastole Relaxation phase of the cardiac cycle.

differentiation *See* cell differentiation.

diffusion Net movement of like molecules (or ions) down their concentration gradient.

digestion The breakdown of food particles into nutrient molecules small enough to be absorbed.

digestive system Organ system with specialized regions where food is ingested, digested, and absorbed and undigested residues are stored, then eliminated.

diploid (DIP-loyd) Having two chromosomes of each type (that is, pairs of homologous chromosomes). Compare *haploid*.

disaccharide (die-SAK-uh-ride) [Gk. *di*, two, and *sakcharon*, sugar] A type of simple carbohydrate, of the class called oligosaccharides; two monosaccharides covalently bonded.

disease Condition that develops when the body's defenses cannot prevent a pathogen's activities from interfering with normal body functions.

disease vector Something (such as an insect) that carries a pathogen from an infected person or contaminated material to new hosts.

disjunction The separation of each homologue from its partner during anaphase I of meiosis.

distal tubule The tubular section of a nephron most distant from the Bowman's capsule; a major site of water and sodium reabsorption.

divergence Accumulation of differences in allele frequencies between populations that have become reproductively isolated from one another.

DNA Deoxyribonucleic acid (dee-ox-ee-rye-bow-new-CLAY-ik). For all cells (and many viruses), the molecule of inheritance. A category of nucleic acids, each usually consisting of two nucleotide strands twisted together helically and held together by hydrogen bonds. The nucleotide sequence encodes the instructions for assembling proteins, and, ultimately, a new individual.

DNA chip A microarray of thousands of DNA sequences that are stamped onto a glass plate; can help identify mutations and diagnose diseases by pinpointing which genes are silent and which are being expressed in a body tissue.

DNA clone An identical copy of foreign DNA that was inserted into plasmids (typically, bacteria).

DNA fingerprint A DNA profile used in criminal investigations.

DNA polymerase (poe-LIM-uh-rase) Enzyme that assembles a new strand on a parent DNA strand during replication; also takes part in DNA repair.

DNA profile Of each individual, a unique array of DNA sequences inherited from each parent.

DNA repair Following an alteration in the base sequence of a DNA strand, a process that restores the original sequence, as carried out by DNA polymerases and other enzymes.

DNA replication The process by which the hereditary material in a cell is duplicated for distribution to daughter nuclei.

DNA sequencing A process that provides information about genes, including their size, their location on chromosomes, and the order of their nucleotides.

dominant allele In a diploid cell, an allele that masks the expression of its partner on the homologous chromosome.

duodenum (doo-oh-DEE-num) The first section of the small intestine.

duplication A change in a chromosome's structure resulting in the repeated appearance of a gene sequence.

dysplasia An abnormal change in the sizes, shapes, and organization of cells in a tissue.

ecological footprint The sum total of resources used by a population or an individual, together with the resulting waste products.

ecological pyramid A way to represent the energy relationships of an ecosystem.

ecology [Gk. *oikos*, home, and *logos*, reason] Study of the interactions of organisms with one another and with their physical and chemical environment.

ecosystem [Gk. *oikos*, home] An array of organisms and their physical environment, all of which interact through a flow of energy and a cycling of materials.

ectoderm [Gk. *ecto*, outside, and *derma*, skin] The outermost primary tissue layer (germ layer) of an embryo, which gives rise to the outer layer of the integument and to tissues of the nervous system.

effector A muscle (or gland) that responds to signals from an integrator (such as the brain) by producing movement (or chemical change) that helps adjust the body to changing conditions.

effector cell Of the differentiated subpopulations of lymphocytes that form during an immune response, the type of cell that engages and destroys the antigen-bearing agent that triggered the response.

efferent arteriole In the urinary system, the arteriole that carries filtered blood from the nephron.

egg *See* ovum.

elastic connective tissue A form of dense connective tissue found in organs that must stretch; made up mostly of the protein elastin, it is quite flexible.

electrocardiogram (ECG) A recording of the electrical activity of the heart's cardiac cycle.

electrolyte Any chemical substance, such as a salt, that ionizes and dissociates in water and is capable of conducting an electrical current.

electron Negatively charged unit of matter, with both particulate and wavelike properties, that occupies one of the orbitals around the atomic nucleus. Atoms can gain, lose, or share electrons with other atoms.

electron transfer When a molecule donates one or more electrons to another molecule.

electron transport system An organized array of enzymes and cofactors, bound in a cell membrane, that accept and donate electrons in sequence. When such systems operate, hydrogen ions (H^+) flow across the membrane, and the flow drives ATP formation and other reactions.

element Any substance that cannot be decomposed into substances with different properties.

embryo (EM-bree-oh) [Gk. *en*, in, and probably *bryein*, to swell] Of animals, a new individual that forms by cleavage, gastrulation, and other early developmental events.

embryonic disk In early development, the oval, flattened cell mass that gives rise to the embryo shortly after implantation.

emerging disease Disease caused by a new strain of an existing pathogen or one that is now exploiting an increased availability of human hosts.

emulsification In digestion, the breaking of large fat globules into a suspension of fat droplets coated with bile salts.

encapsulated receptor Receptor surrounded by a capsule of epithelial or connective tissue; common near the body surface.

endangered species Endemic (native) species highly vulnerable to extinction.

endemic disease A disease that occurs more or less continuously in a region.

endergonic reaction A reaction in which there is a net increase in energy in the product compound.

endocrine gland A ductless gland that secretes hormones, which usually enter interstitial fluid and then the bloodstream.

endocrine system System of cells, tissues, and organs that is functionally linked to the nervous system and that exerts control by way of hormones and other chemical secretions.

endocytosis (en-doe-sigh-TOE-sis) Movement of a substance into cells in which the substance becomes enclosed by a patch of plasma membrane that sinks into the cytoplasm, then forms a vesicle around it. Phagocytic cells also engulf pathogens this way.

endoderm [Gk. *endon*, within, and *derma*, skin] The inner primary tissue layer, or germ layer, of an embryo, which gives rise to the inner lining of the gut and organs derived from it.

endomembrane system System in cells that includes the endoplasmic reticulum, Golgi bodies, and various kinds of vesicles, and in which new proteins are modified into final form and lipids are assembled.

endometrium (en-doh-MEET-ree-um) [Gk. *metrios*, of the womb] Inner lining of the uterus consisting of connective tissue, glands, and blood vessels.

endoplasmic reticulum (ER) (en-doe-PLAZ-mik reh-TIK-yoo-lum) An organelle that begins at the nucleus and curves through the cytoplasm. In rough ER (which has many ribosomes on its cytoplasmic side), new polypeptide chains acquire specialized side chains. In many cells, smooth ER (with no attached ribosomes) is the main site of lipid synthesis.

endotherm Organism such as a human that maintains body temperature from within, generally by metabolic activity and controls over heat conservation and dissipation.

energy The capacity to do work.

energy carrier A molecule that delivers energy from one metabolic reaction site to another. ATP is the premier energy carrier; it readily donates energy to nearly all metabolic reactions.

energy pyramid A pyramid-shaped representation of an ecosystem's trophic structure (feeding levels), illustrating the energy losses at each transfer to a different feeding level.

enzyme (EN-zime) One of a class of proteins that greatly speed up (catalyze) reactions between specific substances. The substances that each type of enzyme acts upon are called its substrates.

eosinophil Fast-acting, phagocytic white blood cell that targets worms, fungi, and other large pathogens.

epidemic A disease outbreak in an area or population that occurs above predicted or normal levels.

epidermis The outermost tissue layer of skin.

epididymis Duct where sperm mature and are stored.

epiglottis A flaplike structure at the start of the larynx, positioned to direct the movement of air into the trachea or food into the esophagus.

epinephrine (ep-ih-NEF-rin) Adrenal hormone that raises blood levels of glucose and fatty acids; also increases the heart's rate and force of contraction.

epiphyseal plate Cartilage that covers either end of a growing long bone, permitting the bone to lengthen. The epiphyseal plate is replaced by bone when growth stops in late adolescence.

epiphysis Each end of a long bone.

epithelium (ep-ih-THEE-lee-um) A tissue consisting of one or more layers of cells that covers the body's external surfaces and lines its internal cavities and tubes. Epithelium has one free surface; the opposite surface rests on a basement membrane between it and an underlying connective tissue. Epidermis is an example.

erythrocyte (eh-RITH-row-site) [Gk. *erythros*, red, and *kytos*, vessel] Red blood cell.

esophagus (ee-SOF-uh-gus) Tubular portion of the digestive system that receives swallowed food and leads to the stomach.

essential amino acid Any of eight amino acids from protein that the body cannot synthesize and must be obtained from food.

essential fatty acid Any of the fatty acids that the body cannot synthesize and must be obtained from food.

estrogens (ESS-tro-jens) Sex hormones that help oocytes mature, trigger changes in the uterine lining during the menstrual cycle and pregnancy, and maintain secondary sexual traits; also influence body growth and development.

eukaryotic cell (yoo-carry-AH-tic) [Gk. *eu,* good, and *karyon,* kernel] A cell that has a "true nucleus," which contains its DNA, and other membrane-bound organelles. Compare *prokaryotic cell.*

evolution [L. *evolutio,* act of unrolling] Genetic change within a line of descent over time.

excitatory postsynaptic potential (EPSP) One of two competing signals at an input zone of a neuron; a graded potential that brings the neuron's plasma membrane closer to the threshold required for an action potential to fire.

excretion *See* urinary excretion.

exergonic reaction Chemical reaction that shows a net loss in energy.

exocrine gland (EK-suh-krin) [Gk. *ex,* out of, and *krinein,* to separate] Glandular structure that secretes products, usually through ducts or tubes, to a free epithelial surface.

exocytosis (ek-so-sigh-TOE-sis) Movement of a substance out of a cell by means of a transport vesicle that fuses with the plasma membrane and releases its contents to the outside.

exon Any of the nucleotide sequences of a pre-mRNA molecule that are spliced together to form the mature mRNA transcript and are ultimately translated into protein.

experiment A test in which some phenomenon in the natural world is manipulated in controlled ways to gain insight into its function, structure, operation, or behavior.

expiration Expelling air from the lungs; exhaling.

exponential growth (ex-po-NEN-shul) Pattern of population growth in which greater and greater numbers of individuals are produced during the successive doubling times; the pattern that emerges when the per capita birth rate remains even slightly above the per capita death rate, putting aside the effects of immigration and emigration.

extinction Permanent loss of a species.

extracellular fluid All the fluid not inside cells; includes plasma (the liquid portion of blood) and tissue fluid (which occupies the spaces between cells and tissues).

extraembryonic membranes Membranes that form along with a developing embryo, including the yolk sac, amnion, allantois, and chorion.

eyes Sensory organs that allow vision; they contain tissue with a dense array of photoreceptors.

facilitated diffusion A form of passive transport where transport proteins provide a channel through which solutes cross a cell membrane.

fact Verifiable information, not opinion or speculation.

fat A lipid with a glycerol head and one, two, or three fatty acid tails. The tails of saturated fats have only single bonds between carbon atoms and hydrogen atoms attached to all other bonding sites. Tails of unsaturated fats additionally have one or more double bonds between certain carbon atoms.

fatty acid A long, flexible hydrocarbon chain with a —COOH group at one end.

femur Thighbone; longest bone of the body.

fermentation [L. *fermentum,* yeast] A type of anaerobic pathway of ATP formation; it starts with glycolysis, ends when electrons are transferred back to one of the breakdown products or intermediates, and regenerates the NAD⁺ required for the reaction. Its net yield is two ATP per glucose molecule broken down.

fertilization [L. *fertilis,* to carry, to bear] Fusion of a sperm nucleus with the nucleus of an egg, which thereupon becomes a zygote.

fetus An embryo after it reaches the age of eight weeks.

fever Body temperature that has climbed above the normal set point, usually in response to infection. Mild fever promotes an increase in body defense activities.

fibrous connective tissue A specialized form of connective tissue that is strong and stretchy; the three types are loose, dense, and elastic.

fibrous joint Type of joint in which fibrous connective tissue unites the adjoining bones and no cavity is present.

fight–flight response The combination of sympathetic and parasympathetic nerve responses that prompt the body to react quickly to intense arousal.

filtration In urine formation, the process by which blood pressure forces water and solutes out of glomerular capillaries and into the Bowman's capsule.

first law of thermodynamics The total amount of energy in the universe remains constant from 14.5°C to 15.5°C at standard pressure.

flagella (fluh-JELL-uh), singular: flagellum [L., *whip*] Tail-like motile structures of many eukaryotic cells; they have a distinctive 9 + 2 array of microtubules. In humans only sperm have a flagellum.

follicle (FOLL-ih-kul) In an ovary, a primary oocyte (immature egg) together with the surrounding layer of cells.

food chain A straight-line sequence of who eats whom in an ecosystem.

food web A network of cross-connecting, interlinked food chains encompassing primary producers and an array of consumers, detritivores, and decomposers.

forebrain Brain region that includes the cerebrum and cerebral cortex, the olfactory lobes, and the hypothalamus.

fossil Physical remains or other evidence of an organism that lived in the distant past. Most fossils are skeletons, shells, leaves, seeds, and tracks that were buried in rock layers before they decomposed.

fossil fuels The fossilized remains of ancient forests. Examples include oil, coal, and natural gas. Fossil fuels are nonrenewable resources.

fossilization How fossils form. An organism or traces of it become buried in sediments or volcanic ash. Water and dissolved inorganic compounds infiltrate the remains. Accumulating sediments exert pressure above the burial site. Over time, the pressure and chemical changes transform the remains to stony hardness.

fovea Funnel-shaped depression in the center of the retina where photoreceptors are densely arrayed and visual acuity is the greatest.

free nerve endings Thinly myelinated or unmyelinated branched endings of sensory neurons in skin and internal tissues. They serve as mechanoreceptors, thermoreceptors, or pain receptors.

free radical Any highly reactive molecule or molecule fragment having an unpaired electron.

FSH Follicle-stimulating hormone. The name comes from its function in females, in whom FSH helps stimulate follicle development in ovaries. In males, it acts in the testes as part of a sequence of events that trigger sperm production.

functional group An atom or group of atoms that is covalently bonded to the carbon backbone of an organic compound and that influences its behavior.

gallbladder Organ of the digestive system that stores bile secreted from the liver.

gamete (GAM-eet) A haploid cell that functions in sexual reproduction. Sperm and eggs are examples.

ganglion (GANG-lee-un), plural: ganglia [Gk. *ganglion*, a swelling] A clustering of cell bodies of neurons in regions other than the brain or spinal cord.

gap junctions Channels that connect the cytoplasm of adjacent cells and help cells communicate by promoting the rapid transfer of ions and small molecules between them.

gastric juice Highly acidic mix of water and secretions from the stomach's glandular epithelium (HCl, mucus, pepsinogens, etc.) that kills ingested microbes and begins food breakdown.

gastrointestinal (GI) tract The digestive tube, extending from the mouth to the anus and including the stomach, small and large intestines, and other specialized regions with roles in food transport and digestion.

gastrulation (gas-tru-LAY-shun) The stage of embryonic development in which cells become arranged into primary tissue layers (germ layers); in humans, the layers are an inner endoderm, an intermediate mesoderm, and a surface ectoderm.

gene A unit of information about a heritable trait that is passed on from parents to offspring. Each gene has a specific location on a chromosome. Chemically, a gene is a sequence of nucleotides in a DNA molecule.

gene flow A microevolutionary process; a physical movement of alleles out of or into a population as individuals leave (emigrate) or enter (immigrate); allele frequencies change as a result.

gene mutation Small-scale change in the nucleotide sequence of a gene.

gene pair In diploid cells, the two alleles at a given locus on a pair of homologous chromosomes.

gene pool Sum total of all genotypes in a population. More accurately, allele pool.

gene therapy Generally, the transfer of one or more normal genes into body cells in order to correct a genetic defect.

genetic abnormality An uncommon version of an inherited trait.

genetic code [After L. *genesis*, to be born] The correspondence between nucleotide triplets in DNA (then in mRNA) and specific sequences of amino acids in the resulting polypeptide chains; the basic language of protein synthesis.

genetic disorder An inherited condition that results in mild to severe medical problems.

genetic drift A microevolutionary process; a change in allele frequencies over the generations due to chance events alone.

genetic engineering Altering the information content of DNA through use of recombinant DNA technology.

genetic recombination Presence of a new combination of alleles in a DNA molecule compared to the parental genotype; the result of processes such as crossing over at meiosis, chromosome rearrangements, gene mutation, and recombinant DNA technology.

genome All the DNA in a haploid number of chromosomes of a species.

genomics The study of whole genomes.

genotype (JEEN-oh-type) Genetic constitution of an individual. Can mean a single gene pair or the sum total of the individual's genes. Compare *phenotype*.

genus (JEEN-us, JEN-er-ah), plural: genera [L. *genus*, race, origin] A grouping of species more closely related to one another in body form, ecology, and history than to others at the same level of classification.

germ cell The cell of sexual reproduction; germ cells give rise to gametes. Compare *somatic cell*.

germ layer One of three primary tissue layers that forms during gastrulation and that gives rise to certain tissues of the adult body. Compare *ectoderm; endoderm; mesoderm*.

gland A secretory cell or multicellular structure derived from epithelium and often connected to it.

glial cells Any of the large number of cells in the nervous system that support neurons physically or in other ways.

glioma Cancer of the glial cells of the brain.

global climate change Major shifts in weather patterns worldwide.

global warming A long-term increase in the temperature of Earth's lower atmosphere.

glomerular capillaries The set of blood capillaries inside the Bowman's capsule of a nephron.

glomerulus (glow-MARE-you-luss), plural: glomeruli [L. *glomus,* ball] The first portion of the nephron, where water and solutes are filtered from blood.

glucagon (GLUE-kuh-gone) Hormone that stimulates conversion of glycogen and amino acids to glucose; secreted by alpha cells of the pancreas when the flow of glucose decreases.

glucocorticoid Hormone secreted by the adrenal cortex that influences metabolic reactions that help maintain the blood glucose level.

gluconeogenesis The process by which liver cells synthesize glucose.

glycemic index A list that ranks a food according to its effect on blood sugar (glucose).

glyceride (GLISS-er-eyed) One of the molecules, commonly called fats and oils, that has one, two, or three fatty acid tails attached to a glycerol backbone. They are the body's most abundant lipids and its richest source of energy.

glycerol (GLISS-er-all) [Gk. *glykys,* sweet, and L. *oleum,* oil] A three-carbon molecule with three hydroxyl groups attached; together with fatty acids, a component of fats and oils.

glycogen (GLY-kuh-jen) A storage polysaccharide that can be readily broken down into glucose subunits.

glycolysis (gly-CALL-ih-sis) [Gk. *glykys,* sweet, and *lysis,* loosening or breaking apart] In first stage of cellular respiration, process by which glucose (or some other organic compound) is partially broken down to pyruvate with a net yield of two ATP. Glycolysis occurs in the cell cytoplasm and oxygen has no role in it.

glycoprotein A protein having oligosaccharides covalently bonded to it. Most human cell surface proteins and many proteins circulating in blood are glycoproteins.

Golgi body (GOHL-gee) Organelle in which newly synthesized polypeptide chains as well as lipids are modified and packaged in vesicles for export or for transport to specific locations within the cytoplasm.

gonad (GO-nad) Primary reproductive organ in which gametes are produced. Ovaries and testes are gonads.

graded potential Of neurons, a local signal that slightly changes the voltage difference across a small patch of the plasma membrane. Such signals vary in magnitude, depending on the stimulus. With prolonged or intense stimulation, they may spread to a trigger zone of the membrane and initiate an action potential.

granulocyte Class of white blood cells that have a lobed nucleus and various types of granules in the cytoplasm; includes neutrophils, eosinophils, and basophils.

gray matter The dendrites, neuron cell bodies, and neuroglial cells of the spinal cord and cerebral cortex.

greenhouse effect Warming of the lower atmosphere due to the presence of the following greenhouse gases: carbon dioxide, methane, nitrous oxide, ozone, water vapor, and chlorofluorocarbons.

growth factor A type of signaling molecule that can influence growth by regulating the rate at which target cells divide.

guanine A nitrogen-containing base; one of those present in nucleotide building blocks of DNA and RNA.

habitat [L. *habitare,* to live in] The type of place where an organism normally lives, characterized by physical features, chemical features, and the presence of certain other species.

hair cell Type of mechanoreceptor that may give rise to action potentials when bent or tilted.

haploid (HAP-loyd) Having only one of each pair of homologous chromosomes; gametes are haploid. Compare *diploid.*

HCG Human chorionic gonadotropin. A hormone that helps maintain the lining of the uterus during the menstrual cycle and during the first trimester of pregnancy.

HDL A high-density lipoprotein in blood; it transports cholesterol to the liver for further processing.

heart Muscular pump that keeps blood circulating through the body.

heat A form of kinetic energy; also called thermal energy.

helper T cell Type of T lymphocyte that produces and secretes chemicals that promote formation of large effector and memory cell populations.

hemoglobin (HEEM-oh-glow-bin) [Gk. *haima,* blood, and L. *globus,* ball] Iron-containing, oxygen-transporting protein that gives red blood cells their color.

hemostasis (hee-mow-STAY-sis) [Gk. *haima,* blood, and *stasis,* standing] Stopping of blood loss from a damaged blood vessel through coagulation, blood vessel spasm, platelet plug formation, and other mechanisms.

hepatic portal system System of blood vessels that transport blood from the digestive tract to and from the liver.

herbivore [L. *herba,* grass, and *vovare,* to devour] Plant-eating animal.

heterotroph (HET-er-oh-trofe) [Gk. *heteros,* other, and *trophos,* feeder] Organism that cannot synthesize its own organic compounds and must obtain nourishment by feeding on autotrophs, each other, or organic wastes. Animals, fungi, many protists, and most bacteria are heterotrophs. Compare *autotroph.*

heterozygous (het-er-oh-ZYE-guss) [Gk. *zygoun,* join together] For a given trait, having nonidentical alleles at a particular locus on a pair of homologous chromosomes.

hindbrain One of the three divisions of the brain; the medulla oblongata, cerebellum, and pons; includes reflex centers for respiration, blood circulation, and other basic functions; also coordinates motor responses and many complex reflexes.

histamine Local signaling molecule that promotes inflammation; makes arterioles dilate and capillaries more permeable (leaky).

HIV Human immunodeficiency virus, which destroys key cells of the immune system and causes AIDS.

homeostasis (hoe-me-oh-STAY-sis) [Gk. *homo,* same, and *stasis,* standing] A physiological state in which the physical and chemical conditions of the internal environment are being maintained within ranges that enable survival of cells and the whole body.

homeostatic feedback loop An interaction in which an organ (or structure) stimulates or inhibits the output of another organ, then shuts down or increases this activity when it detects that the output has exceeded or fallen below a set point.

hominin [L. *homo,* man] All species on the evolutionary branch leading to modern humans. *Homo sapiens* is the only living representative.

Homo erectus A hominin lineage that emerged between 1.5 million and 300,000 years ago and that may include the direct ancestors of modern humans.

Homo habilis A type of early hominin that may have been the maker of stone tools that date from about 2.5 million years ago.

Homo sapiens The species of modern humans that emerged between 300,000 and 200,000 years ago.

homologous chromosome (huh-MOLL-uh-gus) [Gk. *homologia,* correspondence] (also called a *homologue*) One of a pair of chromosomes that resemble each other in size, shape, and the genes they carry, and that line up with each other at meiosis I. The X and Y chromosomes differ in these respects but still function as homologues.

homologous structure The same body part, modified in different ways, in different lines of descent from a common ancestor.

homozygous (hoe-moe-ZYE-guss) Having two identical alleles at a given locus (on a pair of homologous chromosomes).

homozygous dominant condition Having two dominant alleles at a given gene locus (on a pair of homologous chromosomes).

homozygous recessive condition Having two recessive alleles at a given gene locus (on a pair of homologous chromosomes).

hormone [Gk. *hormon,* to stir up, set in motion] Any of the signaling molecules secreted from endocrine glands, endocrine cells, and some neurons that the bloodstream distributes to nonadjacent target cells (any cell having receptors for that hormone).

host An organism that can be infected by a pathogen.

Human Genome Project A research project in which the estimated 3 billion nucleotides present in the DNA of human chromosomes were sequenced.

human immunodeficiency virus (HIV) The pathogen that causes AIDS (acquired immune deficiency syndrome) by destroying lymphocytes.

human papillomavirus (HPV) Virus that causes genital warts; strains of HPV are found in nearly all cervical cancers.

humerus The long bone of the upper arm.

hydrocarbon A molecule having only hydrogen atoms attached to a carbon backbone.

hydrogen bond A weak attraction between an electronegative atom and a hydrogen atom that is already taking part in a polar covalent bond.

hydrogen ion A free (unbound) proton; a hydrogen atom that has lost its electron and so bears a positive charge (H^+).

hydrolysis reaction (high-DRAWL-ih-sis) [L. *hydro,* water, and Gk. *lysis,* loosening or breaking apart] Enzyme-driven reaction in which covalent bonds break, splitting a molecule into two or more parts, and H^+ and OH^- (derived from a water molecule) become attached to the exposed bonding sites.

hydrophilic [Gk. *philos,* loving] Characteristic of a polar substance that is attracted to the polar water molecule and so dissolves easily in water. Sugars are examples.

hydrophobic [Gk. *phobos,* dreading] Characteristic of a nonpolar substance that is repelled by the polar water molecule and so does not readily dissolve in water. Oil is an example.

hydroxide ion Ionized compound of one oxygen and one hydrogen atom (OH^-).

hyperplasia An abnormal enlargement of tissue that leads to a tumor.

hypertension Chronically elevated blood pressure.

hyperthermia Condition in which the body core temperature rises above the normal range.

hypertonic Having a greater concentration of solutes relative to another fluid.

hypodermis A subcutaneous layer having stored fat that helps insulate the body; although not part of skin, it anchors skin while allowing it some freedom of movement.

hypothalamus [Gk. *hypo,* under, and *thalamos,* inner chamber] A brain center that monitors visceral activities (such as salt–water balance and temperature control) and that influences related forms of behavior (as in hunger, thirst, and sex).

hypothermia Condition in which the body core temperature falls below the normal range.

hypothesis A possible explanation of a specific phenomenon.

hypotonic Having a lower concentration of solutes relative to another fluid.

ileum Final section of the small intestine, where absorption is completed and residues move toward the large intestine.

immune response A series of events by which B and T cells recognize a specific antigen, undergo repeated cell divisions that form huge lymphocyte populations, and differentiate into subpopulations of effector and memory cells. Effector cells engage and destroy antigen-bearing agents. Memory cells enter a resting phase and are activated during subsequent encounters with the same antigen.

immune system Interacting white blood cells that defend the body through self/nonself recognition, specificity, and memory. T and B cell antigen receptors ignore the body's own cells yet collectively recognize at least a billion specific threats. Some B and T cells formed in a primary response are set aside as memory cells for future battles with the same antigen.

immunity The body's overall ability to resist and combat any substance foreign to itself.

immunization Various processes, including vaccination, that promote increased immunity against specific diseases.

immunodeficiency Disorder in which a person's immune system is weakened or absent.

immunoglobulin Ig; any of the five classes of antibodies that participate in defense and immune responses. Examples are IgM antibodies (first to be secreted during immune responses) and IgG antibodies (which activate complement proteins and neutralize many toxins).

immunological tolerance The lack of an immune response against normal body cells.

immunotherapy Procedures that enhance a person's immunological defenses against tumors or certain pathogens.

implantation Series of events in which a blastocyst (pre-embryo) invades the endometrium (lining of the uterus) and becomes embedded there.

independent assortment Genetic principle that each gene pair tends to assort into gametes independently of other gene pairs located on nonhomologous chromosomes.

induced-fit model According to this model, a surface region of each substrate has chemical groups that are almost but not quite complementary to chemical groups in an active site.

infection Invasion and multiplication of a pathogen in a host. *Disease* follows if defenses are not mobilized fast enough; the pathogen's activities interfere with normal body functions.

inflammation Process in which, in response to tissue damage or irritation, phagocytes and plasma proteins, including complement proteins, leave the bloodstream, then defend and help repair the tissue. Occurs during both nonspecific and specific (immune) defense responses.

inheritance The transmission, from parents to offspring, of body structures and functions that have a genetic basis.

inhibiting hormone A signaling molecule produced and released by the hypothalamus that controls secretions by the anterior lobe of the pituitary gland.

inhibitory postsynaptic potential (IPSP) Of neurons, one of two competing types of graded potentials at an input zone; tends to drive the resting membrane potential away from the threshold required to trigger a nerve impulse.

innate immunity The body's inborn, preset immune responses, which act quickly when tissue is damaged or microbes have invaded.

inner cell mass In early development, a clump of cells in the blastocyst that will give rise to the embryonic disk.

insertion The end of a muscle that is attached to the bone that moves most when the muscle contracts.

inspiration The drawing of air into the lungs; inhaling.

insulin Pancreatic hormone that lowers the level of glucose in blood by causing cells to take up glucose; also promotes the synthesis of fat and protein and inhibits the conversion of protein to glucose.

integration, neural *See* synaptic integration.

integrator Of homeostatic systems, a control point where different bits of information are pulled together in the selection of a response. The brain is an example.

integument [L. *integere,* to cover] The organ system that provides a protective body covering; in humans, the skin, oil and sweat glands, hair, and nails.

interferon Protein produced by T cells that interferes with viral replication. Some interferons also stimulate the tumor-killing activity of macrophages.

interleukin One of a variety of chemical communication signals—secreted by macrophages and helper T cells—that drive immune responses.

intermediate Substance that forms between the start and end of a metabolic pathway.

intermediate filament A ropelike element of the cytoskeleton that mechanically strengthens cells.

internal environment The fluid bathing body cells and tissues; it consists of blood plus interstitial fluid.

internal respiration Movement of oxygen into tissues from the blood, and of carbon dioxide from tissues into the blood.

interneuron Any of the neurons in the brain and spinal cord that integrate information arriving from sensory neurons and that influence other neurons in turn.

interphase Of cell cycles, the time interval between nuclear divisions in which a cell increases its mass, roughly doubles the number of its cytoplasmic components, and finally duplicates its chromosomes (replicates its DNA).

interstitial fluid (in-ter-STISH-ul) [L. *interstitus,* to stand in the middle of something] The extracellular fluid in spaces between cells and tissues.

intervertebral disk One of a number of disk-shaped structures containing cartilage that serve as shock absorbers and flex points between vertebrae.

intron A noncoding portion of a newly formed mRNA molecule.

invertebral disk Fibrocartilage pad between vertebrae.

in vitro fertilization Conception outside the body ("in glass" petri dishes or test tubes).

ion (EYE-on) An atom (or a compound) that has gained or lost one or more electrons and hence has acquired an overall negative or positive charge.

ionic bond An association between ions of opposite charges.

iris Of the eye, a circular pigmented region behind the cornea with a "hole" in its center (the pupil) through which incoming light enters.

isotonic Having the same solute concentration as a fluid against which it is being compared.

isotope (EYE-so-tope) For a given element, an atom with the same number of protons as the other atoms but with a different number of neutrons.

jejunum Middle section of the small intestine, where most nutrients are digested and absorbed.

joint An area of contact or near-contact between bones.

juxtaglomerular apparatus In kidney nephrons, a place where the arterioles of the glomerulus come into contact with the distal tubule. Cells in this region secrete renin, which triggers hormonal events that stimulate increased reabsorption of sodium.

karyotype (CARRY-oh-type) A preparation of an individual's of metaphase chromosomes arranged by length, shape, and the location on the centromere.

keratin A tough, water-insoluble protein manufactured by most epidermal cells.

keratinocytes Cells of the epidermis that make keratin.

kidney One of a pair of organs that filter organic wastes, toxins, and other substances from the blood and help regulate the volume and solute concentrations of extracellular fluid.

kilocalorie 1,000 calories of heat energy, or the amount of energy needed to raise the temperature of 1 kilogram of water by 1°C; the unit of measure for the caloric value of foods.

kinetic energy The energy of motion.

Krebs cycle With a few conversion steps that precede it, the stage of aerobic respiration in which pyruvate is completely broken down to carbon dioxide and water and 2 ATP form. Coenzymes accept the protons (H^+) and electrons removed from intermediates during the reactions and deliver them to the next stage.

lactate fermentation Anaerobic pathway of ATP formation in which pyruvate from glycolysis is converted to the three-carbon compound lactate, and NAD^+ (a coenzyme used in the reactions) is regenerated. Its net yield is two ATP.

lactation The production of milk by hormone-primed mammary glands.

lacteal Small lymph vessel in villi of the small intestine that receives absorbed triglycerides. Triglycerides move from the lymphatic system to the general circulation.

large intestine The colon; a region of the GI tract that receives unabsorbed food residues from the small intestine and concentrates and stores feces until they are expelled from the body.

larynx (LARE-inks) A tubular airway that leads to the lungs. It contains vocal cords, where sound waves used in speech are produced.

LDL Low-density lipoprotein that transports cholesterol; excess amounts contribute to atherosclerosis.

lens Of the eye, a saucer-shaped region behind the iris containing multiple layers of transparent proteins. Ligaments can move the lens, which focuses incoming light onto photoreceptors in the retina.

leukocytes White blood cells.

Leydig cell In testes, cells in connective tissue around the seminiferous tubules that secrete testosterone and other signaling molecules.

LH Luteinizing hormone, secreted by the anterior lobe of the pituitary gland. In males it acts on Leydig cells of the testes and prompts them to secrete testosterone. In females, LH stimulates follicle development in the ovaries.

life cycle Recurring series of genetically programmed events from the time individuals are produced until they themselves reproduce.

ligament A strap of dense, elastic, regular connective tissue that connects two bones at a joint.

limbic system Brain regions that, along with the cerebral cortex, collectively govern emotions; it includes parts of the thalamus, hypothalamus, amygdala, and hippocampus.

limiting factor Any essential resource that is in short supply and so limits population growth.

linkage The tendency of genes located on the same chromosome to end up in the same gamete. For any two of those genes, the probability that crossing over will disrupt the linkage is proportional to the distance separating them.

lipid A greasy or oily compound of mostly carbon and hydrogen that shows little tendency to dissolve in water, but that dissolves in nonpolar solvents (such as ether). Cells use lipids as energy stores and structural materials, especially in membranes.

lipid bilayer The structural basis of cell membranes, consisting of two layers of mostly phospholipid molecules. Hydrophilic heads force all fatty acid tails of the lipids to become sandwiched between the hydrophilic heads.

lipoprotein A protein that has a lipid attached to it. Molecule that forms when proteins circulating in blood combine with cholesterol, triglycerides, and phospholipids absorbed from the small intestine.

liver Organ with roles in storing and interconverting carbohydrates, lipids, and proteins absorbed from the gut; disposing of nitrogen-containing wastes; and other tasks.

local signaling molecule A molecule that alters chemical conditions in the immediate vicinity where it is secreted, then is swiftly broken down.

locus (LOW-cuss) The location of a particular gene on a chromosome.

logistic growth (low-JIS-tik) Pattern of population growth in which a low-density population slowly increases in size, goes through a rapid growth phase, then levels off once the carrying capacity is reached.

loop of Henle The hairpin-shaped, tubular region of a nephron that functions in reabsorption of water and solutes.

loose connective tissue Flexible fibrous connective tissue with few fibers and cells.

lung One of a pair of sac-shaped organs that provide a moist surface for gas exchange.

lymph (limf) [L. *lympha,* water] Tissue fluid that has moved into the vessels of the lymphatic system.

lymph capillary A small-diameter vessel of the lymph vascular system that has no obvious entrance; tissue fluid moves inward by passing between overlapping endothelial cells at the vessel's tip.

lymph node A lymphoid organ that serves as a battleground of the immune system; each lymph node is packed with macrophages and lymphocytes that cleanse lymph of pathogens before it reaches the blood.

lymph vascular system [L. *vasculum,* a small vessel] The vessels of the lymphatic system, which take up and transport excess tissue fluid and reclaimable solutes as well as fats absorbed from the digestive tract.

lymphatic system An organ system with vessels that take up fluid and solutes from interstitial fluid and deliver them to the bloodstream; its lymphoid organs have roles in immunity.

lymphocyte A T cell or B cell.

lymphoid organ The lymph nodes, spleen, thymus, tonsils, adenoids, and other organs with roles in immunity.

lysosome (LYE-so-sohm) A cell organelle that contains enzymes that can break down polysaccharides, proteins, nucleic acids, and some lipids.

lysozyme Present in mucous membranes that line body surfaces, an infection-fighting enzyme that attacks and destroys various types of bacteria.

macroevolution The large-scale patterns, trends, and rates of change among groups of species.

macrophage A phagocytic white blood cell. It engulfs anything detected as foreign. Some also become antigen-presenting cells that serve as the trigger for immune responses by T and B lymphocytes. Compare *antigen-presenting cell.*

mammal A type of vertebrate; the only animal having offspring that are nourished by milk produced by mammary glands of females.

mandible The lower jaw; the largest single facial bone.

mass extinction An abrupt rise in extinction rates above the background level; a catastrophic, global event in which large groups of organisms are wiped out simultaneously.

mass number The total number of protons and neutrons in an atom's nucleus. The relative masses of atoms are also called atomic weights.

mast cell A type of white blood cell that releases enzymes and histamine during tissue inflammation.

matrix In connective tissue, fiberlike structural proteins together with a "ground substance" of polysaccharides that give each kind of tissue its particular properties.

mechanical processing In digestion, the breaking up and mixing of food by the teeth, tongue, and peristalsis.

mechanoreceptor Sensory cell or cell part that detects mechanical energy associated with changes in pressure, position, or acceleration.

medical imaging Any of several diagnostic methods including magnetic resonance imaging (MRI), X-ray, ultrasound, and CT (computed tomography) scanning.

medulla oblongata Part of the brain stem with reflex centers for respiration, blood circulation, and other vital functions.

meiosis (my-OH-sis) [Gk. *meioun,* to diminish] Two-stage nuclear division process in which the chromosome number of a germ cell is reduced by half, to the haploid number. (Each daughter nucleus ends up with one of each type of chromosome.) Meiosis forms gametes.

melanocytes Cells in the deepest layer of epidermis that produce the brown-black pigment melanin found in keratinocytes.

membrane attack complexes Structures that form pores in the plasma membrane of a pathogen, causing disintegration.

memory The storage and retrieval of information about previous experiences; underlies the capacity for learning.

memory cell Any of the various B or T cells of the immune system that are formed in response to invasion by a foreign agent and that are available to mount a rapid attack if the same type of invader reappears.

menarche A female's first menstruation.

meninges Membranes of connective tissue that are layered between the skull bones and the brain and cover and protect the neurons and blood vessels that service the brain.

menopause (MEN-uh-pozz) [L. *mensis,* month, and *pausa,* stop] End of the reproductive period of a human female's life cycle.

menstrual cycle The cyclic release of oocytes and priming of the endometrium (lining of the uterus) to receive a fertilized egg; the complete cycle averages about 28 days in female humans.

menstruation Periodic sloughing of the blood-enriched lining of the uterus when pregnancy does not occur.

mesoderm (MEH-so-derm) [Gk. *mesos,* middle, and *derm,* skin] In an embryo, a primary tissue layer (germ layer) between ectoderm and endoderm. Gives rise to muscle; organs of circulation, reproduction, and excretion; most of the internal skeleton; and connective tissue layers of the gastrointestinal tract and integument.

messenger RNA (mRNA) A linear sequence of ribonucleotides transcribed from DNA and translated into a polypeptide chain; the only type of RNA that carries protein-building instructions.

metabolic acidosis Lower than optimal blood pH caused by diabetes mellitus.

metabolic alkalosis Higher than optimal pH in blood and other body fluids.

metabolic pathway An orderly sequence of enzyme-driven reactions by which cells maintain, increase, or decrease the concentrations of particular substances.

metabolic syndrome A cluster of symptoms, including slightly elevated blood sugar that increases the risk of developing type 2 diabetes.

metabolism (meh-TAB-oh-lizm) [Gk. *meta*, change] All controlled, enzyme-driven chemical reactions by which cells acquire and use energy. Through these reactions, cells synthesize, store, break apart, and eliminate substances in ways that contribute to growth, survival, and reproduction.

metaphase Of mitosis or meiosis II, the stage when each duplicated chromosome has become positioned at the midpoint of the microtubular spindle, with its two sister chromatids attached to microtubules from opposite spindle poles. Of meiosis I, the stage when all pairs of homologous chromosomes are positioned at the spindle's midpoint, with the two members of each pair attached to opposite spindle poles.

metastasis The process in which cancer cells break away from a primary tumor and migrate (via blood or lymphatic tissues) to other locations, where they establish new cancer sites.

MHC marker Any of a variety of proteins that are self markers. Some occur on all body cells of an individual; others occur only on macrophages and lymphocytes.

micelle (my-CELL) Tiny droplet of bile salts, fatty acids, and monoglycerides; plays a role in fat absorption from the small intestine.

microevolution Changes in allele frequencies brought about by mutation, genetic drift, gene flow, and natural selection.

microfilament [Gk. *mikros*, small, and L. *filum*, thread] One of a variety of cytoskeletal components. Actin and myosin filaments are examples.

micrograph Photograph of an image brought into view with the aid of a microscope.

microorganism Organism, usually single-celled, too small to be observed without a microscope.

microscopy The use of a microscope to view objects, including cells, that are not visible to the naked eye.

microtubule A cytoskeletal element with roles in cell shape, motion, and growth and in the structure of cilia and flagella. The largest element of the cytoskeleton.

microvillus (my-crow-VILL-us) [L. *villus*, shaggy hair] A slender extension of the cell surface that functions in absorption or secretion.

midbrain A brain region that evolved as a coordination center for reflex responses to visual and auditory input; together with the pons and medulla oblongata, part of the brain stem, which includes the reticular formation.

mineral An inorganic substance required for the normal functioning of body cells.

mineralocorticoid Hormone secreted by the adrenal cortex that mainly regulates the concentrations of mineral salts in extracellular fluid.

mitochondrion (my-toe-KON-dree-on), plural: mitochondria. Organelle that specializes in ATP formation; it is the site of the second and third stages of aerobic respiration.

mitosis (my-TOE-sis) [Gk. *mitos*, thread] Type of nuclear division that maintains the parental chromosome number for daughter cells. It is the basis of bodily growth and the repair of tissue damage.

mixture A substance of two or more elements whose proportions can and usually do vary.

molecule A unit of matter in which chemical bonding holds together two or more atoms of the same or different elements.

monoclonal antibody [Gk. *monos*, alone] Antibody produced in the laboratory by a population of genetically identical cells that are clones of a single "parent" antibody-producing cell.

monomer A small molecule that is commonly a subunit of polymers, such as the sugar monomers of starch.

monosaccharide (mon-oh-SAK-ah-ride) [Gk. *sakharon*, sugar] The simplest carbohydrate, with only one sugar monomer. Glucose is an example.

monosomy Condition in which one of the chromosomes in a gamete has no homologue.

morphogenesis (more-foe-JEN-ih-sis) [Gk. *morphe*, form, and *genesis*, origin] Processes by which differentiated cells in an embryo become organized into tissues and organs.

morula A compact ball of sixteen embryonic cells formed after the third round of cleavage.

motility In digestion, the movement of ingested material through the GI tract.

motor neuron A neuron that delivers signals from the brain and spinal cord that can stimulate or inhibit the body's effectors (muscles, glands, or both).

motor unit A motor neuron and the muscle fibers under its control.

mRNA *See* messenger RNA.

mucous membranes (mucosae) Pink, moist membranes that line the tubes and cavities of various body systems; most absorb or secrete substances.

multifactorial trait A trait that is shaped by more than one gene as well as by environmental factors.

multiple allele system A gene that has three or more different molecular forms (alleles).

muscle fatigue A decline in the ability of a muscle to contract; occurs when a muscle has been kept in a state of strong contraction as a result of continuous, high-frequency stimulation.

muscle tension A mechanical force, exerted by a contracting muscle, that resists opposing forces such as gravity and the weight of objects being lifted.

muscle tissue Tissue having cells able to contract in response to stimulation, then passively lengthen and so return to their resting state.

muscle tone In muscles, a steady low-level contracted state that helps stabilize joints and maintain general muscle health.

muscle twitch Muscle response in which the muscle contracts briefly, then relaxes, when a brief stimulus activates a motor unit.

muscular system Skeletal muscles, which attach to bones and pull on them to move the body and its parts.

mutation, gene *See* gene mutation.

myelin sheath Of many sensory and motor neurons, an axonal sheath that affects how fast action potentials travel; formed from the plasma membranes of Schwann cells or oligodendrocytes that wrap repeatedly around the axon and are separated from each other by a small node.

myocardium The cardiac muscle tissue.

myofibril (MY-oh-fy-brill) One of many threadlike structures inside a muscle cell; each is functionally divided into sarcomeres, the basic units of contraction.

myosin (MY-uh-sin) A contractile protein. In muscle cells, it interacts with the protein actin to bring about contraction.

NADP Nicotinamide adenine dinucleotide phosphate; a phosphorylated nucleotide coenzyme. When carrying electrons and unbound protons (H$^+$) between reaction sites, it is abbreviated NADPH$_2$.

nasal cavity The region of the respiratory system in the nosewhere inhaled air is warmed, moistened, and filtered of airborne particles and dust.

natural selection A difference in survival and reproduction among members of a population that vary in one or more traits.

negative feedback A homeostatic mechanism in which an activity changes some condition in the internal environment and so triggers a response that reverses the change.

nephron (NEFF-ron) [Gk. *nephros,* kidney] Of the kidney, a slender tubule in which water and solutes filtered from blood are selectively reabsorbed and in which urine forms.

nerve Cordlike communication line of the peripheral nervous system, composed of axons of sensory neurons, motor neurons, or both, encased in connective tissue. In the brain and spinal cord, similar cordlike bundles are called nerve tracts.

nerve impulse *See* action potential.

nerve tract A bundle of myelinated axons of interneurons inside the spinal cord and brain.

nervous system System of neurons oriented relative to one another in precise message-conducting and information-processing pathways.

nervous tissue Tissue composed of neurons and (in the central nervous system) neuroglia.

neural tube Embryonic forerunner of the brain and spinal cord.

neuroglia (nur-oh-GLEE-uh) Cells that structurally and metabolically support neurons. They make up about half the volume of nervous tissue in the human body.

neuromodulator A signaling molecule that influences the effects of transmitter substances by enhancing or reducing membrane responses in target neurons.

neuromuscular junction Chemical synapse between axon terminals of a motor neuron and a muscle cell.

neuron A nerve cell; the basic unit of communication in the nervous system. Neurons collectively sense environmental change, integrate sensory inputs, then activate muscles or glands that initiate or carry out responses.

neurotransmitter Any of the class of signaling molecules that are secreted from neurons, act on adjacent cells, and are then rapidly degraded or recycled.

neutron Unit of matter, one or more of which occupies the atomic nucleus. Neutrons have mass but no electric charge.

neutrophil Phagocytic white blood cell that takes part in inflammatory responses against bacteria.

niche (nitch) [L. *nidas,* nest] The full range of physical and biological conditions under which members of a species can live and reproduce.

nitrogen cycle Biogeochemical cycle in which gaseous nitrogen is captured by nitrogen-fixing microorganisms and then moves through organisms and ecosystems before being returned to the atmosphere. The atmosphere is the largest reservoir of nitrogen.

nitrogen fixation Process by which a few kinds of bacteria convert gaseous nitrogen (N$_2$) to ammonia.

NK cell Natural killer cell; kills virus-infected cells and some types of cancer cells.

nociceptor A receptor, such as a free nerve ending, that detects stimuli causing tissue damage.

nondisjunction Failure of one or more chromosomes to separate properly during mitosis or meiosis.

nonpoint source A source of pollution not tied to a particular location.

nonrenewable resource A natural resource that exists in a finite amount and cannot be replenished.

nonshivering heat production Response to severe cold; brown adipose tissue releases energy as heat rather than storing it as ATP.

nonsteroid hormone A type of water-soluble hormone, such as a protein hormone, that cannot cross the lipid bilayer of a target cell. These hormones enter the cell by receptor-mediated endocytosis, or they bind to receptors that activate membrane proteins or second messengers within the cell.

nosocomial infection An infection that is acquired in a hospital, usually by direct contact with a microbe.

nuclear envelope A double membrane (two lipid bilayers and associated proteins) that is the outermost portion of a cell nucleus.

nucleic acid (noo-CLAY-ik) A long, single- or double-stranded chain of four different nucleotides joined at their phosphate groups. Nucleic acids differ in which nucleotide base follows the next in the sequence. DNA and RNA are examples.

nucleolus (noo-KLEE-oh-lus) [L. *nucleolus*, a little kernel] Within the nucleus of a nondividing cell, a site where the protein and RNA subunits of ribosomes are assembled.

nucleotide (NOO-klee-oh-tide) A small organic compound having a five-carbon sugar (deoxyribose), nitrogen-containing base, and phosphate group. Nucleotides are the structural units of adenosine phosphates, nucleotide coenzymes, and nucleic acids.

nucleotide sequence The order of nucleotides in a gene; it codes for a specific polypeptide chain.

nucleus (NOO-klee-us) Of atoms, the central core consisting of one or more positively charged protons and (in all but hydrogen) electrically neutral neutrons. In cells, a membranous organelle that physically isolates and organizes the DNA, out of the way of cytoplasmic machinery.

nutrient Element with a direct or indirect role in metabolism that no other element fulfills.

nutrition All those processes by which food is ingested, digested, absorbed, and later converted to the body's own organic compounds.

obesity An excess of fat in the body's adipose tissues, often caused by imbalances between caloric intake and energy output.

olfactory receptors Receptors in the nasal epithelium that detect water-soluble or volatile substances.

oligosaccharide A carbohydrate consisting of a short chain of two or more covalently bonded sugar units. One subclass, disaccharides, has two sugar units. Compare *monosaccharide; polysaccharide.*

omnivore [L. *omnis*, all, and *vovare*, to devour] An organism that feeds on a variety of food types, such as plant and animal tissues. Most humans are omnivores.

oncogene (ON-coe-jeen) A gene that has the potential to induce cancerous transformations in a cell.

oocyte An immature egg.

oogenesis (oo-oh-JEN-uh-sis) Formation of a female gamete, from a germ cell to a mature haploid ovum (egg).

opinion A subjective judgment.

oral cavity The mouth.

orbital Volume of space around the nucleus of an atom in which electrons are likely to be at any instant.

organ A body structure of definite form and function that is composed of more than one kind of tissue.

organ of Corti Region of the inner ear that contains the sensory hair cells involved in hearing.

organ system Two or more organs that interact chemically, physically, or both in performing a common task.

organelle In cells, an internal, membrane-bounded sac or compartment that has a specific, metabolic function.

organic compound A compound having a carbon backbone, often with carbon atoms arranged as a chain or ring structure, and at least one hydrogen atom.

orgasm The culmination of the sex act that involves muscle contractions and sensations of warmth, release, and relaxation.

origin The end of a muscle that is attached to the bone that remains relatively stationary when the muscle contracts.

osmoreceptor Sensory receptor that detects changes in water volume (solute concentration) in the fluid bathing it.

osmosis (oss-MOE-sis) [Gk. *osmos*, act of pushing] The tendency of water to move across a cell membrane in response to a concentration gradient.

osteoblast A cell that forms bone.

osteoclast A bone cell that breaks down the matrix of bone tissue.

osteocyte A living bone cell.

osteon A set of thin, concentric layers of compact bone tissue surrounding a narrow canal carrying blood vessels and nerves; arrays of osteons make up compact bone.

ovarian cycle Cycle during which a primary oocyte matures and is ovulated.

ovary (OH-vuh-ree) The primary female reproductive organ, where eggs form.

oviduct (OH-vih-dukt) Duct through which eggs travel from the ovary to the uterus. Also called Fallopian tube.

ovulation (ahv-you-LAY-shun) During each turn of the menstrual cycle, the release of a secondary oocyte (immature egg) from an ovary.

ovum (OH-vum) A mature female gamete (egg).

oxidation-reduction reaction An electron transfer from one atom or molecule to another. Often hydrogen is transferred along with the electron or electrons.

oxygen debt Lowered O_2 level in blood when muscle cells have used up more ATP than they have formed by aerobic respiration.

oxyhemoglobin A hemoglobin molecule that has oxygen bound to it.

ozone thinning Pronounced seasonal thinning of Earth's ozone layer, as in the lower stratosphere above Antarctica.

palate Structure that separates the nasal cavity from the oral cavity. The bone-reinforced hard palate serves as a hard surface against which the tongue can press food as it mixes it with saliva.

pancreas (PAN-cree-us) Gland that secretes enzymes and bicarbonate into the small intestine during digestion, and that also secretes the hormones insulin and glucagon.

pancreatic islets Any of the two million clusters of endocrine cells in the pancreas, including alpha cells, beta cells, and delta cells.

pandemic A situation in which epidemics of a disease break out in several countries around the world within a given time span.

parasite [Gk. *para*, alongside, and *sitos*, food] An organism that obtains nutrients directly from the tissues of a living host, which it lives on or in and may or may not kill.

parasympathetic nerve Of the autonomic nervous system, any of the nerves carrying signals that tend to slow the body down overall and divert energy to basic tasks; parasympathetic nerves also work continually in opposition with sympathetic nerves to bring about minor adjustments in internal organs.

parathyroid glands (pare-uh-THY-royd) Endocrine glands embedded in the thyroid gland that secrete parathyroid hormone, which helps restore blood calcium levels.

parturition Birth.

passive immunity Temporary immunity conferred by deliberately introducing antibodies into the body.

passive transport Diffusion of a solute through a channel or carrier protein that spans the lipid bilayer of a cell membrane. Its passage does not require an energy input; the protein passively allows the solute to follow its concentration gradient.

pathogen (PATH-oh-jen) [Gk. *pathos*, suffering] An infectious, disease-causing agent, such as a virus or bacterium.

PCR *See* polymerase chain reaction.

pectoral girdle Set of bones, including the scapula (shoulder blade) and clavicle (collarbone), to which the long bone of each arm attaches. The pectoral girdles form the upper part of the appendicular skeleton and are only loosely attached to the rest of the body by muscles.

pedigree Family history of a genetic trait.

pelvic cavity Body cavity in which the reproductive organs, bladder, and rectum are located.

pelvic girdle Set of bones including coxal bones that form the pelvis; the lower part of the appendicular skeleton. The upper portions of the two coxal bones are the hipbones; the thighbones (femurs) join the coxal bones at hip joints. The pelvic girdle bears the body's weight when a person stands.

penetrance In a given population, the percentage of individuals in which a particular genotype is expressed (that is, the percentage of individuals who have the genotype and also exhibit the corresponding phenotype).

penis Male organ that deposits sperm into the female reproductive tract; also houses the urethra.

pepsin Any of several digestive enzymes that are part of gastric fluid in the stomach.

pepsinogen A precursor to the digestive enzyme pepsin.

peptide bond Covalent bond that joins the amino group of one amino acid to the carboxyl group of a second amino acid.

peptide hormone A hormone that consists of a short chain of amino acids.

perception The conscious interpretation of some aspect of the external world created by the brain from nerve impulses generated by sensory receptors.

perforin A type of protein secreted by a natural killer cell of the immune system, and which creates holes (pores) in the plasma membrane of a target cell.

peripheral nervous system (per-IF-ur-uhl) [Gk. *peripherein*, to carry around] The nerves leading into and out from the spinal cord and brain and the ganglia along those communication lines.

peristalsis (pare-ih-STAL-sis) Rhythmic contraction of muscles that moves food forward through the gastrointestinal tract.

peritubular capillaries The set of blood capillaries that threads around the tubular parts of a nephron; they function in reabsorption of water and solutes and in secretion of hydrogen ions and some other substances as urine is formed.

peroxisome Enzyme-filled vesicle in which fatty acids and amino acids are digested first into hydrogen peroxide (which is toxic), then to harmless products.

PGAL Phosphoglyceraldehyde. A key intermediate in glycolysis.

pH scale A scale used to measure the concentration of free hydrogen ions in blood, water, and other solutions; pH 0 is the most acidic, 14 the most basic, and 7, neutral.

phagocyte (FAYG-uh-sight) [Gk. *phagein*, to eat, and *-kytos*, hollow vessel] A macrophage or other white blood cell that engulfs and destroys foreign agents.

phagocytosis (fayg-uh-sigh-TOE-sis) [Gk. *phagein*, to eat, and *-kytos*, hollow vessel] Engulfment of foreign cells or substances by specialized white blood cells by means of endocytosis.

pharynx (FARE-inks) A muscular tube by which food enters the gastrointestinal tract; the dual entrance for the tubular part of the digestive tract and windpipe (trachea).

phenotype (FEE-no-type) [Gk. *phainein*, to show, and *-typos*, image] Observable trait or traits of an individual; arises from interactions between genes, and between genes and the environment.

pheromone (FARE-oh-moan) [Gk. *phero*, to carry, and *-mone*, as in hormone] A type of signaling molecule secreted by exocrine glands that serves as a communication signal between individuals of the same species.

phospholipid A type of lipid that is the main structural component of cell membranes. Each has a hydrophobic tail (of two fatty acids) and a hydrophilic head that incorporates glycerol and a phosphate group.

phosphorylation (foss-for-ih-LAY-shun) The attachment of unbound (inorganic) phosphate to a molecule; also the transfer of a phosphate group from one molecule to another, as when ATP phosphorylates glucose.

photoreceptor A light-sensitive sensory cell.

phytochemicals Plant molecules that may reduce the risk of some disorders, but are not essential.

pigment A light-absorbing molecule.

pilomotor response Contraction of smooth muscle controlling the erection of body hair when outside temperature drops. This creates a layer of still air that reduces heat losses from the body. (It is most effective in mammals that have more body hair than humans do.)

pineal gland (PY-neel) A light-sensitive endocrine gland that secretes melatonin, a hormone that influences reproductive cycles and the development of reproductive organs.

pituitary gland An endocrine gland that interacts with the hypothalamus to coordinate and control many physiological functions, including the activity of many other endocrine glands. Its posterior lobe stores and secretes hypothalamic hormones; the anterior lobe produces and secretes its own hormones.

placenta (pluh-SEN-tuh) Of the uterus, an organ composed of maternal tissues and extraembryonic membranes (the chorion especially); it delivers nutrients to the fetus and accepts wastes from it, yet allows the fetal circulatory system to develop separately from the mother's.

plaque Cholesterol and other lipids that build up in the arterial wall, leaving less room for flowing blood.

plasma (PLAZ-muh) Liquid portion of blood; consists of water, various proteins, ions, sugars, dissolved gases, and other substances.

plasma cell In adaptive immunity, an effector B cell that quickly floods the bloodstream with antibodies.

plasma membrane The outermost cell membrane. Proteins in its lipid bilayer carry out most functions, including transport across the membrane and reception of extracellular signals.

platelet (PLAYT-let) A cell fragment in blood that releases substances necessary for blood clotting.

pleiotropy (ply-AH-trow-pee) [Gk. *pleon*, more, and *trope*, direction] A type of gene interaction in which a single gene exerts multiple effects on seemingly unrelated aspects of an individual's phenotype.

pleura (plural: pleurae) Thin, double membrane surrounding each lung.

point source A single place where a form of pollution begins.

polar body Any of up to three cells that form during the meiotic cell division of an oocyte; the division also forms the mature egg, or ovum.

pollutant Any substance with which an ecosystem has had no prior evolutionary experience in terms of kinds or amounts, and that can accumulate to disruptive or harmful levels. Can be naturally occurring or synthetic.

polygenic trait Trait that results from the combined expression of several genes.

polymer (PAH-lih-mur) [Gk. *polus*, many, and *meris*, part] A molecule composed of three to millions of small subunits that may or may not be identical.

polymerase chain reaction (PCR) DNA amplification method; DNA containing a gene of interest is split into single strands, which enzymes (polymerases) copy; the enzymes also act on the accumulating copies, multiplying the gene sequence by the millions.

polymorphism (poly-MORE-fizz-um) [Gk. *polus*, many, and *morphe*, form] Of a population, the persistence through the generations of two or more forms of a trait.

polypeptide chain Three or more amino acids joined by peptide bonds.

polysaccharide [Gk. *polus*, many, and *sakharon*, sugar] A straight or branched chain of covalently bonded monomers of the same or different kinds of sugars. The most common polysaccharides are cellulose, starch, and glycogen.

pons Hindbrain traffic center for signals between centers of the cerebellum and forebrain.

population A group of individuals of the same species occupying a given area.

population density The number of individuals of a population that are living in a specified area or volume.

population size The number of individuals that make up the gene pool of a population.

positive feedback mechanism Homeostatic mechanism by which a chain of events is set in motion that intensifies a change from an original condition.

potential energy A capacity to do work.

precapillary sphincter A ring of smooth muscle that regulates the flow of blood into a capillary.

prediction A statement about what one can expect to observe in nature if a theory or hypothesis is correct.

preimplantation diagnosis Test for birth defects in an early embryo that was conceived by in vitro fertilization.

primary immune response Activity of white blood cells and their products elicited by a first-time encounter with an antigen; includes both antibody-mediated and cell-mediated responses.

primary productivity Of ecosystems, *gross* primary productivity is the rate at which the producer organisms capture and store a given amount of energy during a specified interval.

primary structure The particular sequence of amino acids that make up a given protein.

primate A type of mammal; primates include monkeys, apes, and humans.

primer A laboratory-made short nucleotide sequence designed to base-pair with any complementary DNA sequence; later, DNA polymerases recognize it as a start tag for replication.

principle of sustainability *See* sustainability.

prion (PREE-on) Small infectious protein that causes rare, fatal degenerative diseases of the nervous system.

probability With respect to any chance event, the most likely number of times it will turn out a certain way, divided by the total number of all possible outcomes.

producer Of ecosystems, any of the organisms that secure energy from the physical environment, as by photosynthesis or chemosynthesis. Green plants are Earth's main primary producers.

product Substance present at the end of a metabolic pathway.

progesterone (pro-JESS-tuh-rown) Female sex hormone secreted by the ovaries.

prokaryotic cell (pro-carry-OH-tic) [L. *pro*, before, and Gk. *karyon*, kernel] A single-celled organism that has no nucleus or any of the other membrane-bound organelles characteristic of eukaryotic cells. Bacteria are prokaryotic.

promoter Of transcription, a base sequence that signals the start of a gene; the site where RNA polymerase initially binds.

prophase Of mitosis, the stage when each duplicated chromosome starts to condense, microtubules form a spindle apparatus, and the nuclear envelope starts to break up.

prophase I Of meiosis, the stage at which the spindle starts to form, the nuclear envelope starts to break up, and each duplicated chromosome condenses and pairs with its homologous partner. At this time, their sister chromatids typically undergo crossing over and genetic recombination.

prophase II Of meiosis, a brief stage during which each chromosome still consists of two chromatids.

prostaglandin Any of various local signaling molecules that typically cause smooth muscle to contract or relax, as in blood vessels, the uterus, and airways.

prostate gland Gland in males that wraps around the urethra and ejaculatory ducts; its secretions become part of semen.

protein A large organic compound composed of one or more chains of amino acids held together by peptide bonds. Proteins have unique sequences of different kinds of amino acids in their polypeptide chains; such sequences are the basis of a protein's three-dimensional structure and chemical behavior.

protein hormone A hormone that consists of a long amino acid chain.

proto-oncogene A gene similar to an oncogene but that codes for a protein required in normal cell function; may trigger cancer, generally when mutations alter its structure or function.

proton Positively charged subatomic particle in the nucleus of all atoms.

proximal tubule The region of a nephron tubule that receives water and solutes filtered from the blood.

psychoactive drug A chemical that acts on the central nervous system, altering the activity of brain neurons and associated mental and physical states.

puberty Period of human development that marks the onset of sexual maturity as the reproductive organs begin to function.

pulmonary circuit Blood circulation route between the heart and lungs.

pulmonary valve Valve in the heart that opens from the right ventricle into the pulmonary artery.

pulse Rhythmic pressure surge of blood flowing in an artery, created during each cardiac cycle when a ventricle contracts.

Punnett square A method to predict the probable outcome of a mating or an experimental cross in a simple diagram.

pyruvate (pie-ROO-vate) A compound with a backbone of three carbon atoms that is the end product of glycolysis.

radiation therapy Cancer treatment that relies on radiation from radioisotopes to damage or destroy cancer cells.

radioisotope An isotope with an unstable nucleus that spontaneously decays to a new, stable atom that is not radioactive.

radiometric dating A method of dating fossils that tracks the radioactive decay of material in the specimen.

radius One of two long bones of the forearm that extend from the humerus (at the elbow joint) to the wrist. The radius runs along the thumb side of the forearm, parallel to the ulna.

receptor, sensory A sensory cell or cell part that may be activated by a specific stimulus.

recessive (allele or trait) [L. *recedere*, to recede] Allele whose expression in heterozygotes is fully or partially masked by expression of its partner; fully expressed only in the homozygous recessive condition.

recognition protein One of a class of glycoproteins that project above the plasma membrane and identify a cell as nonself (foreign) or self (belonging to one's own body tissues).

recombinant DNA A DNA molecule that contains genetic material from more than one organism of the same species or from different species.

recombinant DNA technology Procedures by which DNA (genes) from different species may be isolated, cut, spliced together, and the new recombinant molecules multiplied in quantity in a population of rapidly dividing cells such as bacteria.

rectum Final region of the gastrointestinal tract, which receives and temporarily stores undigested food residues (feces).

red blood cell Erythrocyte; an oxygen-transporting cell in blood.

red marrow A substance in the spongy tissue of many bones that serves as a major site of blood cell formation.

reflex [L. *reflectere*, to bend back] A simple, stereotyped movement in response to a stimulus. Sensory neurons synapse on motor neurons in the simplest reflex pathways.

reflex arc [L. *reflectere,* to bend back] A neural pathway in which signals from sensory neurons directly stimulate or inhibit motor neurons.

regulatory protein A protein that enhances or suppresses transcription of a gene.

renewable resource A natural resource that can, in theory, be tapped indefinitely if replenished.

reproduction In biology, processes by which a new generation of cells or multicellular individuals is produced. Sexual reproduction requires meiosis, formation of gametes, and fertilization. Asexual reproduction refers to the production of new individuals by any mode that does not involve gametes.

reproductive base The number of actually and potentially reproducing individuals in a population.

reproductive cloning Creating a cloned embryo to produce a pregnancy.

reproductive isolation An absence of gene flow between populations.

reproductive system An organ system consisting of a pair of gonads (testes in males, ovaries in females). Its sole function is the continuation of the species.

respiration [L. *respirare,* to breathe] The exchange of oxygen from the environment for carbon dioxide wastes from cells by way of circulating blood. Compare *aerobic cellular respiration; cellular respiration.*

respiratory bronchiole Smallest airway in the respiratory system; opens onto alveoli.

respiratory cycle One inhalation, one exhalation of air into and out of the lungs.

respiratory membrane Two-layer membrane between the walls of lung capillaries and alveoli; blood gases diffuse across it.

respiratory surface In alveoli of the lungs, the thin, moist membrane across which gases diffuse.

respiratory system An organ system specialized for bringing in oxygen and carrying away carbon dioxide wastes; human lungs and airways.

resting membrane potential Of neurons and other excitable cells that are not being stimulated, the steady voltage difference across the plasma membrane.

restriction enzymes Class of bacterial enzymes that cut apart foreign DNA injected into them, as by viruses; also used in recombinant DNA technology.

reticular formation A major network of neurons in the brain stem that helps govern activity of the whole nervous system.

retina A thin layer of neural tissue in the eye that contains densely packed photoreceptors.

retrovirus An RNA virus that infects animal cells and with reverse transcriptase creates an RNA template to synthesize a DNA molecule that integrates itself into the host's DNA.

Rh blood typing A method of characterizing red blood cells on the basis of a protein that serves as a self marker at their surface; Rh⁺ signifies its presence and Rh⁻, its absence.

rhodopsin Substance in rod cells of the eye consisting of the protein opsin and a side group, cis-retinal. When the side group absorbs incoming light energy, a series of chemical events follows that result in action potentials in associated neurons.

rib cage Portion of the axial skeleton in the upper torso, formed by the ribs and sternum, which supports and protects the heart, lungs, and other organs.

ribosome The cell structure at which amino acids are strung together to form the polypeptide chains of proteins. An intact ribosome consists of two subunits, each composed of ribosomal RNA and protein molecules.

ribosomal RNA (rRNA) Type of RNA molecule that combines with proteins to form ribosomes, on which the polypeptide chains of proteins are assembled.

rigor mortis A stiffening of skeletal muscles caused when a person dies and body cells stop making ATP.

RNA Ribonucleic acid. A category of single-stranded nucleic acids that function in processes by which genetic instructions are used to build proteins.

RNA polymerase Enzyme that catalyzes the assembly of RNA strands on DNA templates.

rRNA *See* ribosomal RNA.

rod cell Of the retina, a photoreceptor sensitive to very dim light that contributes to coarse perception of movement.

rugae The crumpled wall folds of an empty stomach.

S-shaped curve A curve characteristic of logistic growth; it is obtained when population size is plotted against time.

salinization A salt buildup in soil as a result of evaporation, poor drainage, and often the importation of mineral salts in irrigation water.

salivary amylase Starch-degrading enzyme in saliva.

salivary gland Any of the glands that secrete saliva, a fluid that initially mixes with food in the mouth and starts the breakdown of starch.

salt Compound that releases ions other than H⁺ and OH⁻ in solution.

saltatory conduction In myelinated neurons, rapid, node-to-node hopping of action potentials.

sampling error Error that develops when an experimenter uses a sample (or subset) of a population for an experimental group that is not large enough to be representative of the whole.

sarcomere (SAR-koe-meer) The basic unit of muscle contraction; a region of myosin and actin filaments organized in parallel between two Z bands of a myofibril inside a muscle cell.

sarcoplasmic reticulum (sar-koe-PLAZ-mik reh-TIK-you-lum) In muscle cells, a membrane system that takes up, stores, and releases the calcium ions required for cross-bridge formation in sarcomeres, hence for contraction.

scapula Flat, triangular bone on either side of the pectoral girdle; the scapulae form the shoulder blades.

Schwann cell A specialized neuroglial cell that grows around a neuron axon, forming a myelin sheath.

scientific method A systematic way of gathering knowledge about the natural world.

scientific theory A thoroughly tested explanation of a broad range of natural events and observations. *See also* theory.

second law of thermodynamics The total amount of energy in the universe is spontaneously flowing from forms rich in energy (such as glucose) to forms having less and less of it.

second messenger A molecule inside a cell that mediates and generally triggers an amplified response to a hormone.

secondary immune response Rapid, prolonged response by white blood cells, memory cells especially, to a previously encountered antigen. *See* memory cell.

secondary oocyte An oocyte (unfertilized egg cell) that has completed meiosis I; it is this haploid cell that is released at ovulation.

secondary sexual trait Trait associated with maleness or femaleness, but not directly involved with reproduction. Beard growth in males and breast development in females are examples.

secretion In general, release of a substance by one or more gland cells. In digestion, the release of enzymes and other substances into the digestive tube.

sedimentary cycle A biogeochemical cycle in which an element having no gaseous phase moves from land, through food webs, to the seafloor, then returns to land through long-term uplifting.

segmentation In the small intestine, an oscillating movement produced by rings of muscle that moved digested food and forces it against the intestinal wall.

segregation The separation of pairs of gametes during meiosis.

selective permeability The capacity of a cell membrane to let some substances but not others cross it at certain times. The property arises as an outcome of the membrane's lipid bilayer structure and its transport proteins.

semen [L. *serere*, to sow] Sperm-bearing fluid expelled from the penis during male orgasm.

semicircular canals Fluid-filled canals positioned at different angles within the vestibular apparatus of the inner ear and that contain sensory receptors that detect head movements, deceleration, and acceleration.

semiconservative replication [Gk. *hemi*, half, and L. *conservare*, to keep] Reproduction of a DNA molecule when a complementary strand forms on each of the unzipping strands of an existing DNA double helix, the outcome being two "half-old, half-new" molecules.

semilunar valve A valve in each half of the heart that opens and closes during each heartbeat in ways that keep blood flowing in one direction, from the ventricle to the arteries leading away from it.

seminal vesicle Part of the male reproductive system; secretes fructose that nourishes sperm.

seminiferous tubules Coiled tubes inside the testes where sperm develop.

senescence (sen-ESS-cents) [L. *senescere*, to grow old] Sum total of processes leading to the natural death of an organism or some of its parts.

sensation The conscious awareness of a stimulus.

sensory adaptation In a sensory system, a state in which the frequency of action potentials eventually slows or stops even when the strength of a stimulus is constant.

sensory neuron Any of the nerve cells that act as sensory receptors, detecting specific stimuli (such as light energy) and relaying signals to the brain and spinal cord.

sensory receptor A sensory cell or specialized cell adjacent to it that can detect a particular stimulus.

sensory system An organ system consisting of sensory receptors (such as photoreceptors), nerve pathways from the receptors to the brain, and brain regions that process sensory information.

septum Of the heart, a thick wall that divides the heart into right and left halves.

serous membranes Membranes that occur in paired sheets and anchor internal organs and reduce friction between organs.

Sertoli cell A type of cells in seminiferous tubules that nourish and otherwise aid the development of sperm.

sex chromosome A chromosome that determines a new individual's gender. Compare *autosomes*.

shell model Model of electron distribution in atoms in which orbitals available to electrons occupy a nested series of shells.

sinoatrial (SA) node Region of conducting cells in the upper wall of the right atrium; the cells generate periodic waves of excitation that stimulate the atria to contract.

sinus In the skull, an air-filled space lined with mucous membrane that reduces the weight of the skull.

sister chromatid Of a duplicated chromosome, one of two DNA molecules (and associated proteins) that remain attached at their centromere during nuclear division. Each ends up in a separate daughter nucleus.

skeletal muscle Type of muscle that interacts with the skeleton to bring about body movements. A skeletal muscle typically consists of bundles of many long cylindrical cells encapsulated by connective tissue.

skeletal system The organ system consisting of bones of the skeleton along with cartilages, joints, and ligaments.

sliding filament mechanism The mechanism by which skeletal muscles contract; sarcomeres contract (shorten) when myosin filaments slide along and pull actin filaments toward the center of the sarcomere.

small intestine The portion of the digestive system where digestion is completed and most nutrients are absorbed.

smog A general term for air pollution; originally the term meant "*fog*" infused with "*smoke*" and other pollutants.

smooth muscle One of the three main muscle types; occurs in the walls of internal organs and generally is not under voluntary control.

sodium–potassium pump A transport protein spanning the lipid bilayer of the plasma membrane. When activated by ATP, its shape changes and it selectively transports sodium ions out of the cell and potassium ions in.

solute (SOL-yoot) [L. *solvere*, to loosen] Any substance dissolved in a solution. In water, this means spheres of hydration surround the charged parts of individual ions or molecules and keep them dispersed.

solvent Fluid in which ions and polar molecules easily dissolve.

somatic cell (so-MAT-ik) [Gk. *soma*, body] Any body cell that is not a germ cell; that is, a body cell that does not give rise to gametes.

somatic nerve Nerves leading from the central nervous system to skeletal muscles.

somatic sensation Awareness of touch, pressure, heat, cold, pain, and limb movement.

somatosensory cortex Part of the gray matter of the cerebral hemispheres that controls somatic sensations.

somatostatin Hormone that inhibits the secretion of insulin in beta cells and glucagen in alpha cells.

somites In a developing embryo, paired blocks of mesoderm that will give rise to most bones and to the skeletal muscles of the neck and trunk.

special senses Vision, hearing, olfaction, or other sensation that arises from a particular location, such as the eyes, ears, or nose.

speciation (spee-cee-AY-shun) The evolutionary process by which species originate. One speciation route starts with divergence of two reproductively isolated populations of a species. They become separate species when accumulated genetic differences prevent them from interbreeding successfully under natural conditions.

species (SPEE-ceez) [L. *species*, a kind] A unit consisting of one or more populations of individuals that can interbreed under natural conditions to produce fertile offspring that are reproductively isolated from other such units.

sperm [Gk. *sperma*, seed] A mature male gamete.

spermatogenesis (sperm-at-oh-JEN-ih-sis) Formation of a male gamete, from a germ cell to a mature sperm.

sphere of hydration A clustering of water molecules around the individual molecules of a substance placed in water. Compare *solute*.

sphincter (SFINK-tur) Ring of smooth muscle between regions of a tubelike system (as between the stomach and small intestine).

spinal cavity Body cavity that houses the spinal cord.

spinal cord The portion of the central nervous system threading through a canal inside the vertebral column. It provides direct reflex connections between sensory and motor neurons, as well as communication lines to and from the brain.

spindle A structure that forms during mitosis or meiosis and that moves the chromosomes. It consists of two sets of microtubules that extend from the opposite poles and that overlap at the spindle's equator.

spleen The largest lymphoid organ; it is a filtering station for blood and a reservoir of lymphocytes, red blood cells, and macrophages.

spongy bone Type of bone tissue in which hard, needle-like struts separate large spaces filled with marrow. Spongy bone occurs inside the shaft and at the ends of long bones and within the breastbone (sternum), pelvis, and bones of the skull.

sporadic disease A disease that breaks out irregularly and affects relatively few people.

start codon Of protein synthesis, a base triplet in a strand of mRNA that serves as the start signal for mRNA translation.

stem cell Unspecialized cell that can give rise to descendants that differentiate into specialized cells.

sternum Elongated flat bone (also called the breastbone) to which the upper ribs attach and so form the rib cage.

steroid (STAIR-oid) A lipid with a backbone of four carbon rings and with no fatty acid tails. Steroids differ in their functional groups. Different types have roles in metabolism, intercellular communication, and cell membranes.

steroid hormone A type of lipid-soluble hormone synthesized from cholesterol. Many steroid hormones move into the nucleus and bind to receptors there; others bind to receptors in the cytoplasm, and the entire complex moves into the nucleus.

sterol A type of lipid with a rigid backbone of four fused carbon rings. Sterols occur in cell membranes; cholesterol is the main type in human tissues.

stimulus [L. *stimulus*, goad] A specific change in the environment, such as a variation in light, heat, or mechanical pressure, that the body can detect through sensory receptors; a form of energy that activates receptor endings of a sensory neuron.

stomach A muscular, stretchable sac that receives ingested food; the organ between the esophagus and intestine in which considerable protein digestion occurs.

stop codon Of protein synthesis, a base triplet in a strand of mRNA that serves as the stop signal for translation, so that no more amino acids are added to the polypeptide chain.

strength training Intense, short-duration exercise that produces larger, stronger muscles.

substrate A reactant or precursor molecule for a metabolic reaction; a specific molecule or molecules that an enzyme can chemically recognize, briefly bind to, and modify in a specific way.

substrate-level phosphorylation A metabolic event; the direct transfer of a phosphate group from a substrate of a reaction to some other molecule.

succession (suk-SESH-un) [L. *succedere*, to follow after] Orderly changes from the time pioneer species colonize a barren habitat through replacements by various species until the climax community, when the composition of species remains steady under prevailing conditions.

surface-to-volume ratio A mathematical relationship in which volume increases with the cube of the diameter, but surface area increases only with the square. In growing cells, the volume of cytoplasm increases more rapidly than the surface area of the plasma membrane that must service the cytoplasm. Because of this constraint, cells generally remain small or elongated, or have elaborately folded membranes.

sustainability An approach for managing human population growth, resource use, and the preservation of wild habitats in ways that will help ensure the long-term survival of the human species.

sympathetic nerve Any of the nerves of the autonomic nervous system; generally concerned with increasing overall body activities during times of heightened awareness, excitement, or danger; sympathetic nerves also work in opposition with parasympathetic nerves to bring about minor adjustments in internal organs.

synaptic integration (sin-AP-tik) The moment-by-moment combining of excitatory and inhibitory signals arriving at a trigger zone of a neuron.

syndrome A set of symptoms that may not individually be notable, but collectively characterize a disorder or disease.

synovial joint Freely movable joint in which adjoining bones are separated by a fluid-filled cavity and stabilized by strap-like ligaments. An example is the ball-and-socket joint at the hip.

synovial membranes Connective tissue membranes that line the cavities of the body's movable joints.

systemic circuit (sis-TEM-ik) Circulation route in which oxygenated blood flows from the lungs to the left half of the heart, through the rest of the body (where it gives up oxygen and takes on carbon dioxide), then back to the right side of the heart.

systole Contraction phase of the cardiac cycle.

target cell Any cell that has receptors for a specific signaling molecule (such as a hormone) and that may alter its behavior in response to the molecule.

taste receptors Chemoreceptors in the taste buds.

T cell T lymphocyte; one of a class of white blood cells that carry out immune responses. The helper T and cytotoxic T cells are examples.

T cell receptor Antigen receptor on a T cell.

tectorial membrane Inner ear structure against which sensory hair cells are bent, producing action potentials that travel to the brain via the auditory nerve.

telophase (TEE-low-faze) Of mitosis, the final stage when chromosomes decondense into threadlike structures and two daughter nuclei form.

tendon A cord or strap of dense, regular connective tissue that attaches a muscle to bone or to another muscle.

teratogens Agents that can cause birth defects.

test An attempt to produce actual observations that match predicted or expected observations.

testcross In genetics, an experimental cross to reveal whether an organism is homozygous dominant or heterozygous for a trait. The organism showing dominance is crossed to an individual known to be homozygous recessive for the same trait.

testes (singular: testis) Male gonads; primary reproductive organs in which male gametes and sex hormones are produced.

testosterone (tess-TOSS-tuh-rown) In males, a major sex hormone that helps control reproductive functions.

tetanus Condition in which a muscle motor unit is maintained in a state of contraction for an extended period.

TFR See total fertility rate.

thalamus Coordinating center in the forebrain for sensory input and a relay station for signals to the cerebrum.

theory See scientific theory.

therapeutic cloning Creating a cloned embryo that will be used as a source of embryonic stem cells.

thermal inversion Situation in which a layer of dense, cool air becomes trapped beneath a layer of warm air; can cause air pollutants to accumulate to dangerous levels close to the ground.

thermoreceptor Sensory cell that can detect radiant energy associated with temperature.

thoracic cavity The chest cavity; holds the heart and lungs.

thirst center Cluster of nerve cells in the hypothalamus that can inhibit saliva production, resulting in mouth dryness that the brain interprets as thirst.

threshold Of neurons and other excitable cells, a certain minimum amount by which the voltage difference across the plasma membrane must change to produce an action potential.

thymine Nitrogen-containing base in some nucleotides; a building block of DNA.

thymus A lymphoid organ with endocrine functions; lymphocytes of the immune system multiply, differentiate, and mature in its tissues, and its hormone affect their functions.

thyroid gland An endocrine gland that produces hormones that affect overall metabolic rates, growth, and development.

tidal volume Volume of air, about 500 milliliters, that enters or leaves the lungs in a normal breath.

tight junction Cell junction where strands of fibrous proteins collectively block leaks between the adjoining cells.

tissue A group of cells and intercellular substances that function together in one or more specialized tasks.

tonicity The relative concentrations of solutes in two fluids, such as inside and outside a cell. When solute concentrations are isotonic (equal in both fluids), water shows no net osmotic movement in either direction. When one fluid is hypotonic (has less solutes than the other), the other is hypertonic (has more solutes) and is the direction in which water tends to move.

total fertility rate (TFR) The average number of children born to the women in a given population during their reproductive years.

trace element Any element that represents less than 0.01 percent of body weight.

tracer A substance with a radioisotope attached to it so that its pathway or destination in a cell, organism, ecosystem, or some other system can be tracked, as by scintillation counters that detect its emissions.

trachea (TRAY-kee-uh) The windpipe, which carries air between the larynx and bronchi.

transcription [L. *trans*, across, and *scribere*, to write] Of protein synthesis, the assembly of an RNA strand on one of the two strands of a DNA double helix; the base sequence of the resulting transcript is complementary to the DNA region on which it was assembled.

transfer RNA (tRNA) Of protein synthesis, any of the type of RNA molecules that bind and deliver specific amino acids to ribosomes and pair with mRNA code words for those amino acids.

translation In protein synthesis, the conversion of the coded sequence of information in mRNA into a particular sequence of amino acids to form a polypeptide chain; depends on interactions of rRNA, tRNA, and mRNA.

translocation A change in a chromosome's structure following the insertion of part of a nonhomologous chromosome into it.

transporter protein One of many kinds of membrane proteins involved in active or passive transport of substances across the lipid bilayer of a plasma membrane. Solutes on one side of the membrane pass through the protein's interior to the other side.

trigger zone The region of a motor neuron where proper stimulation can trigger an action potential (nerve impulse).

triglyceride (neutral fat) A lipid having three fatty acid tails attached to a glycerol backbone. Triglycerides are the body's most abundant lipids and richest energy source.

trisomy (TRY-so-mee) The abnormal presence of three of one type of chromosome in a diploid cell.

tRNA *See* transfer RNA.

T tubules Tubelike extensions of a muscle cell's plasma membrane.

tubular reabsorption In the kidney, the diffusion or active transport of water and usable solutes out of a nephron and

into capillaries leading back to the general circulation; regulated by ADH and aldosterone.

tubular secretion The step of urine formation in which unwanted substances in peritubular capillaries are moved to the urine forming in nephron tubules.

tumor A tissue mass composed of cells that are dividing at an abnormally high rate.

tumor marker A substance that is produced by a specific type of cancer cell or by normal cells in response to cancer.

tumor suppressor gene A gene whose protein product operates to keep cell growth and division within normal bounds, or whose product has a role in keeping cells anchored in place within a tissue.

tympanic membrane The eardrum, which vibrates when struck by sound waves.

ulna One of two long bones of the forearm; the ulna extends along the little finger side of the forearm, parallel to the radius on the thumb side.

umami One of the five primary tastes; the brothy, savory taste associated with aged cheese or meats.

umbilical cord Structure containing blood vessels that connect a fetus to its mother's circulatory system by way of the placenta.

uracil (YUR-uh-sill) Nitrogen-containing base found in RNA molecules; can base-pair with adenine.

urea The main nitrogen-containing waste product when cells break down proteins.

ureter Channel that carries urine from each kidney to the urinary bladder.

urethra Tube that carries urine outside the body from the bladder.

urinary bladder Hollow organ that stores urine.

urinary excretion A mechanism by which excess water and solutes are removed by way of the urinary system.

urinary system An organ system that adjusts the volume and composition of blood and so helps maintain extracellular fluid.

urination Urine flow from the body; a reflex response to tension in the smooth muscle of a full bladder.

urine Fluid formed by filtration, reabsorption, and secretion in kidneys; consists of wastes, excess water, and unneeded solutes.

uterus (YOU-tur-us) [L. *uterus*, womb] Chamber in which the developing embryo is contained and nurtured during pregnancy.

vaccine Antigen-containing preparation injected into the body or taken orally; it elicits an immune response leading to the proliferation of memory cells that offer long-lasting protection against that particular antigen.

vagina Chamber of the female reproductive system that receives the male penis and sperm, forms part of the birth canal, and channels menstrual flow to the exterior.

variable In a scientific experiment, the only factor that is not the same in the experimental group as it is in the control group.

vas deferens Tube leading to the ejaculatory duct; one of several tubes through which sperm move after they leave the testes just prior to ejaculation.

vasoconstriction Decrease in the diameter of an arteriole, so that blood pressure rises; may be triggered by the hormones epinephrine and angiotensin.

vasodilation Enlargement of arteriole diameter, so that blood pressure falls; may be triggered by hormones including epinephrine and angiotensin.

veins Of the circulatory system, the large-diameter vessels that lead back to the heart.

ventricle (VEN-tri-kul) Of the heart, one of two chambers from which blood is pumped out. Compare *atrium*.

venule Small blood vessel that receives blood from tissue capillaries and merges into larger-diameter veins; a limited amount of diffusion occurs across venule walls.

vertebra (plural: vertebrae) One of a series of hard bones arranged with intervertebral disks into a backbone.

vertebrate Animal having a backbone of bony segments, the vertebrae.

vesicle (VESS-ih-kul) [L. *vesicula*, little bladder] One of a variety of small membrane-bound sacs in the cell cytoplasm that function in the transport, storage, or digestion of substances or in some other activity.

vestibular apparatus A closed system of fluid-filled canals and sacs in the inner ear that functions in the sense of balance. Compare *semicircular canals*.

villus (VIL-us), plural: villi. Any of several types of absorptive structures projecting from the free surface of an epithelium.

virulence The relative ability of a pathogen to cause serious disease.

virus A noncellular infectious agent consisting of DNA or RNA and a protein coat; can replicate only after its genetic material enters a host cell and takes over its metabolic machinery.

vision Sensory reception of visual stimuli (especially light) followed by image formation in the brain.

visual cortex Part of the brain that receives signals from the optic nerves.

vital capacity Maximum volume of air that can move out of the lungs after a person inhales as deeply as possible.

vitamin Any of numerous organic substances that the body requires in small amounts for normal cell metabolism but generally cannot synthesize for itself.

vocal cords A pair of elastic ligaments on either side of the larynx wall. Air forced between them causes the cords to vibrate and produce sounds.

water (hydrologic) cycle The movement of water from oceans to the atmosphere, the land, and back to the ocean.

watershed Any region in which all precipitation drains into a single stream or river.

white blood cell Leukocyte; any of the macrophages, eosinophils, neutrophils, and other cells that are the central components of the immune system.

white matter Of the spinal cord, major nerve tracts so named because of the glistening myelin sheaths of their axons.

X chromosome A sex chromosome with genes that cause an embryo to develop into a female, provided that it inherits a pair of these.

X inactivation A compensating phenomenon in females that "switches off" one X chromosome soon after the first cleavages of the zygote.

X-linked gene Any gene on an X chromosome.

X-linked recessive inheritance Recessive condition in which the responsible, mutated gene occurs on the X chromosome.

Y chromosome A sex chromosome with genes that cause the embryo that inherited it to develop into a male.

Y-linked gene Any gene on a Y chromosome.

yellow marrow Bone marrow that consists mainly of fat and hence appears yellow. It can convert to red marrow and produce red blood cells if the need arises.

yolk sac One of four extraembryonic membranes. Part becomes a site of blood cell formation and some of its cells give rise to the forerunners of gametes.

zona pellucida A protein layer around an ovarian follicle.

zoonosis An infectious disease that mainly affects animals other than humans, but can also be passed on to humans.

zygote (ZYE-goat) The first cell of a new individual, formed by the fusion of a sperm nucleus with the nucleus of an egg (fertilization).

Aflatoxin, 433
African Americans
- and sickle-cell anemia, 383
- asthma risk in, 191
- birth defect risk in, 346
African emergence model, of human evolution, 457
- African sleeping sickness, 10f
Afterbirth, 344, 344f
Age structure (of population), 477, 477f
Agglutination, 146–147, 147f
Aging (senescence)
- Alzheimer's disease and, 260, 351, 351f
- and cancer risk, 433
  basal metabolic rate and, 219, 351
  brain and, 351, 351f
  cardiovascular system and, 134, 135t, 350–351
  of cloned animals, 425
  as development stage, 349, 349t
  endocrine system and, 351
- exercise combating effects of, 351
- eye disorders and, 283
  health impacts of, 350–351, 350t, 351f
  and hearing, 277, 351
  heart and, 350–351
  life span, genetic programming of, 350
  muscle health and, 116, 350, 350t
  respiratory system and, 191, 350–351, 350t
  sensory system and, 351
  skeletal system and, 98, 350
  skin and, 350
  vitamins and, 217
Agranulocytes, 143
- Agriculture
  - and land use, 486, 486f
  - and water use, 484, 484f, 486
  - chemicals used in, 37, 37f, 486 (*See also* Pesticides)
  - climate change and, 483
    green revolution, 486
  - runoff from, 484
- AID (artificial insemination by donor), 320
- AIDS (acquired immune deficiency syndrome), 172–173, 172f, 172t, 324t. *See also* HIV
Airline travel
- and deep-vein thrombosis, 140
- and ear popping, 274
- and jet lag, 302
- Air pollution, 190–191, 194, 194f, 480–481, 480f, 481f, 488
Air sacs. *See* Alveolus
- Albinism, 387, 405t
- ALB test, 149t
Albumin, 35, 142, 142f, 205

Alcohol
  absorption across stomach wall, 202
- birth defects caused by, 347, 347f
  and blood-brain barrier, 255
- and cancer risk, 439
- and cirrhosis, 41, 210, 210f
  crossing of placenta by, 339
  drug interactions, 262
  effects of, 262
  functional groups of, 26, 26f
- health impact of, 41, 136, 236
- and heart function, 135
  processing of, 51, 205, 462
- and ulcers, 212
  and unsafe sex, 325
Alcohol dehydrogenase (ADH), 462
- Alcoholism, 385, 386
Aldosterone, 232f, 233, 233f, 289f, 290t, 294t, 298
Ali, Muhammad, 261f
Alkaline fluids, 24
Alkalinity, 24. *See also* Base(s); pH
- Alkalosis, 25
  - metabolic, 234
Alkaptonuria, 411
Allantois, 338
Alleles, 376, 376f, 376t, 377, 377f, 390. *See also* Dominant alleles; Gene(s); Recessive alleles
  expression of, 303, 382–383, 382f, 384–385, 385f, 393, 393f
- genetic disorders and, 394–399, 394f–399f
  independent assortment of, 380, 380f, 381f, 390
  multiple allele system, 383, 384
  penetrance, 384
  probability calculations for, 378–379, 378f, 379f
  SNPs and, 420
  variations in, and evolution, 445, 446
- in X-linked disorders, 398–399, 398f, 399f
Allergens, 170, 170f, 191
- Allergen-specific IgE test, 149t
- Allergic responses/allergies, 157, 165, 165t, 170–171, 170f, 171f, 282, 395
- Allergy desensitization treatments, 171
Alpha carotene, 23
Alpha cells, 300, 300f
Alpha waves (sleep stage), 258f
- ALS (amyotrophic lateral sclerosis), 421, 421f
Altitude
- and blood composition, 149
- and breathing, 183, 183f
Alveolar duct, 180f
Alveolar sac, 180f

Alveolus (alveoli), 180f, 181, 182, 184, 185, 186, 186f, 187, 188f, 189
- Alzheimer's disease (AD), 260, 351, 351f, 354, 421, 421f
Amacrine cells, 281, 281f
- Amazon, deforestation in, 487, 487f
- Amelogenesis imperfecta, 399, 399f
- Amenorrhea, 330
American Cancer Society, 326, 327f, 429, 434, 437, 437t
American Heart Association, 121
Ames, Bruce (Ames test), 434
Amine groups, 26, 26f
Amine hormones, 290, 290t
Amino acids
  absorption in small intestine, 206, 207f
  in blood plasma, 142, 142f
  digestive processing of, 205, 206, 207f, 207t, 215, 227
  essential, 215, 215f
  liver processing, 205
  and origin of life, 458
  in polypeptide chains, 32–33, 32f, 33f
  in proteins, 32–33, 32f, 33f
  in protein synthesis, 414–417, 414f–417f
  taste of, 273
  urinary system processing of, 230–231, 231t
Amino groups, 32, 32f, 33, 33f
Aminopeptidase, 207t
Amish, genetic conditions in, 394f
Ammonia, 205, 227, 234, 472, 472f
- Ammonification, in nitrogen cycle, 472, 472f
Ammonium, 472, 472f
- Amnesia, 259
- Amniocentesis, 348, 348f
Amnion, 338, 338f, 340f–341f
Amniotic cavity, 336f
Amniotic fluid, 338
Amniotic sac, 344
- Amoebic dysentery, 175t
- Amphetamines, 136, 262
Amplitude, of sound wave, 274, 274f
- Amputation, 99
- AMT test, 149t
Amygdala, 257, 257f, 259, 259f
Amyloid protein, 260
Amyloidosis, 421f
Amylose, 29, 29f
- Amyotrophic lateral sclerosis (ALS), 421, 421f
Anabolic steroids, 103
Anabolism, 58
- Anaerobic exercise, 66
Anal canal, 208, 208f
Analogous structures, 450
Anal sphincters, 208, 208f

Bone tissue, 70t, 71, 71f, 85t, 88, 88f, 91f
Bonobos, 3f, 454, 454f
■ Boreal forest, 465f
Borrelia burgdorferi bacteria, 10, 136, 136f
■ Botox injections, 111
■ Botulism, 115
■ Bovine spongiform encephalitis
    (BSE), 260, 261f. *See also*
    Mad cow disease
Bowman's (glomerular) capsule,
    229, 229f, 230, 230f, 233f
Brachial artery, 122f
■ Bradycardia, 135
Bradykinins, 270
Brain
    aging impact on, 351
    ■ aneurysms in, 135
    blood–brain barrier, 255
    cerebral hemispheres, 254, 254f, 256
    cerebrospinal fluid surrounding,
        188, 189f, 254–255, 255f
    cerebrum of, 254, 254f, 256–257, 256f,
        257f, 265t, 270–271, 270f, 272, 281
    development of, 346f, 347
    endocrine system interaction
        with, 292–293, 292f, 293f
    evolution of, 3, 455
    formation of during human
        development, 336, 336f
    functional areas of, 254, 254f
    hemispheres of, 254f
    ■ injury to, 260
    as integrator, 80–81, 80f–82f
    lobes of, 256, 256f
    ■ mind-altering drug impact
        on, 241, 262, 262t
    nervous system role of, 250,
        250f, 254–259, 254f–259f
    neuron interaction in, 249
    peripheral nervous system
        and, 253, 253f
    ■ PET scan of, 256, 256f, 261f
    sensory systems interaction, 268–281
    skull protecting, 92, 92f–93f
    structure and function of, 250–251,
        251f, 254–259, 254f–259f
■ Brain cancer, 260, 260f, 423
Brain case (cranium), 92
Brain stem, 188, 188f, 189f, 254, 257
Brazil, 477f, 487
BRCA1 gene, 326, 433, 437, 437f
BRCA2 gene, 326, 433, 437, 437f
■ Breast, lump in, 326, 326f
Breastbone, 91f
■ Breast cancer, 169, 179, 326, 326f, 421,
    433, 435f, 436t, 437, 437f, 439
Breast milk (breast feeding), 69, 164,
    172, 172f, 293, 345, 345f
■ Breast self-examination,
    326, 326f, 436t

Breath, amount of air in, 185, 185f
Breathing (respiratory cycle), 184–185,
    184f, 185f, 188–189, 188f,
    189f. *See also* Respiration;
    Respiratory system
    nervous system control of, 253, 254
■ Breech birth, 344
British Columbia, 487
Broca's area, 256, 256f
■ Bronchial cancer, 435f
Bronchial trees, 180f, 181
Bronchioles, 180f, 181, 189, 190
Bronchitis, 190–191
Bronchus, 180f, 181
Brown fat, 83, 83f, 116, 301
Brush border, 203, 203f, 206,
    206f, 207f, 211
■ BSE (bovine spongiform
    encephalitis), 260, 261f.
    See also Mad cow disease
■ Bubble babies. *See* SCID (severe
    combined immune deficiency)
Bubonic plague, 9, 9f, 478
Buffers, 25, 187, 200, 234, 234f
Bulbourethral glands, 312f,
    312t, 313, 314f
■ Bulimia nervosa, 221, 221t
Bulk. *See* Fiber (dietary)
Bulk flow, 132–133, 133f
■ BUN test, 149t
Burger, MacKenzie, 287, 287f,
    301, 301f
■ Burns, and skin, artificial, 73
Bursae, 106, 106f
Butyrate, 383

C

■ Caffeine, 135, 233, 255, 262,
    273, 339
Calcitonin, 89, 289f, 290t, 294t,
    296, 297
Calcium
    absorption of, 78, 228, 290
    blood levels, regulation of,
        89, 296–297, 297f
    in body, 16, 16f
    in bones, 88–89, 89f, 90, 90t, 100
    deficiency in, 98
    ■ deposits, in atherosclerotic
        plaque, 134
    ■ in diet, 102, 217t
    in extracellular fluid, 227
    functions of, 296–297
    homeostasis, 89, 100, 111,
        296–297, 297f
    storage of, 50
Calcium ions
    and muscle contraction,
        109–111, 109f–111f, 113
    and nerve impulses, 25, 246

■ Calories (food), 219
cAMP (cyclic adenosine
    monophosphate), 290–291, 291f
■ Camptodactyly, 384, 405t
Canaliculi, 88, 88f
■ Cancer
    ■ aging and, 433
    ■ anal, 324
    ■ bladder, 190f, 236, 434t, 435f, 436
    ■ and blood clots,
        susceptibility to, 154
    ■ bone, 99, 99f
    ■ of bone marrow (leukemia), 86,
        152, 152f, 401, 421f, 422, 423,
        434t, 435, 435f, 436, 438, 439
    ■ brain, 260, 260f, 423
    ■ breast, 169, 179, 326, 326f, 421,
        433, 435f, 436t, 437, 437f, 439
    ■ bronchial, 435f
    ■ carcinogenesis, 432–433, 432f, 433f
    ■ carcinogens, 433, 434,
        434f, 434t, 440t
    ■ cell reproduction in, 374, 431
    ■ cervical, 324, 327, 327f, 431f, 436t
    ■ characteristics of, 430–431,
        430f, 430t, 431f
    ■ chemicals as cause of, 192,
        433, 434, 434f, 434t, 439
    ■ chromosome changes and, 401
    ■ colorectal, 211, 211f, 432, 432f,
        433, 433f, 435f, 436t
    ■ of connective tissue, 434t, 435
    ■ deaths from, 429
    ■ diet and, 211, 217, 439
    ■ endometrial, 326–327
    ■ epithelial, 434t, 435
    ■ of esophagus, 190f, 439
    ■ of eye, 282–283, 433, 433f
    ■ factors leading to, 433, 440t
    ■ gender and, 429
    ■ gene therapy for, 423
    ■ as genetic disease, 326, 385,
        422, 430, 432–433, 432f
    ■ genetic *vs.* environmental
        risk factors, 385
    ■ glial, 260, 260f, 435, 439
    ■ heart, 136
    ■ HIV patient vulnerability
        to, 172, 172f
    ■ immune system response to,
        166–167, 167f, 433, 440t
    ■ immunotherapy for, 169
    ■ incidence of, 429
    ■ inherited susceptibility
        to, 432–433
    ■ of kidney, 236, 236f, 435f
    ■ larynx, 190f, 439
    ■ leukemias, 86, 152, 152f,
        401, 421f, 422, 423, 434t,
        435, 435f, 436, 438, 439

as pH buffer, 187
■ porphyria and, 86
■ sickle-cell anemia and, 382f, 383
structure and function of, 34, 34f, 143, 144, 144f
■ Hemolytic anemias, 152, 152f, 421f
■ Hemolytic disease of the newborn, 148
■ Hemophilia, 151, 398–399, 398f, 399f, 405t, 422, 424, 446
■ Hemorrhagic fevers, 10
■ Hemorrhoids, 210
Hemostasis, 150–151, 150f
Hepatic arteries, 127
Hepatic portal system, 127, 127f, 205, 205f
Hepatic portal vein, 127, 127f, 205, 205f
Hepatic vein, 122f, 127, 127f, 205, 205f
■ Hepatitis, 41, 169, 175t, 210, 324
■ Hepatitis B virus (HBV), 324
■ Hepatitis C virus (HCV), 169, 324
Hepato- (prefix), 435f
HER2 protein, 169
■ Herbicides, 37, 434. *See also* Pesticides
■ Herbivores, 466, 468f
■ Herceptin, 169
■ Hereditary methemoglobinemia, 405t
Heredity. *See* Gene(s); Genetic
■ Herniated disk, 93
■ Heroin, 262
■ Herpes infections (herpes virus), 79, 260, 282, 324, 324f, 324t
Heterotrophs. *See* Consumers
Heterozygous individuals, 376, 376t, 377, 377f, 395, 396, 397, 398–399
Hexosaminidase A, 396
HGH. *See* Human growth hormone
■ High blood pressure (hypertension), 129, 129t
■ and cardiovascular disease, 135, 135t
■ genetic *vs.* environmental risk factors, 385
■ health risks associated with, 129, 135, 135t, 235
■ and licorice, 239
■ medications for, 400
■ in metabolic syndrome, 220
■ risk factors for, 129t, 140, 218, 220, 299
High-density lipoproteins (HDLs), 134, 179, 205
Hindbrain, 254, 254f, 265t
Hip joint, 95, 97f, 98
Hippocampus, 257, 257f, 259, 259f
Histamine, 160–161, 161f, 165, 165t, 170, 170f, 171, 270

Histones, 356
■ HIV (human immunodeficiency virus), 149t, 152, 169, 172–173, 172f, 172t, 173f, 339, 433
■ Hives, 170, 171f
HIV protease, 173
hMG (human menopausal gonadotropin), 320
Homeostasis
blood pressure, 131, 131f, 228, 293, 302
blood role in, 137, 142
of body fluids, 80–81, 226–227, 226t, 231, 232–233, 232f, 233f, 235
body temperature, 82–83, 82f, 83f, 83t
bones and, 89, 90, 100
calcium levels, 89, 111, 296–297, 297f
cancer and, 442
cardiovascular system role in, 123, 123f, 137
circulatory system and, 130
definition of, 2
digestive system in, 213
endocrine system in, 303
feedback controls. *See* Negative feedback loop; Positive feedback loop
glia and, 242
glucose blood levels, 288, 293, 298
hypothalamus and, 255
kidneys and, 145, 145f, 231, 232–233, 232f, 233f, 234, 234f
in living things, 2
maintaining, 80–81, 80f, 81f
muscular system and, 11, 11f, 117
nervous system and, 251, 255, 263
organ systems role in, 80–81, 100
pH balance and, 24, 25, 187, 231, 234, 234f
red blood cell count, 145, 145f
respiratory system in, 188–189, 188f, 189f, 193
skeletal system and, 90, 90t, 100
skin and, 78
sodium blood levels, 298, 302
urinary system and, 226–227, 226t, 231, 232–233, 232f, 233f, 237
water properties and, 22–23
Hominins, 456, 456f, 462
Hominoids, 454f
Homocysteine, 134, 135t
■ Homocystinuria, 421f
Homo erectus, 453, 456, 456f, 457
Homo floresiensis, 457, 457f
Homo habilis, 447, 447f, 453, 456, 456f
Homologous chromosomes, 357, 357f, 368
Homologous structures, 450, 450f

Homologue(s). *See* Homologous chromosomes
Homo neanderthalensis (Neanderthals), 453, 453f, 456, 457f
Homo sapiens, 443, 453, 453f, 454, 454t, 455f, 456–457, 456f, 457f
Homozygous dominant condition, 376, 376t
Homozygous individuals, 376, 376t, 396
Homozygous recessive condition, 376, 376t
■ Hookworm, 175t
Hopf, Colin, 396f
Horizontal cells, 281, 281f
Hormone replacement therapy, 306
Hormones
adrenal, 294t, 298–299, 299f
and appetite, 219
and blood-brain barrier, 255
in blood plasma, 142, 142f
and blood pressure control, 131
blood sugar regulation with, 204, 300, 300f
bones and, 89
cardiac, 294t, 302
as chemical signals, 288, 288t
childbirth triggered by, 344
digestive system and, 202, 206, 209, 209f, 209t, 294t, 302
discovery of, 288
follicle-stimulating, 289f, 290t, 292t, 293, 293f, 310–311, 311f, 315, 315f, 320, 335
and genetic development, 340–341, 341f
gonadal, 289f, 294t, 302, 302f
heart, 294t, 302
human growth. *See* Human growth hormone
hypothalamic, 292, 292f, 293f
interactions ("partnerships"), types of, 288
kidney, 294t
liver, 294f
liver processing of, 204f, 205
as long-term controllers, 294, 294t
luteinizing, 289f, 290t, 292t, 293, 293f, 310–311, 311f, 315, 315f
menstrual cycle and, 308–309
milk production control by, 345, 345f
nonsteroid, 290–291, 290t, 291f
ovarian, 289f, 290, 294t, 302, 302f
pancreatic, 294t, 300, 300f
parathyroid, 89, 289f, 290t, 294t, 296–297, 297f
peptide, 290–291, 290t, 291f
pineal, 289f, 302
pituitary, 289f, 292–293, 292f, 292t, 293f

endocrine system interaction, 117, 303, 344

in homeostasis, 11, 11f, 117

immune system interaction, 117

integumentary system interaction, 117

lymphatic system interaction, 117

nervous system interaction, 110–112, 110f, 111f, 113f, 114, 115, 117, 128, 246, 246f, 248–249, 249f, 251f, 252–253, 252f, 256–258, 256f, 257f, 263

overview of, 76, 77f

reflexes, 248–249, 249f

reproductive system interaction, 117

respiratory system interaction, 117, 180f, 184, 184f, 193, 263

sensory system interaction, 117, 269, 269f, 270, 279, 279f

skeletal system interaction, 100, 117

urinary system interaction, 117, 237

Musick, Matt, 103

Mycobacterium tuberculosis bacteria, 192, 192f

Myelin sheath, 248, 248f, 253, 261

Myelo- (prefix), 435f

Myeloid stem cells, 142, 143f

Myo- (prefix), 435f

■ Myocardial infarction (MI; heart attack), 121, 129, 135, 140

■ Myocarditis, 136

Myocardium, 124, 124f. *See also* Cardiac muscle

Myofibrils, 106, 106f, 108, 108f, 110, 110f, 116

Myoglobin, 107, 107f, 116

Myometrium, 308, 309f

■ Myopathy, 114–115, 114f, 115f

■ Myopia (nearsightedness), 282, 282f, 283

Myosin, 53, 108–109, 108f, 109f, 110, 111f

■ Myotonic muscular dystrophy, 115

N

NAD⁺, 61, 62f

Nader, Matt, 121

NADH, 61, 61f, 62f

Nails, 78

Nanobes, 459, 459f

Nanoparticles, 66

Nasal cavity/nasal passages
olfactory receptors in, 272–273, 273f
in respiratory system, 180, 180f
sinuses and, 92–93, 92f–93f

Nash, John, 404, 404f

National Academy of Engineering, 63

National Academy of Sciences, 434

National Cancer Institute, 439

National Heart, Lung, and Blood Institute, 129, 129t, 149t, 153

National Institute on Drug Abuse (NIDA), 103

National Institutes of Health, 73, 324

National Institutes of Mental Health, 247

National Keratoconus Foundation, 167

National Kidney Foundation, 236

■ Natural gas, as energy source, 488, 488f

Natural killer (NK) cells, 167, 172, 433, 439

Natural selection, 9, 9t, 444, 446, 447

■ Natural world. *See also* Ecosystem(s); Environment
■ elements in, 16, 16f
■ organization of, 4, 4f–5f
■ scientific research on, 6–7, 6f, 7f, 9

Navel, 344

■ NBDs (neurobiological disorders), 404, 404f

Neanderthals (*Homo neanderthalensis*), 453, 453f, 456, 457f

■ Nearsightedness (myopia), 282, 282f, 283

Neck, bones of, 93, 93f. *See also* Spinal cord; Throat

Negative feedback loop
ADH release inhibited by, 232, 233f
for cortisol, 298, 299f
in digestive system, 209
in homeostasis, 81, 81f, 82–83, 82f, 83t, 89
hormonal control through, 288, 296, 315, 315f
norepinephrine production, 298–299
red blood cell count maintained by, 144, 145f
in respiratory cycle, 188
stress response and, 298–299

Neisseria gonorrhoeae bacteria, 322–323, 322f

Neon, 18f

Neonates (newborns)
■ chlamydia infection in, 322
as development stage, 349, 349t
immune system and, 164
skull of, 96

■ Neoplasm, 430, 430f. *See also* Tumor(s)

■ Nephritis, 236

Nephrons, 228–229, 229f
urine formation in, 230–231, 230f, 231t

Nerve(s). *See also* Nervous system; Neuron(s); *other nerve-related entries*
afferent, 250, 251f
auditory, 274f–275f, 275
autonomic, 250, 251f, 252–253, 252f
in bone, 88
cranial, 250, 250f, 251f, 252

efferent, 250, 251f
free endings, 79, 270, 271f
optic, 278–279, 278f, 278t, 281, 281f
parasympathetic, 251f, 252–253, 252f
somatic, 250, 251f, 252
spinal, 250, 250f, 252
structure and function of, 248, 248f, 248t
sympathetic, 251f, 252–253, 252f
thoracic, 250f
ulnar, 250f
vestibular, 276, 276f

Nerve cells, 25, 43f, 73

Nerve endings, free, 79, 270, 271f

Nerve fiber, 248, 248t

■ Nerve growth factors, research on, 260

Nerve impulses, 243–245, 243f–245f, 268–269, 269f. *See also* Action potentials

Nerve tracts, 248, 253

Nervous system. *See also entries under* Nerve
aging and, 351
brain role in, 250, 250f, 254–259, 254f–259f (*See also* Brain)
cardiovascular system interaction, 137, 263
central, 250, 251f, 265t
development of, 336–337, 336f, 337f
digestive system interaction, 209, 209f, 213, 263
■ disorders of, 260–261, 260f, 261f
endocrine system interaction, 263, 292–293, 292f, 293f, 303
in homeostasis, 251, 255, 263
immune system interaction, 263
integumentary system interaction, 263
lymphatic system interaction, 263
muscular system interaction, 110–112, 110f, 111f, 113f, 114, 115, 117, 128, 246, 246f, 248–249, 249f, 251f, 252–253, 252f, 256–258, 256f, 257f, 263
nervous tissue in, 73, 76f, 85t
neuron role in, 242–247, 242f–247f
overview of, 77f, 250–251, 250f, 251f
peripheral, 250, 251f, 252–253, 252f, 253f
reproductive system interaction, 263
respiratory system interaction, 188–189, 188f, 189f, 193, 263
sensory system interaction, 257, 259, 259f, 263, 268–281, 269f–271f, 281f
skeletal system interaction, 100, 263
spinal cord role in, 250, 250f, 251f (*See also* Spinal cord)

Polysomes, 416, 417f
Polyunsaturated fatty acids, 214, 214t
Pons, 254, 254f, 258f, 265t
Population(s)
    adaptive radiation of, 452–453, 453f
    age structure of, 477, 477f
    controls on growth of, 478, 478f
    defined, 445
    demographics, 477, 477f
    genetic drift in, 446, 447
    genetic variations in, 445, 445f
    growth of human, 476–478, 476f–478f
    ■ habitats of, 464, 464f, 465f
    ■ and land use, 486, 486f
    as level of biological organization, 4, 4f–5f
    natural selection in, 444, 446, 447
    total fertility rate and, 476
Population density, 477
Pores
    in capillary walls, 132–133, 132f
    nuclear, 48, 48f, 49f
    in plasma membranes, 48, 48f, 49f, 160, 160f
■ Porphyria, 86, 86f
Positive feedback loop, 81
    breast milk production as, 345, 345f
    in nerve impulses, 244
■ Positron Emission Tomography (PET) scan, 17, 17f, 256, 256f, 260f, 266, 266f, 363, 436
Posterior pituitary lobe, 292–293, 292f, 292t
Postganglionic neurons, 252
Postsynaptic cells, 246
Potassium
    homeostasis, 298
    ■ in diet, 217t
    and muscle function, 114
Potassium ions
    in extracellular fluid, 226, 227
    in neuron signaling, 243–245, 243f–245f
    urinary system processing of, 230f, 230r, 231
■ Power plants, and pollution, 480, 480f, 484
■ Prayer, research on, 14
Precapillary sphincters, 133, 133f
Precipitation
    ■ acid rain, 4, 25, 25f, 480, 480f, 488
    ■ global warming and, 482f, 483
    ■ in water cycle, 469, 469f
■ Prediabetes, 301
Predictions, scientific, 6–7, 7f
Prefrontal cortex, 257, 259, 259f
Preganglionic neurons, 252

Pregnancy. *See also* Development; Prenatal development
    and cardiovascular system, 131
    ■ diet in, 346
    ■ drug use in, 347
    ■ ectopic, 335, 354, 354f
    ■ herpes infection and, 282
    ■ HIV transmission in, 172, 172f
    ■ home tests for, 169, 169f, 335
    ■ miscarriage, 346, 347, 352, 363
    ■ Rh blood typing and, 148, 148f
    sexual intercourse and, 316
    ■ and smoking, 190f
    ■ STDs and, 323
    ■ toxoplasmosis and, 352
    ■ weight gain in, 346
Prehensile hand movements, 454–455, 455f
■ Preimplantation diagnosis, 348
■ Premature birth, 37, 342, 344, 347
■ Premenstrual syndrome (PMS), 330
Premolars, 200, 200f
Premotor cortex, 256
Prenatal development. *See also* Birth; Embryo(s); Fetus; Pregnancy
    comparative, as evidence for evolution, 450, 451f
    ■ disorders of, 346–347, 346f, 347f
    embryo implantation, 334–335, 334f
    embryonic period, 332–341, 332f–341f, 332t
    endocrine system role in, 349
    extraembryonic membrane formation in, 338, 338f, 339f
    fetal period, 342–343, 342f, 343f
    overview of, 332–333, 332f, 332t, 333f
    retina formation in, 281
    sex chromosomes and, 392, 393f
    stages of, 349, 349t
Pressure
    atmospheric, 182, 182f
    blood. *See* Blood pressure
    diastolic, 129, 129f, 129t
    intrapulmonary, 184
    sensory system response to, 270
    systolic, 129, 129f, 129t
Pressure gradients, in respiration, 182–183, 182f, 184, 186–187, 187f
Presynaptic cells, 246
Primary consumers, 466, 467f, 468f
Primary motor cortex, 256, 256f, 257f
Primary oocytes, 310, 310f, 364–365, 365f, 372
Primary productivity, 468, 468f
Primary somatosensory cortex, 256f, 257. *See also* Somatosensory cortex
Primary spermatocytes, 314, 314f, 364, 365f

Primary structure of protein, 33, 33f, 34, 34f
Primary succession, 464, 465f
Primary tissues (germ layers), 332, 332t, 333f
Primary visual cortex, 257. *See also* Visual cortex
Primates
    classification of, 3, 3f
    evolution of, 451, 451f, 452, 454–457, 454f
■ Primers (in PCR), 419, 419f
Primitive streak, 336, 336f
■ Principle of sustainability, 491, 493
Prion, 260, 261f
Probability (in genetics). *See also* Punnett square
    ■ calculation of, 378–379, 378f, 379f
    factors affecting, 384–385, 385f
    independent assortment and, 380, 380f, 381f, 390
Probiotic bacteria, 210
Problem identification, in scientific method, 6
■ Producers (autotrophs), 4, 5f, 466, 466f, 467f, 468, 468f, 472, 472f, 482
Product (of metabolic reaction), 59
■ Progeria, 405t
Progestational phase, of menstrual cycle, 309, 309t
Progesterone, 289f, 290t, 294t, 298, 302, 309, 309f, 311, 311f, 335, 345, 351
■ Progestin injections/ implants, 318t, 319
Prokaryotic cells, 42, 42f, 42t
Prolactin, 288, 289f, 290t, 292t, 293, 293f
Proliferating phase, of menstrual cycle, 309, 309t
Proline, 415
Prometaphase, 360
Promoter (in transcription), 412, 412f–413f
Pronation, 97f
proofreading of DNA, 411
Prophase (mitosis), 358f, 360, 361f, 370f–371f
Prophase I (meiosis), 364, 366, 366f–367f, 368, 370f–371f
Prophase II (meiosis), 364, 366f–367f, 367, 370f–371f
Prostaglandins, 161, 170, 270, 288, 288t, 313, 344
■ Prostate, enlarged, 327
■ Prostate cancer, 37, 169, 327, 435f, 436t
Prostate gland, 312f, 312t, 313, 314f, 316
Prostate-specific antigen (PSA), 436
    ■ blood test for, 327
Protease inhibitors, 173

Stretch receptors, 209, 209f, 269, 269f, 270
Stretch reflex, 248–249, 249f
Striation, of muscle tissue, 72
■ Strip mining, 488
■ Stroke, 111, 129, 151
Structural formula, of molecule, 19t, 20
Structural model, of molecule, 19t
Structural proteins, 32, 32t, 70, 350
Student stress scale, 304
Subatomic particles, 16–17, 16f. *See also* Electron(s)
Subjectivity, 8, 9
Sublingual gland, 200, 200f
Submandibular gland, 200, 200f
Submucosa, in digestive tract wall, 199, 199f
Substance P, 247, 270
Substrates, 59, 59f
■ Succession (ecological), 464, 465f
Sucking reflex, 342
Sucrose, 28, 28f
■ Sudden cardiac arrest (SCA), 121, 135. *See also* Heart attack
Sugars. *See also* Blood sugar; Fructose; Glucose; Sucrose
    absorption of in small intestine, 207f
    in ATP, 58, 58f
    disaccharides, 28, 28f
    in DNA, 408, 408f
    functional groups in, 26, 26f
    in nucleotides, 36
    oligosaccharides, 28–29
    polysaccharides, 29, 29f, 70, 207t, 214
    ■ refined, 214, 215
    in RNA, 412
    simple (monosaccharides), 28, 214
    storage of, 29
    table sugar, 28, 28f
■ Suicide, antidepressant drugs and, 247
Sulfhydryl groups, 26f
Sulfur
    in coal, 488
    ■ in diet, 217t
    disulfide bridges in, 34
Sulfuric acid, 24, 25, 227
Sulfur oxides, 480
Summation, of action potentials, 247, 249
■ Sun
    ■ and air pollution, 480
    ■ and cancer risk, 439
    ■ effect on skin, 79, 79f
    ■ energy from, 4, 5f, 466, 466f, 482f
    ■ exposure, and radiation, 433
    ■ and porphyria, 86, 86f
    ■ role in ecosystem, 466, 466f, 468
    ■ and vitamin D production, 78
    ■ water cycle and, 469

Superior vena cava, 122f, 124f, 125f, 127
Supination, 97f
■ Supplements, dietary
    ■ amino acids as, 33
    ■ antioxidants, 217
    ■ regulation of, 33
    ■ vitamins and minerals as, 217, 346
Support (by skeletal system), 90t, 93
Surface barriers to infection, 79, 156, 160
Surface-to-volume ratio, of cell, 42–43, 42f
Surfactants, 342
■ Surgery
    ■ blood transfusions and, 141
    ■ for cancer, 283, 326, 438
    ■ for cardiovascular disease, 134, 134f
    ■ cataract, 283
    ■ to control fertility, 318, 318t, 319f
    ■ coronary bypass, 134, 134f
    ■ dental, and cardiovascular disease, 136
    ■ for ectopic pregnancy, 335, 354
    ■ gall bladder removal, 205
    ■ gastric bypass/banding, 220, 220f
    ■ hysterectomy, 322
    ■ infections from, 11
    ■ for skeletal disorders, 98, 98f, 99
    ■ and transfusions, autologous, 147
    ■ transplants, 155, 155f, 167, 240, 283
■ Surrogate mothers, 321, 321f
Survival of the fittest, 446. *See also* Natural selection
■ Sustainability, principle of, 491, 493
Sutures (skull), 96
Swallowing, 181, 185, 201, 201f
Sweat and sweating, 22–23, 82–83, 82f, 83t, 226t, 227, 238. *See also* Sweat glands
Sweat glands, 69, 78–79, 78f, 82–83, 82f, 83t, 350
■ Sweeteners, artificial, 273
Swelling. *See* Edema; Inflammation
■ Swine flu (H1N1 virus), 175, 175f, 192
Sympathetic nerves, 251f, 252–253, 252f
Sympto-thermal method. *See* Rhythm method
Synapses, 111, 111f, 246–247, 246f, 249, 249f
Synaptic integration, 247, 249
Synaptic vesicles, 246, 246f
Syndrome, definition of, 394
Synergistic interaction, of hormones, 288
Synergist muscles, 107
Synovial fluids, 96
Synovial joints, 96, 96f, 97f, 98–99

Synovial membranes, 75, 75f, 96, 98
■ Syphilis, 11, 322f, 323, 323f, 324t
Systemic circuit, 126–127, 126f, 127f
■ Systemic lupus erythematosus (SLE), 157, 171
Systems of organs. *See* Organ systems
Systole, 124–125, 125f
Systolic blood pressure, 129, 129f, 129t

T

Table salt. *See* Salt
Table sugar, 28, 28f
■ Tachycardia, 135
Tail, of embryo, 340–341, 340f–341f
Tailbone (coccyx), 93, 93f
■ Tanning, 79, 385, 433, 433f, 439, 442
Target cells, 288, 288t, 290–291, 291f, 292t, 294t
Tarsal bones, 91f, 95, 95f
Taste (gustation), 257, 272, 272f, 273, 351
Taste buds, 272, 272f, 273
Taste receptors, 272, 272f
■ Tau, 351
■ Tay-Sachs disease, 396, 396f
T cell(s)
    cytotoxic, 163, 166–167, 166f, 167f, 439
    effector, 162, 162f
    formation of, 159
    helper, 163, 164, 165f, 166f, 167, 172
    HIV and, 172–173, 172f
    immune system role, 157, 162–163, 162f, 163f, 164, 165f, 166–167, 166f, 167f, 351, 423, 433, 439
    ■ in immune disorders, 170f, 171
    ■ in immunotherapy, 169
    memory, 162, 162f, 164, 165f, 166, 166f
    origin of, 143f
T cell receptors, 162
■ Tear (in muscle), 114, 114f
Tear glands (eye), 180
Tears (eye), 160, 164
Tectorial membrane, 274f–275f, 275
Teeth, 200, 200f
    ■ amelogenesis imperfecta, 399, 399f
    ■ dental caries, bacteria causing, 212, 212f
    ■ dental surgery, and cardiovascular disease, 136
    evolution of, 455
    in food processing, 200
    joints anchoring to sockets, 96
    sockets for, 92
Telangiectasia, 132
Telomeres, 350
Telophase (mitosis), 358f, 360, 361f, 370f–371f
Telophase I (meiosis), 364, 366, 366f–367f, 369f, 370f–371f
Telophase II (meiosis), 67, 364, 366f–367f, 369f, 370f–371f